discovering advanced mathematics

mechanics

Ted Graham
Centre for Teaching Mathematics,
University of Plymouth

series editor
John Berry
Centre for Teaching Mathematics,
University of Plymouth

contributory authors
Rob Lincoln
Stuart Rowlands
Stewart Townend

project contributors
Steve Dobbs, Bob Francis,
Roger Fentem, Howard Hampson,
Penny Howe, Sue de Pomerai,
Claire Rowland, John White

Collins Educational
An imprint of HarperCollinsPublishers

BRIDGWATER COLLEGE
B0054034
CANCELLED

531
B0054037

Ted Graham is Senior Examiner for AEB Mechanics. He is a senior lecturer at the School of Mathematics and Statistics, University of Plymouth, and a member of the Centre for Teaching Mathematics at the University.

Rob Lincoln is Head of Mathematics at Warminster School in Wiltshire.

Stuart Rowlands taught for 13 years at Calverhay High School, Bath, before joining the Centre for Teaching Mathematics, University of Plymouth.

M. Stewart Townend is a Principal Lecturer in mathematics at the Liverpool John Moores University.

John Berry is Professor of Mathematics Education and Director of the Centre for Teaching Mathematics at the University of Plymouth. John Berry worked with SCAA on the development of the new A-level syllabuses, and the applications of new technology on the teaching and assessment of A-level mathematics.

Published by Collins Educational
An imprint of HarperCollinsPublishers
77–85 Fulham Palace Road
Hammersmith
London W6 8JB

© Ted Graham/HarperCollinsPublishers 1995
First published in 1995
ISBN 0 00 322372 8

All rights reserved. No part of this publication may be reproduced, stored in a retrieval system or transmitted in any form or by any means, electronic, mechanical, photocopying or otherwise, without the prior permission of the publishers.

Designed and illustrated by: Ken Vail Graphic Design, Cambridge, UK
Cover design by: Moondisks, Cambridge, UK
Project editor: Joan Miller
Additional illustration: Tom Cross
Printed and bound in the UK by: Scotprint Ltd., Musselburgh

Contents

Preface

discovering advanced mathematics

Mathematics is not just an important subject in its own right, but also a tool for solving problems. Mathematics at A-level is changing to reflect this: during your A-level course, you must study at least one area of the *application* of mathematics. This is what we mean by 'mathematical modelling'. Mathematicians have been applying mathematics to problems in mechanics and statistics for many years. But now, modelling is included throughout A-level maths.

This book focuses on the application known as 'mechanics'. It draws on many real examples and includes activities that are specifically designed to tackle the demands of the Common Core for mathematics A-level. Calculus is not used until quite late in the book so that you can begin your work in mechanics early in your A-level course, without having covered a lot of pure mathematics.

Technology is advancing as well. Hand-held calculators that can produce graphs and even do simple algebra are revolutionising the subject. The Common Core for A-level expects you to know how to use appropriate technology in mathematics and be aware that this technology has limitations.

We have written *discovering advanced mathematics* to meet the needs of the new A- and AS-level syllabuses and the Common Core for mathematics. The books provide opportunities to study advanced mathematics while learning about modelling and problem-solving. We show you how to make best use of new technology, including graphics calculators (but you don't need more than a good scientific calculator to work through the book).

In every chapter in this book, you will find:

- an introduction that explains a new idea or technique in a helpful context;
- plenty of worked examples to show you how the techniques are used, starting with a question and then a 'Commentary' that leads you to the solution;
- exercises in two sets, A and B (the B exercises 'mirror' the A set so that you can practise the same work with different questions);
- consolidation exercises that test you in the same way as real exam questions.

And then, once you have finished the chapter:

- modelling and problem-solving exercises to help you pull together all of the ideas.

We hope that you will enjoy advanced mathematics by working through this book.

We thank the many people involved in developing *discovering advanced mathematics*. In particular, we thank the students and staff of Lipson School and Warminster School where much of the material in this book was trialled, Jon Reeves for checking the exercises and Karen Eccles for typing the manuscript. Also to Veronica, Rachel, Ben and Emma for their inspiration and patience.

Ted Graham *January 1995*

Mechanics and the real world

- *Mechanics is the study of structures and objects that move.*
- *Real situations are often very complex and simplifications need to be made to allow solutions to be found to them.*
- *Mathematical modelling is the process of using mathematics to solve real problems.*

WHAT IS MECHANICS?

The world around us is full of examples of the applications of mechanics. Whenever anything moves the laws of mechanics describe its motion and what is needed to cause and maintain that motion.

The study of mechanics in the context of objects that move is called **dynamics**. Some examples of problems in dynamics that you will be able to solve by the time you reach the end of this book are:

- How fast can a car go round a bend without spinning out of control?
- How long a piece of elastic should a bungee jumper use?
- How fast will a parachutist be moving when she hits the ground?
- How far will the chairs on a chairoplane ride swing out?
- How fast must the cars on a roller coaster travel to get through a corkscrew?

When someone sits on a chair it is important that the legs are strong enough to support them. When a structure such as a suspension bridge is built it is important that the towers and cables can support the deck and the traffic it carries. It is often necessary to make predictions about the strengths of structures that remain at rest. The study of mechanics in this context is known as **statics**.

Take a few moments to identify some examples of mechanics that you meet in your own life and classify them as examples of dynamics or statics.

Exploration 1.1

Mechanics in action

This diagram below shows a cut-away picture of a car.

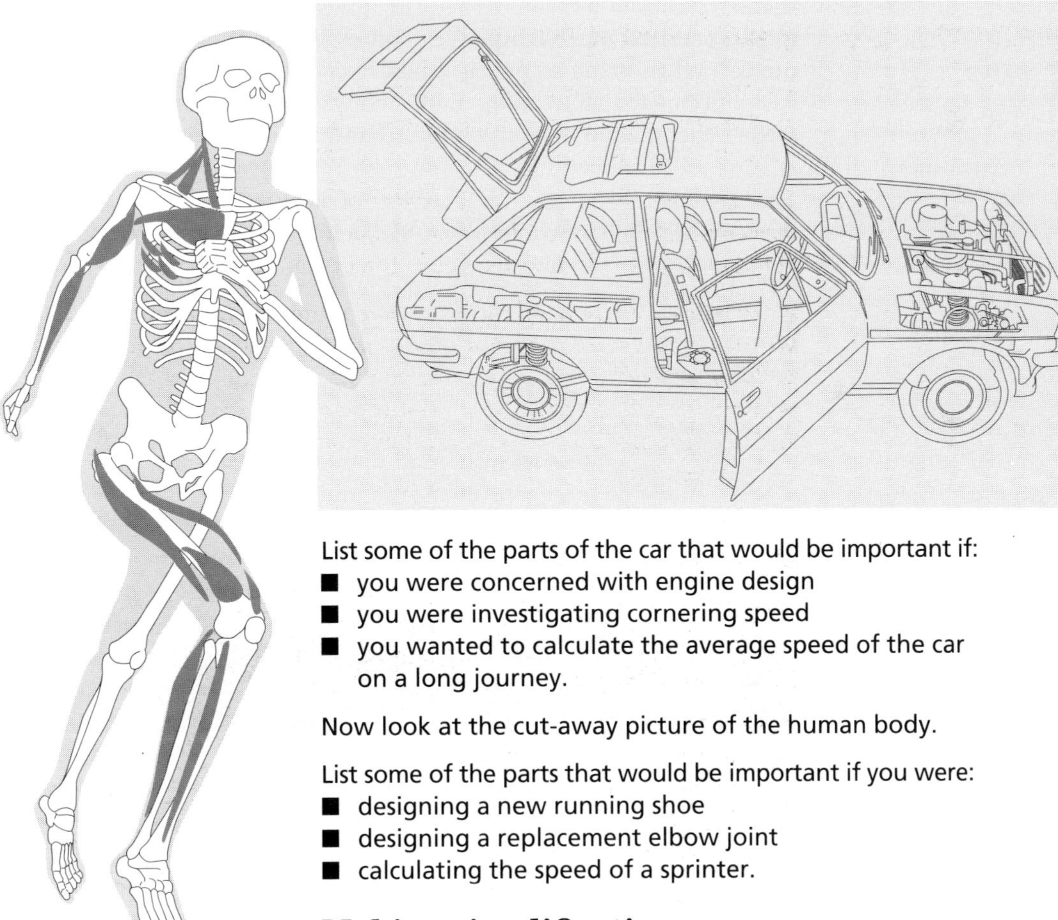

List some of the parts of the car that would be important if:

- you were concerned with engine design
- you were investigating cornering speed
- you wanted to calculate the average speed of the car on a long journey.

Now look at the cut-away picture of the human body.

List some of the parts that would be important if you were:

- designing a new running shoe
- designing a replacement elbow joint
- calculating the speed of a sprinter.

Making simplifications

Both a motor car and a human body are incredibly complex, with many different moving parts. To consider every part of either of them would be a major task. However, it may be that we are concerned only with a single part of them or with their movement as a whole.

BRIDGWATER COLLEGE LIBRARY

In mechanics it is often useful to make simplifications, so that solutions can be obtained to problems that otherwise would be too sophisticated to solve.

When describing the motion of a car we can consider it as a point moving along a line or curve, very much like an aircraft seen on a radar screen. We know how it is moving, but assume that it is at a single point or it is a particle, ignoring its size and the complex motions within it.

When describing the motion of a sprinter we may think of the sprinter as a point or particle that moves along a straight line, ignoring the motion of the arms, legs and internal structures.

In mechanics it is often very useful to assume that objects are points that have some of the same characteristics as the object, such as mass and speed, but ignore other characteristics such as rotation, size and the movement of internal parts. This process is an early stage of **mathematical modelling**, and is often described as using a **particle model**. When using a particle model it is important to be aware of the assumptions that have been made and whether or not there are other factors that should be taken into account.

MATHEMATICAL MODELLING

Mathematical modelling is the process by which real situations can be described mathematically and thereby problems can be solved. The process contains a number of phases or stages which are shown on the diagram.

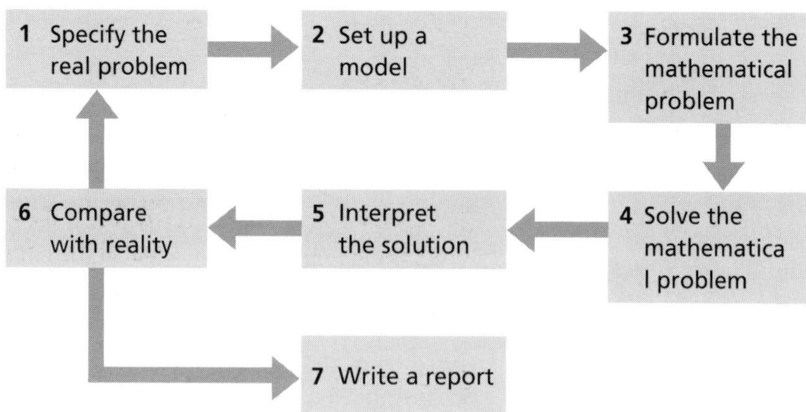

Each of these stages is now discussed briefly to give an overview of the whole process.

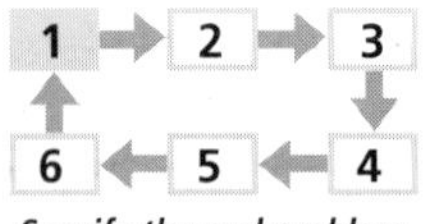

Specify the real problem

1 Problem statement

It is important to decide exactly what problem is to be solved. A clear ***problem statement*** can be an important first step if we are asked to tackle a vague problem that is not well defined.

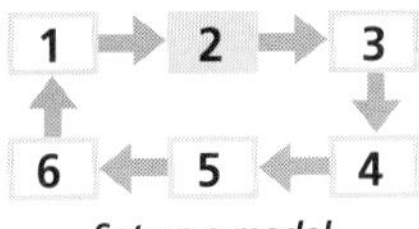

Set up a model

2 Set up a model

This is a transfer from the problem statement to a model of that situation that can be analysed mathematically. Involved in this may be making a list of all the features that could or should be taken into account and then using this to specify a set of assumptions or simplifications to create a model of the situation. This model then provides a simplified situation from which to formulate and solve a mathematical problem.

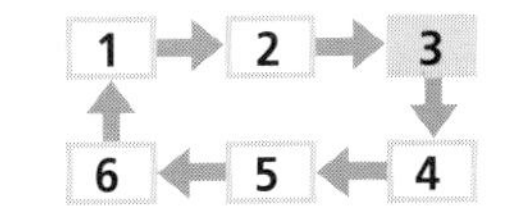

Formulate the mathematical problem

3 Formulate a mathematical problem

Once a simplified situation has been established it is necessary to define a mathematical problem to solve. This will consist of a highly specified mathematical problem, very much like the sort of problems that you will find in a student text book exercise.

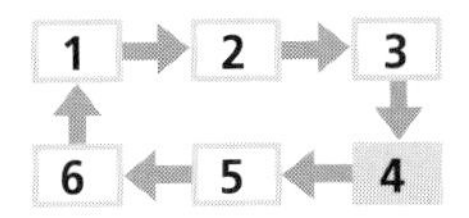

Solve the mathematical problem

4 Mathematical solution

At this stage mathematical techniques are used to solve the mathematical problem defined at the previous stage. In this book you will learn a number of techniques for solving problems that involve mechanics.

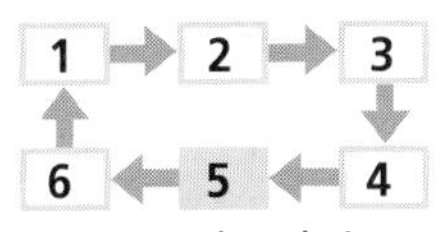

Interpret the solution

5 Interpretation

It is important to return to the original situation and interpret the solution in that context. It may well be that the original problem leads to a solution that is either impossible or unsafe, for example, or that the solution is quite reasonable.

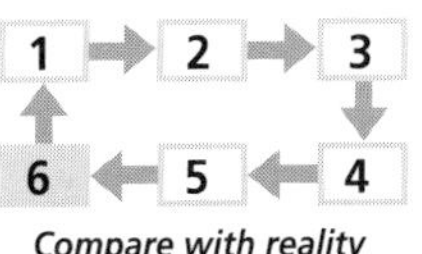

Compare with reality

6 Compare with reality

This stage develops naturally from the interpretation of the solution. Whenever predictions have been made it is important to try to validate them in some way. This can often be done by comparison with real situations. Sometimes an experiment can be used to do this.

The comparison will either confirm that the solutions are reasonable or throw some doubt on their value. In the latter case, it is necessary to return to the original problem and modify the assumptions, so that a better outcome is produced.

7 Report – refine the model

Once a model that produces good solutions has been established, then the modelling process is completed with a report on the findings.

In this chapter you have begun to think about the first two stages of this process. As you study further you will learn how to solve different types of mathematical problems and to interpret the results. Almost all of the chapters contain a mathematical modelling activity that will help you to develop the sort of skills that you need if you are to apply the mechanics that you learn to real life problems. The following exercises help you to begin to think about making assumptions that form an essential part of the mathematical modelling process.

EXERCISES

1.1 A

1 A car can be modelled as a particle. State whether or not this is a reasonable assumption for each problem given below and if not state what other factors may need to be considered.
a) Deciding how fast a car can go round a bend.
b) Deciding how far apart to put sleeping policemen.
c) Calculating the average speed for a long journey.
d) Designing a car park.

2 Identify three sporting activities where it is:
a) acceptable to model the athletes as particles,
b) not acceptable to model the athletes as particles.

3 The Earth rotates about its axis. Is this factor important when considering:
a) the orbit of the Earth round the sun,
b) a child on a swing in the park,
c) the launch of a space shuttle?

4 The radius of the Earth is 6400 km. Is it reasonable to model the Earth as a particle, when considering:
a) its orbit round the sun,
b) the flight of an aeroplane,
c) the placing of a satellite in orbit around the Earth?

5 Select some objects that move and decide whether it is reasonable to model them as particles.
Give some different examples of situations where this model may not be reasonable.

6 Make a list of the features that you would consider when deciding whether or not to install a pedestrian crossing.

EXERCISES

1.1B

1 A football could be modelled as a particle. Decide if this is an appropriate assumption for each situation described below.
a) Predicting the speed at which it should be kicked to travel 20 m before bouncing.
b) Deciding if the ball will go into the goal.
c) Deciding how to kick it, so that it swerves past a line of defenders and into the goal.
d) When describing its position on a football pitch.

2 When objects are modelled as particles, any rotation of the object as it moves is ignored. For example if a football is modelled as a particle, the rotation as it rolls is ignored.

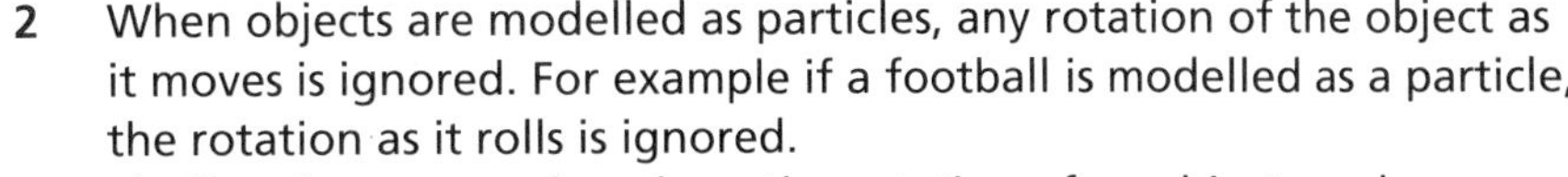

a) Give three examples where the rotation of an object can be ignored, because its effects are likely to be insignificant.

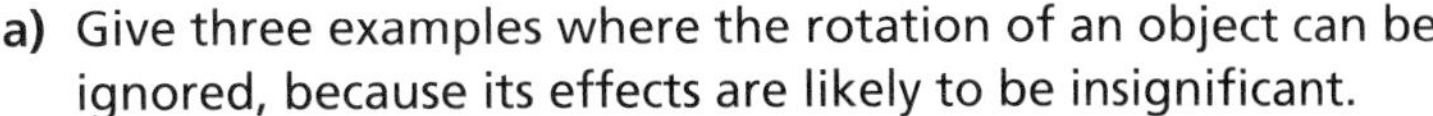

b) Give three examples where the rotation of an object is important and should not be ignored.

3 The child on the swing shown here is to be modelled as a particle. Discuss what the position of the particle should be on the diagram.

4 A motorcycle could be modelled as a particle. Is this assumption appropriate when:

a) calculating the maximum speed at which to round a bend,

b) trying to predict the top speed on a straight road,

c) estimating the difference in top speeds for riders in upright or prone positions?

5 Make a list of the features that you think should be taken into account when trying to predict the maximum height that could be attained in a high jump contest.

6 When the size and shape of an object is important a rigid body model is used. This assumes that the object does not change shape in any way. Select some objects that could be modelled as rigid bodies and some which are not appropriate to model as a rigid body.

Summary

After working through this chapter you should:

- be aware of the need to make simplifications
- be aware of some of the advantages and disadvantages of the particle model
- have had an introduction to mathematical modelling.

Force

- *A force causes a change in the motion of an object.*
- *There are many different types of force. The most common examples are explored in this chapter.*

INTRODUCTION TO FORCE

Exploration 2.1

What do we understand by 'force'?

The word force is often used in an everyday sense, but has a more precise scientific meaning. What do you understand by the word force?

Consider the questions below, bearing in mind your answers to the question above.

- If you drop a ball, gravity exerts a downward force on the ball. What happens to the ball? What is the effect of this force?
- You apply the brakes on a bicycle that you are riding. The result of this action is a backward force. What is the effect of this force?
- A tennis ball is travelling towards you, and you hit it with a racquet. The racquet exerts a force on the ball. What is the effect of this force?
- You are travelling on a well-oiled skateboard on a level surface. What happens to your speed? Why? In a perfect environment how far would the skateboard move before stopping?

The effect of forces

The effect of a force is to change the motion of an object. It can cause it to start moving, stop moving, change speed or change direction. If no force acts the motion of an object will not change, it will either stay still or move at a constant speed in a straight line for ever! Newton formally stated this in his **first law**:

> *Any body will remain at rest or move at constant speed in a straight line unless acted on by a force.*

Balanced or unbalanced forces

In reality there will be more than one force acting on any object and so we must consider the combined effect of these forces. In some cases the forces will balance each other out, so that the motion of the object will not change. In other cases the forces will not balance each other out, there will be an overall or resultant force, and so there will be a change in motion.

Example 2.1

A parachutist falls with an open parachute. Choose a suitable model and decide if the forces acting are in balance when:

a) *the parachute has just been opened,*
b) *terminal speed has been reached.*

Solution

The picture shows a parachutist and parachute. They can be modelled as a particle with two forces acting on it: gravity downwards and air resistance upwards.

a) *When the parachute is first opened the parachutist starts to lose speed, and so the overall force must be upwards. This is illustrated in the diagram where the air resistance force is greater, so the resultant force is upwards.*

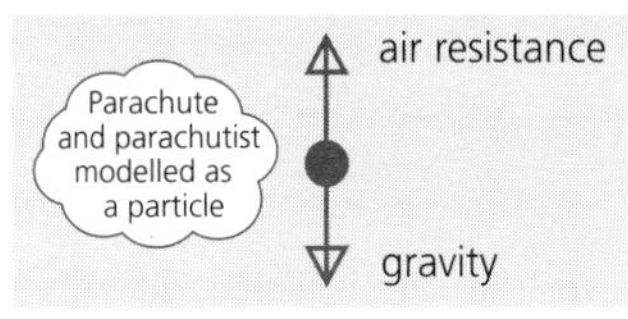

b) *At a later stage the parachute and parachutist are moving at a constant (terminal) speed, so the forces are in balance. This is shown in the diagram where both forces cancel each other out.*

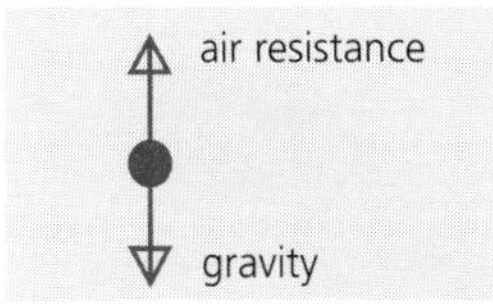

Example 2.2

A car travels along a straight horizontal road. Are the forces on the car balanced or unbalanced? What happens if it goes round a bend?

Car modelled as a particle

Solution

Consider now the forces acting on a car as it moves along a straight horizontal road. If we think of the car as a particle this is much easier. Whatever the motion, gravity will always pull down on the car and this will be balanced by upward forces on the wheels, so that the vertical forces are in balance.

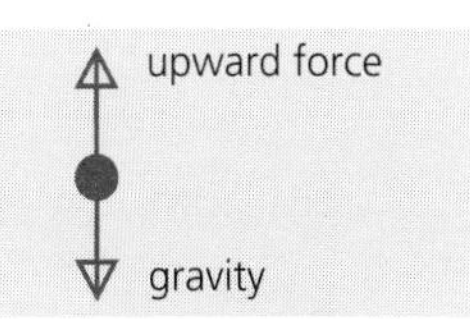

Vertical forces in balance

If the car is moving forward there will also be resistance forces acting to try to slow down the car. If the car is gaining speed the forward force must be greater than the resistance forces, so the overall or resultant force is forwards.

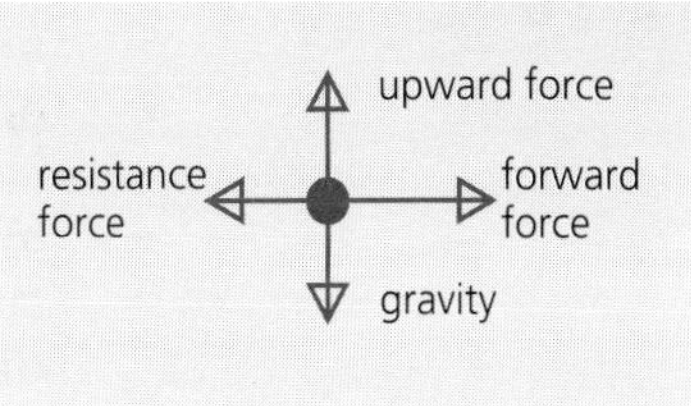

Car gaining speed

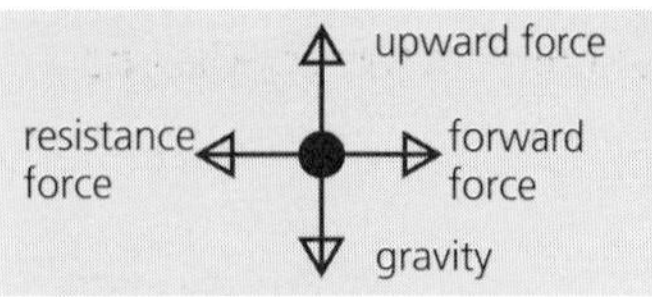

Car at constant speed

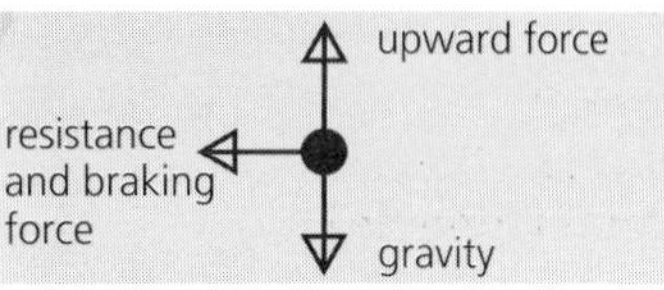

Car slowing down

If the car is moving at a constant speed such as when cruising at its top speed, then as the motion is constant, the forces on the car must be in balance.

If the car driver applies the brakes, then the forward force will be removed and a large backward force will act. Note that even though the force acts backwards the car continues to move forward until it stops.

If the car changes direction as it moves, for example as it goes round a bend, then there must also be a sideways force acting on the car to cause the change in motion.

EXERCISES

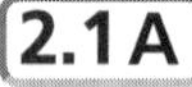
2.1 A

1 For each situation described below state whether or not the forces acting are in balance.
- **a)** A book resting on a table in a train that is travelling at constant speed along a straight set of tracks.
- **b)** A mass that is moving up and down on the end of a spring.
- **c)** A pendulum in a clock as it swings.
- **d)** A car going round a roundabout at a constant speed.
- **e)** A skateboard travelling down a slope at a constant speed.
- **f)** A stone falling through a fluid at its terminal speed.
- **g)** A bouncing ball while it is in contact with the ground.

2 Describe situations where each of the following are true.
- **a)** The forces acting on a moving object are in balance.
- **b)** The resultant force on an object acts in the direction in which it is moving.
- **c)** The resultant force on an object acts in the opposite direction to which it is moving.
- **d)** There is a force acting at an angle to the path of the object.

3 A cyclist free wheels down a hill.
- **a)** Describe what happens to his speed.
- **b)** Describe what happens to the resultant or overall force on him.

EXERCISES

2.1 B

1 For each situation described below, decide if the forces acting are in equilibrium.
- **a)** A sail boarder travelling at a constant speed in a straight line on a smooth lake.
- **b)** A water-skier describing an arc of a circle.
- **c)** A sledge travelling at a constant speed down a slope.
- **d)** The Earth as it orbits the sun.
- **e)** A bungee jumper, when she reaches her lowest position.
- **f)** A cup of coffee on a table in a train moving at a constant speed in a straight line up a hill.
- **g)** A car parked on a hill.

2 For each example below, describe when or under what conditions the forces acting could be in equilibrium.
a) A crate on the back of a lorry.
b) A bungee jumper.
c) An ice puck that is sliding on ice.
d) A diver ascending from the sea bottom.

3 A pebble is dropped into a deep pool of water.
a) What happens to the speed of the pebble?
b) What happens to the resultant force on the pebble?

4 Repeat question 3 for a car that freewheels to rest.

GRAVITY ON EARTH

Gravity is a force that the Earth exerts on all objects. On Earth the magnitude of this force is given by mg, where m is the mass in kilograms and g is a constant known as the **acceleration due to gravity**. On Earth g has the value of approximately $9.8\,\text{m s}^{-2}$. In fact the value of g does have slightly different values at different places, but in this book we shall take g as $9.8\,\text{m s}^{-2}$. The force of gravity always acts towards the centre of the Earth.

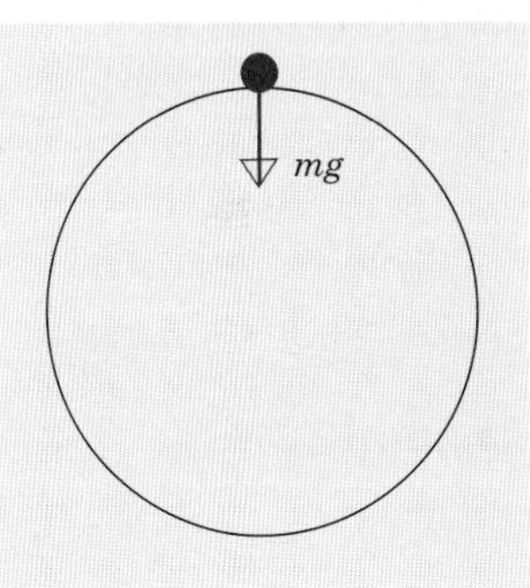

Example 2.3

Calculate the magnitude of the force of gravity on:

a) *a crate of mass 55 kg,*
b) *a car of mass 1.2 tonnes.*

Illustrate these forces on a suitable diagram.

mg = 539 N

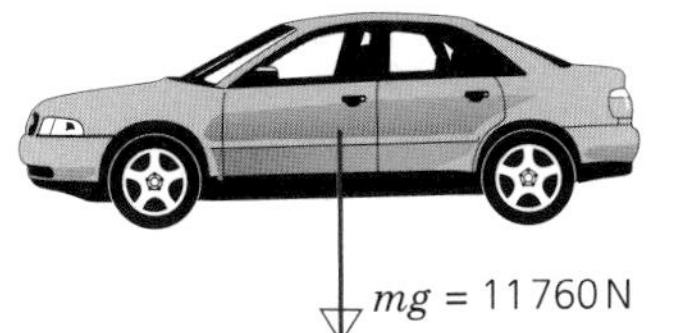

Solution

a) *Force of gravity given by* $mg = 55 \times 9.8 = 539$ *N*
b) *Note 1 tonne = 1000 kg.*
So $mg = 1200 \times 9.8 = 11\,760$ *N*

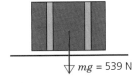

The units of force are newtons, denoted by N. These forces are illustrated in the diagrams. Note that the force has been assumed to act at the centre of each object.

Note on weight

The force on a body exerted by gravity is often referred to as **weight**. In everyday language, weight and mass are often confused or interchanged. For example people may talk about the weight of a person being 66 kg, when they mean their mass is 66 kg. Their weight is $66 \times 9.8 = 647$ N.

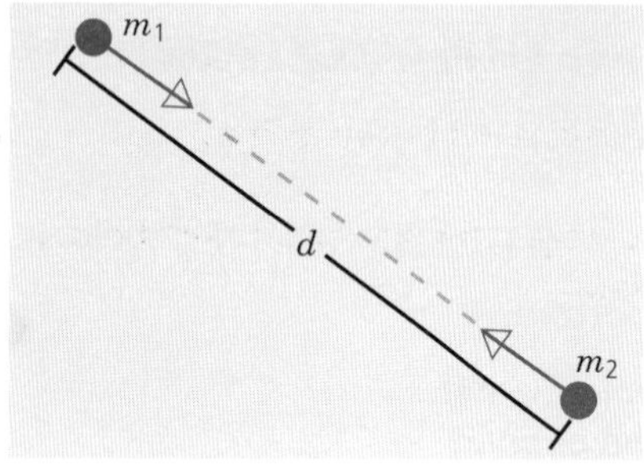

The universal law of gravitation

There is always a force of gravitational attraction between any two objects. For example there will be gravitational attraction between any two people, but fortunately the magnitude of this attraction is small enough to be insignificant. There are strong gravitational forces between the sun and other planets in the solar system. Without these forces the planets would not continue in their orbits.

Newton formulated a **law of universal gravitation**, based on his observations of the motion of the planets. This law states that the force of attraction between any two objects is:

$$\frac{Gm_1m_2}{d^2}$$

where m_1 and m_2 are the masses of the objects, in kilograms, d is the distance between their centres, in metres, and G is a constant which has a value $6.67 \times 10^{-11}\,\text{kg}^{-1}\,\text{m}^3\,\text{s}^{-1}$.

Note that forces act on both objects as shown and both have the same magnitude.

Example 2.4

Use the data given below to calculate the force that the sun exerts on the Earth.

Mass of Earth $= 5.98 \times 10^{24}$ *kg*
Mass of sun $= 1.99 \times 10^{30}$ *kg*

Average distance between Earth and sun $= 1.47 \times 10^{10}$ *m*

Solution
The force is calculated using:

$$\frac{Gm_1m_2}{d^2} = \frac{6.67 \times 10^{-11} \times 5.98 \times 10^{24} \times 1.99 \times 10^{30}}{\left(1.47 \times 10^{10}\right)^2} = 3.67 \times 10^{24}\ N$$

Example 2.5

Use the universal law of gravitation and the data below to find the value of g on the moon.

Radius of moon $= 1.73 \times 10^{6}$ *m*
Mass of moon $= 7.38 \times 10^{22}$ *kg*

Solution
Consider a particle of mass M on the surface of the moon. Then the force acting on this mass is given by:

$$\frac{Gm_1m_2}{d^2} = \frac{6.67 \times 10^{-11} \times 7.38 \times 10^{22} \times M}{\left(1.73 \times 10^{6}\right)^2} = 1.6M$$

As the mass is M then the acceleration due to gravity is 1.6 m s^{-2}.

EXERCISES

2.2 A

Questions marked * involve the universal law of gravitation.

1 Calculate the magnitude of the force of gravity on each object below and illustrate these forces on a diagram:
a) a box of mass 25 kg,
b) a table tennis ball of mass 3 grams,
c) a van of mass 1.8 tonnes.

2 The force exerted on a car by gravity is 8820 N. What is the mass of the car?

3 Estimate the size of the force that gravity exerts on:
a) a tennis ball, **b)** an elephant, **c)** a bicycle.

4 Two objects of different masses are released together and allowed to fall. Does the mass of the objects affect the time that they take to fall? Try a simple experiment to confirm your answer.

5 On the moon the acceleration due to gravity is $1.6\,\text{m s}^{-2}$. Find the force of gravity exerted by the moon on an astronaut of mass 65 kg.

*6 Determine the magnitude of the gravitational attraction between a man of mass 70 kg and a woman of mass 60 kg if they are 1 m apart.

*7 The radius of the planet Venus is 4.59×10^{6} m and it has a mass of 4.78×10^{24} kg. Find the value of g on the planet.

*8 Calculate the gravitational force that a person of mass 75 kg would experience when standing on the surface of the planet Mars. The mass of Mars is 6.55×10^{23} kg and the radius is 3.36×10^{6} m.

*9 An astronaut of mass 75 kg is space walking at a height of 3×10^{5} m above the surface of the Earth. What is the magnitude of the force of gravity on the astronaut? (The radius of the Earth is 6.37×10^{6} m.)

*10 The distance between the centres of the Earth and the moon is 3.84×10^{8} m. A spaceship wants to position itself so that the force of gravity of the moon exactly balances the force of gravity of the Earth. How far should the ship position itself from the centre of the Earth?

EXERCISES

2.2 B

Questions marked * involve the universal law of gravitation.

1 Calculate the magnitude of the force of gravity acting on each object below and illustrate the forces on a diagram:
a) a student of mass 68 kg,
b) a letter of mass 50 grams,
c) a train of mass 200 tonnes.

2 As a ball falls, it experiences a force of 2.8 N, due to gravity. What is the mass of the ball?

3 Estimate the magnitude of the force that gravity exerts on:
a) a copy of this book, **b)** a horse, **c)** a tennis racquet.

4 Two similar objects of different masses are held at the same height and released together. What differences would you notice if this experiment were done on both Earth and the moon and the results were compared?

5 An astronaut of mass 74 kg steps onto a planet where $g = 5.2 \text{ m s}^{-2}$. Find the force that gravity exerts on the astronaut.

6 A space probe of mass 200 kg lands on an unexplored planet. The probe's sensor detects that a force due to gravity of 3000 N is acting on the probe. Find the acceleration due to gravity on this planet.

*7 Find the magnitude of the gravitational attraction between two cars both of mass 1000 kg, parked so that their centres are 5 m apart.

*8 An astronaut of mass 68 kg is standing on a planet of radius 4×10^8 m. If the force of gravity on the astronaut is 500 N, find the mass of the planet.

*9 A satellite of mass 700 kg is placed into orbit at a height of 8×10^7 m above the surface of the Earth. The radius of the Earth is 6.37×10^6 m. Find the force that gravity exerts on the satellite.

*10 It is estimated that an unexplored planet has twice the mass of the Earth and that its radius is 1.8 times bigger. What would be the value of g on this planet?

CONTACT FORCES

Normal reaction

Think about a book resting on a shelf. Gravity exerts a **downward** force on the book, but the shelf exerts an **upward** force on it, to balance the force due to gravity.

The upward force is at right angles to the surface of the shelf and is called the **normal reaction** force, usually labelled R on diagrams.

The book also exerts a downward force on the shelf. This force has the same size or **magnitude** as the normal reaction, R, but acts in the opposite direction.

Example 2.6

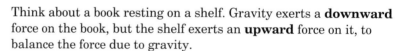

Three boxes, each of mass 5 kg, are stacked one on top of another. Draw a diagram to show the forces acting on each box and find the magnitude of each force.

Solution

Start with the top box and work down. The forces on the top box are gravity and a normal reaction R_1. The force of gravity is 49 N and to balance this R_1 must be 49 N.

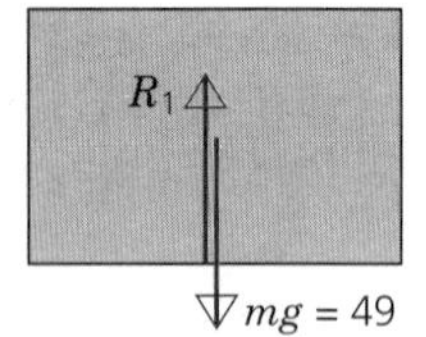

The middle box has a downward normal reaction force of magnitude R_1, the force of gravity and an upward normal reaction force, R_2. The normal reaction, R_2, must balance both R_1 and mg and so must have magnitude 98 N.

The bottom box again has two normal reactions and the force of gravity. The upward normal reaction force, R_3, must balance mg and R_2 and so has magnitude 147 N.

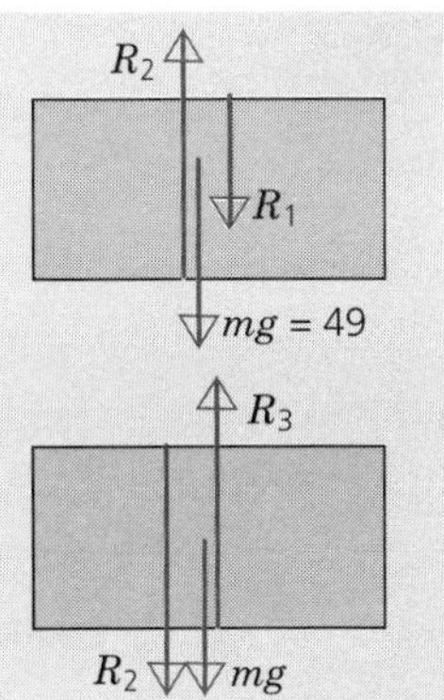

Example 2.7

At a fairground two dodgems are heading straight towards each other. Draw a diagram to show the forces on the dodgems when they collide.

Solution

Gravity acts on both cars and in both cases is balanced by a normal reaction, R_1.

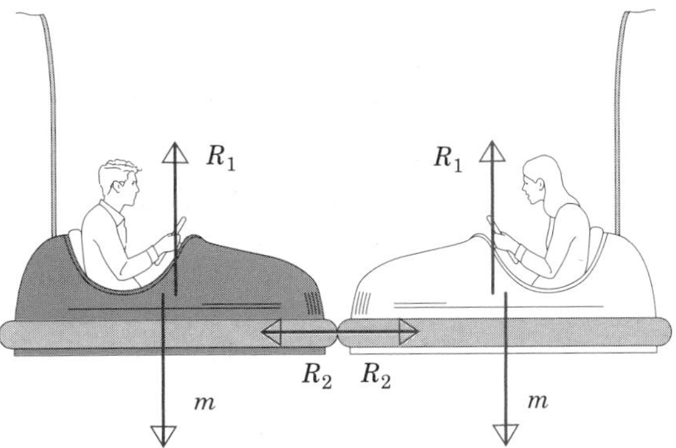

When the cars collide they exert normal reaction forces, R_2, of equal size but opposite directions at the point of contact.

Newton's third law

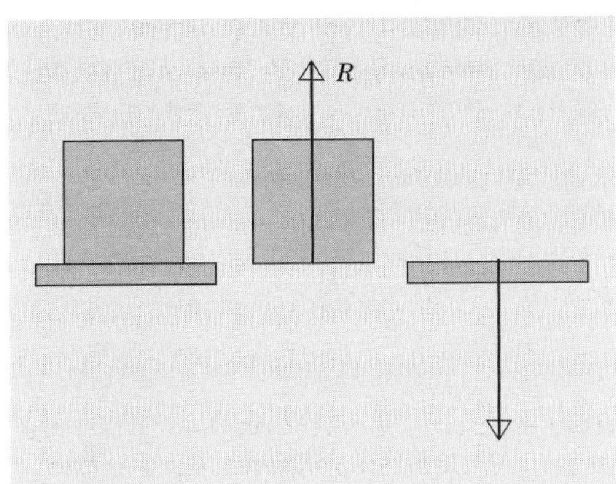

Whenever two surfaces are in contact, they exert forces on each other that are equal in magnitude, but acting in opposite directions.

Newton stated this formally in his **third law** of motion, as:

> *For every action there is an equal and opposite reaction.*

This law is easily seen to apply when considering normal reaction forces that act on surfaces in contact. However, it is important to remember that it is true when any force acts.

Friction

Imagine a man trying to push a heavy box, but it doesn't move. If he exerts a horizontal force on the box, then there must be another force acting on the box to balance the force he applies.

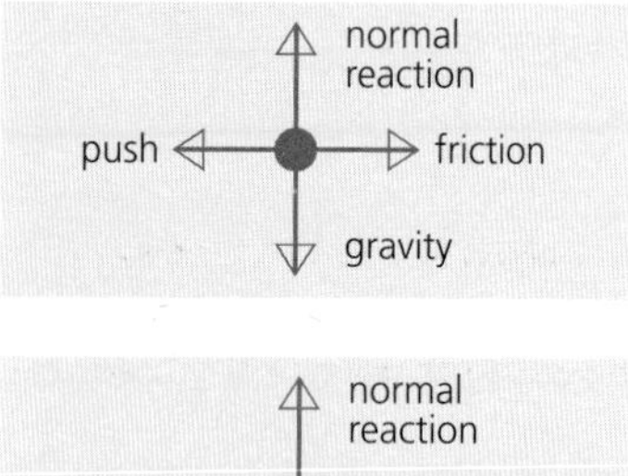

Modelling the box as a particle allows us to draw the force diagram (left).

The force that balances the push is friction, which acts because of the rough surfaces that are in contact. If the man pushes harder (bottom left) and the box stays still, the friction must have increased to match the greater force he exerts.

There is, however, a limit to the size or magnitude of the friction force and when the force applied exceeds this, then the box will begin to move, gaining speed.

Take a few moments to think about what factors affect the size or magnitude of the friction force.

The friction model

A model for the magnitude of a friction force is:

$$F \leq \mu R$$

where R is the magnitude of the normal reaction and μ is the **coefficient of friction**, a constant which depends on the surfaces in contact. When an object is at rest, F will be just sufficient to keep the forces in balance. When the object is about to move, or on the point of slipping, or actually moving, F takes its maximum value.

Exploration 2.2 *Finding the value of μ*

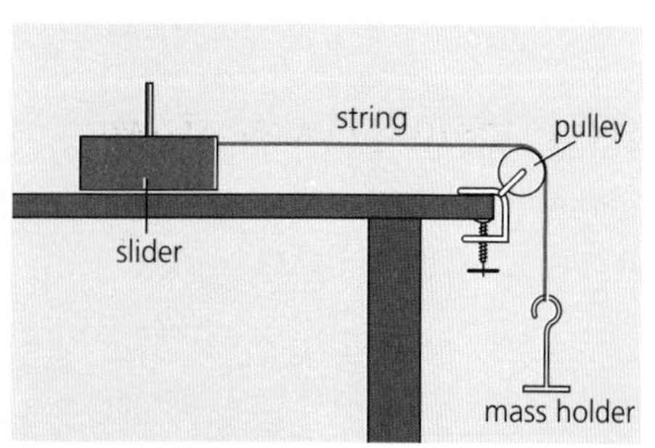

- *Set up the apparatus as shown in the diagram. Add 10 gram masses until the slider just begins to move. Assume that the force the string exerts on the slider is then equal to the maximum friction force. Record the mass needed to achieve this.*
- *Add a 100 gram mass to the slider and repeat the last step. Continue this procedure until you have six or seven results.*
- *Produce a table like this, which contains a sample set of results. (Use your own results.)*

Total mass of slider, M	$R = Mg$	Mass on string, m	Force exerted by string, mg
70 grams	0.686 N	30 grams	0.294 N
170 grams	1.666 N	70 grams	0.686 N

- Plot a graph of the friction force, F, which is assumed to be equal to the force exerted by the string, against R the normal reaction on the slider. The gradient of the line of best fit gives a value for μ.
- Comment on the accuracy of the friction model in the light of your experimental results.

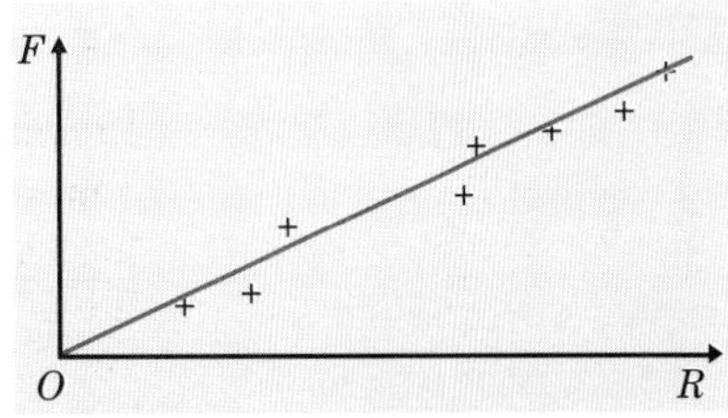

Example 2.8

A horse pulls a large sledge of mass 250 kg across a snow-covered horizontal surface at a constant speed. If the coefficient of friction between the snow and the sledge is 0.3, find the force exerted by the horse.

Solution

As the sledge moves at constant speed the forces acting must be in balance. Modelling the sledge as a particle gives the force diagram shown.

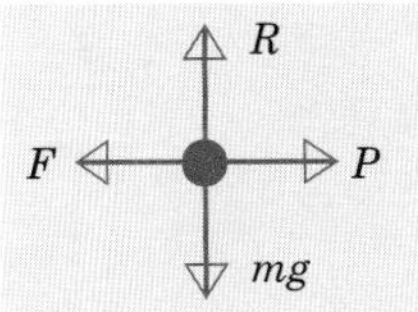

For the forces to be in balance:

$R = mg$

and

$P = F$

Since the sledge is sliding, the friction model gives:

$F = \mu R = \mu mg = 0.3 \times 250 \times 9.8 = 735$ N

So to keep the forces in balance the horse exerts a force of 735 N.

Example 2.9

An accident investigator drags a car tyre along a road surface in an experiment to determine the value of μ for the tyre and the road. The mass of the tyre is 8 kg and he needs to exert a force of 60 N to move it at constant speed. Find the value of μ.

Solution

Modelling the tyre as a particle leads to the force diagram shown, where the forces are in balance.

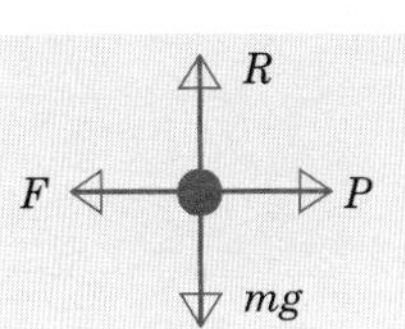

As the forces are in balance:

$R = mg = 8 \times 9.8 = 78.4$ N

and

$F = P = 60$ N.

Using the friction model, $F = \mu R$, gives:

$60 = \mu \times 78.4$

$$\mu = \frac{60}{78.4} = 0.765$$

Example 2.10

Draw a diagram to show the forces acting on a child sliding down a slide in a park.

Solution

Modelling the child as a particle allows us to represent her by a point.

Gravity pulls down on her. There is a normal reaction, R, at right angles to the slide. Friction acts parallel to the surface of the slide.

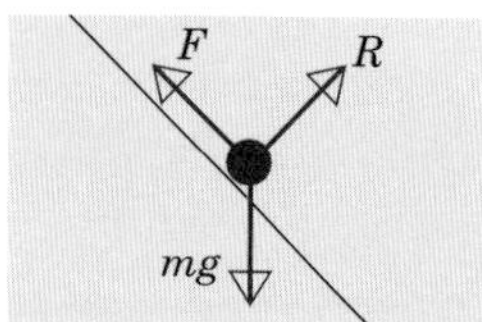

EXERCISES

2.3A

Questions marked * involve the use of the coefficient of friction.

1 Calculate the magnitude of the normal reaction forces on:

a) a man of mass 75 kg standing on a horizontal surface,
b) a book of mass 0.6 kg resting on a table,
c) one leg of a three-legged stool of mass 2 kg,
d) one leg of a bed of mass 20 kg.

2 For each situation illustrated below draw a diagram to show the force of gravity, normal reactions and friction forces that are acting on the object that has been underlined.

a) A <u>box</u> on a table.

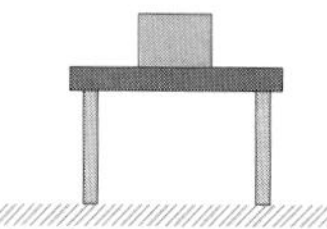

b) A <u>sledge</u> moving on a slope.

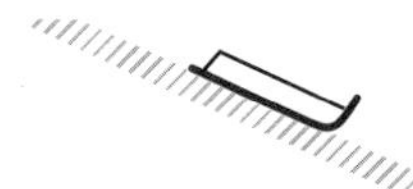

c) A <u>ball</u> as it hits the ground.

d) A <u>ladder</u> resting against a wall.

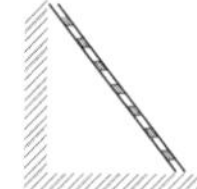

3 A book of mass 2 kg is placed at the centre of a desk of mass 30 kg. Draw a diagram to show the forces acting on the desk. Describe each force and find its magnitude.

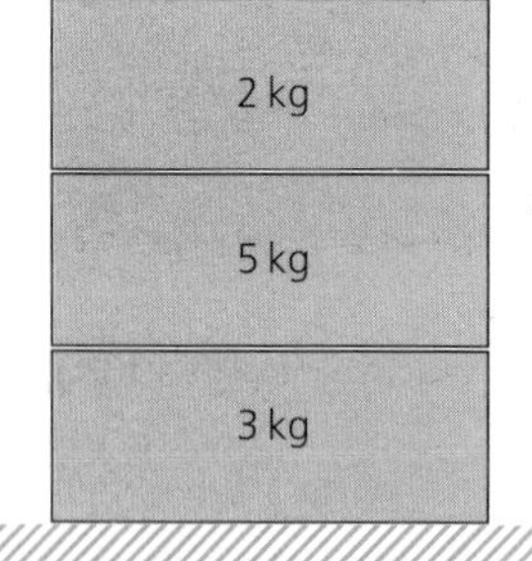

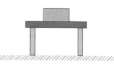

4 Three boxes are stacked as shown in the diagram. Draw diagrams to show the forces acting on each box and find their magnitudes.

5 Bottles of cooking oil are packed in boxes that have a mass of 12 kg when full. The boxes must not be subjected to reaction forces greater than 650 N or their contents will be damaged. How many boxes of cooking oil can be stacked in a single pile?

6 Each diagram shows a system that is at rest. In each case find the magnitude of the friction force acting and the direction of the force.

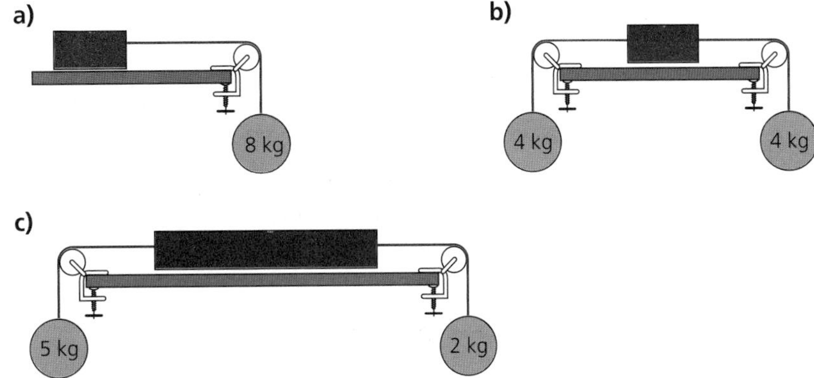

7 A crate of mass 50 kg rests on a horizontal surface and a horizontal force is applied to it. What is the magnitude of the friction force if the crate remains at rest and the applied force is:
a) 280 N, **b)** 200 N?

*8 The coefficient of friction between a crate of mass 50 kg and the ground is 0.6. What is the least horizontal force that can be applied for the crate to move on level ground?

*9 A car locks its brakes on a horizontal road. What is the magnitude of the frictional force on the car if its mass is 1150 kg and the coefficient of friction between the tyres and the road is:
a) 0.9 on a dry road, **b)** 0.65 on a wet road?

*10 A box of mass 10 kg is placed on a sledge of mass 11 kg, on horizontal (level) ground.
a) Draw diagrams to show the forces acting on the box and the sledge if a horizontal force is applied to the box.
b) Calculate the magnitude of all the normal reaction forces present.
c) The coefficient of friction between the sledge and the snow is 0.2 and between the box and the sledge it is 0.6. Does the sledge move if a horizontal force of 40 N is applied to the box?

*11 An empty skip which has a mass of 500 kg is resting on a horizontal surface. The coefficient of friction between the skip and the ground is 0.68.
a) Calculate the magnitude of the normal reaction force, R, on the skip.
b) The magnitude of the maximum friction force is $0.68R$. Find the maximum value of the friction force, using the value of R found above.
c) Some boys try to move the skip. If each boy can exert a force of 400 N, how many are needed to move the skip?
d) In fact there are three boys unsuccessfully trying to move the skip. What is the magnitude of the friction force?

EXERCISES

2.3 B

Questions marked * involve the use of coefficient of friction.

1 Find the magnitude of the normal reaction forces on:
a) a woman of mass 50 kg standing on a horizontal surface,
b) a book of mass 500 grams on a shelf,
c) a crate of mass 2 tonnes, resting on the back of a lorry.

2 Find the magnitude of the normal reactions described below. In each case state clearly any assumptions that you make.
a) The reaction force on each leg of a table of mass 20 kg.
b) The reaction force on each leg of a chair of mass 8 kg.
c) The reaction force on each wheel of a bus of mass 5 tonnes.

3 For each situation illustrated below, draw a diagram to show the force of gravity, normal reactions and friction forces that are acting on the object that has been underlined.
a) A <u>plank</u> resting on 2 concrete blocks.

b) A <u>skier</u> travelling down a slope.

c) A <u>plank</u> leaning against a wall.

d) A <u>pole</u> stuck between two bars of a climbing frame.

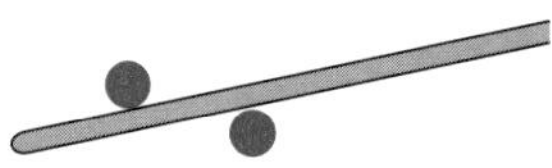

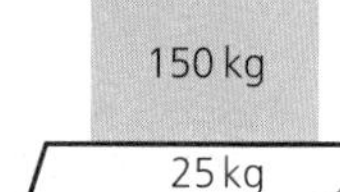

4 A box of supplies of mass 150 kg is loaded onto a sledge of mass 25 kg. Draw a diagram to show the forces acting on the sledge when it is at rest on horizontal ground. Find the magnitude of each force.

5 Boxes that each have a mass of 30 kg are stacked in a pile eight boxes high. Find the reaction forces acting on the bottom box and illustrate them on a diagram.

6 If the forces shown in each diagram act, find the magnitude of the force that friction must exert if the object stays at rest in each case. Also state the direction of the friction force.

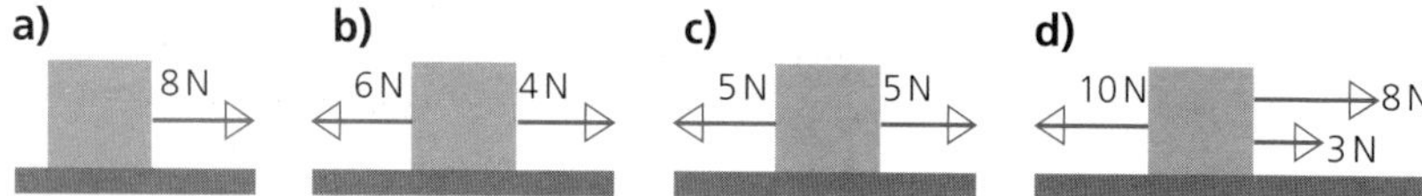

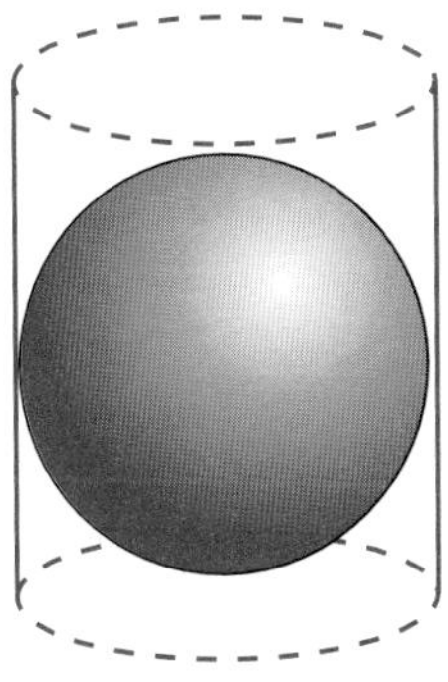

7 A ball of mass 100 grams is dropped into a vertical tube which has the same diameter. Find the magnitude of the friction force if:
a) the ball remains at rest, **b)** the ball moves at a constant speed.

What information about the friction force can be deduced if:
c) the ball is slowing down, **d)** the ball is gaining speed?

*8 The coefficient of friction between a set of skis and the snow is 0.21. A skier stands on horizontal snow wearing these skis.
a) The skier has a mass of 71 kg, find the normal reaction force on the skier.
b) What forward force must be applied to the skier if he is to begin to move?

*9 Tests carried out on a car of mass 1200 kg by police officers suggest that the friction force experienced when the car skids varies between 10 200 N and 6830 N, depending on the road conditions. What is the corresponding range of values for the coefficient of friction?

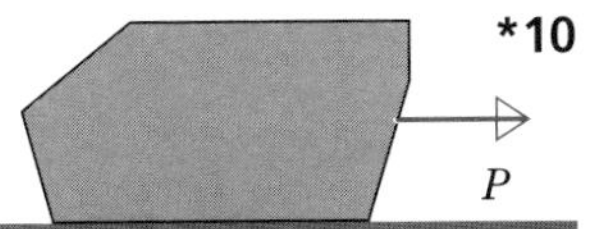

*10 The diagram shows a force, P, acting on a skip of mass 500 kg. The friction between the skip and the ground is $0.8R$ where R is the normal reaction on the skip.
a) Find the magnitude of the friction when the skip is moving.
b) Find the magnitude of the friction if the skip is at rest.

*11 A car of mass 1000 kg is parked on a horizontal road with the hand brake on. (***Note:*** The hand brake only applies to the rear wheels.) The coefficient of friction between the tyres and the road is 0.8.
a) Assuming that the normal reaction forces on all the wheels are equal, calculate the magnitude of each force.
b) An attempt is made to move the car by applying a horizontal forward force. How big must this force be if the back wheels are to slide?
c) In fact the car will move with the back wheels sliding when a horizontal force of 800 N is applied. Calculate the reaction forces on the front and rear wheels of the car.

TENSION AND THRUST

Exploration 2.3

Stretching springs

- Think of some examples where a spring is stretched or compressed. What happens in these situations? What forces are exerted?
- What happens if you try to stretch a rod or a string? Are there similarities with the spring?

Stretching and compressing a spring

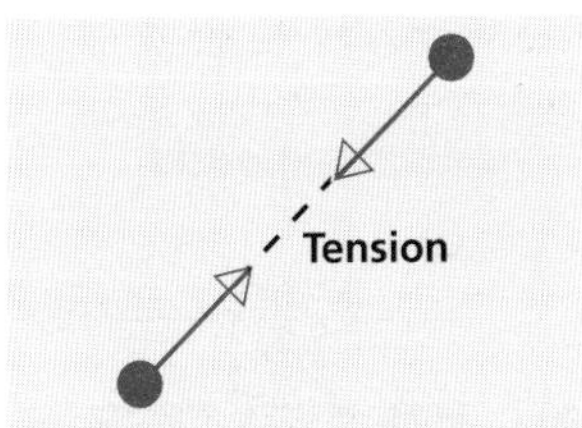

When a spring is **stretched** it exerts forces on the objects to which it is attached. The forces act along the spring towards its centre and have the same magnitude. These forces are known as **tensions**. When any other object is pulled, for example a normal string or a rod, there are also tension forces present in the same way.

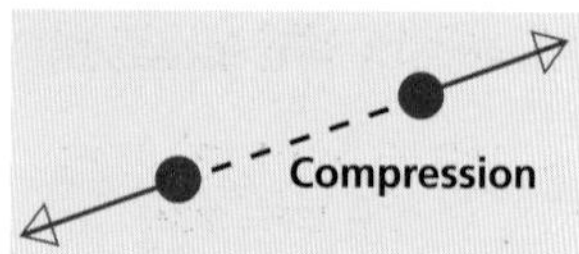

When a spring is **compressed** it exerts forces that act away from its centre. These forces are known as **thrusts**. A rod can be in compression and exerts a thrust in the same way that a spring does. However, it is impossible for a string to exert a thrust. Why?

Example 2.11

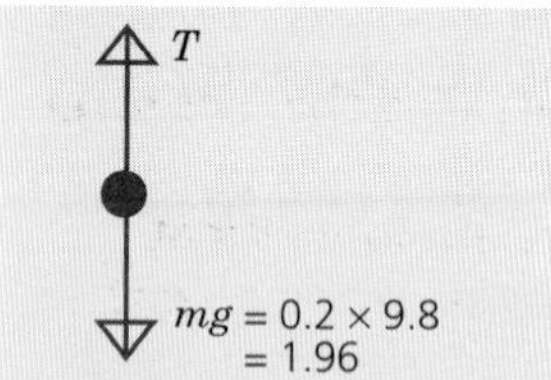

A spring stretches when it supports a mass of 200 grams. Find the tension in this spring.

Solution

The diagram shows the forces acting on the mass.
Gravity acts down and the tension, T, in the spring upwards.
The tension, T, balances the force of gravity so the tension in the spring must be 1.96 N.

Example 2.12

Three identical objects are suspended using a spring, an inelastic string (which does not stretch) and a metal rod.

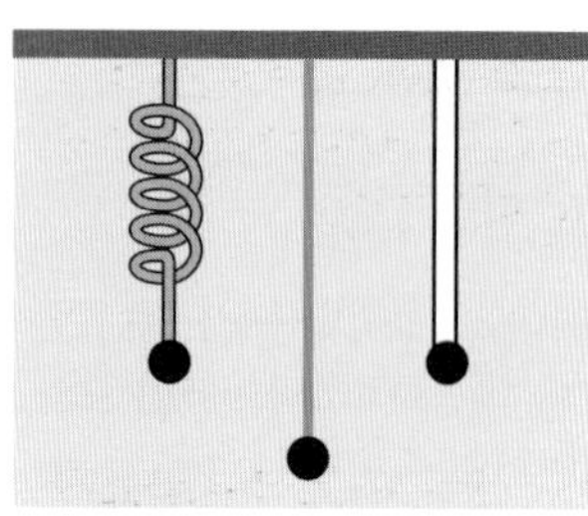

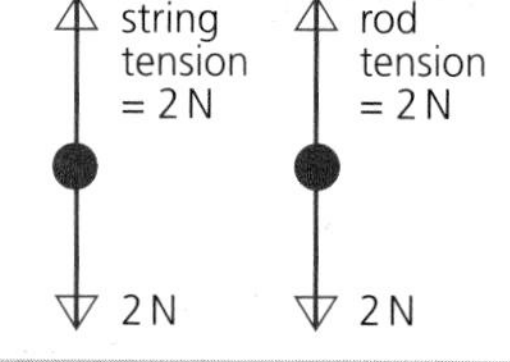

The spring stretches so that it exerts a force of 2 N.

a) *Find the mass of the objects.*

b) *What forces act in the other two cases?*

Solution

a) *The tension in the spring balances the force of gravity, so:*

$$T = mg \Rightarrow 2 = m \times 9.8 \Rightarrow m = \frac{2}{9.8} = 0.204 \text{ kg}$$

b) *In each case, the force of gravity on the objects is 2 N so the force of gravity, 2 N, must be balanced by an upward force of 2 N, as illustrated. In each of the other two cases the force of gravity, 2 N, must be balanced by an upward force of 2 N, as illustrated. Both the rod and the string also exert tension forces with the same magnitude and direction as the spring.*

Example 2.13

The diagram shows two inelastic strings supporting a bar of mass 100 grams, that in turn supports a spring and a further mass of 200 grams. Find the tension in the strings and the spring.

Solution

The diagram shows the forces acting on the 200 gram mass. The force T_1 acts upwards to balance the force of gravity and so has magnitude 1.96 N.

The diagram shows the forces acting on the bar. The downward forces exerted by gravity and the spring have a total magnitude of 2.94 N, which must be balanced by an upward force of 2.94 N. This is exerted by the two strings, so they must each have a tension of 1.47 N.

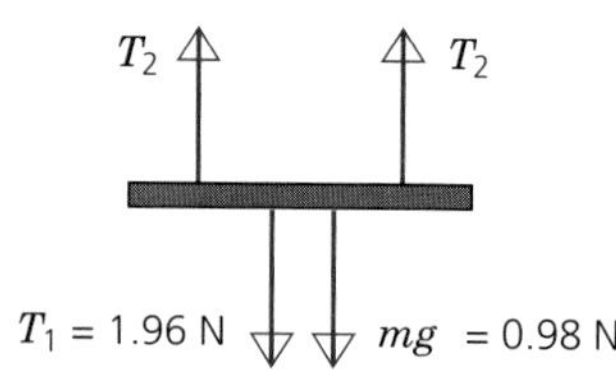

EXERCISES

2.4A

1 The pointer on an angler's spring balance moves when a fish of mass 1.5 kg is hung on it. Find the tension in the spring in the balance.

2 A 1 kg bag of sugar causes the spring in a set of kitchen scales to be compressed. Find the tension in this spring.

3 A cuckoo clock is controlled by a 50 gram mass that vibrates on the end of a spring.

a) When the clock stops the mass remains at rest. Find the tension in the spring.

b) The mass moves up and down from its central position. What happens to the tension in the spring as it moves up and down?

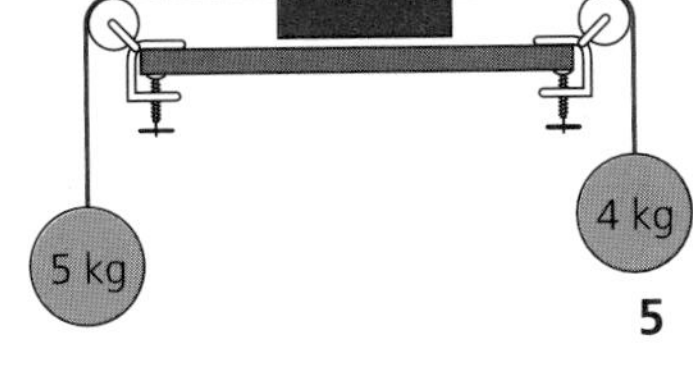

4 The diagram shows a system that is at rest.

a) Find the tension in each string.

b) Draw a diagram to show the forces acting on the block.

c) What is the magnitude of the friction between the block and the plane?

5 The diagram shows a block of mass 500 grams attached to a spring. The friction between the block and the plane is $0.7R$, where R is the normal reaction, when the block is on the point of sliding.

a) What is the magnitude of the normal reaction force on the block?

b) What is the magnitude of the friction force if the block is on the point of sliding?

c) What is the tension in the spring?

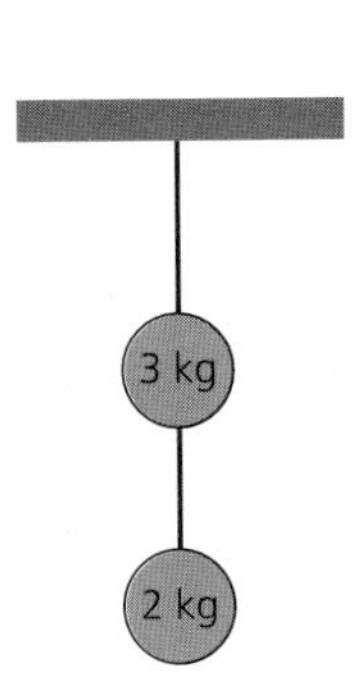

6 Consider the system of masses and strings, which is at rest.

a) Draw a diagram to show the forces acting on the 3 kg mass.

b) Find the tension in each string, assuming that the system is on Earth.

c) What would be the tension in the strings if the system were on the moon?

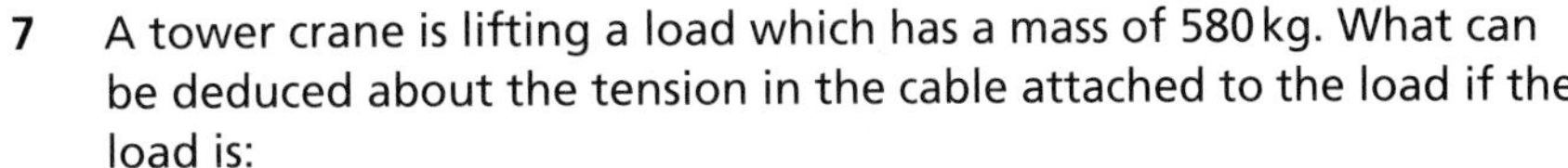

7 A tower crane is lifting a load which has a mass of 580 kg. What can be deduced about the tension in the cable attached to the load if the load is:

a) at rest, **b)** rising at a constant 4 m s^{-1},

c) gaining speed as it rises?

8 A child of mass 25 kg sits on a swing in the park. There is no motion.

a) Estimate the tension in each supporting rope.

b) What assumptions have you made to obtain this answer? How do these effect your answer?

9 Two people are using a rope and a pulley to lift the engine out of a car. If the mass of the engine is 250 kg, find the force that they must exert on the rope, to hold the engine in any one position.

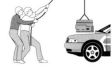

10 For each system shown, calculate the tension in each rope if the system is at rest.

a) 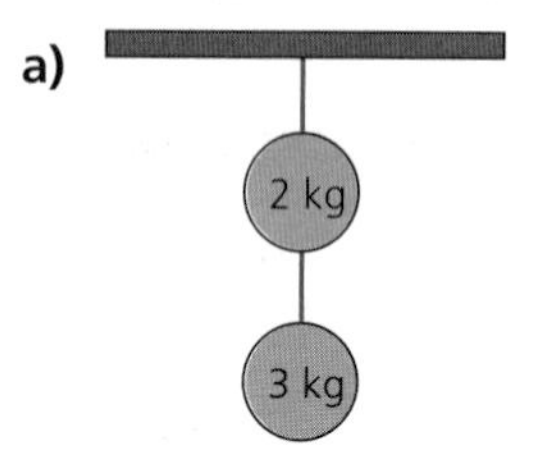

b) 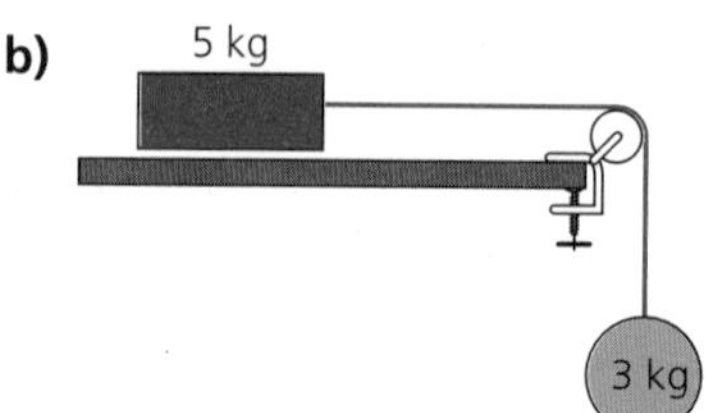

c) 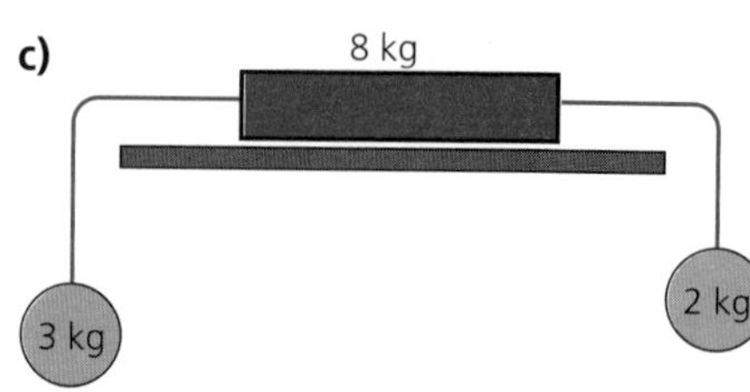

Find the friction between the block and the surface in **b)** and **c)**.

EXERCISES

2.4B

1 The spring in a spring balance stretches when supporting a load of mass 10 kg. What is the tension in the spring?

2 A spring stretches when supporting a load.
a) What is the tension for a load mass of 9 kg?
b) What is the mass of the load if the tension is 100 N?

3 The tension in a spring varies between 8 N and 12 N as the mass on the end moves up and down.
a) What would you expect the tension in the spring to be if the object were at rest?
b) Use you answer to **a)** to estimate the mass of the object.

4 The diagram shows a system that is at rest. The spring shown is stretched. Find the tension in the spring if:
a) there is no friction,
b) there is a friction force of 18 N between the block and the plane.

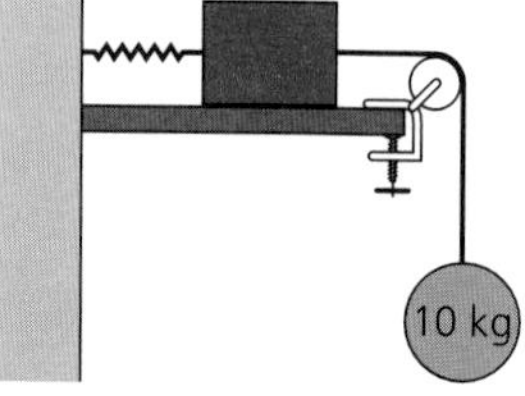

5 A dinghy of mass 280 kg is being hoisted out of the water. Find the tension in the rope if it is rising at a constant speed.

6 The diagram shows a system of masses and strings that is at rest.
a) Draw diagrams to show the forces acting on each mass.
b) Find the tension in each string.
c) If the strings were replaced by springs, what would be the tension in each spring?

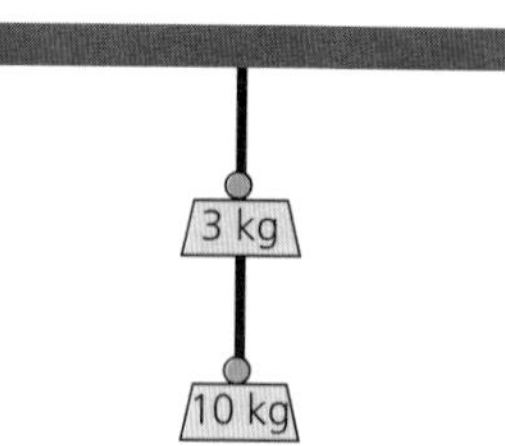

7 For each system shown, find the tension in the string if there is no movement. Find the friction present between the block and the plane in each case.

a) **b)**

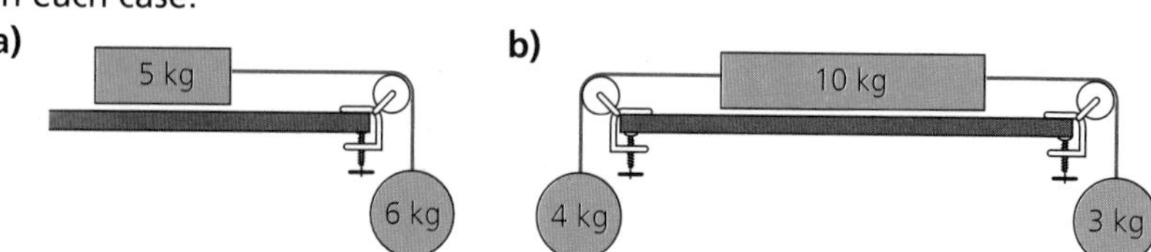

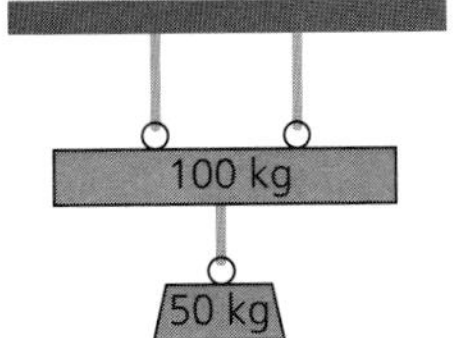

8 The diagram shows a steel beam of mass 100 kg that supports a load of mass 50 kg. Find the tensions in each cable, stating clearly any assumptions that you make.

9 The tension, T, in a spring is given by $T = 50e$, where e is the extension measured in metres. Three identical springs and three masses are arranged as shown.

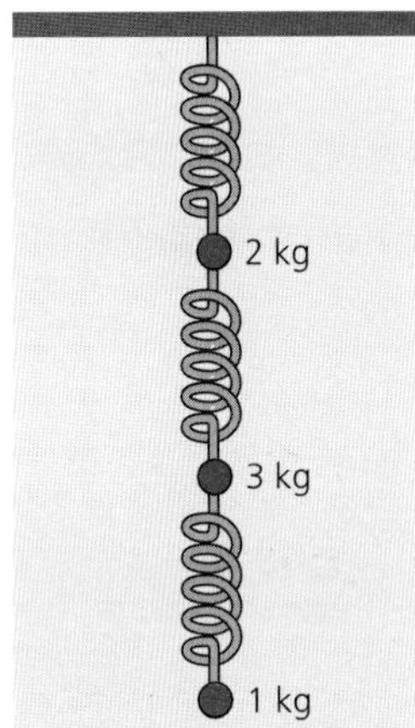

a) Find the tension in each spring.
b) Find the extension of each spring.

10 Draw diagrams to show the forces acting on each object that is underlined below.

a) A <u>pendulum bob</u> as it swings.

b) A <u>sign</u> supported by two cables.

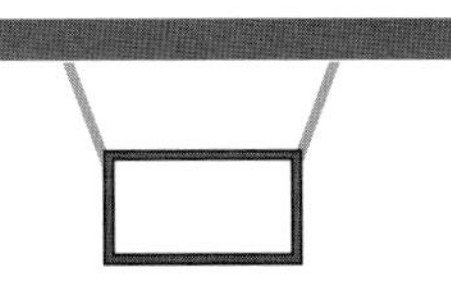

c) A <u>concrete slab</u> being lifted.

d) A <u>sledge</u> pulled as shown.

OTHER FORCES

In this chapter we have discussed the main types of force that we meet in this mechanics book. However there are a few others that we need to mention and these are described below.

Air resistance

This always acts in opposite direction to any motion. It increases as the speed increases.

Upthrust or buoying

If an object is placed in a fluid then there will be an upward force due to the displacement of the fluid.

Electromagnetic forces

These are found in electric motors, for example. They are beyond the scope of this book.

Aerodynamic forces

An example of this type of force is the lift force on an aeroplane or a hydrofoil.

Note: The first and most crucial step in solving any mechanics problem is to identify the forces that are acting on the object under consideration.

Example 2.14

A ball is dropped into a pond. It falls to a lowest point and then rises up out of the water. Draw diagrams to show the forces acting on the ball at X where it is going down, Y at the lowest point and Z on the way back up.

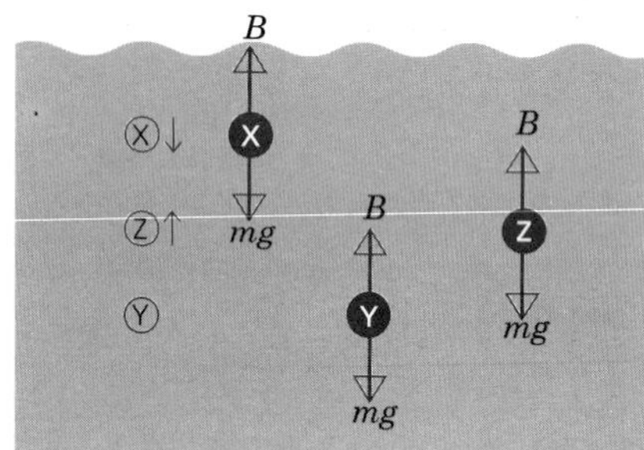

Solution

At X, there is gravity, but also an upward buoyancy force, B.
As the ball must be losing speed, $B > mg$.
At the lowest point, Y, the same two forces act. As the motion is changing from downwards to upwards again $B > mg$.
At Z the same two forces act. As the ball is gaining speed as it rises, $B > mg$.

Example 2.15

Draw a diagram to show the forces acting on the aeroplane. What can be deduced about these forces if:

a) *the aeroplane is travelling at a constant speed,*
b) *the aeroplane is rising at a constant speed,*
c) *the aeroplane is gaining speed, in horizontal flight?*

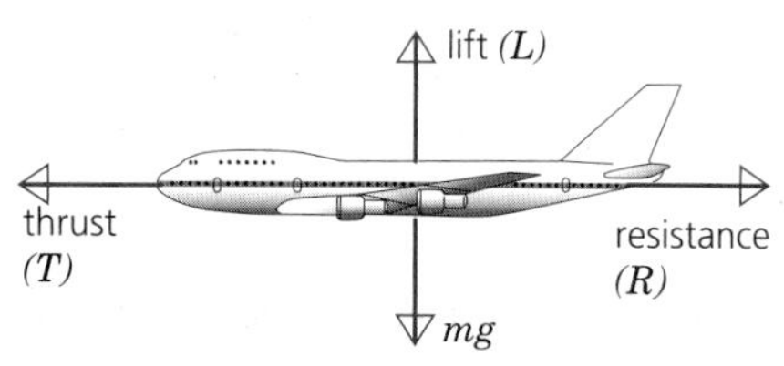

Solution

The diagram shows the forces present.

a) *The motion of the aeroplane does not change and so all the forces must be in balance giving $R = T$ and $L = mg$.*
b) *Here again the motion is not changing, so $R = T$ and $L = mg$.*
c) *As the aeroplane is gaining speed there must be an overall forward force, so $T > R$, but $L = mg$.*

CONSOLIDATION EXERCISES FOR CHAPTER 2

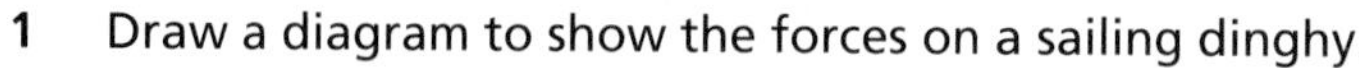

1 Draw a diagram to show the forces on a sailing dinghy.

2 A bottle is floating in a tank of water.

a) If the bottle is at rest show the forces acting on the bottle. What can be stated about the magnitude of each force?
b) The bottle is pushed down so that it bobs up and down in the water. Draw diagrams to show the forces on the bottle when it is at its highest and lowest positions.
c) Are the forces on the moving bottle ever in balance? If so when?

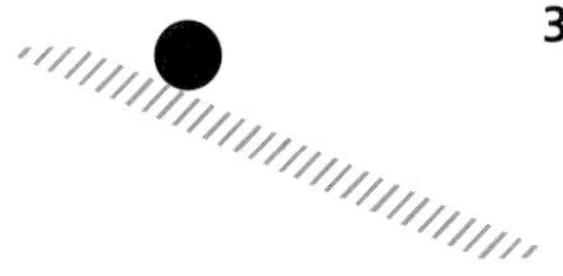
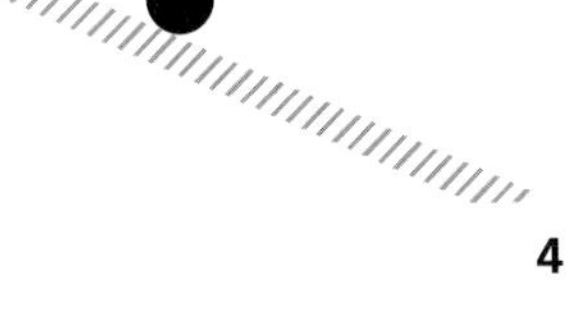

3 A marble is placed on a slope and once released it rolls down the slope.
  a) Draw a diagram to show the forces acting on the marble.
  b) What do you think the effect of the friction is on the motion of the marble?

4 A box is placed on a conveyor belt. Draw diagrams to show the forces acting if the box is:
  a) at rest,
  b) moving at constant speed,
  c) gaining speed,
  d) losing speed.

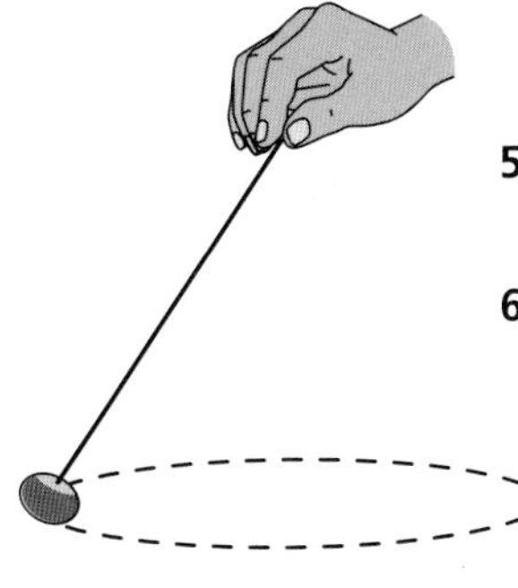

5 A conker on the end of a length of string is swinging in a circle. Show the forces that act on the conker. Is it possible for these forces to be in balance?

6 During a game of tennis the ball is hit by the racquet. Draw a diagram to show the forces on the ball:
  a) before it is hit,
  b) while in contact with the racquet,
  c) after it has been hit.

7 A golf ball is at rest on the ground when it is hit by a club. Draw diagrams to show the forces acting on the ball when it is:
  a) at rest on the ground,
  b) in contact with the club on the ground,
  c) in contact with the club off the ground,
  d) moving but not in contact with the club or the ground.

POSSIBLE MODELLING TASKS

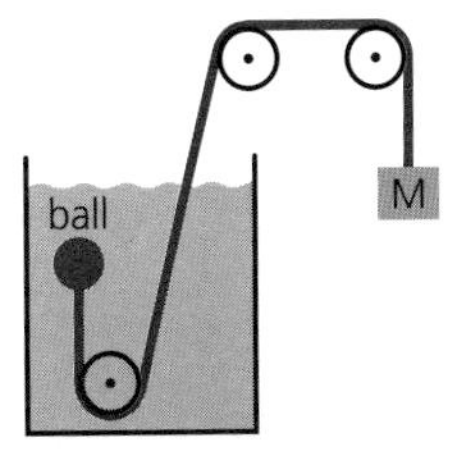

1 Devise and conduct an experiment to discover how the tension in a spring or string depends on its extension.

2 The diagram (left) shows an experiment that can be set up to explore how the buoyant force on an object varies with depth. Conduct your own experiment using this or another approach to find a relationship between the buoyant force and depth.

Summary

After working through this chapter you should:

- be able to identify different types of forces and draw force diagrams
- be able to calculate the force of gravity on an object on Earth
- be aware of the contact forces: friction and normal reactions
- be aware of the tension or compression that can be found in a spring or rod
- be able to determine whether or not forces are in balance by examining the motion of the object on which they act.

MECHANICS

3

Vectors and forces

- *It is easy to see if horizontal or vertical forces are in balance, this idea can be extended to any forces.*
- *In many real problems, solutions can be obtained because all the forces are in equilibrium or balance.*

INTRODUCTION

Many of the quantities that we discuss in this book will be defined in terms of a size and a direction. A force has a size or magnitude, given in newtons, and a direction in which it acts. For example a force may have magnitude 80 N and act straight down. Quantities which have a magnitude and a direction are known as **vectors**.

Exploration 3.1

Recognising vectors

Think about each of the following quantities.

distance force speed mass velocity weight

- If they *are* vectors give a simple example of each one.
- If they *are not* vectors, explain why not.

Exploration 3.2

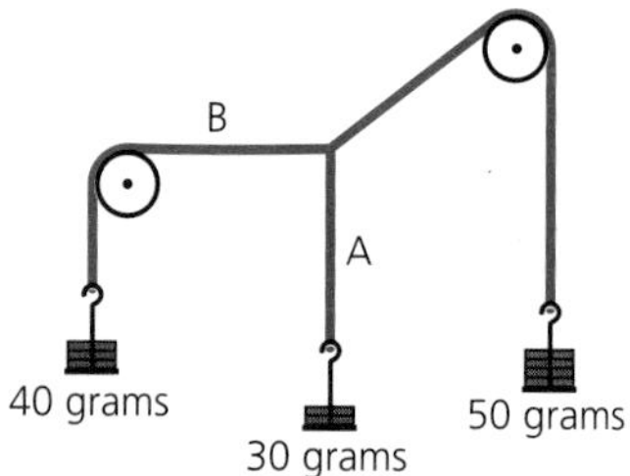

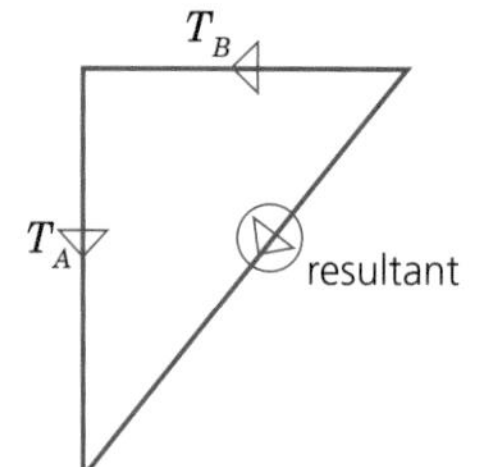

Force and vectors

Using pulleys and strings set up the system shown.

- Move one of the masses and check that the system returns to the position shown, when you release it.
- Verify that string A is vertical and string B is horizontal. Swap the 30 and 40 gram masses. What happens?
- Use three masses of 130 grams, 120 grams and 50 grams. Show that it is again possible to have one string vertical and one horizontal.
- Now use three different masses, suspending one at the end of each string. Decrease or increase the masses at the ends of the strings until one string is vertical and another horizontal.
  Find the tension in each string, by considering the forces acting where the masses are attached, and draw a diagram like the one shown. The horizontal and vertical force are equivalent to a single resultant. Find the size and direction of this resultant. How does it compare with the tension in the third string?
- Repeat the last step for another set of masses.

EXPRESSING FORCES AS VECTOR

Exploration 3.2 suggests that any force can be exp resultant of a horizontal and vertical force.

In Chapter 2 we saw how easy it is to determine whether or n are in balance by looking at the horizontal and vertical forces separately. A more formal approach involves the use of **unit vectors i** and **j** that are perpendicular to each other, in the same two-dimensional plane. It is usual to take the vector **i** as horizontal, directed to the right. A force of 4**i** is a force of 4 N that acts horizontally to the right, and a force of –2**i** is a force of 2 N that acts horizontally to the left.

The vector **j** is usually assumed to act in a vertical direction, either up or down.

The diagram on the right shows the case when **j** is upwards. The vector 3**j** represents an upward force of 3 N, while –2**j** represents a downward force of 2 N.

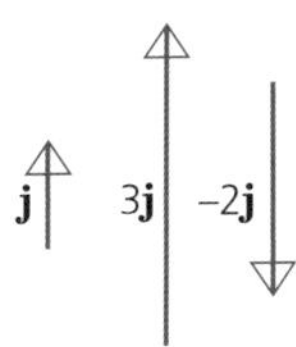

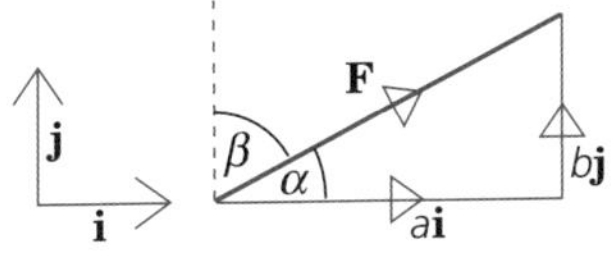

Forces that act in other directions can be expressed in the form $a\mathbf{i} + b\mathbf{j}$. The numbers a and b are known as the horizontal and vertical **components** of the force. If the force has magnitude F, then $a = F\cos\alpha$ and $b = F\cos\beta$ where α is the angle between the force and the **i** direction and β is the angle between the force and the **j** direction.

Example 3.1

*A force of magnitude 6 N acts at 30° to the horizontal as shown. Express this force in the form $a\mathbf{i} + b\mathbf{j}$, where **i** and **j** are horizontal and vertical unit vectors.*

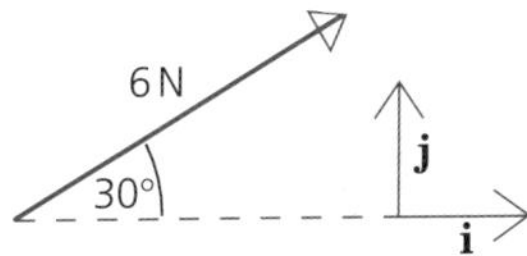

Solution

*Note that the angle between the force and the unit vector **i** is 30°, and the angle between the force and the unit vector **j** is 60°.*

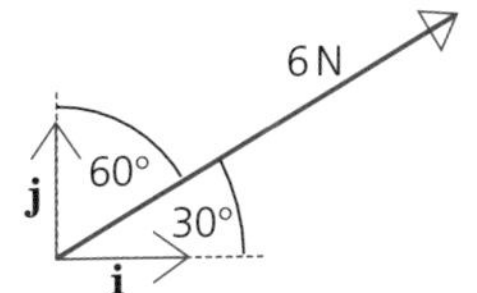

So the horizontal component is $6\cos30°$ or 5.20 N and the vertical component is $6\cos60°$ or 3.00 N.

Now the force can be expressed as:

$6\cos30°\mathbf{i} + 6\cos60°\mathbf{j} = 5.20\mathbf{i} + 3.00\mathbf{j}$

The equivalence of these two forces to the original can be seen by considering the diagram on the right.

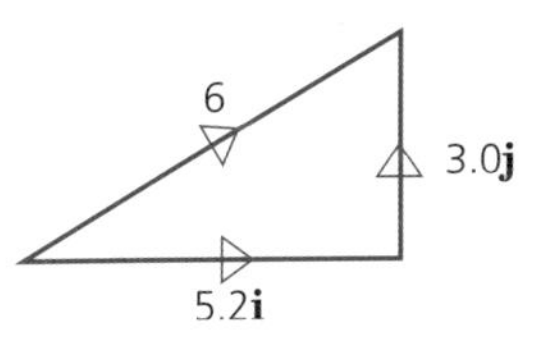

Example 3.2

*Three forces act at a point as shown in the diagram. Express each force in terms of the unit vectors **i** and **j**. Also find the resultant or combined effect of these forces.*

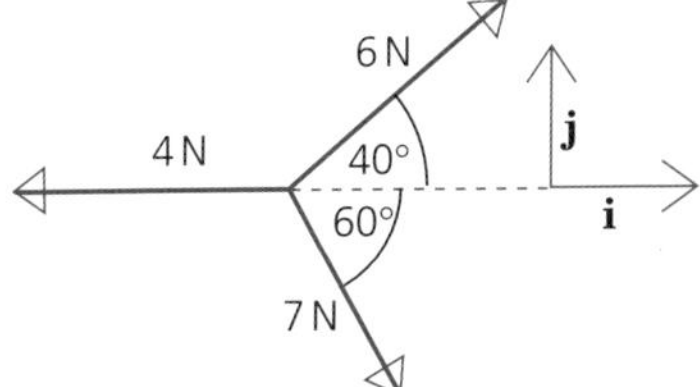

Solution

Considering each force in turn, gives the results in the table below.

Force	Vector form
4 N	$-4\mathbf{i}$
6 N	$6\cos 40°\mathbf{i} + 6\cos 50°\mathbf{j} = 4.60\mathbf{i} + 3.86\mathbf{j}$
7 N	$7\cos 60°\mathbf{i} + 7\cos 150°\mathbf{j}$ or $7\cos 60°\mathbf{i} - 7\cos 30°\mathbf{j} = 3.50\mathbf{i} - 6.06\mathbf{j}$

The resultant force, $\mathbf{F}$, is the sum of all the individual forces.

$$\mathbf{F} = (-4\mathbf{i}) + (4.60\mathbf{i} + 3.86\mathbf{j}) + (3.50\mathbf{i} - 6.06\mathbf{j})$$
$$= (-4 + 4.60 + 3.50)\mathbf{i} + (3.86 - 6.06)\mathbf{j}$$
$$= 4.10\mathbf{i} - 2.20\mathbf{j}$$

The force is illustrated in the diagram. The magnitude of $\mathbf{F}$ is:

$$F = \sqrt{4.10^2 + 2.20^2} = 4.65\ N$$

The angle θ to the $\mathbf{i}$ direction is:

$$\theta = \tan^{-1}\left(\frac{2.20}{4.10}\right) = 28.2°$$

Example 3.3

The forces shown in the diagram act at a point. Find the magnitude and direction of the force $\mathbf{F}$, if the resultant force is zero.

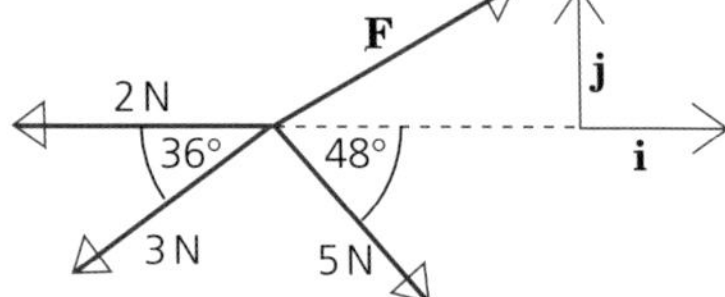

Solution

Assume that the force $\mathbf{F}$ is $a\mathbf{i} + b\mathbf{j}$. Each force is expressed in terms of the unit vectors $\mathbf{i}$ and $\mathbf{j}$ in the following table.

Force	Vector form
$\mathbf{F}$	$a\mathbf{i} + b\mathbf{j}$
2 N	$-2\mathbf{i}$
3 N	$-3\cos 36°\mathbf{i} - 3\cos 54°\mathbf{j} = -2.43\mathbf{i} - 1.76\mathbf{j}$
5 N	$5\cos 48°\mathbf{i} - 5\cos 52°\mathbf{j} = 3.35\mathbf{i} - 3.72\mathbf{j}$

The resultant force is:

$$(a\mathbf{i} + b\mathbf{j}) + (-2\mathbf{i}) + (-2.43\mathbf{i} - 1.76\mathbf{j}) + (3.35\mathbf{i} - 3.72\mathbf{j})$$
$$= (a - 2 - 2.43 + 3.35)\mathbf{i} + (b - 1.76 - 3.72)\mathbf{j}$$
$$= (a - 1.08)\mathbf{i} + (b - 5.48)\mathbf{j}$$

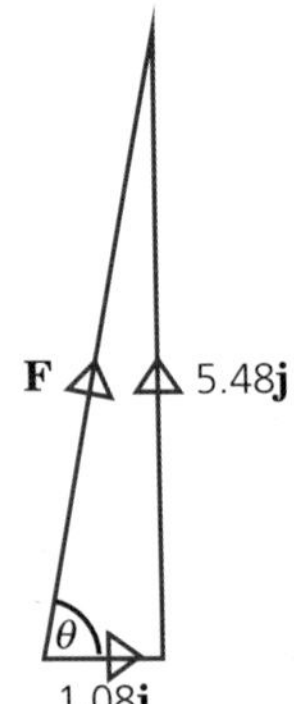

For the resultant to be zero, $a = 1.08$ and $b = 5.48$, so the force $\mathbf{F}$ can be expressed as:

$$\mathbf{F} = 1.08\mathbf{i} + 5.48\mathbf{j}$$

This is shown in the diagram on the left.

The magnitude can be found using Pythagoras' theorem.

$F = \sqrt{1.08^2 + 5.48^2} = 5.59\ N$

The angle θ can be found using trigonometry.

$$\tan\theta = \frac{5.48}{1.08} \Rightarrow \theta = \tan^{-1}\frac{5.48}{1.08} = 78.9°$$

Thus the force **F** *has magnitude 5.59 N and acts at 78.9° above the unit vector* **i**.

Vector form and algebraic notation

The force **F** shown in the diagram can be expressed in vector form:

$$\mathbf{F} = F\cos\alpha\mathbf{i} + F\cos(90° - \alpha)\mathbf{j}$$

As $\cos(90° - \alpha) = \sin\alpha$ this can be expressed as:

$$\mathbf{F} = F\cos\alpha\mathbf{i} + F\sin\alpha\mathbf{j}$$

Note that $F\cos\alpha$ is often referred to as the **horizontal component** and $F\cos(90° - \alpha)$ or $F\sin\alpha$ as the **vertical component** of the force.

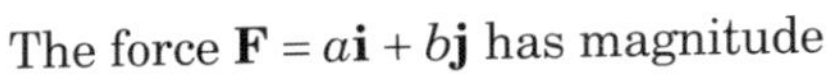

The force $\mathbf{F} = a\mathbf{i} + b\mathbf{j}$ has magnitude:

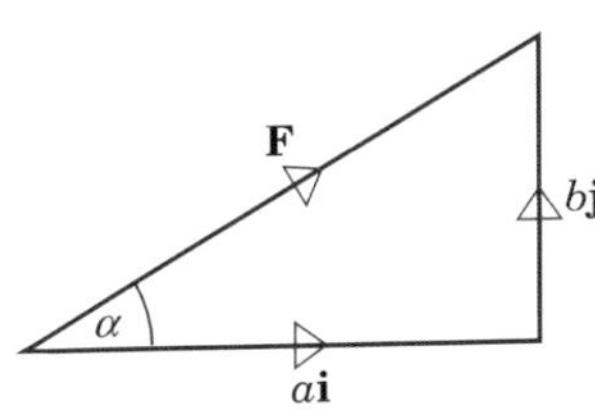

$$F = \sqrt{a^2 + b^2}$$

and direction:

$$\alpha = \tan^{-1}\frac{b}{a}$$

Column vectors are sometimes used as an alternative notation. The force $\mathbf{F} = a\mathbf{i} + b\mathbf{j}$ would be written as $\begin{pmatrix} a \\ b \end{pmatrix}$ in column vector form.

The force F shown would be expressed as $\mathbf{F} = \begin{pmatrix} F\cos\alpha \\ F\cos(90° - \alpha) \end{pmatrix}$.

EXERCISES

3.1A

1 Express each force illustrated below in terms of the unit vectors **i** and **j**.

a) 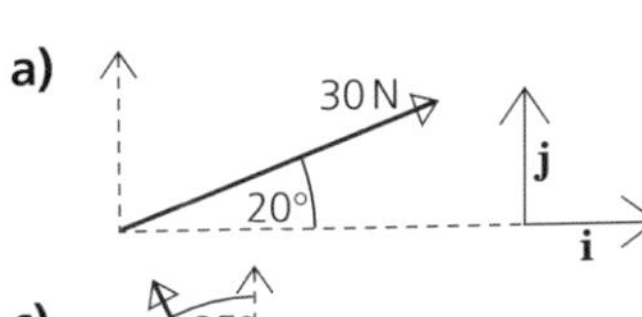

b) 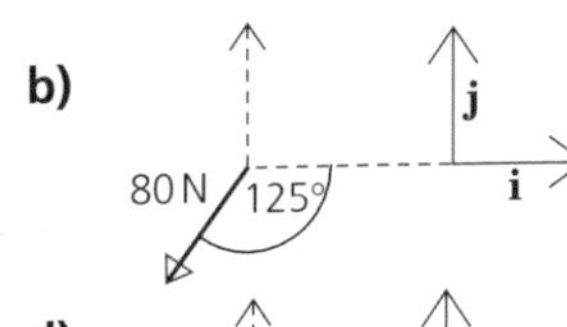

c) 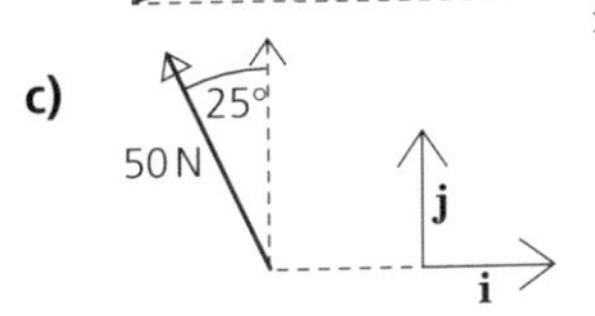

d) 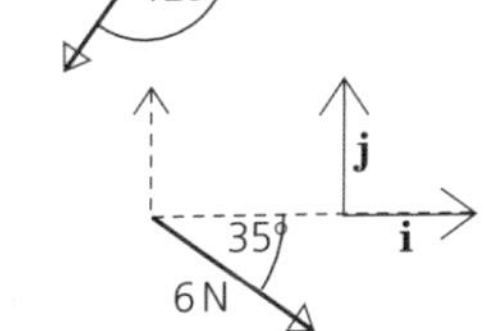

2 Express each force shown below as a column vector.

a) 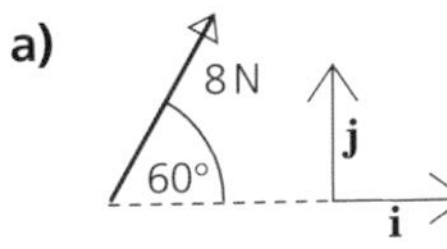

b) 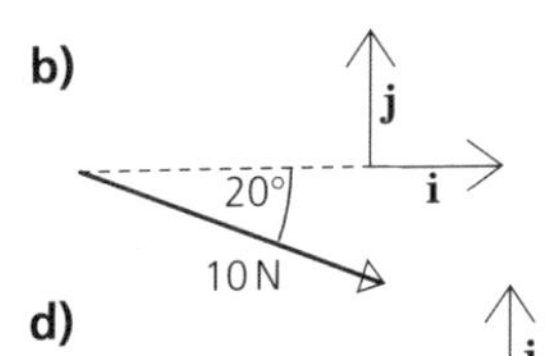

c) 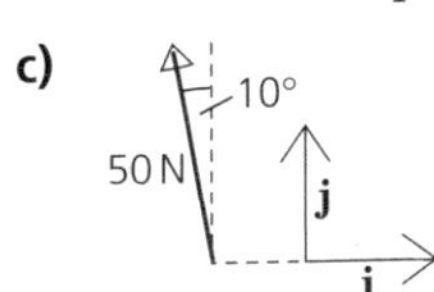

d) 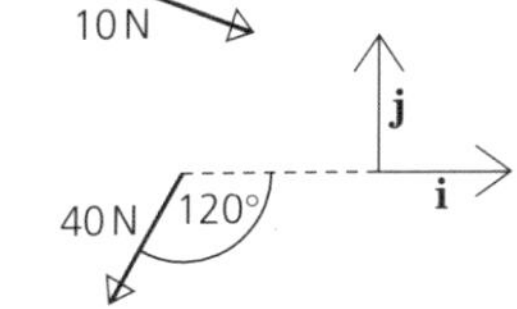

3 For each set of forces shown below, find in the resultant force in terms of the unit vectors **i** and **j**. Find also the magnitude and direction of the resultant force.

a) 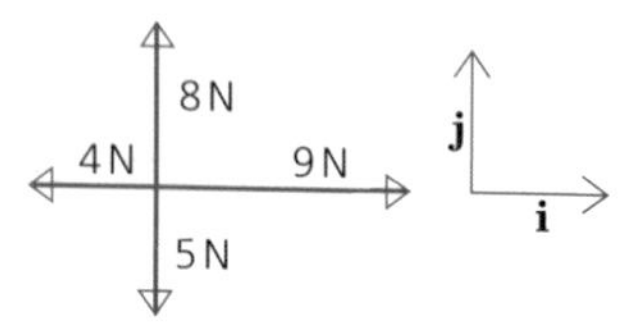

b) 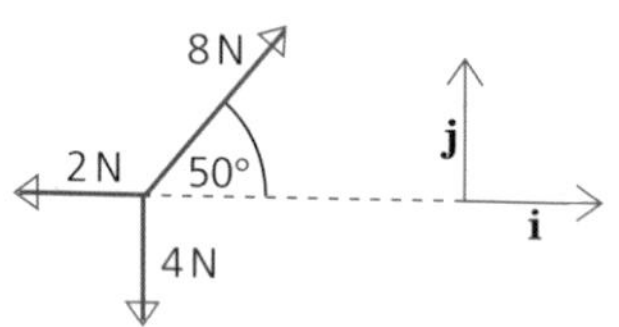

c) 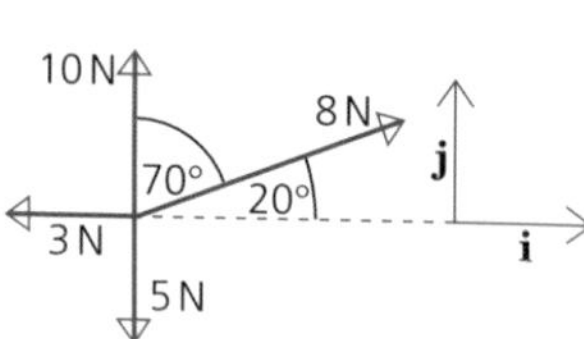

d) 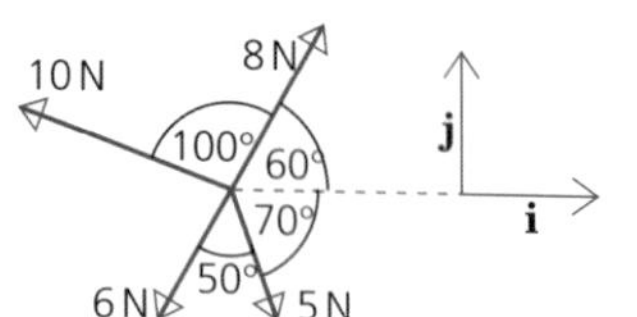

e) 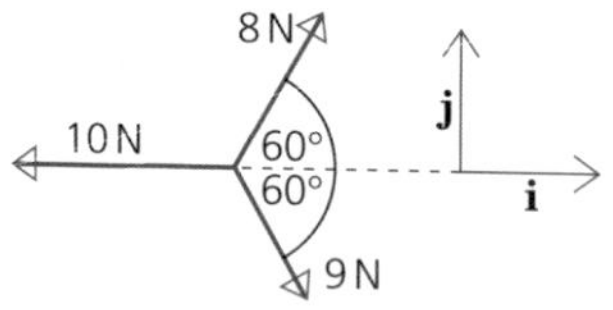

f) 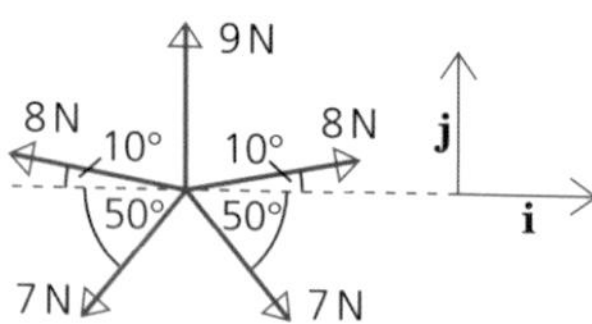

4 Find the magnitude of each force below and the angle it makes with the positive **i** direction.

a) $10\mathbf{i} + 12\mathbf{j}$ b) $16\mathbf{i} + 32\mathbf{j}$ c) $11\mathbf{i} + 8\mathbf{j}$
d) $22\mathbf{i} - 100\mathbf{j}$ e) $-16\mathbf{i} - 12\mathbf{j}$ f) $-3\mathbf{i} + 4\mathbf{j}$

5 Find the magnitude of each force below. Describe the direction in which the force acts.

a) $\begin{pmatrix} 3 \\ 2 \end{pmatrix}$ b) $\begin{pmatrix} -6 \\ 7 \end{pmatrix}$ c) $\begin{pmatrix} -4 \\ -2 \end{pmatrix}$ d) $\begin{pmatrix} -2 \\ 1 \end{pmatrix}$

6 Express each force illustrated in terms of perpendicular unit vectors **i** and **j**.

a) 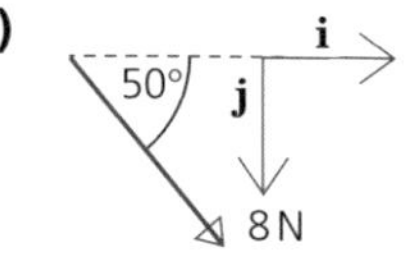

b) 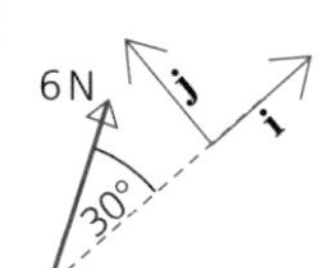

c) 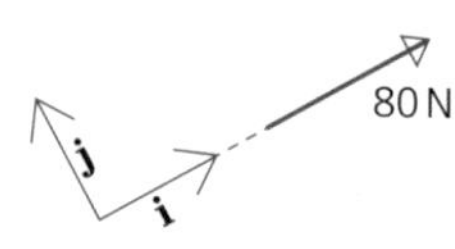

d 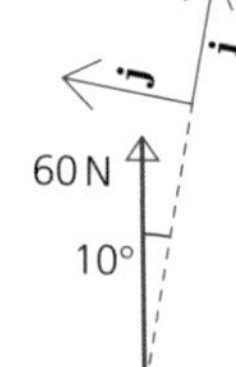

7 Find the resultant force in terms of the unit vectors **i** and **j**.

a) 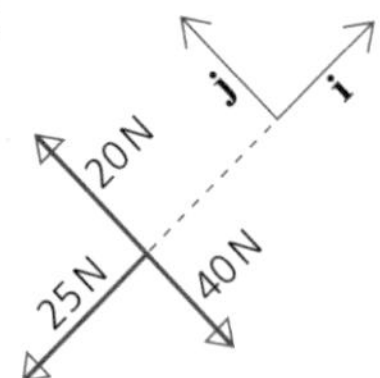

b) 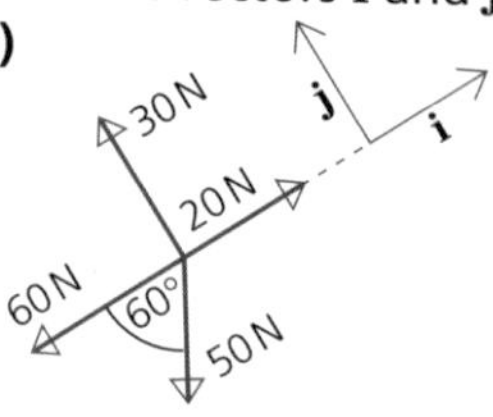

EXERCISES

3.1B

1 For each force shown below, express it as a vector in terms of **i** and **j**.

a) 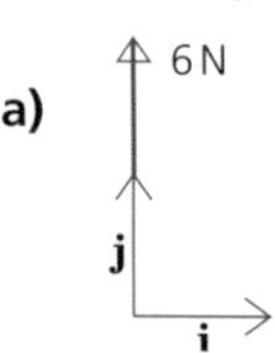

b) 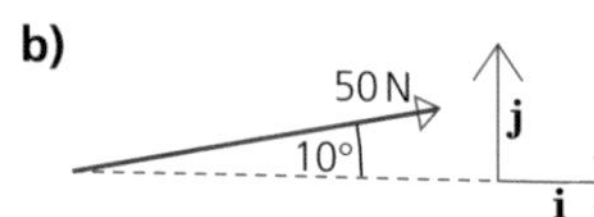

c) 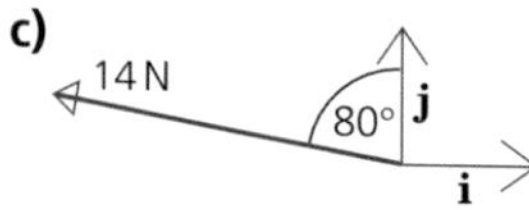

d)

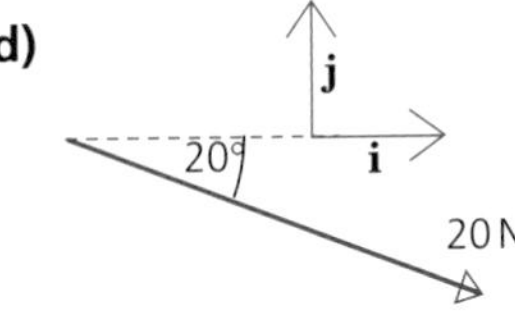

e)

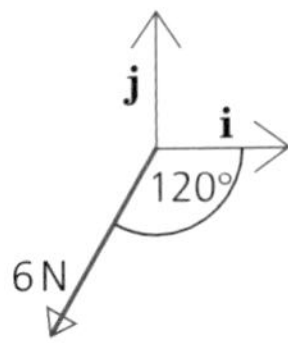

f)

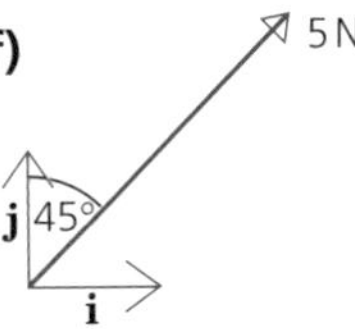

2 In each example below, find the resultant force in terms of the unit vectors $\mathbf{i}$ and $\mathbf{j}$.

a)

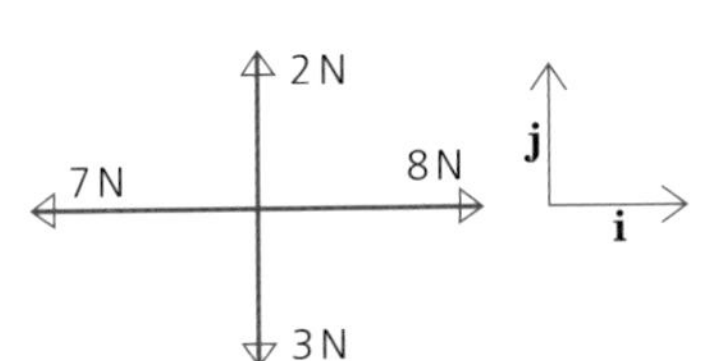

b)

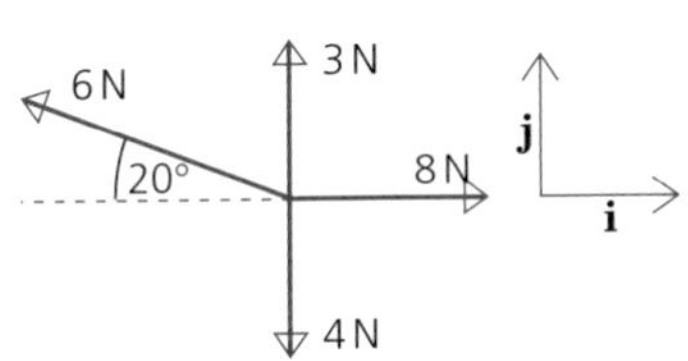

c)

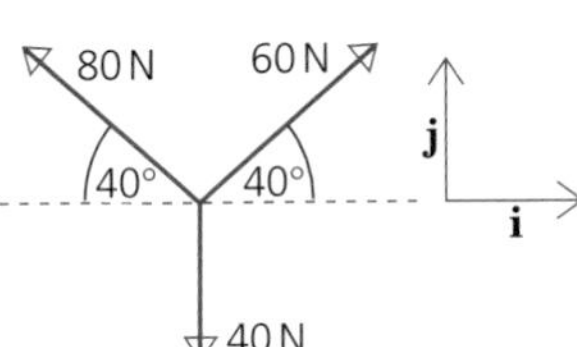

d)

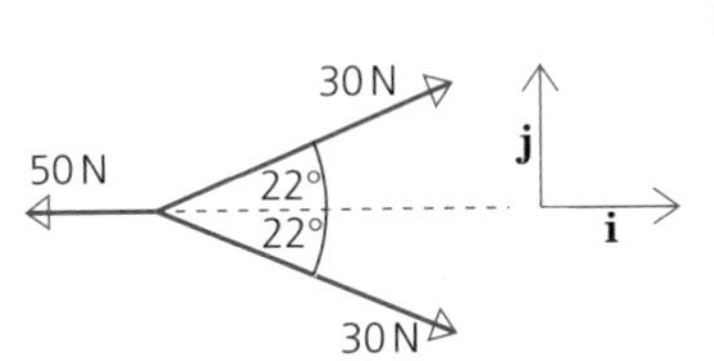

e)

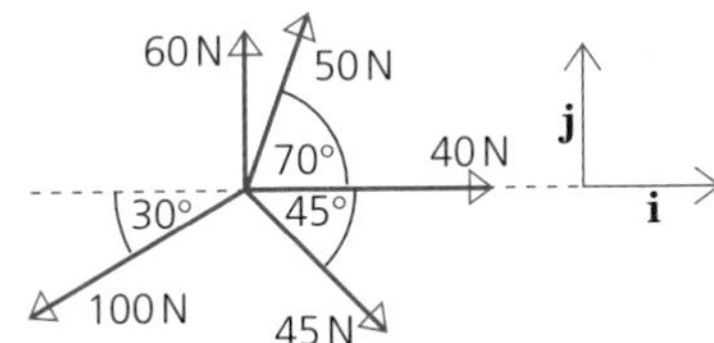

3 Express each force below as a column vector.

a)

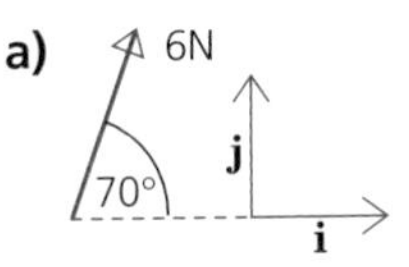

b)

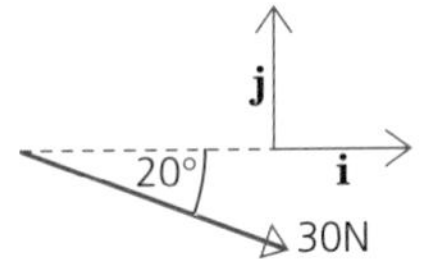

c)

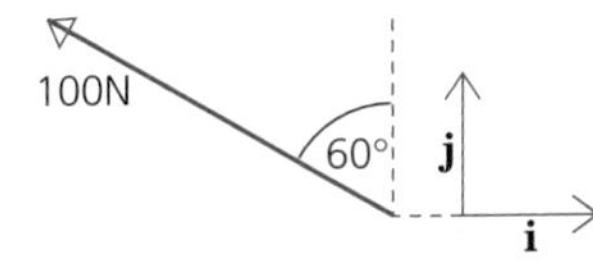

d)

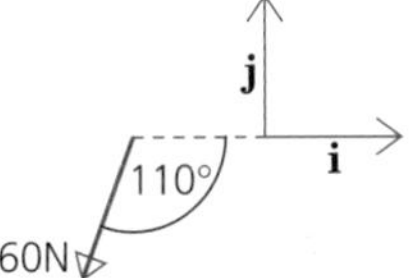

4 Find the magnitude and direction of each force below. In each case draw a diagram, indicating the angle that you specify.

a) $3\mathbf{i} + 10\mathbf{j}$ **b)** $60\mathbf{i} - 110\mathbf{j}$ **c)** $-84\mathbf{i} - 30\mathbf{j}$
d) $-40\mathbf{i} + 20\mathbf{j}$ **e)** $6\mathbf{i} - 8\mathbf{j}$ **f)** $61\mathbf{i} + 12\mathbf{j}$

5 For each force below, find the magnitude and illustrate the direction clearly on a diagram, specifying one angle.

a) $\begin{pmatrix} 7 \\ 4 \end{pmatrix}$ **b)** $\begin{pmatrix} 6 \\ -2 \end{pmatrix}$ **c)** $\begin{pmatrix} 10 \\ -2 \end{pmatrix}$ **d)** $\begin{pmatrix} -6 \\ 8 \end{pmatrix}$

6 Express each force illustrated below in terms of the perpendicular unit vectors $\mathbf{i}$ and $\mathbf{j}$.

a)

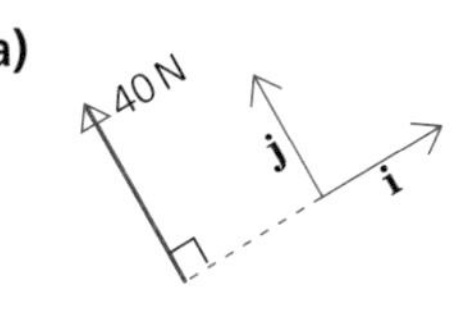

b)

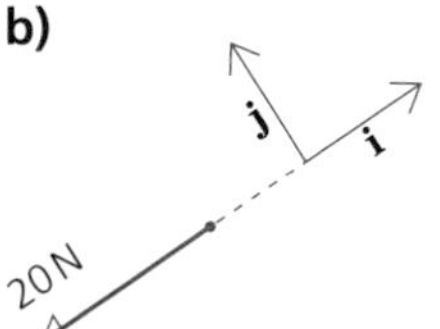

c)

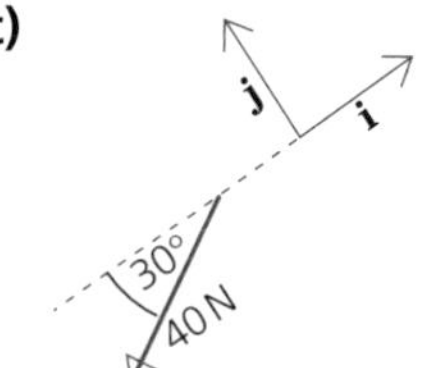

d)

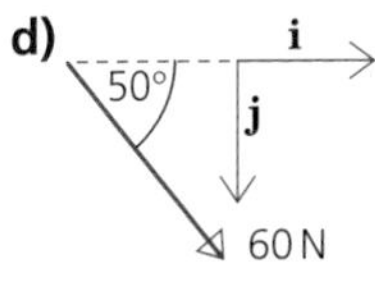

7 For each set of forces, find the resultant in terms of $\mathbf{i}$ and $\mathbf{j}$. Give the magnitude of the resultant and the angle it makes with the $\mathbf{i}$ direction.

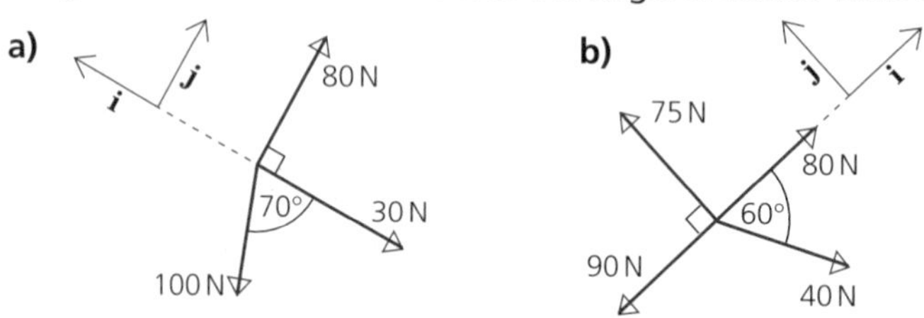

FORCES AND EQUILIBRIUM

From Newton's first law, if the resultant force on an object is zero, then the object will either remain at rest or move with a constant speed in a straight line. In our everyday life, we see many real situations where the resultant force is zero. We can now look more closely at problems involving this phenomenon, using the ideas of vectors.

Example 3.4

A car of mass 700 kg is travelling up a slope making an angle of 4° above the horizontal at a constant speed. Calculate the forward force that acts on the car, assuming that there are no resistance forces present. Comment on your result.

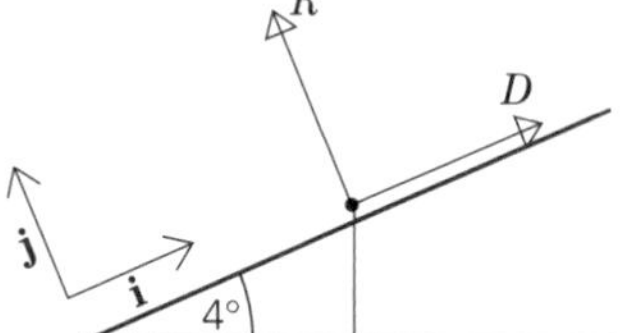

Solution

If the car is modelled as a particle which experiences no resistance, then there are three forces acting, namely gravity, the normal reaction, R, and a forward force, D.

When considering objects on a slope it is often helpful to use unit vectors that are parallel and perpendicular to the slope. In this case $\mathbf{i}$ and $\mathbf{j}$ are taken as shown in the diagram. In problems like this, it is normal to talk about component of forces perpendicular and parallel to the slope rather than horizontal or vertical.

Now each force can be expressed in vector form.

Force	Vector form
R	$R\mathbf{j}$
D	$D\mathbf{i}$
mg	$-6860\cos 86°\mathbf{i} - 6860\cos 4°\mathbf{j}$

The resultant force is:
$(D - 6860\cos 86°)\mathbf{i} - (R - 6860\cos 4°)\mathbf{j}$
As the car moves at a constant speed in a straight line the forces must be in equilibrium, so:

$$D - 6860\cos 86° = 0 \quad \textit{and} \quad R - 6860\cos 4° = 0$$

This gives $R = 6843$ N and $D = 479$ N. So the car must exert a forward force of 479 N if it is to move up the slope. As in reality the car will experience a resistance force, the magnitude of D must actually be greater than 479 N.

Example 3.5

A hopper of cement is lifted by a crane and then pulled sideways until it is in the position shown. If the cables attached to the hopper are positioned as in the diagram and the mass of the hopper and its contents are 500 kg, find the tension in each cable.

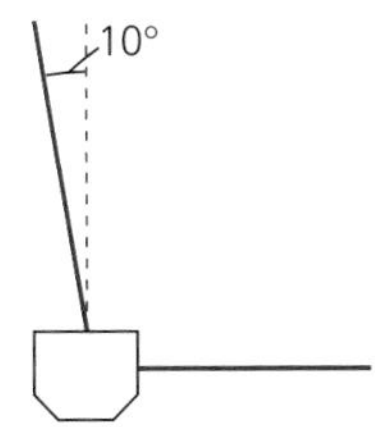

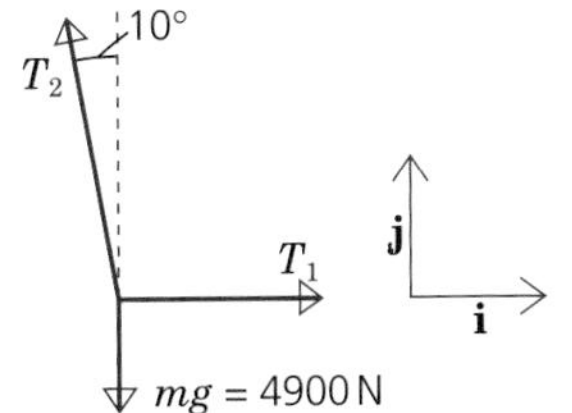

Solution

The forces acting on the hopper are the two tensions and the force of gravity, as shown in the diagram, where the hopper is modelled as a particle.

The resultant force on the hopper is:

$$(T_1 - T_2\cos 80°)\mathbf{i} + (T_2\cos 10° - 4900)\mathbf{j}$$

For equilibrium the resultant force is zero, so:

$$T_1 - T_2\cos 80° = 0 \qquad (1)$$

and

$$T_2\cos 80° - 4900 = 0 \qquad (2)$$

From equation (2):

$$T_2 = \frac{4900}{\cos 10°} = 4976\ N$$

Substituting this value back into equation (1) gives:

$$T_1 = 4976\cos 80° = 864\ N$$

Example 3.6

A break-down truck is being used to pull a skip of mass 2 tonnes over some rough ground. The coefficient of friction between the skip and the ground is 0.6. Find the tension in the rope, if the skip is on the point of sliding.

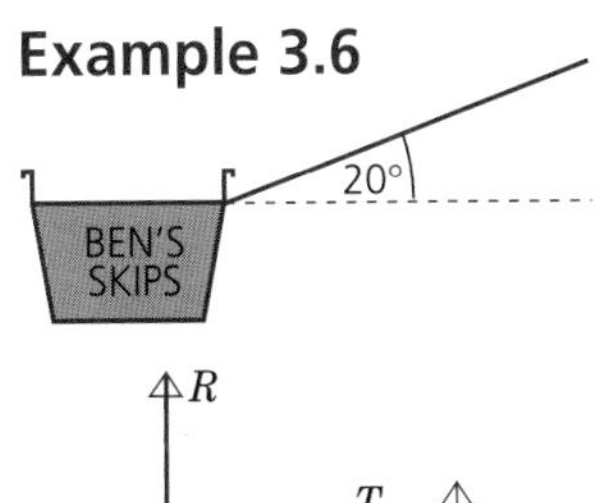

Solution

If the skip is modelled as a particle, then there are four forces acting, the force of gravity, $m\mathbf{g}$, a normal reaction, R, friction, F, and the tension on the rope, T.

Each force can be expressed in vector form as below.

Force	Vector form
R	$R\mathbf{j}$
F	$-F\mathbf{i}$
T	$T\cos 20°\mathbf{i} + T\cos 70°\mathbf{j}$
mg	$-19\,600\mathbf{j}$

The resultant force is:
$(T\cos 20° - F)\mathbf{i} + (R + T\cos 70° - 19\,600)\mathbf{j}$

As the skip is on the point of sliding the forces acting must be in equilibrium, so both the horizontal and vertical components of the resultant force must be zero:

$T\cos 20° - F = 0$ *(1)*
and
$R + T\cos 70° - 19\,600 = 0$ *(2)*

Also because the skip is on the point of sliding, F *takes its maximum value of* μR *or 0.6*R *in this case. So equation (1) becomes:*

$T\cos 20° - 0.6R = 0$
or

$$R = \frac{T\cos 20°}{0.6}$$

Substituting this value of R *in equation (2) to eliminate* R *gives:*

$$\frac{T\cos 20°}{0.6} + T\cos 70° - 19\,600 = 0$$

Now this can be solved for T.

$$T\left(\frac{\cos 20°}{0.6} + \cos 70°\right) = 19\,600$$

$$T = \frac{19\,600}{\dfrac{\cos 20°}{0.6} + \cos 70°} = 10\,272\ N$$

Example 3.7

A block is placed on an inclined plane at an angle α *to the horizontal. The coefficient of friction between the block and the plane is* μ. *Show that if the block is on the point of sliding:*

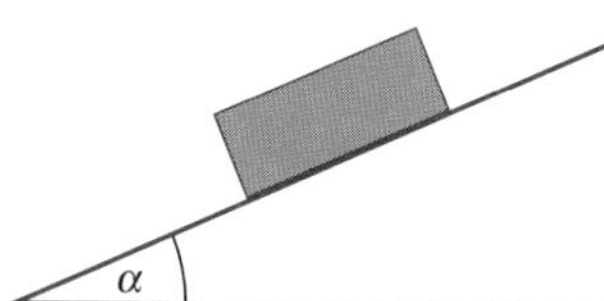

$\tan\alpha = \mu$

Solution

The diagram shows the forces acting on the block, which has been modelled as a particle.

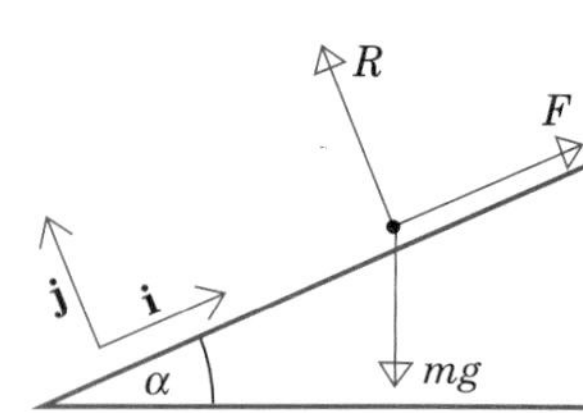

Using the unit vectors as shown, the resultant force is:
$(F - mg\sin\alpha)\mathbf{i} + (R - mg\cos\alpha)\mathbf{j}$

As the block is at rest the resultant must be zero, so:

$F - mg\sin\alpha = 0$ *(1)*
and
$R - mg\cos\alpha = 0$ *(2)*

As the block is on the point of sliding, $F = \mu R$, *so equation (1) becomes:*
$\mu R - mg\sin\alpha = 0$

Using equation (2) to eliminate R gives: $\mu mg\cos\alpha - mg\sin\alpha = 0$
so that:

$$\mu mg\cos\alpha = mg\sin\alpha$$

$$\mu = \frac{\sin\alpha}{\cos\alpha} = \tan\alpha$$

*This angle α is known as the **angle of friction**.*

Example 3.8

The diagram shows two cables that support a sign of mass 3 kg.

a) Draw a diagram to show the forces acting on the sign.
b) The resultant of these three forces is zero. Draw a diagram to illustrate this.
c) Use the sine rule to find the tension in each string.

Solution

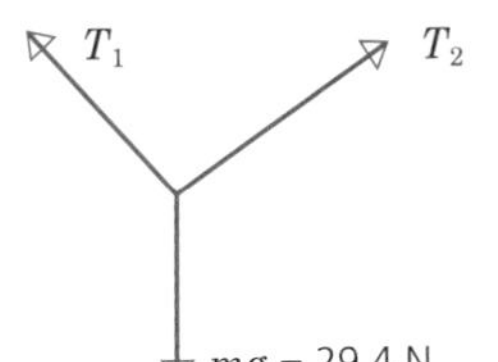

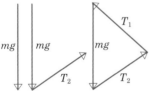

a) The diagram shows the three forces acting, the force of gravity and the two tensions.
b) Because the resultant force is zero the three vectors representing the forces must combine to form a triangle. We begin by selecting one force, for example the force of gravity. This can be represented by a vector that points straight down. Then we add the tension T_2, beginning at the end of the gravity vector. Finally add T_1, beginning at the end of T_2, to complete the triangle.
c) Note that T_2 makes an angle of 30° with the horizontal and so is at 60° to the vertical. Also T_1 is at 40° to the horizontal and so is at 50° to the vertical. So the third angle in the triangle is 180° – (50° + 60°) = 70°.

Applying the sine rule gives:

$$\frac{29.4}{\sin 70^\circ} = \frac{T_1}{\sin 60^\circ} = \frac{T_2}{\sin 50^\circ}$$

so

$$T_1 = \frac{29.4\sin 60^\circ}{\sin 70^\circ} = 27\ N$$

and

$$T_2 = \frac{29.4\sin 50^\circ}{\sin 70^\circ} = 24\ N$$

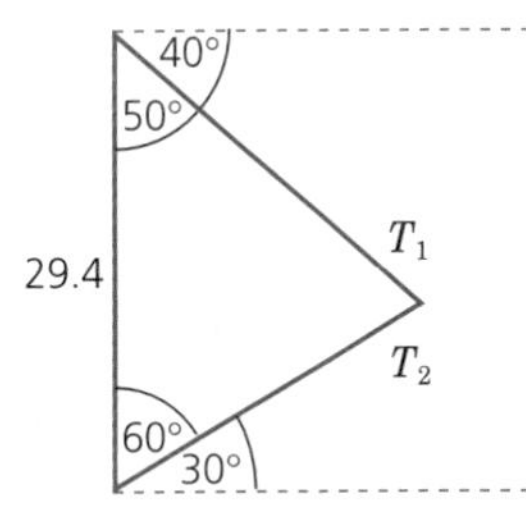

Three forces in equilibrium

When only three forces act on a body which is in equilibrium it can often be easier to use the approach of Example 3.8. This is because the forces will always form a triangle, and then the sine rule can be used.

If three forces act as shown, then this approach can be used to demonstrate that:

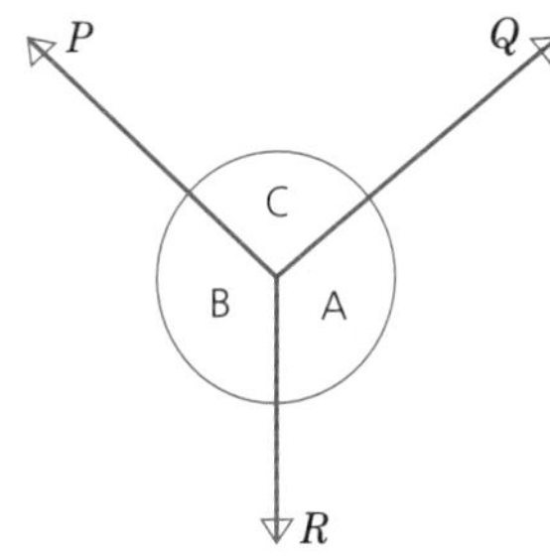

$$\frac{P}{\sin A} = \frac{Q}{\sin B} = \frac{R}{\sin C}$$

This result is known as **Lami's theorem**.

EXERCISES

3.2A

*Questions marked * involve the coefficient of friction.*

1 The four forces shown act on the top of a telegraph pole.

a) Express the resultant force acting on the top of the pole, in terms of the unit vectors $\mathbf{i}$ and $\mathbf{j}$.

b) Find, in terms of $\mathbf{i}$ and $\mathbf{j}$, the additional force that must also act if the forces are to be in equilibrium.

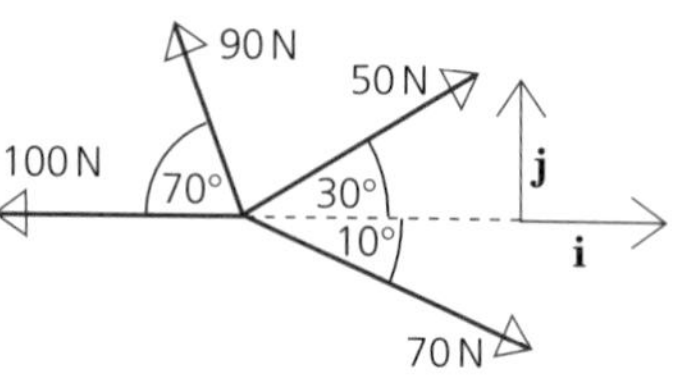

c) Calculate the magnitude of this force and draw a diagram to show the direction in which it acts.

2 The diagram shows two cables and a light rod that support a mass of 50 kg.

a) State the tension in cable 2.

b) Write each of the forces that act at the point A, in terms of the unit vectors $\mathbf{i}$ and $\mathbf{j}$.

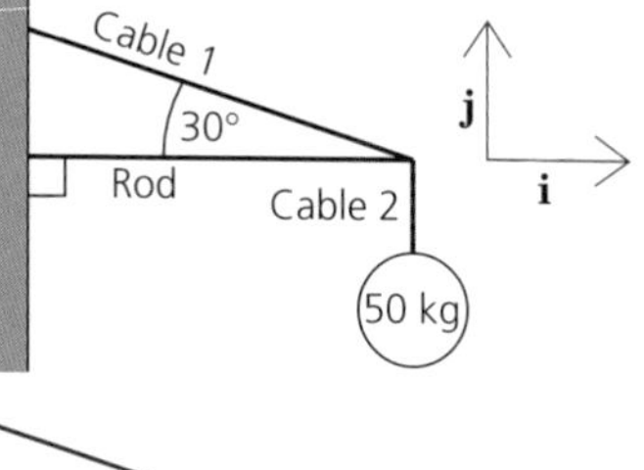

c) Calculate the tension in cable 1 and the thrust in the rod.

d) A third cable is attached to the mass and pulled horizontally so that cable 2 makes an angle of 10° to the vertical as shown. Calculate the new tensions in cables 1 and 2. What happens to the force exerted by the rod?

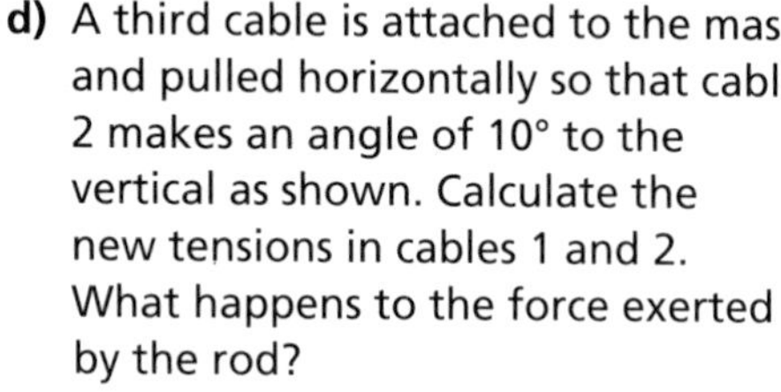

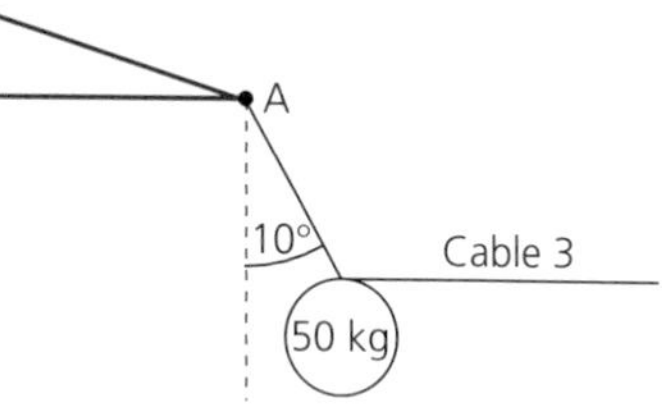

3 The diagram shows a heavy object suspended at rest by two strings.

a) Draw a diagram to show the forces acting on the mass.

b) Express each force in terms of the unit vectors $\mathbf{i}$ and $\mathbf{j}$, which act horizontally and vertically respectively.

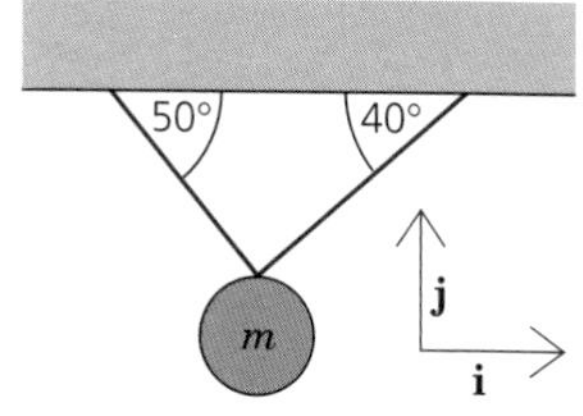

c) If the tension in the left hand string is 45 N, find the tension in the other string.

d) Find the mass of the object.

4 A child of mass 40 kg is trying to use a plastic tray as a sledge.

a) The child sits on the tray on a slope at 10° to the horizontal. Model the child and tray as a particle. Draw a diagram to show the forces acting on them.

b) Express each force in terms of unit vectors $\mathbf{i}$ and $\mathbf{j}$, which are parallel and perpendicular to the slope respectively. Then find the resultant force.

c) If the child remains at rest find the magnitude of each force.

d) The child tries a steeper slope at 30° to the horizontal, and finds that she gains speed as she slides. State what you can deduce about the magnitude of the friction force.

5 **a)** Draw a diagram to show the forces acting on a car of mass 1000 kg as it is driven along a straight horizontal road.

b) What is the relationship between the forward forces and the resistance forces when the car is travelling at top speed?

c) The maximum forward force that can be produced by the car is 2925 N. The resistance forces on the cars are given by $R = v^2 + 20v$, where v is the speed of the car in m s^{-1}. Find the top speed for the car.

d) Show that, when travelling down a slope inclined at 5° to the horizontal, the car experiences a forward force parallel to the slope of 854 N. What is the maximum speed of the car if it begins to freewheel from rest down the slope?

6 Three cables exert forces that act in a horizontal plane as shown, on the top of a telegraph pole.

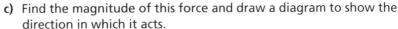

a) Find the resultant of these three forces, in terms of the unit vectors **i** and **j**.

b) A fourth force acts so that the forces on the pole are in equilibrium. Express this force in terms of **i** and **j**.

c) Find the magnitude of this force and draw a diagram to show the direction in which it acts.

d) The fourth cable does not lie in the same horizontal plane as the other cables and the tension in it is in fact 420 N. Find the angle between this cable and the horizontal.

e) Is it likely that the three original forces all act in the horizontal plane?

***7** **a)** Find the tension in the rope shown in the diagram, if there is assumed to be no friction on the slope.

b) The coefficient of friction between the block and the plane is 0.4. Find the tension in the rope, if the block is on the point of sliding.

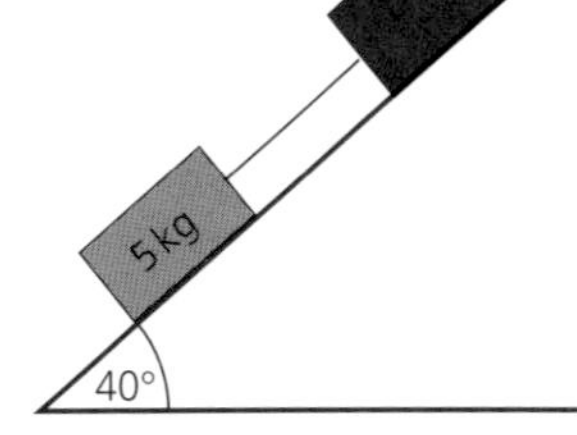

8 The diagram shows two smooth balls inside a smooth cylinder.

Draw diagrams to show:

a) the forces on the large ball,

b) the forces on the small ball.

The radius of the small ball is 4 cm and its mass is 100 grams. The radius of the large ball is 6 cm and its mass is 200 grams. If the radius of the cylinder is 7 cm, find the magnitude of the reaction forces acting on each ball.

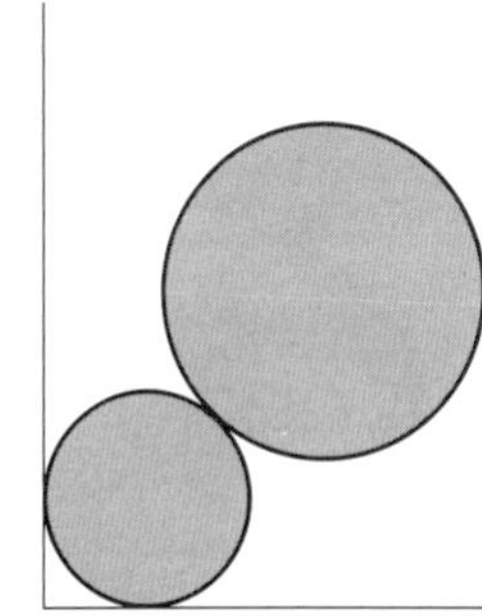

9 a) Two cables support an object of mass 50 kg as shown in part (a) of the diagram. Find the tension in each cable.

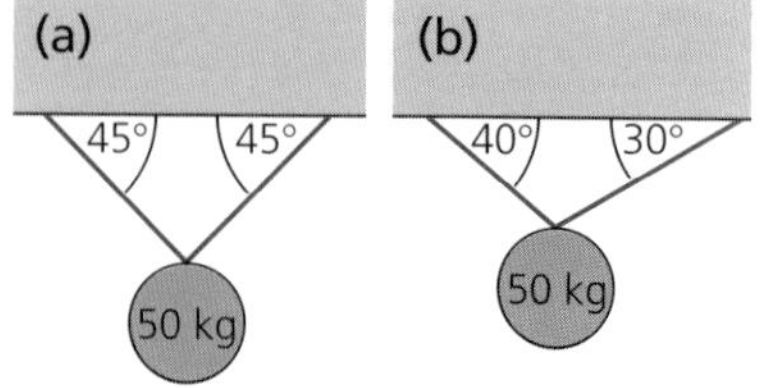

b) Repeat (a) for the second system, shown in part (b) of the diagram.

10 A car of mass 1.2 tonnes is parked on a slope that makes an angle of 8° to the horizontal.

a) Draw a diagram to show the forces acting on the car.
b) Using appropriate unit vectors, express each force in component form and find the resultant.
c) What is the magnitude of the friction forces acting on the car?

EXERCISES

3.2B

*Questions marked * involve the coefficient of friction.*

1 The forces shown in the diagram act on a ship that is being pulled by two tugs at a constant speed.

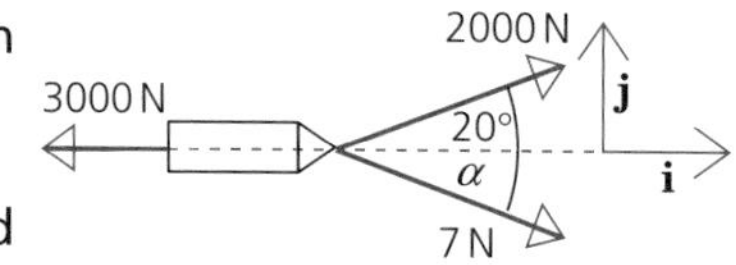

a) Find the resultant of the 3000 N and 2000 N forces, in terms of the unit vectors **i** and **j**.
b) Find the other force marked, in terms of **i** and **j**, if all three are in equilibrium.
c) Find the magnitude of the third force and the size of the angle, α

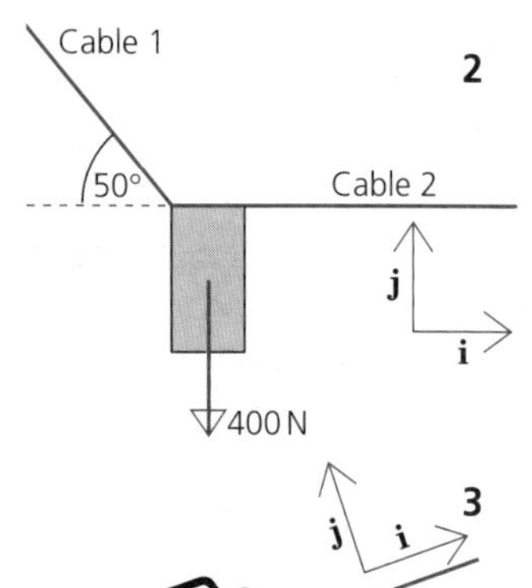

2 The diagram shows a crate that is being lowered into position with two cables. In the position shown the crate is at rest.

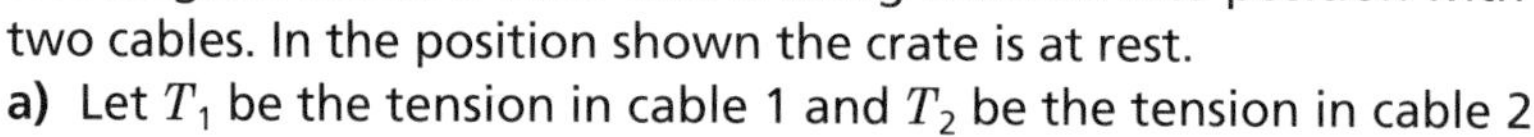

a) Let T_1 be the tension in cable 1 and T_2 be the tension in cable 2. Write down the resultant force in terms of T_1, T_2, **i** and **j**.
b) Use the vertical component of the resultant to find the tension in cable 1.
c) Use the horizontal component of the resultant to find the tension in cable 2.

3 A car of mass 950 kg is parked on a hill as shown in the diagram.

a) Draw a diagram to show the forces acting on the car, if it is modelled as a particle.
b) Express the force of gravity, in terms of the unit vectors **i** and **j**.
c) Find the magnitude of the friction and normal reaction forces.

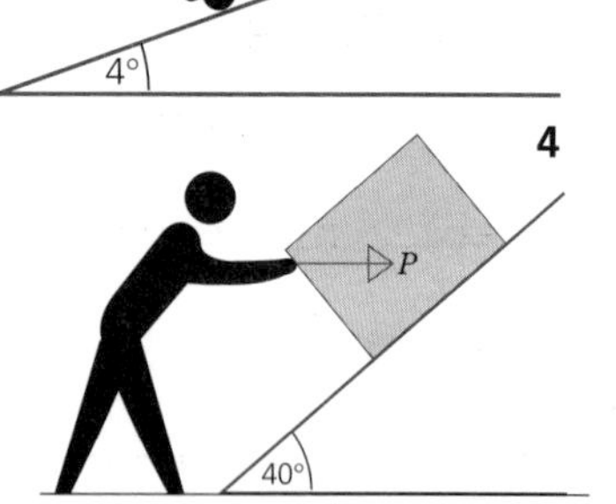

4 A man is trying to stop a box of mass 25 kg from sliding down a slope. He exerts a horizontal force P.

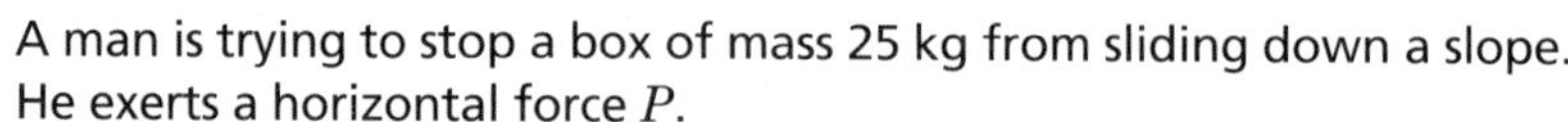

a) Draw a diagram to show the forces acting on the box.
b) Assume that there is no friction and find the magnitude of the force, P.

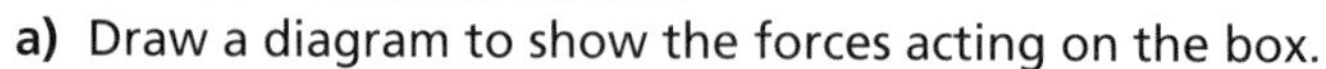

c) If, due to friction, P is, in fact, 10% less than your prediction, find the magnitude of the friction force between the slope and the box.

5 A child hauls a sledge of mass 18 kg up a slope at 7° to the horizontal.

a) Assume that the rope is parallel to the slope and there is no friction between the sledge and the slope. Find the tension in the rope.

b) Explain which assumptions in (a) are unrealistic, and suggest any alternatives.

c) The rope in fact makes an angle of 10° to the slope. Find the tension in the rope and compare this to your solution for a horizontal rope.

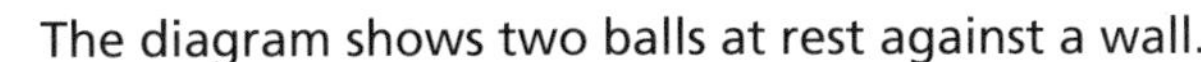

radius 4 cm
mass 500 grams
P
radius 10 cm
mass 2 kg

6 The diagram shows two balls at rest against a wall.

A horizontal force is applied by a child to keep them in place, as shown in the diagram.

a) Draw diagrams to show the forces acting on each ball.

b) Find the magnitude of the force exerted by the child, if the reaction force beteen the wall and the larger ball is zero.

*7 A skier of mass 72 kg is travelling, at a constant speed, down a slope at 13° to the horizontal.

a) Find the friction force acting on the skier.

b) What value of the coefficient friction does your answer to (a) suggest should be used when modelling the motion of the skier?

c) Criticise this model for friction and suggest how it could be modified.

8 Two removal men are carrying a bookcase up a flight of stairs. The mass of the bookcase is 30 kg. The higher man exerts a horizontal force P on the bookcase. The lower man exerts a force Q at 80° to the horizontal. Find the magnitude of the forces P and Q.

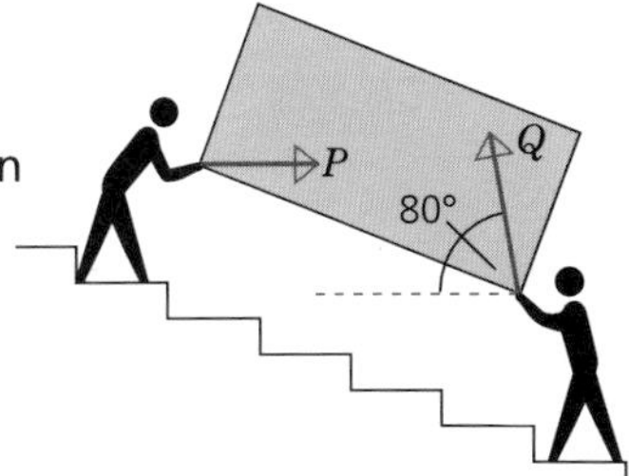

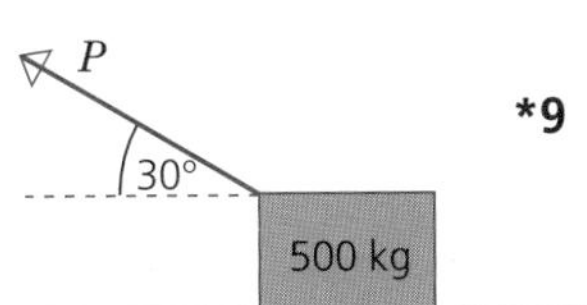

*9 A force P is applied as shown, to try to move a large concrete block of mass 500 kg. The coefficient of friction between the block and the surface is 0.6. Find P when the block is on the point of sliding.

10 The system shown is in equilibrium.

a) Find the tension in each string.

b) Find the mass of the middle object.

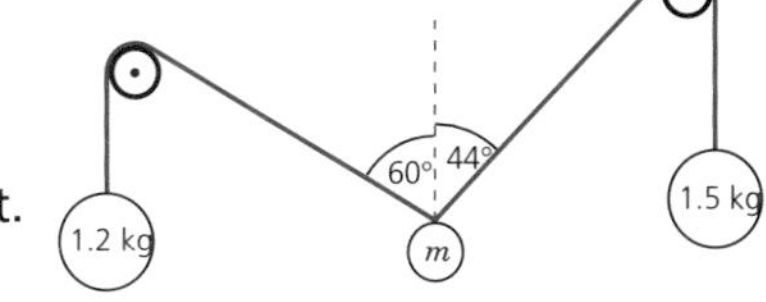

CONSOLIDATION EXERCISES FOR CHAPTER 3

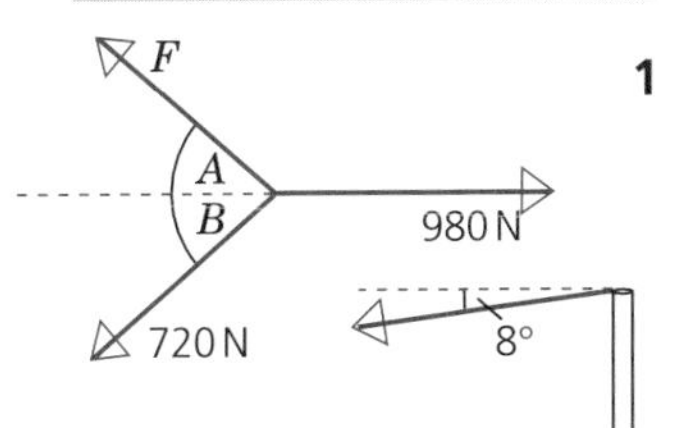

1 The diagram shows three horizontal forces that act on the top of a telegraph pole.

a) Find F and the angle A if B is 20° and the forces on the pole are in equilibrium.

b) If each force in fact acts along a direction at 8° to the horizontal, find the force exerted by the top of the pole.

c) Explain why it is unlikely that the forces actually act in a horizontal plane.

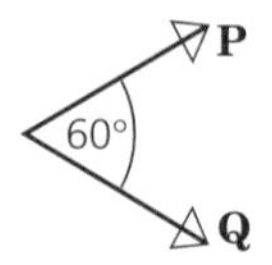

2 Two forces $\mathbf{P}$ and $\mathbf{Q}$, each of magnitude 100 N, are inclined to each other at an angle of 60°, and act on an object. Find the magnitude of the resultant $\mathbf{P} + \mathbf{Q}$.

(UCLES (Linear) Question 6, Specimen Paper 2, 1994)

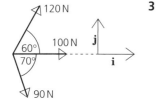

3 Three cables exert forces that act in a horizontal plane on the top of a telegraph pole.

a) Find the resultant of these three forces, in terms of the unit vectors $\mathbf{i}$ and $\mathbf{j}$.

b) A fourth cable is attached to the top of the telegraph pole to keep the pole in equilibrium. Find the force, exerted by this fourth cable, in terms of $\mathbf{i}$ and $\mathbf{j}$.

c) Show that the magnitude of the fourth force is 192 N, correct to three significant figures.

d) On a copy of the diagram, show clearly the direction in which the fourth force acts.

e) The fourth cable does not lie in the same horizontal plane as the other three cables and the tension in this cable is in fact 200 N. Find the angle between this cable and the horizontal plane.

(AEB Question 4, Specimen Paper 3, 1994)

4 Three forces $\mathbf{F}_1$, $\mathbf{F}_2$ and $\mathbf{F}_3$ act on a particle and $\mathbf{F}_1 = (-3\mathbf{i} + 7\mathbf{j})$ newtons, $\mathbf{F}_2 = (\mathbf{i} - \mathbf{j})$ newtons and $\mathbf{F}_3 = (p\mathbf{i} + q\mathbf{j})$ newtons.

a) Given that this particle is in equilibrium, determine the value of p and the value of q.

The resultant of the forces $\mathbf{F}_1$ and $\mathbf{F}_2$ is $\mathbf{R}$.

b) Calculate, in N, the magnitude of $\mathbf{R}$.

c) Calculate, to the nearest degree, the angle between the line of action of $\mathbf{R}$ and the vector $\mathbf{j}$.

(ULEAC Question 3, M1 Specimen Paper, 1994)

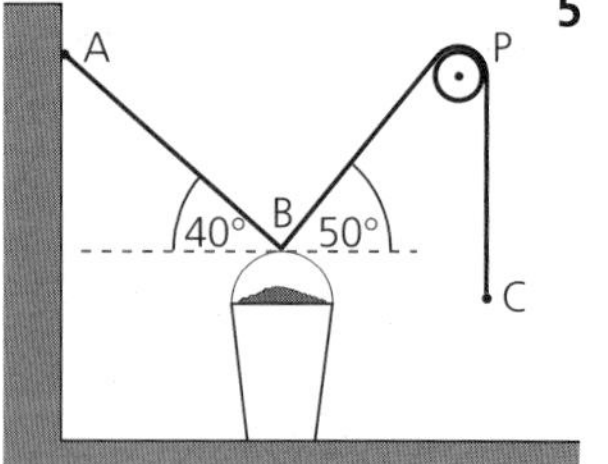

5 A basket of earth, of total mass 40 kg, is attached to two light inextensible ropes, BA and BC. The end A is attached to a fixed point and BC passes over a smooth pulley, P. A downward pull is applied to the end C. The parts AB and BP of the rope make angles of 40° and 50° respectively with the horizontal (see diagram). The basket is at rest on the ground; the force exerted on the basket by the ground is vertical and has magnitude 150 N. Find the tensions in AB and BP and find the magnitude and direction of the resultant force on the pulley due to the rope.

(UCLES Question 6, Specimen Paper M1, 1994)

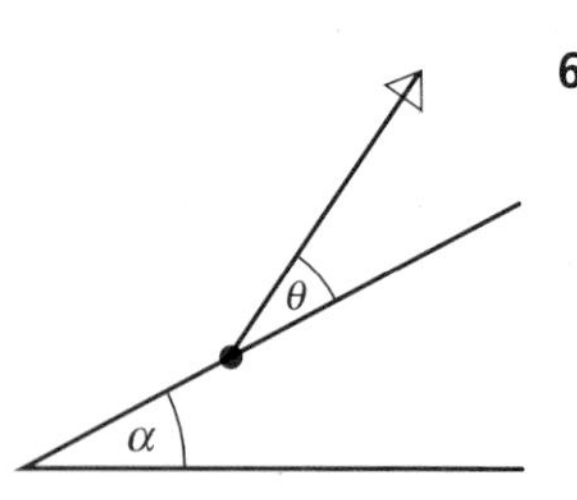

6 A particle of mass m lies on a smooth plane inclined as an angle α to the horizontal. The particle is held in equilibrium by a string which lies in a vertical plane through a line of greatest slope and makes an angle θ with the plane, as shown in the diagram. The tension in the string is of magnitude T, and the force exerted by the plane on the particle is of magnitude R.

a) Find, in terms of m, g, α and θ, an expression for T.

b) Show that $R = mg\cos\alpha(1 - \tan\alpha\tan\theta)$.

(ULEAC Question 2, M1 January, 1993)

MATHEMATICAL MODELLING ACTIVITY

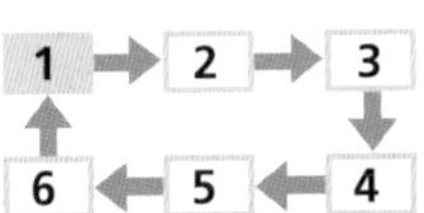

Specify the real problem

Problem statement

How steep should an artificial ski-slope be?

The first stage in solving a modelling problem is to identify the important features.

Make a list of the important features that you think should be considered in this problem.

Set up a model

Set up a model

Having considered the features of a situation it is then necessary to draw up a set of assumptions that will allow a mathematical problem to be formulated. The assumptions set out below have been prepared as a possible way of starting the problem.

- The skier is modelled as a particle.
- The skier skis straight down the slope.
- The skier maintains a constant speed, unless pushing with the sticks.
- No air resistance acts on the skier.
- The magnitude of the friction force is constant on the slope and can be modelled as $0.2R$, where R is the magnitude of the normal reaction between the skier and the slope.
- The mass of the skier is m kg.
- The slope makes a constant angle α with the horizontal.

The assumptions allow a first simple model to be formulated.

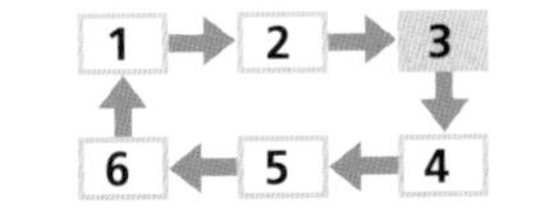

Formulate the mathematical problem

Formulate a mathematical problem

The skier skis down on a slope subject to a resistance force of $0.2R$. What angle should the slope make to the horizontal, if the skier is to travel with a constant speed?

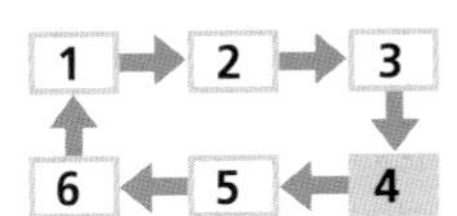

Solve the mathematical problem

Mathematical solution

The diagram shows the forces acting.

The resultant force can be expressed as:
$(mg\sin\alpha - F)\mathbf{i} + (R - mg\cos\alpha)\mathbf{j}$.

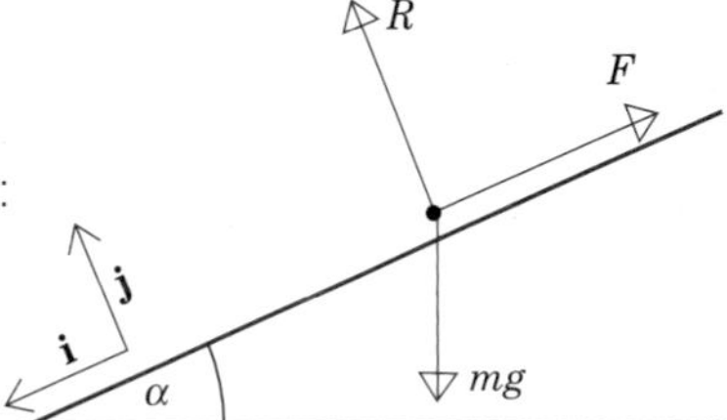

As the skier travels at a constant speed the resultant force is zero so,

$mg\sin\alpha - F = 0$ (1)

and

$R - mg\cos\alpha = 0$ (2)

But $F = 0.2R$ so equation (1) becomes:

$mg \ \sin\alpha - 0.2R = 0$ (3)

Eliminating R from equations (2) and (3) gives:

$mg\cos\alpha = 5mg\sin\alpha$

or

$\frac{\sin\alpha}{\cos\alpha} = \frac{1}{5} \Rightarrow \tan\alpha = \frac{1}{5}$

so $\alpha = 11.3°$.

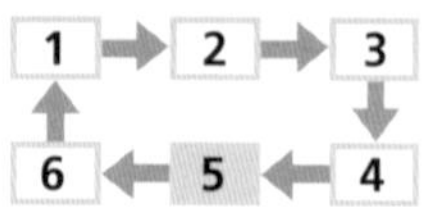

Interpret the solution

Interpretation

If the slope makes an angle of 11.3°, skiers will be able to travel down it at a constant speed, but will need to push off to start moving.

Two problems that arise are that skiers may not travel straight down the slope and the air resistance on the skiers may be significant.

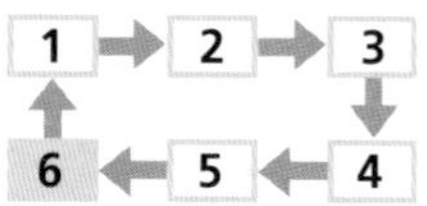

Compare with reality

Compare with reality

It may be possible for you to visit a local ski slope or obtain some data. Does the angle of 11.3° degrees seem about right? Is the model suggested for the friction reasonable?

Refining the model

Make an additional assumption about the path of the skier or the air resistance acting. Keep this refinement simple and use it to obtain a revised solution to the problem.

POSSIBLE MODELLING TASKS

1 What is the steepest hill that it would be possible to park a car on safely? (**Note**: The hand brake normally acts only on the back wheels.)

2 A 'button lift' is often used at artificial ski slopes to take skiers to the top of the slope. Investigate the forces that act on a skier who is using such a device.

3 A winch is to be installed at the top of a slipway to help sailors get their dingys out of the water. How strong a rope would be needed for the winch?

Summary

After working through this chapter you should:

- be able to express forces in the form $a\mathbf{i} + b\mathbf{j}$
- be able to find the magnitude and direction of a force
- know that in equilibrium the resultant force is zero
- be able to find the magnitude of the forces that act on an object that is at rest or moving with constant velocity.

Kinematics in one dimension

- *There are many objects that move in only one dimension, for example, a car moving along a straight road or a ball that is dropped and allowed to fall.*
- *Simple formulae can be developed to deal with these cases.*

UNIFORM MOTION

Exploration 4.1

Motion of a dropped ball

Imagine that you drop a ball from the top of a tall building. You check the ball every second. Sketch a diagram to show the position of the ball every second while it is falling.

Example 4.1

A cyclist is travelling at a constant speed of 4 m s^{-1}. Sketch a graph of speed against time. What is the gradient of the line that you have drawn? What does it represent? Use the graph to find the distance travelled between:

a) $t = 0$ *and* $t = 2$,
b) $t = 2$ *and* $t = 5$,
c) $t = 0$ *and* $t = T$.

Sketch a graph of distance travelled against time. What is the relationship between the gradient of this graph and the speed?

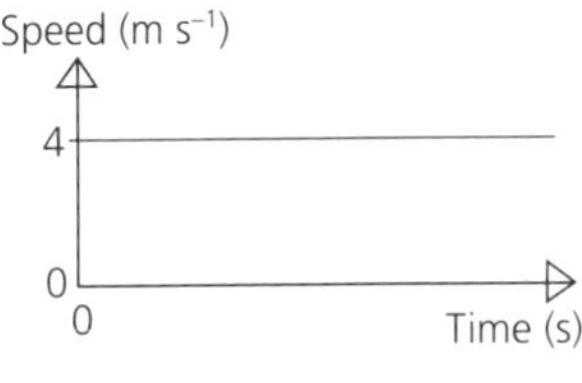

Solution

As the speed of the cyclist does not change this can be simply represented by a horizontal line, as shown. The gradient of the line is 0, which simply means that the speed is not changing.

The distance travelled is represented by the area under the line, between the required time points.

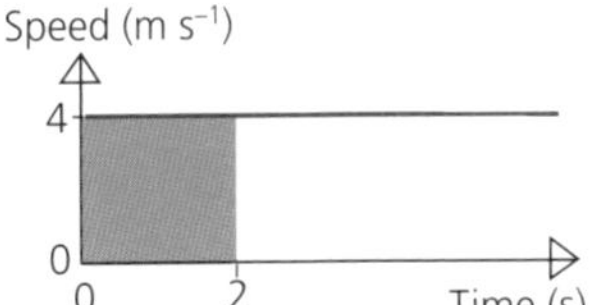

a) *For the distance between* $t = 0$ *and* $t = 2$ *we require the area of the shaded region shown. This gives distance* $4 \times 2 = 8$ *m.*

b) *For the distance between $t = 2$ and $t = 5$, we require the area of the shaded region shown.*
This gives distance $= 4 \times (5 - 2) = 12$ m.

c) *The distance between $t = 0$ and $t = T$ is illustrated by the area of the shaded region in the diagram.*
This gives distance $= 4 \times T = 4T$
The distance–time graph is simply a straight line because the distance travelled, s, is $s = 4t$. The gradient of the line is 4 which is the same as the speed of the cyclist.

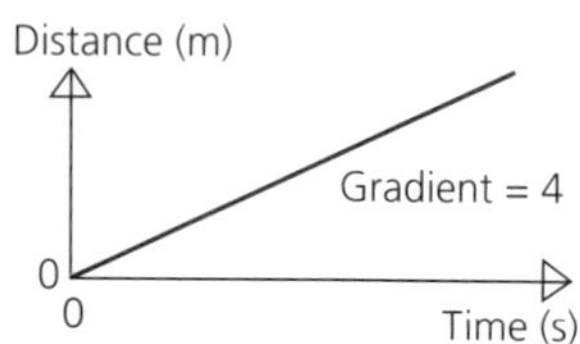

Example 4.2

The speed of a car increases at a constant rate from 0 to 15 m s^{-1} in a ten-second interval. Sketch a graph to show the speed of the car against the time. Find the gradient of the line and describe what information this gives about the motion of the car. Find the distance travelled by the car between:

a) *$t = 0$ and $t = 5$,*
b) *$t = 0$ and $t = T$.*

Use the answer to **b)** *to sketch a graph of distance against time. What is the relationship between the gradient of this graph and the speed?*

Solution

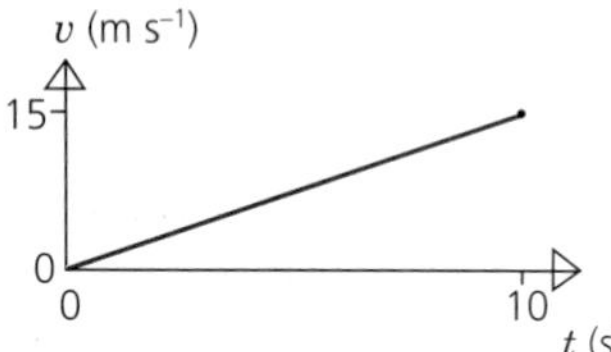

The sketch shows how the speed increases. The gradient of the line is given by $\frac{15}{10} = 1.5$. This value describes how the speed of the car increases, in this case by 1.5 m s^{-1} per second. This is known as the acceleration of the car. In this case it is 1.5 m s^{-2}. The speed of the car is then $v = 1.5t$.

a) *The distance travelled between $t = 0$ and $t = 5$ is represented by the area of the shaded region.*
So the distance is:
$\frac{1}{2} \times 5 \times (1.5 \times 5) = 18.75$ *m*

b) *The distance travelled between $t = 0$ and $t = T$ is represented by the area of the shaded region.*
This gives distance $= \frac{1}{2} \times T \times 1.5T = 0.75T^2$

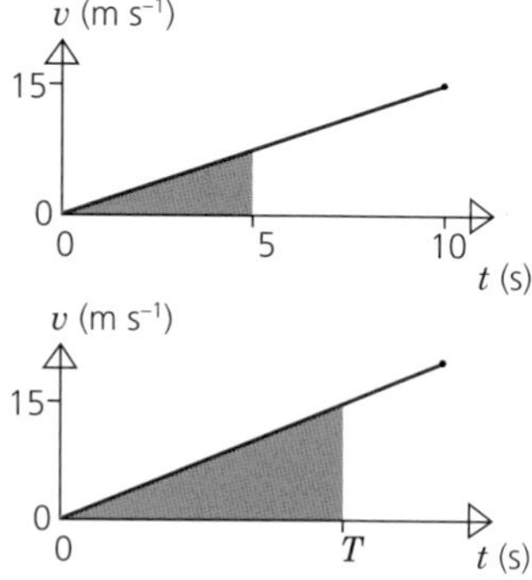

Using this result, we can draw a graph of distance against time as shown. Note that the gradient of this graph increases with time and that the speed also increases with time. In fact the gradient of the curve gives the speed at that time.

Gradient and area

The **gradient** of the distance–time graph shown gives the **speed**.

The **area under** the speed–time graph shown gives the **distance travelled**. The gradient of this speed–time graph gives the **acceleration**.

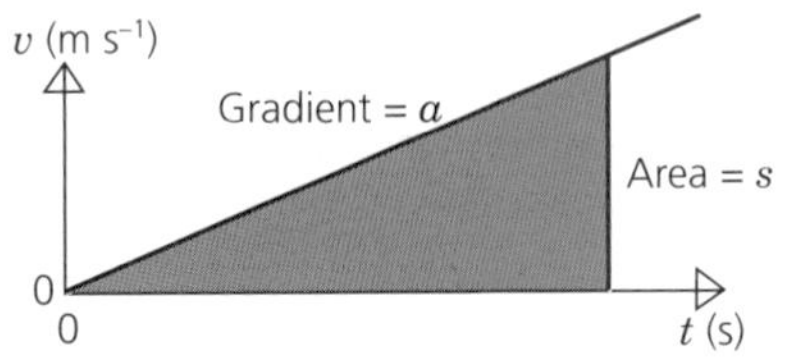

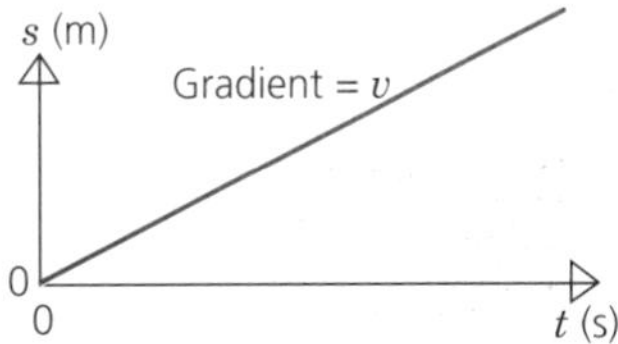

EXERCISES

4.1 A

1 An aeroplane is travelling at a constant speed of 140 m s^{-1}. Sketch a graph of speed against time for this motion. Use this graph to find the distance travelled by the aeroplane in:
a) 5 seconds, **b)** T seconds.
What is the gradient of the line and what is the acceleration of the aircraft?

2 A hot air balloon rises vertically with a speed that is assumed to increase at a constant rate from 0 to 2 m s^{-1} over a 30-second period.
a) What is the acceleration of the balloon? Use this to find an expression for the speed of the balloon.
b) Assuming that the balloon's speed continues to increase at the same rate, find the distance that the balloon rises in:
i) 10 seconds, **ii)** 40 seconds, **iii)** T seconds.
c) Sketch a graph of the distance the balloon rises against the time.
d) Suggest a more realistic way for the speed of the balloon to increase.

3 A ball is dropped from a height of 18 m. As it falls, its speed is given by $v = 9.8t$.
a) Sketch a graph of speed against time for the ball, and find its acceleration.
b) Find a formula for the distance travelled by the ball in t seconds.
c) How long does it take for the ball to reach the ground?

4 A car is travelling at 10 m s^{-1} when it begins to increase its speed at a constant rate of 1.2 m s^{-2}.
a) Sketch a graph to show how the speed increases. Assume the speed starts to increase when $t = 0$.
b) Find the speed of the car when $t = T$.
c) Find the distance travelled by the car between:
i) $t = 0$ and $t = 5$, **ii)** $t = 0$ and $t = T$.
d) Sketch a graph of distance against time for the car.

5 An ice hockey puck is hit so that it has an initial speed of 18 m s^{-1}, but as it travels over the ice it loses speed at a rate of 1.5 m s^{-2}.
a) Sketch a graph of speed against time for the puck.
b) What is the acceleration of the puck?

c) Find an expression for the speed of the puck.
d) How far does the puck travel between:
i) $t = 0$ and $t = 3$, ii) $t = 0$ and $t = T$?
e) Sketch a graph of distance against time for the puck.
f) How far does the puck travel before it stops?

EXERCISES

4.1 B

1 A car travels at a constant speed of $20\,\text{m}\,\text{s}^{-1}$. Sketch a graph of speed against time for the car. Use the graph to find the distance travelled by the car in:
a) 10 seconds, b) t seconds.

2 A cyclist increases her speed from 0 to $5\,\text{m}\,\text{s}^{-1}$ over a 15-second period. Assume that the speed increases at a constant rate.
a) Sketch a graph of speed against time for the cyclist. Find the gradient of the line and state the acceleration of the cyclist.
b) Write down an expression for the speed of cyclist at time t.
c) Find the distance the cyclist travels in:
i) 5 seconds, ii) 15 seconds, iii) t seconds.
d) This question has assumed that the speed increases at a constant rate. Suggest an alternative model for the way the speed increases, illustrating this with a sketch graph. Would the total distance travelled in the 15 seconds be more or less than for the original model?

3 As a stone falls in a tank of water its speed is modelled by $v = 0.8t$.
a) Sketch a speed–time graph and find the acceleration of the stone.
b) Find an expression for the distance travelled by the stone.
c) How long does it take the stone to travel the 1.2 m to the bottom of the tank and what is its speed when it hits the bottom?

4 An aeroplane is travelling at $70\,\text{m}\,\text{s}^{-1}$. After 40 seconds its speed has increased to $90\,\text{m}\,\text{s}^{-1}$.
a) Sketch a graph of speed against time. What is the acceleration of the aeroplane? What assumptions have you made to reach this conclusion?
b) Find an expression for the speed of the aeroplane t seconds after its speed began to increase.
c) Find the distance travelled by the aeroplane in the 40 seconds.
d) Find an expression for the total distance travelled by the aeroplane t seconds after it began to increase speed.

5 A train is travelling at $30\,\text{m}\,\text{s}^{-1}$ when it begins to slow down at a rate of $2\,\text{m}\,\text{s}^{-1}$.
a) Sketch a graph of speed against time for the train.
b) How long does the train take to stop?
c) How far has it travelled before it stops?
d) What was the acceleration of the train?

EQUATIONS OF MOTION FOR CONSTANT ACCELERATION

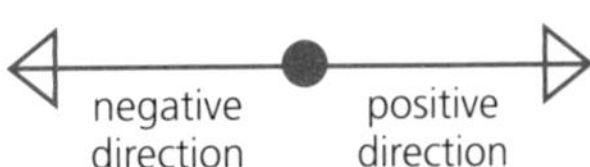

When motion is restricted to one dimension, the objects can move in one of two directions, one defined as **positive** and the other as **negative**. The **speed** of an object does not give any indication of its direction of motion. The **velocity** of an object describes the speed and direction of the motion. A velocity of $+4\,\text{m s}^{-1}$ describes motion with speed of $4\,\text{m s}^{-1}$ in the positive direction, while a velocity of $-4\,\text{m s}^{-1}$ describe a motion with a speed of $4\,\text{m s}^{-1}$ in the negative direction.

Accelerations can also be either positive or negative when motion is restricted to one dimension. Similarly an object can **move** in either a positive or negative direction. The distance an object moves simply tells us how far the object has travelled, so instead of calling it distance we shall now use the vector term, **displacement**, which describes direction as well, The magnitude of the displacement describes how far the object is from the origin and the sign defines whether this is in the positive or negative direction.

Example 4.3

A car has a velocity u and experiences an acceleration a.

a) *Sketch a graph of velocity against time for the car.*
b) *Find an expression for the velocity of the car.*
c) *Find an expression for the distance travelled by the car.*

Solution

a) *The graph shows how the velocity changes with time.*
b) *The acceleration is a, so the gradient of the line will be a.*
This gives: $v = u + at$
c) *The distance travelled is represented by the area shaded in the graph. The area of the rectangle is ut and the area of the triangle is $\frac{1}{2}at^2$, so the distance travelled is given by: $s = ut + \frac{1}{2}at^2$*

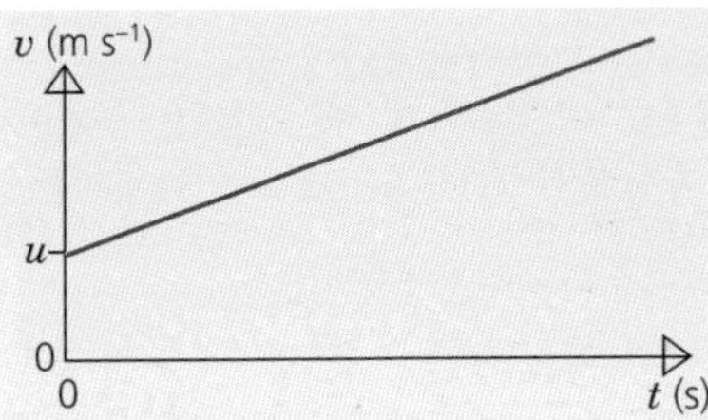

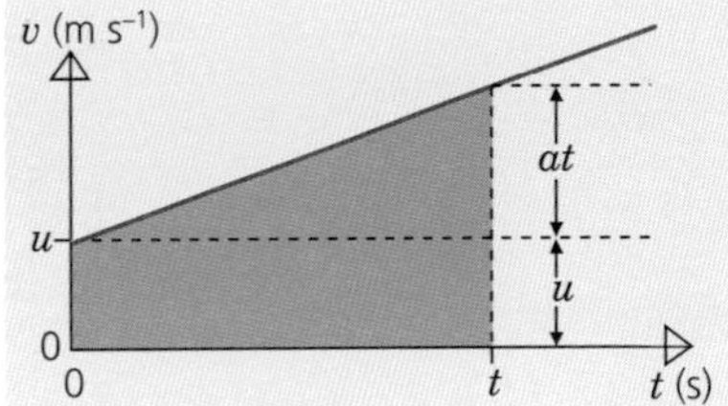

Note: *It is usual to denote the distance travelled by s.*

Example 4.4

Use the equations $v = u + at$ and $s = ut + \frac{1}{2}at^2$ to show that $v^2 = u^2 + 2as$.

Solution

Rearranging $v = u + at$ gives: $t = \dfrac{v-u}{a}$

Substituting this value for v into $s = ut + \frac{1}{2}at^2$ gives:

$$s = \frac{u(v-u)}{a} + \frac{a(v-u)^2}{2a^2}$$

$$\Rightarrow 2as = 2u(v-u) + (v-u)^2$$

$$\Rightarrow 2as = 2uv - 2u^2 + v^2 - 2uv + u^2$$

$$\Rightarrow 2as = v^2 - u^2$$

$$\Rightarrow v^2 = u^2 + 2as$$

Constant acceleration equations

The equations derived in the above examples are known as the constant acceleration equations. They can be used whenever an acceleration is constant. These equations are listed below.

$$v = u + at$$

$$s = ut + \tfrac{1}{2}at^2$$

$$v^2 = u^2 + 2as$$

Note that s is displacement, v is final velocity, u is initial velocity, a is acceleration and t is time.

Example 4.5

A ball is dropped from a height of 5 m and falls under the influence of gravity so that its acceleration is 9.8 m s^{-2}.

a) *When does the ball hit the ground?*
b) *How fast is it moving when it hits the ground?*

Solution

Assume 'down' is the positive direction in this problem.

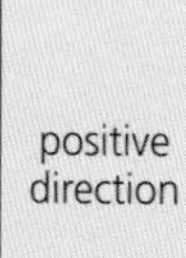

a) *Using $s = ut + \frac{1}{2}at^2$ with $a = 9.8$ and $u = 0$ gives:*

$$s = 0 \times t + \tfrac{1}{2} \times 9.8t^2$$

When the ball hits the ground $s = 5$, so $5 = 4.9t^2$

so $t^2 = \sqrt{\dfrac{5}{4.9}} = 1.01$ *seconds*

b) *The velocity of the ball is given by:*

$$v = u + at$$

Using $u = 0$, $a = 9.8$ and $t = 1.01$ gives:

$$v = 0 + 9.8 \times 1.01 = 9.90 \text{ m s}^{-1}$$

Example 4.6

A hot air balloon is rising at 0.5 m s^{-1} at a height of 40 m, when it begins to experience a downward acceleration of 0.25 m s^{-2}.

a) *Find the height that the balloon reaches before it stops rising.*
b) *How long does it take to return to the ground?*
c) *At what speed is the balloon travelling when it reaches the ground?*

Solution

First note $a = -0.25$ and $u = 0.5$.

a) *Using $v^2 = u^2 + 2as$ we have:*

$$0^2 = 0.5^2 + 2 \times (-0.25)s$$

because at the maximum height, $v = 0$, which gives:

$$s = \frac{0.5^2}{2 \times 0.25} = 0.5 \text{ m}$$

so the balloon rises a further 0.5 m to 40.5 m.

b) *Relative to the original position, we require $s = -40$.*
Using $s = ut + \frac{1}{2}at^2$ we obtain:

$$-40 = 0.5t + \tfrac{1}{2} \times (-0.25)t^2$$

Rearranging and multiplying by 8 gives:

$$t^2 - 4t - 320 = 0$$

and solving this quadratic equation gives $t = -16$ or $t = 20$.
We require the positive value, $t = 20$, which means that it takes 20 seconds for the balloon to reach the ground.

c) *Using $v = u + at$, with $a = -0.25$, $u = 0.5$ and $t = 20$ gives:*

$$v = 0.5 + (-0.25) \times 20 = -4.5 \text{ m s}^{-1}$$

positive direction

EXERCISES

1 A ball is dropped from a height of 10 m so that it has an acceleration of 9.8 m s^{-2}.

a) How long does it take to reach the ground?
b) How fast is it moving when it hits the ground?

2 Find the acceleration of each object below.

a) A skateboarder whose velocity increases from 0.5 m s^{-1} to 1.3 m s^{-1} in seven seconds.
b) A car that increases in velocity from 0 to 25 m s^{-1} in five seconds.
c) A skier whose velocity increases from 0.5 m s^{-1} to 7 m s^{-1} while travelling 20 m.
What assumptions have you had to make to answer this question?

3 A ball is thrown straight up into the air with an initial speed of 8 m s^{-1} and has a downward acceleration of 9.8 m s^{-2}.

a) Find the maximum height that the ball reaches if it is released 1.5 m above ground level.
b) Find the speed of the ball when it hits the ground.

4 A cyclist was travelling at 6 m s^{-1} when she slowed down and stopped in 5 m. Assume that the acceleration of the cyclist was constant.

a) Find the acceleration of the cyclist.
b) How long did it take the cyclist to stop?

5 A car travelling at 20 m s^{-1} was 200 m from a set of traffic lights, which were red. The driver slowed down, until he was 50 m from the lights and travelling at 8 m s^{-1}, when the lights changed to green. He immediately began to accelerate, passing the lights at a speed of 15 m s^{-1}.

a) What was the acceleration of the car while it was slowing down?
b) If the driver had maintained the same acceleration would he have stopped before he reached the lights?

c) How long was it before the lights changed to green?
d) What was the acceleration of the car after the lights changed?
e) How long did it take the driver to increase the speed of the car from $8\,m\,s^{-1}$ to $15\,m\,s^{-1}$?
f) What was the total time taken to travel the 200 m?
g) What was the average speed of the car?
h) What assumptions have you had to make to solve this problem?

6 A balloon was at a height of 80 m and ascending at a constant speed of $3\,m\,s^{-1}$, when the crew released some hot air so that the balloon experienced a downward acceleration of $0.25\,m\,s^{-2}$.
a) How much further did the balloon ascend?
b) How long did it take the balloon to return to its original position?
c) How long did it take the balloon to reach the ground?
d) How fast was the balloon moving when it hit the ground?

7 A car accelerated from rest at $3\,m\,s^{-2}$ for ten seconds, travelled at a constant speed for two minutes and then took six seconds to stop. How far did the car travel in total?

8 A lift accelerated at $0.2\,m\,s^{-2}$ from rest for ten seconds, then travelled at constant speed for 20 seconds, and then took 15 seconds to stop. Find the distance travelled by the lift if it was always moving upwards.

9 It has been estimated that sprinters reach their top speed after two seconds and maintain this constant speed for the remainder of the sprint.
a) Assume that the acceleration of the sprinter during the first two seconds is a. Find expressions for the distance travelled and the speed reached during this two-second period, in terms of a.
b) It takes the sprinter a further ten seconds to complete the sprint. Express the distance travelled by the sprinter in this ten-second period, in terms of a.
c) Find a if it was a 100 m race.

10 A ball is dropped from a height of 1 m and bounces, rebounding to a height of 70 cm.
a) Find the speed of the ball when it hits the ground.
b) Find the speed of the ball when it leaves the ground.
c) If the ratio of the speeds before and after bouncing are in a constant ratio, find the height after the second bounce.

EXERCISES

4.2B

1 Find the acceleration of each object described below.
a) A train that increases its velocity from rest to $8\,m\,s^{-1}$ in 40 seconds.
b) A ball whose velocity increases from $3\,m\,s^{-1}$ to $11\,m\,s^{-1}$ in four seconds.
c) A car whose velocity decreases from $10\,m\,s^{-1}$ to $2\,m\,s^{-1}$ over 40 m.

2 A ball experiences an acceleration of $-9.8\,m\,s^{-2}$ after it has been thrown straight upwards with an initial speed of $2\,m\,s^{-1}$ from a height of 1 m.

a) When is the velocity of the ball zero?
b) What is the maximum height reached by the ball?
c) Find the speed of the ball when it hits the ground.

3 A ball hits the ground travelling at $5\,\mathrm{m\,s^{-1}}$ and rebounds in the opposite direction at $4\,\mathrm{m\,s^{-1}}$. Find the acceleration of the ball if it is in contact with the ground for 0.5 seconds.

4 The *Highway Code* states that the braking distance for a car travelling at 30 mph ($13.4\,\mathrm{m\,s^{-1}}$) is 45 m.
a) Find the acceleration of a car that stops in this way, and the time it takes for the car to stop.
b) What would be the stopping distance for a car travelling at 100 mph ($44.7\,\mathrm{m\,s^{-1}}$)?
c) How long would the car take to stop?

5 As a ball falls in a pond of water it experiences an acceleration of $-2\,\mathrm{m\,s^{-2}}$. Assume that the ball enters the pond travelling downwards with a velocity of $10\,\mathrm{m\,s^{-1}}$.
a) Find the time it takes the ball to come to rest.
b) Find the greatest depth in the water to which the ball travels.
c) What is the speed of the ball when it resurfaces?
d) Comment on the validity of this model of the ball in motion.

6 A train accelerates at $0.4\,\mathrm{m\,s^{-2}}$ from rest, until it reaches a velocity of $20\,\mathrm{m\,s^{-1}}$. It then slows down, stopping in 100 m.
a) What is the total distance travelled by the train?
b) Find the time that the train is in motion.

7 A lift travels 5 m in 20 seconds. The graph shows how the velocity changes.
a) Express the maximum speed of the lift in terms of T and a, the acceleration of the lift during its first stage of motion.
b) Express the total distance travelled in terms of a and T.
c) If $T = 5$, find a and the maximum speed of the lift.

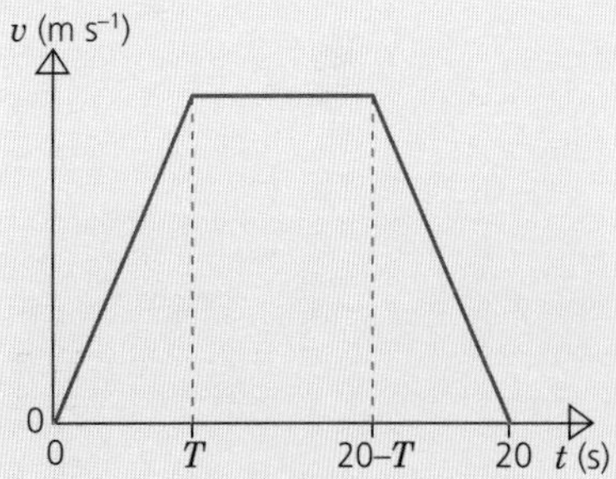

8 Two cars are level for an instant at a set of traffic lights, one is at rest and the other travelling at a constant $10\,\mathrm{m\,s^{-1}}$. The car that was at rest catches up with the other car when they have both travelled 100 m.
a) Find the acceleration of the car that was at rest and the time that it took to travel the 100 m.
b) Find the speed of this car when it catches up with the other car.
c) Once the cars are level they both continue at a constant speed. How far apart are they after a further 20 seconds?

9 A car is 25 m from a set of traffic lights when they change to red. The speed of the car is $15\,\mathrm{m\,s^{-1}}$. The driver brakes for two seconds and travels a further 20 m.
a) Find the acceleration of the car.

b) Can the driver stop before the car reaches the lights?
c) The driver panics and accelerates forwards at $4\,\mathrm{m\,s^{-2}}$. Find the position of the car when the traffic from the lights on the other road at the junction turn to green, eight seconds after the first set turned to red.
d) Comment on the driver's actions.

10 A cyclist travels 10 km in one hour. A possible model for the velocity of the cyclist is shown in the diagram.

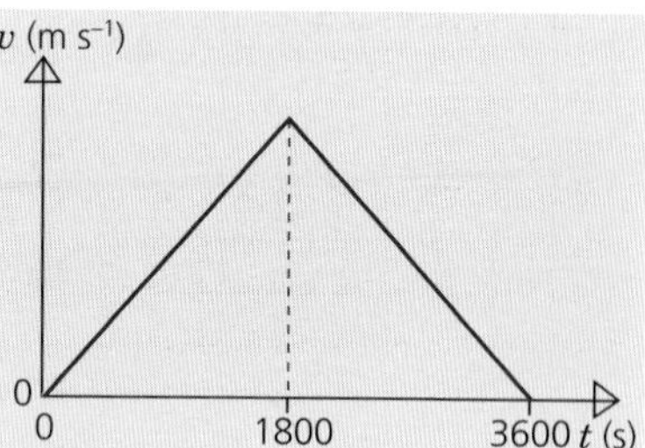

a) Find the acceleration of the cyclist based on this model and the maximum speed reached.
b) Criticise this model and suggest a more suitable alternative.

CONSOLIDATION EXERCISES FOR CHAPTER 4

1 A train starts from rest at a station A and moves with uniform acceleration along a straight track. The train passes a signal box B with speed $15.5\,\mathrm{m\,s^{-1}}$. From B, the train then covers a 900 m stretch of level track in 50 s, with the same uniform acceleration. By considering a speed–time sketch, or otherwise, find:
a) the acceleration, in $\mathrm{m\,s^{-2}}$, of the train,
b) the time, in seconds, taken by the train to move from A to B.

(ULEAC Question 2, Old Specimen M1)

2 A train was scheduled to travel at $50\,\mathrm{m\,s^{-1}}$ for 15 minutes on part of its journey. The velocity–time graph illustrates the actual progress of the train which was forced to stop because of signals.

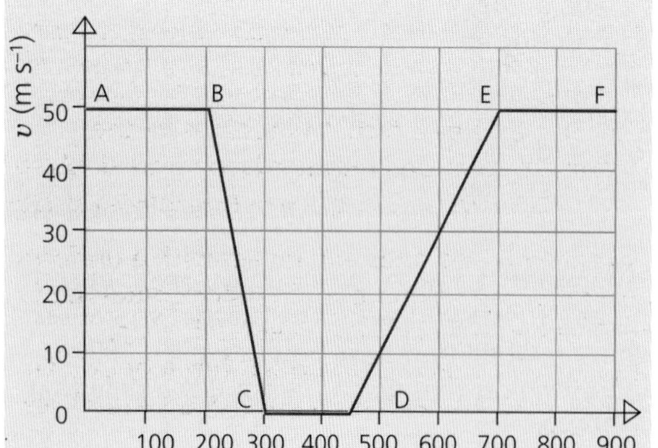

a) Without carrying out any calculations, describe what was happening to the train in each of the stages BC, CD and DE.
b) Find the deceleration of the train while it was slowing down and the distance travelled during this stage.
c) Find the acceleration of the train when it starts off again and the distance travelled during this stage.
d) Calculate how long the stop will have delayed the train.
e) Sketch the distance–time graph for the journey from A to F, marking the stages A, B, C, D, E and F.

(MEI Question 1, Mechanics 1, June 1992)

3 An airport has a straight runway of length 3000 m. During take-off, a jet aircraft, starting from rest, moves with constant acceleration along the runway and reaches its take-off speed of $270\,\mathrm{km\,h^{-1}}$ after 40 seconds.
a) Express the take-off speed in $\mathrm{m\,s^{-1}}$.

b) Find the acceleration of the jet during the take-off, in m s^{-2}.
c) Find the fraction of the length of the runway used by the jet during its take-off.

(NEAB Question 4, Specimen Paper 2, 1994)

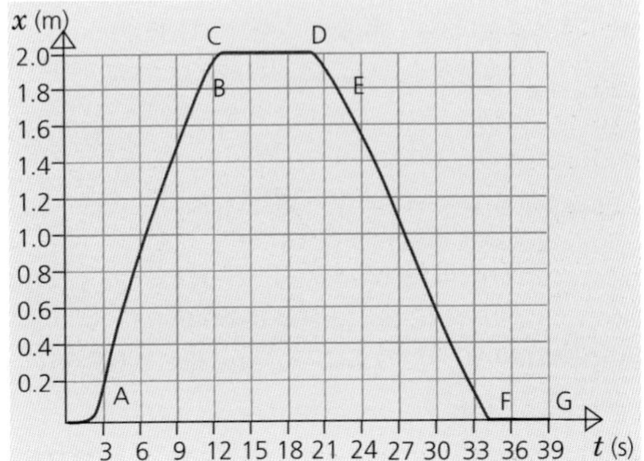

4 The (t, x) graph of a toy train moving on straight rails is as shown in the diagram. The distance x is measured in metres and the time t in seconds.

The coordinates of the points marked are: A(3, 0.2), B(11, 1.8), C(13, 2.0), D(20, 2.0), E(22, 1.8), F(34, 0), G(40, 0). The lines AB, CD, EF and FG are all straight line segments. Find the speeds when $t = 5$, $t = 15$ and $t = 25$.
Sketch the (t,v) graph and describe, in words, the motion of the train for $0 \leq t \leq 40$.

(UCLES Question 7, Specimen Paper M1, 1994)

5 A hovercraft which has a maximum speed of 80 km h^{-1} makes journeys across the English Channel, a distance of 52 km, in 40 minutes.

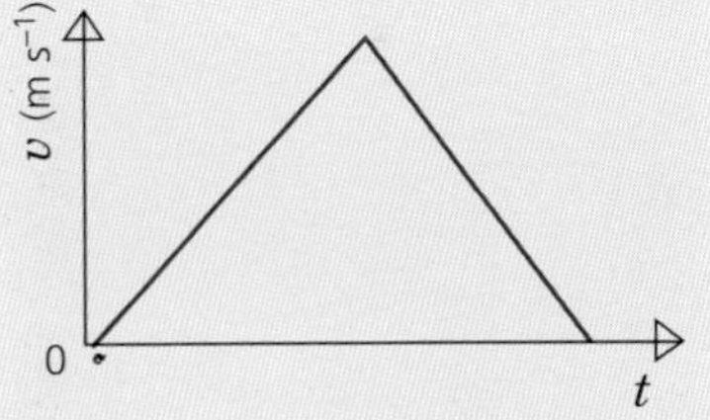

a) Show that the velocity–time (v–t) graph for a 40-minute journey could *not* be as shown.

b) The v–t graph for a 40-minute crossing in which the hovercraft reaches its maximum speed is of the form shown.

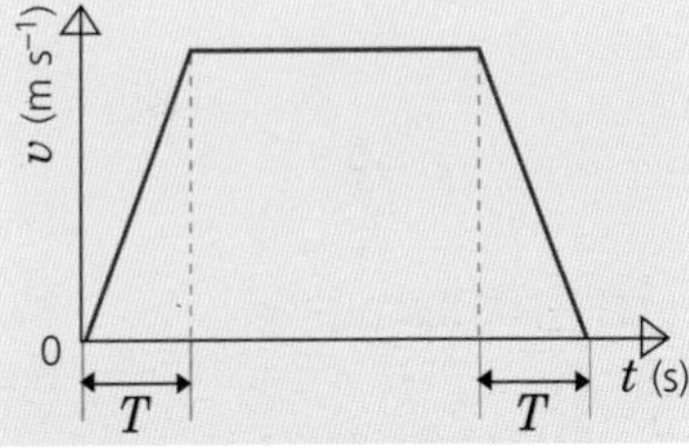

i) Find the time T.
ii) Calculate the acceleration of the hovercraft during the first part of the crossing.
iii) How would the crossing time be affected if the times for the acceleration to maximum speed and the retardation were doubled?

(Oxford Question 8, Specimen Paper Module 2, 1994)

6 A car starts from rest at time $t = 0$ seconds and moves with a uniform acceleration of magnitude 2.3 m s^{-2} along a straight horizontal road. After T seconds, when its speed is V m s^{-1}, it immediately stops accelerating and maintains this steady speed until it hits a brick wall when it comes instantly to rest. The car has then travelled a distance of 776.25 m in 30 s.

a) Sketch a speed–time graph to illustrate this information.
b) Write down an expression for V in terms of T.
c) Show that $T^2 - 60T + 675 = 0$.

(ULEAC Question 2, M1, January 1992)

MATHEMATICAL MODELLING ACTIVITY

Problem statement

In many towns traffic calming schemes have now been introduced in residential areas. One of the most unpopular but effective methods is to use humps in the road called 'sleeping policemen'. How far apart should sleeping policemen be placed to ensure that vehicles do not exceed a speed of 30 mph (13.4 m s^{-1})?

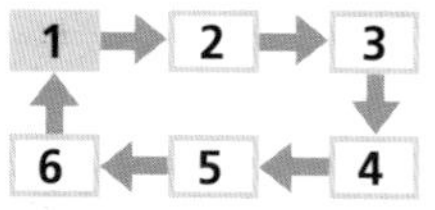

Specify the real problem

Set up a model

The first stage in setting up a model is to identify the important features. The list below includes the main features in this problem.

- The speed at which vehicles can cross the bumps.
- The acceleration of vehicles once they have crossed the bumps.
- The way in which vehicles slow down when approaching the bumps.

Are there any other features that you feel are important and should be included in the list?

Using this list of features it is now possible to draw up a set of assumptions, which is listed below.

- The cars cross the bumps at a speed of zero.
- The cars are treated as particles with no length.
- The cars stop according to the *Highway Code*, so that at 30 mph (13.4 m s^{-1}) it takes 14 m to stop.
- The acceleration of the cars once over the first bump is such that it can accelerate from 0 to 60 mph (26.8 m s^{-1}) in ten seconds.

Show that the acceleration of the car would be 2.7 m s^{-2} while gaining speed and –6.42 m s^{-2} while slowing down.

These assumptions allow a first model to be formulated.

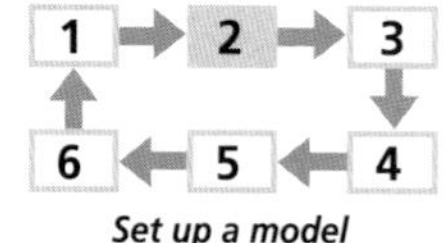

Set up a model

Formulate a mathematical problem

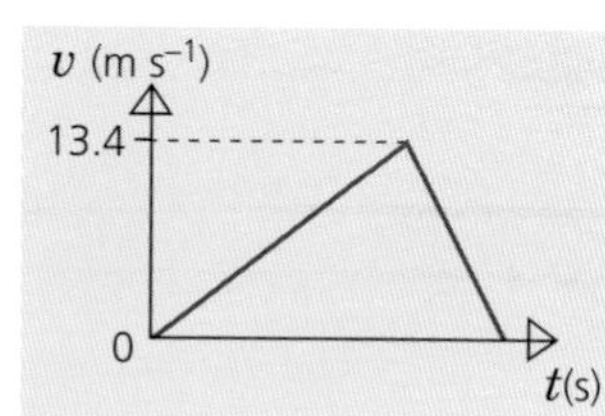

The diagram shows a sketch of a velocity time graph for the vehicle. Find the distance travelled by the car, as it is gaining speed and total distance covered.

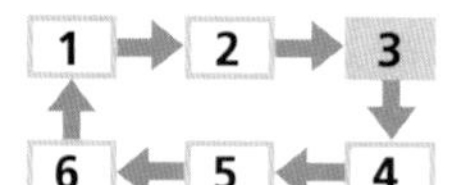

Formulate the mathematical problem

Mathematical solution

The distance travelled while the car accelerates can be found by first using the equation $v = u + at$ with v = 13.4 m s^{-1} (30 mph), u = 0 and a = 2.7 m s^{-2}. Show that s = 33 m.

The distance travelled by the car as it accelerates is 33 m, and it travels a further 14 m while braking. The total distance travelled by the car is 47 m.

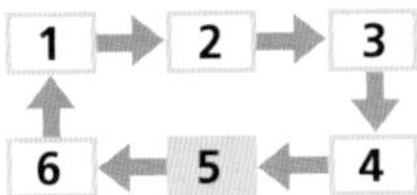

Interpret the solution

Interpretation

The speed bumps or sleeping policemen should be placed 47 m apart. What do you think will be the effects of the humps on the drivers?

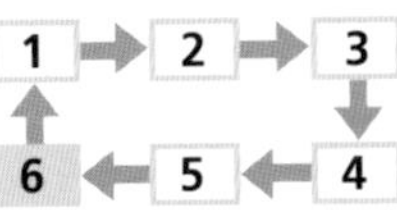

Compare with reality

Compare with reality

It is very likely that placing humps 47 m apart will cause a great deal of irritation for drivers and may result in other problems on the road.

As the distance between the humps is relatively small, it may be important to consider the length of the cars likely to travel on the stretch of road.

Very few drivers will actually accelerate from 0 to 60 mph in ten seconds under normal conditions. What other criticisms, if any, can be made of the results obtained? Can you find an existing traffic calming scheme to compare the results with?

Refining the model

In order to respond to the criticisms above define a new problem where the lengths of the vehicles are taken into account, and a more realistic figure is used for their acceleration. Find a solution to your revised problem.

POSSIBLE MODELLING TASKS

1 On Guernsey at many junctions a 'filter in turn' system operates. Vehicles take it in turn to leave from the roads joining the junction, one at a time from each road, taking each road in turn. Investigate how traffic would flow through a junction like this.

2 Consider a pedestrian crossing with traffic lights. Devise a timing schedule for the changing of the lights.

Summary

After working through this chapter you should:

- be able to use the constant acceleration equations:

$$s = ut + \frac{1}{2}at^2$$
$$v = u + at$$
$$v^2 = u^2 + 2as$$

- be aware of the conditions when these equations can be applied.

Motion and vectors

- *In real life motion is rarely restricted to one dimension.*
- *The ideas of motion developed in Chapter 4 can be extended into two dimensions.*

POSITION VECTORS

Exploration 5.1

Describing position and motion

Imagine a yacht sailing from a port at A to another port at B. You are at O, where you have a clear view of the yacht.

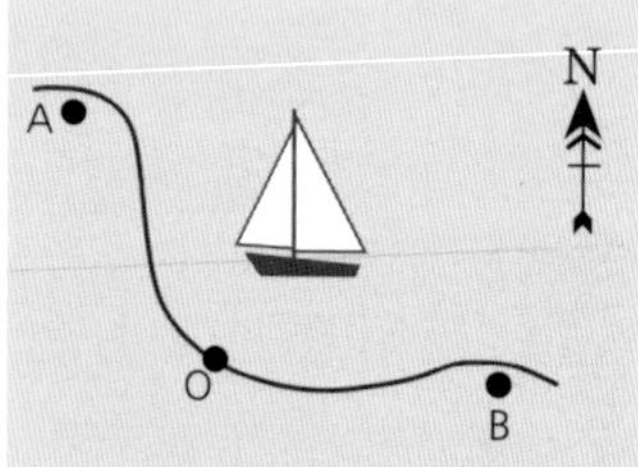

- How could you describe the **position** of the yacht?
- How could you describe the **motion** of the yacht?

Relative position

In the exploration, O is a **fixed position** or **reference point**. The positions of the yacht as it moves can be described **relative to** that fixed point. In Mechanics, a **position vector** describes where an object is, relative to a **fixed origin** or reference point. It gives information about the distance of the object from the reference point, and its direction from that point.

This diagram again shows the yacht, but this time a position vector has been drawn in. The vector could be described 'six kilometres on a bearing of 068°', or alternatively as $5.56\mathbf{i} + 2.25\mathbf{j}$, where $\mathbf{i}$ is a unit vector pointing east and $\mathbf{j}$ is a unit vector pointing north.

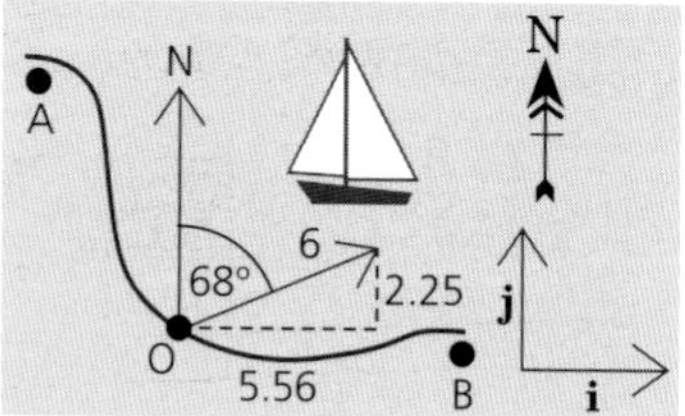

It is important to remember that every time the yacht, or any other object, moves, then its position vector changes. The diagram has now been revised to show the position vectors at four different positions, and the path followed by the yacht. It is a common convention that $\mathbf{r}$ is used for the position vector of an object.

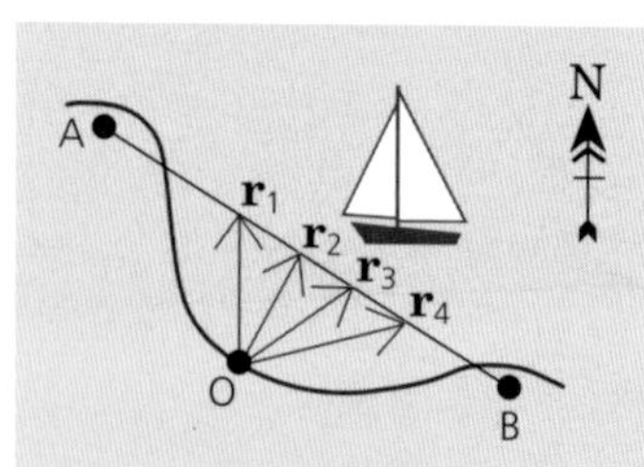

Example 5.1

The path of a rocket launched at a fireworks display is given by:

$$\mathbf{r} = 10t\,\mathbf{i} + \left[2(t-3)^3 + 54\right]\mathbf{j}$$

until the rocket explodes when $t = 3$. Find the position of the rocket when $t = 0$, 1, 2 and 3. Use this information to sketch the path of the rocket.

Solution

To find the position vectors, substitute the values $t = 0$, 1, 2 and 3 into

$$\mathbf{r} = 10t\,\mathbf{i} + \left[2(t-3)^3 + 54\right]\mathbf{j}$$

For $t = 0$:

$$\mathbf{r} = 10 \times 0\,\mathbf{i} + \left[2(-3)^3 + 54\right]\mathbf{j} = 0\mathbf{i} + 0\mathbf{j}$$

For $t = 1$:

$$\mathbf{r} = 10 \times 1\,\mathbf{i} + \left[2(-2)^3 + 54\right]\mathbf{j} = 10\mathbf{i} + 38\mathbf{j}$$

For $t = 2$:

$$\mathbf{r} = 10 \times 2\,\mathbf{i} + \left[2(-1)^3 + 54\right]\mathbf{j} = 20\mathbf{i} + 52\mathbf{j}$$

For $t = 3$:

$$\mathbf{r} = 10 \times 3\,\mathbf{i} + \left[2(0)^3 + 54\right]\mathbf{j} = 30\mathbf{i} + 54\mathbf{j}$$

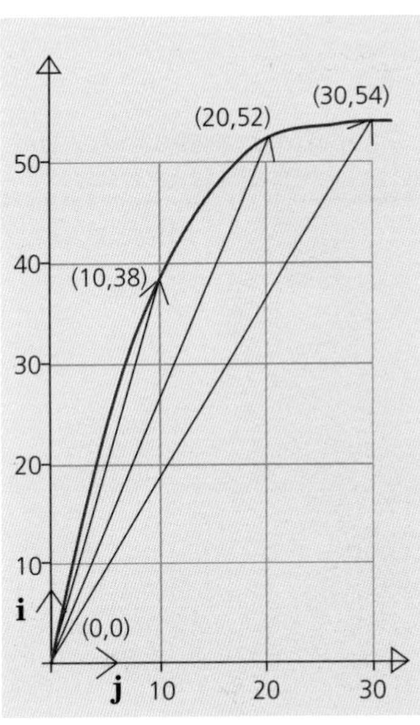

The diagram shows the position vectors that were calculated. The changing position of the rocket has been drawn as a smooth curve through the known positions.

GRAPHICS CALCULATOR ACTIVITY 5.1

Plotting paths

For this activity your calculator must be set into parametric mode.

The position vector of an object that moves in two dimensions will be of the form:

$$\mathbf{r} = x\mathbf{i} + y\mathbf{j}$$

where x and y are functions of time, t. In parametric mode it is possible to enter the expressions for x and y and plot the path of the object.

Consider:

$$\mathbf{r} = 4t\,\mathbf{i} + \left(8 - 5t^2\right)\mathbf{j}$$

where

$x = 4t$ and $y = 8t - 5t^2$

To enter these follow the instructions below.

Texas

Press `Y=`

Type `4T` then `ENTER`

Type `8T-5T²` then `ENTER`

Press `WINDOW` and enter the values below:

$X_{min} = 0$; $X_{max} = 9$
$Y_{min} = 0$; $Y_{max} = 6$; $Y_{scl} = 1$
$T_{min} = 0$; $T_{max} = 1.6$; $T_{step} = 0.1$

Then press `GRAPH` to see the path.

Casio

Press `GRAPH`

Type `4T, 8T-5T²` and `EXE`

Press `RANGE` and enter the values below:

$X_{min} = 0$; $X_{max} = 9$; scl = 1
$Y_{min} = 0$; $Y_{max} = 6$; scl = 1

T, q min : 0 max : 1.6 ptch : 0.1

Then press `EXE`

1 T_{max} gives the maximum time value for the plotting process. Try changing this value from 1.6 to 1.0 and see what happens to the graph.

2 To produce a plot that gives the true shape of the path the value of X_{max} must be 1.5 times bigger than the value of Y_{max}. Try changing these values so that the plot fills the whole screen.

3 Try to describe situations where the motion of an object would follow the path you have plotted.

4 Now plot the path defined by:
$\mathbf{r} = 4t\mathbf{i} + (4t - 2.5t^2)\mathbf{j}$

How does it compare with the path you plotted above?

5 The path of the rocket in the worked example was defined by:

$\mathbf{r} = 10t\mathbf{i} + \left[2(t-3)^3 + 54\right]\mathbf{j}$ for $0 \le t \le 3$

Plot the path on your calculator. (Use $T_{max} = 3$, $X_{max} = 90$, and $Y_{max} = 60$)

6 Plot the path of objects that move with the position vectors given below: You will need to choose suitable values of X_{max}, Y_{max}, etc.

a) $\mathbf{r} = 4t\mathbf{i} + 3t\mathbf{j}$ for $0 \le t \le 5$

b) $\mathbf{r} = (3t - t^2)\mathbf{i} + 4t\mathbf{j}$ for $0 \le t \le 3$

c) $\mathbf{r} = \left(4 - \frac{1}{t}\right)\mathbf{i} + 3t\mathbf{j}$ for $1 \le t \le 10$

d) $\mathbf{r} = \left(t + \frac{1}{t}\right)\mathbf{i} + t\mathbf{j}$ for $1 \le t \le 5$

EXERCISES

5.1A

1 A hockey ball is struck so that it begins to move on a pitch, so that its position vector is given by $\mathbf{r} = 4t\mathbf{i} + 0.1t^2\mathbf{j}$ where $\mathbf{i}$ and $\mathbf{j}$ are unit vectors parallel and perpendicular to the side of the pitch.

a) Find the position vector when, $t = 0$, 1, 2, 4 and 6 seconds.

b) Sketch the path of the ball.

2 The yacht considered in the discussion at the start of this section moves with position vector relative to O:

$$\mathbf{r} = 2t\mathbf{i} + (10 - 3t)\mathbf{j}$$

where distances are in kilometres and time is in hours.

a) The yacht leaves A when $t = 0$ and arrives at B when $t = 4$. Find the position vectors of A and B relative to O.

b) Find the position of the yacht when $t = 2$. Express this position relative to O using a distance and a bearing.

c) When is the position vector of the ship $3\mathbf{i} + 5.5\mathbf{j}$?

3 A child throws a ball and its position vector, relative to an origin at ground level is:

$$\mathbf{r} = 8t\mathbf{i} + (15t - 5t^2 + 1)\mathbf{j}$$

where $\mathbf{i}$ and $\mathbf{j}$ are horizontal and vertical unit vectors.

a) Find the position of the ball when $t = 0$, 0.5, 1.0, 1.5, 2.0, 2.5 and 3.0.

b) Plot the path of the ball.

c) From the path, estimate the maximum height of the ball and the distance from O to where it lands.

4 Two boats A and B move so that their position vectors are:

$$\mathbf{r}_A = (8 + 2t)\mathbf{i} + (4 - 3t)\mathbf{j} \text{ and } \mathbf{r}_B = 4t\mathbf{i} + 2t\mathbf{j}$$

where $\mathbf{i}$ and $\mathbf{j}$ are unit vectors north and east respectively.

a) Find the positions of A and B, when $t = 0$, 2 and 3.

b) Sketch the path of each boat.

c) When does B reach the initial position of A? How far apart are the boats at this time?

5 The position vector of a javelin is modelled by:

$$\mathbf{r} = 15t\mathbf{i} + (2 + 10t - 5t^2)\mathbf{j}$$

where $\mathbf{i}$ and $\mathbf{j}$ are horizontal and vertical unit vectors respectively.

a) Find the position of the javelin when $t = 0$, 0.5, 1, 1.5 and 2.0.

b) Draw the path of the javelin.

c) Find the range of values of t for which the model is valid.

d) Find the length of the throw.

e) What problems may arise in modelling a javelin as a particle?

6 A snooker ball is struck by a cue and moves so that its position, relative to its starting point is:

$$\mathbf{r} = 0.5t\mathbf{i} + (0.4t + 0.08t^2)\mathbf{j}$$

where $\mathbf{i}$ and $\mathbf{j}$ are unit vectors parallel to the sides of the table.

a) Find $\mathbf{r}$ when $t = 0$, 1, 2 and 3.

b) Plot the path of the ball.
c) The table is not quite level. Draw an arrow on your diagram to show the direction of the slope.

7 The path of a rocket launched at a fireworks display was:

$$\mathbf{r} = 20t\mathbf{i} + \left[(t-4)^3 + 64\right]\mathbf{j}$$

where i and $\mathbf{j}$ are horizontal and vertical unit vectors respectively. The rocket exploded when $t = 4$.

a) Find the position of the rocket when $t = 0$, 1, 3, 3.5 and 4.
b) Draw the path of the rocket.
c) In what direction was the rocket moving when it exploded?

8 A snooker table has the dimensions shown in the diagram. A ball has position vector $\mathbf{r} = 0.2\mathbf{i} + 0.3\mathbf{j}$ when it is struck by the cue. It then moves so that it drops into the pocket marked X.

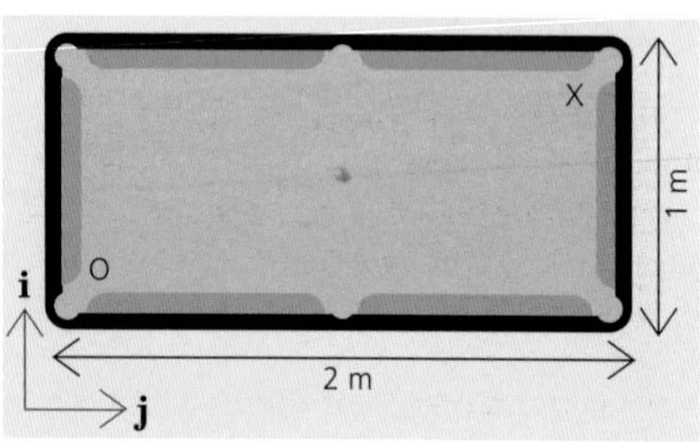

a) State the position vector of the ball when it drops into the pocket.
b) The position vector of the ball is given by:

$$\mathbf{r} = (a + bt)\mathbf{i} + (c + dt)\mathbf{j}$$

Use the initial position of the ball to determine the values of the constants a and c.

c) The ball drops into the pocket two seconds after it was hit. Use this to find the value of the constants b and d.

EXERCISES

5.1 B

1 The motion of a car as it travels round a bend is described by the position vector:

$$\mathbf{r} = \left(0.5t^2 + 4t\right)\mathbf{i} + 3t\mathbf{j}$$

where $\mathbf{i}$ and $\mathbf{j}$ are perpendicular unit vectors.

a) Find the position of the car when $t = 0$, 1, 2 and 3.
b) Plot the path of the car.

2 A ship moves with position vector:

$$\mathbf{r} = (-3 + 2t)\mathbf{i} + 3t\mathbf{j}$$

where $\mathbf{i}$ and $\mathbf{j}$ are horizontal and vertical unit vectors respectively.

a) Find the position of the ship when $t = 0$, 2, 4, and 6.
b) Plot the path of the ship.
c) Find the time when the ship is due north of the origin.

3 A basket ball moves with position vector:

$$\mathbf{r} = 3t\mathbf{i} + \left(2 + 12t - 5t^2\right)\mathbf{j}$$

where $\mathbf{i}$ and $\mathbf{j}$ are horizontal and vertical unit vectors.

a) Find the positions of the basket ball when $t = 0$, 1, 2 and 3.
b) Draw the path of the basket ball.
c) What is the maximum height of the ball?

d) Find the time for which the ball is in the air.
e) How far does the ball land from the point where it was thrown?

4 A football rolls on a sloping pitch so that its position vector is:

$$\mathbf{r} = 3t\mathbf{i} + 0.1t^2\mathbf{j}$$

where **i** and **j** are perpendicular unit vectors.
a) Find the position of the football when $t = 0$, 1, 2, 3 and 4.
b) Plot the path of the ball.
c) On your plot draw an arrow to indicate the direction of the slope.

5 A cyclist is moving down a hill, so that her position vector is given by:

$$\mathbf{r} = 3t\mathbf{i} + (16 - 2t)\mathbf{j} \quad \text{for } 0 \le t \le 8$$

where **i** and **j** are horizontal and vertical unit vectors.
a) Plot the path and describe the slope of the hill.

Another cyclist moves down another hill with position vector:

$$\mathbf{r} = 2t\mathbf{i} + \left(16 - \frac{t^2}{9}\right)\mathbf{j}$$

b) Plot the path of this cyclist and describe the slope of the hill.
c) If the second cyclist descends the same vertical distance as the first cyclist, find the range of values of t for which the second model is valid.

6 A ball moves with position vector:

$$\mathbf{r}_A = 4.7t\mathbf{i} + (1.7t - 5t^2)\mathbf{j}$$

and a second ball moves with position vector:

$$\mathbf{r}_B = 1.7t\mathbf{i} + (4.7t - 5t^2)\mathbf{j}$$

where **i** and **j** are horizontal and vertical unit vectors.
a) Show that both balls land in the same place. Describe the differences between the two paths.
b) Which ball is in the air for the longer time?

7 The diagram shows part of a blow football table. The ball moves with position vector:

$$\mathbf{r} = (50 - 20t)\mathbf{i} + 5t^2\mathbf{j}$$

a) When does the ball hit the end of the pitch?
b) Is a goal scored? Sketch the path of the ball.

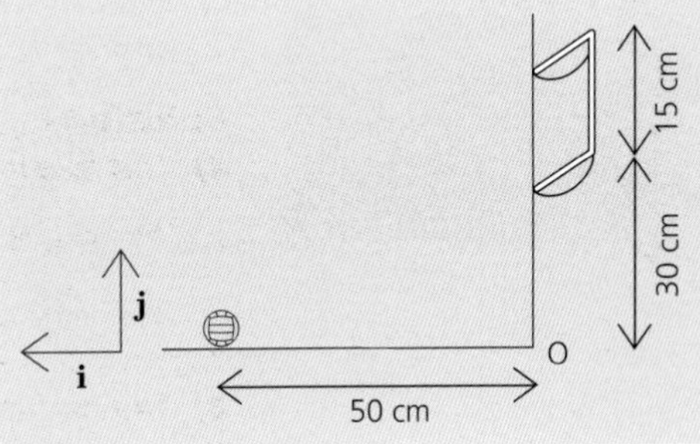

8 Two boats, A and B, move with position vectors:

$$\mathbf{r}_A = 3t\mathbf{i} + 2.5t\mathbf{j} \text{ and } \mathbf{r}_B = 2t\mathbf{i} + (t^2 - 4t + 8)\mathbf{j}$$

a) Show that the paths of both boats have one point in common.
b) Find how far apart the boats are when the second boat reaches this point.

SPEED AND VELOCITY

Exploration 5.2

Motion of athletes as they run

Imagine a group of athletes on a running track competing in a middle distance race. What happens to the motion of the athletes as they run round the track?

The difference between speed and velocity

During a race, the **speed** of the athletes will almost certainly change from time to time, but overall, for the majority of the race, it will probably be reasonable to assume that the speed is constant. However, the **direction of motion** will change during the race. On the straight it will be constant, but on the bends it will change continually, always being **along the tangents** to the curves.

The speed of an object simply describes how fast an object moves.

The **average velocity** of an object is defined as the change in position or displacement divided by the time taken.

$$\text{Average velocity} = \frac{\text{displacement}}{\text{time taken}}$$

The velocity of an object describes both how fast it moves and in which direction. The velocity is in fact the **rate of change of position**.

Example 5.2

A jet-ski moves from position A to position B in 30 seconds. The position vector of A is $\mathbf{r}_A = 50\mathbf{i} + 100\mathbf{j}$ *and the position vector of B is* $\mathbf{r}_B = 200\mathbf{i} - 20\mathbf{j}$.

a) *Find the average velocity of the jet-ski, in terms of* $\mathbf{i}$ *and* $\mathbf{j}$.
b) *Find the magnitude of the average velocity.*
c) *Explain why the average speed of the jet-ski is greater than the answer to* ***b)***.

Solution

a) *The displacement of the jet-ski will be:*

$$\mathbf{r}_B - \mathbf{r}_A = (200\mathbf{i} - 20\mathbf{j}) - (50\mathbf{i} + 100\mathbf{j}) = 150\mathbf{i} - 120\mathbf{j}$$

$$\text{Average velocity} = \frac{\text{displacement}}{\text{time taken}} = \frac{150\mathbf{i} - 120\mathbf{j}}{30} = 5\mathbf{i} - 4\mathbf{j}$$

b) *The magnitude of the average velocity is* $\sqrt{5^2 + 4^2} = 6.4$ *m s*$^{-1}$.
c) *If the jet-ski travelled from A to B directly at a constant speed, its average speed would be 6.4 m s*$^{-1}$*. However, it cannot do this as there is some land in the way, so its speed must be greater than 6.4 m s*$^{-1}$.

Example 5.3

A cyclist rides over the small hill shown in the diagram. Draw arrows to represent the velocity of the cyclist at each position shown.

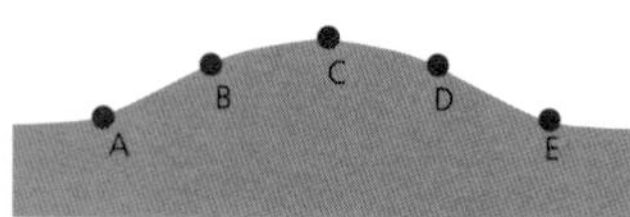

Solution

The direction will always be at a tangent to the curve of the hill. The cyclist will probably lose speed between A and C and gain speed between C and E, so shorter arrows indicate a lower speed near to the top of the hill.

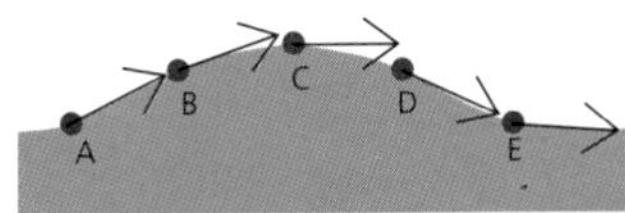

Example 5.4

A motor boat is travelling at a speed of 3 m s^{-1} on a bearing of 230°. Express the velocity of the boat in terms of unit vectors **i** *and* **j** *that are directed east and north respectively.*

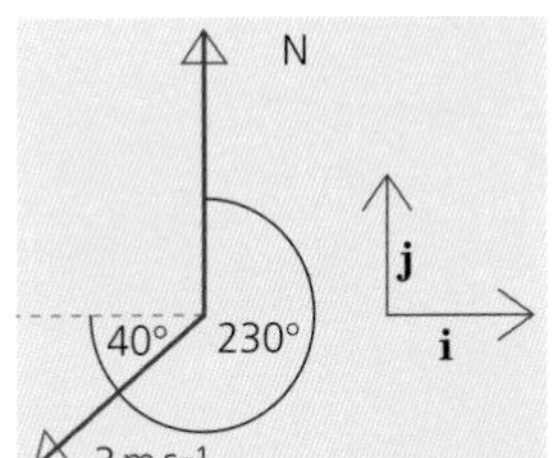

Solution

The velocity vector can be treated in exactly the same way as force vectors were in the previous chapter.

$\mathbf{v} = -3\cos 40°\,\mathbf{i} - 3\cos 50°\,\mathbf{j} = -2.30\mathbf{i} - 1.93\mathbf{j}$

EXERCISES

5.2A

1 Express each velocity given below in terms of the unit vectors **i** and **j** which are east and north respectively.

a) A jet-ski moving at 5 m s^{-1} on a bearing of 050°.

b) An aeroplane flying at 85 m s^{-1} on a bearing of 302°.

c) A competitor in an orienteering competition running at 3 m s^{-1} on a bearing of 188°.

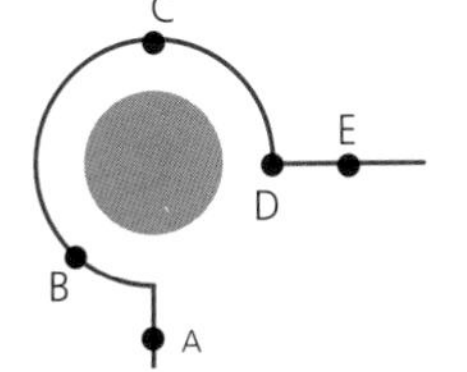

2 A car maintains a constant speed while negotiating a roundabout. Draw arrows to show the velocity of the car at each point shown on the diagram.

3 Two children sit on the roundabout as shown in the diagram. How do their velocities compare if they sit at positions:

a) A and B,

b) A and C,

c) B and C?

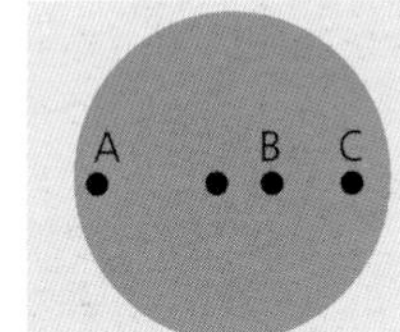

4 Describe situations in which the motion of an object is likely to have:

a) constant velocity,

b) velocity with constant direction only,

c) constant speed, but variable velocity.

5 The positions of three boats are recorded when $t = 0$ and when $t = 20$, and are listed in the table below.

	$\mathbf{r}(0)$	$\mathbf{r}(20)$
A	$4\mathbf{i} + 10\mathbf{j}$	$4\mathbf{i} + 110\mathbf{j}$
B	$100\mathbf{i} + 100\mathbf{j}$	$80\mathbf{i} + 20\mathbf{j}$
C	$38\mathbf{i} + 110\mathbf{j}$	$46\mathbf{i} + 10\mathbf{j}$

a) Find the average velocity of each boat, in terms of $\mathbf{i}$ and $\mathbf{j}$, and then find the magnitude of each average velocity.
b) Sketch the path of each boat.
c) If all the boats move with constant speed what can be deduced about the speed of A, B and C?

6 Three aeroplanes, A, B and C, are approaching an airport. The velocities of the planes are given as:

$$\mathbf{v}_A = 100\mathbf{i} + 20\mathbf{j} \quad \mathbf{v}_B = 60\mathbf{i} - 150\mathbf{j} \quad \mathbf{v}_C = -40\mathbf{i} - 160\mathbf{j}$$

where $\mathbf{i}$ and $\mathbf{j}$ are unit vectors east and north respectively.

Find the speed of each aeroplane and the bearing on which they are flying.

7 A football is kicked so that it is initially at $20\,\mathrm{m\,s^{-1}}$, in a direction 20° above the horizontal. Express the initial velocity of the ball in terms of the unit vectors, $\mathbf{i}$ and $\mathbf{j}$, as illustrated in the diagram.

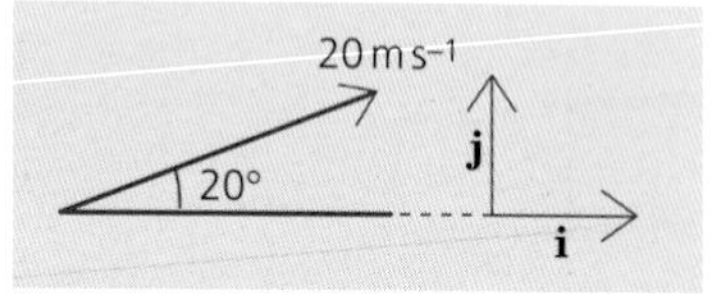

8 A train travelling up a steady incline has a velocity of $40\mathbf{i} + 2\mathbf{j}$, where $\mathbf{i}$ and $\mathbf{j}$ are horizontal and vertical unit vectors respectively. Find the speed of the train and the slope of the incline.

9 A child sits on a roundabout that rotates anti-clockwise. The child has speed $2\,\mathrm{m\,s^{-1}}$ and describes a circle of radius 1.8 m.

a) Express the velocity of the child at positions A and B in terms of the unit vectors $\mathbf{i}$ and $\mathbf{j}$.
b) Find the time it takes for the child to get from A to B.
c) Find the average velocity of the child.
d) When the child returns to position A, what is her average velocity?

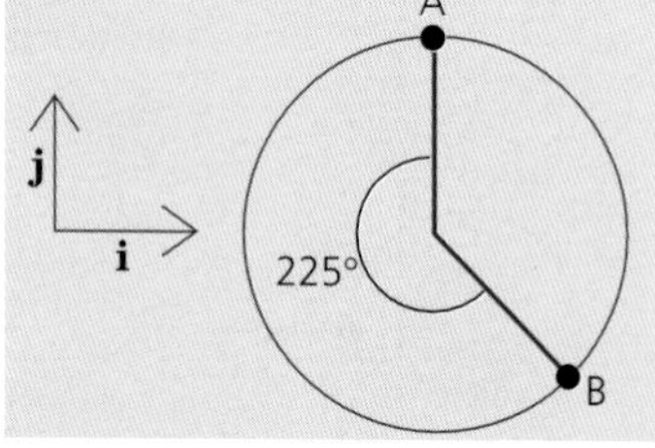

10 A bird flies between two trees. The position vector of the second tree is $40\mathbf{i} + 80\mathbf{j}$ relative to the first tree, where $\mathbf{i}$ and $\mathbf{j}$ are unit vectors east and north respectively. It takes the bird ten seconds to complete the flight.

a) Show that it is possible to model the velocity of the bird as $\mathbf{v} = 4\mathbf{i} + 8\mathbf{j}$. Explain the assumptions on which this model is based.
b) What criticisms can be made of this model?
c) A revised model is:

$$\mathbf{v} = \frac{6}{25}t(10-t)\mathbf{i} + \frac{12}{25}t(10-t)\mathbf{j} \quad \text{for } 0 \le t \le 10$$

Show that the bird travels directly from A to B.
d) Find an expression for the speed of the bird, at time t.
e) Find the initial and final speeds of the bird and the maximum speed reached.
f) What advantages or disadvantages does the revised model have compared to the original model?

EXERCISES

5.2 B

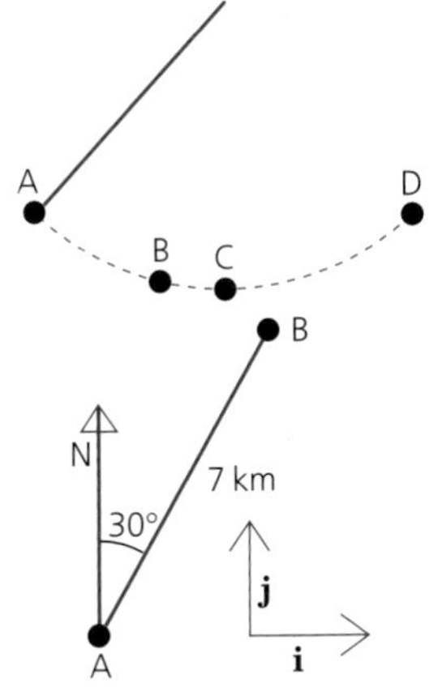

1 Express each velocity given below in terms of the unit vectors $\mathbf{i}$ and $\mathbf{j}$, east and north respectively.

a) A hill walker moving at 1 m s^{-1} on a bearing of 074°.
b) A ship sailing on a bearing of 250° at 6 m s^{-1}.
c) A hovercraft travelling at 30 m s^{-1} on a bearing of 125°.

2 The diagram shows the path of a pendulum that swings between A and D. Draw vectors to represent the velocity of the bob on the end of the pendulum at each point as:

a) it swings from A to D, **b)** it swings from D to A.

3 The bearing of town B from town A is 030° and they are 7 km apart. A car travels from A to B in eight minutes.

a) Find the position vector of B relative to A, using units of metres.
b) Find the average velocity of the car and its magnitude, in m s^{-1}.
c) What can you deduce about the average speed of the car?

4 As part of a display two aeroplanes fly in a circle centred at O, but so that they are always both in line with O.

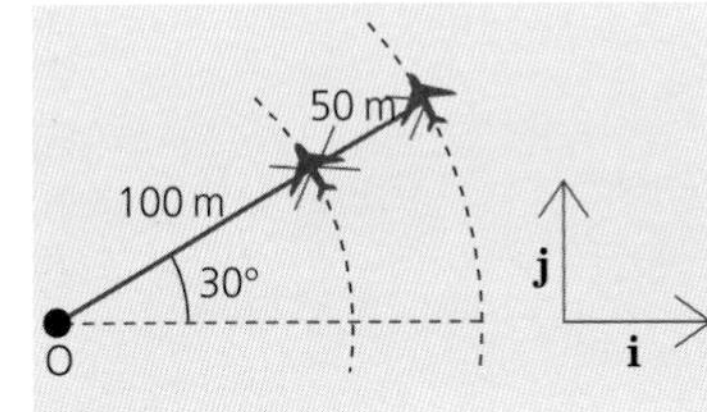

a) How do the speeds of the aeroplanes compare?
b) Find the average speed of each plane if they complete the circles in four seconds.

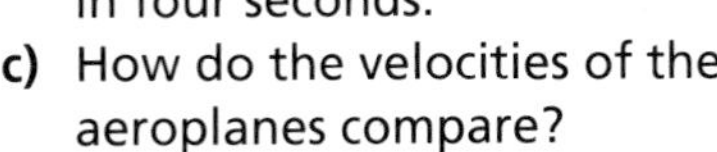

c) How do the velocities of the aeroplanes compare?
d) Find the velocity of each plane in the position shown.

5 The velocity of a ship, in m s^{-1}, at three positions, A, B and C is expressed in terms of unit vectors $\mathbf{i}$ and $\mathbf{j}$, which are east and north respectively.

$$\mathbf{v}_A = 8\mathbf{i} + 5\mathbf{j} \quad \mathbf{v}_B = 3\mathbf{i} - 4\mathbf{j} \quad \mathbf{v}_C = -2\mathbf{i} - 6\mathbf{j}$$

a) Express each velocity as a bearing and a speed.
b) If the ship travels for 120 seconds with each velocity, find its final position, relative to its starting point.
c) Sketch the path of the ship and criticise your answer to **b)**.

6 The radar screen shows the position of a helicopter at two-second intervals. It is initially at A, it then moves to B and then to C.

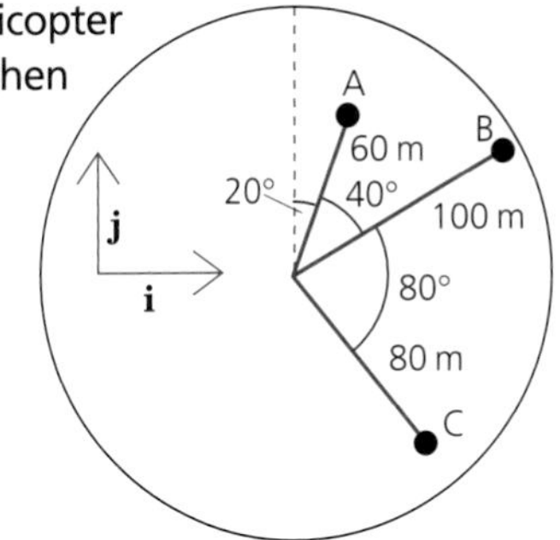

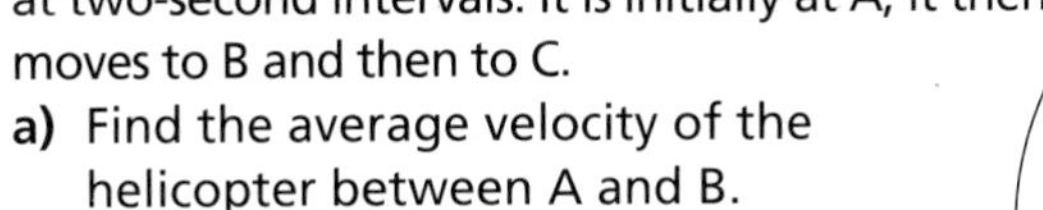

a) Find the average velocity of the helicopter between A and B.
b) Find the average velocity of the helicopter between B and C.
c) Find the speed and bearing of the helicopter, assuming it moves with the average velocity between B and C.

7 A bicycle is moving up a hill with velocity $2\mathbf{i} + 0.2\mathbf{j}$, where $\mathbf{i}$ and $\mathbf{j}$ are horizontal and vertical unit vectors respectively.

a) Find the speed of the cyclist.

b) What angle does the hill make with the horizontal?
c) If the cyclist travels at the same speed up a slope at 2° to the horizontal, find her velocity.

8 When an aeroplane takes off it initially travels at $50\,\mathrm{m\,s^{-1}}$ at 25° to the horizontal.
a) Express this speed as a vector, defining appropriate unit vectors.
b) Express the velocity using an alternative pair of unit vectors.
c) How long does it take the aeroplane to reach an altitude of 100 m? Explain any assumptions that you need to make.

9 Two children, Anil and Ben, amuse themselves on a long train journey by playing with a ball. The train is moving at $40\,\mathrm{m\,s^{-1}}$ and the children remain seated.

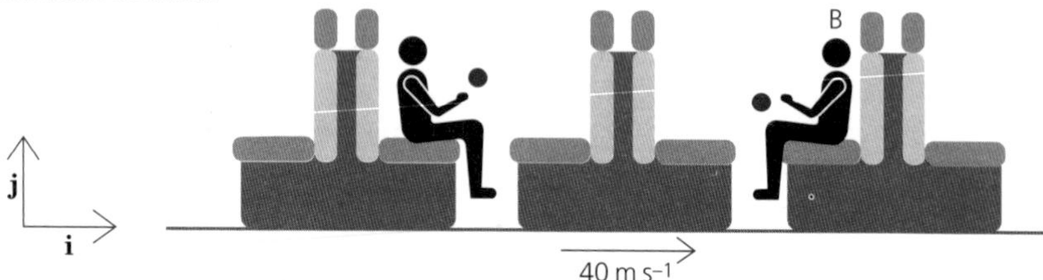

Find the velocity of the ball, in terms of the unit vectors $\mathbf{i}$ and $\mathbf{j}$, if:
a) Anil is holding the ball,
b) the ball is travelling horizontally away from Ben at $5\,\mathrm{m\,s^{-1}}$,
c) Anil throws it towards Ben at $10\,\mathrm{m\,s^{-1}}$, at an angle of 10° above the horizontal.

10 The velocity of a ball that is thrown into the air is modelled as:

$$\mathbf{v} = 10\mathbf{i} + (8 - 5t)\mathbf{j}$$

where $\mathbf{i}$ and $\mathbf{j}$ are horizontal and vertical unit vectors respectively.
a) Find the initial speed and direction of motion of the ball.
b) When the ball reaches its highest position it is travelling horizontally. Find the time when this takes place and the horizontal distance travelled at this time. (In practice the ball is travelling at 8 $\mathrm{m\,s^{-1}}$ horizontally when the ball reaches its highest point.)
c) Devise a revised model for the horizontal component of velocity, which does not alter the initial speed, but gives a speed of $8\,\mathrm{m\,s^{-1}}$, when the ball is moving horizontally.
d) Criticise your revised model.

ACCELERATION

Exploration 5.3

What is acceleration?

Which of the objects described below are accelerating?

- A car slowing down to stop at a pedestrian crossing.
- A teenager on a fairground ride that rotates at a constant rate.
- A cyclist travelling along a country lane at a constant speed.
- A planet orbiting the sun.
- A parachutist who has reached terminal speed.

What we mean by acceleration

The term **acceleration** is used to describe how the velocity of an object is changing. If the velocity is constant, then the acceleration is zero. If the velocity changes in any way, then the object is accelerating. The acceleration is defined as the **rate of change of velocity**.

In this section only objects that have constant accelerations will be considered.

The constant acceleration formulae developed in Chapter 3 can also be expressed in vector form.

$\mathbf{v} = \mathbf{u} + \mathbf{a}t$

$\mathbf{r} = \mathbf{u}t + \frac{1}{2}\mathbf{a}t^2 + \mathbf{r}_0$

Note that $\mathbf{r}_0$ is the initial position.

Example 5.5

A bullet leaves the barrel of a gun 1.5 m above ground level, travelling horizontally at 70 m s⁻¹. It experiences a downward acceleration of 10 m s⁻². When does the bullet hit the ground and how far has it travelled horizontally?

Solution

The diagram shows the bullet modelled as a particle at its initial position, with vectors to show its initial velocity and acceleration. The initial velocity is $\mathbf{u} = 70\mathbf{i}$, *the acceleration is* $\mathbf{a} = -10\mathbf{j}$ *and the initial position* $\mathbf{r}_0 = 1.5\mathbf{j}$.

Using the constant acceleration equation:

$\mathbf{r} = \mathbf{u}t + \frac{1}{2}\mathbf{a}t^2 + \mathbf{r}_0$

gives:

$\mathbf{r} = 70\mathbf{i}t + \frac{1}{2}\times -10\mathbf{j}t^2 + 1.5\mathbf{j} = 70t\mathbf{i} + \left(1.5 - 5t^2\right)\mathbf{j}$

The bullet will hit the ground when the $\mathbf{j}$ *component of* $\mathbf{r}$ *is zero. This occurs when:*

$1.5 - 5t^2 = 0$

$\Rightarrow \quad t^2 = \frac{1.5}{5} = 0.3$

$\Rightarrow t = 0.548 \text{ s}$

When $t = 0.548$ s *the position vector is:*

$\mathbf{r} = 70 \times 0.548\mathbf{i} + 0\mathbf{j} = 38.36\mathbf{i}$

So the horizontal distance travelled by the bullet is 38.36 m.

Example 5.6

A boat leaves a marker buoy where it was at rest. It experiences an acceleration of 1.2 m s⁻² on a bearing of 070° for ten seconds and then travels with a constant velocity for a further 20 seconds.

Find the position and speed of the boat after:

a) *10 seconds,* ***b)*** *30 seconds.*

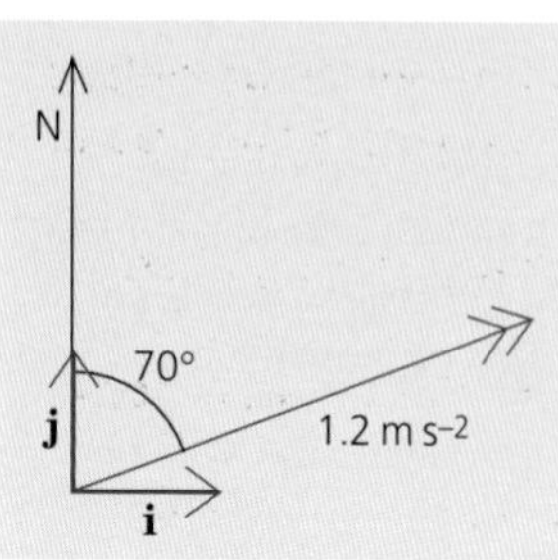

Solution

The diagram shows the acceleration of the boat at its initial position. The unit vectors $\mathbf{i}$ *and* $\mathbf{j}$ *are east and north respectively.*
The acceleration can be expressed as:

$$\mathbf{a} = 1.2\cos 20°\,\mathbf{i} + 1.2\cos 70°\,\mathbf{j} = 1.13\mathbf{i} + 0.41\mathbf{j}$$

a) *In this case both* $\mathbf{u} = 0\mathbf{i} + 0\mathbf{j}$ *and* $\mathbf{r}_0 = 0\mathbf{i} + 0\mathbf{j}$ *so using:*

$$\mathbf{r} = \mathbf{u}t + \tfrac{1}{2}\mathbf{a}t^2 + \mathbf{r}_0$$

gives:

$$\mathbf{r} = \tfrac{1}{2}(1.13\mathbf{i} + 0.41\mathbf{j})t^2 = 0.565t^2\mathbf{i} + 0.205t^2\mathbf{j}$$

and using:

$$\mathbf{v} = \mathbf{u} + \mathbf{a}t$$

gives:

$$\mathbf{v} = (1.13\mathbf{i} + 0.41\mathbf{j})t = 1.13t\mathbf{i} + 0.41t\,\mathbf{j}$$

So when $t = 10$*:*

$$\mathbf{r} = 0.565 \times 10^2\,\mathbf{i} + 0.205 \times 10^2\,\mathbf{j} = 56.5\mathbf{i} + 20.5\mathbf{j}$$

and

$$\mathbf{v} = 1.13 \times 10\mathbf{i} + 0.41 \times 10\mathbf{j} = 11.3\mathbf{i} + 4.1\mathbf{j}$$

b) *The boat now moves with a constant speed of* $11.3\mathbf{i} + 4.1\mathbf{j}$*, so in the next 20 seconds its position changes by:*

$$20(11.3\mathbf{i} + 4.1\mathbf{j}) = 226\mathbf{i} + 82\mathbf{j}$$

So its final position is given by:

$$\mathbf{r} = (56.5\mathbf{i} + 20.5\mathbf{j}) + (226\mathbf{i} + 82\mathbf{j}) = 282.5\mathbf{i} + 102.5\mathbf{j}$$

EXERCISES

1 A boat is initially moving at $2\,\text{m s}^{-1}$ on a bearing of 045°.

a) Express this initial speed in terms of the unit vectors $\mathbf{i}$ and $\mathbf{j}$, which are east and north respectively.

b) The boat then experiences an acceleration of $0.1\mathbf{i} + 0.3\mathbf{j}$.
Find the velocity of the boat when it has been accelerating for ten seconds.

c) Find the position of the boat at the end of the ten-second period, taking its starting point as the origin.

2 A bullet is fired with an initial velocity of $180\mathbf{i}$, from a point with position vector $1.5\mathbf{j}$. The bullet experiences an acceleration of $-10\mathbf{j}$. The unit vectors $\mathbf{i}$ and $\mathbf{j}$ are horizontal and vertical respectively.

a) Find expressions for the velocity and position of the bullet, t seconds after it is fired.

b) Find the time taken for the bullet to reach the ground and the horizontal distance travelled.

c) What is the speed of the bullet when it hits the ground?

3 For the first five seconds of its motion down a runway, an aeroplane experiences an acceleration of $16\,\text{m s}^{-2}$. Use unit vectors **i** and **j** parallel and perpendicular to the runway respectively.

a) Find the position and velocity of the aeroplane after the first five seconds.

The acceleration then changes to $4\mathbf{i} + 6\mathbf{j}$ and remains constant for a further 20 seconds.

b) Find the position and velocity of the plane at the end of this 20-second period.

4 A golf ball is hit so that its initial velocity is $20\,\text{m s}^{-1}$ at 60° above the horizontal.

a) Express the initial velocity of the golf ball in terms of the unit vectors, **i** and **j**, which are horizontal and vertical respectively.

b) The ball is initially at the origin and experiences an acceleration of $-10\mathbf{j}$. Find an expression for the position of the ball, at time t.

c) Sketch the path of the ball.

d) When does the ball hit the ground and how far is it from the point where it was hit?

5 A child on a slide experiences an acceleration of $2\mathbf{i} - 1.5\mathbf{j}$. Assume that the child begins from rest, and is at the origin.

a) If the child starts at rest find an expression for the position of the child, at time t.

b) Sketch the path of the child for $0 \leq t \leq 3$.

c) If the child takes three seconds to reach the bottom of the slide, find the length of the slide.

d) What is the speed of the child at the bottom of the slide?

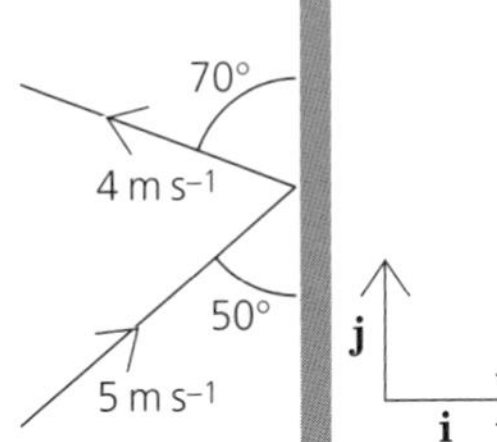

6 A squash ball hits the front wall, moving at $5\,\text{m s}^{-1}$, and rebounds at $4\,\text{m s}^{-1}$. The diagram shows the change in direction.

a) Express both velocities in terms of the unit vectors **i** and **j**.

b) Find the acceleration of the ball while in contact with the wall, if this contact is maintained for 0.1 seconds.

7 A submarine at a depth of 100 m jettisons a buoyant package to be collected by a contact waiting in a boat on the surface. The submarine is travelling horizontally at $2\,\text{m s}^{-1}$ and the package accelerates at $0.5\,\text{m s}^{-2}$ vertically as it rises.

a) Find an expression for the velocity and position of the package relative to its point of release.

b) Sketch the path of the package.

c) Find the speed of the package when it reaches the surface, and its distance from the boat, if the boat was above the submarine when the package was jettisoned.

d) Why is the acceleration given likely to be unrealistic?

8 A football is kicked so that it leaves the ground with a speed of $30\,\text{m s}^{-1}$, travelling at 40° above the horizontal. The unit vectors **i** and **j** are horizontal and vertical respectively.

a) Express the initial velocity of the ball in terms of **i** and **j**.

b) Explain why the acceleration of the ball is $-10\mathbf{j}$. What assumptions have been made in order to obtain this?
c) Find expressions for the velocity and position of the ball if it leaves the ground at the origin.
d) When does the ball hit the ground again and what horizontal distance has it travelled?
e) Comment on your answer to **d)** in the light of your answer to **b)**.

9 When an aeroplane leaves a runway its velocity, in m s^{-1}, is $100\mathbf{i} + 50\mathbf{j}$, where $\mathbf{i}$ and $\mathbf{j}$ are horizontal and vertical unit vectors respectively. Ten seconds after take-off, its position, relative to where it leaves the runway, is $1200\mathbf{i} + 600\mathbf{j}$.
a) Find the acceleration of the aeroplane, stating any assumptions that you make.
b) Find the speed of the aeroplane after ten seconds.

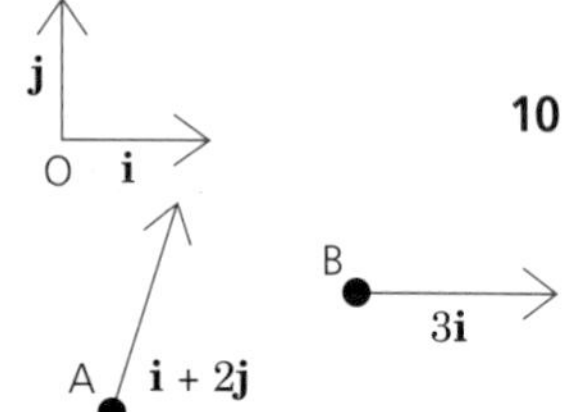

10 The points A and B in the diagram represent the positions of a bird. A has position vector $2\mathbf{i} + 3\mathbf{j}$ and B has position vector $6\mathbf{i} + 5\mathbf{j}$. At each point the bird has the velocity shown.
a) Find the time it takes the bird to get from A to B.
b) Find the acceleration of the bird.

EXERCISES

5.3 B

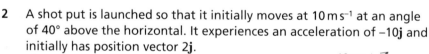

1 The cue ball on a snooker table is struck so that it moves at $1.2\,\text{m s}^{-1}$ parallel to the long side of the table.
a) Find the initial velocity of the ball in terms of $\mathbf{i}$ and $\mathbf{j}$.
b) As the table slopes slightly it experiences an acceleration $0.1\mathbf{j}$. Find an expression for the velocity of the ball.
c) After 1.5 seconds the ball has travelled the distance required by the player. Find its velocity and position relative to its starting point.
d) Do you think the cue ball hits the ball it was aimed at?

2 A shot put is launched so that it initially moves at $10\,\text{m s}^{-1}$ at an angle of 40° above the horizontal. It experiences an acceleration of $-10\mathbf{j}$ and initially has position vector $2\mathbf{j}$.

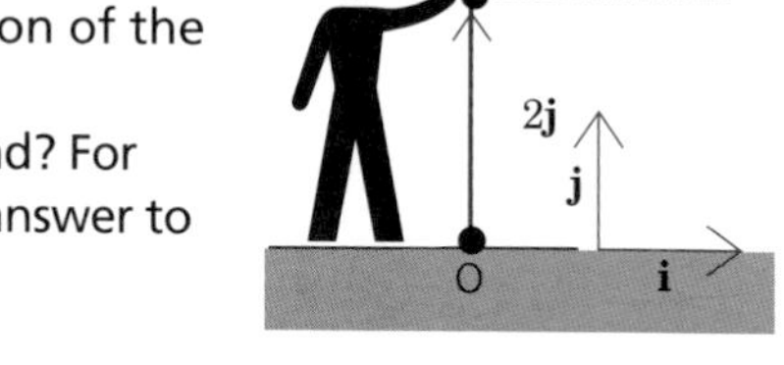

a) Express the initial velocity of the shot put in terms of $\mathbf{i}$ and $\mathbf{j}$.
b) Find an expression for the position of the shot at any time.
c) When does the shot hit the ground? For what range of values of t is your answer to **b)** valid?
d) Find the length of the throw.

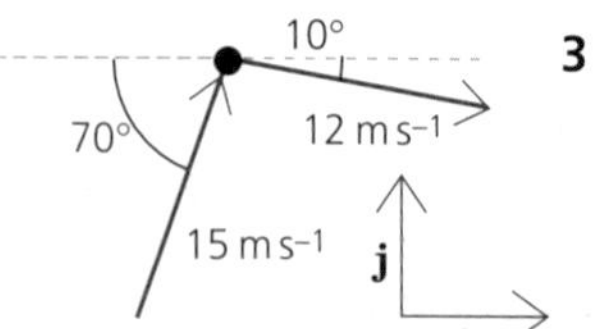

3 The diagram shows the velocity of a football, before and after it hits a goal post.
a) Express both velocities in terms of the unit vectors $\mathbf{i}$ and $\mathbf{j}$.
b) Find the acceleration of the ball if it is in contact with the post for 0.5 seconds.

c) State any assumptions that you made to obtain your answer to **b)**. Suggest a possible revised assumption that may be more realistic.

4 A sky diver jumps out of a moving aeroplane, that is travelling horizontally at $50\,\text{m s}^{-1}$. Assume that **i** and **j** are unit vectors, horizontal and vertical respectively. The sky diver's acceleration is $-10\mathbf{j}$.

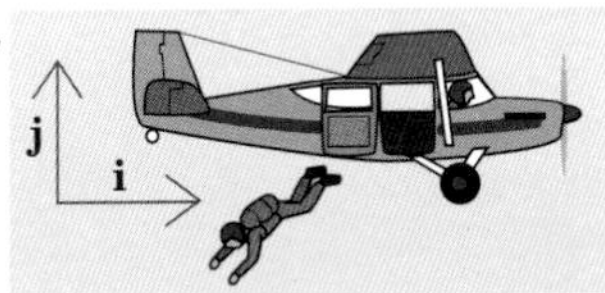

a) Explain why the initial velocity of the sky diver is $50\mathbf{i}$.
b) Find the velocity of the sky diver after 20 seconds freefall.
c) Calculate the speed of the sky diver and comment on your result.
d) If the terminal speed of a sky diver is $80\,\text{m s}^{-1}$, suggest the range of values for which your model is valid.

5 A hockey ball is moving at $3.4\,\text{m s}^{-1}$. It experiences an acceleration of $25\,\text{m s}^{-2}$ at right angles to its path, while it is in contact with a hockey stick. Assume that the acceleration is present for 0.6 seconds.
a) If the velocity of the ball is initially $3.4\mathbf{i}$, express the acceleration in terms of the perpendicular unit vectors **i** and **j**.
b) Find the velocity of the ball after it has been hit.
c) Find the angle through which the motion of the ball has been deflected.

6 As a skateboarder travels down a slope, he experiences an acceleration $\mathbf{i} - 0.05\mathbf{j}$. Assume that the skateboarder starts at rest at the top of the slope. The unit vectors **i** and **j** are horizontal and vertical respectively.
a) Find an expression for the position vector of the skateboarder at time t.
b) Find an expression for the distance travelled.
c) How long does it take the skateboarder to travel 20 m down the slope? What are the velocity and speed at this time?

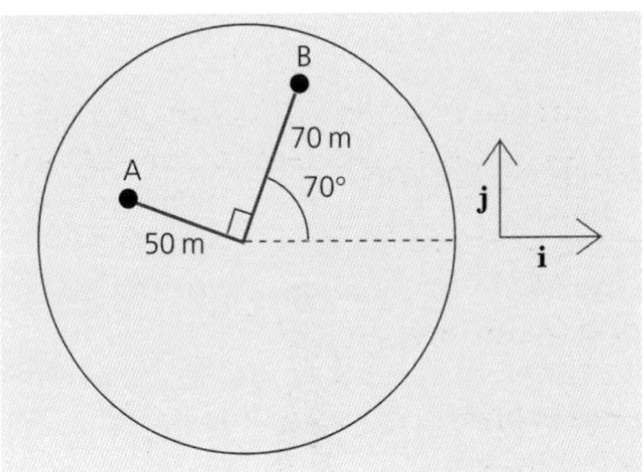

7 The radar screen shows the positions of a power boat that travels directly between the points A and B. At A it is at rest, and it accelerates until it is at B two seconds later.
a) Find the acceleration of the boat.
b) Find the velocity of the boat at B.
c) Find the position of the boat two seconds after it leaves B, if it maintains a constant velocity. Express the position as a distance at a bearing, given that **j** points north.

8 A hot air balloon rises from a point on the ground. After ten seconds its position vector, relative to its starting point, is $5\mathbf{i} + 20\mathbf{j}$, where **i** and **j** are horizontal and vertical unit vectors respectively.
a) Find expressions for the acceleration and velocity of the balloon.
b) The horizontal movement of the balloon is due to the wind. After the first ten seconds the horizontal component of the velocity remains constant. Find the position of the balloon ten seconds later.

9 A firework starts at rest and experiences an acceleration of $4.8\,\text{m s}^{-2}$ at an angle of 80° to the horizontal for four seconds before it explodes. Find the height of the rocket when it explodes, and the distance it has travelled.

10 A jet-ski is initially moving at $10\,\text{m s}^{-1}$. It experiences an acceleration of $1.8\,\text{m s}^{-2}$ at right angles to its direction of motion. Find the speed of the jet-ski after five seconds and its change in position in this time.

CONSOLIDATION EXERCISES FOR CHAPTER 5

1 A car travels at steady speed on a horizontal winding road.

a) State whether the velocity is:
- **i)** always zero,
- **ii)** always constant but not zero,
- **iii)** variable.

Give a reason for your choice of answer.

b) State whether the acceleration is:
- **i)** always constant,
- **ii)** always parallel to the velocity,
- **iii)** variable but not always parallel to the velocity.

Give a reason for your choice of answer.

(WJEC Question 3, Specimen Paper A2, 1994)

2 A marksman is shooting on a rifle range. The target is 400 m away and at the same horizontal level as the rifle. The bullet is fired at a speed of $350\,\text{m s}^{-1}$ and air resistance may be neglected.

Let the vertical distance the bullet would fall over this range if it were fired horizontally be d.

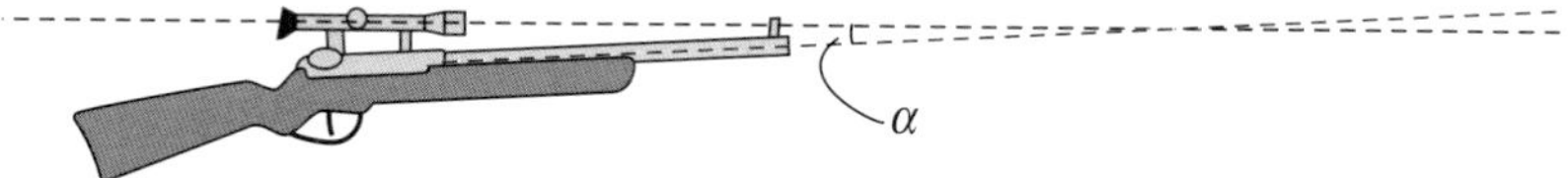

a) Given that $g = 9.8\ \text{m s}^{-2}$, show that d is 6.4 m.

In practice, a rifle is aimed by using sights which are adjustable. The line of the sights is inclined to the barrel to compensate for the bullet falling under gravity, as indicated in the diagram.
The marksman attempts to adjust the sights for a range of 400 m. This results in the line of the sights being horizontal when the barrel is directed at a point 6.4 m above the target.

b) Calculate the angle α between the sights and the barrel.

c) Show that the actual distance which the bullet drops relative to the rifle is less than 0.1 m.

The rifle is now fired horizontally at a target 800 m away (ie. the range is doubled).

d) Write down the approximate vertical distance the bullet will fall over this range.

(MEI Question 4, MI, June 1992)

MATHEMATICAL MODELLING ACTIVITY

Specify the real problem

Problem statement

Will an umbrella keep you dry when it is raining?

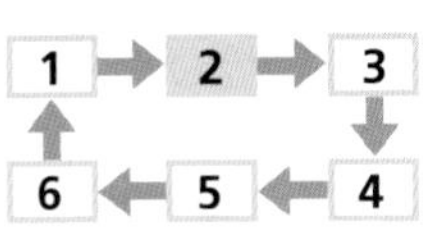

Set up a model

Set up a model

The first stage in setting up a model is to identify the important features. The list below includes the main features.

- The size and position of the umbrella.
- The speed at which the person holding the umbrella is walking.
- The speed of the falling rain and the angle at which it falls.

Are there other important factors that you feel should be included? Having considered these features a set of assumptions can now be set up.

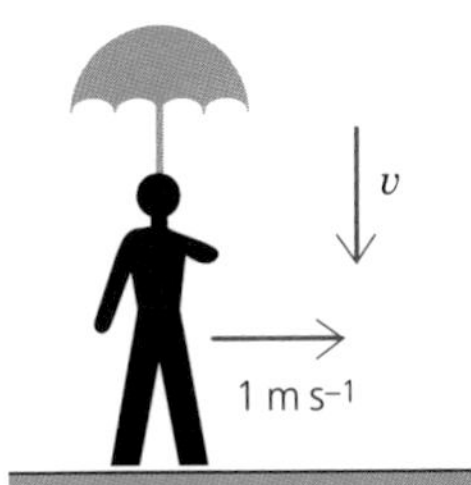

- The rain falls vertically at a constant v m s^{-1}.
- The person walks at 1 m s^{-1}.
- The umbrella is at a height of 2 m and has a radius of 50 cm.
- The person is modelled as a vertical rod and the umbrella handle is also vertical and an extension of the person.

These assumptions allow a simple first model to be formulated.

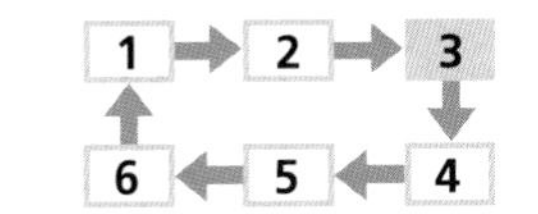

Formulate the mathematical problem

Formulate a mathematical problem

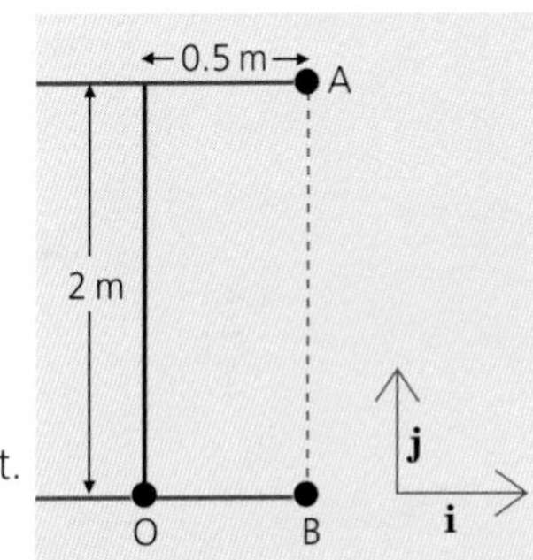

Consider a drop of rain at A. It has a velocity $-v\mathbf{j}$ and its initial position is $0.5\mathbf{i} + 2\mathbf{j}$.

The velocity of the person is $1\mathbf{i}$.

Find v if the rain drop lands at the person's feet.

Solve the mathematical problem

Mathematical solution

The position of the rain drop is:

$$\mathbf{r} = 0.5\mathbf{i} + (2 - vt)\mathbf{j}$$

When the raindrop hits the ground at B the time taken to fall from A is:

$$t = \frac{2}{v}$$

The time taken for the person's feet to get to B is 0.5 seconds, so their feet will be dry if $t < 0.5$ or:

$$\frac{2}{v} < t \quad \Rightarrow v > 4 \text{ m s}^{-1}$$

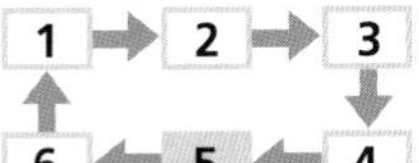

Interpret the solution

Interpretation

The person will not get wet unless the rain falls at a speed of less than 4 m s^{-1}. Is it reasonable to assume that the rain falls a speed greater than 4 m s^{-1}?

If your answer is yes, then your umbrella will keep you dry.

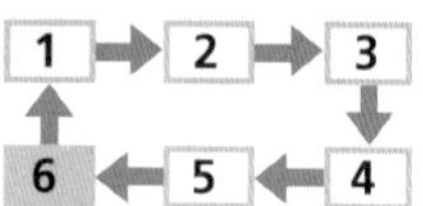

Compare with reality

Compare with reality

The rain does not always fall straight down and often descends at an angle.

There are very few people who are seen walking with their umbrellas vertical. They will often position them at an angle to provide greater protection.

The two reasons outlined above illustrate some of the inadequacies of the original model. Can you find other criticisms of the original model?

Refining the model

Select one or both of the two criticisms outlined above, define a revised set of assumptions and solve the revised problem.

POSSIBLE MODELLING TASKS

Hockey players are often taught to stop a ball before hitting it. Explain why this is a good technique and develop a strategy for hitting moving balls.

Summary

After working through this chapter you should:

- be able to use vectors to describe positions, velocities and accelerations in two dimensions
- use the constant acceleration equations

  $\mathbf{v} = \mathbf{u} + \mathbf{a}t$

  and

  $\mathbf{r} = \mathbf{u}t + \frac{1}{2}\mathbf{a}t^2 + \mathbf{r}_0$

  in two dimensional situations.

Newton's first and second laws

- *An applied force will cause a change in the motion of a particle.*
- *The change in motion depends on the mass, initial velocity and force applied.*

Exploration 6.1

Motion

- Think about an ice hockey puck that is hit and moves across the ice. What happens to the motion of the puck? What difference would it make if the two surfaces in contact were perfectly smooth?
- Consider a parachutist falling through the air. What happens to the motion as the parachutist falls?

NEWTON'S FIRST LAW

During his study of mechanics, Newton formulated his first law of motion which states:

All bodies will continue to remain at rest, or move in a straight line at constant speed, unless acted on by an external force.

In fact there are very rarely, if ever, any situations where no external forces act on a body. However if all the forces acting on an object are in ***equilibrium*** *or balance, the object can be treated as if there were no external force acting.*

An alternative way of stating Newton's first law is:

All bodies will continue to remain at rest, or move in a straight line with constant speed, if the forces acting are in equilibrium.

Example 6.1

A parachutist of mass 71 kg is falling at a constant speed. Find the upward force acting on the parachutist.

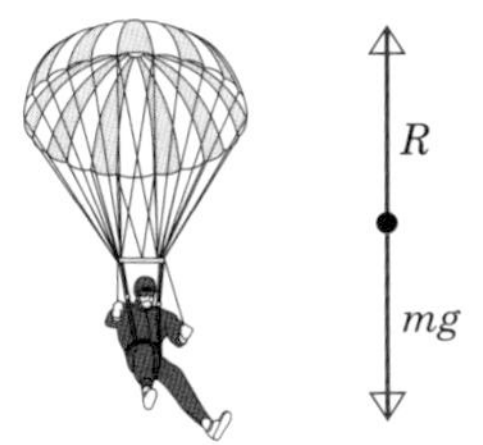

Solution

Consider the parachute and parachutist together as a particle. There are two forces acting, air resistance, R, which acts up, and gravity, mg, which acts down. As the speed of the parachutist is constant, these forces must be in balance, so that their resultant is zero. So:

$R = mg = 71 \times 9.8 = 695.8$ *N*

Example 6.2

A lorry of mass 20 tonnes is travelling up a hill inclined at 4° to the horizontal. If there are assumed to be no resistance forces on the lorry, and it travels at a constant speed, find the forward force on the lorry.

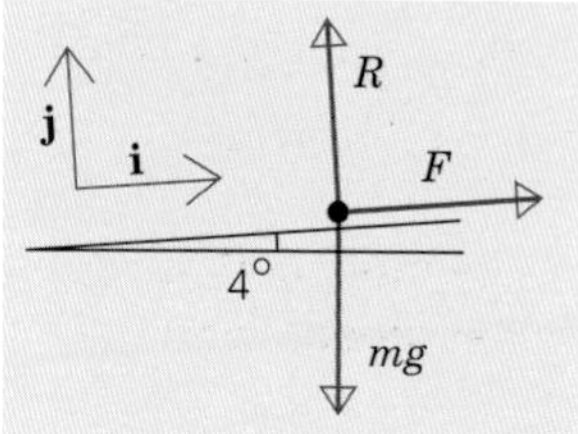

Solution

The diagram show the forces acting on the lorry, when it is modelled as a particle. Gravity acts downward, F is the forward force on the lorry and R the normal reaction.

Using the unit vectors as shown in the diagram gives the resultant force as:

$$(F - mg\cos 86°)\mathbf{i} + (R - mg\cos 4°)\mathbf{j} \doteq (F - 13\,672)\mathbf{i} + (R - 195\,523)\mathbf{j}$$

As the lorry is moving at a constant speed the resultant force must be zero, so,

$F - 13\,672 = 0$

and

$R - 195\,523 = 0$

The forward force F is then 13672 N.

EXERCISES

1 A parachutist of mass 80 kg is falling at a constant speed. Find the upward force acting on him.

2 A motorboat is travelling at a constant speed of 8 m s^{-1}. If a forward force of 780 N acts on the boat, find the magnitude of the resistance forces. What assumptions have you made to reach your answer?

3 The resistance forces on a car are assumed to be proportional to the speed of the car. When the car is travelling at 40 m s^{-1} the resistance force is 1800 N.

a) Find the constant of proportionality.

b) Find the forward force needed to keep the car moving at 30 m s^{-1}. Further investigation suggests that at 10 m s^{-1} the resistance forces are 600 N.

c) Suggest an alternative linear model for the resistance force and use this to find the forward force required to maintain a constant speed of 30 m s^{-1}.

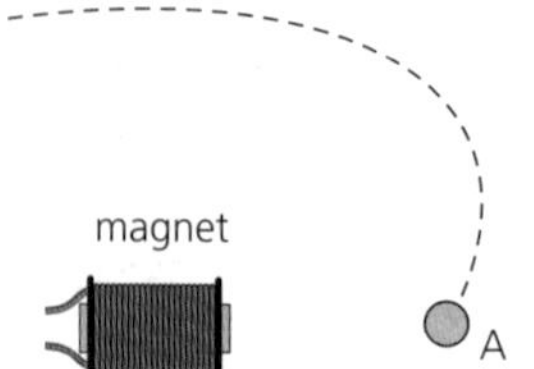

4 The diagram shows the path of a ball bearing that is rolling on a horizontal surface close to an electro-magnet. When the ball gets to the point A the magnet is switched off. Sketch the path of the ball bearing beyond A.

5 A puck, attached to a length of string fixed at O, describes a circle on a smooth horizontal table top, as shown in the diagram. Sketch the path of the puck if the string breaks at A. What happens to the speed of the puck?

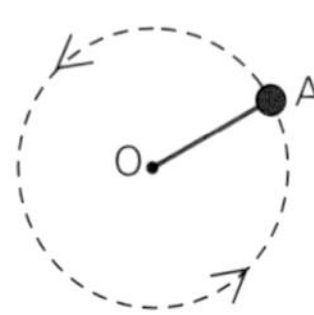

6 A car of mass 900 kg is travelling up a slope inclined at 3° to the horizontal. Find the forward force exerted by the car if it is travelling at a constant speed, and there is no resistance to motion.

7 A cyclist of mass 73 kg is free-wheeling down a slope inclined at 5° to the horizontal. If there are assumed to be no resistance forces, and the cyclist travels at a constant speed, find the force exerted by the brakes if the speed of the cyclist is:
a) $10\,\mathrm{m\,s^{-1}}$, **b)** $5\,\mathrm{m\,s^{-1}}$.

8 The resistance force exerted by a parachute can be modelled as $R = 196v$, where v is the velocity. Find the terminal speeds for a parachutist of mass:
a) 68 kg, **b)** 78 kg.

9 A car of mass 1000 kg travels down a slope inclined at 5° to the horizontal, at a constant speed. If a resistance force of 150 N acts, find the force exerted by the brakes on the car.

10 A slide in a park is at 40° to the horizontal. Children reach a maximum speed partway down the slide. Find the magnitude of the resistance forces acting on a child of mass 45 kg.

EXERCISES

6.1 B

1 A diver of mass 69 kg is rising from the sea bed at a constant speed. Find the upward force acting on the diver.

2 An aeroplane is travelling at a constant velocity. If a forward force of 2080 N is acting on the plane, find the magnitude of the resistance forces on the plane.

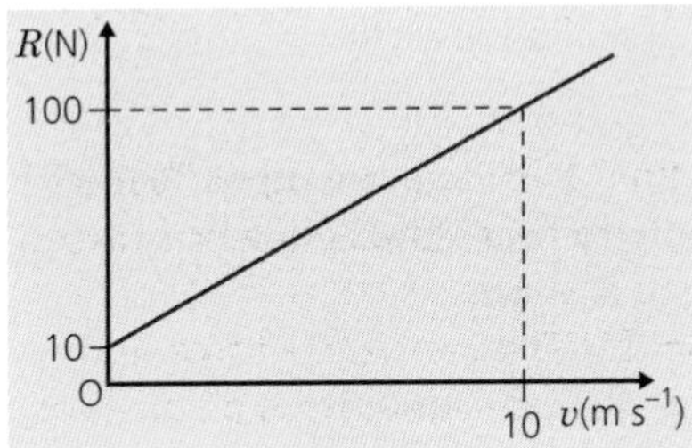

3 The graph shows how the resistance forces on a skateboard vary with speed.
a) Find an algebraic model for the resistance force.
b) What forward force must be acting on the skateboard if it is to move at a constant velocity of $5\,\mathrm{m\,s^{-1}}$?
c) Criticise the model for the resistance forces and sketch a graph of an alternative.

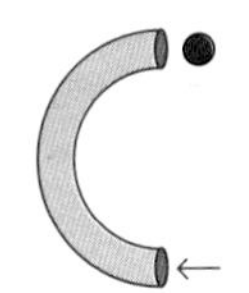

4 A semi-circular length of tube is placed on a horizontal table. A small ball is fired into one end and leaves from the other. Sketch the path of the ball after it leaves the tube.

5 The diagram shows a car travelling round a bend. A child drops a coin out of the window in the position shown. Sketch the path of the coin.

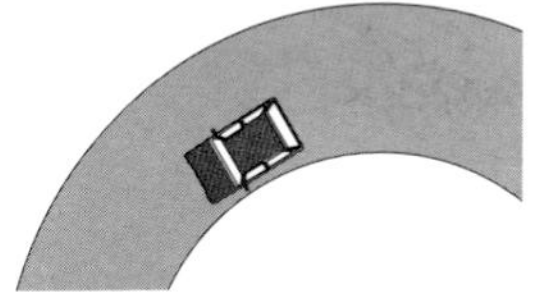

6 An aeroplane of mass 2000 kg is rising at a constant rate of $3\,\mathrm{m\,s^{-1}}$. What is the magnitude of the vertical lift force on the aeroplane?

7 A child of mass 40 kg is travelling down a slope on a sledge of mass 10 kg at a constant speed. If there is a friction force of 471 N acting on the sledge, find the angle between the slope and the horizontal.

8 A lorry is ascending a hill inclined at 6° to the horizontal at a constant speed. The mass of the lorry is 13 000 kg. Find the magnitude of the forward force on the lorry, if no resistance forces are present.

9 The resistance force, R, on a car is related to the speed, v, of the car by $R = 20v$. The mass of the car is 980 kg.
   a) Find the forward force acting on a car that is moving at a constant speed of 32 m s^{-1}.
   b) Find the forward force acting on a car that is travelling up a hill, inclined at 3° to the horizontal. The speed of the car is a constant 20 m s^{-1}.

10 The terminal speed of a parachutist of mass 60 kg is 6 m s^{-1}.
   a) Assume that the air resistance force, R, on the parachutist is proportional to her speed. Find a model of the air resistance.
   b) Which would be the terminal speed for a parachutist of mass 80 kg?

NEWTON'S SECOND LAW

Newton's first law states that if no force acts on an object it will remain at rest or move with a constant velocity. Newton's second law describes how the motion of an object changes when a force acts. The law states:

When a resultant force acts on a body it produces an acceleration that is proportional to the applied force.

In this chapter the use of the law will be restricted to bodies with constant masses. In this case Newton's second law can be stated as:

$$\mathbf{F} = m\mathbf{a}$$

The equation $\mathbf{F} = m\mathbf{a}$ is a vector equation, but if we are concerned with motion in a straight line, or wish simply to consider the magnitudes of $\mathbf{F}$ and $\mathbf{a}$, then it can be used in the form $F = ma$.

Exploration 6.2

When is $F = ma$ not appropriate?

Try to identify some situations where it would not be appropriate to use Newton's second law in the form $F = ma$.

Example 6.3

Find the force required to make a car of mass 1050 kg accelerate at 2 m s^{-2}.

Solution

Using $F = ma$ with $m = 1050$ and $a = 2$ m s^{-2} gives:

$F = ma = 1050 \times 2 = 2100$ *N*

Example 6.4

As a crane lifts a crate of mass 700 kg off the ground, the crate experiences an acceleration of 0.6 m s^{-2}. Find the tension in the cable.

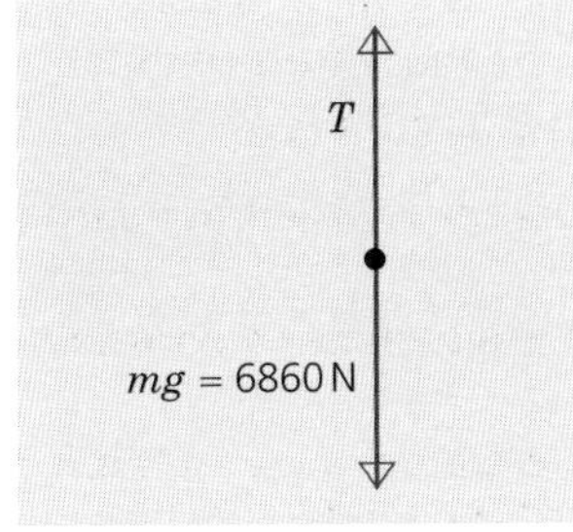

Solution

Modelling the crate as a particle, and assuming that the tension and gravity are the only forces acting on the crate, gives the force diagram shown.

As the acceleration is upward we shall take upwards as the positive sense in this case. So the resultant force is $T - 6860$. Using $F = ma$ gives:

$$T - 6860 = 700 \times 0.6 = 420 \Rightarrow T = 7280 \text{ N}$$

So the tension in the cable is 7420 N.

Example 6.5

A skier of mass 65 kg is on a ski slope that is inclined at 20° to the horizontal. If the skier experiences a friction force of 120 N, find the acceleration of the skier. What factor may be important in reality, but has not been included in your solution?

Solution

Assuming that there is no air resistance, and modelling the skier as a particle gives the force diagram shown. If unit vectors are introduced parallel and perpendicular to the slope, then the resultant force is:

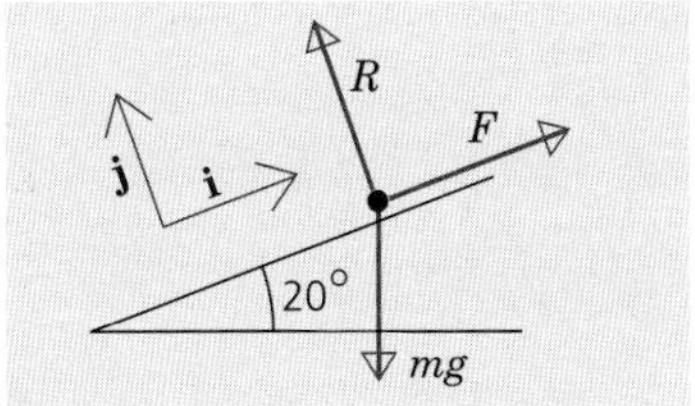

$$(F - mg\cos 70°)\mathbf{i} + (R - mg\cos 20°)\mathbf{j}$$

As there is no change in the component of motion perpendicular to the slope:

$$R - mg\cos 20° = 0$$

and so the resultant force is:

$$(F - mg\cos 70°)\mathbf{i} = (120 - 218)\mathbf{i} = -98\mathbf{i}$$

To find the acceleration, this force must be divided by the mass to give:

$$\mathbf{a} = \frac{-98\mathbf{i}}{65} = -1.51\mathbf{i}$$

So the skier experiences an acceleration of 1.51 m s^{-2} down the slope. The factor that has been ignored in this solution is air resistance which would reduce the acceleration of the skier.

Example 6.6

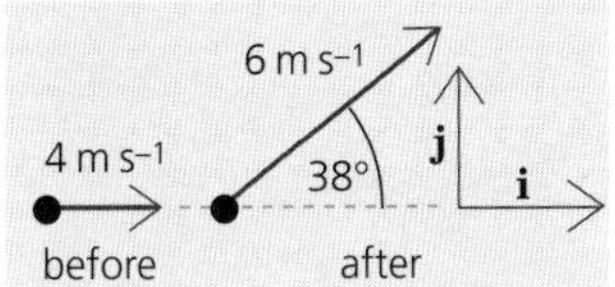

The diagram shows the path of a hockey ball of mass 0.25 kg, before and after it has been hit by a stick.

a) *Find the change in velocity of the ball.*

b) *If the ball is in contact with the stick for 0.2 seconds, calculate the magnitude of the average force on the ball.*

c) *What aspect of the ball's motion has not been considered?*

Solution

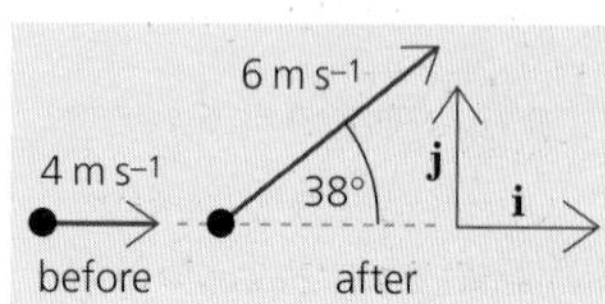

a) *The initial velocity of the ball is:*

$\mathbf{u} = 4\mathbf{i}$

while the final velocity is:

$\mathbf{v} = 6\cos 38°\,\mathbf{i} + 6\cos 52°\,\mathbf{j} = 4.73\mathbf{i} + 3.69\mathbf{j}$

So the change in velocity is:

$\mathbf{v} - \mathbf{u} = 4.73\mathbf{i} + 3.69\mathbf{j} - 4\mathbf{i} = 0.73\mathbf{i} + 3.69\mathbf{j}$

b) *The average acceleration is:*

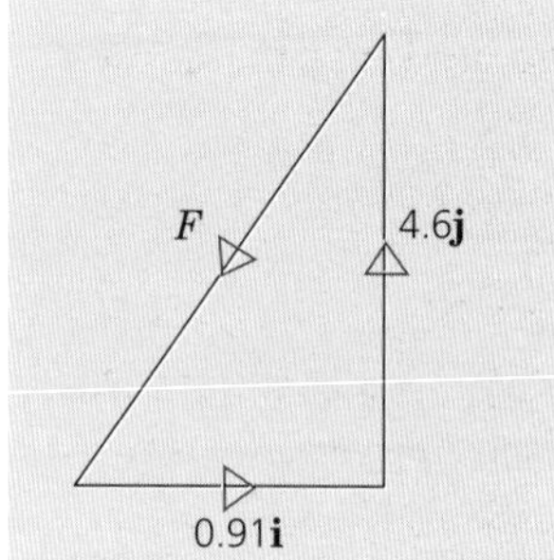

$$\frac{\mathbf{v} - \mathbf{u}}{t} = \frac{0.73\mathbf{i} + 3.69\mathbf{j}}{0.2} = 3.65\mathbf{i} + 18.45\mathbf{j}$$

The average force can now be found using $\mathbf{F} = m\mathbf{a}$*, as:*

$\mathbf{F} = 0.25(3.65\mathbf{i} + 18.45\mathbf{j}) = 0.91\mathbf{i} + 4.6\mathbf{j}$

The magnitude of this force is:

$F = \sqrt{0.91^2 + 4.6^2} = 4.69\ N$

c) *The rotation of the ball as it rolls is one factor that has been ignored in the solution of this problem.*

EXERCISES

6.2A

1 Calculate the magnitude of the force required to make each object experience the acceleration stated.
a) A lorry of mass 30 tonnes accelerating at 0.5 m s^{-2}.
b) A boat of mass 350 kg accelerating at 1.2 m s^{-2}.
c) A ball of mass 200 grams accelerating at 9.8 m s^{-2} while it falls.

2 Calculate the acceleration that each object would experience when acted on by the force stated.
a) A force of 300 N acting on a child of mass 45 kg on a slide.
b) A force of 450 N acting on a cycle and cyclist of combined mass 70 kg.
c) A force of 80 N acting on a table-tennis ball of mass 3 grams.

3 A lift has mass 400 kg and holds three people of masses 60 kg, 70 kg and 52 kg. Find the tension in the lift cable if it is moving upwards and:
a) accelerating at 0.02 m s^{-2},
b) travelling at constant speed,
c) has an acceleration of –0.15 m s^{-2}.

4 A car of total mass 1200 kg is travelling at 30 m s^{-1} when the brakes are applied. It is assumed that the brakes cause a constant force of 9000 N to act.
a) What is the acceleration of the car?
b) How far does the car travel before it stops?
c) How long does it take the car to stop?

5 A motorcycle skids 24 m before stopping on a horizontal surface.

a) By assuming that the coefficient of friction between the tyre and the road is 0.8, find the acceleration of the motorcycle.

b) Find the speed of the motorcycle when it began to skid.

c) What forces, that you have not considered above, could also act on the motorcycle? How would they effect your answer to **b)**?

6 A ball of mass 200 grams is moving at $2.5\,\text{m s}^{-1}$ when it is hit by a bat. It then moves at $3\,\text{m s}^{-1}$ and has been deflected by 30° from its original path.

a) Find the change in velocity of the ball.

b) Find the magnitude of the average force exerted on the ball by the bat, if they are in contact for 0.5 seconds.

7 The forces shown act on a 5 kg mass.

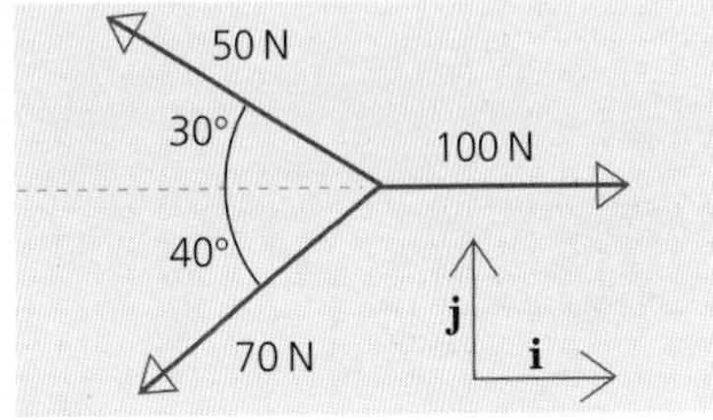

a) Find the resultant force on the mass in terms of the unit vectors **i** and **j**.

b) Find the acceleration of the mass.

c) What are the magnitude and direction of the acceleration?

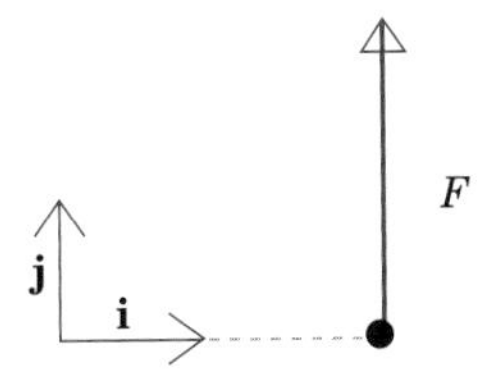

8 A hockey ball of mass 200 grams is rolling along the path shown by the dotted line when it is hit by a stick. The stick exerts a force in the direction shown for a short period of time.

a) Sketch the path of the ball after it has been hit.

b) The force has magnitude 50 N and acts for 0.4 seconds. Find the acceleration of the ball.

c) The ball was initially moving at $4\,\text{m s}^{-1}$. Find the velocity of the ball after it has been hit.

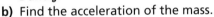

9 An archaeologist discovers the remains of an ancient catapult that would have been used to launch rocks. Elastic ropes were used to join A to B and C.

a) Find the force needed to give a rock of mass 20 kg an initial acceleration of $20\,\text{m s}^{-2}$.

b) Find the angle between the two ropes and hence find the tension in each rope.

c) Before release the catapult was held in position by soldiers. The soldiers would have had an average mass of 75 kg and the coefficient of friction between their feet and the ground was 0.4. How many soldiers were needed to hold the catapult in position?

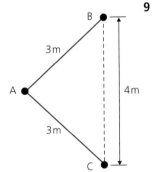

10 A car of mass 900 kg is at rest on a slope that makes an angle of 5° to the horizontal. Assume that there are no resistance forces present.

a) Find the resultant force on the car.

b) Find the speed that the car reaches if it is allowed to free wheel for five seconds.

c) If there are resistance forces present, and their effect is to reduce the resultant force by 20%, revise your answer to part **b)**.

EXERCISES

6.2 B

1 Calculate the magnitude of the force required to make each object accelerate as described.
- **a)** A table tennis ball of mass 3 grams accelerating at $9.8\,\text{m s}^{-2}$ as it falls.
- **b)** An aeroplane of mass 3000 kg accelerating at $5\,\text{m s}^{-2}$.
- **c)** A train of mass 20 tonnes accelerating at $0.6\,\text{m s}^{-2}$.

2 Find the magnitude of the acceleration that would be produced in each example below, assuming that any other forces are in balance.
- **a)** A force of 3000 N acting on a car of mass 800 kg.
- **b)** A force of 200 N acting on a dinghy of mass 175 kg.
- **c)** A force of 500 N acting on a bullet of mass 50 grams.

3 A crane is lifting, vertically, a load of mass 700 kg. Find the tension in the lifting cable if the load:
- **a)** is accelerating upwards at $0.01\,\text{m s}^{-2}$,
- **b)** is accelerating downwards at $-0.2\,\text{m s}^{-2}$,
- **c)** is travelling at constant speed.

4 A bicycle and rider of mass 74 kg are travelling at $10\,\text{m s}^{-1}$, when the brakes are applied and a retarding force of 200 N acts on the bicycle.
- **a)** Find the acceleration of the bicycle.
- **b)** How far does the bicycle travel before it stops?
- **c)** How long does it take for the bicycle to stop?

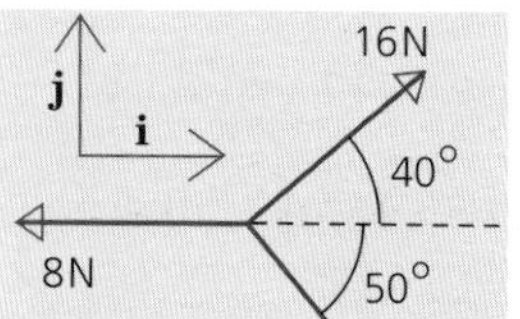

5 The three forces shown act on a 2 kg mass.
- **a)** Find the resultant force on the mass in terms of $\mathbf{i}$ and $\mathbf{j}$.
- **b)** Find the acceleration of the mass in terms of $\mathbf{i}$ and $\mathbf{j}$.
- **c)** What are the magnitude and direction of the acceleration?

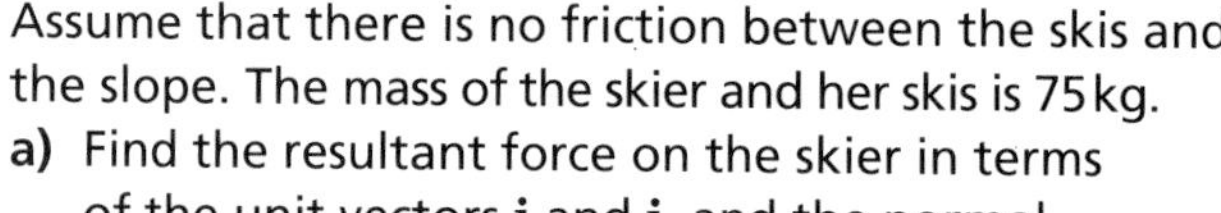

6 A skier is on a slope inclined at 15° to the horizontal. Assume that there is no friction between the skis and the slope. The mass of the skier and her skis is 75 kg.

- **a)** Find the resultant force on the skier in terms of the unit vectors $\mathbf{i}$ and $\mathbf{j}$, and the normal reaction R.
- **b)** Explain why the acceleration of the skier is $-a\mathbf{i}$.
- **c)** Find the acceleration of the skier.
- **d)** If the acceleration of the skier is in fact $-1.54\,\text{m s}^{-2}$, find the friction force acting on the skier.

7 A cyclist of mass 74 kg free-wheels down a hill inclined at 8° to the horizontal.
- **a)** Find the component of the force of gravity parallel to the slope.
- **b)** What is the acceleration of the cyclist if no resistance forces act?
- **c)** What is her speed at the bottom of the hill, 50 m from where she started to move?

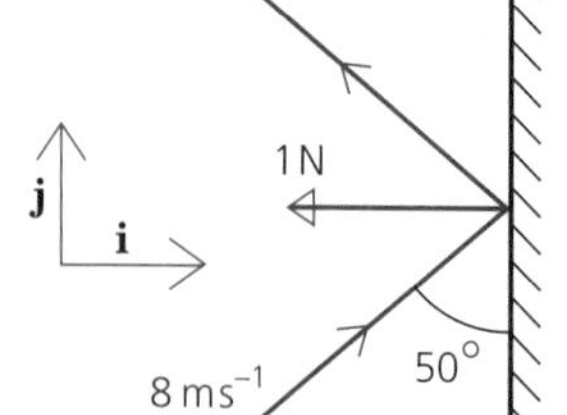

8 A squash ball of mass 24 grams hits a wall as shown in the diagram. The wall exerts a force of 1 N on the ball for 0.3 seconds.
- **a)** Find the acceleration of the ball in terms of unit vectors $\mathbf{i}$ and $\mathbf{j}$.
- **b)** Find the velocity of the ball as it leaves contact with the wall.

9 A car of mass 1000 kg is accelerating up a slope at 2° to the horizontal. If the acceleration of the car is $0.8\,\text{m}\,\text{s}^{-2}$ and there are no resistance forces, find the forward force acting on the car.

10 A spaceship of mass 8000 kg is drifting sideways in space at $0.5\,\text{m}\,\text{s}^{-1}$. At O it fires its engines which exert a constant forward force of 1000 N.

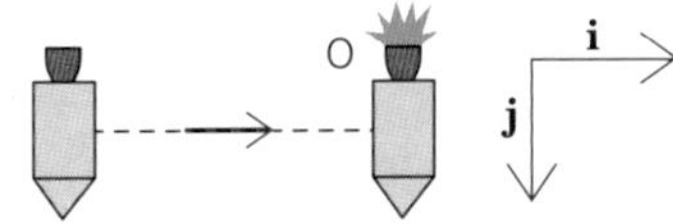

a) Draw a diagram to show how you would expect the ship to move.
b) Express the acceleration of the spaceship in terms of the unit vectors **i** and **j** shown.
c) Find expressions for the velocity and position of the ship.
d) The engines fire for ten seconds. Sketch the path of the spaceship.
e) Find the velocity of the spaceship after the engines are turned off and its position relative to O after a further 20 seconds.

CONNECTED PARTICLES

Exploration 6.3

Investigating pulleys

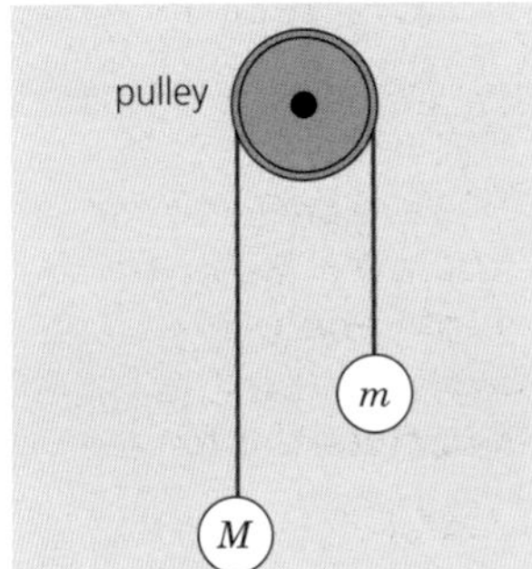

Consider the situation shown in the diagram, where two objects are connected by a string that passes over a pulley. They are held in the position shown and released.

- What happens if:
  a) $M > m$,
  b) $M < m$,
  c) $M = m$?
- Consider the forces acting on each mass. How do the two tension forces compare?
- How do the velocity and acceleration of each mass compare?

Forces in strings

When objects are connected by a string, provided that the string remains taut:
a) the tension forces acting on each object have the same magnitude.
b) the velocity and acceleration of both objects have the same magnitudes.

Example 6.7

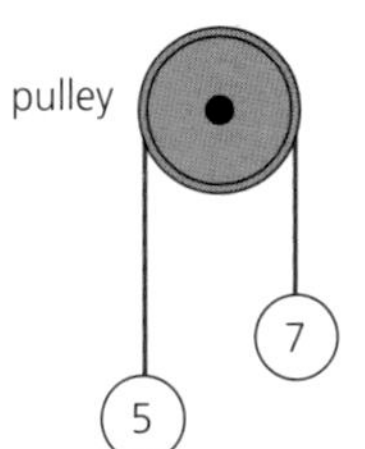

A 5 kg mass is joined to a 7 kg mass by a string as shown in the diagram. Find the acceleration of each mass.

Solution

Both masses will have accelerations with the same magnitudes, a, but opposite directions. The diagrams show the forces on each mass and the direction of the acceleration.

For the 5 kg mass the acceleration is upward, as is the resultant force which has magnitude $T - 49$, so:

$F = ma$

$T - 49 = 5a \qquad (1)$

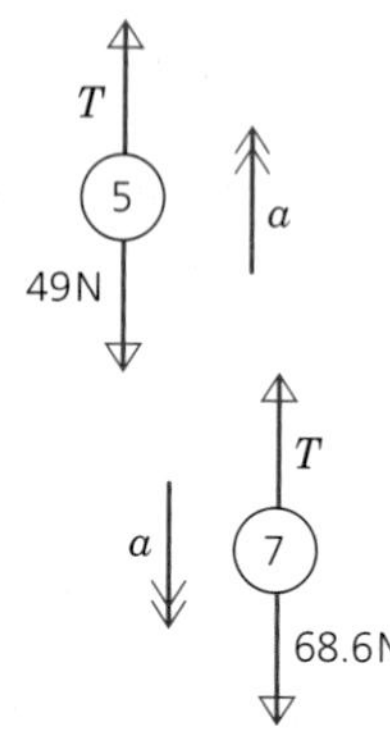

Similarly for the 7 kg mass, but noting that the resultant force and acceleration are downward give:

$F = ma$

$68.6 - T = 7a$ *(2)*

From equation (1):

$T = 49 + 5a$

and substituting into equation (2) gives

$7a = 68.6 - (49 + 5a)$

$7a = 19.6 - 5a$

$12a = 19.6$

$a = \frac{19.6}{12} = 1.63\ m\ s^{-1}$

Example 6.8

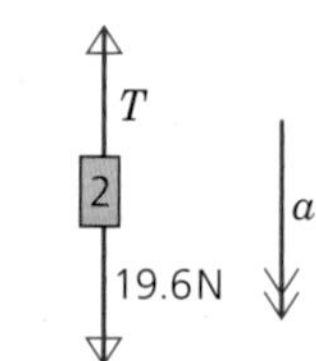

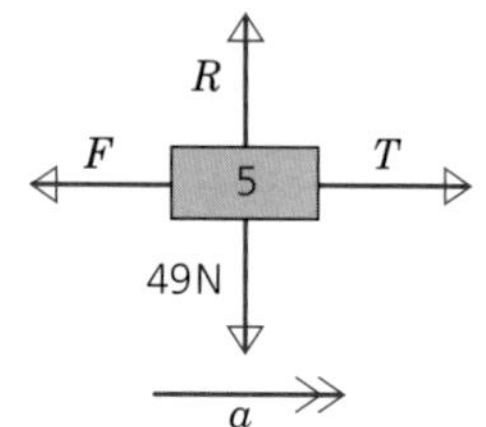

The diagram shows a block of mass 5 kg resting on a horizontal surface and attached by a string to a 2 kg mass. The coefficient of friction between the block and the surface is 0.15. Find the acceleration of the block.

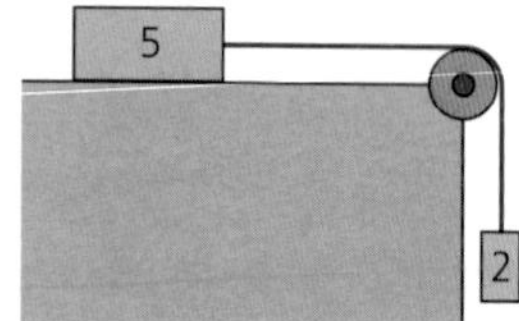

Solution

Note that the acceleration of both the block and the mass will have the same magnitude, a.

Consider first the mass. The acceleration is downward and so is the resultant force which has magnitude 19.6 – T. Using $F = ma$ gives:

$19.6 - T = 2a$

$\Rightarrow T = 19.6 - 2a$ *(1)*

Now consider the block. The reaction force will balance the force of gravity, so $R = 49$. The friction can then be found using:

$F = \mu R = 0.15 \times 49 = 7.35$

The resultant force on the block will act to the right so:

$T - F = 5a$

$\Rightarrow T - 7.35 = 5a$ *(2)*

Substituting for T in equation (2) from equation (1) gives:

$19.6 - 2a - 17.35 = 5a$

$7a = 12.25$

$a = 1.75\ m\ s^{-2}$

Exploration 6.4

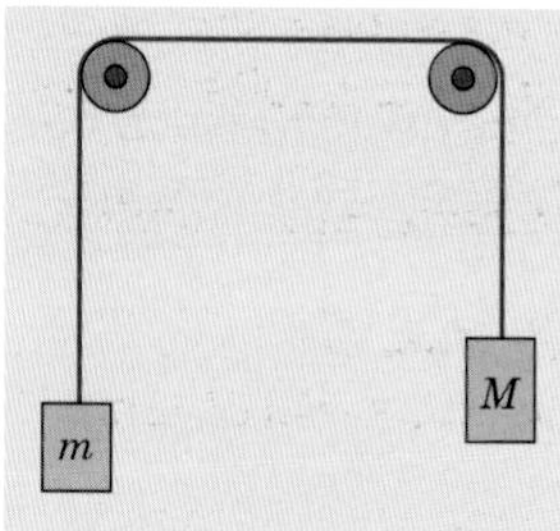

Practical activity

- Set up the apparatus shown, where two masses M and m are connected by a string passing over two pulleys. Arrange the pulleys as high as possible and choose the length of the string so that when one mass is on the floor the other is just below the pulley.
- Select masses so that there is a small difference between m and M, for example 150 grams and 170 grams. Observe the motion that takes place, when the system is released with the lighter mass at floor level. Record the distance travelled by each mass and the time taken. Using your results calculate the acceleration of each mass. (Hint: $s = ut + \frac{1}{2}at^2$)

- Use the approach of Example 6.7 to predict the acceleration of the masses.
- Compare your observed and predicted acceleration. What factors have been ignored in your prediction? Would they explain any difference that you have observed?
- Repeat the last three steps for some different pairs of masses.

Example 6.9

A van of mass 1100 kg is towing a car of mass 750 kg. Normally the van has a maximum acceleration of 1.9 m s^{-2}. Find the maximum acceleration of the van when towing and the tension in the tow rope under these conditions, if there is a resistance force of 300 N on the car.

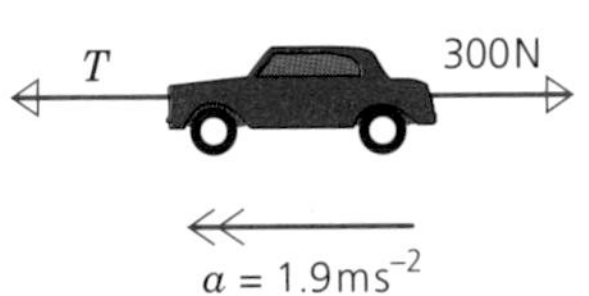

Solution

To produce an acceleration of 1.9 m s^{-2} requires a resultant force of magnitude:

$F = ma = 1100 \times 1.9 = 2090\,N$

Assuming that there is a forward force of 2090 N on the van, the resultant force is 2090 – T which produces an acceleration a. So using Newton's second law, $F = ma$, gives:

$2090 - T = 1100a$ *(1)*

Considering the towed car, there will be a resultant force of T – 300 which gives an acceleration a. So using Newton's second law, $F = ma$, gives:

$T - 300 = 750a$ *(2)*

From equation (1):

$T = 2090 - 1100a$

This can be substituted into equation (2) to give:

$(2090 - 1100a) - 300 = 750a$

$1790 = 1850a$

$$a = \frac{1790}{1850} = 0.9676\ m\ s^{-2}$$

This value can now be substituted into:

$T = 2090 - 1100a$

to give:

$T = 2090 - 1100 \times 0.9676 = 1025.6\,N$

EXERCISES

6.3A

1 A car of mass 1000 kg tows a caravan of 700 kg. Both experience resistance forces of 250 N, and the car exerts a forward force of 2000 N.

a) If T is the magnitude of the force that the car exerts on the caravan, find the resultant force on the car and the resultant force on the caravan.

b) Use Newton's second law to write down two equations, if both car and caravan have acceleration a.

c) Solve the two equations to find the values of T and a.

d) How long does it take the car and caravan to reach a speed of 10 m s^{-1}?

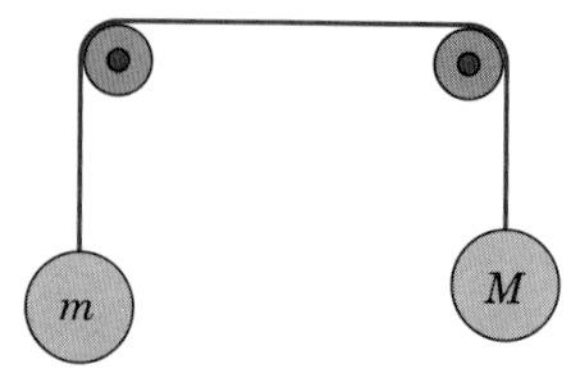

2 a) Find the acceleration of the masses shown in the diagram if:
i) $m = 5$ kg and $M = 3$ kg,
ii) $m = 400$ grams and $M = 500$ grams,
iii) $m = 1$ kg and $M = 1$ kg.
b) Express the acceleration of the system in terms of m, M and g.

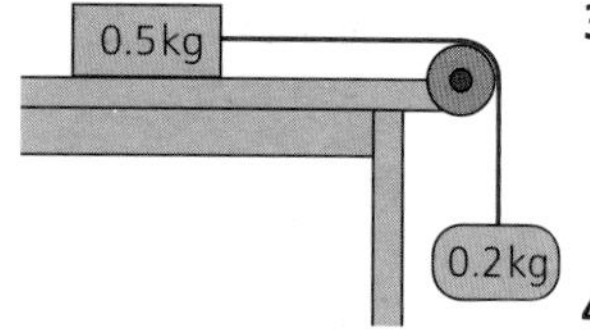

3 The diagram shows a block resting on a horizontal surface and attached to a mass by a string. Find the acceleration of each mass if:
a) there is no friction between the block and the surface,
b) the coefficient of friction between the block and the surface is 0.3.

4 The diagram shows a train that pulls two carriages. The engine is able to exert a forward force of 10 000 N. Each carriage is subjected to resistance force of 4000 N. The engine has a mass of 5 tonnes and each carriage has a mass of 2 tonnes.

a) By modelling the whole train as a single particle, find the acceleration of the train.
b) By considering the last carriage, find the tension in the coupling between the two carriages.
c) Find the tension in the coupling between the engine and the first carriage.

5 Two blocks of masses 3 kg and 5 kg are connected by a string as shown in the diagram.

a) If there is no friction between the blocks and the surfaces, find the acceleration of each block.
b) If the coefficient of friction between either mass and the slopes is 0.1, find the tension in the string and the acceleration of each mass.

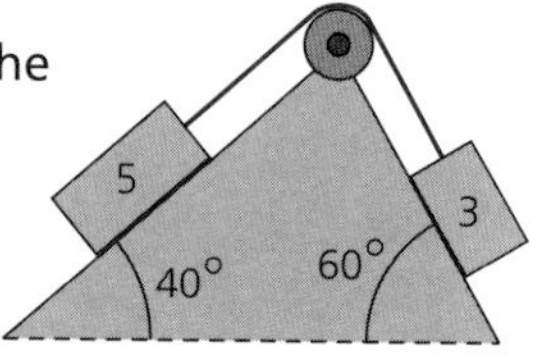

6 A free-fall simulator consists of a rope passing over a pulley attached to a 200 kg mass. Those using the simulator fall for 10 m before the rope becomes taut.

a) Calculate the speed reached by a person using the simulator when he has fallen 10 m and the rope is about to become taut.
b) Find the acceleration of the person using the simulator, when the rope is taut. Assume the person has a mass of 70 kg.
c) How far will the person fall before he stops moving?

7 A tow truck of mass 1500 kg is towing a car of mass 800 kg. The car experiences a resistance force of 200 N. The truck exerts a forward force of 1500 N.

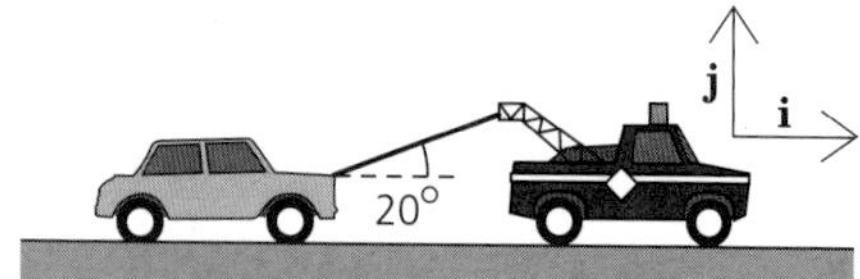

a) Express the resultant force on the truck and the car in terms of the unit vectors $\mathbf{i}$ and $\mathbf{j}$.
b) Find the acceleration of both vehicles.
c) Calculate the tension in the tow rope.

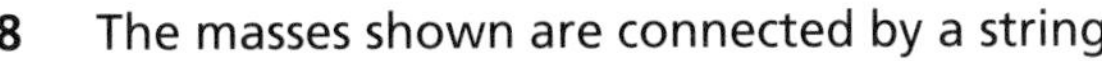

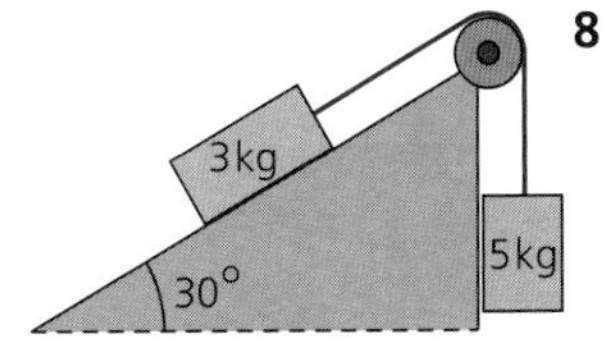

8 The masses shown are connected by a string.
a) If there is no friction between the slope and the mass, find the acceleration of both masses.
b) If the coefficient of friction between the mass and the slope is 0.7, find the tension in the string and the acceleration of each mass.

EXERCISES

6.3B

1 The diagram shows a block on a plane that is attached to a mass.
A frictional force of 10 N acts on the block.

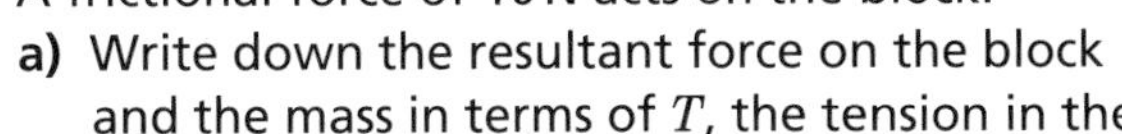

a) Write down the resultant force on the block and the mass in terms of T, the tension in the string.

b) Find expressions for the acceleration of the block and the mass.
c) Find the tension in the string and the acceleration of each mass.

10N
5kg
2kg

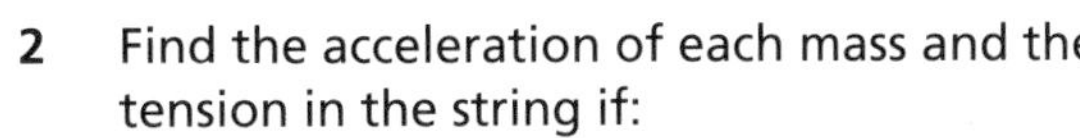

2 Find the acceleration of each mass and the tension in the string if:

a) $F = 20$ N, $M = 10$ kg and $m = 4$ kg,
b) $F = 0$, $M = 1$ kg and $m = 0.5$ kg,
c) $F = 10$ N, $M = 6$ kg and $m = 2$ kg.

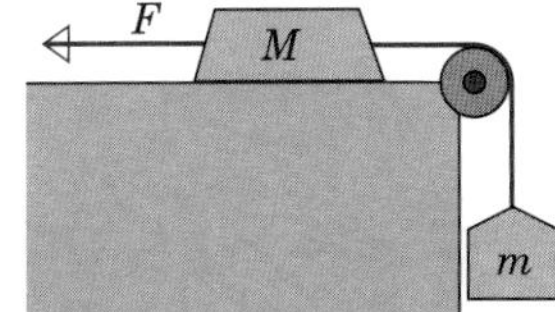

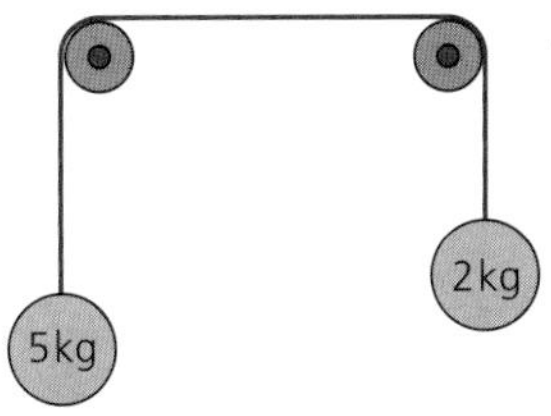

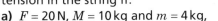

3 The diagram shows two masses connected by a string passing over two pulleys.
a) Express the resultant force on each mass in terms of T, the tension in the string.
b) Find expressions for the acceleration of each mass.
c) Find the tension in the string and then the acceleration of each mass.
d) Which assumptions have you made in order to solve this problem?

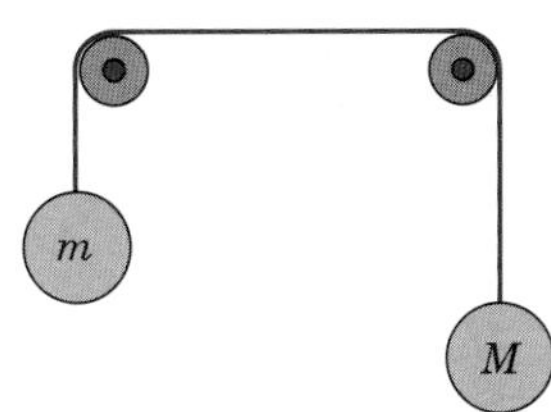

4 The diagram shows two masses connected by a string that passes over two smooth pulleys.
a) Find the acceleration of each mass if $m = 10$ kg and $M = 8$ kg.
b) Find the tension in the string in terms of M, m and g.

5 A car towing a caravan exerts a force of 5400 N. The mass of the car is 1100 kg and the mass of the caravan is 700 kg.
a) The acceleration of the car and caravan is 2.5 m s^{-2}. Find the total resistance on the car and caravan.
b) If this total resistance is divided equally between the car and caravan, find the force that the car exerts on the caravan.

6 A car tows an unbraked trailer down a hill that is inclined at 4° to the horizontal. The mass of the car is 1000 kg and the mass of the trailer is 200 kg.
a) If the cars brakes exert a force of 200 N find the acceleration of the car and trailer.
b) Find the force that the car exerts on the trailer.

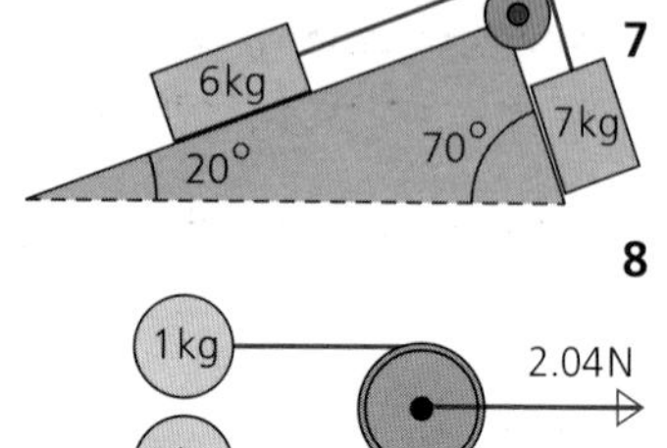

7 The diagram shows two blocks connected by a string passing over a smooth pulley. Find the acceleration of each mass and the tension in the string if there is no friction present.

8 Two objects with masses as shown are connected by a string that passes over a pulley on a smooth horizontal surface. A force of 2.04 N is applied to the pulley as shown.

1 kg
2 kg
2.04 N

a) If the pulley has mass 0.2 kg and is accelerating at 0.2 m s^{-2}, show that the tension in the string joining the two masses is 1 N.

b) Find the acceleration of each mass.

CONSOLIDATION EXERCISES FOR CHAPTER 6

(In this exercise assume $g = 9.8$ m s^{-2} unless stated otherwise.)

1 A lift when empty is of mass 1200 kg. A person of mass 65 kg stands in the lift holding a parcel of mass 4 kg.

a) If the lift is descending with an acceleration of 0.5 m s^{-2}, find the tension in the cable supporting the lift and the force exerted on the parcel by the person holding it.

b) The parcel is dropped from a height of 1.5 m. It takes one second for the person to react and grab the parcel. Does the parcel hit the floor of the lift?

2 A car of mass 1300 kg is travelling along a horizontal road. The resistance force on the car is $70v$, and the maximum forward force on the car is 3000 N.

a) Find the maximum initial acceleration of the car.

b) Find the maximum speed of the car on a horizontal surface.

c) Find the maximum speed of the car if it drives down a slope at 3° to the horizontal.

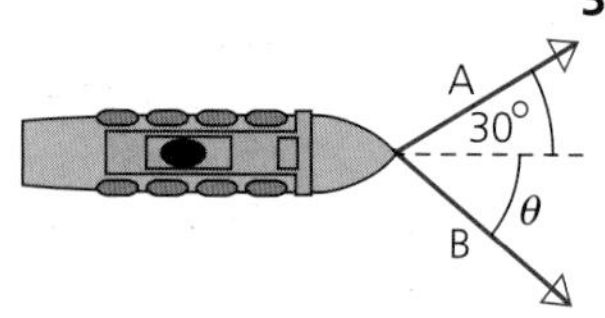

3 Two tugs are towing a large ship of mass 10 000 tonnes into harbour. Tug A exerts a force of 40 000 N and Tug B exerts a force of 35 000 N.

a) Find the angle θ if the ship is to move directly forwards.

b) If the acceleration of the ship is 0.05 m s^{-2}, find the magnitude of the resistance forces on the ship.

c) How long does it take the ship to travel 400 m?

4 Two men are moving a heavy crate of mass 50 kg. They both exert forces of 140 N at 20° to the direction of motion. There is a resistance force of 108 N on the crate. Find the speed of the crate when it has:

a) moved 3 m,

b) been pushed for five seconds.

5 The diagram shows two masses connected by a string. The coefficient of friction between the surface and the 5 kg mass is 0.5.

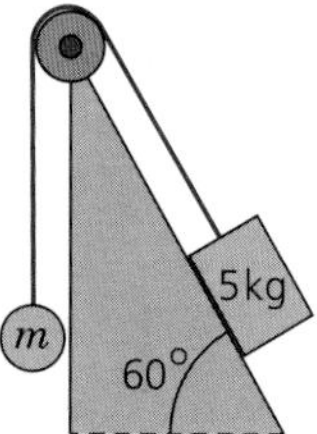

a) Show that the friction force on the 5 kg mass as it moves is 12.25 N.

b) Find the unknown mass if it has a downward acceleration of 1.2 m s^{-2}.

c) What assumptions have you made about the string?

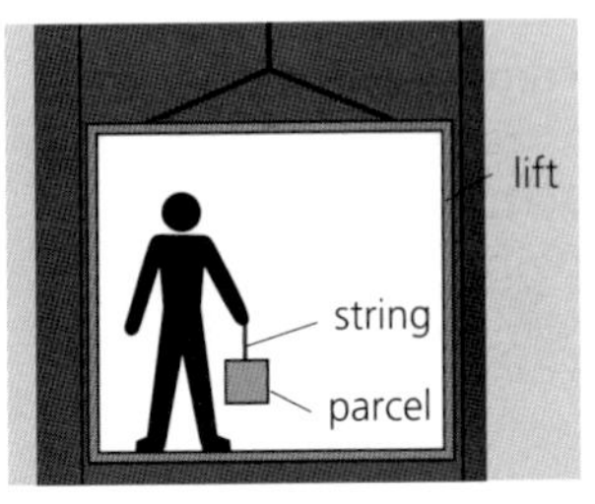

6 A man of mass 70 kg is standing in a lift which, at a particular time, has an acceleration of 1.6 m s^{-2} upwards. He is holding a parcel of mass 5 kg by a single string.

a) Draw a diagram marking the forces acting on the parcel and the direction of the acceleration.

b) Show that the tension in the string is 57 N.

c) Calculate the reaction of the lift floor on the man.

During the first two seconds after starting from rest, the lift has acceleration in m s^{-2} modelled by $3t(2 - t)$, where t is in seconds. The maximum tension the string can withstand is 60 N.

d) By investigating the maximum acceleration of the system, or otherwise, determine whether the string will break during this time.

(MEI Question 2, Mechanics 1, June 1992)

7 A toboggan of mass 15 kg carries a child of mass 25 kg. It starts from rest on a snow slope of inclination 10°. Given that the acceleration is 1.2 m s^{-2} and that air resistance may be ignored:

a) find the coefficient of friction,

b) find the speed when it has moved 50 m from rest and find the time taken.

Having reached the bottom of the slope, the toboggan and child are pulled slowly back up the slope by a light rope which is parallel to a line of greatest slope.

c) Find the tension in the rope.

(UCLES Question 8, Specimen Paper M1, 1994)

8 A particle A, of mass $4m$, lies on a smooth plane inclined at 30° to the horizontal. A light inextensible string is attached to A and passes over the small smooth pulley P fixed at the top of the inclined plane. To the other end of the string is attached a particle B, of mass m, which hangs freely. The particles are released from rest with the string taut and with the portion AP parallel to a line of greatest slope of the inclined plane.

a) Write down the equation of motion for particle A down the plane and the equation of motion of particle B.

b) Hence calculate, in terms of m and g, the magnitude of the tension in the string and the acceleration of the particle A down the plane.

(ULEAC Question 10, Old Specimen M1)

MATHEMATICAL MODELLING ACTIVITY

1 → 2 → 3 → 4 → 5 → 6

Specify the real problem

Problem statement

When is a hill too steep to drive up?

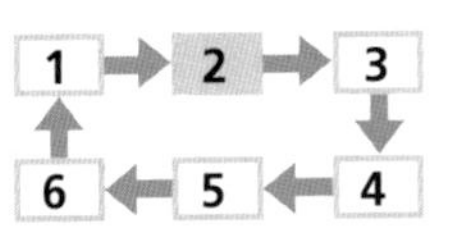

Set up a model

Set up a model

The list below gives the main features of the problem.

- the forward force that can be exerted by the car
- the slope of the hill

Are there any other factors which you feel are important in this problem? The assumptions listed below will allow the model to be formulated.

- The car can be modelled as a particle of mass m kg.
- The hill is at a constant angle of α to the horizontal.
- The car can exert a maximum forward force of $2.5m$ N.

Justify using a value of $2.5m$ N for the maximum forward force.

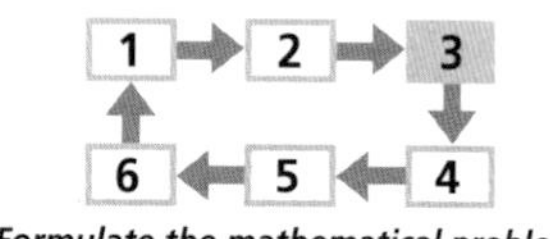

Formulate the mathematical problem

Formulate a mathematical problem

Find the angle, α, between the slope and the horizontal when the forces on the car are in equilibrium.

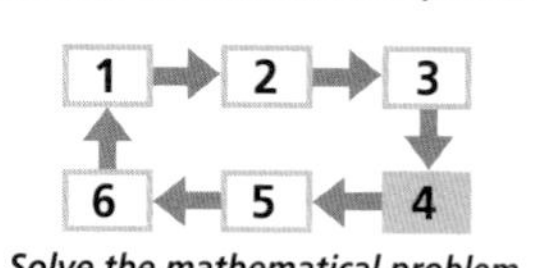

Solve the mathematical problem

Mathematical solution

The diagram shows the forces acting. The resultant force can be expressed in terms of the unit vectors **i** and **j** as:

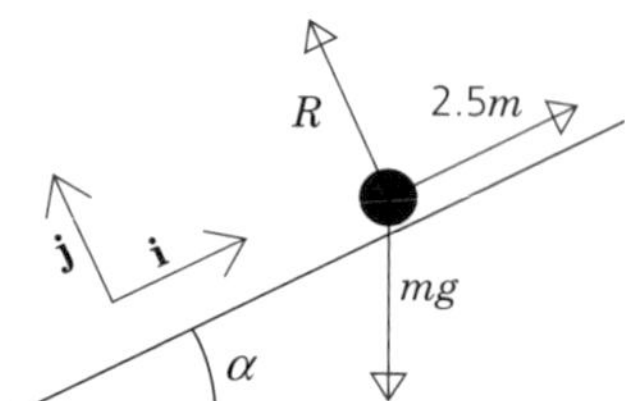

$(2.5m - mg\sin\alpha)\mathbf{i} + (R - mg\cos\alpha)\mathbf{j}$

When in equilibrium, both components of the resultant force will be zero, so:

$2.5m - mg\sin\alpha = 0$

Rearranging gives:

$$\sin\alpha = \frac{2.5}{g} \Rightarrow \alpha = \sin^{-1}\left(\frac{2.5}{9.8}\right) = 15°$$

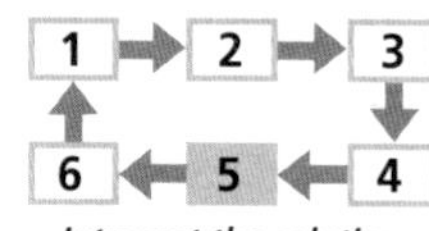

Interpret the solution

Interpretation

A car can continue to travel up a slope at 15° to the vertical if driven flat out, if it is moving at the bottom of the hill. The car would only be able to maintain speed, and not accelerate.

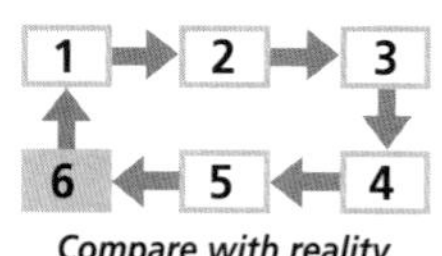

Compare with reality

Compare with reality

Find the angle that a hill with 1:5 gradient makes with the horizontal. Comment on the solution in the light of this.

One criticism that could be made of this model is that the wheels of the car may spin on a steep slope.

What other criticisms can you make of this model?

Refining the model

A typical car will have two driving wheels that provide the forward force on the car, using friction between them and the road.

From the mathematical solution the total reaction force on the car is $mg\cos\alpha$ which is equal to $9.5m$ N for the 15° slope. Assume that the reaction on each wheel is a quarter of this figure, that is $2.4m$ N.

The friction for each driving wheel is $1.25m$ N.

So using $F \le \mu R$ gives:

$$1.25m \le \mu \times 2.4m \;\Rightarrow\; \mu \ge \frac{1.25}{2.4} = 0.52$$

if the wheels do not slip.
Do you think it is likely that the wheels will slip?

POSSIBLE MODELLING TASKS

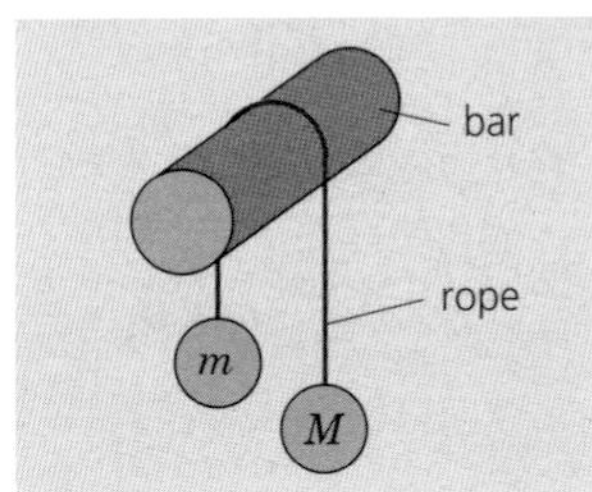

1 Use a rope passed over a peg or bar with masses attached as shown to investigate the friction between the rope and the bar.

2 Drop a ball and record how high it bounces each time and the total time for a number of bounces. Use your results to estimate the value of g.

3 Drop a ball from a high building. Record the time taken to fall. Use this to estimate g and investigate the effects of air resistance on the ball.

4 The *Highway Code* contains a table of stopping distances. Some assumptions are given in the table. What other assumptions do you think have been made? Investigate the effects of hills and towing unbraked trailers or caravans on the stopping distances given.

5 Find a gentle constant slope. Allow a cyclist or skateboarder to free-wheel down the slope. Record the time taken to travel a range of distances. Find the acceleration of the skateboarder or cyclist at different times and try to find a model for the resistance forces acting on him.

6 A solicitor writes to you asking for assistance. His client was stationary in a car when it was hit by a motorcycle. The motorcycle left a 78 ft (24 m) skid on the road as it skidded down the hill into the car. The solicitor would like you to prepare a report on the speed that the motorcycle was travelling when it hit the car.

Summary

After working through this chapter you should:

- be able to apply Newton's first law
- be able to make predictions about motion by using Newton's second law
- solve problems involving connected particles.

Motion with variable forces and acceleration

- *The equations used in earlier chapters apply to constant accelerations only.*
- *In reality some forces vary in magnitude and direction and so produce variable accelerations.*

Exploration 7.1

Forces acting on an object

Describe what happens to the forces acting on each object described below.

- A mass vibrating on the end of a spring.
- A ball that is hit by a racquet or bat.
- A pendulum.

Would it be reasonable to model any of the forces above as constant forces?

Try to describe some situations where variable forces act. Is it reasonable to model any of these forces as constant forces?

VELOCITY AND ACCELERATION

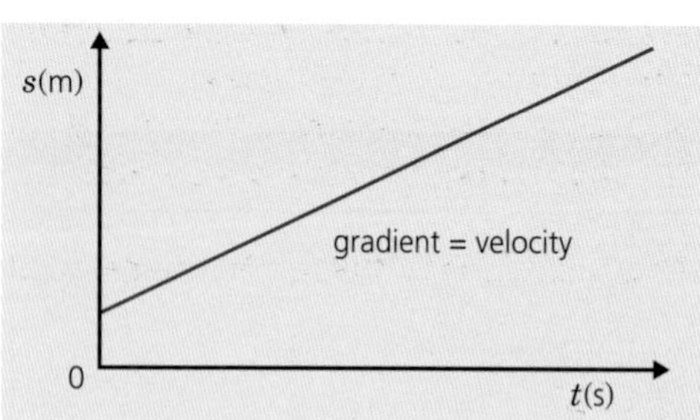

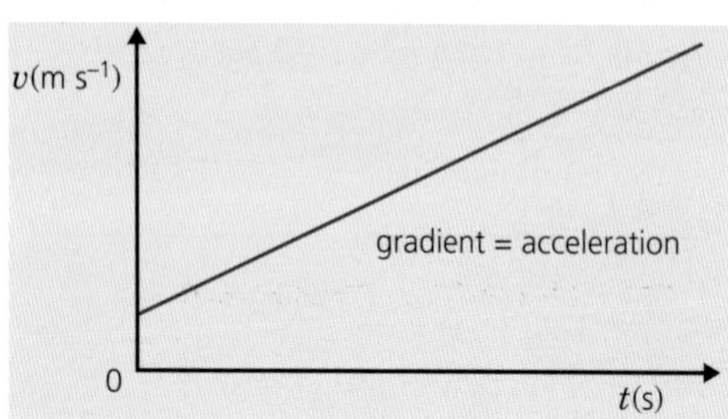

In Chapter 4 we studied velocity and acceleration by looking at the gradients of distance–time and speed–time graphs. All the examples that we considered in that chapter were restricted to the special case where acceleration was constant.

Using the notation of calculus, we can extend these ideas to a more general format that is not restricted to constant acceleration.

As the gradient of a distance–time graph is $\frac{\mathrm{d}s}{\mathrm{d}t}$, we can write $v = \frac{\mathrm{d}s}{\mathrm{d}t}$.

Similarly the acceleration can be expressed as:

$$a = \frac{\mathrm{d}v}{\mathrm{d}t} \text{ or } a = \frac{\mathrm{d}^2 s}{\mathrm{d}t^2}.$$

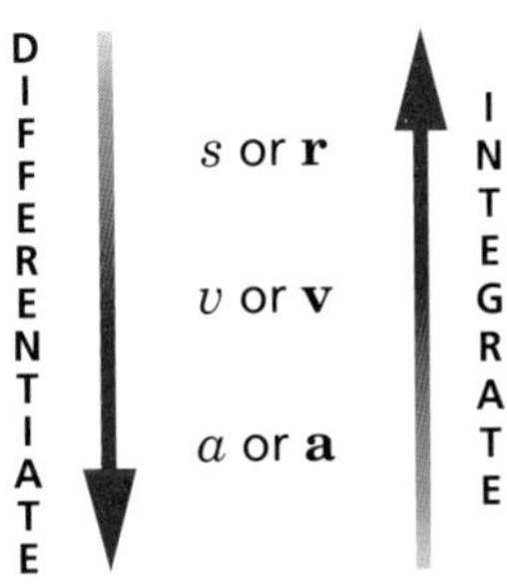

We can then find speeds and accelerations by **differentiating** expressions for distance. Distances and speeds can also be found by **integrating** expressions for acceleration.

Two and three dimensions

The principle of using calculus in one dimension can be extended to two or three dimensions. If the position vector is defined as:

$$\mathbf{r} = x\mathbf{i} + y\mathbf{j} + z\mathbf{k}$$

then:

$$\mathbf{v} = \frac{d\mathbf{r}}{dt} = \frac{dx}{dt}\mathbf{i} + \frac{dy}{dt}\mathbf{j} + \frac{dz}{dt}\mathbf{k}$$

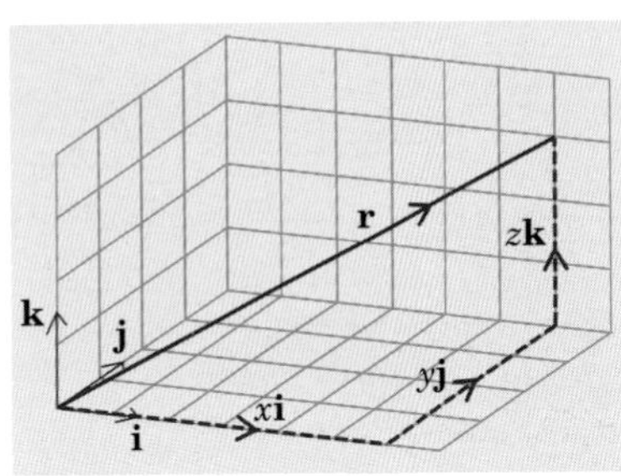

Note: Each component of the position vector is differentiated in turn. To find the acceleration, the velocity is differentiated in the same way.

$$\mathbf{a} = \frac{d\mathbf{v}}{dt} = \frac{d^2x}{dt^2}\mathbf{i} + \frac{d^2y}{dt^2}\mathbf{j} + \frac{d^2z}{dt^2}\mathbf{k}$$

In two dimensions the **k** term is simply omitted.

Example 7.1

As a car travels round a bend its position vector is modelled by:

$$\mathbf{r} = (2t + 4)\mathbf{i} + \tfrac{1}{4}t^3\mathbf{j}$$

After four seconds the car leaves the bend and the expression above no longer applies.

a) *If you have a graphics calculator, or software, plot the path of the car.*
b) *Calculate the position of the car when t = 0, 1, 2, 3 and 4. Use this information to plot the path of the car.*
c) *Find an expression for the velocity of the car. Calculate the velocity of the car when t = 0, 1, 2, 3 and 4. Draw vectors on your diagram to represent the velocity of the car at t = 0, 1, 2, 3 and 4.*
d) *Find an expression for the acceleration of the car. Calculate the acceleration when t = 0, 1, 2, 3 and 4 and draw vectors on the diagram to illustrate these.*
e) *Describe what happens to the acceleration of the car. Is this realistic?*

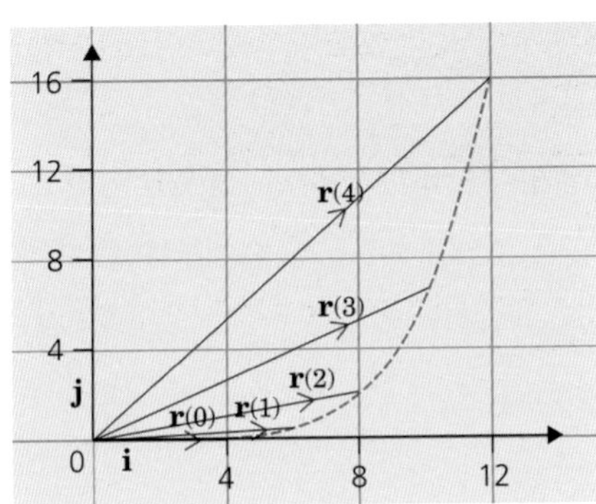

Solution

a) *See the graph on the left.*
b) *Substituting the values of t into the position vector gives:*

$t = 0 \Rightarrow \mathbf{r} = 4\mathbf{i} + 0\mathbf{j} = 4\mathbf{i}$
$t = 1 \Rightarrow \mathbf{r} = 6\mathbf{i} + 0.25\mathbf{j}$
$t = 2 \Rightarrow \mathbf{r} = 8\mathbf{i} + 2\mathbf{j}$
$t = 3 \Rightarrow \mathbf{r} = 10\mathbf{i} + 6.75\mathbf{j}$
$t = 4 \Rightarrow \mathbf{r} = 12\mathbf{i} + 16\mathbf{j}$

These points have been plotted on the diagram.

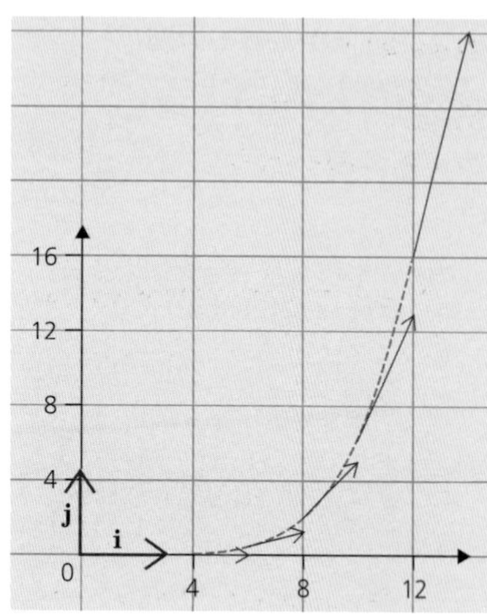

c) *The velocity of the car is given by:*

$$\mathbf{v} = \frac{d\mathbf{r}}{dt} = 2\mathbf{i} + \tfrac{3}{4}t^2\,\mathbf{j}$$

Substituting the values of t into the expression for the velocity gives:

$t = 0 \Rightarrow \quad \mathbf{v} = 2\mathbf{i} + 0\mathbf{j} = 2\mathbf{i}$

$t = 1 \Rightarrow \quad \mathbf{v} = 2\mathbf{i} + 0.75\mathbf{j}$

$t = 2 \Rightarrow \quad \mathbf{v} = 2\mathbf{i} + 3\mathbf{j}$

$t = 3 \Rightarrow \quad \mathbf{v} = 2\mathbf{i} + 6.75\mathbf{j}$

$t = 4 \Rightarrow \quad \mathbf{v} = 2\mathbf{i} + 12\mathbf{j}$

The diagram shows the path with the velocity vectors superimposed, note that the velocity vectors are at tangents to the path.

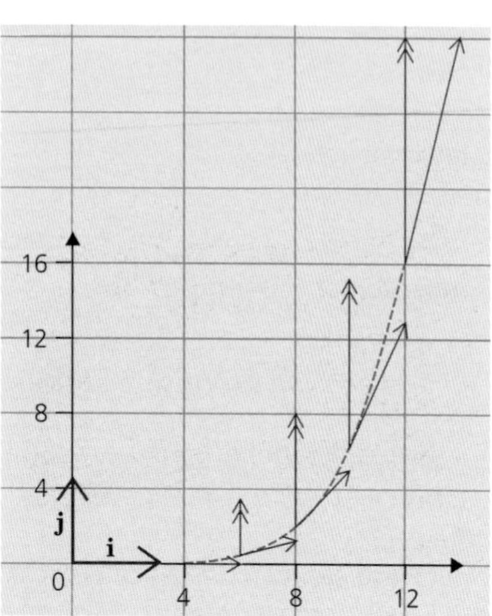

d) *The acceleration is obtained by differentiating the velocity.*

$$\mathbf{a} = \frac{d\mathbf{v}}{dt} = 0\,\mathbf{i} + \frac{3}{2}t\,\mathbf{j} = \frac{3}{2}t\,\mathbf{j}$$

Substituting values for t gives:

$t = 0 \Rightarrow \ \mathbf{a} = 0\mathbf{j} = 0 \qquad t = 1 \Rightarrow \ \mathbf{a} = 1.5\mathbf{j} \qquad t = 2 \Rightarrow \ \mathbf{a} = 3\mathbf{j}$

$t = 3 \Rightarrow \ \mathbf{a} = 4.5\mathbf{j} \qquad t = 4 \Rightarrow \ \mathbf{a} = 6\mathbf{j}$

The diagram shows the path with both the velocity and acceleration vectors superimposed.

e) *The acceleration is always in the **j**-direction and increases in magnitude from 0 to 6 m s^{-2}. It is very unlikely in reality that a normal car can produce an acceleration of 6 m s^{-2}, a car that accelerates from 0 to 60 mph in 6 seconds has an acceleration of 4.5 m s^{-2}.*

Example 7.2

The distance travelled by a cyclist is modelled by:

$$s = \frac{t^2}{2} - \frac{t^4}{1200} \text{ for } 0 \le t \le 17.3\text{ s}$$

a) *Find expressions for the velocity and acceleration of the cyclist.*

b) *When does the cyclist begin to slow down?*

Solution

a) *The velocity and acceleration can be found by differentiating the expression for s with respect to t.*

$$v = \frac{ds}{dt} = t - \frac{t^3}{300}$$

and

$$a = \frac{dv}{dt} = 1 - \frac{t^2}{100}$$

b) *For the cyclist to be slowing down $a < 0$ so:*

$$1-\frac{t^2}{100}<0 \quad\Rightarrow\quad 1<\frac{t^2}{100} \quad\Rightarrow\quad 100<t^2 \quad\Rightarrow\quad 10<t$$

So the cyclist begins to slow down after 10 seconds.

Motion in three dimensions

If the position vector of any object that moves in three dimensions is defined as: $\mathbf{r} = x\mathbf{i} + y\mathbf{j} + z\mathbf{k}$
then its velocity is:

$$\mathbf{v} = \frac{d\mathbf{r}}{dt} = \frac{dx}{dt}\mathbf{i} + \frac{dy}{dt}\mathbf{j} + \frac{dz}{dt}\mathbf{k}$$

and its acceleration is:

$$\mathbf{a} = \frac{d\mathbf{v}}{dt} = \frac{d^2x}{dt^2}\mathbf{i} + \frac{d^2y}{dt^2}\mathbf{j} + \frac{d^2z}{dt^2}\mathbf{k}$$

If an object moves in one dimension with displacement s then its velocity is:

$$v = \frac{ds}{dt}$$

and its acceleration is: $a = \dfrac{dv}{dt} = \dfrac{d^2s}{dt^2}$

EXERCISES

7.1 A

1 The position vector for a ball thrown by a child is given by:

$$\mathbf{r} = 5t\mathbf{i} + \left(1 + 6t - 4.9t^2\right)\mathbf{j}$$

a) Find an expression for the velocity of the ball.
b) Show that the acceleration of the ball is constant and state its magnitude and direction.

2 An aeroplane has position vector:

$$\mathbf{r} = \frac{t^3}{30}\mathbf{i} + \frac{t^4}{120}\mathbf{j} + 5t\mathbf{k}$$

a) Find expressions for the velocity and acceleration of the aeroplane.
b) At what rate is the aeroplane gaining height?
c) What can be deduced about the direction of the acceleration of the plane?

3 The height of a ball thrown up into the air is modelled by:

$$h = 16t - 4.9t^2 + 2$$

a) Find an expression for the velocity of the ball.
b) Find the maximum height reached by the ball.
c) Find the acceleration of the ball.

4 The distance s travelled by a dragster is modelled by:

$s = 4t^2 - \dfrac{t^3}{15}$ while the dragster is accelerating.

a) Find when the acceleration of the dragster becomes zero.
b) Hence find the maximum speed attained by the dragster and the distance it travels while accelerating.

5 The path of a javelin is modelled by:

$$\mathbf{r} = \left(18t - 3t^2\right)\mathbf{i} + \left(14t - 4.9t^2\right)\mathbf{j}$$

where $\mathbf{i}$ and $\mathbf{j}$ are horizontal and vertical unit vectors respectively.

Find the length of the throw and the velocity of the javelin when it hits the ground.

6 The speed of a car that starts from rest is modelled by:

$v = 2\sqrt{t}$ for $0 \le t \le 1$

a) Find the speed of the car when $t = 100$.
b) Find an expression for the acceleration of the car.
c) Find the acceleration of the car when $t = 100$.

7 The height of a rocket fired vertically upwards is given by:

$$h = 10t^2 - \frac{5t^3}{3}$$

a) Find expressions for the velocity and acceleration of the rocket.
b) The expressions above are only valid while $a \ge 0$. Find the time when the acceleration of the rocket becomes zero. Calculate the velocity and height at this time.
c) The rocket then moves under the influence of gravity with an acceleration of $-10\,\mathrm{m\,s^{-2}}$. Find the maximum height reached by the rocket.

8 As a car travels round a bend its position vector is:

$\mathbf{r} = 5t\mathbf{i} + \frac{t^2}{6}\mathbf{j}$ for $0 \le t \le 4$

a) Find expressions for the velocity and acceleration of the car while it is on the bend.
b) Find the speed of the car as it enters the bend when $t = 0$ and leaves the bend when $t = 4$.
c) Describe what happens to the acceleration of the car as it rounds the bend.

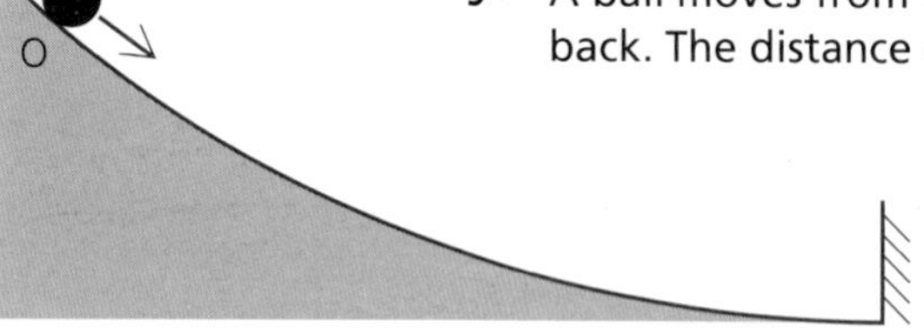

9 A ball moves from O down a curved slope, hits a barrier and bounces back. The distance of the ball from O is modelled as:

$x = \dfrac{t^2(t-5)^2}{20}$ until the ball returns to O.

a) Find expressions for the velocity and acceleration of the ball.
b) What is the acceleration of the ball when it is instantaneously at rest and in contact with the barrier?
c) What is the speed of the ball at O?

10 The position of a rider on a helter-skelter is given by:

$$\mathbf{r} = 2\sin t\,\mathbf{i} + 2\cos t\,\mathbf{j} + (10 - 0.5t)\mathbf{k}$$

a) Find expressions for the velocity and acceleration of the rider.
b) Show that the speed of the rider is constant and find its magnitude.
c) Describe the acceleration of the rider.

EXERCISES

7.1B

1 In each case below, find an expression for the velocity and the acceleration.

a) $\mathbf{r} = (6t^2 - 2)\mathbf{i} + (5t^2 - 3t + 2)\mathbf{j}$ **b)** $\mathbf{r} = \frac{t^3}{100}\mathbf{i} + \frac{t^3}{50}\mathbf{j}$

c) $\mathbf{r} = (6t^4 - 3t^2)\mathbf{i} + (4t - 8)\mathbf{j}$

2 A diver dives from a high board into a pool of water. The depth of the diver in the water is modelled as:

$$d = t^3 - 4t^2 + 4t$$

a) Find an expression for the velocity of the diver while she is in the water.

b) What is the greatest depth that the diver reaches?

c) What is the velocity of the diver when she returns to the surface?

d) An alternative model for d is:

$$d = \frac{32t(2-t)}{27}$$

Show that the diver still reaches the same maximum depth and find the speed at which the diver returns to the surface.

e) Explain which is the more realistic model.

3 The distance travelled by a cyclist is modelled as: $s = 0.8\,t^2$ for $0 \le t \le 5$

a) Find the velocity and acceleration of the cyclist.

b) Describe what happens to the velocity and acceleration, using sketch graphs to help.

c) Suggest how the cycle moves for $t > 5$.

4 When a large rocket takes off it is suggested that the height of the rocket is given by:

$$h = kt^6$$

where k is a constant. If the acceleration of the rocket is 30 m s^{-2} after 10 seconds, find the value of k.

5 A football is kicked so that its motion is modelled by:

$$\mathbf{r} = 10t\mathbf{i} + (12t - 4.9t^2)\mathbf{j}$$

where $\mathbf{i}$ and $\mathbf{j}$ are horizontal and vertical unit vectors respectively. Show that the acceleration of the football is constant, stating clearly its direction and magnitude.

6 A hockey ball is hit on a sloping pitch, so that its motion is modelled as:

$$\mathbf{r} = 5t\mathbf{i} + (12t - 0.1t^3)\mathbf{j}$$

a) Find the velocity and acceleration of the ball.

b) Sketch the path of the ball and indicate the direction of the acceleration of the ball.

c) Describe the slope of the pitch.

7 A 'demon-drop' slide at an adventure centre is such that users move with position vector:

$$\mathbf{r} = \frac{t^4}{5}\mathbf{i} + \left(8 - 2t^2\right)\mathbf{j}$$

where $\mathbf{i}$ and $\mathbf{j}$ are horizontal and vertical unit vectors respectively. The model is only valid while the vertical component is greater than zero. After this, the riders all travel on a horizontal surface.

a) Find the range of values for which the model is valid.

b) What are the velocity and acceleration of the riders when they are about to reach the level surface?

c) Do you think that this ride is safe?

8 The path of an object is modelled by:

$$\mathbf{r} = \sqrt{t}\,\mathbf{i} + \frac{1}{\sqrt{t}}\mathbf{j}$$

where $\mathbf{i}$ and $\mathbf{j}$ are perpendicular unit vectors. Find the velocity and acceleration of the object when $t = 4$.

9 A child descends a slide, with displacement, s, modelled by:

$$s = 0.6t + 0.52t^2$$

a) Find the velocity and acceleration of the child.

b) Describe the initial motion of the child.

c) The slide is at 45° to the horizontal and straight. What assumptions have been made about the friction acting on the child?

10 A car on a bend moves so that: $\mathbf{r} = 4t\,\mathbf{i} + 0.2t^3\mathbf{j}$ for $0 \le t \le 5$ where $\mathbf{i}$ and $\mathbf{j}$ are perpendicular unit vectors. The car moves with a constant velocity once it has left the bend.

a) Sketch the path of the car and indicate the direction of the acceleration of the car.

b) Find the speed of the car when it enters the bend at $t = 0$ and when it leaves the bend at $t = 5$.

c) Find the position of the car when $t = 10$.

VARIABLE ACCELERATION

So far we have discussed situations in which the acceleration was taken to be constant. In real life, this rarely happens, so now we need to consider what is more likely to happen in 'everyday' situations.

Example 7.3

A spaceship experiences an acceleration that increases with time so that:

$$a = \frac{3t}{400} \quad \text{for} \quad 0 \le t \le 20$$

Find the speed reached by the spaceship in the first 20 seconds of its motion and the distance it travels if it initially has velocity 0.3 m s^{-1}.

Solution

The velocity can be found by integrating the expression for the acceleration.

$$v = \int a\,dt = \int \frac{3t}{400}\,dt = \frac{3t^2}{800} + c$$

We now need to find the value of the constant of integration. When $t = 0$, $v = 0.3$, so:

$$0.3 = 0 + c$$

giving the constant $c = 0.3$. In this case, the constant represents the initial velocity of the spaceship. Thus the velocity is:

$$v = \frac{3t^2}{800} + 0.3$$

so at the end of the 20 second period:

$$v = \frac{3 \times 20^2}{800} + 0.3 = 1.8 \text{ m s}^{-1}$$

The distance travelled by the ship is given by:

$$s = \int v\,dt = \int\left(\frac{3t^2}{800} + 0.3\right)dt = \frac{t^3}{800} + 0.3t + c$$

Assuming that the motion begins at the origin, then $s = 0$ when $t = 0$, so $c = 0$, giving:

$$s = \frac{t^3}{800} + 0.3t$$

After 20 seconds, the distance travelled is given by:

$$s = \frac{20^3}{800} + 0.3 \times 20 = 16 \text{ m}$$

Example 7.4

When a ball bounces on the ground it is in contact with the ground for 0.5 seconds and experiences an acceleration that is modelled by:

$$a = 200t - 400t^2 - 10 \quad \text{for} \quad 0 \le t \le 0.5$$

a) *Calculate the magnitude of the acceleration when $t = 0$, 0.25 and 0.5. Use this to sketch a graph of the acceleration against time.*
b) *If initially the velocity of the ball is $-\frac{5}{3}$ m s^{-1} find an expression for v, the velocity of the ball.*
c) *Find the velocity of the ball when it leaves the ground at $t = 0.5$.*
d) *What criticisms could be made about this model?*

Solution

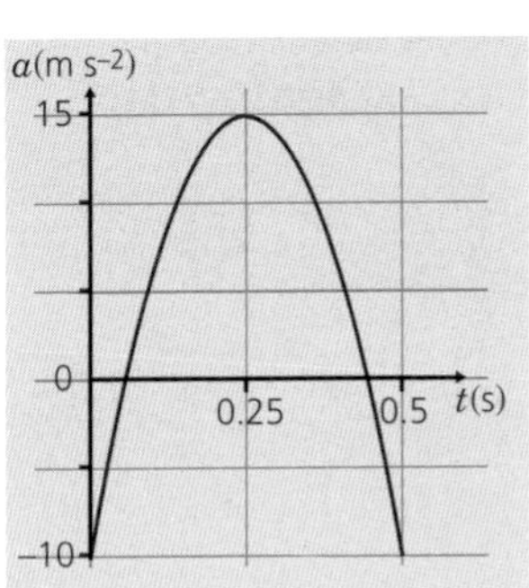

a) *The acceleration can be found by substituting the values of t into the expression to give:*

$t = 0$ $\quad a = -10$
$t = 0.25$ $\quad a = 15$
$t = 0.5$ $\quad a = -10$

These three points are represented by the points on the diagram. As the acceleration is a quadratic, it must have the shape shown.

b) *The velocity can be found by integrating the acceleration.*

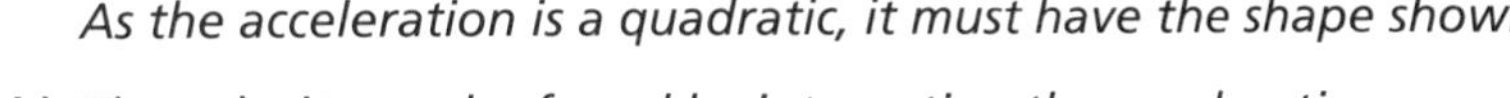

$$v = \int a\,dt = \int(200t - 400t^2 - 10)dt = 100t^2 - \frac{400t^3}{3} - 10t + c$$

As the velocity is $-\frac{5}{3}$ when $t = 0$, we have: $-\dfrac{5}{3} = 0 + c \Rightarrow c = -\dfrac{5}{3}$

thus the velocity is given by: $v = 100t^2 - \dfrac{400t^3}{3} - 10t - \dfrac{5}{3}$

c) The velocity of the ball when $t = 0.5$ is:

$$v = 100 \times 0.5^2 - \frac{400 \times 0.5^3}{3} - 10 \times 0.5 - \frac{5}{3} = \frac{5}{3}$$

d) The ball rebounds at the same speed as it hits the ground. It is very unlikely that this will happen in reality.

Example 7.5

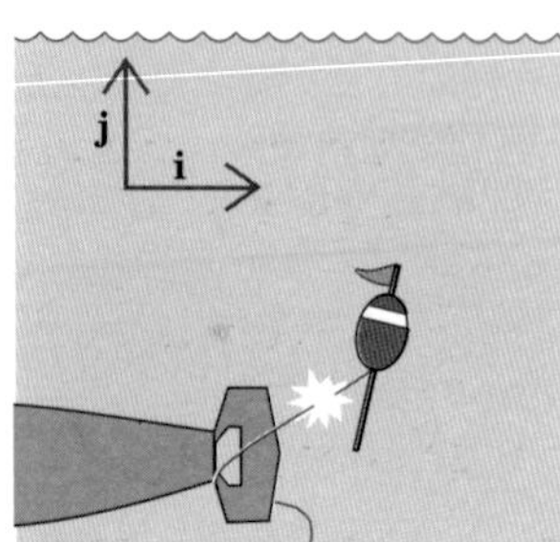

A buoy is snagged by a submarine and dragged down under the water. It becomes detached from the submarine which is moving horizontally at 2 m s^{-1}. As the buoy rises, it experiences an upward acceleration of $5 - \frac{1}{2}t$. Find expressions for the velocity and position of the buoy relative to its point of release.

If the acceleration of the buoy is zero when it reaches the surface, find the depth to which it was dragged.

Solution

In terms of the unit vectors shown on the diagram, the acceleration of the buoy is: $\mathbf{a} = \left(5 - \frac{1}{2}t\right)\mathbf{j}$

The velocity can be found by integrating this expression.

$$\mathbf{v} = \int \mathbf{a}\,\mathrm{d}t = \int\left(5 - \frac{t}{2}\right)\mathrm{d}t\,\mathbf{j} = c_1\mathbf{i} + \left(5t - \frac{t^2}{4} + c_2\right)\mathbf{j}$$

Note that it is necessary to introduce the horizontal component of **velocity**. The initial velocity of the ball was $-2\mathbf{i} + 0\mathbf{j}$, so the values of the two constants are $c_1 = -2$ and $c_2 = 0$, which gives:

$$\mathbf{v} = -2\mathbf{i} + \left(5t - \frac{t^2}{4}\right)\mathbf{j}$$

This expression can now be integrated to give the position vector.

$$\mathbf{r} = \int \mathbf{v}\,\mathrm{d}t = \int -2\mathrm{d}t\,\mathbf{i} + \int\left(5t - \frac{t^2}{4}\right)\mathrm{d}t\,\mathbf{j} = (-2t + c_3)\mathbf{i} + \left(\frac{5t^2}{2} - \frac{t^3}{12} + c_4\right)\mathbf{j}$$

If the initial position of the buoy is $0\mathbf{i} + 0\mathbf{j}$ then $c_3 = 0$ and $c_4 = 0$ so that the position vector is:

$$\mathbf{r} = -2t\,\mathbf{i} + \left(\frac{5t^2}{2} - \frac{t^3}{12}\right)\mathbf{j}$$

At the surface, the acceleration is zero so:

$$5 - \frac{2}{t} = 0 \quad \text{or} \quad t = 10\text{ s}$$

When t = 0, the position is:

$$\mathbf{r} = (-2 \times 10)\mathbf{i} + \left(\frac{5 \times 10^2}{2} - \frac{10^3}{12}\right)\mathbf{j} = -20\mathbf{i} + 167\mathbf{j}$$

So the buoy must have been dragged to a depth of 167 m.

EXERCISES

7.2A

1 In each case below find an expression for the position of the object using the given acceleration, $\mathbf{a}$, initial velocity, $\mathbf{u}$, and initial position $\mathbf{r}_0$.

a) $\mathbf{a} = -10\mathbf{j}$ $\mathbf{u} = 4\mathbf{i} + 2\mathbf{j}$ $\mathbf{r}_0 = 2\mathbf{j}$

b) $\mathbf{a} = 4\mathbf{i} + t\mathbf{j}$ $\mathbf{u} = 3\mathbf{i} - 2\mathbf{j}$ $\mathbf{r}_0 = 0\mathbf{i} + 0\mathbf{j}$

c) $\mathbf{a} = t^2\mathbf{i} - t^3\mathbf{j}$ $\mathbf{u} = 0\mathbf{i} + 0\mathbf{j}$ $\mathbf{r}_0 = 10\mathbf{i} - 20\mathbf{j}$

2 As a ship begins to move, it experiences an acceleration that decreases linearly from $0.5\,\text{m s}^{-2}$ to 0 over a 30-second period.

a) Write down an expression for the acceleration of the ship in terms of time, t, that is valid for $0 \le t \le 30$.

b) Find an expression for the speed of the ship if it starts at rest and the speed reached at the end of the 30 seconds.

c) Find an expression for the distance travelled by the ship and the distance covered in the 30-second period.

3 A cyclist is initially at rest and accelerates at $0.4\,\text{m s}^{-2}$ but this decreases uniformly to 0 over 20 seconds. The cyclist then continues at a constant speed.

a) Write down an expression for the acceleration of the cyclist.

b) Find the maximum speed reached by the cyclist.

c) How far does the cyclist travel in the 20-second period?

4 A rocket launched at a fireworks display moves with position vector:

$$\mathbf{r} = 10t\mathbf{i} + \left(2t^3 - 18t^2 + 54t\right)\mathbf{j} \quad \text{for} \quad 0 \le t \le 3$$

a) By differentiating the expression, find the velocity of the rocket.

b) Find the acceleration of the rocket.

c) Find the velocity and position of the rocket when $t = 3$.

d) When $t = 3$, the rocket explodes and the shell of the rocket returns to Earth with a downward acceleration of $9.8\,\text{m s}^{-2}$. Find, by integrating, expressions for the velocity and position of the shell of the rocket after it has exploded. (Let t be the time since the explosion.)

e) How far does the shell land from the point where the rocket was launched?

5 An object moves so that its position vector is given by:

$$\mathbf{r} = \left(6t^2 - 5\right)\mathbf{i} + \left(5t^2 - 3t + 8\right)\mathbf{j}$$

a) Find the acceleration of the object.

b) Find the velocity and speed of the object when $t = 3$.

c) When does the object have position vector $\mathbf{r} = 145\mathbf{i} + 118\mathbf{j}$?

6 A car experiences an initial acceleration of $3\,\text{m s}^{-2}$ but this decreases uniformly. The car reaches a maximum speed of $30\,\text{m s}^{-1}$ after 20 seconds. Find the distance travelled by the car while it is accelerating.

7 A car has velocity $30\mathbf{i}$ when it enters a bend. While on the bend its acceleration is $0.2\mathbf{i} + 0.1t\mathbf{j}$.

a) Find expressions for the velocity and position of the car while on the bend.

b) Sketch the path of the car for $0 \le t \le 5$.

8 The acceleration of a piston in an engine is modelled as $a = -10t$. The piston is initially at the mid-point of its path and moving at $20\,\text{m s}^{-1}$.

a) Find the time taken for the piston to come to rest.
b) Sketch a graph of the displacement of the piston against time.
c) State the range of values for which the model could be considered reasonable.
d) Suggest how the model for the acceleration could be refined.

9 The acceleration of a jet-ski is modelled by: $\mathbf{a} = 2\mathbf{i} - \frac{t}{4}\mathbf{j}$

where $\mathbf{i}$ and $\mathbf{j}$ are perpendicular unit vectors. The initial velocity of the jet-ski is $4\mathbf{j}$. Describe the path of the jet-ski for the 10 seconds that the acceleration is present.

10 The graph shows how the magnitude of the acceleration of the car varies during an emergency stop.

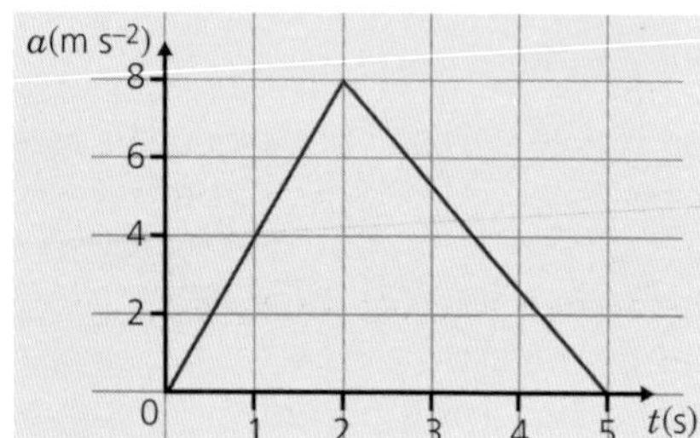

a) What is the direction of the acceleration?
b) Find $\int_0^5 a\,dt$.
c) What was the initial speed of the car?
d) Sketch a graph of speed against time.
e) What was the speed of the car after 2 seconds?
f) Give reasons why the acceleration graph has the shape illustrated.

EXERCISES

7.2 B

1 In each case below, the acceleration, $\mathbf{a}$, and initial velocity, $\mathbf{u}$, of an object are given. Initially the object was at the origin. Find an expression for the position vector of the object after t seconds.

a) $\mathbf{a} = \frac{t}{2}\mathbf{i} + \frac{t}{3}\mathbf{j}$ $\quad \mathbf{u} = 4\mathbf{i}$

b) $\mathbf{a} = 3t\mathbf{i} - 2t\mathbf{j}$ $\quad \mathbf{u} = 20\mathbf{i} - 18\mathbf{j}$

c) $\mathbf{a} = \sqrt{t}\mathbf{i} + 2\sqrt{t}\mathbf{j}$ $\quad \mathbf{u} = 10\mathbf{i} - 2\mathbf{j}$

2 As a car moves away from rest it accelerates to its maximum speed in 10 s. A model for its acceleration during this period is $a = 10 - t$.

a) Find the maximum speed of the car.
b) Find the distance the car travels while accelerating.

3 A cyclist who starts from rest accelerates for 20 seconds. A model for the acceleration of the cyclist is:

$$a = \frac{20 - t}{20}$$

a) Find the speed of the cyclist at the end of the 20 seconds.
b) Find the total distance travelled.

4 The acceleration of an aeroplane on a runway is modelled as: $a = \frac{\sqrt{t}}{2}$

The aeroplane takes off when it reaches a speed of $72\,\text{m s}^{-1}$. Find the time that the aeroplane accelerates before take-off and the distance travelled before take-off.

5 As a car accelerates from rest its acceleration is $\mathbf{a} = \frac{20-t}{5}\mathbf{i} + \frac{20-t}{10}\mathbf{j}$ where $\mathbf{i}$ and $\mathbf{j}$ are perpendicular unit vectors.

a) Find the velocity of the car.

b) Show that when $t = 20$, the speed is a maximum.

c) When does the car come to rest?

d) Find the displacement of the car from its starting point when it comes to rest.

6 A ball is thrown so that its acceleration, including some of the effects of air resistance, has been modelled as:

$$\mathbf{a} = -\frac{t}{20}\mathbf{i} - 10\mathbf{j}$$

where $\mathbf{i}$ and $\mathbf{j}$ are horizontal and vertical unit vectors respectively.

a) Find an expression for the velocity of the ball if its initial velocity is $8\mathbf{i} + 2\mathbf{j}$.

b) Find the position vector of the ball.

c) Where does the ball hit the ground?

7 The acceleration of a jet-ski is modelled as $\mathbf{a} = \frac{\sqrt{t}}{2}\mathbf{i} - \frac{\sqrt{t}}{4}\mathbf{j}$ where $\mathbf{i}$ and $\mathbf{j}$ are perpendicular unit vectors.

a) If the jet-ski is initially at rest, find its position vector after 16 seconds.

b) Describe the path of the jet-ski.

8 The mass attached to a spring is given an initial velocity of $2\,\text{m s}^{-1}$. The acceleration of the mass is modelled as $a = -\frac{1}{2}t$

a) Find an expression for the displacement of the mass from its initial position.

b) When does the mass come to rest for the first time?

c) For what range of values of t could the model be reasonable?

d) Suggest an alternative model for the acceleration.

$2\,\text{m s}^{-1}$

9 The acceleration of a person on a free-fall slide is modelled as:

$$\mathbf{a} = 0.5\mathbf{i} - \frac{(t-4)^2}{2}\mathbf{j}$$

where $\mathbf{i}$ and $\mathbf{j}$ are horizontal and vertical unit vectors respectively.

a) Find the expressions for the velocity and position of the person on the slide.

b) When does the motion become horizontal?

c) The acceleration from this point on is $-5\mathbf{i}$. How far does the person travel on the horizontal section at the bottom of the slide?

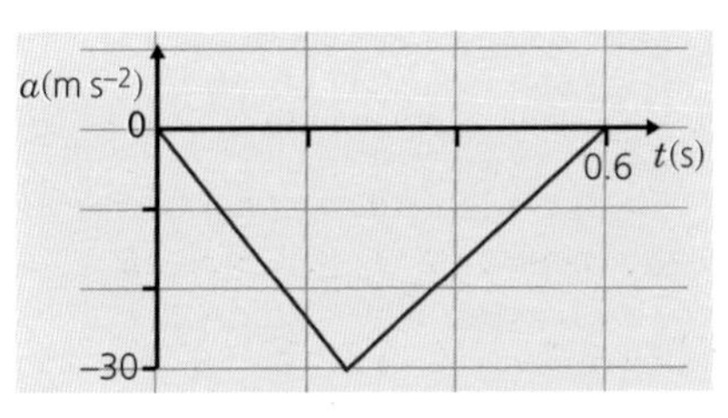

10 A ball hits a wall travelling at $5\,\text{m s}^{-1}$. It experiences an acceleration that can be modelled as shown in the graph.

a) Use the graph to find $\int_0^{0.6} a\,dt$.

b) If u is the impact speed and v the rebound speed, explain why $u - v = \int_0^{0.6} a\,dt$

c) Find the rebound speed.

CONSOLIDATION EXERCISES FOR CHAPTER 7

1 A particle P moves from rest, at a point O at time $t = 0$ seconds, along a straight line. At any subsequent time t seconds the acceleration of P is proportional to $(7 - t^2)$ m s^{-2} and the displacement of P from O is s metres. The speed of P is 6 m s^{-1} when $t = 3$.

a) Show that: $s = \frac{1}{24}t^2(42 - t^2)$

b) Find the total distance, in metres, that P moves before returning to O.

(ULEAC Question 7, M1, January 1992)

2 With respect to a fixed origin O, the position vector, $\mathbf{r}$ metres, of a particle P, of mass 0.15 kg at time t seconds is given by:

$\mathbf{r} = t^3\mathbf{i} - 4t^2\mathbf{j} \quad t \geq 0$

a) Find, in m s^{-1}, the speed of P when $t = 2$.

b) Find, in N to two decimal places, the magnitude of the resultant force acting on P when $t = 2$.

(ULEAC Question 3, M1, June 1992)

3 At time t seconds, where $t \geq 0$, the acceleration of a particle P, moving in a straight line, is $(2t - 8)$ m s^{-2}.

Given that the initial speed of P is 12 m s^{-1}, find the times at which P comes instantaneously to rest.

(ULEAC, Question 1, M1, January 1993)

4 A dot moves on the screen of an oscilloscope so that its position relative to a fixed origin is given by:

$$\mathbf{r} = 2t\mathbf{i} + \sin\left(\frac{\pi t}{2}\right)\mathbf{j}$$

a) Sketch the path of the dot for $0 \leq t \leq 4$.

b) Find the velocity and acceleration of the dot when $t = 3$. Draw vectors on your diagram to show these two quantities.

(AEB Question 1, Specimen Mechanics Paper, 1994)

5 Over the first two seconds as it accelerates from rest, the velocity v m s^{-1} of an aeroplane is modelled by the equation $v = 10\sqrt{t}$ where t is measured in seconds from the moment that the aeroplane starts to move.

a) According to the model, how far does the aeroplane travel in the first two seconds?

b) Calculate the acceleration of the aeroplane when $t = 2$.

c) Why does the model not give a satisfactory value for the acceleration at $t = 0$?

(Nuffield Question 6, Specimen Paper 1, 1994)

6 A particle of mass 3 kg is placed on a rough horizontal surface and a horizontal force of gradually increasing magnitude P is applied to it.

a) Draw a diagram and show the force of friction $\mathbf{F}$ acting on the particle, together with all the other forces.

b) After t seconds P is given by:

$P = 2t \quad 0 \le t \le 7$

$P = 0 \quad 7 < t$

i) Find the value of the coefficient of friction μ, if the body is in limiting equilibrium when $t = 4$.

ii) Show that the distance travelled by the body in t seconds, where $4 \le t \le 7$ is:

$$x = \frac{1}{9}\left(t^3 - 12t^2 + 48t - 64\right)$$

(Oxford Question 7, Specimen Paper M5, 1994)

MATHEMATICAL MODELLING ACTIVITY

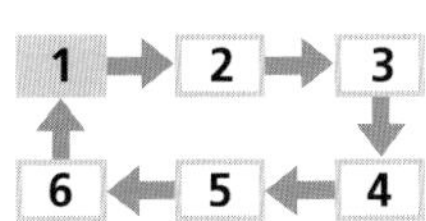

Specify the real problem

Problem statement

It has been suggested that a sprinter will reach their top speed in 4 s and maintain that speed for the rest of the sprint. Develop a model that describes how a sprinter moves during the first four seconds.

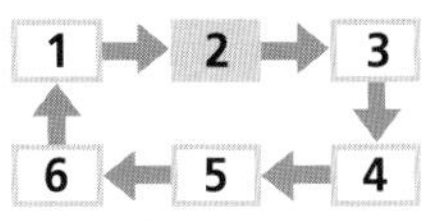

Set up a model

Set up a model

A simple first model can be developed from the assumptions listed below.

- The sprinter is modelled as a particle travelling along a straight line.
- The acceleration decreases linearly during the first four seconds and is then zero,
  i.e. $a = p - qt$ for $t \le 4$.
- The sprinter reaches a speed of 6.5 m s^{-1}. (This figure has been chosen as it relates to data that can be used later.)

Formulate the mathematical problem

Formulate a mathematical problem

If the sprinter has acceleration $a = p - qt$, for $0 \le t \le 4$, and reaches a maximum speed of 6.5 m s^{-1} find the values of p and q.

Solve the mathematical problem

Mathematical solution

First consider the fact that the acceleration is zero when $t = 4$.
This gives the equation:

$$0 = p - (q \times 4) \Rightarrow p = 4q$$

Now to consider the velocity of the sprinter, which can be obtained by integration.

$$v = \int a\,\mathrm{d}t = \int (p - qt)\,\mathrm{d}t = pt - \frac{qt^2}{2} + c$$

When $t = 0$, $v = 0$ so the constant c is zero and $v = pt - \dfrac{qt^2}{2}$.

When $t = 4$, $v = 6.5$ m s^{-1}, and so $6.5 = 4p - 8q$.
But $p = 4q$, so

$$6.5 = 16q - 8q = 8q$$

$$q = \frac{6.5}{8} = 0.8125$$

Using this gives $p = 3.25$.

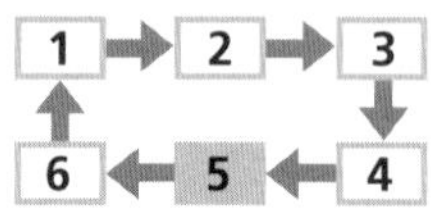

Interpret the solution

Interpretation

The acceleration can be modelled as:

$$a = 3.25 - 0.8125t$$

and the velocity can be modelled as:

$$v = 3.25t - 0.40625t^2$$

Show that the distance travelled by the sprinter can be modelled as:

$$s = 1.625t^2 - 0.135\,416t^3$$

Also find the distance travelled by the sprinter in the four seconds.

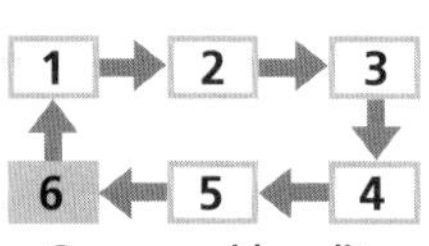

Compare with reality

Compare with reality

The table below gives a set of observed velocities recorded with an electronic measuring device.

t (s)	0	0.5	1.0	1.5	2.0	2.5	3.0	3.5	4.0
v (m s^{-1})	0	1.52	2.87	3.90	4.69	5.24	5.67	6.10	6.50

Compare the actual results with the models predictions and comment on the quality of the model.

Refinement of the model

Suggest possible ways in which the model could be refined.

POSSIBLE MODELLING TASK

Many real situations involve variable forces. Choose a situation where an object begins to move from rest, such as a car or cyclist. Record how far the object travels over short periods of time. Use this to investigate how the resultant force on the object varies with time.

Summary

After working through this chapter you should:

- be able to use differentiation to find velocity and acceleration from a given position vector
- be able to use integration to find velocity and position from an acceleration
- be able to work in two or three dimensions.

Projectiles

- *In mechanics, any unpowered object that is launched into the air, for example a tennis ball hit by a racquet is called a projectile.*
- *In this chapter, equations modelling the motion of a projectile are developed.*
- *In order to keep things simple, we have assumed that there is no air resistance to the motion of a projectile.*

THE MOTION OF A PROJECTILE

When a particle is launched into the air, its motion follows mathematical rules and patterns. In this chapter we shall explore these patterns and establish some of the basic rules.

Exploration 8.1

Simplifications and assumptions

When trying to describe the path of a ball in flight mathematically it is necessary for us to ignore certain aspects of its motion, so that we can formulate a simple model.

- Air resistance could be ignored. Are there any other assumptions that it may be advisable to make? You may wish to identify specific assumptions for balls used in different sports.
- How do you think the speed of a projectile, for example a golf ball, varies during its flight? Looking at this diagram may give some clues.

Assumptions

In some ball games, the spin of the ball affects its flight by generating a lift force. In this chapter we shall model projectiles as particles that do not spin, and are influenced only by gravity. This allows to draw up a very simple model.

The speed of a cricket ball decreases as it gains height, reaching a minimum at the ball's highest point, before gaining speed again as the ball returns to the ground. It is useful to compare this situation with a ball that is thrown straight up and then falls straight down again in the air. The speed decreases to zero on the way up and increases again on the way down.

BRIDGWATER COLLEGE LIBRARY

Exploration 8.2

The path of a projectile

- Set up the apparatus shown in the diagram. The plastic tube should be fixed to a retort stand with a clamp. It is used to provide a consistent way of launching the ball or marble. Provided the angle of the tube to the vertical is not altered, the speed at which the ball leaves the tube will always be the same. The sloping table provides a means of slowing down the motion of the projectile. Imagine that the table top is in a vertical plane, and that you are seeing a ball move in slow motion.

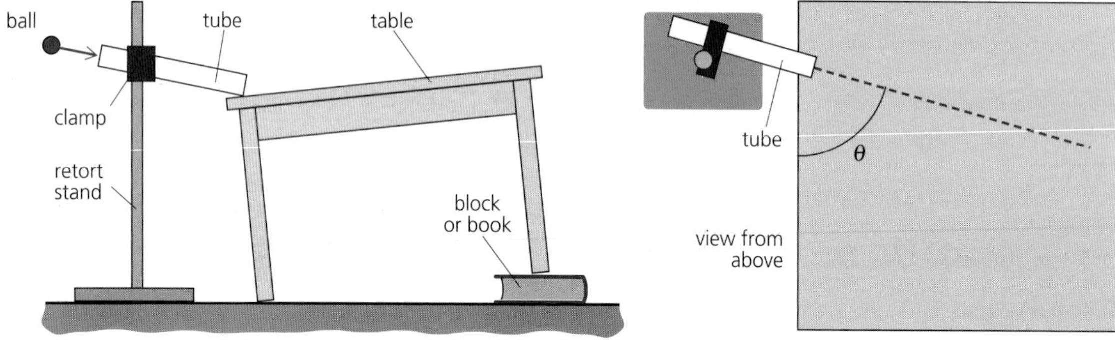

- Begin by launching the ball straight up the table. It should return to the tube! Now vary the angle the tube makes with the edge of the table. What angle will give you the maximum range?
- Change the launch speed and repeat the last step. How does the path of the ball change?
- Place a target somewhere on the table top (a pencil sharpener or small object that the ball can knock over is ideal). By changing the angle of launch and the launch speed, try to find four different ways of hitting the target.

The path of a projectile

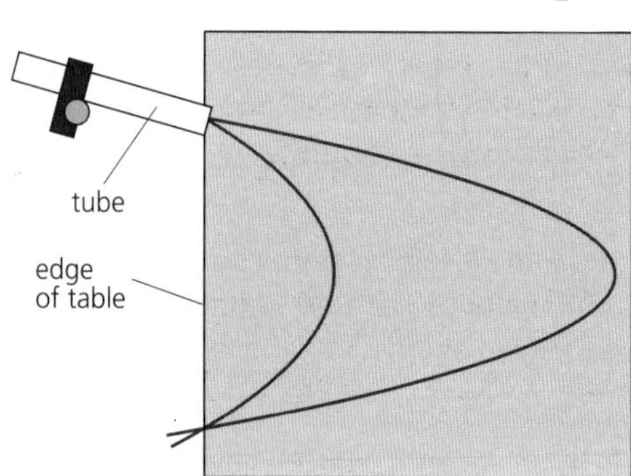

The distance travelled by the ball on the table top increases as the launch speed is increased. The angle that the table makes with the edge of the tube is also very important. When it is 45° the ball will travel the greatest distance. The same point on the edge of the table can be reached by two different paths, one launched with a small angle, less than 45°, and the other with a large angle, greater than 45°, as shown in the diagram. The path that the ball describes is a **parabola**.

The worked examples that follow show how to find the position vector for a projectile and to determine some of the features of the motion.

CALCULATOR ACTIVITY 8.1

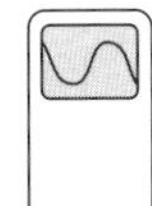

1 First select parametric plotting mode and degree mode. Then set the range or window values to be
$x_{min} = 0$; $x_{max} = 5$; $x_{scl} = 1$
$y_{min} = 0$; $y_{max} = 2$; $y_{scl} = 1$
$T_{min} = 0$ $T_{max} = 2$ $T_{step/pitch} = 0.1$
2 The path of a particular projectile is modelled by:
$\mathbf{r} = (6t\cos 30°)\mathbf{i} + (6t\sin 30° - 5t^2)\mathbf{j}$

where **i** and **j** are the unit vectors in the horizontal and vertical directions, as usual. Enter this in the form:
$x = (6t\cos 30°)$ and $y = (6t\sin 30° - 5t^2)$

Is the path plotted what you may expect for a projectile?
3 In the equations above, 30° is the **angle of projection**. Use some different values for this angle. Sketch a graph to show that the **range** (horizontal distance travelled) varies with the angle. What angle gives the maximum range?
4 In the equations above, 6 represents the speed of projection. Use some different values for this speed, in the equations. How does the speed affect the range?

Modelling the path of a projectile

The path of a projectile can be modelled as: $\mathbf{r} = ut\mathbf{i} + \left(vt - \frac{1}{2}gt^2\right)\mathbf{j}$

where u and v are the horizontal and vertical components of the initial velocity. Using the equations of motion from earlier work, we can say that the horizontal displacement is given by ut. The displacement increases at a constant rate and so the projectile is moving at a constant speed. The equation for the vertical displacement is exactly the same as would apply to an object projected vertically upwards. Combining the vertical and horizontal components, the parabolic path is produced.

As we have assumed that the only force that acts on a projectile is gravity, the resultant force will simply be $-mg\mathbf{j}$. Using Newton's second law ($\mathbf{F} = m\mathbf{a}$), we can see that the acceleration of the projectile will be $-g\mathbf{j}$. We shall use this expression for the acceleration of a projectile as a starting point for the following examples.

Example 8.1

A golf ball is struck so that its initial velocity is 20 m s^{-1} at an angle of 40° above the horizontal.

a) Express the acceleration of the ball in terms of the unit vectors **i** *and* **j**.
b) Express the initial velocity in terms of the unit vectors **i** *and* **j**.
c) Find an expression for the position of the ball, assuming that it starts at the origin.
d) Find how far the ball is from the origin when it hits the ground.
e) What factors about the ball's motion have been ignored?

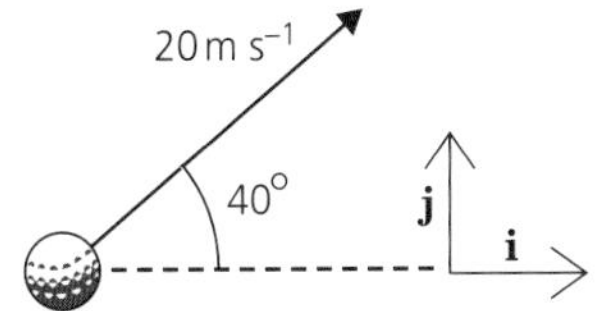

Solution

a) *If the ball is modelled as a particle that is acted on by only the force of gravity, then the acceleration will simply be the acceleration due to gravity, so:*

$\mathbf{a} = -g\mathbf{j} = -9.8\mathbf{j}$

b) *The initial velocity of the ball is given by:*

$\mathbf{v}(0) = 20\cos 40°\,\mathbf{i} + 20\cos 50°\,\mathbf{j} = 15.3\mathbf{i} + 12.9\mathbf{j}$

20
20 cos 50° or 20 sin 40°
40°
20 cos 40°

c) *The first step is to integrate the acceleration to find the velocity. This gives:* $\mathbf{v} = \int 0\,\mathrm{d}t\,\mathbf{i} + \int -9.8\,\mathrm{d}t\,\mathbf{j} = c\mathbf{i} + (-9.8t + d)\mathbf{j}$
where c *and* d *are constants that depend on the initial velocity.*
*As the initial velocity is 15.3***i** *+ 12.9***j***, the value of* c *is 15.3 and* d *is 12.9, giving* $\mathbf{v} = 15.3\mathbf{i} + (12.9 - 9.8t)\mathbf{j}$
Integrating again gives the position vector:

$$\mathbf{r} = \int 15.3\,\mathrm{d}t\,\mathbf{i} + \int (12.9 - 9.8t)\,\mathrm{d}t\,\mathbf{j}$$

$$= (15.3t + e)\mathbf{i} + \left(12.9t - 4.9t^2 + f\right)\mathbf{j}$$

where e *and* f *are constants that depend on the initial position.*
*As the projectile begins at the origin, 0***i** *+ 0***j***, both* e *and* f *must be zero, so* $\mathbf{r} = 15.3t\,\mathbf{i} + \left(12.9t - 4.9t^2\right)\mathbf{j}$

d) *When the ball hits the ground the vertical component of the position vector,* $(12.9t - 4.9t^2)$*, will be zero, giving:*

$12.9t - 4.9t^2 = 0 \Rightarrow t(12.9 - 4.9t) = 0$

This equation has two solutions:

$t = 0$ *and* $t = \dfrac{12.9}{4.9} = 2.63$

The first solution, $t = 0$ *corresponds to the instant that the ball leaves the ground and* $t = 2.63$ *corresponds to the instant when the ball hits the ground. The time that the ball is in the air is often referred to as the* ***time of flight****.*
Substituting $t = 2.63$ *into the position vector gives:*

$\mathbf{r} = 15.3 \times 2.63\,\mathbf{i} + \left(12.9 \times 2.63 - 4.9 \times 2.63^2\right)\mathbf{j} = 40.2\mathbf{i}$

So the ball hits the ground 40.2 m from the point where it began its flight. This distance is known as the ***range*** *of the projectile.*

e) *Golf balls are designed so that as they spin they experience an aerodynamic lift force that increases their range. This factor has been completely ignored, as has any air resistance. It has also been assumed that there is no wind present.*

Example 8.2

An archer shoots an arrow so that its initial velocity is 40 m s^{-1} *at an angle of 60° to the horizontal.*

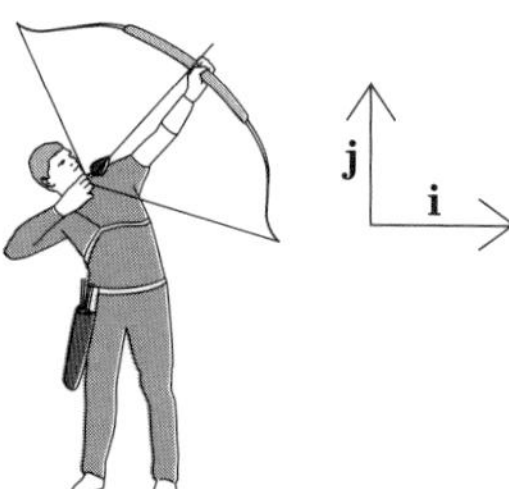

a) *Find the initial velocity of the arrow in terms of the unit vectors* **i** *and* **j***.*

b) *Find the position vector that describes the position of the arrow.*

c) *Find the maximum height reached by the arrow.*

40
40 cos 30°
or
40 sin 60°
60°
40 cos 60°

Solution

a) *The initial velocity is:*

$$40\cos 60^\circ\,\mathbf{i} + 40\cos 30^\circ\,\mathbf{j} = 20\mathbf{i} + 34.6\mathbf{j}$$

b) *Assuming that the only force acting is gravity gives the acceleration as:*

$$\mathbf{a} = -g\mathbf{j} = -9.8\mathbf{j}$$

This can be integrated to give the velocity vector as:

$$\mathbf{v} = \int 0\,\mathrm{d}t\,\mathbf{i} + \int -9.8\,\mathrm{d}t\,\mathbf{j} = c\,\mathbf{i} + (-9.8t + d)\mathbf{j}$$

where c and d are constants that depend on the initial velocity. As the initial velocity is 20$\mathbf{i}$ + 34.6$\mathbf{j}$, then c = 20 and d = 34.6, giving:

$$\mathbf{v} = 20\mathbf{i} + (34.6 - 9.8t)\mathbf{j}$$

Integrating again gives the position vector as:

$$\mathbf{r} = \int 20\,\mathrm{d}t\,\mathbf{i} + \int (34.6 - 9.8t)\,\mathrm{d}t\,\mathbf{j} = (20t + e)\mathbf{i} + \left(34.6t - 4.9t^2 + f\right)\mathbf{j}$$

where e and f are constants that depend on the initial position. If it is assumed that the arrow begins at 0$\mathbf{i}$ + 0$\mathbf{j}$, then e and f are both zero, to give:

$$\mathbf{r} = 20t\,\mathbf{i} + \left(34.6t - 4.9t^2\right)\mathbf{j}$$

c)

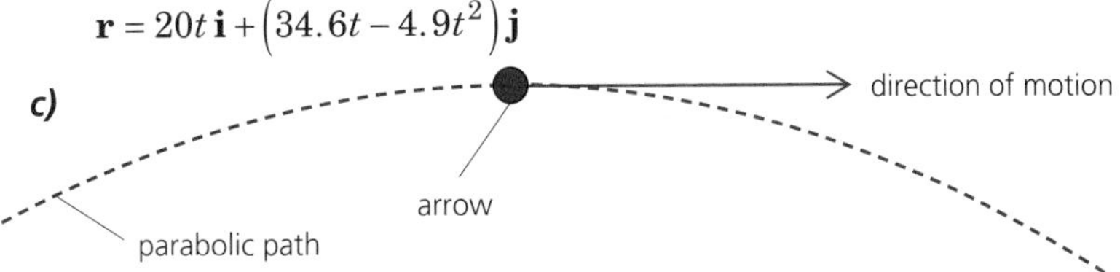

At the highest position of the arrow it is for an instant moving horizontally, so the vertical component of its velocity, $(34.6 - 9.8t)$, is zero. This gives the equation:

$$34.6 - 9.8t = 0 \text{ or } t = \frac{34.6}{9.8} = 3.53$$

So the arrow attains its maximum height when t = 3.53. Substituting this into the position vector gives:

$$\mathbf{r} = 20 \times 3.53\mathbf{i} + \left(34.6 \times 3.53 - 4.9 \times 3.53^2\right)\mathbf{j} = 70.6\mathbf{i} + 61.1\mathbf{j}$$

So the maximum height reached by the arrow is 61.1 m.

The examples above have considered specific cases of projectile motion. The next example involves a general projectile. This approach allows results for the range, height and time of flight to be formulated.

Example 8.3

A projectile is launched with an initial speed V at an angle α above the horizontal.

a) *Find the initial velocity of the projectile.*
b) *Find the position vector of the projectile.*
c) *Find the range of the projectile.*
d) *Find the maximum height reached by the projectile.*

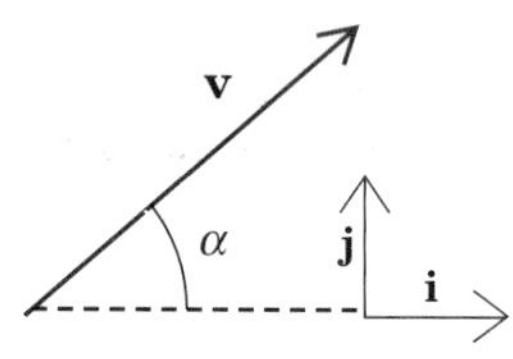

Solution

a) *The initial velocity is:*

$V\cos\alpha\,\mathbf{i} + V\cos(90° - \alpha)\,\mathbf{j}$

As α *is unknown it is easier to use sin* α*, so the initial velocity is:*

$V\cos\alpha\,\mathbf{i} + V\sin\alpha\,\mathbf{j}$

b) *If gravity is assumed to be the only force then the acceleration will be:* $\mathbf{a} = -g\mathbf{j}$

This can be integrated to give the velocity as:

$\mathbf{v} = \int 0\,\mathrm{d}t\,\mathbf{i} + \int -g\,\mathrm{d}t\,\mathbf{j} = c\mathbf{i} + (-gt + d)\mathbf{j}$

As the initial velocity is $V\cos\alpha\,\mathbf{i} + V\sin\alpha\,\mathbf{j}$*, the constant* c *is* $V\cos\alpha$ *and* d *is* $V\sin\alpha$*, giving:*

$\mathbf{v} = V\cos\alpha\,\mathbf{i} + (V\sin\alpha - gt)\,\mathbf{j}$

Integrating again will give the position vector:

$$\mathbf{r} = \int V\cos\alpha\,\mathrm{d}t\,\mathbf{i} + \int (V\sin\alpha - gt)\,\mathrm{d}t\,\mathbf{j}$$

$$= (Vt\cos\alpha + e)\mathbf{i} + \left(Vt\sin\alpha - \tfrac{1}{2}gt^2 + f\right)\mathbf{j}$$

If the projectile is launched from $0\mathbf{i} + 0\mathbf{j}$*, then* e *and* f *are zero, so that:*

$\mathbf{r} = Vt\cos\alpha\,\mathbf{i} + (Vt\sin\alpha - \tfrac{1}{2}gt^2)\mathbf{j}$

c) *The time of flight is found by considering when the projectile returns to the level from which it was launched. When this is the case the vertical component of the position vector,* $Vt\sin\alpha - \tfrac{1}{2}gt^2$*, is zero, giving:*

$Vt\sin\alpha - \tfrac{1}{2}gt^2 = 0$

As t *is a factor, this can be expressed as:*

$t\left(V\sin\alpha - \tfrac{1}{2}gt\right) = 0$

Solving this equation gives:

$$t = 0 \text{ or } \left(V\sin\alpha - \tfrac{1}{2}gt\right) = 0 \Rightarrow \quad t = 0 \text{ or } t = \frac{2V\sin\alpha}{g}$$

and the time of flight is given by $\dfrac{2V\sin\alpha}{g}$.

d) *The range is given by the horizontal component of the position vector,* $Vt\cos\alpha$*, when the projectile hits the ground.*

Substituting $t = \dfrac{2V\sin\alpha}{g}$ *and using the trigonometric identity,* $2\sin\alpha\cos\alpha = \sin 2\alpha$ *gives:*

$$\text{range} = V\cos\alpha\frac{2V\sin\alpha}{g} = \frac{V^2 2\sin\alpha\cos\alpha}{g} = \frac{V^2\sin 2\alpha}{g}$$

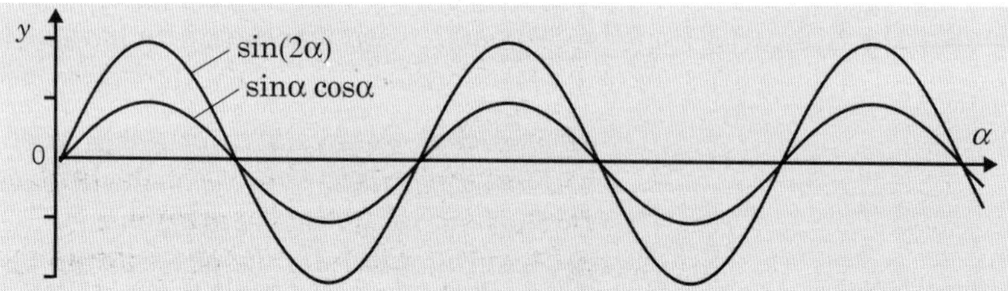

Note: If you are studying Pure Mathematics as well as Mechanics, you will recognise the identity $2\sin\alpha\cos\alpha = \sin 2\alpha$. If you do not, try plotting, $\sin\alpha\cos\alpha$ and $\sin 2\alpha$ and comparing the curves.

e) *The maximum height is reached when the vertical component of the velocity, $V\sin\alpha - gt$ is zero. At this point:*

$$V\sin\alpha - gt = 0$$

so that

$$t = \frac{V\sin\alpha}{g}$$

Note that this is half the time of flight.
Substituting the time into the vertical component of the position vector, $V\sin\alpha\, t - \frac{1}{2}gt^2$, gives:

$$\text{maximum height} = V\sin\alpha\frac{V\sin\alpha}{g} - \frac{g}{2}\left(\frac{V\sin\alpha}{g}\right)^2$$

$$= \frac{V^2\sin^2\alpha}{g} - \frac{V^2\sin^2\alpha}{2g} = \frac{V^2\sin^2\alpha}{2g}$$

Summarising the results so far

The results produced in Example 8.3 can be very useful for dealing with projectile problems. They are summarised below.

$$\mathbf{r} = Vt\cos\alpha\mathbf{i} + \left(Vt\sin\alpha - \tfrac{1}{2}gt^2\right)\mathbf{j}$$

$$\text{Range} = \frac{V^2\sin 2\alpha}{g}$$

$$\text{Time of flight} = \frac{2V\sin\alpha}{g}$$

$$\text{Maximum height} = \frac{V^2\sin^2\alpha}{2g}$$

However, it is important that problems can be solved by finding the position vector for that situation, as well as using standard formulae.

EXERCISES

8.1 A

1 A cricket ball is hit so that it leaves ground level with a speed of $4\,\text{m s}^{-1}$ and moving at an angle of 20° above the horizontal.

a) Express the initial velocity of the ball in terms of the unit vectors $\mathbf{i}$ and $\mathbf{j}$ which are horizontal and vertical respectively.

b) State the acceleration of the ball, in terms of $\mathbf{i}$ and $\mathbf{j}$.

c) If the ball has been in motion for t seconds, find the velocity of the ball in terms of t.

d) Find the position vector of the ball, in terms of t, stating a suitable origin.

2 A kangaroo hops so that it leaves the ground moving at $5\,\text{m s}^{-1}$ and at an angle of 15° to the horizontal.

a) Express the initial velocity of the kangaroo in terms of the unit vectors $\mathbf{i}$ and $\mathbf{j}$ which are horizontal and vertical respectively.

b) Find the position vector of the kangaroo in terms of t, where t is the time, in seconds, the kangaroo has been in motion.
c) For how long is the kangaroo in the air during one hop?
d) How much ground does the kangaroo cover during one hop?
e) What is the greatest height reached by the kangaroo?

3 An athlete competing in a long jump event claims that he can take off at $10\,\text{m}\,\text{s}^{-1}$. Assume that he initially moves at 25° to the horizontal.
a) Find the distance that the athlete would jump.
b) Do you think that the claim is reasonable? Explain why.

4 A tennis ball is launched from ground level so that it initially moves at $15\,\text{m}\,\text{s}^{-1}$ at 39° to the horizontal.
a) Find the range of the ball.
b) Would it make any difference to your answer if it was a cricket ball?
c) If you were to include air resistance in the problem, which ball would be affected more?

5 An archer shoots an arrow at a target, hitting it dead centre. Assume that the arrow is launched from the same height as the centre of the target, at an initial speed of $25\,\text{m}\,\text{s}^{-1}$ and at 25° to the horizontal.
a) Find the distance of the target from the archer.
b) Show that if the arrow were fired at the same speed but at 65° to the horizontal it would hit the target in the same place.
c) Find the speeds of the arrows when they hit the target in each case.
d) Comment on which angle the archer would prefer to use, giving reasons to support your conclusion.

6 In a school sports event a student jumps 1.8 m in the long jump event.
a) Assuming that the student initially moves at 18° to the horizontal find their initial speed.
b) What angle of projection would give the greatest range?
c) If the angle of the initial velocity was unknown, find the minimum initial speed that would be needed to travel 1.8 m.

7 A shot putt is thrown with an initial speed of $8\,\text{m}\,\text{s}^{-1}$ and travels 4.8 m.
a) Find the angle at which the shot putt was projected.
b) If a lighter shot putt were launched with the same initial velocity, how would its range compare to the heavier one?
c) Explain why an athlete can achieve a greater range with a lighter shot putt.

8 A crate of relief supplies is dropped from an aeroplane travelling horizontally at $75\,\text{m}\,\text{s}^{-1}$. It falls for three seconds before a parachute attached to it opens automatically. The unit vectors $\mathbf{i}$ and $\mathbf{j}$ are horizontal and vertical respectively.
a) Explain why the initial velocity of the crate is $75\mathbf{i}$.
b) Find an expression for the position vector of the crate relative to its point of release.
c) Find the position of the crate when the parachute opens.
d) Find the speed of the crate when the parachute opens.
e) Draw a diagram to show the path of the crate and the aeroplane, assuming that the aeroplane maintains a constant velocity.

9 A footballer heads a ball giving it an initial velocity of $6\,\text{m}\,\text{s}^{-1}$ at 70° to the horizontal. The ball is initially at a height of 2 m.

a) Using a suitable origin, find a position vector for the ball.

b) Find the maximum height reached by the ball.

c) Find the time that the ball is in the air and the distance it travels horizontally.

10 A basketball player throws a ball from a height of 1.8 m with an initial velocity of $8\,\text{m}\,\text{s}^{-1}$ at 38° to the horizontal. The basket is 3 m above ground level.

a) Find the position vector of the ball relative to the player's feet.

b) The basket is situated 4 m from the player. Find the time it takes the ball to travel a horizontal distance of 4 m.

c) What is the height of the ball when it has travelled 4 m horizontally?

d) Suggest how the player could improve the chance of the ball going in the basket.

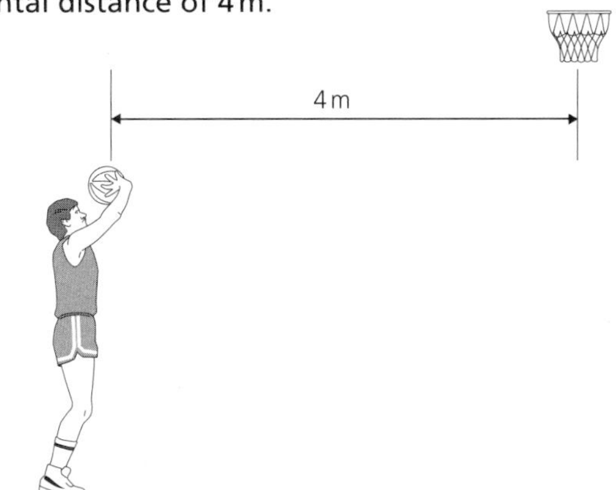

EXERCISES

8.1 B

1 A football is kicked from the ground so that it leaves the ground with a speed of $7\,\text{m}\,\text{s}^{-1}$ at an angle of 30° above the horizontal.

a) Express the initial velocity of the ball in terms of the unit vectors $\mathbf{i}$ and $\mathbf{j}$ which are horizontal and vertical respectively.

b) State the acceleration of the ball in terms of $\mathbf{i}$ and $\mathbf{j}$.

c) Find the velocity of the ball in terms of t, if the ball has been in motion for t seconds.

d) Find the position vector of the ball in terms of t, using a suitable origin.

2 During a sack race, a girl jumps so that she leaves the ground moving at $2\,\text{m}\,\text{s}^{-1}$ and at an angle of 20° to the horizontal.

a) Express the initial velocity of the girl in terms of the unit vectors $\mathbf{i}$ and $\mathbf{j}$ which are horizontal and vertical respectively.

b) Find a position vector to describe the position of the girl in terms of t, if t is the amount of time in seconds she has been in motion.

c) How much time passes between each jump?

d) How much ground does she cover during each jump?

e) What is the greatest height reached by the girl?

3 A rugby player claims to be able to kick a rugby ball from the ground at $12\,\text{m}\,\text{s}^{-1}$. Assume that the ball initially moves in a direction 35° above the horizontal.

a) Find the distance that the rugby player can kick the ball.
b) Do you think that the claim is reasonable? Explain why.

4 A tennis ball is launched from ground level so that initially moves at $14\,\mathrm{m\,s^{-1}}$ at 42° to the horizontal.
a) Find the range of the ball.
b) Would it make any difference to your answer if it were a table tennis ball?
c) If you were to consider the effects of air-resistance in the problem, how would the motion of the balls compare with your earlier answers?

5 During a darts match a darts player hits the bulls eye. Assume that the dart is thrown from the same height as the bulls eye at an initial speed of $8\,\mathrm{m\,s^{-1}}$and at 10° above the horizontal.
a) Find how far from the dart board the player is standing.
b) Show that if the dart had been thrown with the same speed but at 80° above the horizontal it would also have hit the bulls eye.
c) Find the speed of the dart when it hits the target in each case.
d) Which would be the best angle to throw the dart? Give reasons for your choice.

6 A motorcycle stunt rider jumps a distance of 24 metres.
a) Assume the rider and bike initially move at 20° to the horizontal, find the initial speed.
b) What angle of projection would give the greatest range?
c) If the angle of the initial velocity was unknown, find the minimum initial speed that would be needed to travel 24 metres.

7 During a game of tiddly-winks, counters initially travel at $1\,\mathrm{m\,s^{-1}}$ but the angle to the horizontal at which they can be projected does not exceed 60°.
a) What is the maximum range of a counter?
b) If a circular cup of radius 4 cm and height 3 cm is placed on the table to collect the counters, what is the maximum distance from the centre of the cup a counter can be projected to land in the cup?
c) What is the minimum distance?
d) How would the radius and height of the circular cup affect your results?

8 A child drops an apple core from the open window of a moving car, travelling on a horizontal road at a constant velocity V.
a) Explain why, when the apple core hits the road, it is still below the child's hand vertically.
b) If t is the time in seconds from the moment the apple core is released, find an expression for the position vector of the apple core relative to its point of release.
c) If the apple core hits the road when $t = 0.6$, find the speed of the apple core at this time if the speed of the car is a constant $20\,\mathrm{m\,s^{-1}}$.

9 A shot-putter is able to project a shot with a speed of $6\,\mathrm{m\,s^{-1}}$ at an angle above the horizontal of 35° from a point 1.7 m vertically above the ground.

a) Using a suitable origin, find a position vector for the shot.
b) Find the time that the shot is in the air and the distance that it travels horizontally.
c) Draw a sketch of the trajectory of the shot as seen from one side.
d) Explain why considering air resistance would not change the path of the ball appreciably.
e) On the same diagram sketch the path of a table-tennis ball projected in exactly the same way but subject to air resistance.

10 An enterprising golf enthusiast decides to stake a claim to build the first golf course on the moon to provide a facility for those who will eventually live and work there. The length of a typical golf drive on Earth is 100 m. Assume that an average golfer hits the ball at 35° to the horizontal.
a) Find the initial speed of the ball.
b) What range would be reached using the same initial velocity on the moon? (Assume $g = 1.6\,\text{m}\,\text{s}^{-2}$ on the moon.)
c) Give brief advice to the enthusiast about the design of his new course.

THE EQUATION OF THE PATH OF A PROJECTILE

Sometimes, when solving problems involving projectiles, it is helpful to find an equation for the path of a projectile that does not involve time. The height of the projectile is then expressed in terms of the horizontal distance travelled. This is useful, for example, when a rugby player is trying to decide how to kick the ball, to obtain a conversion.
The following exploration demonstrates that the path of a projectile can be described in two different ways, one of which is independent of time.

CALCULATOR ACTIVITY 8.2

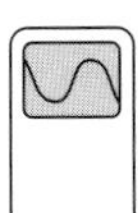

1 A projectile has a position vector:
$\mathbf{r} = 15t\mathbf{i} + (10t - 5t^2)\mathbf{j}$
Set your calculator to parametric plotting mode and to the range/window values below.
$x_{\text{min}} = 0$; $x_{\text{max}} = 40$; $x_{\text{scl}} = 10$
$y_{\text{min}} = 0$; $y_{\text{max}} = 10$; $y_{\text{scl}} = 10$
$T_{\text{min}} = 0$; $T_{\text{max}} = 2$; $T_{\text{step/pitch}} = 0.1$
Find the range and maximum height of the projectile.

2 Repeat **1** for the projectile with position vector
$\mathbf{r} = 12t\mathbf{i} + (12t - 5t^2)\mathbf{j}$

3 Change the Function mode (Texas) or Rec mode (Casio).
Plot $y = \dfrac{x(30 - x)}{45}$

Is the shape similar to the path of the projectile drawn in **1**? Use Trace to verify that the range and the maximum height are the same.

4 Now plot $y = \dfrac{x(28.8 - x)}{28.8}$ and compare with the projectile path in **2**.

5 The equation of the path is in fact given by:

$$y = \frac{4hx(R-x)}{R^2}$$

where h is the maximum height and R the range. Verify that this gives the equation of the path for some other projectiles.

The path of a projectile as a quadratic equation

It is possible to express the path of the projectile where

$$\mathbf{r} = x\mathbf{i} + y\mathbf{j}$$

in the form

$$y = ax^2 + bx + c.$$

The method of obtaining this result, without the range or maximum height, is demonstrated in the following examples.

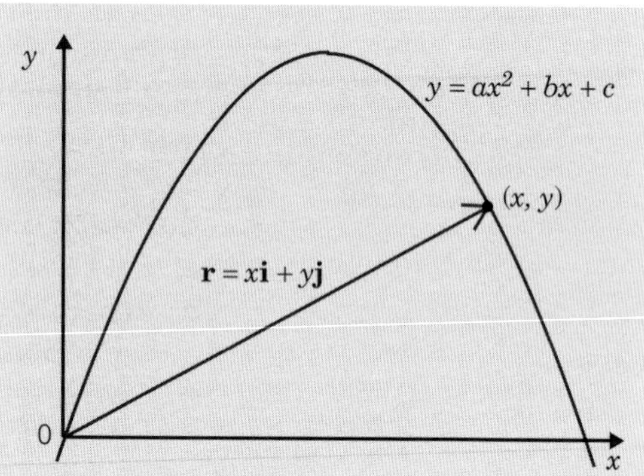

Example 8.4

A projectile moves with position vector $\mathbf{r} = 8t\mathbf{i} + (4t - 4.9t^2)\mathbf{j}$.

a) *If $\mathbf{r}$ is considered as $\mathbf{r} = x\mathbf{i} + y\mathbf{j}$, express y in terms of x.*

b) *Factorise the result and hence find the range.*

Solution

a) *Comparing* $\mathbf{r} = 8t\mathbf{i} + (4t - 4.9t^2)\mathbf{j}$
and $\mathbf{r} = x\mathbf{i} + y\mathbf{j}$
gives $x = 8t$ *and* $y = 4t - 4.9t^2$.

From the first of these equations, $t = \frac{x}{8}$*. This can then be substituted into the second equation to give:*

$$y = 4t - 4.9t^2 = 4 \times \left(\frac{x}{8}\right) - 4.9 \times \left(\frac{x}{8}\right)^2 = \frac{x}{2} - \frac{4.9x^2}{64} = 0.5x - 0.077x^2$$

b) *The result above can be factorised to give* $y = x(0.5 - 0.077x)$.
To find the range we require $y = 0$, *so:*

$0 = x(0.5 - 0.077x)$ so $x = 0$ or $0.5 - 0.077x = 0$

$\Rightarrow x = 0$ or $x = \frac{0.5}{0.077} = 6.5\,\text{m}$

So the range is 6.5 m.

Example 8.5

A footballer is 10 m from an open goal when he kicks the football with a speed of 20 m s^{-1} at 38° to the horizontal.

a) *Express the path of the ball in the form* $y = \mathrm{f}(x)$.

b) *Find the height of the ball when it crosses the goal line. Do you think a goal is scored?*

Solution

a) *The position vector of the ball is:*

$$\mathbf{r} = 20t\cos 38°\mathbf{i} + \left(20t\cos 52° - 4.9t^2\right)\mathbf{j}$$

This could be expressed as $\mathbf{r} = x\mathbf{i} + y\mathbf{i}$ *so that:*

$x = 20t\cos 38°$ *and* $y = 20t\cos 52° - 4.9t^2$

From the equation for x*:*

$$t = \frac{x}{20\cos 38°}$$

This can be substituted into the expression for y *to give:*

$$y = 20 \times \frac{x}{20\cos 38°}\cos 52° - 4.9t\left(\frac{x}{20\cos 38°}\right)^2 = 0.7813x - 0.0197x^2$$

b) *The height is given by* y*, so when* $x = 10$*:*

$y = 0.7813 \times 10 - 0.0197 \times 10^2 = 7.813 - 1.97 = 5.843$ *m*

The ball goes way over the top of the goal.

In the last two examples, projectiles were launched from ground level. The next example shows how to obtain an equation for all projectiles launched in this way.

Example 8.6

A projectile moves with position vector $\mathbf{r} = Vt\cos\alpha\ \mathbf{i} + (Vt\sin\alpha - \frac{1}{2}gt^2)\mathbf{j}$.
If $\mathbf{r} = x\mathbf{i} + y\mathbf{i}$, *express* y *in terms of* x.

Solution

Comparing $\mathbf{r} = Vt\cos\alpha\ \mathbf{i} + (Vt\sin\alpha - \frac{1}{2}gt^2)\mathbf{j}$ *and* $\mathbf{r} = x\mathbf{i} + y\mathbf{j}$ *gives:*

$x = Vt\cos\alpha$ *and* $y = Vt\sin\alpha - \frac{1}{2}gt^2$

From the expression for x, t *can be expressed as:*

$$t = \frac{x}{V\cos\alpha}$$

This can be substituted into the above expression for y *to give:*

$$y = V \times \frac{x}{V\cos\alpha}\sin\alpha - \frac{1}{2}g\left(\frac{x}{V\cos\alpha}\right)^2$$

$$= x\tan\alpha - \frac{gx^2}{2V^2\cos^2\alpha} = x\tan\alpha - \frac{gx^2\sec^2\alpha}{2V^2}$$

Projectiles launched from points not at ground level

Not all projectiles are launched from ground level. In a game of football the ball is often kicked from ground level, so the ball can be modelled as a particle that is projected from ground level and returns to ground level. However, if the ball is headed rather than kicked, then the height at which it is projected becomes significant and must be taken into account. The next example illustrates one approach to dealing with this problem by taking the point of release as the origin.

Example 8.7

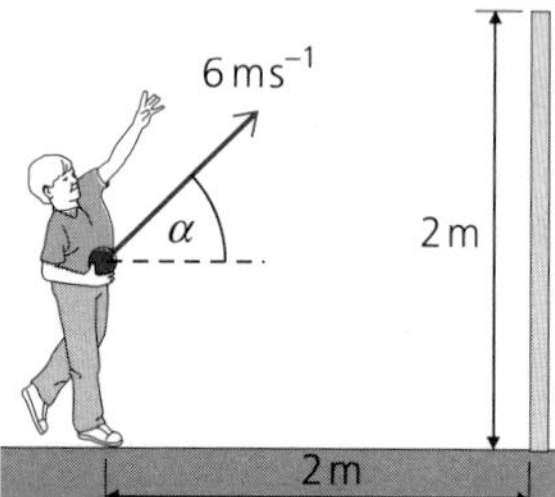

A disruptive pupil is hoping to throw a tennis ball over the school fence, as shown in the diagram. He can throw the ball at 6 m s⁻¹. At what angle α should he throw the ball to get it over the fence?

Solution

The position vector for the ball is given by:

$$\mathbf{r} = 6t\cos\alpha\mathbf{i} + \left(6t\sin\alpha - 4.9t^2\right)\mathbf{j}$$

where the point of release is taken as the origin.
The path can then be found since:

$$x = 6t\cos\alpha \quad \textit{and} \quad y = 6t\sin\alpha - 4.9t^2$$

Eliminating t gives:

$$y = x\tan\alpha - 0.136x^2\sec^2\alpha$$

The ball just passes over the fence if $x = 2$ and $y = 1$, so substituting these values gives:

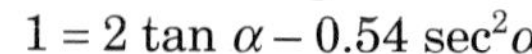

$$1 = 2\tan\alpha - 0.54\sec^2\alpha$$

but $\sec^2\alpha = 1 + \tan^2\alpha$ giving:

$$1 = 2\tan\alpha - 0.54(1 + \tan^2\alpha)$$

Rearranging gives a quadratic equation:

$$0.54\tan^2\alpha - 2\tan\alpha + 1.54 = 0$$

and solving this quadratic for $\tan\alpha$ gives:

$$\tan\alpha = \frac{2 \pm \sqrt{2^2 - 4 \times 1.54 \times 0.54}}{2 \times 0.54} = \frac{2 \pm 0.821}{1.08} = 2.61 \;\textit{ or }\; 1.08$$

Now $\tan\alpha = 2.61$ gives $\alpha = 69°$ and $\tan\alpha = 1.08$ gives $\alpha = 47°$.
Using either angle will have the required result. The diagram shows the two different paths.

EXERCISES

1 For each position vector given below, express the path followed in the form $y = f(x)$ and use this to find the range.

a) $\mathbf{r} = 40t\mathbf{i} + (40t - 4.9t^2)\mathbf{j}$ b) $\mathbf{r} = 10t\mathbf{i} + (20t - 4.9t^2)\mathbf{j}$
c) $\mathbf{r} = 12t\mathbf{i} + (18t - 4.9t^2)\mathbf{j}$ d) $\mathbf{r} = 6t\mathbf{i} + (4t - 4.9t^2 + 3)\mathbf{j}$

2 A projectile on the moon has position vector $\mathbf{r} = 4t\mathbf{i} + (2t - 0.8t^2)\mathbf{j}$
Express the path of the projectile in the form $y = f(x)$.

3 A golf ball is 20 m from some trees. The player then hits the ball with an initial speed of 45 m s⁻¹ at an angle of 30° to the horizontal.

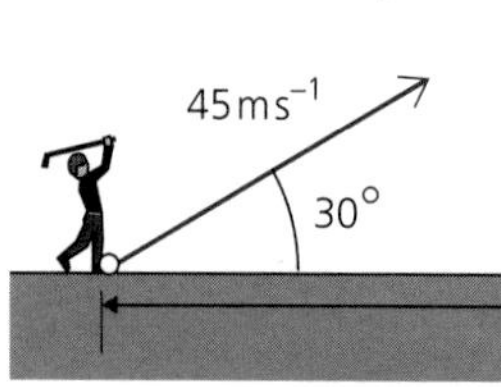

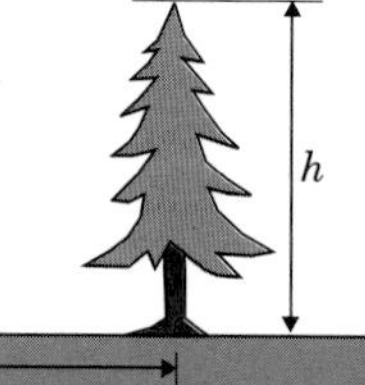

a) Find the position vector of the golf ball.
b) Express the path of the ball in the form $y = f(x)$.
c) If the ball just clears the trees, find the height of the trees.

4 A footballer is about to take a free kick. He kicks the ball so that it initially moves at 18 m s⁻¹ at 30° to the horizontal.

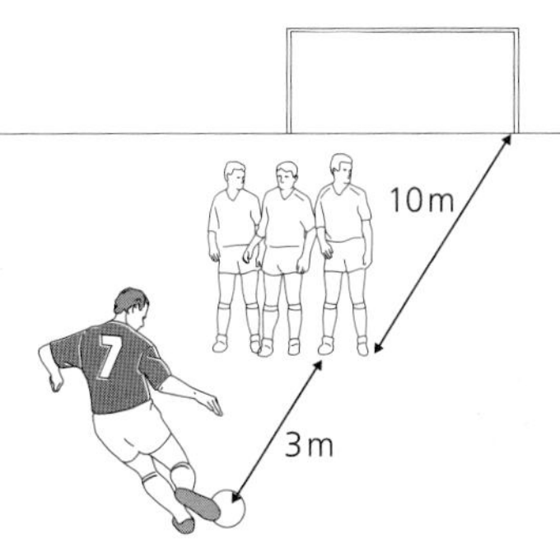

a) Write down an expression for the position vector of the ball.
b) If this can be expressed as $\mathbf{r} = x\mathbf{i} + y\mathbf{j}$ find y in terms of x.
c) What is the height of the ball when it passes over the wall of defenders?
d) Is it possible for a goal to be scored?

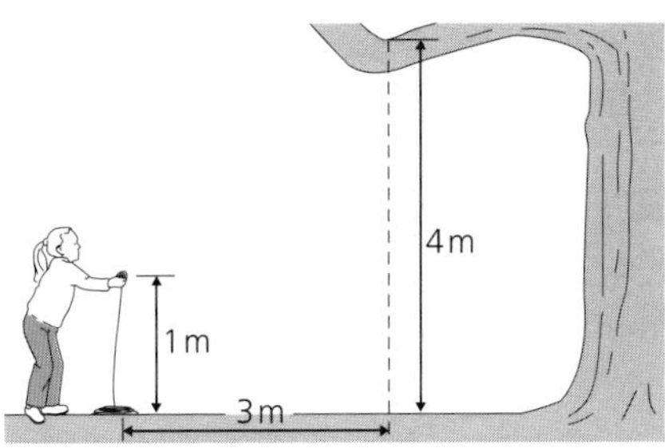

5 A child is trying to throw a stone, tied to a piece of string, over a branch, to make a rope swing. The diagram shows the positions of the child and the branch. Assume that the stone is thrown with speed v at 70° to the horizontal.

a) Write down the position vector of the stone.
b) Find the equation of the path of the stone in terms of x and v.
c) At what speed must the stone be thrown if it is to go over the branch?

6 A shuttlecock is hit when it is in the position shown in the diagram. It is hit so that it initially moves at 60° to the horizontal.

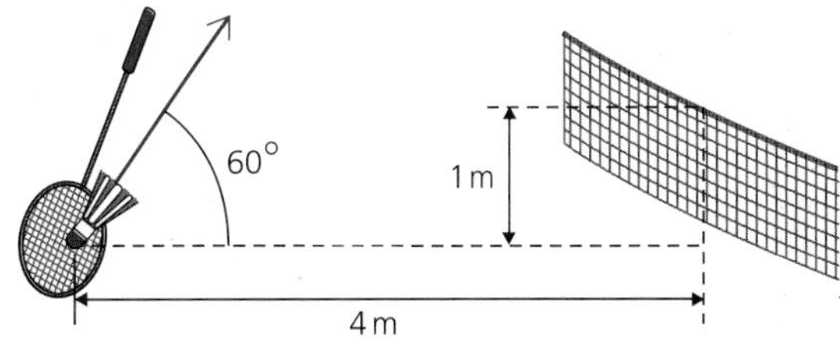

a) Find the equation of the path of the shuttlecock in terms of x and v.
b) For what values of v will the shuttlecock just clear the net?
c) Is it reasonable to model a shuttlecock as a particle moving under gravity alone? Explain why.

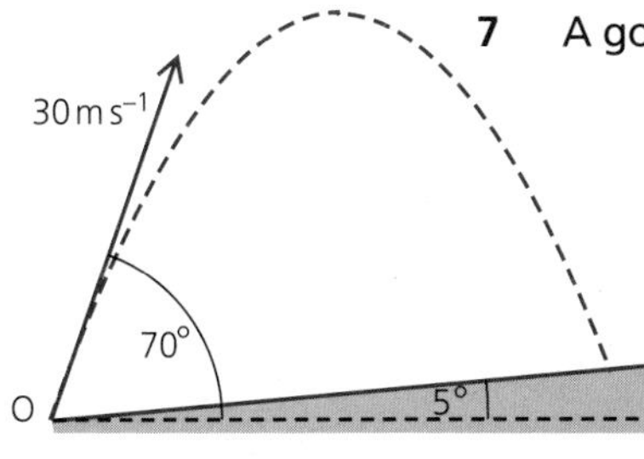

7 A golf ball is hit at a speed of $30\,\text{m}\,\text{s}^{-1}$ at 70° to the horizontal on a slope.

a) Find the equation of the path of the ball.
b) Find the equation of the slope.
c) Use the two equations above to find where the ball lands by eliminating y from the equations.
d) How far from O does the ball land?

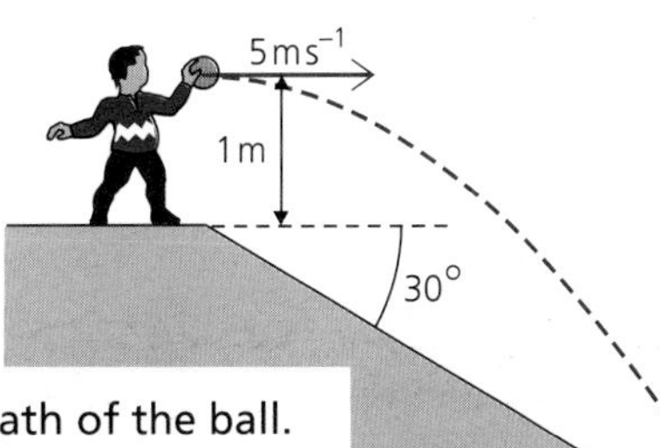

8 A child stands at the top of a bank and throws a ball horizontally at $5\,\text{m}\,\text{s}^{-1}$ from the position shown.

a) Find the equation of the path of the ball.
b) Find the equation of the bank.
c) Find when the ball first hits the bank.

9 A motorcyclist intends to ride off a ramp at A and attempt to clear a wall at B.

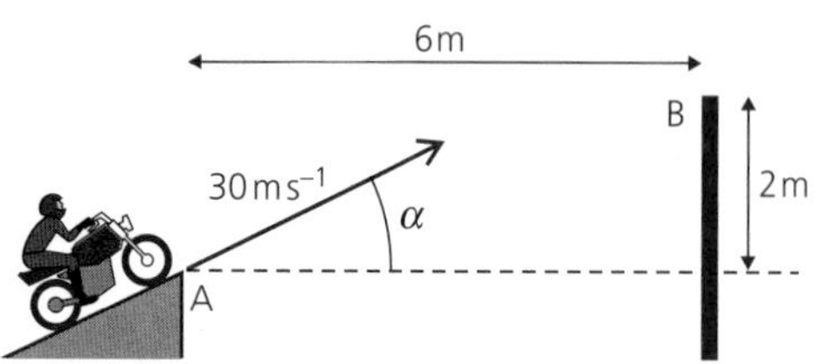

a) Find the equation of the path of the motorcyclist, in terms of x and α, if the take-off speed is 30 m s^{-1}.
b) Use the result $\sec^2\alpha = 1 + \tan^2\alpha$ to produce a quadratic equation in tan α, if he just clears the wall.
c) Solve this equation to find two possible values of α.
d) What would you recommend and why?

10 A tennis player can serve a ball at a speed of 20 m s^{-1}, with the ball at an initial height of 2 m. The diagram shows the dimensions of the net and court.

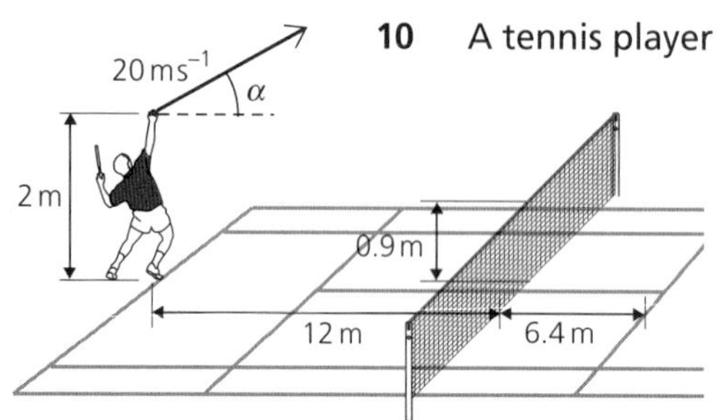

a) Find the equation of the path of the ball, assuming that it starts at the origin, in terms of x and α.
b) Find the possible values of α if the ball just clears the net.
c) Which would you recommend and why?

EXERCISES

8.2 B

1 For each position vector given below, express the path followed in the form $y = f(x)$ and use this to find the range.

a) $\mathbf{r} = 21.21t\mathbf{i} + (21.21t - 4.9t^2)\mathbf{j}$ b) $\mathbf{r} = 60t\mathbf{i} + (80t - 4.9t^2)\mathbf{j}$

c) $\mathbf{r} = 0.87t\mathbf{i} + (4.92t - 4.9t^2)\mathbf{j}$ d) $\mathbf{r} = ut\cos\beta\mathbf{i} + \left(ut\sin\beta - \frac{gt^2}{2}\right)\mathbf{j}$

Sketch the trajectory of the projectile in each case.

2 A projectile on the moon has position vector $\mathbf{r} = 5\sqrt{3}t\mathbf{i} + \left(5t - \frac{4}{5}t^2\right)\mathbf{j}$. Express the path of the projectile in the form $y = f(x)$.

3 A rugby ball is 26 m from the goal post. It is kicked with an initial speed of 17 m s^{-1} at an angle of 45° to the horizontal.
a) Find the position vector of the rugby ball relative to its starting point, using suitable unit vectors.
b) Express the path of the ball in the form $y = f(x)$.
c) If the ball just clears the bar, find the height of bar.

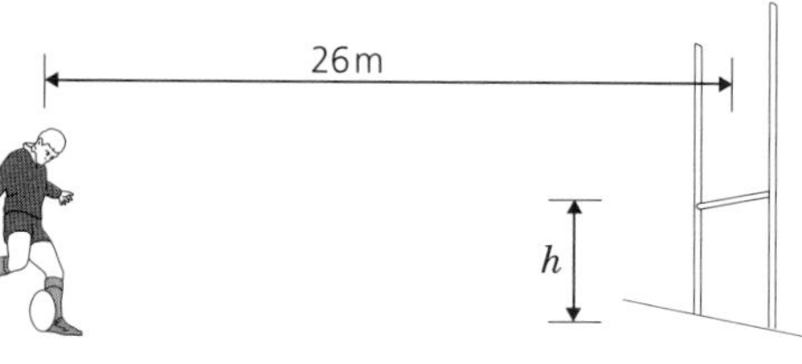

4 A golf ball is 5 m from some bushes. The player hits the ball with an initial speed of 10 m s^{-1} at an angle of 30° to the horizontal.

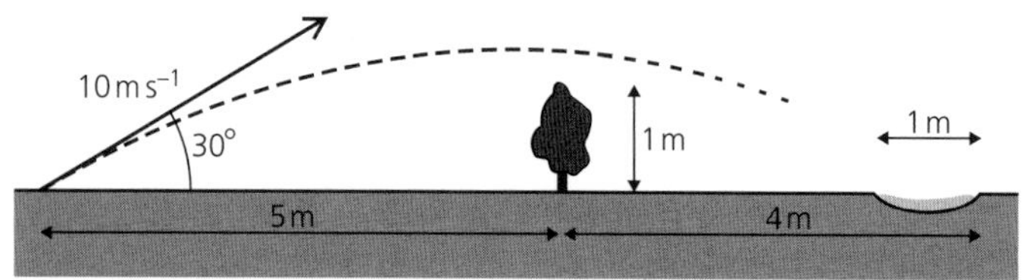

a) Write down an expression for the position vector of the ball, relative to its original position in terms of t, $\mathbf{i}$ and $\mathbf{j}$, where $\mathbf{i}$ and $\mathbf{j}$ are horizontal and vertical unit vectors respectively.

b) If this can be expressed as $\mathbf{r} = x\mathbf{i} + y\mathbf{j}$ find y in terms of x.

c) How much clearance is there when the ball passes over the bushes, if they have a height of 1 m?

d) Does the ball land in the sand pit?

5 A netball is thrown with an initial speed of $V\,\text{m s}^{-1}$ at an angle of 60° to the horizontal. The netball leaves the hands of the player 2 m from the ground at a horizontal distance of 4 m from the net.

a) Find the position vector of the ball, using appropriate unit vectors and origin.

b) Find the equation of the path of the ball in terms of x and y.

c) Find the value of V if the ball is to go into the net.

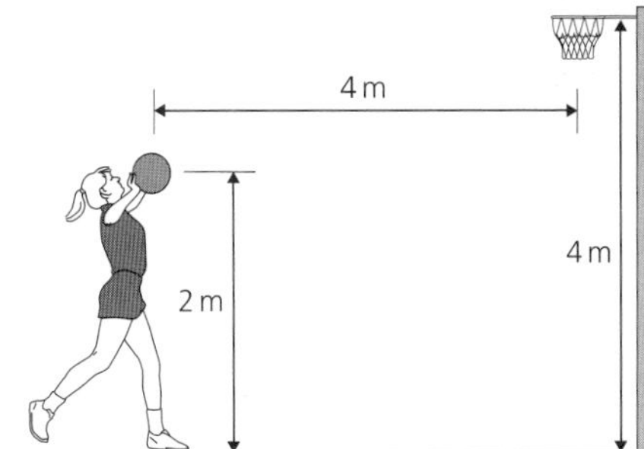

6 An office worker lobs a screwed up piece of paper at an angle of 50° to the horizontal so as to clear a partition and hit the office worker behind, as shown in the diagram.

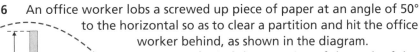

a) Find the equation of the path of the projectile in terms of x and V, where the position of the ball is $x\mathbf{i} + y\mathbf{j}$.

b) For what values of V will the projectile just clear the partition?

c) Is it reasonable to model the screwed up paper ball as a particle moving under gravity alone? Explain why.

7 A football is kicked up a bank with a speed of $20\,\text{m s}^{-1}$ at 60° to the horizontal as shown.

a) Find the equation of the path of the ball.

b) Find the equation of the slope.

c) Use the two equations above to find where the ball lands, by eliminating y from the equations.

d) How far from O does the ball land?

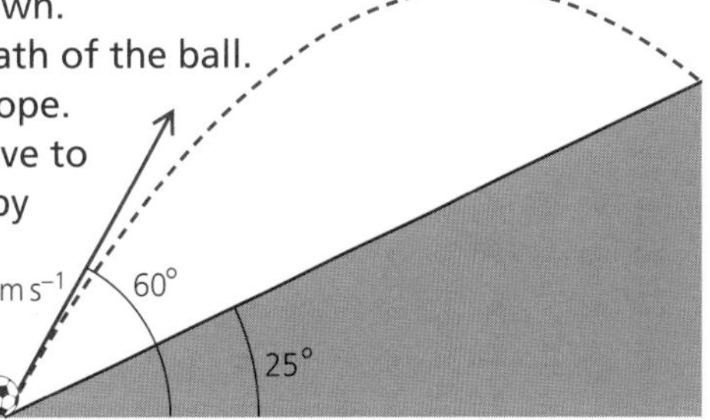

8 A stunt car speeds off a horizontal ramp at $30\,\text{m s}^{-1}$ as shown.

a) Find the equation of the path of the car.

b) Find the equation of the inclined plane.

c) Find where the car hits the inclined plane.

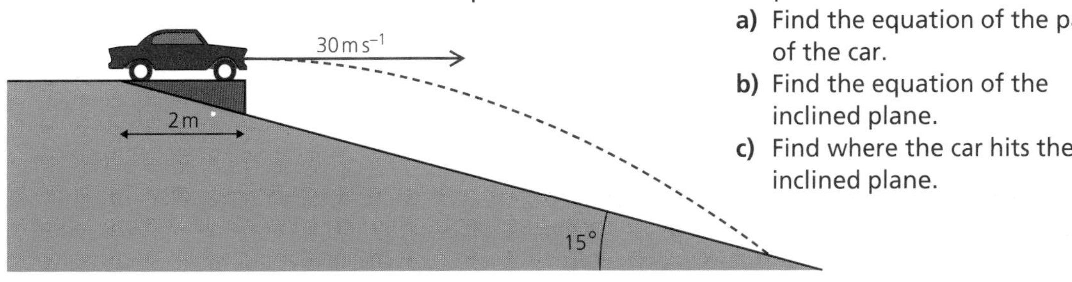

9 A motorcyclist intends to clear a row of buses by riding off a ramp as shown.

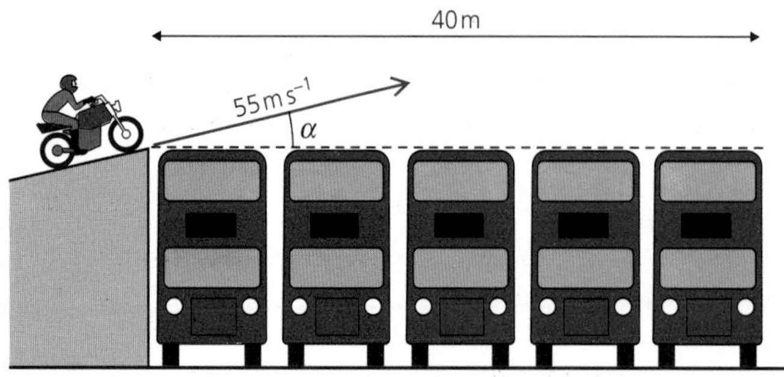

a) Find the equation of the path of the motorcyclist in terms of x and α, if the take-off speed is 55 m s^{-1}.

b) Use the result $\sec^2\alpha = 1 + \tan^2\alpha$ to produce a quadratic equation in $\tan\alpha$ if the rider just clears the buses.

c) Solve the equation to find two possible values of α.

d) Which would you recommend and why?

10 A tennis player serves with a speed V at an angle α to the horizontal. The diagram shows the dimensions of the net and court.

a) Find the equation of the path of the ball, assuming that it starts at the origin, in terms of x, y and V. (Take g to be 10 m s^{-2}.)

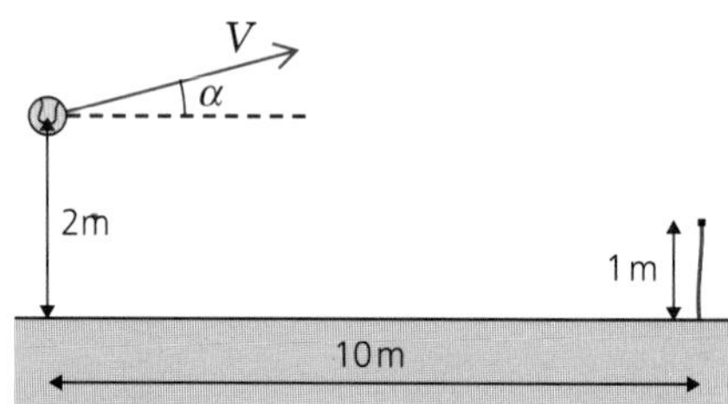

b) If the ball just clears the net, write down a quadratic equation in tan α.

c) From your equation in tanα, prove that if the ball just clears the net then $V \geq 9.51$ m s^{-1}.

CONSOLIDATION EXERCISES FOR CHAPTER 8

1 A particle of mass 5 grams is fired from a tower 25 metres above the ground with a speed of 40 m s^{-1} at an angle of 30° with the horizontal. Find:

a) the time taken to reach its highest point,

b) the velocity of the particle when it hits the ground.

(SMP 16–19 Question B2, Modelling with Force and Motion Specimen Paper, 1994)

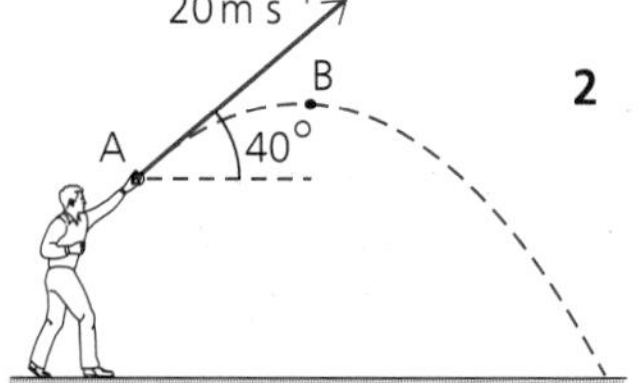

2 A ball is thrown with a speed of 20 m s^{-1} at an angle of 40° to the horizontal. Write down the component of the ball's velocity in the horizontal direction when it is at:

a) A, the point from which it is thrown.

b) B, the highest point of its path.

(Nuffield Question 3, Specimen Paper 4a, 1994)

3 A cannonball fired at an angle of elevation of 10° has a range, on a horizontal plane, of 1.25 km. Ignoring air resistance, find the muzzle velocity. (If a general formula for the range of a projectile is used then it should be proved.) If air resistance is taken into account, what can be said about the muzzle velocity?

(UCLES Question 3, M1 Specimen Paper, 1994)

4 An archer with a cross-bow projects an arrow from the firing position A which hits the target B. The line joining A and B is horizontal and of length 78 m. The initial direction of the arrow makes an angle of 24°

with the horizontal. By modelling the arrow as a particle as it moves through the air from A to B, find:

a) the magnitude of the velocity of projection,

b) the time taken by the arrow to reach B from A,

c) the greatest height of the arrow above AB during its flight.

Write down two assumptions which you have made about the forces acting on the arrow during its flight.

(ULEAC Question 7, M1 Specimen Paper, 1994)

5 In a football match, a goalkeeper takes and kicks the ball from a stationary position on the ground with a velocity whose horizontal and vertical components are respectively $20\,\text{m s}^{-1}$ and $15\,\text{m s}^{-1}$. If a forward whose height is 1.75 m is to head the ball and his distance measured along the horizontal pitch from the kick is 56 m, how high does he need to jump?

(You may model the ball as a particle and its motion should be considered to be in a vacuum.)

Give reasons why particle motion in a vacuum is inappropriate for the modelling of the motion of a football.

If air resistance is taken into account, which would you expect to be affected more:

a) the greatest height the ball reaches or

b) the horizontal distance it travels?

Give reasons for your answer.

(Oxford Question 6, Paper 2, January 1990)

6 A golf ball is projected with speed $49\,\text{m s}^{-1}$ at an angle of elevation from a point A on the first floor of a golf driving range. Point A is at a height of $3\frac{4}{15}$ m above horizontal ground. The ball first strikes the ground at a point Q which is at a horizontal distance of 98 m from the point A as shown below.

a) Show that $6\tan^2\alpha - 30\tan\alpha + 5 = 0$.

b) Hence find, to the nearest degree, the two possible angles of elevation.

c) Find, to the nearest second, the smallest possible time of direct flight from A to Q.

(ULEAC Question 9, Paper M1, January 1992)

7 In this question take $g = 10\,\text{m s}^{-2}$ and neglect air resistance.

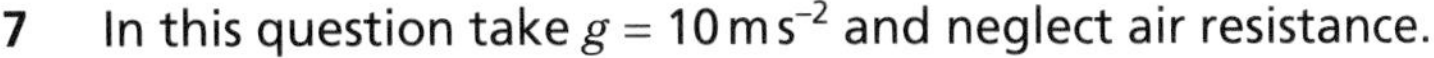

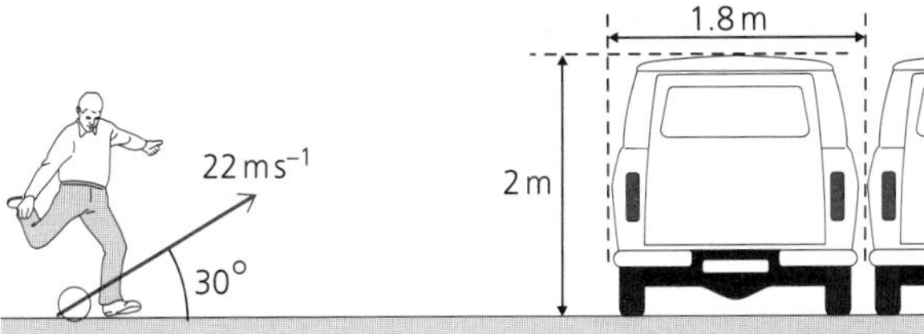

In an attempt to raise money for a charity, participants are sponsored to kick a ball over some small vans. The vans are each 2 m high and 1.8 m wide and stand on horizontal ground.

One participant kicks the ball at an initial speed of $22\,\text{m s}^{-1}$ inclined at 30° to the horizontal.

a) What are the initial values of the vertical and horizontal components of velocity?

b) Show that while in flight the vertical height y metres at time t seconds satisfies the equation $y = 11t - 5t^2$ and calculate at what times the ball is at least 2 m above the ground. The ball should pass over as many vans as possible.

c) Deduce that the ball should be placed about 3.8 m from the first van and find how many vans the ball will clear.

(MEI Question 3, Paper M1, January 1992)

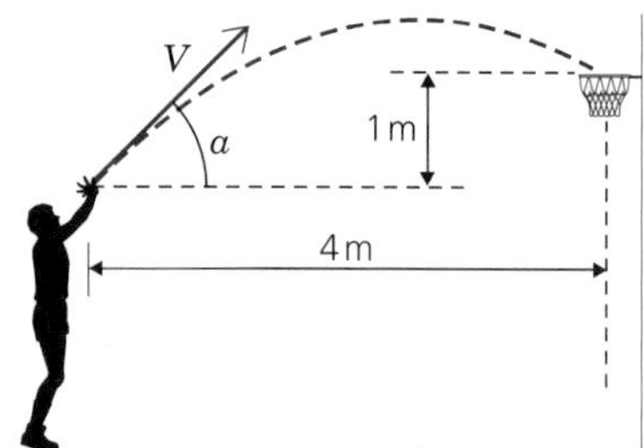

8 The motion of the ball in a free shot in basketball is illustrated in the diagram.
The model assumes that the ball is a point particle, acted on by constant gravity, g. The ball is projected from a position, distance 4 m horizontally and 1 m vertically from the basket, with speed $V\,\mathrm{m\,s^{-1}}$ at an angle of $a°$ with the horizontal.

a) Show that, taking $g = 9.8\,\mathrm{m\,s^{-2}}$, V and a must satisfy

$$1 = 4\tan a - \frac{7.84}{V^2}\sec^2 a.$$

b) Use this equation to find the required speed of projection when angle a equals 45°.

c) Also use this equation to find the two possible trajectories when $V = 8.0\,\mathrm{m\,s^{-2}}$.
For the ball to fall through the basket, the angle made with the vertical at the basket should be as small as possible. Which of your two solutions above would be preferred?

(AEB Question 8, Mechanics Specimen Paper, 1994)

MATHEMATICAL MODELLING ACTIVITY

1 → 2 → 3 ↓ 4 ← 5 ← 6 ↑

Specify the real problem

Problem statement

Many sports involve hitting, kicking or throwing balls. It is often easy to measure the distance that a ball travels and the time it is in the air, but it is more difficult to find the speed at which it was projected. Find a method for determining the speed at which a football has been kicked.

1 → 2 → 3 ↓ 4 ← 5 ← 6 ↑

Set up a model

Set up a model

Make a list of the key features that should be considered in solving this problem. In order to solve the problem the following set of assumptions will be made.

- The ball is a particle.
- The only force acting is gravity.
- The ball is launched from ground level.
- The ball is projected at 45° to the horizontal.
- It is possible to record the distance travelled by the ball at the time of flight.

Comment on this set of assumptions, in particular add any further assumptions that you feel are appropriate.

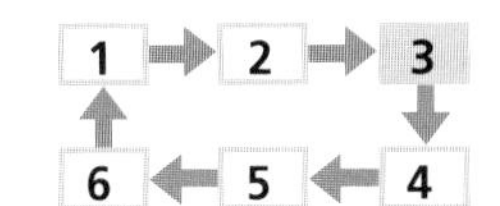

Formulate the mathematical problem

Formulate a mathematical problem

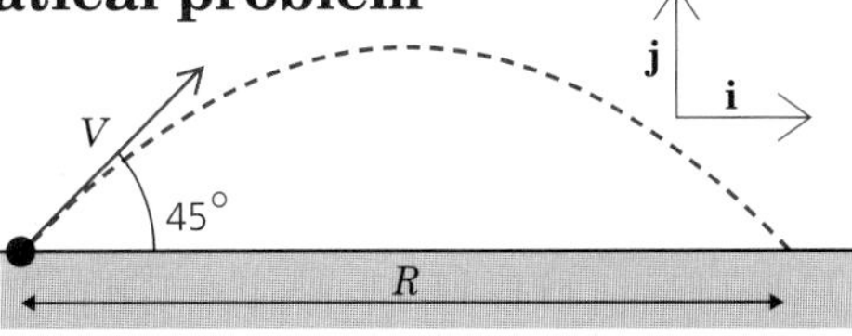

Assume a ball is projected at speed V at 45° to the horizontal and travels a distance R. Express V in terms of R the range of the ball.

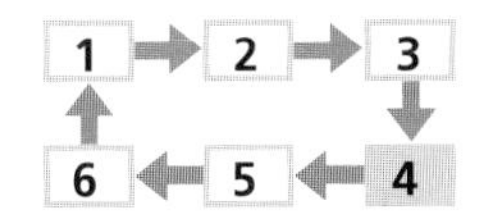

Solve the mathematical problem

Mathematical solution

The position vector $\mathbf{r}$ of the ball at time t, is $\mathbf{r} = \frac{Vt}{\sqrt{2}}\mathbf{i} + \left(\frac{Vt}{\sqrt{2}} - 4.9t^2\right)\mathbf{j}$.

Show that $V = \sqrt{9.8R}$.

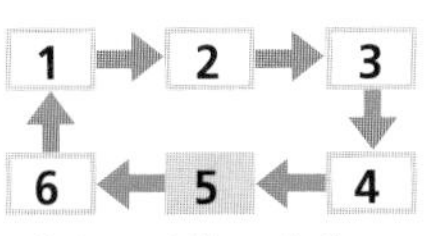

Interpret the solution

Interpretation

The result above shows how the speed of projection is related to the range. It is interesting to note that $R \propto V^2$ so that an increase in V gives a greater increase in R provided $V > 1$.

The speed of projection of a projectile can be found by measuring the range and using the result $V = \sqrt{9.8R}$.

The three factors that his solution does not address are that the ball may not have been launched at 45°, it may not have been launched at ground level and no account has been taken of air resistance.

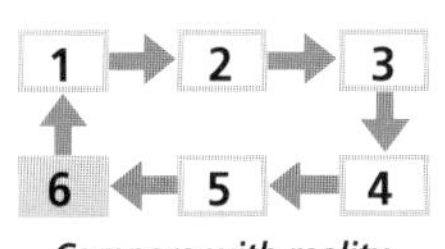

Compare with reality

Compare with reality

Conduct an experiment with a football to determine the speed of projection. Comment on your results. If you are unable to conduct the experiment find the speed at which a football must be kicked if it travels a distance of between 60 to 70 m, which could be done by a good player.

Refining the model

Apply one or both of the revisions outlined below and conduct a further experiment. Also state which sport you are considering and why that revision is appropriate.
Three criticisms have been made of the model. Make at least one further criticism of this model.
The criticisms outlined above are now considered.

1 The effect of the air resistance is that the ball will be slowed down during its flight, reducing the range. So the air resistance will cause the initial speed of the football to be under estimated.
2 Assume that the angle of projection is now α, and not fixed at 45°. Show that:

$$\tan\alpha = \frac{gT^2}{2R}$$

where T is the time of flight and that:

$$V = \frac{R}{T\cos\alpha}$$

Using the time of flight the angle of projection can be calculated and then the speed of projection found.
Comment on the value of the original model based on the 45° angle of projection.

3 Assume that the ball is launched from a height h, at 45° to the horizontal, show that:

$$V = \sqrt{\frac{R^2 g}{R+h}}$$

Consider two cases, first where a drop kick takes place and secondly where a ball is headed rather than being kicked.
Calculate the speed of projection with and without taking account of the height at which the ball is projected. Is it important to take account of the height of projection?

POSSIBLE MODELLING TASKS

1. By measuring the range and time of flight of a golf ball (or a ball used in some other sport), estimate its speed of projection.
2. Find the optimum angle of projection for a shot putt, taking into account the height of release. The speed of launch is about 13 m s^{-1} for a strong athlete. Should athletes concentrate on gaining the maximum speed of release or the most accurate angle of projection?
3. Estimate the speed of a tennis ball that has been served by one of your fellow students. (Investigate how they should hit the ball when they serve.)
4. Investigate how the wind speed may affect the path of a projectile.
5. Consider a set of golf clubs. Investigate how the different clubs have been designed to achieve different ranges when used.

Summary

After working through this chapter you should:

- be able to find the position vector, time of flight, maximum height and range of a projectile
- be aware of the key results listed below:

$$\text{range} = \frac{V^2 \sin 2\alpha}{g}$$

$$\text{time of flight} = \frac{2V \sin \alpha}{g}$$

$$\text{maximum height} = \frac{V^2 \sin^2 \alpha}{2g}$$

$$\text{position vector } \mathbf{r} = Vt\cos\alpha\,\mathbf{i} + \left(Vt\sin\alpha - \tfrac{1}{2}gt^2\right)\mathbf{j}$$

- be able to express the path of a projectile as an equation in terms of x and y and, in particular, obtain the result:

$$y = x\tan\alpha - \frac{gx^2 \sec^2\alpha}{2V^2}$$

for a projectile launched from ground level.

MECHANICS

9

Momentum and collisions

- *The mass of an object and its velocity are important factors in a collision.*
- *Momentum is the product of the mass of an object and its velocity.*
- *Momentum is a fundamental quantity of motion.*

MOMENTUM

When we want to change or cancel the motion of an object, we need to exert a force to do so. Two factors that will influence the size of the force needed are the **velocity** of the object concerned, and its **mass**. Objects with greater masses and velocities will require greater forces, or forces that act for longer, than will smaller, slower objects. In this chapter we explore what happens when there are changes in motion, and in particular in the case of collisions.

Which of the objects below, would it be possible for a human being, of average strength and fitness, to stop?

- A football travelling at 70 mph.
- A small car moving at 2 mph.
- A 70 kg rugby player moving at $10\,\text{m}\,\text{s}^{-1}$.

You may find it helpful to express all the velocities in the same unit. Remember,

$$1\text{ mph} \approx \frac{1600}{3600} = \frac{4}{9}\text{ m s}^{-1}$$

In all these situations, both the mass and speed of the objects are important factors. If the motion of an object is to be changed, the force required to produce the change will depend on the **mass** of the object and the **acceleration**, or **change in velocity**. Newton realised that both mass and velocity were important and defined a fundamental quantity of motion, now known as **momentum**.

The momentum of an object is defined as:

$$m\mathbf{v}$$

where m is its mass and $\mathbf{v}$ its velocity.

As velocity is a vector, then momentum must also be a vector quantity. However in the same way that it is often useful to talk about the speed of an object, it is useful to talk about the magnitude of the momentum.

Example 9.1

Calculate the magnitude of the momentum of:
a) *a sprinter of mass 65 kg travelling as 8 m s^{-1},*
b) *a lorry of mass 30 tonnes travelling at 20 m s^{-1}.*

Solution

The magnitude of the momentum is given by mv, where m is the mass and v the speed.

a) *Here $m = 65$ kg and $v = 8$ m s^{-1} so:*
$mv = 65 \times 8 = 520$ kg m s^{-1}

b) *Here $m = 30\,000$ kg and $v = 20$ m s^{-1} so:*
$mv = 30\,000 \times 20 = 600\,000$ kg m s^{-1}

Example 9.2

A boat of mass 300 kg is travelling at 2 m s^{-1} on a bearing of 070°. Express the momentum in terms of the unit vectors $\mathbf{i}$ and $\mathbf{j}$ which are east and north respectively.

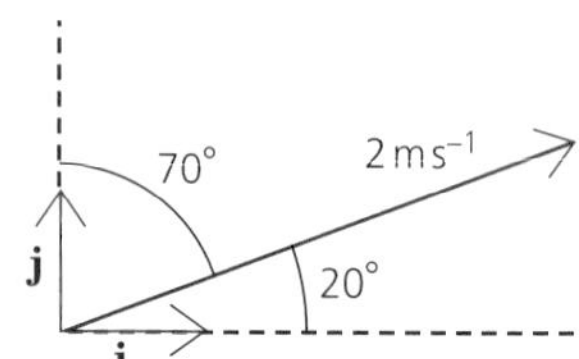

Solution

The diagram shows the velocity of the boat, which can be expressed as:

$$\mathbf{v} = 2\cos 20°\,\mathbf{i} + 2\cos 70°\,\mathbf{j} = 1.88\mathbf{i} + 0.68\mathbf{j}$$

then the momentum of the boat is given by:

$$m\mathbf{v} = 300(1.88\mathbf{i} + 0.68\mathbf{j}) = 564\mathbf{i} + 204\mathbf{j}$$

EXERCISES

9.1A

1 Calculate the magnitude of the momentum of each object described below.
a) A ball of mass 0.2 kg travelling at 20 m s^{-1}.
b) A ferry of mass 800 tonnes travelling at 5 m s^{-1}.
c) A bullet of mass 10 grams travelling at 950 m s^{-1}.

2 Estimate the magnitude of the momentum of each object described below.
a) A table tennis ball in flight.
b) A car on the motorway.
c) A tennis ball that has just been served.

3 Two pucks of masses 8 grams and 16 grams are fired along straight tracks at constant speeds. The diagram shows their positions after a period of time. How does the momentum of each puck compare?

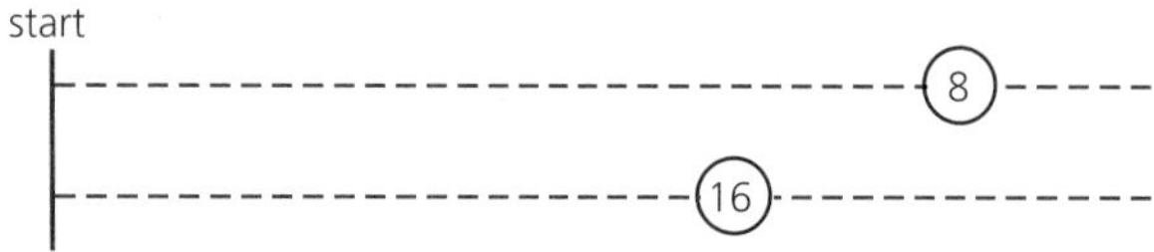

4 A skater follows a curved path, moving at a constant speed. What happens to the skater's momentum?

5 Express the momentum of each of the following objects in terms of unit vectors $\mathbf{i}$ and $\mathbf{j}$, that are east and north respectively.

a) A boat of mass 750 kg travelling at 8 m s^{-1} on a bearing of 027°.
b) An aeroplane of mass 1.5 tonnes travelling at 60 m s^{-1} on a bearing of 220°.
c) A walker of mass 70 kg travelling at 1.1 m s^{-1} on a bearing of 110°.

6 Express the momentum of each object below in terms of unit vectors **i** and **j**, that are horizontal and vertically upwards respectively.
a) A tennis ball of mass 70 grams hit so that it moves at 15 m s^{-1} at an angle of 5° below the horizontal.
b) A shot putt of mass 5 kg launched at 13 m s^{-1} at 42° above the horizontal.
c) A cyclist and cycle of combined mass 80 kg travelling down a slope at 9° to the horizontal with a speed of 8.5 m s^{-1}.

EXERCISES

9.1 B

1 Calculate the magnitude of the momentum of each object that moves as described below.
a) A sprinter of mass 68 kg travelling at 7 m s^{-1}.
b) A lorry of mass 10 tonnes travelling at 20 m s^{-1}.
c) A table tennis ball of mass 3 grams moving at 4 m s^{-1}.

2 Estimate the magnitude of the momentum of each object described below.
a) A superball dropped from a height of 2 m, just before it hits the ground.
b) A cyclist on a level road.
c) A football that has been kicked.

3 An athlete claims that her momentum remains constant as she runs round a track. Is this claim justified?

4 When a rocket takes off its mass decreases as fuel is used up. What happens to the momentum of the rocket if the speed increases at the same rate that the mass is decreasing?

5 Express the momentum of each object described below in terms of unit vectors **i** and **j**, that are east and north respectively.
a) A dinghy of mass 300 kg travelling at 2.1 m s^{-1} on a bearing of 070°.
b) A ship of mass 2000 tonnes travelling at 5 m s^{-1} on a bearing of 200°.
c) A competitor of mass 60 kg at an orienteering contest, who runs at 4.3 m s^{-1} on a bearing of 190°.

6 Express the momentum of each object below in terms of the unit vectors **i** and **j**, that are horizontal and vertical respectively.
a) A basket ball of mass 900 grams travelling at 5 m s^{-1} at 40° to the horizontal.
b) A golf ball of mass 45 grams that is hit so it initially moves at 15 m s^{-1} at an angle of 50° to the horizontal.
c) A girl of mass 45 kg running up a ramp at 30° to the horizontal with a speed of 1.8 m s^{-1}.

IMPULSE AND CHANGE OF MOMENTUM

When the motion of any object changes, it is because there is a force acting. The force could act for any length of time. In this section we explore the relationship between the force, the time for which it acts and the resulting change in momentum. This can be useful in many situations. For example in a car crash it may be easy to estimate the change of momentum and the time taken for a vehicle to come to rest. The forces that acted during the collision can then be estimated.

Exploration 9.1

When a ball bounces on the ground, the force acts on the ball for a very short period of time. It would be very difficult to measure either the time of contact or the force acting on the ball. However, it is relatively easy to describe the change in momentum.

- Try to describe other situations where it would be hard to measure forces and contact times, but where the change in momentum is easier to observe.

Impulse

When a force acts for a short time, we often refer to its effects as an **impulse**. For a constant force, Newton's second law states that

$\mathbf{F} = m\mathbf{a}$ but as $\mathbf{a} = \frac{(\mathbf{v} - \mathbf{u})}{t}$ this can be expressed as:

$$\mathbf{F}t = m\mathbf{v} - m\mathbf{u}.$$

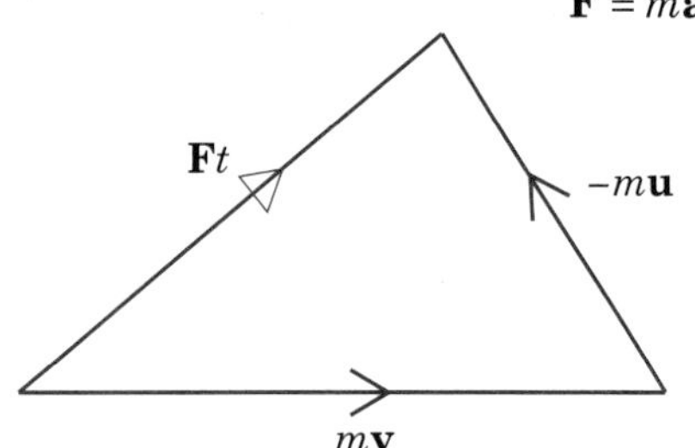

This is a vector equation and can be represented as shown in the diagram.

Impulse, **I** is defined as the change in momentum, so:

$$\mathbf{I} = m\mathbf{v} - m\mathbf{u} \text{ and also } \mathbf{I} = \mathbf{F}t$$

While the idea of impulse is very useful for forces that act for a very short time, the results above are valid for any situation, such as one in which the time for which the force acts can be measured.

We have expressed these equations in vector form, but they can also be expressed as $I = Ft$

and: $I = mv - mu$

for use in one-dimensional situations. The units of I are N s, or newton seconds. **Note:** The units N s and kg m s^{-1} are the same.

It is also interesting to note that Newton's second law can be expressed in terms of momentum. Consider a force F. Then by Newton's second law:

$$F = ma \text{ or } F = m\frac{dv}{dt}$$

If m is a constant, then $F = \frac{d(mv)}{dt}$

This shows that the resultant force is equal to the rate of change of momentum. It was in this form that Newton originally stated his second law.

Example 9.3

A car involved in a collision has a mass of 900 kg and is brought to rest from a speed of 10 m s^{-1}.

a) *Find the impulse on the car.*

b) *If the car is brought to rest in a period of 1.5 seconds, find the average force acting on the car.*

Solution

a) *Here the initial speed of the car is 10 m s^{-1} so, $u = 10$, the final speed is 0, so $v = 0$, and $m = 900$.*

Using $I = mv - mu$ gives:

$I = 900 \times 0 - 900 \times 10 = -9000$ *N s*

Note that I is negative because the force involved acts to oppose the motion of the car.

b) *Using $I = Ft$ with $t = 1.5$ gives:*

$-9000 = F \times 1.5$

$$F = \frac{-9000}{1.5} = -6000 \text{ N}$$

Example 9.4

A hockey ball is travelling at 7 m s^{-1} when it is hit by a stick. After being hit it is deflected through 60° and its speed increases to 9 m s^{-1}.

a) *Explain which aspects of the ball's motion will be ignored if the ball is treated as a particle.*

b) *Find the impulse on the ball, in terms of suitable perpendicular unit vectors* $\mathbf{i}$ *and* $\mathbf{j}$.

c) *If the stick is in contact with the ball for 0.1 seconds, find the average force on the ball. Illustrate this force on a diagram and find its magnitude.*

Solution

a) *If the ball is to be modelled as a particle its rotation must be ignored.*

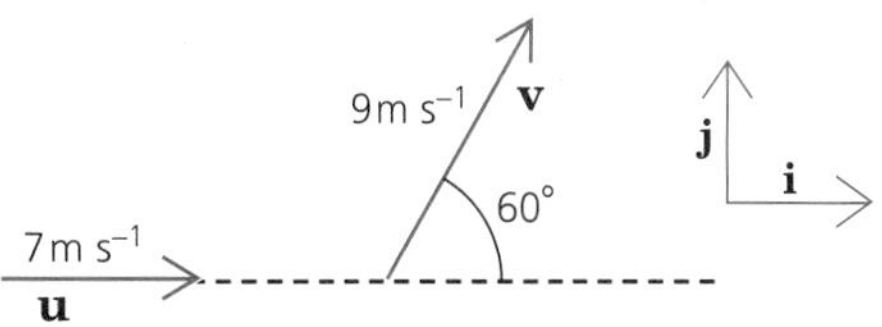

b) *The diagram shows the initial and final velocities of the ball. The initial velocity, is clearly:*

$\mathbf{u} = 7\mathbf{i}$

The final velocity $\mathbf{v}$ *is given by:* $\mathbf{v} = 9\cos 60°\mathbf{i} + 9\cos 30°\mathbf{j} = 4.5\mathbf{i} + 7.8\mathbf{j}$

Now the impulse can be calculated.

$\mathbf{I} = m\mathbf{v} - m\mathbf{u} = 0.2(4.5\mathbf{i} + 7.8\mathbf{j}) - 0.2 \times 7\mathbf{i}$

$= (0.9 - 1.4)\mathbf{i} + 1.6\mathbf{j} = -0.5\mathbf{i} + 1.6\mathbf{j}$

This is illustrated in the diagram.

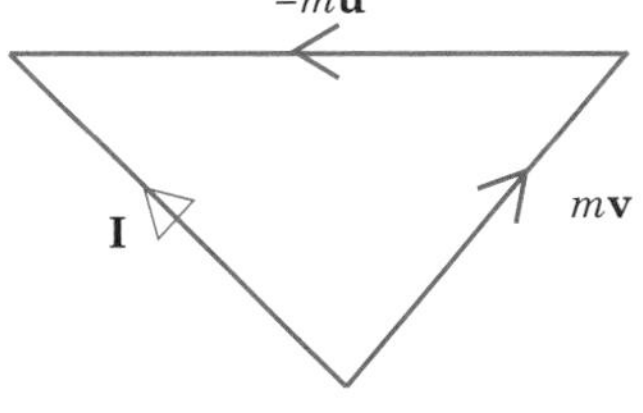

c) *The average force can now be found using,* $\mathbf{I} = \mathbf{F}t$.

$-0.5\mathbf{i} + 1.6\mathbf{j} = 0.1\mathbf{F} \quad \Rightarrow \mathbf{F} = -5\mathbf{i} + 16\mathbf{j}$

The diagram shows this force which has magnitude given by:

$$F = \sqrt{5^2 + 16^2} = \sqrt{281} = 16.8 \text{ N}$$

EXERCISES

9.2 A

1 When a superball of mass 75 grams bounces on the floor, a downward velocity of $3\,m\,s^{-1}$ is converted to an upward velocity of $2.5\,m\,s^{-1}$.
 a) Find the magnitude of the impulse on the ball.
 b) If a lighter superball of mass 50 grams experiences the same change in velocity, find the magnitude of the impulse on this ball.

2 At a fairground a dodgem with a mass of 300 kg including passengers hits the side of the driving area at a speed of $2\,m\,s^{-1}$ and bounces back at $0.5\,m\,s^{-1}$.
 a) Find the magnitude of the impulse on the car.
 b) If the car is in contact with the side of the area for 1.2 seconds, find the average force on the car.

3 In a road traffic accident a car of mass 1500 kg, travelling at $5\,m\,s^{-1}$, hits a stationary vehicle and is brought to rest in two seconds.
 a) Find the magnitude of the impulse on the car.
 b) Find the average force on the car.

4 A diver of mass 55 kg dives from a board at a height of 5 m into a pool.
 a) Show that, if air resistance is ignored, the diver is travelling at $9.9\,m\,s^{-1}$ when she hits the water.
 b) When she hits the water she moves for a further 1.2 seconds before stopping. Find the magnitude of the impulse on the diver.
 c) What is the average force exerted on her by the water?

5 Two roller skaters collide head on at a skating rink. They have initial velocities of $3\,m\,s^{-1}$ and $-4\,m\,s^{-1}$. Their masses are 45 kg and 50 kg respectively. After the collision the lighter skater has velocity $-1\,m\,s^{-1}$.
 a) Find the impulse on this skater.
 b) The other skater experiences an impulse of the same magnitude, but opposite direction. Find the velocity of this skater after the collision.

6 A cricket ball of mass 200 grams hits the ground and rebounds with velocities as shown.

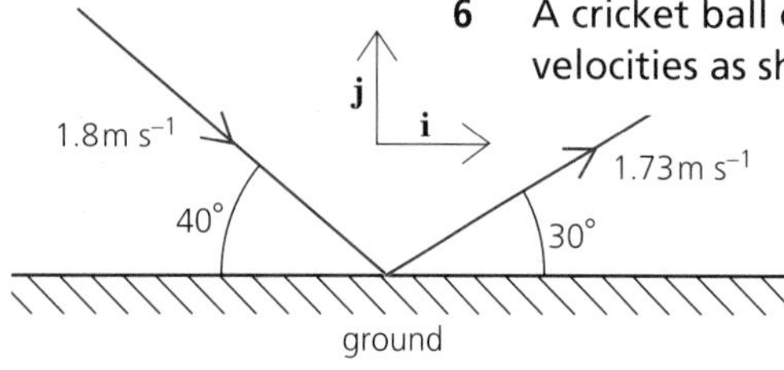

 a) Express both velocities in terms of the unit vectors **i** and **j**.
 b) Find the impulse on the ball.
 c) If the average force on the ball during contact is 9 N, find the time of contact.

7 A skateboarder of mass 65 kg is heading towards a wall and tries to take evasive action by pushing on the wall. During the impact the velocity of the skateboard is changed as shown in the diagram.

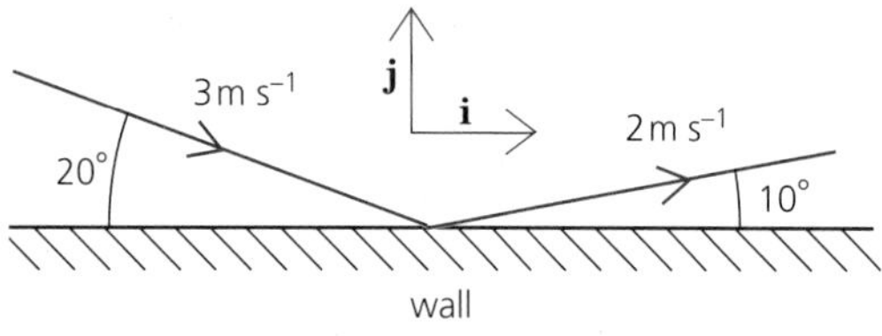

 a) Find the impulse on the skateboarder.
 b) Draw a diagram to show the impulse, the initial and final momentum of the skateboarder and their relationship.

8 A soldier of mass 78 kg, on an assault course, jumps off a ledge at A holding a rope and falls vertical to B, where the rope becomes tight.

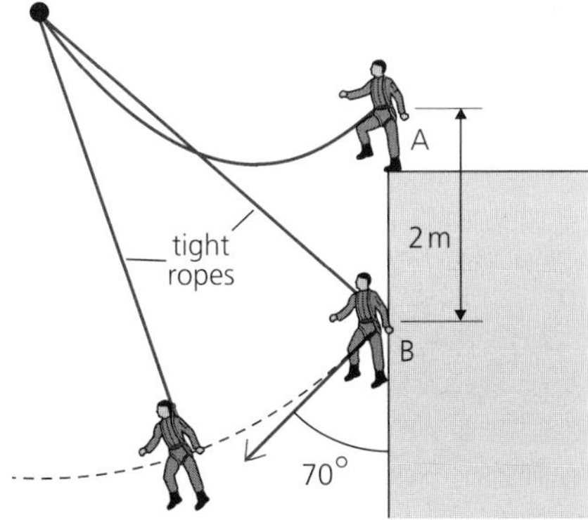

At B the direction of the velocity of the soldier changes.

a) Show that the speed of the soldier at B is $6.26\,\text{m s}^{-1}$.

b) Find the impulse on the soldier as he changes direction. Assume that his speed does not change.

c) Draw a diagram to show the impulse and change of momentum of the soldier.

d) Estimate the time taken for the direction of the velocity to change and use this to find the average tension in the rope while the soldier changes direction.

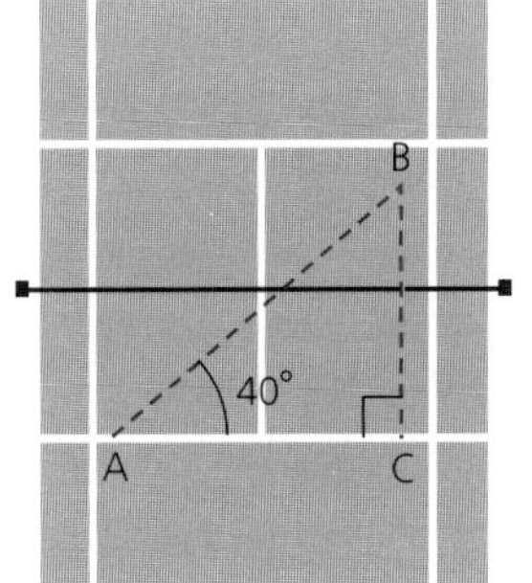

9 A shuttlecock of mass 30 grams travels from A to B. At B it has a speed of $5\,\text{m s}^{-1}$ and is travelling at an angle of 20° below the horizontal. After it has been hit it travels towards C with a speed of $6\,\text{m s}^{-1}$ and at 30° above the horizontal.

a) Express the velocities of the shuttlecock before and after it is hit in terms of the unit vectors $\mathbf{i}$ and $\mathbf{j}$ as shown, and $\mathbf{k}$ which is vertical.

b) Find the impulse on the shuttlecock.

c) If the time of contact is 0.5 seconds, find the magnitude of the average force on the shuttlecock.

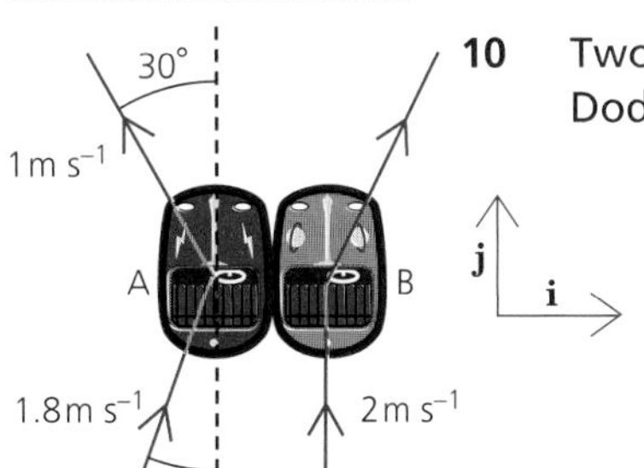

10 Two dodgems at a fair collide giving each other a glancing blow. Dodgem A has mass 320 kg and dodgem B has mass 290 kg. The initial velocities are as shown and A moves with a speed of $1\,\text{m s}^{-1}$ after changing direction as shown.

a) Express the initial and final speeds of A in terms of the unit vectors $\mathbf{i}$ and $\mathbf{j}$.

b) Find the impulse on A.

c) Assume that the impulse on B has the same magnitude as the impulse on A, but acts in the opposite direction. Use this to find the final velocity of B.

EXERCISES

1 A croquet ball of mass 300 grams is hit by a mallet and moves at $0.8\,\text{m s}^{-1}$.

a) Find the magnitude of the impulse that the ball received when it was hit.

b) If the same impulse were given to a ball of mass 500 grams, what would its speed be?

2 A ball of mass 60 grams hits the ground travelling at $4\,\text{m s}^{-1}$ and rebounds at $2\,\text{m s}^{-1}$.

a) Find the momentum of the ball when it hits the ground and when it rebounds.

b) Find the impulse on the ball.

c) If the ball was in contact with the ground for 0.5 seconds, find the average force on the ball.

3 A ball of mass 500 grams is travelling at $6\,\text{m s}^{-1}$. It is caught and brought to rest in 0.6 seconds.

a) Find the magnitude of the impulse on the ball as it is brought to rest.

b) Find the force on the ball, if this is assumed to be constant.
c) Describe how the force on the ball may change as it is caught.

4 A parachutist of mass 75 kg is travelling vertically at 5 m s^{-1} when he lands on horizontal ground and comes to rest.
a) What is the impulse of the ground on the man?
b) Explain why he should bend his legs upon landing.
c) He bends his legs and becomes stationary after 1.8 seconds. Calculate the average force the ground exerts on the man.
d) If he doesn't bend his legs, and the impulse lasts for only 0.3 seconds, calculate the average force the ground exerts on the man.

5 Experiments suggest that a car of mass 1000 kg moving at 13 m s^{-1} can be brought to rest in 1.2 s, when in collision with a solid wall. The use of crumple zones can extend this time to 1.9 s. Find the average force on a car with and a car without crumple zones.

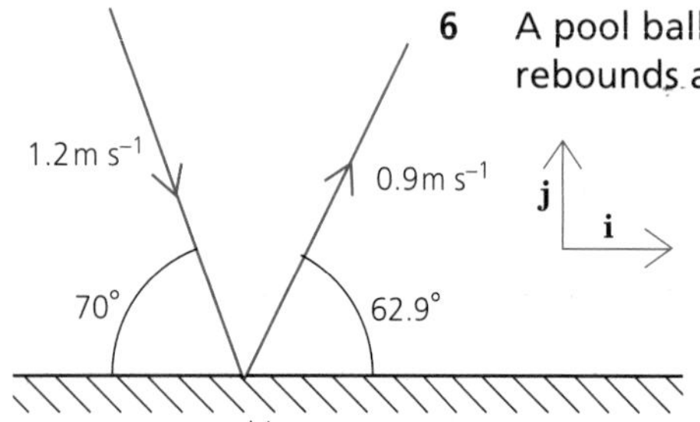

6 A pool ball of mass 120 grams hits the side cushion of a table and rebounds as shown in the diagram.
a) Express the initial and final velocities of the ball in terms of the unit vectors $\mathbf{i}$ and $\mathbf{j}$.
b) Find the impulse on the ball.
c) If the ball is in contact with the cushion for 0.8 s, find the magnitude of the average force on the ball.
d) Suggest how the force on the ball may vary during the collision.

7 A stone of mass 80 grams strikes a glancing blow on a glass window pane as shown in the diagram.

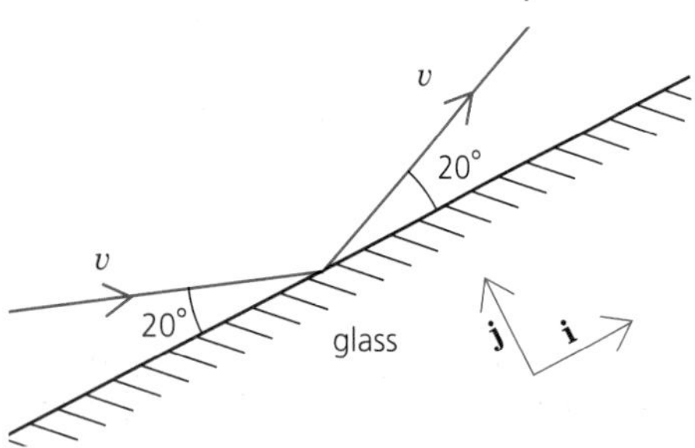

a) Express the velocities illustrated in terms of $\mathbf{i}$ and $\mathbf{j}$, if $v = 5$ m s^{-1}.
b) Find the impulse on the stone and the impulse on the window.
c) If the stone had hit the window at right angles to its path and stopped, what would have been the impulse on the window if the speed of the stone was 5 m s^{-1}?
d) If a glancing blow at 20° as illustrated causes the same impulse as a direct hit, find v.

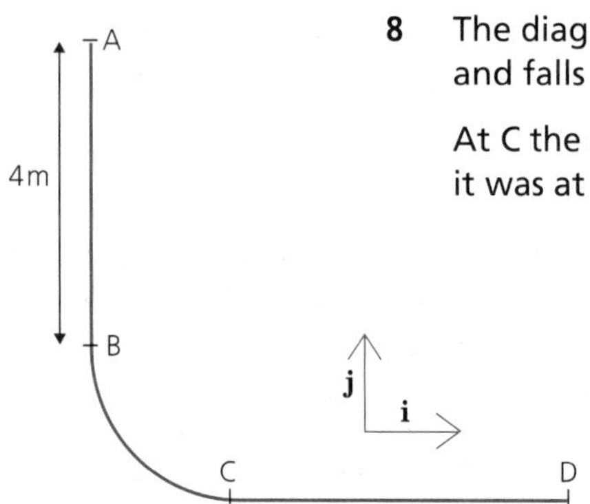

8 The diagram shows a freefall slide. A child of mass 50 kg jumps from A and falls freely to B.

At C the child is travelling horizontally and her speed is 10% less than it was at B.
a) Assume that $g = 9.8$ m s^{-2} and find the speed of the child at B.
b) Express the velocities of the child at B and C in terms of $\mathbf{i}$ and $\mathbf{j}$.
c) Find the impulse on the child between B and C.
d) If it takes 0.8 s to get from B to C find the average force on the child.

9 A racing car travels at a constant speed of $40\,\text{m s}^{-1}$ around a bend. The direction of motion changes through 20° as shown. The mass of the driver is 66 kg.

a) Find the impulse upon the driver in terms of the unit vectors $\mathbf{i}$ and $\mathbf{j}$.

b) If the change takes place in three seconds find the magnitude of the average force on the driver and of his average acceleration.

10 The diagram shows a collision between two pool balls, one of which was initially at rest. Assume that the mass of each ball is m.

a) Find the impulse on the moving ball, in terms of m and the unit vectors $\mathbf{i}$ and $\mathbf{j}$.

b) Assume that the stationary ball experiences an impulse of the same magnitude, but opposite direction. Find the velocity of the ball after the collision.

c) At what speed and in what direction does the ball travel after the collision?

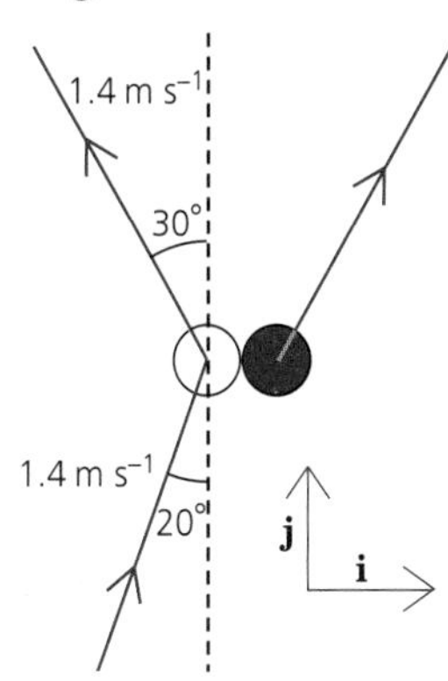

IMPULSE AND VARIABLE FORCES

The relationship $Ft = mv - mu$ can be useful, but in many cases we must take into account the fact the forces acting during the period of time are *not* constant. When a ball is squashed as it bounces the reaction force between the ball and the ground increases to a maximum and then decreases as the ball begins to rise. In this section we explore how to deal with the variable forces which are encountered much more often, in reality, than the constant forces that we have considered so far.

Exploration 9.2

Forces acting on a ball

- When a ball bounces on the ground or is hit by a bat or racquet, a force acts on the ball. How does this force change during the time of contact? Sketch possible graphs of force against time.
- Now consider the resultant force on the ball, i.e. including gravity. How does the resultant force change during the bounce? Try sketching graphs of this.

Variable reaction forces

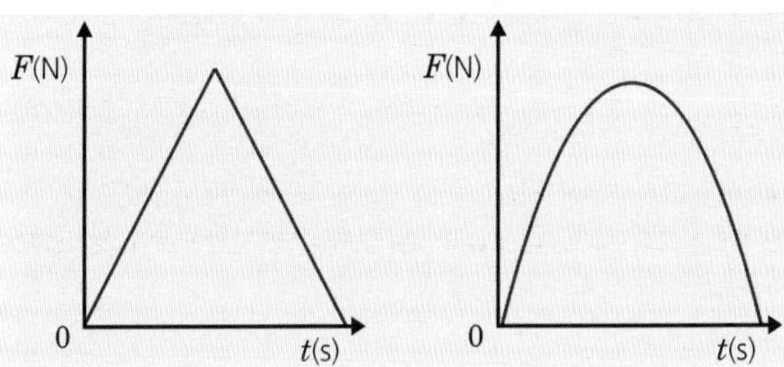

A reaction force begins to act as soon as the ball makes contact with the ground. The way in which the force increases as the ball is deformed or squashed depends on the type of ball and the material from which it is made. The force may be quite complex. Some simple models for the behaviour of this force are shown in the graphs.

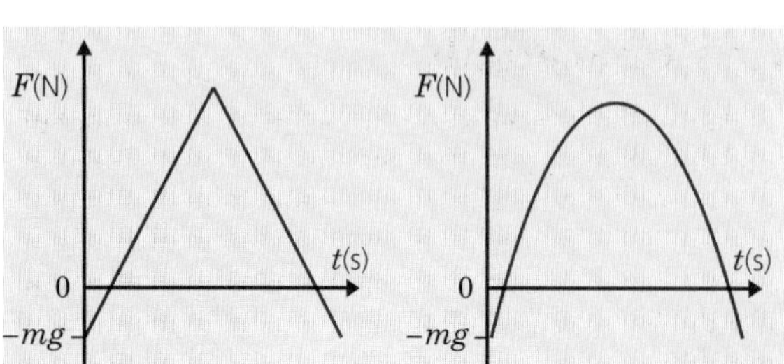

The resultant force on the ball is the resultant of the reaction force R and the force of gravity mg. As they act in opposite directions the resultant is $R - mg$, taking the upward direction as positive. The graph shows the resultant force corresponding to each model illustrated above.

Note that as the ball is squashed the resultant force remains negative, or downward, until the reaction becomes greater than the force of gravity.

Example 9.5

A ball of mass 250 grams bounces on the ground, hitting the ground at 5 m s⁻¹ and rebounding at 4 m s⁻¹. It is in contact with the ground for 0.25 s.

a) *Find the impulse on the ball.*

b) *Assume that the ground exerts a constant force on the ball during contact. Sketch a graph of force against time and find the area under the curve.*

c) *Assume that the force on the ball varies as shown in the graph. Find the area under the graph and compare with the impulse on the ball.*

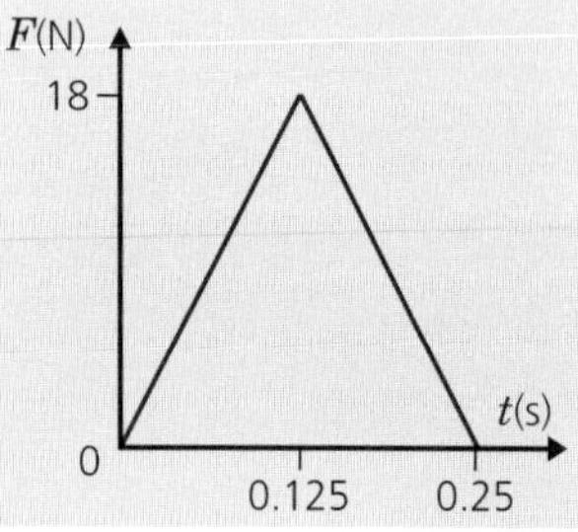

Solution

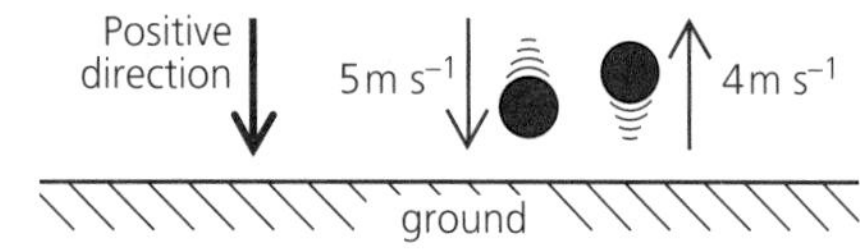

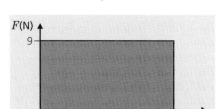

a) *Defining down as positive, gives $u = 5$ and $v = -4$ with $m = 0.25$.*

Using: $I = mv - mu$

gives: $I = 0.25 \times 5 - 0.25 \times (-4) = 2.25\,Ns$

b) *For a constant force $I = Ft$, so with $t = 0.25$ and $I = 2.25$ this gives:*

$2.25 = 0.25\,F$

$\Rightarrow F = \dfrac{2.25}{0.25} = 9\,N$

The graphs shows a sketch of F against t during the collision. The shaded area is:

$9 \times 0.25 = 2.25$

F(N)
9
0
0.25 t(s)

c) *The area under the graph can be found easily as the area of a triangle is $\frac{1}{2} \times$ base $\times$ height*

So here:

area $= \frac{1}{2} \times 0.25 \times 18 = 2.25$

*The area here is the same as the area under the graph in **b)**, which is the same as the impulse on the ball.*

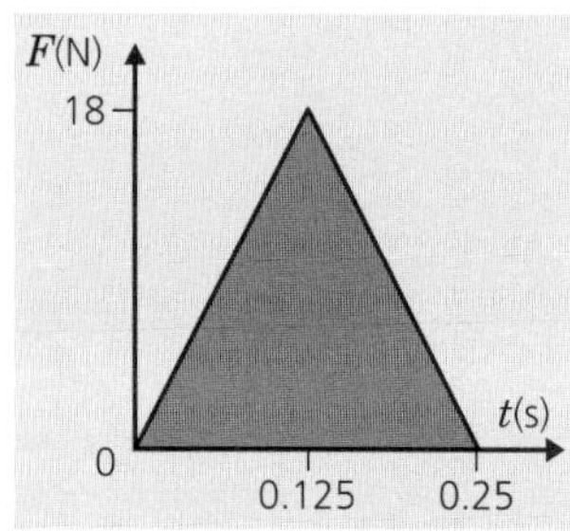

Area under a force–time graph

There seems to be a relationship between the area under a force–time graph and impulse. Consider the area under the graph shown.

This area is given by:

$$\int_{t_1}^{t_2} F(t)\mathrm{d}t$$

Using Newton's second law this can be expressed as:

$$\int_{t_1}^{t_2} F(t)\mathrm{d}t = m\int_{t_1}^{t_2} a(t)\mathrm{d}t$$

where m is assumed to be constant.

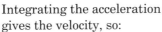

Integrating the acceleration gives the velocity, so:

$$m\int_{t_1}^{t_2} a(t)\mathrm{d}t = m\big[v(t)\big]_{t_1}^{t_2} = m\big(v(t_2) - v(t_1)\big)$$

If the velocity at $t = t_1$ is v and at $t = t_2$ is u, then:

$$m\big(v(t_2) - v(t_1)\big) = mv - mu$$

So combining these steps gives:

$$\int_{t_1}^{t_2} F(t)\mathrm{d}t = mv - mu$$

or $I = \int_{t_1}^{t_2} F(t)\mathrm{d}t$

This result allows the impulse to be found when a variable force acts. When the force is constant it reduces to $Ft = I$ as used in the previous section.

CALCULATOR ACTIVITY 9.1

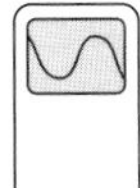

(Assume $g = 10\,\mathrm{m\,s^{-2}}$.)

1 Assume that the resultant force acting on a ball of mass 100 grams is modelled by:

$$F(t) = 200t - 1000t^2 - 1$$

Set the plotting range to the values given below.

$x_{\min} = 0$; $x_{\max} = 0.3$; $x_{\mathrm{scl}} = 0.1$

$y_{\min} = -5$; $y_{\max} = 20$; $y_{\mathrm{scl}} = 5$

Use your graphics calculator to plot a graph of F against t using the range settings given.

a) What is the force of gravity acting on the ball before it hits the ground?

b) The ball first makes contact with the ground when $t = 0$. Explain why $F(0) = -1$.

c) What is F when the ball leaves the ground? When does the ball leave the ground?

d) Use your graphics calculator to find

$$\int_0^{0.2}(200t - 1000t^2 - 1)\mathrm{d}t$$

and state the impulse on the ball.

e) If the ball hits the ground at a speed of $8\,\mathrm{m\,s^{-1}}$, find the speed at which it rebounds.

2 The force that the ground exerts on a bouncing ball of mass 50 grams is modelled as:

$$F(t) = 10\sin 10\pi t$$

Set the plotting range to the values given below.

$x_{min} = 0$; $x_{max} = 0.2$; $x_{scl} = 0.1$;
$y_{min} = -5$ $y_{max} = 10$ $y_{scl} = 1$

a) Sketch a graph of $F(t)$ using the range settings given.
b) For what range of values of t is the model above valid?
c) Explain why the resultant force on the ball during this period of time is $F(t) - 0.5$.
d) Use your calculator to find the impulse on the ball.
e) If the ball hits the ground with a speed of $7\,\mathrm{m\,s^{-1}}$, find the speed at which it rebounds.

Example 9.6

When a train hits a buffer it is brought to rest in 0.8 seconds. The mass of the train is 10 tonnes and the force exerted by the buffer is modelled by:

$$F(t) = 6400t^2 - 10\,000t^4$$

a) *Find the impulse on the train.*
b) *Find the initial speed of the train.*

Solution

a) *The impulse is given by:*

$$I = \int_0^{0.8}(6400t^2 - 10\,000t^4)\mathrm{d}t$$

$$= \left[\frac{6400t^3}{3} - \frac{10\,000t^5}{5}\right]_0^{0.8}$$

$$= \frac{6400}{3} \times 0.8^3 - 2000 \times 0.8^5 = 437\ N\ s$$

b) *As the train stops $v = 0$, so using:*

$$I = mv - mu$$

and noting that $I = -437$ since the force opposes the motion of the train gives:

$$-437 = -10\,000u$$

$$u = \frac{437}{10\,000} = 0.044\ m\,s^{-1}$$

EXERCISES

9.3 A

1 A squash ball of mass 24 grams is travelling horizontally at 6 m s^{-1} when it hits a wall and rebounds at a speed of 4.5 m s^{-1}.

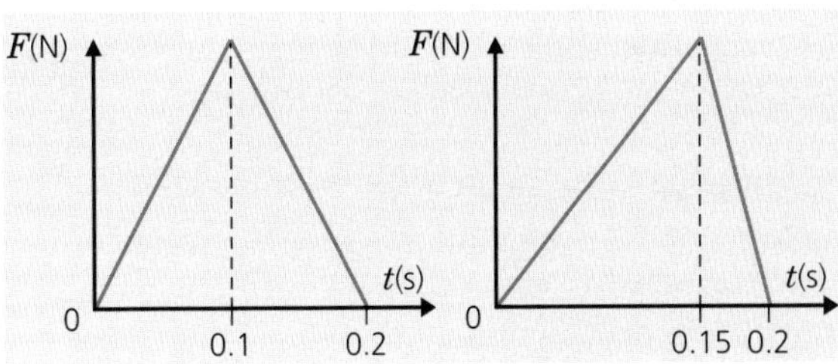

a) Find the magnitude of the impulse on the ball.

b) If the force on the ball increases and decreases as shown in the first graph, find the magnitude of the maximum force on the ball.

c) If the force varies as shown in the second graph, find the magnitude of the maximum force on the ball.

2 A golf ball hits a tree and rebounds along its original path. The speed of the ball when it hit the tree was 10 m s^{-1} and the mass of the ball is 0.046 kg. Assume that the force on the ball can be modelled as in the graph.

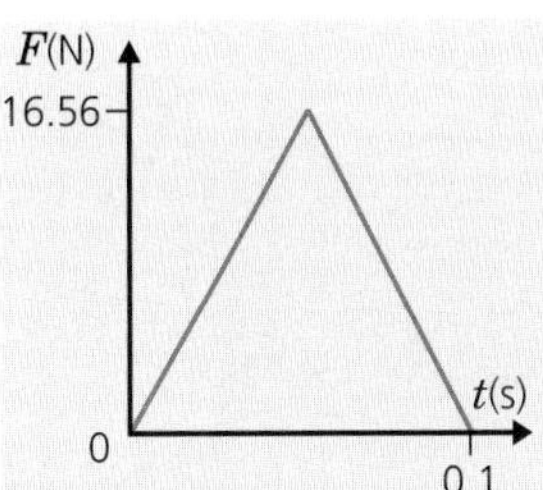

a) Find the magnitude of the impulse on the ball.

b) Find the speed at which it rebounds.

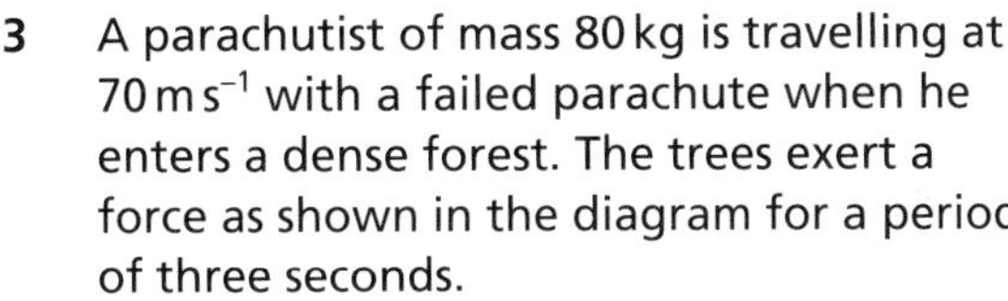

3 A parachutist of mass 80 kg is travelling at 70 m s^{-1} with a failed parachute when he enters a dense forest. The trees exert a force as shown in the diagram for a period of three seconds.

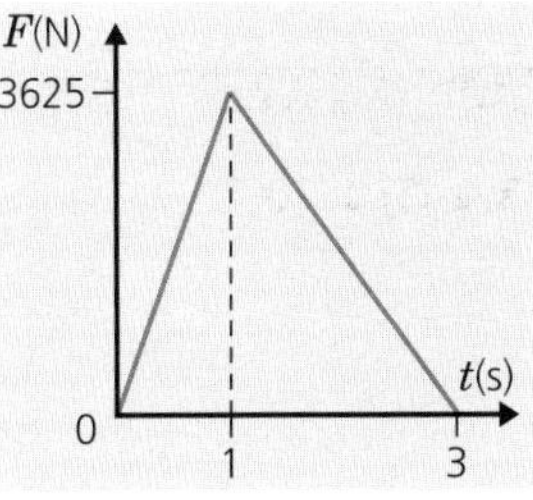

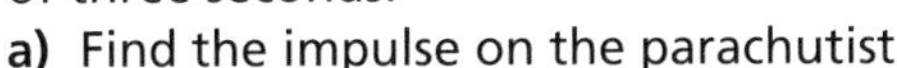

a) Find the impulse on the parachutist.

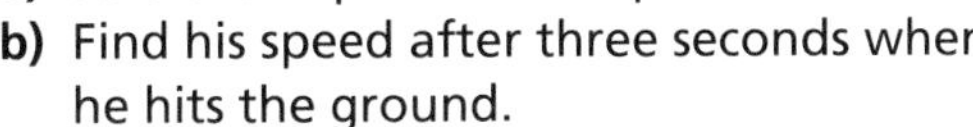

b) Find his speed after three seconds when he hits the ground.

4 When a squash ball of mass 24 grams bounces on a wall the force exerted on the ball can be modelled by:

$F(t) = 10\sin 20\pi t$ for $0 \le t \le 0.05$ s

a) Find the impulse on the ball during the 0.05 seconds that the ball is in contact with the wall.

b) If the ball rebounds at 6 m s^{-1}, find the speed of impact with the wall.

c) An alternative model for $F(t)$ is: $F(t) = 750t - 15\,000t^2$
What speed of impact does this model give?

5 A skateboarder of mass 50 kg collides with a bush, which brings her to rest after two seconds. The force exerted by the bush on the skateboarder can be modelled by: $F(t) = 1000t - 500t^2$

a) Find the impulse on the skateboarder.

b) Find the initial speed of the skateboarder.

c) An alternative model for the force on the skateboarder is:

$F(t) = A\sin \omega t$

Find the values of A and ω if the skateboarder is still stopped after two seconds, has the same initial speed and the force becomes zero when the skateboarder stops.

6 When a superball of mass 170 grams bounces off a vertical wall, at right angles to the surface, with an impact speed of $5\,\text{m}\,\text{s}^{-1}$, it experiences a force that can be modelled as:

$$F(t) = 25\sin 15\pi t$$

a) Sketch a graph of $F(t)$ and state the range of values of t for which you think the model is valid. How long is the ball in contact with the wall?

b) Find the impulse on the ball and the speed at which it leaves the wall.

c) The same ball is dropped onto the ground at the same speed. Assume that the ground exerts the same force on the ball as the wall. Find the resultant force on the ball.

d) Find the speed at which the ball rebounds.

7 A simple model for the resultant force acting on a ball of mass 250 grams as it bounces on a horizontal surface is:

$$F(t) = 120t - 300t^2$$

a) The ball is in contact with the ground for 0.4 s. Find the rebound speed, if it hits the ground moving at $8\,\text{m}\,\text{s}^{-1}$.

b) By considering $F(0)$ and $F(0.4)$ criticise the model.

c) A revised model for $F(t)$ is $F(t) = 120t - 300t^2 - a$ where a is a constant. Select a suitable value for a and find the revised rebound speed, that takes account of your criticism. (Assume $g = 10\,\text{m}\,\text{s}^{-2}$)

8 The resultant force on a bouncing ball is to be modelled as:

$$F(t) = at - bt^2 - c$$

Assume that the ball has mass 200 grams, only travels vertically, hits the ground at $10\,\text{m}\,\text{s}^{-1}$ and rebounds at $7\,\text{m}\,\text{s}^{-1}$, and that $g = 10\,\text{m}\,\text{s}^{-2}$.

a) The ball first comes into contact with the ground when $t = 0$. Explain why $F(0) = -2$ and find c.

b) If the ball leaves contact with the ground when $t = 0.5$, state the value of $F(0.5)$ and show that $b = 2a$.

c) By integrating $F(t)$, find the impulse on the ball in terms of a.

d) Find the values of a and b.

EXERCISES

9.3 B

1 A ball of mass 120 grams hits the ground travelling at $6\,\text{m}\,\text{s}^{-1}$ and rebounds at $4\,\text{m}\,\text{s}^{-1}$ in the opposite direction.

a) Find the magnitude of the impulse on the ball.

b) The graph shows a model for how the force on the ball could vary with time. Find the impulse on the ball in terms of T, the time for which the ball is in contact with the ground.

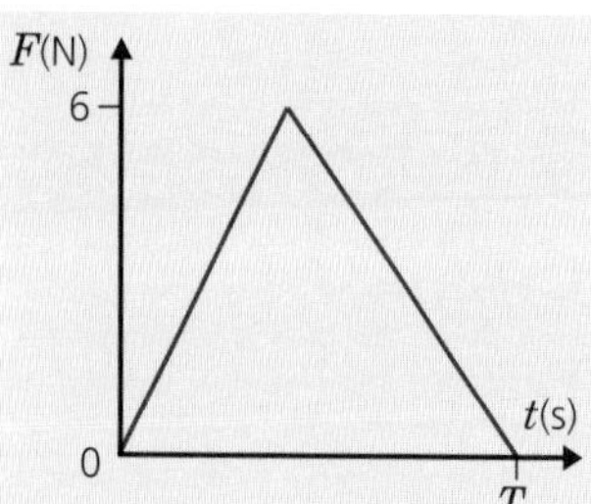

c) Find T.

d) Comment on the shape of the graph, suggest an alternative model for the force.

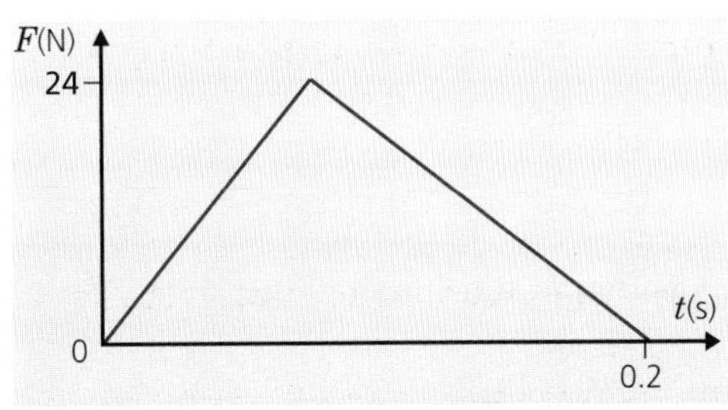

2 A ball of mass 300 grams, travelling horizontally at $5\ \text{m s}^{-1}$, hits a wall.
Assume that the force exerted on the ball by the wall can be modelled by the graph shown.

a) Find the impulse on the ball.
b) Find the speed at which the ball rebounds.
c) What difference would it make to your answer if the force peaked earlier (for example when $t = 0.05$)?

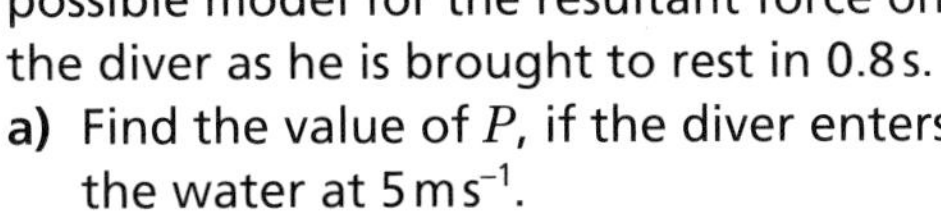

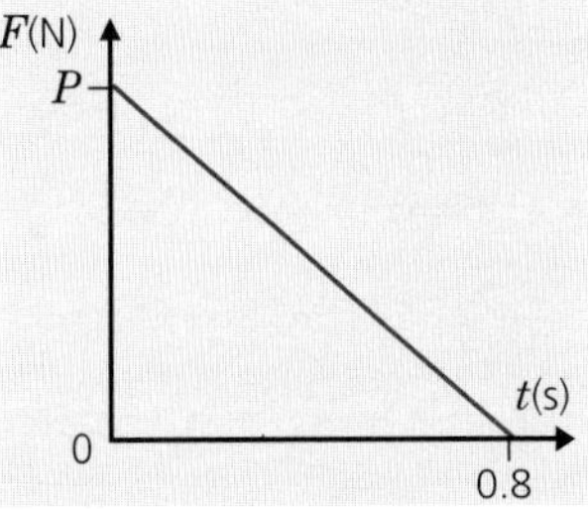

3 A diver of mass 64 kg enters the water travelling vertically. The graph shows a possible model for the resultant force on the diver as he is brought to rest in 0.8 s.

a) Find the value of P, if the diver enters the water at $5\ \text{m s}^{-1}$.
b) An alternative model is:

$$F(t) = 400 - 625t^2$$

Show that $F(0) = \frac{1}{2}p$ and $F(0.8) = 0$. Find the speed at which the diver would have entered the water if this model is used.

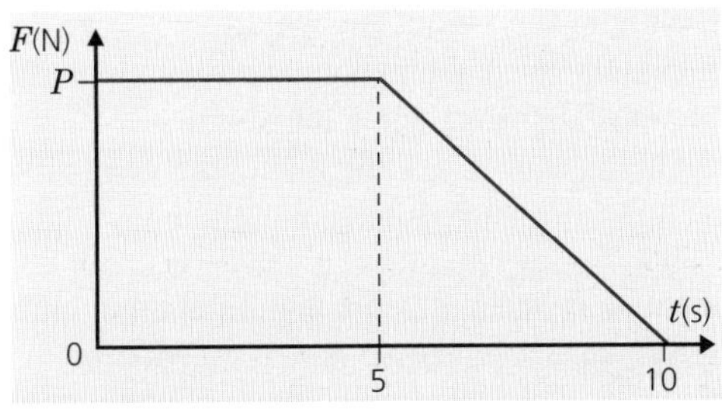

4 A car of mass 1020 kg is travelling at $20\ \text{m s}^{-1}$ when the driver applies the brakes and brings the car to rest in 10 seconds. Assume the magnitude of the resistance force on the car is modelled as illustrated in the diagram.

a) Find the initial momentum of the car.
b) By considering the area under the force–time graph find the magnitude of the impulse on the car, in terms of P.
c) Find the value of P.

5 As a light aeroplane of mass 1500 kg travels down a runway the resultant force on the plane can be modelled as:

$$F(t) = 5000 - \frac{50}{9}t^2$$

The aeroplane takes off when $t = 30$ s.

a) Find the impulse on the plane between $t = 0$ and $t = 30$.
b) Find the speed of the plane when it takes off.

6 When a ball bounces against a vertical wall, the force that the wall exerts on the ball can be modelled by: $F(t) = A\sin 10\pi t$

a) Find when $F(t)$ is zero and explain why the ball is in contact with the wall for 0.1 s.
b) Find the impulse on the ball during the bounce in terms of A.
c) If a ball of mass 50 grams hits the wall at $6\ \text{m s}^{-1}$ and rebounds at $5\ \text{m s}^{-1}$ find the value of A.
d) When the ball bounces against the ground the model for the resultant force, is:

$$F(t) = A\sin 10\pi t - mg$$

Assume $g = 10\ \text{m s}^{-2}$ and find the speed at which a ball would rebound if it hits the ground at $6\ \text{m s}^{-1}$.

7 When a ball of mass 100 grams bounces off a vertical wall, the wall exerts a force given by:

$$F(t) = 120t - 400t^2$$

a) The ball hits the wall when $t = 0$. Find when the force is zero and state when the ball leaves the wall.

b) If the ball hits the wall at 10 m s^{-1} find the speed at which it bounces back off the wall.

c) If the ball were to bounce off a horizontal surface, with the same speed of impact, what would happen to the speed at which it rebounds?

8 A ball, of mass m, hits a wall at a speed u, and rebounds with a speed v, having been in contact with the wall for time T. The force on the ball is modelled by:

$$F(t) = A \sin \alpha t$$

Find the values of A and α in terms of v, u, m and T, stating any assumptions you make.

COLLISIONS AND CONSERVATION OF MOMENTUM

During a collision it is very hard to monitor the forces acting, or even the direction in which they act. This is true whether we are considering a collision between two balls on a snooker table, or two cars in an accident. However, we can still find out a lot about a collision by comparing the momentum before and after impact. In this section we examine the change in momentum during a collision.

Exploration 9.3

Collision of two vehicles

What happens when two vehicles collide? Consider the situations below, commenting on which objects gain speed and which lose speed as a result of the collision.

- A lorry colliding with a stationary car.
- A car colliding with a stationary lorry.
- An Intercity 125 train colliding with a fly.

Exploration 9.4

Collisions

This activity can be carried out using equipment that is available in most mathematics or science departments, such as the 'Unilab Collisions Kit' or an air track. It is very simple to set up. The instructions here refer to the 'Collisions Kit'.

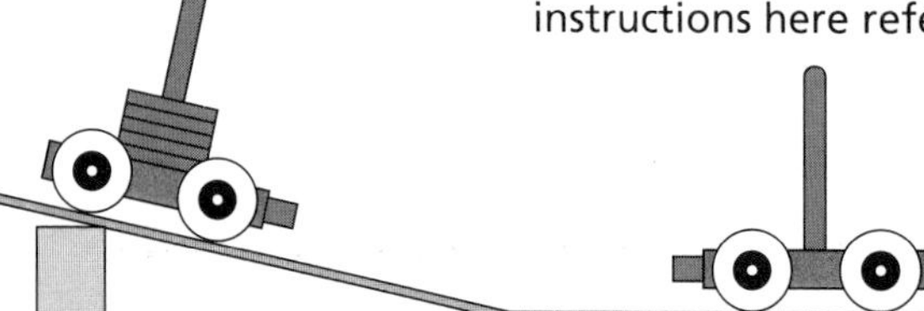

- Set up the apparatus as shown in the diagram. Let one buggy roll down the slope so that it collides with the stationary buggy. What happens?

- Attach Velcro to the buggies, so that they stick together after collision. What happens?
- In each situation, describe what happens to the momentum of each buggy.
- Vary the masses of the buggies so that heavy buggies collide with light buggies, and vice versa. What happens to the momentum of the buggies in each case?

Momentum in collisions

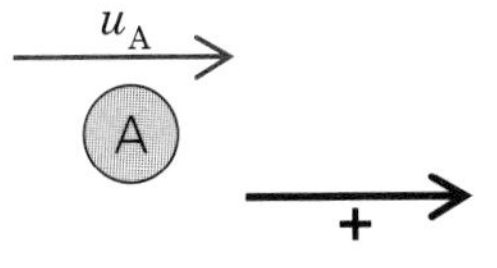

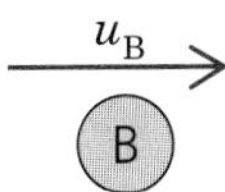

Consider the simple case of two masses A and B, where A has mass m_A and velocity u_A and B has mass m_B and velocity u_B.

Assume that A is travelling faster than B so that they collide, and that they travel along a straight path.

It is likely that their velocities will be changed by the collision so the diagram can be completed by letting v_A and v_B represent their respective velocities after the collision.

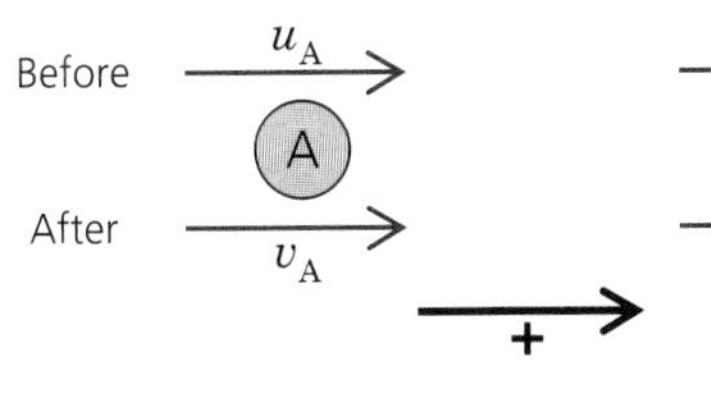

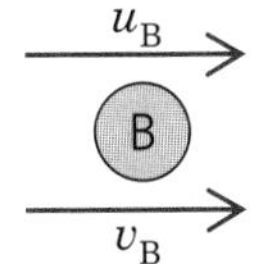

During the impact both masses experience an impulse (i.e. Ft where F is the force exerted and t is the time for which the force is acting).

Impulse $I = mv - mu$
so the impulse on each mass is:

$$I_A = m_A v_A - m_A u_A$$

$$I_B = m_B v_B - m_B u_B$$

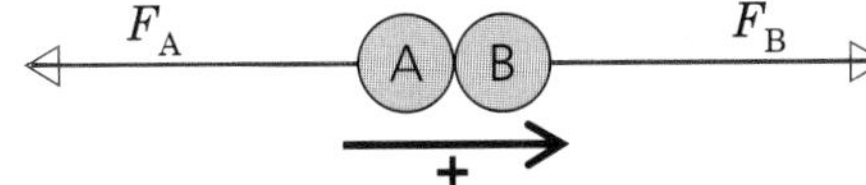

Since the *size* of the force is the same for both masses, we can see that the forces F_A and F_B have the relationship:

$$F_A = -F_B$$

Since they both act for the same time:

$$F_A t = -F_B t$$

or:

$$I_A = -I_B$$

Using $I = mv - mu$ gives:

$$m_A v_A - m_A u_A = -(m_B v_B - m_B u_B)$$

and rearranging gives:

$$m_A v_A + m_B v_B = m_A u_A + m_B u_B$$

This is an extremely important equation. It can also be expressed as follows:

total momentum after the impact = total momentum before the impact

This important result is known as **conservation of momentum**.

Example 9.7

On the track of a model railway, a truck with a mass of 0.2 kg is moving at 2 m s^{-1} towards a stationary truck of mass 0.3 kg. The trucks collide and join together during the collision. Find how fast they are moving after the collision.

Solution

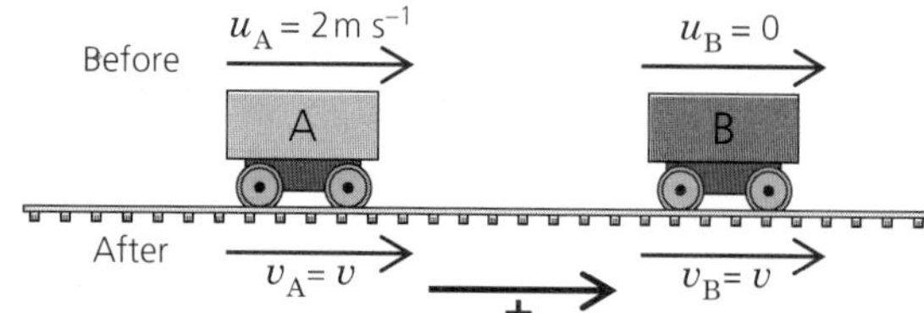

The diagram summarises the collision. Note that $v_A = v_B$ because the trucks join together during the collision so both v_A and v_B can simply be referred to as v.

Using conservation of momentum:

$m_A v_A + m_B v_B = m_A u_A + m_B u_B$

gives:

$0.2v + 0.3v = 0.2 \times 2 + 0.3 \times 0 \quad \Rightarrow 0.5v = 0.4$

$v = 0.8\, m\, s^{-1}$

So the trucks collide and move together after the collision with speed 0.8 m s^{-1}.

Example 9.8

Two balls are moving directly towards each other. Ball A has mass 3 kg and is moving with speed 5 m s^{-1} and ball B has mass 2 kg and is moving with speed 2 m s^{-1}. After the collision ball A continues to move in its original direction but at the reduced speed of 1 m s^{-1}. Find how ball B is moving after the collision.

Solution

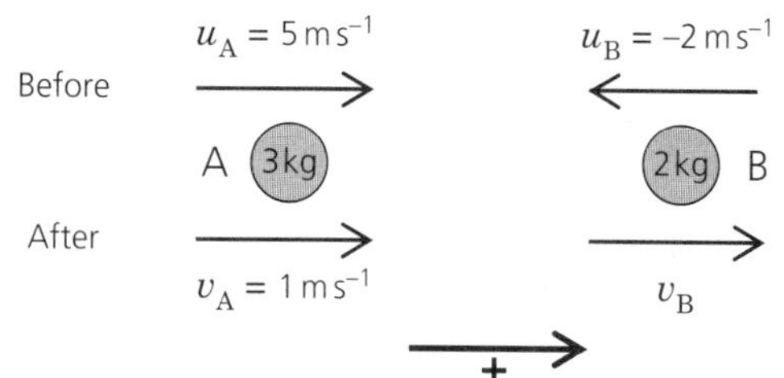

Defining the positive direction to the right as shown in the diagrams gives

$u_A = 5$, $u_B = -2$ and $v_A = 1$.

Using conservation of momentum:

$m_A v_A + m_B v_B = m_A u_A + m_B u_B$

gives:

$3 \times 1 + 2v_B = 3 \times 5 + 2 \times -2 \quad \Rightarrow 2v_B = 8$

$v_B = 4\, m\, s^{-1}$

The value for v_B tells us that B moves away from the collision at 4 m s^{-1}. Since it is also positive it moves to the right.

Exploration 9.5

Assumption about collisions

In the above example we have ignored several features of the motion of the balls. What are these features?

Rotation and impact

In the examples we have considered so far, the balls are moving directly towards each other with constant speed. This means they must be moving on a horizontal surface or guide of some sort. If they are moving on a smooth horizontal wire then the assumptions used in the example are probably quite good. However, if they are moving on a horizontal surface, the balls will almost certainly rotate as they move. The effects of this on the collision could well be very significant. Snooker players make a great deal of use of the rotation of the balls to enable them to gain advantages in the position of the balls on the table, and to achieve those impossible shots!

The rotation of the balls may well effect the collision but the model of two particles considered above gives a good prediction, if the effects of the rotation are not significant.

Example 9.9

A boy of mass 50 kg stands on a skateboard of mass 4 kg. He is not moving when he decides to jump off the skateboard. If he jumps forward with speed 1 m s^{-1}, find the velocity of the skateboard.

Solution

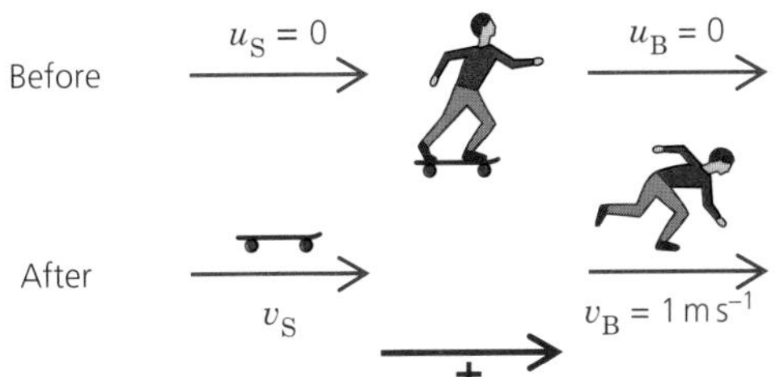

Defining the positive direction to be to the right allows the known velocities to be specified. The initial velocity of the boy is $u_B = 0$ and his final velocity $v_B = 1$. The initial velocity of the skateboard is $u_S = 0$ and its final velocity is v_S.

Using conservation of momentum:

$m_A v_A + m_B v_B = m_A u_A + m_B u_B$

gives:

$4v_S + 50 \times 1 = 4 \times 0 + 50 \times 0 \Rightarrow 4v_S = -50$

$v_S = -12.5\,m\,s^{-1}$

It can now be seen that the skateboard moves at 12.5 m s^{-1}. The negative sign indicates that the velocity of the skateboard is in the direction to the left on the diagram.

It is helpful to draw any unknown vectors on diagrams pointing in the positive direction. Then if we obtain a negative value for any of them, we know it represents a quantity in the negative direction.

EXERCISES

9.4 A

1 Two balls of mass 100 grams from a Velcro catching game collide. Before the collision they are both travelling in the same direction with speeds of 5 m s^{-1} and 2 m s^{-1}. During the collision they stick together and then move with the same velocity.

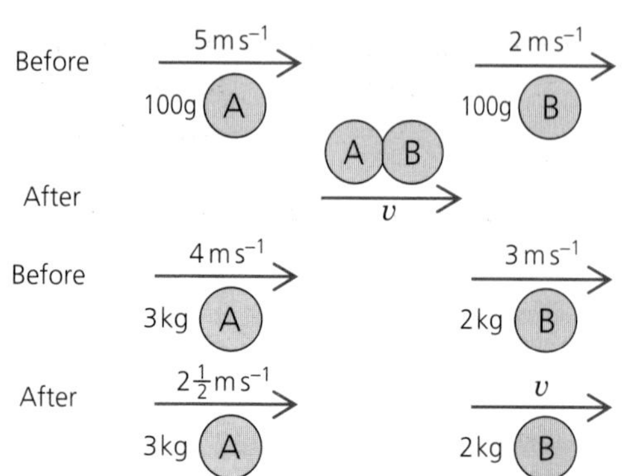

a) What is the total momentum of both balls before the collision?
b) What is the total momentum after the collision?
c) Find the speed of the balls after the collision.

2 The two particles shown in the diagram collide on a smooth horizontal surface. After the collision, the lighter particle moves with speed v.
a) What is the total momentum of the particles before the collision?
b) Find the value of v.

3 Two bodies, with masses of 250 grams and 450 grams, approach each other from opposite directions, on a smooth horizontal surface with speeds of $8\,\text{m s}^{-1}$ and $2\,\text{m s}^{-1}$ respectively. The heavier body has its direction of motion reversed by the collision and moves away at $1\,\text{m s}^{-1}$. Find the speed, v, of the lighter body after the collision.

4 Two bodies, A and B, of masses 2 kg and m kg respectively, are travelling in the same direction. A is travelling at $4.50\,\text{m s}^{-1}$ and B at $2\frac{1}{3}\,\text{m s}^{-1}$ before they collide. After the collision A and B continue to travel in their original directions but with speeds $2\,\text{m s}^{-1}$ and $4\,\text{m s}^{-1}$ respectively. Find the value of m, the mass of B.

5 A lorry of mass 8 tonnes is travelling at $10\,\text{m s}^{-1}$ when it hits a stationary car of mass 1020 kg. If the two vehicles move together after the collision, what is their initial combined speed?

6 Particle A, with mass 3 kg and speed $5\,\text{m s}^{-1}$, collides with a particle B, which is a 2 kg mass initially at rest. If particle A continues to travel in the same direction but with a speed of $1\,\text{m s}^{-1}$, find the speed with which B is moving after the collision.

7 If, in question **6**, the two masses join together, find the speed with which they move after the collision.

8 A firework display includes a rocket which sends a 0.3 kg mass into the air. When travelling horizontally at the top of its trajectory it explodes into two pieces of masses 0.1 kg and 0.2 kg. If the speed of the 0.3 kg mass is $8\,\text{m s}^{-1}$ before the explosion and the 0.2 kg mass is propelled at a speed of $12\,\text{m s}^{-1}$ in the same direction after the explosion, find the speed of the 0.1 kg mass.

9 Two boys, Paul and Mark, are on the bank beside a lake when they see a rowing boat floating a short distance from the bank. The boat is not moving and has a mass of 100 kg.

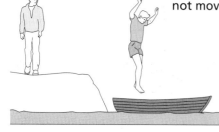

Paul, who has a mass of 50 kg, decides to jump from the bank into the boat and to do this he first runs at right angles to the bank towards the water. On the edge of the lake he jumps and lands in the boat. If he has a horizontal speed of $5\,\text{m s}^{-1}$ just before landing in the boat, calculate the speed with which he and the boat move immediately after he lands in the boat.

10 The boat in question **9** moves further away from the bank with Paul in it, so Mark decides to try to jump to the boat as well. Mark fails to reach the boat and lands in the water near the boat, which is now at rest. Paul throws an oar of mass 10 kg with a horizontal speed of 4 m s^{-1} towards Mark. Find how fast Paul and the boat move away from Mark as a result of this.

EXERCISES

9.4 B

1 An air gun pellet of mass 1 gram and travelling at 30 m s^{-1} hits the centre of an apple of mass 120 grams. The apple is suspended by a string and is at rest before the pellet hits and becomes embedded within it.

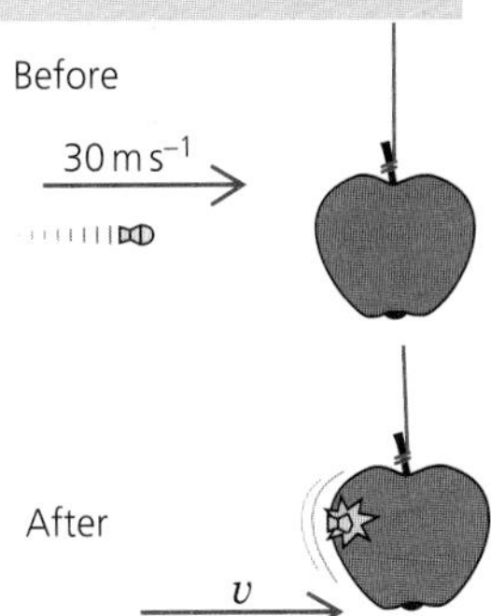

a) What is the total momentum of the two objects before the collision?

b) What is the total momentum after the collision?

c) Find the speed of the apple after the pellet has become embedded within it.

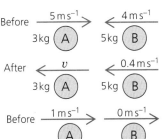

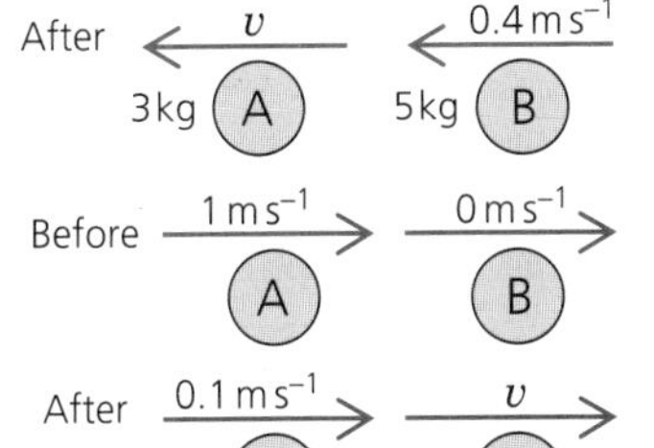

2 The two bodies shown collide on a smooth horizontal surface.

a) What is the total momentum of the two bodies before the collision?

b) What is the total momentum after the collision?

c) Find the value of v, the speed of the 3 kg mass after the collision.

3 A snooker ball collides directly with an identical ball which is initially at rest. The first ball has its speed reduced from 1 m s^{-1} to 0.1 m s^{-1} but continues to move in the same direction. Find the speed with which the second ball moves after the collision.

4 The collision in question **3** is repeated but the first ball is replaced by a plastic ball which has half the mass of a snooker ball. If the snooker ball moves away from the collision with speed 0.4 m s^{-1}, find the speed and direction of motion the plastic ball has after the collision.

5 Particle A of mass m travels on a smooth horizontal surface at 6 m s^{-1}. It collides with particle B which has a mass of 3 kg and travels at 4 m s^{-1} directly towards A. Both particles have their direction of motion reversed by the collision. A now has a speed of 2 m s^{-1} and B has a speed of $1\frac{1}{3}$ m s^{-1}. Find the mass of particle A.

6 A bullet of mass m is fired from a gun of mass 2.4 kg. After the bullet is fired at a speed of 400 m s^{-1} the gun recoils at 3 m s^{-1}.

Find the value of m, the mass of the bullet in grams.

7 A model railway truck of mass 120 grams travels on a smooth, horizontal track at speed v. It strikes a stationary truck of mass 180 grams and the trucks become coupled together. The trucks move with speed 12 cm s^{-1} after the collision. Find the value of v (to the nearest cm s^{-1}).

8 A bullet of mass 6 grams is fired at a wooden block of mass 36 grams which is initially at rest on a smooth horizontal surface. The bullet approaches the block with a horizontal speed v and becomes embedded in the block. Find the value of v to the nearest m s^{-1}, if the block moves at a speed of 7 m s^{-1} after the bullet has hit it.

9 A ping pong ball has a mass of 10 grams and collides directly with a football which has a mass of 1 kilogram. Just before they collide the balls are travelling towards each other, the ping pong ball with a speed of 6 m s^{-1} and the football with a speed of 1 m s^{-1}. After the collision, the ping pong ball has its direction of motion reversed and a speed of 4 m s^{-1}. How is the motion of the football affected?

10 A 5 kg bowling ball collides directly with a stationary skittle of mass 1 kg. Immediately after the collision, the ball and the skittle move in the original direction of the ball with speeds of 3 m s^{-1} and 4 m s^{-1} respectively. Find the original speed of the bowling ball.

COLLISIONS IN TWO DIMENSIONS

Sometimes collisions do take place in one dimension, often where the motion is constrained in some way, as with trucks that run on tracks. However, as the result of a good many collisions, there is motion in two or even three dimensions. When a footballer heads a ball there is motion in three dimensions. When two snooker balls collide there is motion in two dimensions.

Exploration 9.6

Is momentum conserved?

We know momentum is conserved in one dimension, but is momentum conserved in two dimensions? Consider the example below.

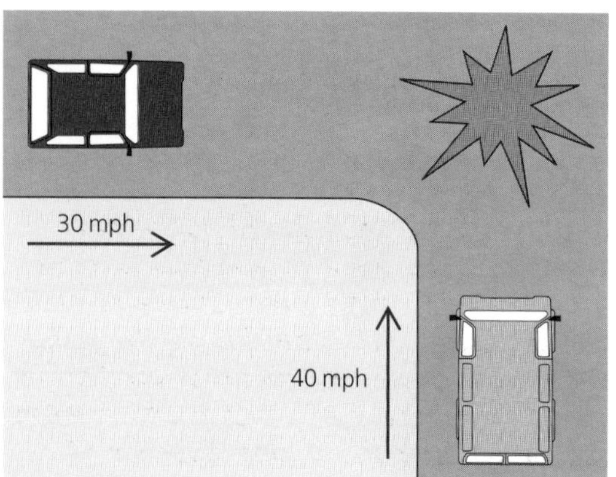

A car travelling at 50 km h^{-1} (30 mph) hits a van travelling at 65 km h^{-1} (40 mph). If they are travelling at right angles when the collision takes place, how do you think the two vehicles could move if:

- they became entangled
- they bounce apart?

Do you think momentum would be conserved in this situation?

What other collisions would support the suggestion that momentum would be conserved in two dimensions?

Comparing momentum

To investigate whether or not momentum is conserved, we can consider two particles that collide with each other. We can study the velocities of two particles A and B before and after impact and the impulse during impact.

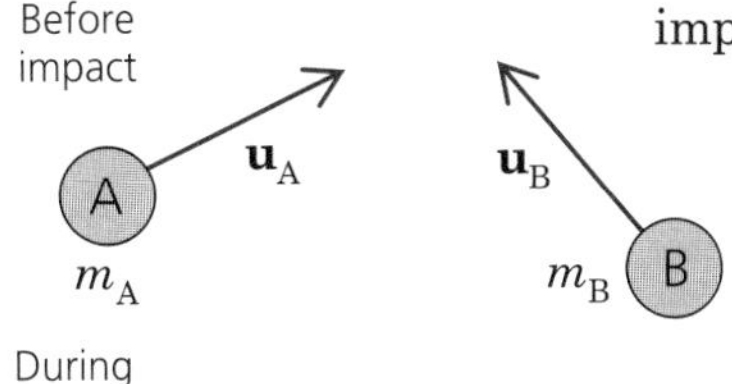

Before the impact A and B have velocities $\mathbf{u}_A$ and $\mathbf{u}_B$ respectively so the total momentum before the collision is $m_A\,\mathbf{u}_A + m_B\,\mathbf{u}_B$.

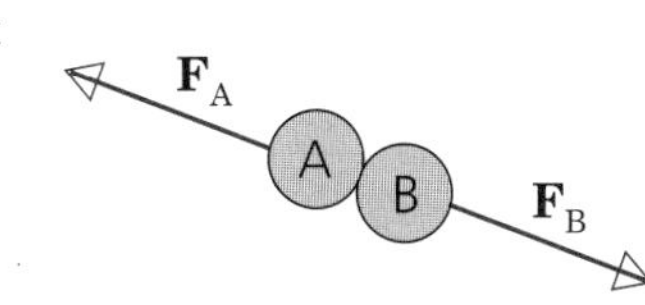

When the balls are in contact they exert forces on each other that are equal in magnitude, but acting in opposite directions. So if $\mathbf{F}_A$ is the force on A and $\mathbf{F}_B$ the force on B, $\mathbf{F}_A = -\mathbf{F}_B$.

If the balls are in contact for t seconds, then:

$$\mathbf{F}_A t = -\mathbf{F}_B\, t$$

or:

$$\mathbf{I}_A = -\mathbf{I}_B$$

So the impulse on A is of equal magnitude, but in the opposite direction to the impulse on B.

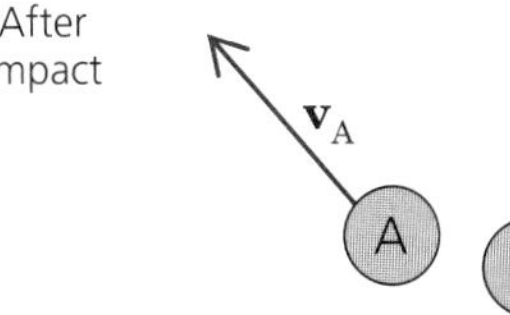

After impact A and B have velocities $\mathbf{v}_A$ and $\mathbf{v}_B$ respectively, so the total momentum after the collision is $m_A\,\mathbf{v}_A + m_B\,\mathbf{v}_B$.

$\mathbf{I}_A$ is the impulse on A and equal to the change in momentum or

$$m_A\,\mathbf{v}_A - m_A\,\mathbf{u}_A.$$

Similarly $\mathbf{I}_B = m_B\,\mathbf{v}_B - m_B\,\mathbf{u}_B$.

As $\mathbf{I}_A = -\mathbf{I}_B$,

this can be expressed as:

$$m_A\,\mathbf{v}_A - m_A\,\mathbf{u}_A = -(m_B\,\mathbf{v}_B - m_B\,\mathbf{u}_B)$$

Bringing the terms with $\mathbf{v}_A$ and $\mathbf{v}_B$ to the left of the equation and those with $\mathbf{u}_A$ and $\mathbf{u}_B$ to the right gives:

$$m_A\,\mathbf{v}_A + m_B\,\mathbf{v}_B = m_A\,\mathbf{u}_A + m_B\,\mathbf{u}_B$$

or:

$$\text{total momentum after collision} = \text{total momentum before collision}$$

So momentum is still conserved when collisions take place in two or three dimensions, as well as in one dimension.

Example 9.10

The driver of a car of mass 900 kg assumed he had right of way as he came out of a junction. He collided at right angles with a car of mass 1200 kg. In defence the driver claimed that the other car was travelling at a speed far in excess of the 30 mph (13 m s^{-1}) speed limit. The police estimated that immediately after impact the 900 kg car glanced off at 30 m s^{-1} and at 90° from the original direction of travel and the 1200 kg car glanced off at 12 m s^{-1} and at 60° from the original direction of travel. Was the first driver right?

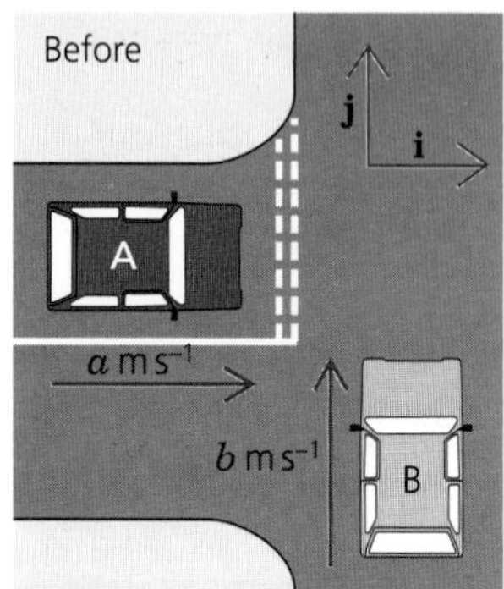

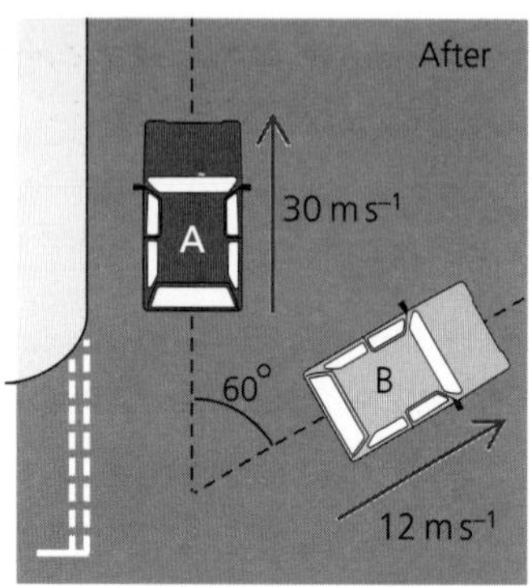

Solution

The first diagram shows the velocities of the cars before impact. The vehicles have been labelled A and B.
The initial velocity of A is $\mathbf{u}_A = a\,\mathbf{i}$ *and the initial velocity of B is* $\mathbf{u}_B = b\,\mathbf{j}$.
The second diagram shows the velocities of the cars after the collision.
The final velocity of car A is $\mathbf{v}_A = 30\mathbf{j}$. *The final velocity of B is:*
$\mathbf{v}_A = 12\cos 30°\mathbf{i} + 12\cos 60°\mathbf{j} = 10.39\mathbf{i} + 6\mathbf{j}$

Now the equation for conservation of momentum can be applied.
$m_A\,\mathbf{v}_A + m_B\,\mathbf{v}_B = m_A\,\mathbf{u}_A + m_B\,\mathbf{u}_B$
so that:
$900 \times 30\mathbf{j} + 1200(10.39\mathbf{i} + 6\mathbf{j}) = 900a\mathbf{i} + 1200b\mathbf{j}$
or $27\,000\mathbf{j} + 12\,468\mathbf{i} + 7200\mathbf{j} = 900a\mathbf{i} + 1200b\mathbf{j}$
Collecting together the $\mathbf{i}$ *and* $\mathbf{j}$ *components gives:*
$(12\,468 - 900a)\mathbf{i} + (27\,000 + 7200 - 1200b)\mathbf{j} = 0$
Consider first the $\mathbf{j}$ *components.*
$34\,200 - 1200b = 0$
$$\Rightarrow b = \frac{34\,200}{1200} = 28.5\ m\ s^{-1}$$
So the car B was travelling well in excess of the 13 m s⁻¹ or 30 mph speed limit.
Consider also the $\mathbf{i}$ *components.*
$12\,468 - 900a = 0$
$$\Rightarrow a = \frac{12\,468}{900} = 13.86\ m\ s^{-1}$$
Car A was also exceeding the speed limit, but only just!

Example 9.11

The white ball travels parallel to the longest side of the table at 10 m s⁻¹. The white ball collides with the black ball in such a way that the black ball goes into the right hand pocket and (to the dismay of the player) the white ball goes into the left hand pocket as shown (game lost!). Calculate the speeds of the two balls after impact.

Solution

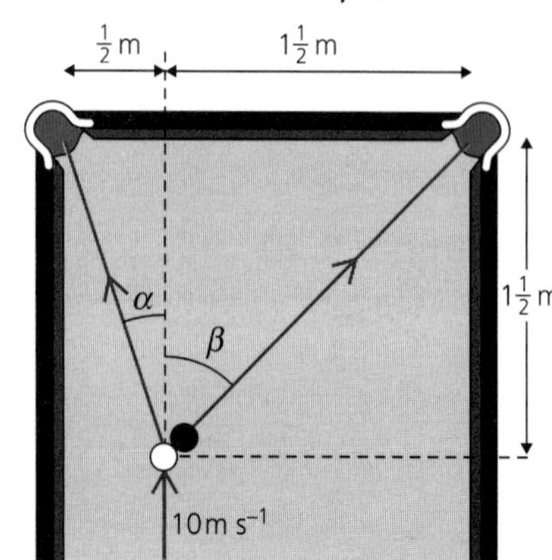

Assume that the masses of the balls are all equal. In this case the conservation of momentum equation, with B for black and W for white,

$m_W\,\mathbf{u}_W + m_B\,\mathbf{u}_B = m_W\,\mathbf{v}_W + m_B\,\mathbf{v}_B$
simplifies to:
$\mathbf{u}_W + \mathbf{u}_B = \mathbf{v}_W + \mathbf{v}_B$
Initially $\mathbf{u}_B = 0$ *and* $\mathbf{u}_W = 10\mathbf{j}$

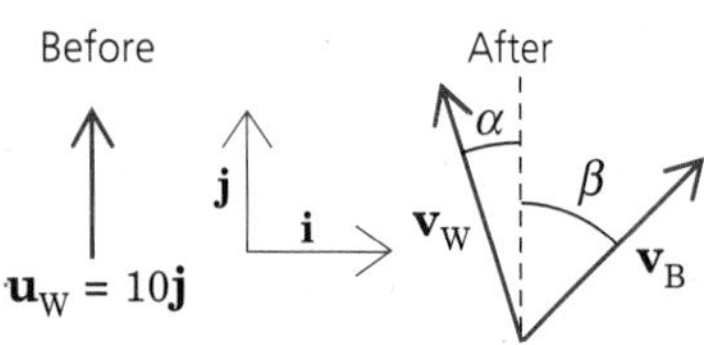

After the collision the balls move as shown in the diagram. First it is necessary to calculate the angles α and β.

$$\tan\alpha = \left(\frac{\frac{1}{2}}{1\frac{1}{2}}\right)$$

so $\alpha = 18.4°$

Clearly $\beta = 45°$.

So the final velocities can now be expressed as vectors.

$$\mathbf{v}_B = v_B\cos 45°\,\mathbf{i} + v_B\cos 45°\,\mathbf{j}$$

and:

$$\mathbf{v}_W = -v_W\cos 71.6°\mathbf{i} + v_W\cos 19.4°\mathbf{j}$$

Applying the conservation of momentum gives:

$$10\mathbf{j} = \left(v_B\cos 45° - v_W\cos 71.6°\right)\mathbf{i} + \left(v_B\cos 45° + v_W\cos 18.4°\right)\mathbf{j}$$

Considering the $\mathbf{i}$ *components gives:*

$$v_B\cos 45° = v_W\cos 71.6°$$

so $v_B = \dfrac{v_W\cos 71.6°}{\cos 45°} = 0.446v_W$

Considering the $\mathbf{j}$ *components gives:*

$$10 = v_B\cos 45° + v_W\cos 18.4°$$

or by substituting for v_B*:*

$$10 = 0.446v_W\cos 45° + v_W\cos 18.4°$$

and solving for v_W *gives:*

$$v_W = \frac{10}{0.446\cos 45° + \cos 18.4°} = 7.91\,ms^{-1}$$

Now as $v_B = 0.446v_W$

$$v_B = 0.446 \times 7.91 = 3.53\,ms^{-1}$$

EXERCISES

9.5 A

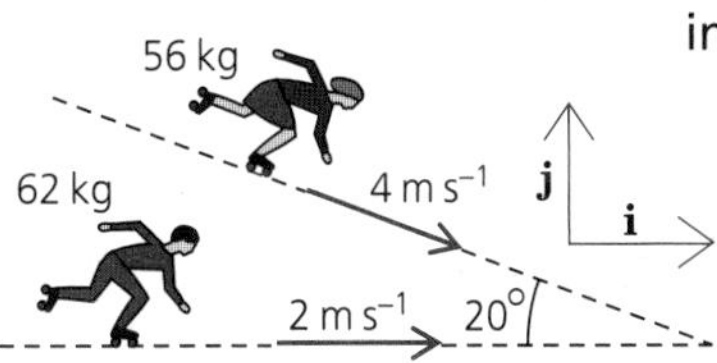

1 Two roller skaters collide and move together after the collision. Their masses are 62 kg and 56 kg and they were moving as shown in the diagram before the collision.

a) Find the momentum of each skater before the collision, in terms of $\mathbf{i}$ and $\mathbf{j}$.

b) What is the momentum of the two skaters after the collision?

c) Find their speed and the direction in which they move after the collision.

2 A small bird of mass 0.4 kg flies horizontally at a speed of $10\,\text{m s}^{-1}$. An owl of mass 2 kg swoops down onto the smaller bird with a speed of $20\,\text{m s}^{-1}$, travelling at 30° to the vertical.

a) Find the momentum of each bird just before impact, using suitable unit vectors.

b) The owl grabs the smaller bird with its claws and they move together. Find their speed just after the impact.

c) Are there any reasons why momentum should not be conserved in this case? If so, explain why.

3 A goalkeeper of mass 70 kg moves forward to catch a football of mass 2 kg. The goalkeeper moves at 1 m s^{-1} and the football travels at 40 m s^{-1} at an angle of 25° above the horizontal.

a) Find the total momentum of the ball and the goalkeeper before the collision, in terms of horizontal and vertical unit vectors.

b) Find the velocity of the goalkeeper after he has caught the ball, in terms of horizontal and vertical unit vectors.

c) If the goalkeeper keeps hold of the ball, in which direction would he move after catching the ball?

4 Two pucks are placed on an air table, where they hover on a cushion of air. Both pucks have the same mass. One puck is fired at 4 m s^{-1} into the other puck, which is stationary. After the collision the two pucks move as shown in the diagram.

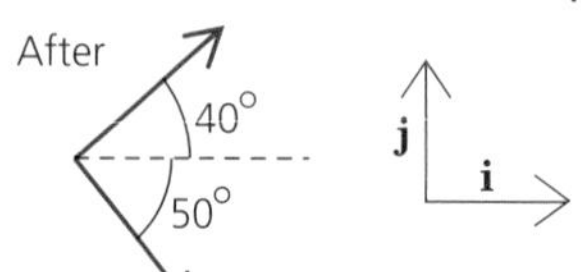

a) Explain the advantages of using unit vectors as illustrated. Why is this possible in this case?

b) Find the speed of each puck after the collision.

5 A car of mass 1000 kg that is being chased by a police car hits a stationary vehicle of mass 1200 kg. The motion of the car is shown in the diagram.

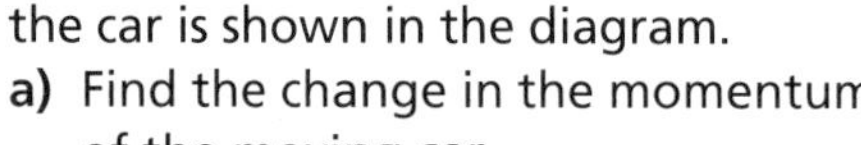

a) Find the change in the momentum of the moving car.

b) What is the speed of the stationary car after the collision?

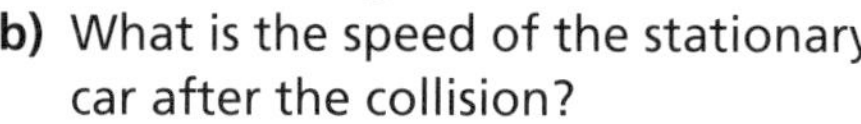

c) Explain why the car that was stationary does not appear to move very much.

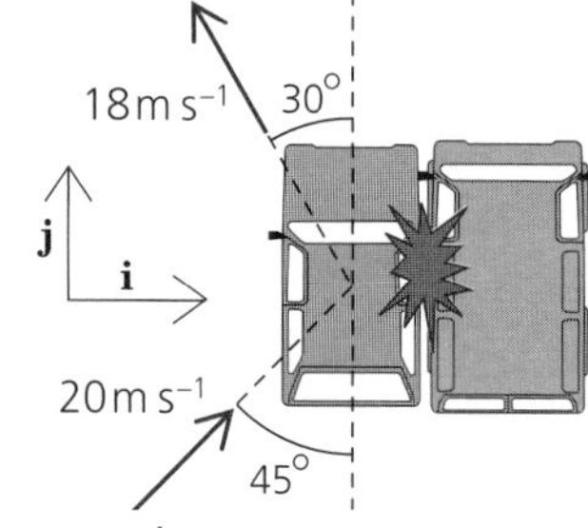

6 A stone of mass 80 grams is fired from a catapult so that it hits a tin can of mass 200 grams. After the collision the can and stone move as shown in the diagram.

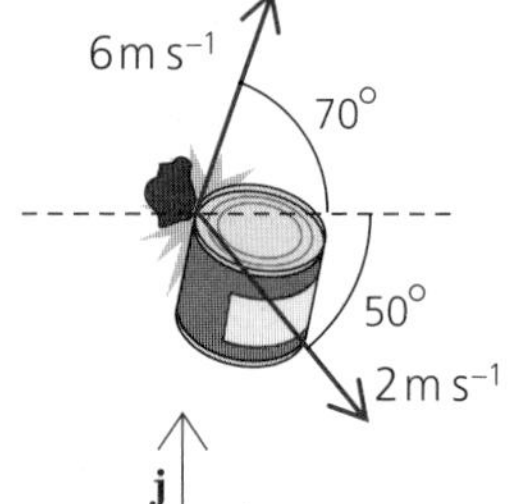

a) Find the total momentum of the can and the stone after the collision, in terms of the unit vectors $\mathbf{i}$ and $\mathbf{j}$.

b) Find the initial velocity of the stone.

c) Use a scale drawing to confirm the result obtained above.

7 A car of mass $3m$, and a motorcycle of mass m, are involved in a collision. Initially, they are travelling at right angles, both with speed u. After the collision both travel together with speed v.

a) Show that the magnitude of the total momentum of both vehicles before the collision is $\sqrt{10}mu$.

b) Express v in terms of u and find the angle through which each vehicle is deflected.

8 A scientist conducts an experiment to find the mass of a helium atom. A proton of mass 1.67×10^{-27} kg is fired with a velocity of 10^7 m s^{-1} towards a stationary helium atom. The collision leaves the

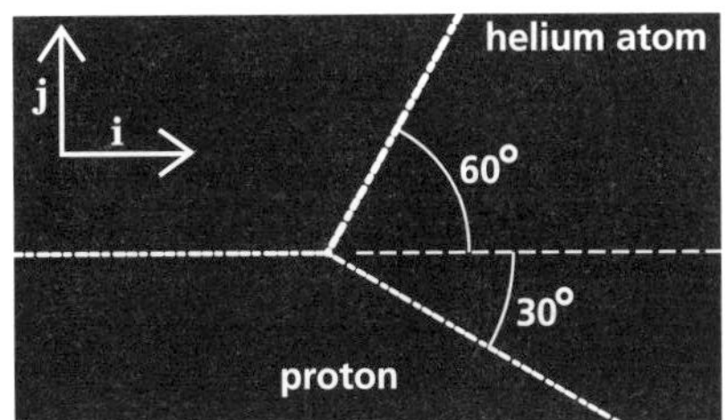

trace shown in a photographic plate. After the collision the speed of the proton is $5 \times 10^6\,\text{m s}^{-1}$ and the speed of the helium atom is $2.17 \times 10^6\,\text{m s}^{-1}$.

a) Find the total momentum of the atom and the proton after the collision in terms of m, the mass of the helium atom, and the unit vectors $\mathbf{i}$ and $\mathbf{j}$.

b) Use conservation of momentum to find m.

9 A sailor of mass 64 kg jumps off a quay into a rubber dinghy of mass 40 kg. He hits the dinghy travelling at $8\,\text{m s}^{-1}$ and at 10° to the horizontal. Assume the dinghy travels horizontally.

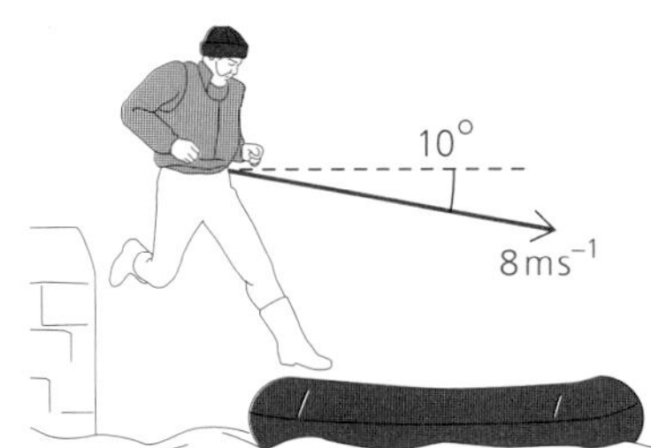

a) Find the vertical impulse on the sailor.

b) If the horizontal component of momentum is conserved, find the speed of the dinghy.

10 A gun of mass 4000 kg fires a shell of mass 100 kg at a speed of $300\,\text{m s}^{-1}$ at an angle of 40° above the horizontal. The gun moves backwards horizontally. Describe what happens to the gun, calculating the magnitude of any quantities you use in your description.

EXERCISES

1 A car of mass 1100 kg is travelling at $14\,\text{m s}^{-1}$, as shown in the diagram, when it is hit by a lorry of mass 4 tonnes travelling as shown.

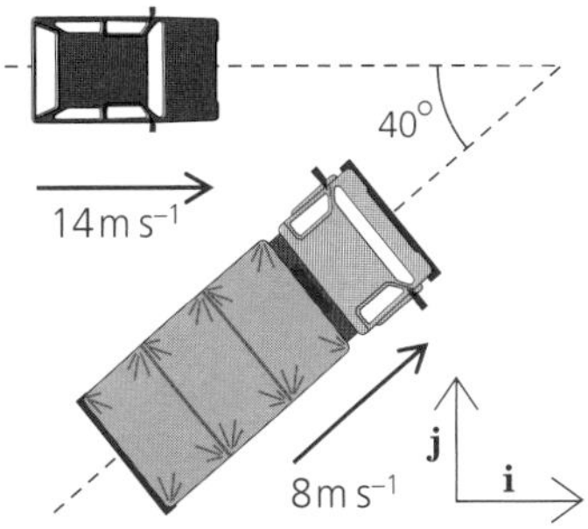

a) Find the momentum of each vehicle before the collision in terms of the unit vectors $\mathbf{i}$ and $\mathbf{j}$.

b) If, after the collision, the two vehicles move together, find their speed and the direction in which they move, just after the collision.

c) What do you expect to happen to the velocity of the vehicles after the collision?

2 A 5000 kg space capsule travelling at $400\,\text{m s}^{-1}$ is struck by a lump of debris of mass 200 kg, as shown in the diagram. The debris becomes embedded in the capsule.

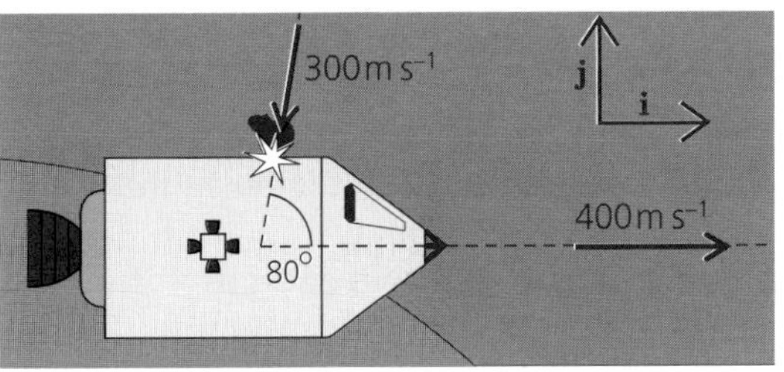

a) Find the momentum of the capsule and the lump of debris before the collision, in terms of the unit vectors $\mathbf{i}$ and $\mathbf{j}$.

b) Find the speed of the capsule after the collision.

c) Find the deflection of the capsule from its original path.

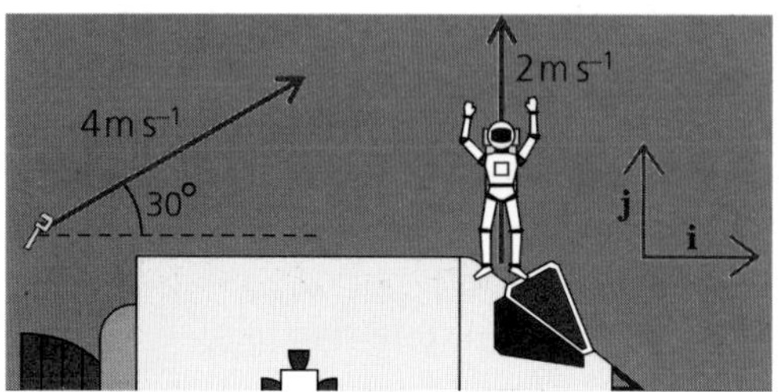

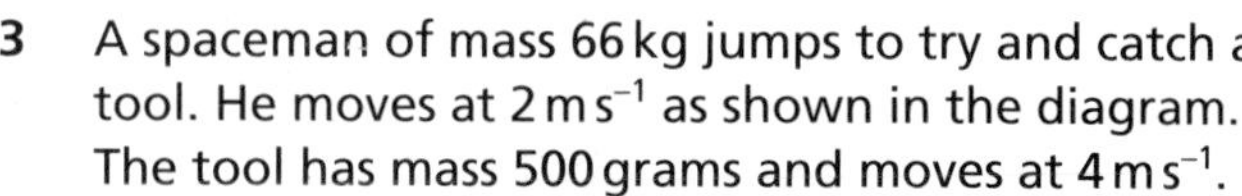

3 A spaceman of mass 66 kg jumps to try and catch a tool. He moves at $2\,m\,s^{-1}$ as shown in the diagram. The tool has mass 500 grams and moves at $4\,m\,s^{-1}$.

a) Calculate the total momentum of the tool and the spaceman before the collision.

b) Find the speed at which they move and the direction in which they move after the collision.

c) Should the spaceman be worried about the effect of catching the tool?

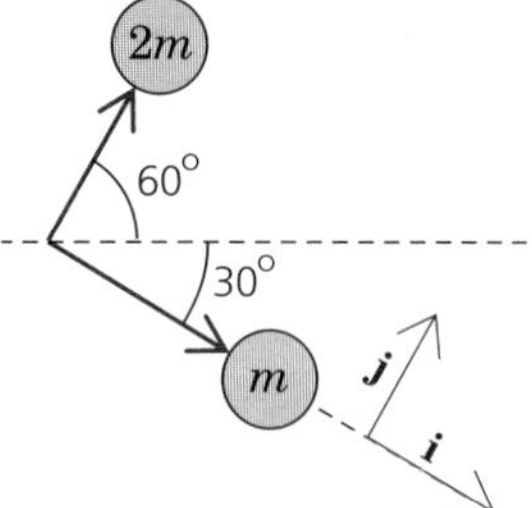

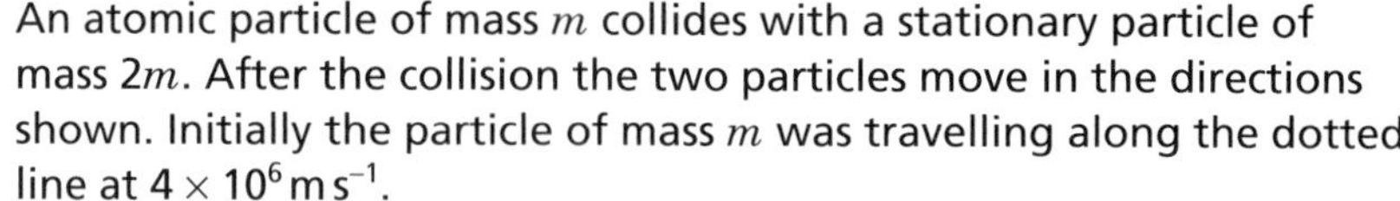

4 An atomic particle of mass m collides with a stationary particle of mass $2m$. After the collision the two particles move in the directions shown. Initially the particle of mass m was travelling along the dotted line at $4 \times 10^6\,m\,s^{-1}$.

a) Express the initial momentum of the particle of mass m in terms of the vectors $\mathbf{i}$ and $\mathbf{j}$.

b) Find the speed of each particle after the collision.

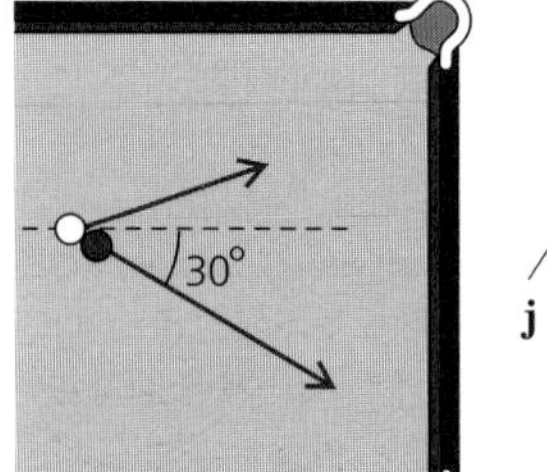

5 A white snooker ball travels straight down a table at $2\,m\,s^{-1}$ parallel to the side and strikes a stationary red ball, which then moves in the direction shown at $1.5\,m\,s^{-1}$.

a) Find the total momentum of the two balls before the collision, in terms of the unit vectors $\mathbf{i}$ and $\mathbf{j}$ and the mass, m, of each ball.

b) Find the direction in which the white ball travels after the collision.

c) State any assumptions that you have made.

6 The diagram show how two snooker balls, of mass m, move after a collision. Assume that the masses of the balls are equal and that one was initially at rest.

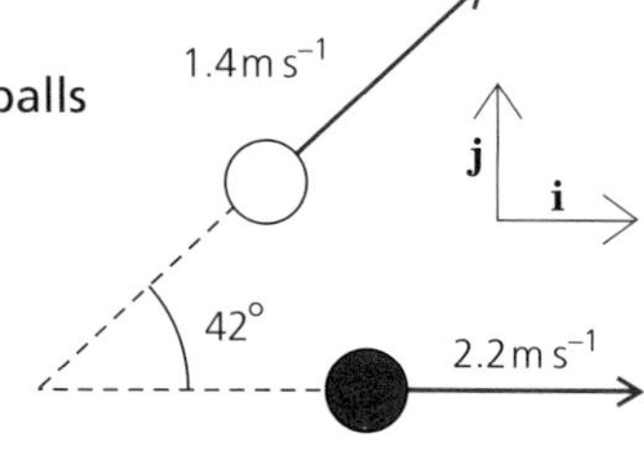

a) Find the total momentum of the two balls after the collision.

b) Find the velocity of the moving ball before the collision.

c) Sketch a diagram that could be used to find the velocity by scale drawing.

7 Two atomic particles A and B collide and join together. Both particles were initially moving at right angles. A has mass m and initial speed u, while B has mass $3m$ and initial speed $2u$.

a) Find the magnitude of the total momentum before the collision.

b) Find the speed of the two combined particles after the collision.

c) Find the deflection of each particle.

8 Two air pucks that hover on a glass surface have masses M and m. The puck of mass M is fired and hits the puck of mass m which was stationary. After the collision the pucks move as shown in the diagram.

a) Find the total momentum of the two pucks after the collision.

b) Find M in terms of m.

c) Find v.

9 A child of mass 40 kg jumps from a wall onto a stationary skateboard of mass 5 kg. The child travels at 5 m s^{-1}, at 20° to the horizontal, before impact.

a) If the skateboard begins to move horizontally, find the magnitude of the vertical impulse on the child.

b) If the horizontal component of momentum is conserved, find the speed of the child and the skateboard.

10 A trolley of mass 80 kg is freewheeling at a constant speed of 4 m s^{-1}. A heavy sack of mass 30 kg is dropped onto the trolley, so that it is moving at right angles to the trolley when it impacts. What happens to the motion of the trolley?

THE COEFFICIENT OF RESTITUTION

Exploration 9.7

Comparing heights of bounce

If you drop different types of balls onto the floor and let them bounce, they will rebound to different heights. Why is there this difference, and what do you think causes it?

Materials that bounce

What happens depends on the material from which the ball is made and the surface onto which it is dropped. Soft, spongy materials will not bounce well. To demonstrate how the nature of a ball affects its bounce, take a table tennis ball and make a small hole in it. First observe how it bounces. Then push some elastic bands in through the hole and see how this affects the bounce. Try to produce a ball that does not bounce. Then push in more elastic bands and observe what happens to the bounce.

Exploration 9.8

Bouncing balls

- Collect several different sports balls (for example, football, basketball, tennis ball, squash ball, cricket ball, etc.).
- Drop each ball, in turn (from a variety of heights), onto the floor and record how high each rebounds. What happens?
- Calculate the speed at which the ball hits the ground and the speed at which it rebounds in each case. (**Note**: $v = \sqrt{2gh}$.) What is the ratio of impact to rebound speeds?
- Next, find a floor with a different covering and repeat the above experiment. What happens this time?

Your results should indicate that the rebound height is dependent on the two materials involved in the collision, and the height from which the balls are released.

Newton's experimental law

Newton proposed an experimental law that describes how the impact and rebound velocities are related. He stated that:

$$v = -eu$$

where v is the rebound velocity, u the impact velocity and e is a constant called the **coefficient of restitution**. The constant e depends on the type of ball and the surface it bounces on.

If $e = 1$ the collision is said to be **perfectly elastic**.
If $0 < e < 1$ the collision is said to be **elastic**.
If $e = 0$ the collision is said to be **inelastic**.

Example 9.12

A ball is dropped from a height of 1.5 m, onto the floor where it bounces. The coefficient of restitution is 0.6.

a) *Find the velocity of the ball on impact.*
b) *Find the velocity of the ball on rebound.*
c) *Find the height to which the ball bounces.*

Solution

Note first that the downward direction is defined as positive.

a) *As the ball is subject to a constant acceleration of 9.8 m s⁻², the constant acceleration formula:*

$v^2 = u^2 + 2as$

can be applied. In this case $u = 0$, $a = 9.8$ *and* $s = 1.5$, *so:*

$v^2 = 0^2 + 2 \times 9.8 \times 1.5 = 29.4$

and:

$v = 5.42\,m\,s^{-1}$

b) *The rebound velocity can be found using:*

$v = -eu$

In this case $e = 0.6$ *and* $u = 5.42$, *so:*

$v = -0.6 \times 5.42 = -3.252\,m\,s^{-1}$

c) *The constant acceleration formula:*

$v^2 = u^2 + 2as$

can be used to find the rebound height. Here $u = -3.252$, $v = 0$ *and* $a = 9.8$, *so:*

$0^2 = (-3.252)^2 + 2 \times 9.8s$

$$s = \frac{-(-3.252)^2}{2 \times 9.8} = -0.54m$$

So the ball rebounds to a height of 54 cm. The answer for s *is negative as the ball has moved upwards.*

Example 9.13

Show that if a ball is dropped from a height H *and rebounds to a height* h, *then the coefficient of restitution is:*

$$e = \sqrt{\frac{h}{H}}$$

Solution

If the ball is allowed to fall from a height H, *under gravity the equation:*

$v^2 = u^2 + 2as$

can be applied with $u = 0$ *and* $s = H$ *to give:*

$v^2 = 2gH$

so the impact velocity is $\sqrt{2gH}$.

Similarly the rebound velocity is $-\sqrt{2gh}$.

Using $v = -eu$ *gives:*

$$-\sqrt{2gh} = -e\sqrt{2gH}$$

$$\Rightarrow \sqrt{h} = e\sqrt{H} \text{ so } e = \sqrt{\frac{h}{H}}$$

Exploration 9.9

Momentum and restitution

When a collision takes place involving two superballs, momentum is conserved. Is momentum conserved when two spongy balls collide? Does conservation of momentum depend on the value of e?

Momentum for any collision

Before

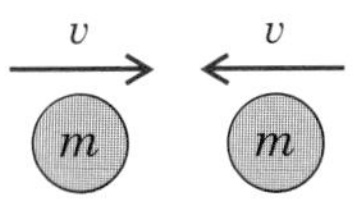

After

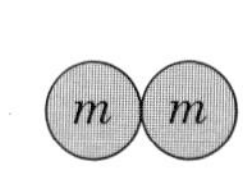

The law of conservation of momentum *does* hold for any collision provided no external forces act. If two balls of the same mass are travelling towards each other at the same speed, then the total momentum before collision is zero. If the collision brings them both to rest, then the total momentum after the collision is also zero. This extreme case illustrates conservation of momentum taking place, but more information is needed about what happens in a collision. This is where the coefficient of restitution is important.

Before

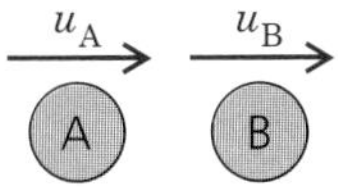

After

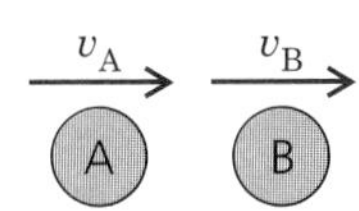

Consider two balls A and B moving, with velocities u_A and u_B respectively, along a straight line. As they collide the velocity of impact is $u_A - u_B$. After the collision, the velocity of separation will be $v_A - v_B$.

Newton's experimental law applies in this situation, with $u = u_A - u_B$ and $v = v_A - v_B$ to give:

$$v = -eu$$

or:

$$v_A - v_B = -e(v_A - v_B)$$

This result can be used in conjunction with conservation of momentum to predict what will happen after a collision.

Example 9.14

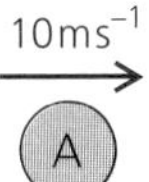

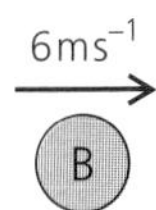

Two identical superballs are moving as shown in the diagram. The coefficient of restitution between the two balls is 0.8. Find what happens to each ball after the collision.

Solution

First define the positive direction to the right, so the initial velocities are $u_A = 10$ *and* $u_B = 6$, *with* v_A *and* v_B *as the final velocities.*

As the balls have the same mass the conservation of momentum equation:

$$m_A u_A + m_A u_B = m_A v_A + m_A v_B$$

reduces to:

$$u_A + u_B = v_A + v_B$$

and substituting for u_A and u_B gives:
$10 + 6 = v_A + v_B$
so $v_A + v_B = 16$ *(1)*

Using the experimental law:
$v_A - v_B = -e(u_A - u_B)$

gives:
$v_A + v_B = -0.8(10 - 8)$
so $v_A + v_B = -1.6$ *(2)*

Adding equations (1) and (2) gives:
$2v_A = 12.8$ *so* $v_A = 6.4\,m\,s^{-1}$
Substituting this into equation (1) gives:
$6.4 + v_B = 16$ *so* $v_B = 9.6\,m\,s^{-1}$
So in the collision A loses speed and B gains speed.

Example 9.15

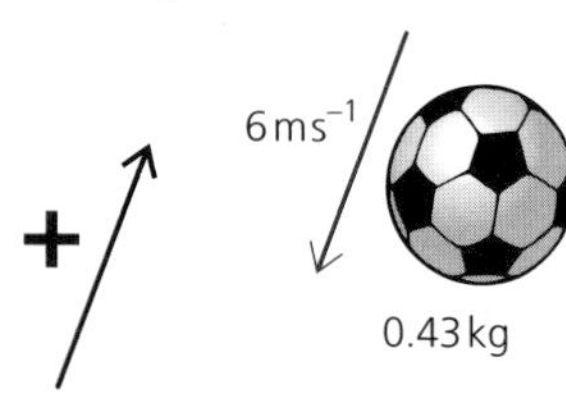

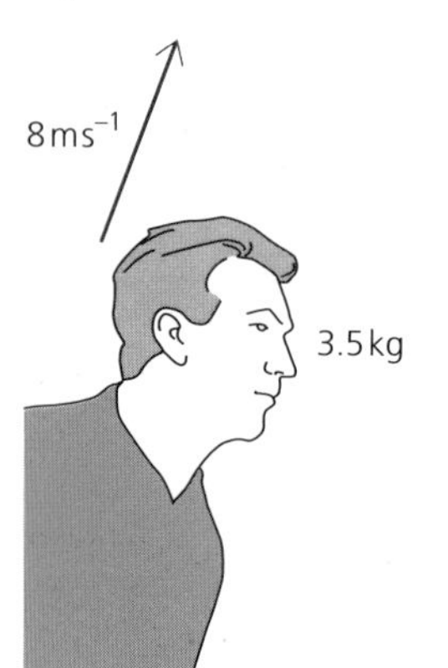

When a footballer heads a football, it can be modelled as a collision between two spheres. Assume that the head and the ball are travelling towards each other along a straight line as shown in the diagram, and that the coefficient of restitution is 0.3.

a) *Find the velocity of the ball after the collision.*
b) *If the contact time is 0.005 s find the force acting on the head and its acceleration.*
c) *What factor has been ignored and do you think that this is reasonable in this situation?*

Solution

The diagram shows the positive direction, the initial velocities and the masses of the two spheres.

a) *Using the conservation of momentum:*

$m_H u_H + m_B u_B = m_H v_H + m_B v_B$

gives:

$3.5 \times 8 + 0.43 \times (-6) = 3.5v_H + 0.43v_B$

or:

$3.5v_H + 0.43v_B = 25.42$ *(1)*

Using the experimental law:

$v_H - v_B = -e(u_H - u_B)$

gives:

$v_H - v_B = -0.3(8 - (-6)) = -4.2$ *(2)*

Multiplying equation (2) by 0.43 gives:

$0.43v_H - 0.43v_B = -1.806$ *(3)*

Now adding equations (1) and (3) gives:

$3.93v_H = 23.614$

so $v_H = 6.01\,m\,s^{-1}$

Substituting this into equation (2) gives:

$6.01 - v_B = -4.2$

so $v_B = 10.21\,ms^{-1}$

b) *The force on the players head can be found using:*

$Ft = mv - mu$

in this case:

$F \times 0.005 = 3.5 \times 6.01 - 3.5 \times 8$

so that $F = -1393\,N$

If the force is assumed to be constant, then the acceleration can be found using:

$F = ma$

$-1393 = 3.5a$

so that $a = -398\,ms^{-2}$

While this figure is very large it should be noted that it is only experienced for a very short period of time. Boxers can experience accelerations of up to about 900 m s⁻².

c) *The effects of gravity have been ignored, but this is justifiable as the time of the collision is so short the effects of gravity would be negligible.*

EXERCISES

9.6 A

1 In each situation below find the unknown quantity.

a)

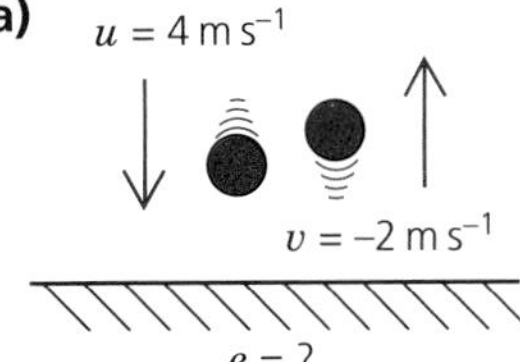

b)

c)

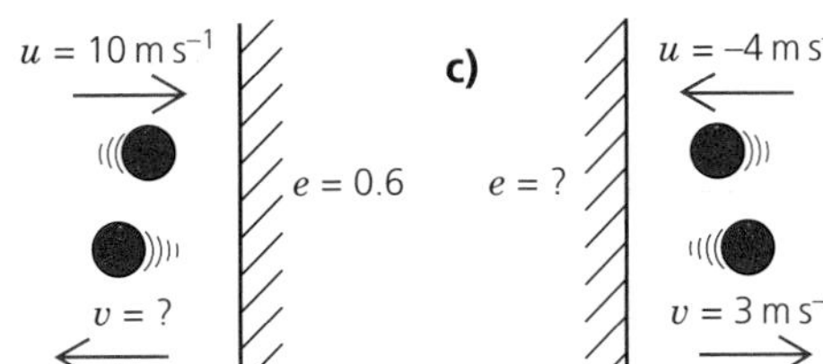

d)

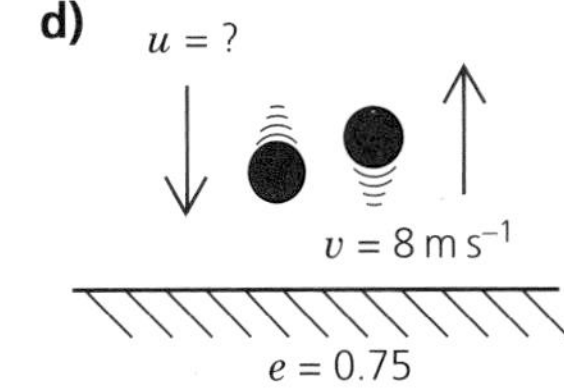

2 A ball is dropped from a height of 2 m and rebounds to a height of 1.2 m.

a) Find the velocity of the ball when it hits the ground.

b) Find the velocity of the ball when it leaves the ground.

c) What is the coefficient of restitution between the ball and the ground?

3 In each situation below find the unknown quantity, when a ball bounces on the ground.

a)

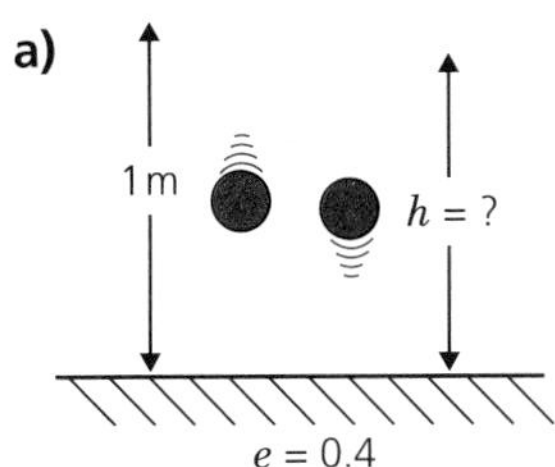

b)

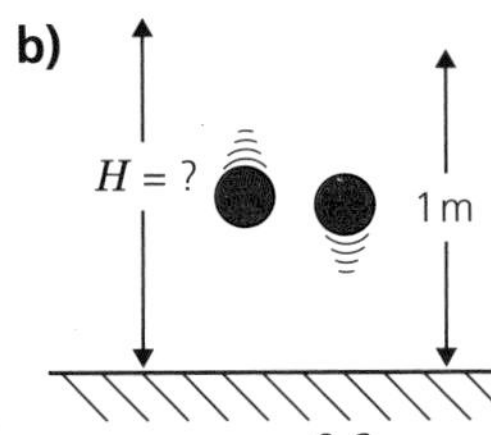

c)

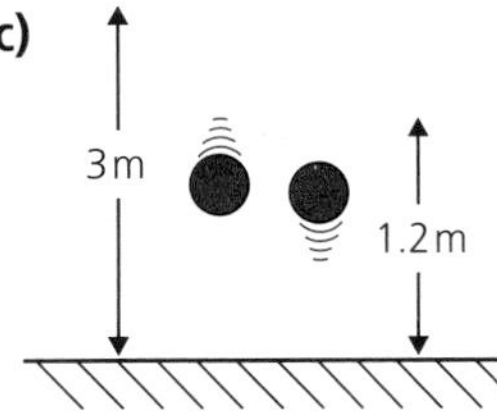

d)

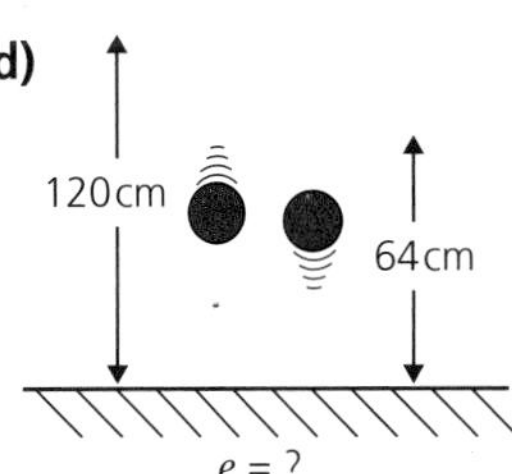

4 Two identical balls are moving directly towards each other at $10\,m\,s^{-1}$. Find the velocities of the two balls after the collision if:

a) $e = 0.95$, **b)** $e = 0.1$.

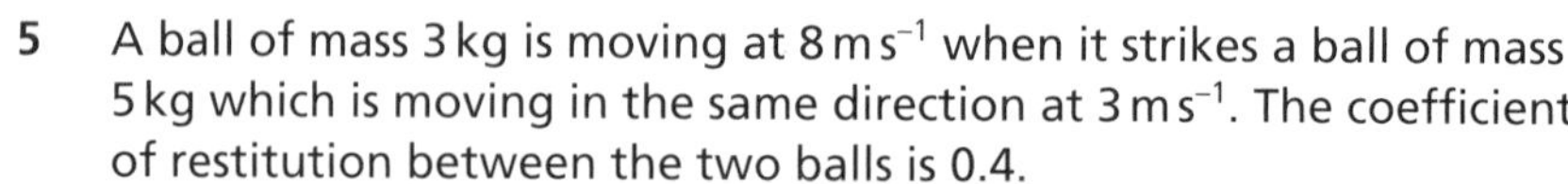

5 A ball of mass 3 kg is moving at $8\,\text{m s}^{-1}$ when it strikes a ball of mass 5 kg which is moving in the same direction at $3\,\text{m s}^{-1}$. The coefficient of restitution between the two balls is 0.4.

a) Find the velocities of the two balls after the impact.

b) What important factor has been ignored?

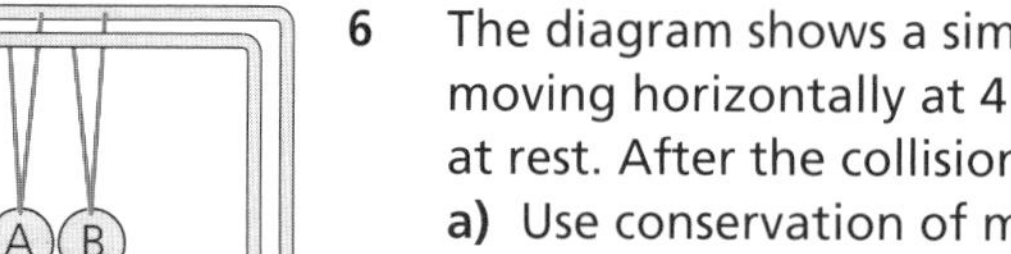

6 The diagram shows a simplified Newton's Cradle. The ball A is moving horizontally at $4\,\text{m s}^{-1}$ when it strikes ball B, which was at rest. After the collision ball B moves at $3.95\,\text{m s}^{-1}$.

a) Use conservation of momentum to find out what happens to ball A after the collision.

b) Find the coefficient of restitution between the balls.

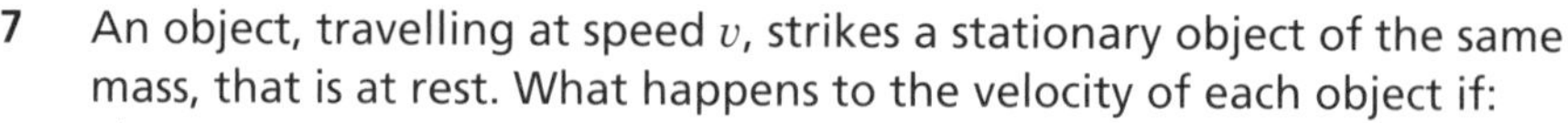

7 An object, travelling at speed v, strikes a stationary object of the same mass, that is at rest. What happens to the velocity of each object if:

a) $e = 1$, **b)** $e = 0.5$, **c)** $e = 0$?

8 Three balls A, B and C are moving horizontally with the speeds shown in the diagram. The coefficient of restitution between the balls is 0.5.

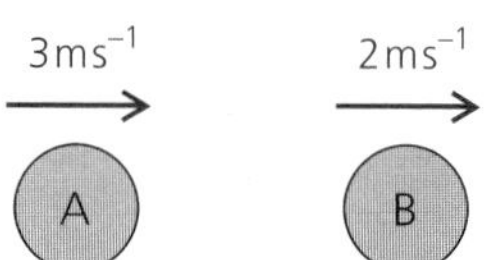

a) The first collision that takes place is between A and B. Find the velocities of these two balls after the collision.

b) The next collision is between B and C. Find the velocities of these two balls after this collision.

c) What collision takes place next?

9 Three identical spheres A, B and C lie with their centres on a straight line. Initially A moves with speed u towards B and C, which are at rest. The coefficient of restitution between the spheres is e.

a) Find the velocities of A and B after the first collision, in terms of e.

b) Find the velocities of B and C after the second collision, in terms of e.

10 A ball is dropped from a height h onto a horizontal floor. The coefficient of restitution between the ball and the floor is e. If the ball is allowed to bounce continually until it comes to rest, find the total distance travelled by the ball.

EXERCISES

9.6 B

1 Find the unknown quantity in each situation described below, where a ball bounces against another surface.

a)

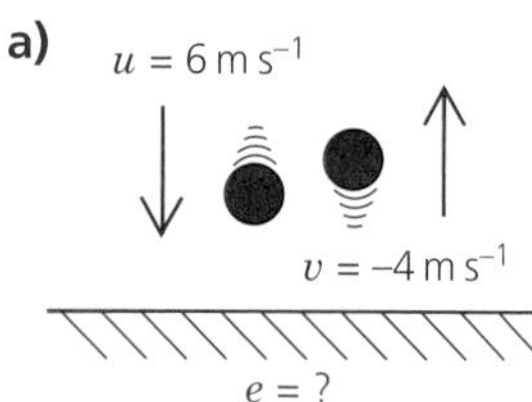

b)

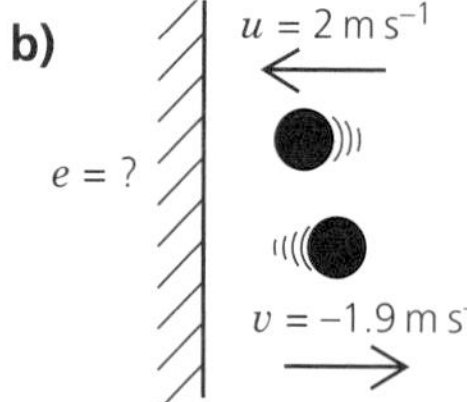

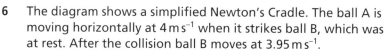

c)

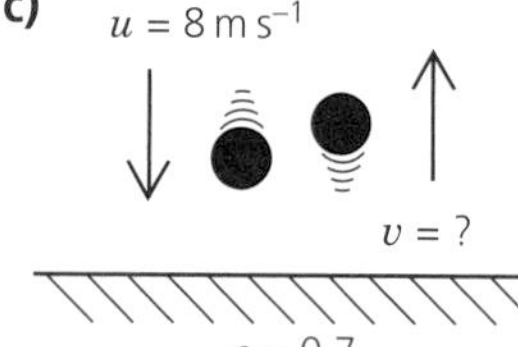

d)

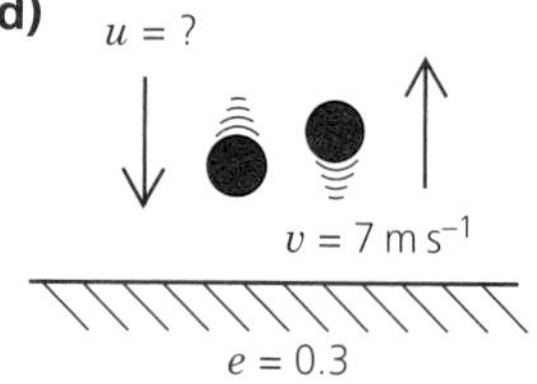

2 A ball is dropped from a height of 5 m. The coefficient of restitution between the ball and the ground is 0.4.

a) Find the velocity of the ball when it hits the ground.

b) Find the rebound velocity of the ball.

c) To what height does the ball rebound?

3 For each situation illustrated below, where a ball is allowed to bounce on the ground, find the unknown quantity.

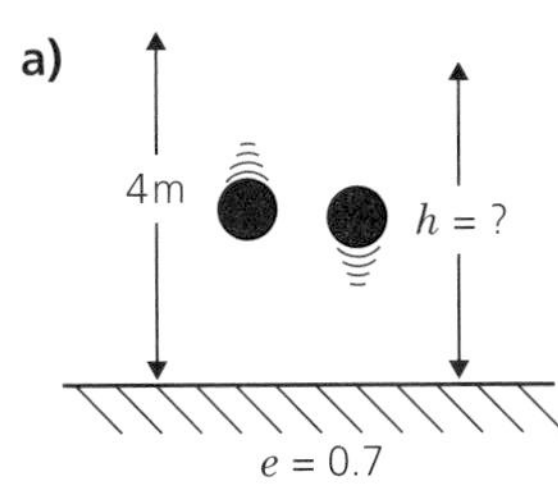

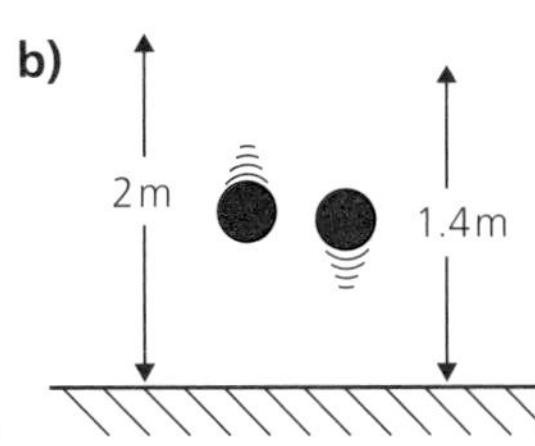

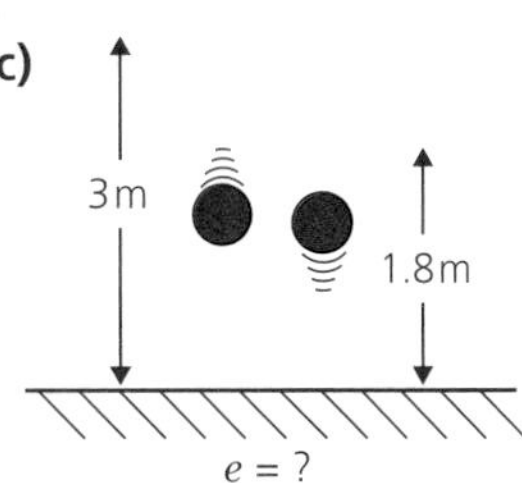

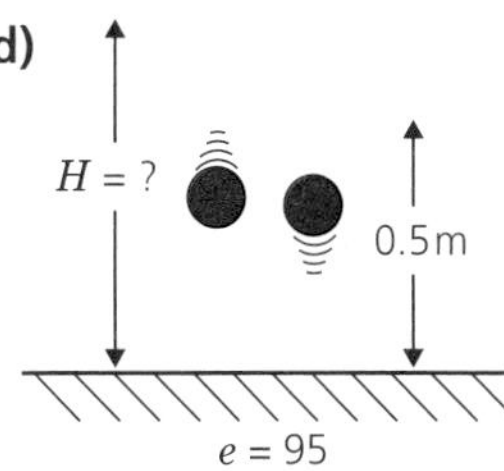

4 Two identical balls are moving in the same direction, one at $5\,\mathrm{m\,s^{-1}}$ and the other at $10\,\mathrm{m\,s^{-1}}$. Find the velocities of the balls after a collision if:

a) $e = 0.8$, **b)** $e = 0.4$.

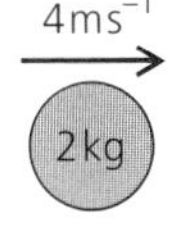

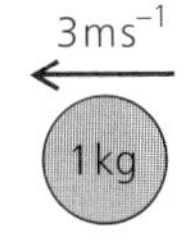

5 Two balls are moving towards each other as shown in the diagram. The coefficient of restitution between the two balls is 0.2. Find the velocity of each ball after the collision.

6 An object travelling at $2\,\mathrm{m\,s^{-1}}$ and with a mass of 1 kg strikes a stationary object of mass 2 kg. What happens after the collision if:

a) $e = 1$, **b)** $e = 0.2$?

7 A car of mass 1000 kg travelling at $18\,\mathrm{m\,s^{-1}}$ drives into a stationary car of mass 900 kg. The speed of the moving car is halved during the collision.

a) Use conservation of momentum to find what happens to the other car.

b) Find the coefficient of restitution between the two cars.

8 Three identical spheres are travelling along a straight line with the speeds shown. The coefficient of restitution between each sphere is 0.4.

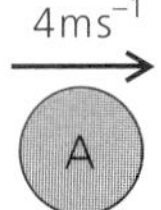

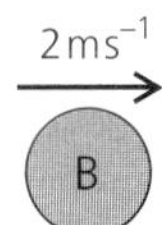

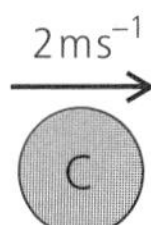

a) Find the speeds of A and B after they collide.

b) Find the speeds of B and C after they collide.

c) Are there any further collisions? If so which spheres are involved in the next collision?

9 Three identical spheres are moving as shown in the diagram. The coefficient of restitution is 0.4.

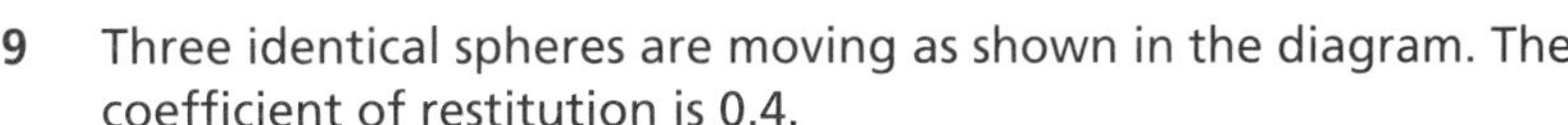

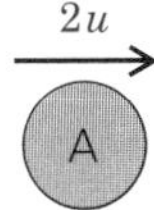

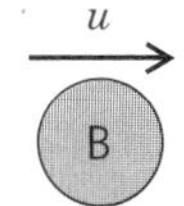

a) If A and B collide first, what are their velocities after the collision?

b) Next, B hits C which is stationary. What are the velocities of B and C after the collision?

c) Do any further collisions take place? Find the final velocities of each sphere.

10 A sphere of mass m, travelling at speed V, hits another sphere of the same mass that is travelling in the same direction, with speed v. The coefficient of restitution between the two spheres is e.

a) Find the velocity of each sphere after the collision.

b) Show that the magnitude of the impulse on each sphere is:

$$\tfrac{1}{2}m(1+e)(V-v)$$

CONSOLIDATION EXERCISES FOR CHAPTER 9

1 A block of wood, of mass 2 kg, is at rest on a smooth horizontal table. A bullet of mass 0.1 kg, moving horizontally at a speed of $420\,\text{m s}^{-1}$, strikes the block and becomes embedded in it. Find the speed of the block after the impact.

(UCLES (Linear) Question 5, Specimen Paper 2, 1994)

2 A large box of mass 2 kg rests on a smooth horizontal surface. A boy decides to propel the box along the surface by throwing a ball against the box. The ball, when travelling horizontally with speed $10\,\text{m s}^{-1}$, strikes the box normally at the centre of a plane face, and then returns to the boy.

a) The ball is of mass 0.4 kg and the coefficient of restitution between the ball and the box is 0.5.

i) Show that after the ball has been thrown once against the box, the box's speed will be $2.5\,\text{m s}^{-1}$.

ii) The ball is thrown a second time against the box, so that it strikes it as before. Show that the resulting speed of the box will then be greater than $4\,\text{m s}^{-1}$.

b) Give a reason why, if the boy uses a ball of mass 2 kg, throwing it as before, he will not be able to use the same ball more than once if he remains in the same position.

(Oxford Question 4, Specimen Paper M5, 1994)

3 Two small uniform smooth spheres A and B, of equal size and of mass m and $4m$ respectively, are moving directly towards each other with speed $2u$ and $6u$ respectively. The coefficient of restitution between the spheres is $\frac{1}{2}$. Find the speed of B immediately after the spheres collide.

(AEB Question 3, Paper 3, June 1991)

4 A stone, A, is sliding across a smooth rink with momentum $\begin{pmatrix}4\\3\end{pmatrix}\text{kg m s}^{-1}$. It hits another stone, B, and its momentum becomes $\begin{pmatrix}2\\-17\end{pmatrix}\text{kg m s}^{-1}$.

a) What is the change in momentum of stone A?

b) What does this tell you about the change in momentum of stone B?

(SMP 16–19 Question B1, Newton's Laws of Motion Specimen Paper, 1994)

5 State the law of conservation of momentum for two interacting particles which move in the same straight line. Derive this law from Newton's laws of motion. Two particles A and B, of mass m and $3m$ respectively, lie at rest on a smooth horizontal table. The particle A is projected horizontally with speed $5u$ and collides directly with B. After the collision A is moving in the same direction with speed u. Find the speed of B. Find also the magnitude of the impulse of A on B.

(NEAB Question 10, Specimen Paper 3, 1994)

6 A small sphere R, of mass 0.08 kg, moving with speed $1.5\,\text{m s}^{-1}$, collides directly with another small sphere S, of mass 0.12 kg, moving in the same direction with speed $1\,\text{m s}^{-1}$. Immediately after the collision R and S continue to move in the same direction with

speeds $U\,\mathrm{m\,s^{-1}}$ and $V\,\mathrm{m\,s^{-1}}$ respectively. Given that:
$U : V = 21 : 26$:

a) show that $V = 1.3$,

b) find the magnitude of the impulse, in N s, received by R as a result of the collision.

(ULEAC Question 3, January 1992)

7 Two cars are being driven on a level 'skid pan', on which resistances to motion, acceleration and braking may all be neglected.

Car A, of mass 1200 kg, is travelling at $15\,\mathrm{m\,s^{-1}}$ when it collides directly with car B, of mass 800 kg, travelling at $10\,\mathrm{m\,s^{-1}}$ in the same direction.

a) Let the speeds of cars A and B after the collision be v_A and v_B respectively. Draw a diagram on which you mark the velocities of the cars before and after collision.

b) The coefficient of restitution between the cars is 0.8. Write down equations involving the momentum of each of the cars and their relative speeds before and after impact. Solve these equations to show that car B has speed $15.4\,\mathrm{m\,s^{-1}}$ after the impact and find the new speed of car A.

Car B now collides directly with another small car of mass 740 kg which is initially at rest and becomes entangled with it.

c) Find the speed of car B (and the entangled car) after this impact.

d) There is now a final direct collision between car A and car B (and the entangled car) after which they separate at a speed of $2.55\,\mathrm{m\,s^{-1}}$. Calculate the coefficient of restitution in this impact.

(MEI Question 1 M2 Specimen Paper, 1994)

8 A particle A of mass 0.7 kg is moving with speed $3\,\mathrm{m\,s^{-1}}$ on a horizontal smooth table of height 0.9 m above a horizontal floor. Another particle B, of mass M kg, is at rest on the edge of the table top. Particle A strikes particle B, and they coalesce into a single particle C. The particle C then falls from the table. From the point of leaving the table to the point of hitting the floor, the horizontal displacement of C is 0.6 m.

a) Show that C takes $\frac{3}{7}$ s to fall to the floor.

b) Find the value of M.

(London Question 4, Paper M1, January 1993)

MATHEMATICAL MODELLING ACTIVITY

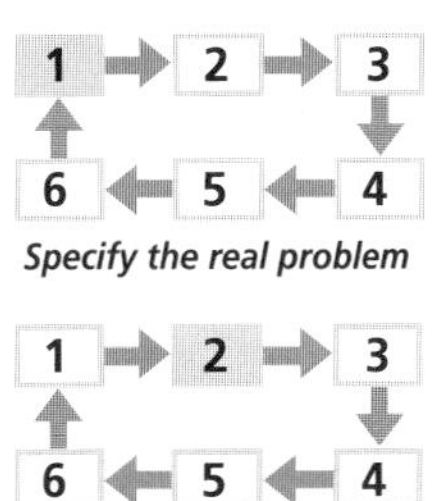

Specify the real problem

Problem statement

Many safety conscious groups have called for the introduction of seat belts in coaches. Develop a mathematical argument to support this call.

Set up a model

Set up a model

Discuss the important features that should be considered in solving this problem. The problem will be developed by considering head on collisions between two vehicles. The assumptions below have been made:

- The maximum speed of impact will be 70 mph (31 m s^{-1}).
- The mass of a loaded lorry will not exceed 38 tonnes.
- The mass of a loaded coach will not exceed 10 tonnes.
- The mass of a car will not exceed 1.5 tonnes.
- Momentum is conserved in all collisions.
- Vehicles move together after collisions.

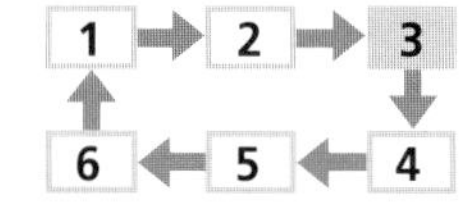

Formulate the mathematical problem

Formulate a mathematical problem

Find an expression for the combined speed, V, of the vehicles after a collision, if they are initially moving towards each other at speed u and have masses m and M.

Before

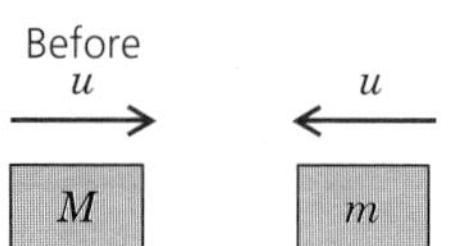

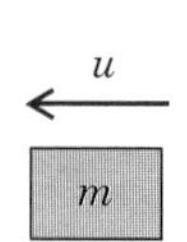

After

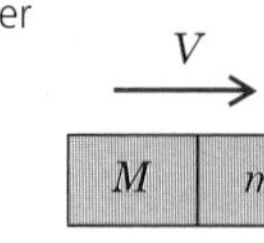

Estimate the acceleration experienced by the passengers in the vehicles if the collision lasts three seconds.

1 2 3 6 5 4

Solve the mathematical problem

Solve the mathematical problem

Using conservation of momentum gives:

$$Mu - mu = V(M + m)$$

so that:

$$V = u\left(\frac{M - m}{M + m}\right)$$

Show that the change in velocity of a passenger in the vehicle of mass M is:

$$\frac{2mu}{m + M}$$

and for a passenger in the vehicle of mass m is:

$$\frac{2Mu}{m + M}$$

If the two vehicles are brought to rest in a time of T seconds then the average accelerations in the experiment will be:

$$\frac{2mu}{T(m + M)} \text{ and } \frac{2Mu}{T(m + M)}$$

Interpretation

Consider a head-on collision between a car and a coach at 70 mph, when the collision lasts three seconds.

Here u = 31 m s^{-1}, M = 10 000 and m = 1500; so the acceleration experienced by the passengers in the car is:

$$\frac{2 \times 10\,000.31}{3(10\,000 + 1500)} = 18.0 \text{ m s}^{-2}$$

The acceleration experienced by the passengers in the coach is:

$$\frac{2 \times 1500}{3(10\,000 + 1500)} = 2.7 \text{ m s}^{-2}$$

There is clearly a major difference between the magnitudes of these accelerations. If the time of the collision was reduced, then both these figures would increase, but the ratios would be similar.

Consider head on collisions between two coaches and a coach and a lorry. Finally give your conclusions.

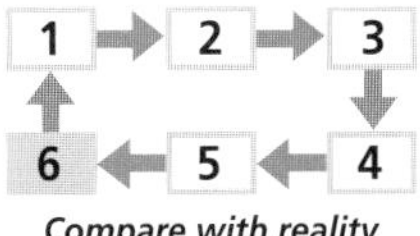

Compare with reality

Compare with reality

The *Highway Code* stopping distances are based on an acceleration of 6.6 m s^{-2}. Comment on the experiences of the passengers in the collisions you have considered.

Suggest any weaknesses of the model considered above.

POSSIBLE MODELLING TASKS

1 A snooker ball is stationary on a table, and is hit by a ball that moves parallel to the side of the table. The balls make contact on a line of symmetry of the table. Investigate which ball reaches the side of the table first.

2 By considering the path of a ball hit by a bat or racquet, try to find the impulse on the ball when it is hit.

3 A Mini and its occupants of total mass 760 kg pulled out of a side road into the path of an Escort of mass 1040 kg including its passengers. The diagram shows the skid marks measured by the police at the scene.
Advise the police who to prosecute bearing in mind that the Mini did not give way and the speed limit was 30 mph.

Summary

After working through this chapter you should:

- be aware that momentum is a vector $m\mathbf{v}$
- know that the change in momentum is equal to the impulse:
  $$\mathbf{I} = m\mathbf{v} - m\mathbf{u}$$
  and that for a constant force:
  $$\mathbf{I} = \mathbf{F}t = m\mathbf{v} - m\mathbf{u}$$
- know that the impulse of a variable force is:
  $$\mathbf{I} = \int F\,\mathrm{d}t$$
- know that momentum is conserved in any collision:
  $$m_A\mathbf{v}_A + m_B\mathbf{v}_B = m_A\mathbf{u}_A + m_B\mathbf{u}_B$$
- know Newton's experimental law of restitution in the forms:
  $$v = -eu$$
  and: $v_A - v_B = -e(u_A - u_B)$

Energy, work and power

- *The ideas of energy can be derived from the basic equations of motion, to provide an alternative approach to solving problems involving energy, work and power.*
- *This chapter concentrates on the energy associated with making objects.*

FORMS OF ENERGY

Exploration 10.1

What is energy?

What do you understand by the word **energy**? Try to identify some situations that involve or are associated with energy. Decide where that energy may have come from and where it may go.

For example:

- a car that starts at rest, gains speed and then stops, using the brakes
- a carriage moving on a ride such as the famous Alton Towers corkscrew
- an athlete competing in a pole vault event.

Kinetic energy and potential energy

There are basically two kinds of energy: **kinetic** and **potential**. Kinetic energy is the energy associated with a moving object. The kinetic energy of an object is determined by the speed or velocity with which it is moving, and its mass. The faster an object moves, the more kinetic energy it has. A heavy moving object will have more kinetic energy than a lighter object moving at the same speed.

Potential energy, considered in its simplest form, is energy that has been stored in some way, but can be released and converted into potential energy. Potential energy can be stored in a compressed spring or in an object held above the ground. If the spring expands or the object falls, the stored potential energy is converted to kinetic energy.

Other 'forms' of energy are often talked about, but can always be related to these first simple – and more accurate – origins. For example, 'heat' energy (the hotter something is, the faster the atoms move about within the object) is kinetic energy. Heat is often released when energy is converted from one form to another. For example, when the brakes are applied in a car, some of the kinetic energy of the car is lost as heat in the brake pads. 'Chemical' energy can most accurately be thought of as potential energy. The atoms within a substance can be said to have energy because of their positions within the electrical fields. Any energy that an object has because of a field effect is called potential energy.

A basic rule of science is that 'energy cannot be created or destroyed', This is called **conservation of energy**. When energy is transferred from one form to another, it may manifest itself in different ways: heat, light, sound for example, but the total amount present stays the same.

EXERCISES

10.1 A

1 Describe examples of situations where:
   a) electricity is converted to kinetic energy or vice versa,
   b) potential energy stored in a spring is converted to kinetic energy.

2 Describe some situations where unwanted heat is converted from one form of energy to another.

3 Describe the energy changes that take place when an athlete takes part in a high jump contest at an athletics event.

4 Electricity is produced at a hydroelectric power station. From what form of energy is electricity obtained?

5 Describe the energy changes that take place during a bungee jump.

EXERCISES

10.1 B

1 Describe examples of situations where:
   a) kinetic energy is converted to potential energy,
   b) chemical potential energy is converted to kinetic energy.

2 Describe the energy changes that take place as a pendulum swings in a grandfather clock. Why does a grandfather clock need to be wound up?

3 Describe the energy changes that take place during a pole vault.

4 A wind farm has several windmills that are used to produce electricity. Describe the energy changes that take place as the electricity is produced. What is the original source of the energy that is produced?

5 Describe the energy changes that take place as a squash ball is hit.

KINETIC ENERGY

All moving objects have some energy associated with their motion, the faster they move the greater the energy. For example much more energy would be associated with the motion of a racing car than an average family car. Similarly, more energy would be associated with the motion of a heavy goods vehicle than a car moving at the same speed. The kinetic energy of an object is the energy that it has due to its motion and depends both on its mass and its speed.

The kinetic energy (KE or E_k) of an object is defined as:

$$E_k = \frac{1}{2}mv^2$$

where m is the mass and v the speed of the object in question. Note this does not depend on the direction of motion, only the speed. The units of energy are joules, J, when $\mathrm{m\,s^{-1}}$ and kg are used for v and m respectively.

Example 10.1

A car of mass 1200 kg is initially moving at 13 m s^{-1} (30 mph). Later, its speed has increased to 15 m s^{-1} (34 mph).

a) *Find the initial kinetic energy of the car.*

b) *Find the gain in kinetic energy of the car when the speed has increased.*

Solution

a) *Here m = 1200 kg and v = 13 m s^{-1}, so:*

$$E_k = \tfrac{1}{2}mv^2 = \tfrac{1}{2} \times 1200 \times 13^2 = 101\,400\,J$$

b) *When the speed has increased:*

$$E_k = \tfrac{1}{2} \times 1200 \times 15^2 = 135\,000\,J$$

Now the gain in kinetic energy can be calculated.

Gain in KE $= 135\,000 - 101\,400 = 13\,600\,J$

Example 10.2

A middle-distance runner of mass 68 kg can run at a steady 5 m s^{-1} or sprint for short distances at 10 m s^{-1}.

a) *Find the KE at steady 5 m s^{-1}.*

b) *Find the KE at sprint speed of 10 m s^{-1}.*

c) *Compare your answers and comment on them.*

Solution

a) *Here m = 68 kg and v = 5 m s^{-1}, so:*

$$E_k = \tfrac{1}{2}mv^2 = \tfrac{1}{2} \times 68 \times 5^2 = 850\,J$$

b) *Now v has increased to 10 m s^{-1} and so:*

$$E_k = \tfrac{1}{2} \times 68 \times 10^2 = 3400\,J$$

c) *Much more energy is needed for a speed of 10 m s^{-1}. If the speed is doubled, the energy increases by a factor of 4.*

EXERCISES

10.2 A

1 Find the kinetic energy of each object described below.

a) A car of mass 900 kg moving at 30 m s^{-1}.

b) An iron of mass 3 kg moving at 0.5 m s^{-1}.

c) A squash ball of mass 24 grams moving at 5 m s^{-1}.

d) A swimmer of mass 64 kg moving at 0.8 m s^{-1}.

2 A cyclist of mass 60 kg on a cycle of mass 10 kg increases speed from 6 m s^{-1} to 9 m s^{-1}. Find the gain in kinetic energy.

3 A train of mass 50 tonnes is moving at a speed of 0.1 m s^{-1} when it is brought to rest by a spring buffer.

a) Find the initial kinetic energy of the train.

b) Find the energy that is stored in the buffer when the train comes to rest.

c) The buffer in fact returns 85% of the energy it gains to the train. Find the speed of the train as it bounces back off the buffer.

4 When a squash ball bounces off a wall it is estimated that it loses 20% of its kinetic energy. A squash ball has a mass of 24 grams and hits a wall at $4\,\text{m}\,\text{s}^{-1}$.

a) Find the kinetic energy of the ball when it hits the wall.

b) What is the speed of the ball when it rebounds?

c) Where could the energy that has been lost have gone?

5 The diagram shows a sprung device that can be used for launching balls into the air. When compressed, the energy stored in the spring is 80 J, which is all transferred to the ball.

a) Find the speed at which a 50 gram ball will be launched.

b) Find the speed at which a 120 gram ball would be launched.

c) A ball is launched at $8\,\text{m}\,\text{s}^{-1}$. What is its mass?

6 The diagram shows a spring-powered launcher that is part of a pin ball machine. The plunger has mass 200 grams and the ball has mass

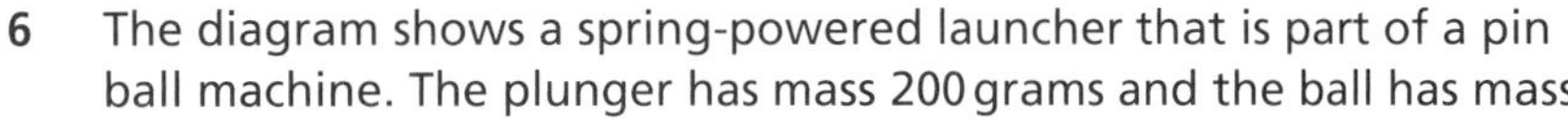

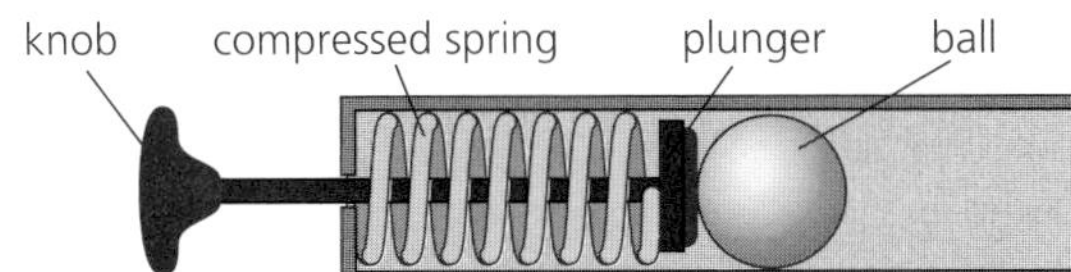

60 grams. When compressed, there is 20 J of energy stored in the spring. As the spring expands, all its energy is transferred to the ball and the plunger. Find their speeds when the spring is fully expanded.

7 A 100 gram ball is dropped and hits the ground travelling at $4\,\text{m}\,\text{s}^{-1}$.

a) What is the kinetic energy of the ball when it hits the ground?

b) How much potential energy did the ball have when it was released?

8 A cyclist of mass 58 kg starts from rest and free-wheels down a gentle slope with constant acceleration of $0.5\,\text{m}\,\text{s}^{-2}$.(**Hint**: Use $v^2 = u^2 + 2as$)

a) Find the speed of the cyclist after travelling 5 m and 10 m.

b) Find the kinetic energy of the cyclist at each position.

c) How much kinetic energy do you think the cyclist will have after 20 m?

9 A ball of mass 150 grams is dropped from a height of 2 m.

a) By assuming that it has a constant acceleration of $9.8\,\text{m}\,\text{s}^{-2}$, find the speed at which it hits the ground.

b) Find the kinetic energy of the ball when it hits the ground. Hence state the amount of potential energy lost as the ball falls to the ground.

c) State clearly any assumptions that you have made to obtain the final answer to **b)**.

10 A ball of mass 80 grams is dropped from a height of 1 m and rebounds to a height of 80 cm.

a) If the acceleration of the ball is a constant $9.8\,\text{m}\,\text{s}^{-2}$, find the speed at which it hits the ground and hence the kinetic energy on impact.

b) By considering the acceleration of the ball and the rebound height, find the speed at which it leaves the ground and the kinetic energy on rebound.
c) What percentage of the ball's energy was lost while bouncing?
d) Comment on your results.

EXERCISES

10.2 B

1 Find the kinetic energy of each object described below.
a) A child of mass 45 kg running at 5 m s^{-1}.
b) A train of mass 50 tonnes travelling at 10 m s^{-1}.
c) A raindrop of mass 0.5 grams falling at 4.8 m s^{-1}.
d) A motorcycle and rider of combined mass 400 kg travelling at 40 m s^{-1}.

2 The speed of an aeroplane of mass 3000 kg increases from 20 m s^{-1} to 40 m s^{-1}. Find the gain in kinetic energy.

3 A squash ball of mass 24 grams is travelling at 8 m s^{-1} when it hits a wall, at right angles to its path.
a) Find the kinetic energy of the ball when it hits the wall.
b) Describe what happens to the kinetic energy of the ball during the bounce.
c) Assume that after the bounce the ball has 80% of its original kinetic energy. Find the speed of the ball after the bounce.

4 A catapult is used to launch a stone of mass 50 g at a speed of 6 m s^{-1}.
a) Find the kinetic energy of the stone when it is launched.
b) Find the energy that was initially stored in the stretched elastic of the catapult.
c) At what speed would a 40 gram stone be launched? Explain what assumptions you have made.

5 An air rifle can be used to fire either pellets of mass 2 grams or darts of mass 4 grams. The rifle transfers 16 J of energy to the pellet or dart.
a) Find the speed at which the pellet is fired.
b) Find the speed at which the dart is fired.
c) A new type of dart is launched at a speed of 80 m s^{-1}.
What is its mass?

6 An air track has a vehicle of mass 1 kg which slides smoothly on the track. It is set into motion by a stretched elastic band.
a) If 5 J of energy are stored in the elastic band, find the speed at which the vehicle will move.
b) Explain why, in practice, the vehicle does not reach this speed.

7 In an advertisement for a car, the car is shown falling to the ground. Assume the mass of the car is 1000 kg and that it hits the ground moving at 5 m s^{-1}.
a) Find the kinetic energy of the car when it hits the ground.
b) How much potential energy did the car have originally? (Assume it was all converted to kinetic energy.)
c) Comment on why your answer to **b)** is unlikely to be true.

8 A car of mass 1200 kg has an acceleration of 1.2 m s^{-2}, and is at rest.
  a) Find the speed of the car when it has travelled 20 m and 40 m.
  b) Find the kinetic energy of the car at each position.
  c) How much kinetic energy would you expect the car to have when it has travelled 80 m? Verify your answer.

9 A parachutist of mass 70 kg falls 30 m before opening his parachute.
  a) If there were no air resistance find the speed that would be reached by the parachutist and the kinetic energy of the parachutist.
  b) How much potential energy has the parachutist lost while falling the 30 m?
  c) In fact the parachutist only reaches a speed of 10 m s^{-1}. How much of the potential energy has not been converted to kinetic energy? Explain what has happened to the lost energy.

10 A ball of mass 100 grams hits the ground travelling at 4 m s^{-1} and rebounds at 3 m s^{-1}. Assume $g = 10$ m s^{-2} in this question.
  a) By assuming that the ball is subject only to the effects of gravity, find the height from which the ball was dropped and the height to which it rebounds.
  b) Find the kinetic energy of the ball immediately before and after the bounce.
  c) Comment on your results.

WORK

The kinetic energy of a moving object has now been used in a range of examples, but how can an object be given kinetic energy? If an object is to gain speed a force must be applied to it. For example a force is required to make a car accelerate and gain kinetic energy. In this section we explore how the gain in kinetic energy is related to the force or forces that are applied.

Spend a few moments thinking about what you understand by the word 'work'. Is there a link between the work done and the kinetic energy of something that has been worked on? For example, consider:

- pushing a car that is initially at rest.
- an object falling under gravity.

By pushing the car, you exert a force on it. You certainly feel as if you are doing work and as the car begins to move it gains energy. The work that you have done gives the car its energy.

Now think about a catapult. As the elastic is stretched, work is done and the resulting energy is stored in the elastic. This is then transferred to the missile that is fired. The elastic in turn does work on the missile.

When there is a change in the energy of a system or object, then work has been done to cause this change. Consider as an example the case where the speed of an object increases from u to v. The change in kinetic energy is:

$$\text{change (gain) in KE} = \tfrac{1}{2}mv^2 - \tfrac{1}{2}mu^2$$

From the constant acceleration equation:

$$v^2 = u^2 + 2as$$

it can be seen that:

$$v^2 - u^2 = 2as$$

Returning to the change in kinetic energy gives:

$$\text{change in KE} = \tfrac{1}{2}m\left(v^2 - u^2\right) = \tfrac{1}{2}m \times 2as = mas$$

The resultant force, F, on the object is equal to ma, by Newton's second law; so the change in KE is:

$$\text{change in KE} = mas = Fs$$

The quantity Fs which is the product of the force and distance is the work done by the force and it is equal to the change in KE. This can be expressed as:

change in KE = work done

or:

$$\tfrac{1}{2}mv^2 - \tfrac{1}{2}mu^2 = Fs$$

The equation $v^2 = u^2 + 2as$ is an equation for straight line motion and so this definition of work done will also be restricted to motion in one dimension and as it is a constant acceleration equation, to problems involving constant forces.

Example 10.3

A child exerts a horizontal force of 80 N on a sledge carrying a friend, the combined mass of friend and sledge being 55 kg.

a) *Calculate the work done by this force as the sledge moves 10 m.*

b) *Find the final speed of the sledge if it was initially moving at 0.75 m s⁻¹.*

c) *What assumption has been made in solving this problem? How does this affect the final speed?*

Solution

a) *The work done is found from:*

work done $= Fs = 80 \times 10 = 800$ *J*

b) *To find the speed, use:*

work done = change in KE

In this case:

$$800 = \tfrac{1}{2} \times 55v^2 - \tfrac{1}{2} \times 55 \times 0.75^2$$

or:

$$800 = 27.5v^2 - 15.5$$

$$v^2 = \frac{800 + 15.5}{27.5}$$

$$v = 5.45 \text{ m s}^{-1}$$

c) We have assumed that there is no resistance to the motion. There would almost certainly be friction and so the final speed would actually be less than that predicted above. Work has to be done against the friction forces that are present.

Example 10.4

A shot putt of mass 5 kg is held 1.2 m above ground level and released from rest.

a) Find the work done by gravity on the mass as it falls.
b) Find the speed of the mass when it hits the ground.

Solution

a) As the shot putt has a mass of 5 kg, the force of gravity acting is 49 N; so the work done is:
work done $= 49 \times 1.2 = 58.8\,J$
This is often referred to as the **gravitational potential energy**.
b) To find the speed, use:
work done $= \frac{1}{2}mv^2 - \frac{1}{2}mu^2$

$$58.8 = \frac{1}{2} \times 5v^2 - 0$$

Solving for v gives:

$$v^2 = \frac{58.5}{2.5}$$

$$v = 4.85\,m\,s^{-1}$$

Example 10.5

A filing cabinet of mass 60 kg is being moved across the floor of an office by a single person pushing and exerting a force of 220 N. There is a friction force of 140 N resisting the motion.

a) Find the work done on the filing cabinet by the person as it is moved 2 m.
b) Find the work done by the friction force.
c) Find the overall amount of work done on the filing cabinet.
d) Find the speed it reaches.

Solution

a) Work done by the person $= 220 \times 2 = 440\,J$
b) Work done to overcome friction $= -140 \times 2 = -280\,J$
Note: This is negative because this force acts in the opposite direction to the motion. This is usually described as the ***work done* against *friction***.
c) The overall amount of work done $= 440 - 280 = 160\,J$
This is the work done by the person minus the work done against friction.
d) Using work done = change in KE gives:

$$160 = \frac{1}{2} \times 60v^2 - \frac{1}{2} \times 60 \times 0^2$$

$$160 = 30v^2$$

$$v = \sqrt{\frac{160}{30}} = 2.3\,m\,s^{-1}$$

Example 10.6

A crane is lifting a heavy load of mass 1080 kg. The lifting cable is vertical and has a constant tension of 10 700 N.

a) *Find the work done by the tension as the load is raised 5 m.*
b) *Find the work done by gravity.*
c) *Find the overall work done on the load.*
d) *Find the speed reached by the load.*

Solution

a) *The work done by the tension* $= 10\,700 \times 5 = 53\,500\,J$

b) *The force of gravity on the load is 10 584 N, and so the work done* $= -10\,584 \times 5 = -52\,920\ J$

Note: *The negative value means that the force acts in the opposite direction to the motion. We say that the work done against gravity is 52 920 J.*

c) *The overall work done* $= 53\ 500 - 52\ 920 = 580\,J$

d) *Using work done = change in KE gives:*

$$580 = \tfrac{1}{2} \times 1080v^2 - \tfrac{1}{2} \times 1080 \times 0^2$$

$$v = \sqrt{\frac{580}{540}} = 1.04\ m\ s^{-1}$$

EXERCISES

10.3 A

1 Two men are pushing a car of mass 1400 kg. They each exert forces of 120 N.
a) Find the work done on the car by the men as it moves 10 m.
b) Find the speed that the car reaches.

2 A cyclist of mass 70 kg on a cycle of mass 12 kg exerts a constant forward force of 80 N.
a) Find the work done as the cyclist moves forward 50 m.
b) Find the speed of the cyclist after moving this distance.
c) What factors have you ignored?

3 A ball of mass 70 grams is dropped from a height of 1.4 m into soft mud.
a) Find the work done by gravity on the ball as it falls the 1.4 m.
b) Find the speed of the ball when it hits the mud.
c) The ball penetrates 8 cm into the mud before stopping. Find the average force on the ball as it moves into the mud.

4 Calculate the work that must be done to cause changes in motion as described below.
a) A car of mass 1200 kg increasing in speed from 10 m s^{-1} to 15 m s^{-1}.
b) A ship of mass 7000 tonnes as it increases in speed from 1 m s^{-1} to 4 m s^{-1}.
c) A ball of mass 200 grams which changes speed from 2 m s^{-1} to 5 m s^{-1} when hit by a bat.

5 In each case below, calculate the work done by the force and the speed reached.
a) A force of 10 000 N acting on a train of mass 25 000 kg as it travels 50 m, starting at rest.

b) A fly of mass 3 grams acted on by a force of 0.0001 N as it moves 1.1 m from an initial speed of 0.5 m s^{-1}.

c) A force of 50 N acting on a sledge of mass 8 kg as a child pulls it from rest for 2 m.

6 A lift has mass 700 kg including its occupants. As the lift begins to move, the tension in the cable is 7010 N.

a) Find the work done by the tension as the lift rises 1.8 m.

b) Find the work done against gravity as the lift rises 1.8 m.

c) Find the gain in kinetic energy of the lift.

d) What speed does the lift reach?

7 A basketball player throws a ball directly up into the air. While the ball, which has a mass of 2 kg is in contact with his hands, an upward force of 50 N is exerted on the ball.

a) If the hands remain in contact with the ball as it moves 50 cm, find the work done by this force.

b) Find the work done against gravity as the ball rises the 50 cm.

c) Find the gain in kinetic energy of the ball.

d) At what speed does the ball leave the player's hands?

8 A cyclist on a horizontal road experiences a constant forward force of 80 N while peddling 50 m. A resistance force of 60 N also acts on the cyclist.

a) Find the work done by the cyclist.

b) Find the work done against the resistance force.

9 A bungee jumper of mass 72 kg jumps off a bridge and falls 10 m before the elastic rope attached to her legs goes taut.

a) Find the work done by gravity as the jumper falls the first 10 m and the speed reached at this point.

b) Before coming to rest the jumper travels a further 8 m. Find the total energy stored in the elastic at this point.

10 Forensic experts working with the police are investigating the scene of a shooting. They find a bullet of mass 30 grams that has penetrated 5 cm into a wooden doorpost. They test the bullet by firing it through a wooden barrier 2 cm thick. If it hits the barrier at 100 m s^{-1} and leaves at 50 m s^{-1}, estimate the speed at which the bullet hit the doorpost. Criticise your answer.

EXERCISES

10.3 B

1 A child exerts a horizontal force of 30 N on a sledge of mass 8 kg as she pushes it on level ground.

a) Find the work done by the child as she pushes the sledge 5 m.

b) Find the speed that the sledge reaches if there are no resistances.

c) Do you think it is reasonable to assume there is no resistance? Why?

2 A forward force of 8000 N acts on a lorry of mass 4000 kg as it moves forward from rest on a level road.

a) Find the work done as the lorry moves forward 60 m.

b) Find the speed of the lorry when it has travelled the 60 m, if there are assumed to be no resistance forces.

c) If the speed of the lorry is only 8 m s^{-1}, find the work done against the resistance forces.

3 A gymnast of mass 62 kg jumps off a vaulting horse of height 1.2 m onto a rubber mat.

a) Find the work done by gravity as the gymnast jumps.

b) What is the speed of the gymnast when she hits the mat?

c) As the gymnast comes to rest the mat is compressed by 1 cm. Find the average force exerted by the mat as the gymnast is brought to rest. (Note that the gymnast falls a total distance of 1.21 m.)

4 Calculate the work that must be done to cause changes in motion as described below.

a) A cyclist of mass 73 kg whose speed increases from 5 m s^{-1} to 8 m s^{-1}.

b) A dinghy of mass 300 kg as its speed increases from 1.5 m s^{-1} to 2.3 m s^{-1}.

c) A child of mass 54 kg on a roundabout as the speed increases from rest to 3 m s^{-1}.

5 In each case below, calculate the work done by the force and the speed reached, assuming that the force always acts in the direction of motion.

a) A force of 2400 N acting on a car of mass 1400 kg as it travels 40 m from an initial speed of 5 m s^{-1}.

b) A train of mass 20 tonnes acted on by a force of 5000 N as it travels 20 m, from an initial speed of 20 m s^{-1}.

c) A force of 30 N acting on a ball of mass 50 grams while it moves 0.2 m and is in contact with a bat. (Assume the ball is initially at rest.)

6 A crane is lifting a load of mass 750 kg. As the load is lifted the tension on the cable is 7400 N.

a) Find the work done by the tension as the load is lifted 10 m.

b) Find the work done against gravity as the load is lifted 10 m.

c) What is the gain in kinetic energy of the load?

d) What speed does the load reach?

7 A weightlifter is taking part in a competition. A set of weights of mass 250 kg is to be lifted.

a) Find the work done against gravity as the weights are lifted 1.2 m.

b) When the weights have been lifted 1.2 m, they are moving at 0.8 m s^{-1}. Find the total work done by the weightlifter.

c) The weights are eventually lifted to a height of 2.2 m. Find the work done to achieve this.

8 A car of mass 1080 kg is travelling on a straight horizontal road at a steady speed of 18 m s^{-1}. The car then begins to accelerate, exerting a forward force of 2000 N. The car is also subject to a resistance forces that are modelled as a constant of magnitude 800 N.

a) Find the work done by the car as it travels a further 50 m.

b) Find the work done against the resistance forces.

c) What is the speed of the car at the end of the 50 m?

d) Criticise the way in which the resistance forces have been modelled. Suggest how the model could be improved and how this would affect your answer to **c)**.

9 A trampolinist of mass 62 kg is bouncing on a trampoline. She rises 1.2 m above the unstretched trampoline's surface on each bounce.

a) Find the work done by gravity and her speed when she makes contact with the bed of the trampoline.

b) As she is brought to rest, before she bounces again, the bed of the trampoline is pushed down 0.5 m. Find the potential energy stored in the trampoline at this point.

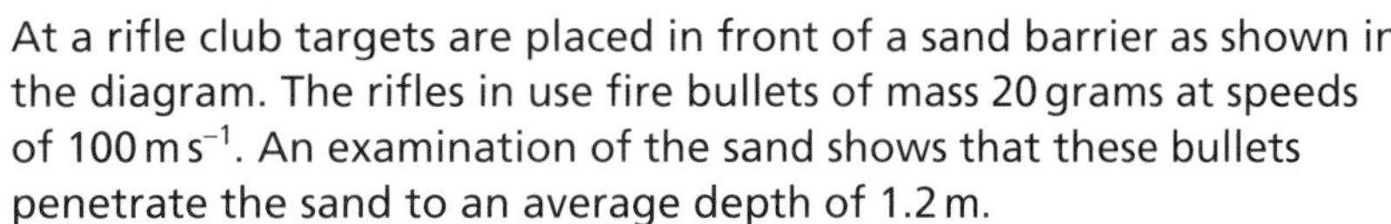

10 At a rifle club targets are placed in front of a sand barrier as shown in the diagram. The rifles in use fire bullets of mass 20 grams at speeds of $100\,\mathrm{m\,s^{-1}}$. An examination of the sand shows that these bullets penetrate the sand to an average depth of 1.2 m.

a) Find a simple model for the force exerted by the sand on the bullets.

b) What is the greatest speed at which a bullet can be safely fired at this range?

c) Comment on the validity of your solution.

WORK IN TWO DIMENSIONS

In the previous section we only looked at forces that act in or directly against the movement of the object. This covers only a small number of the situations that arise in real life. When an athlete runs up a hill gravity acts, but at an angle to the direction of motion. Often, forces act at angles to the direction of motion of an object, rather than in the same direction. There are also forces that do no work. When you stand still a normal reaction acts on you but does not change your kinetic energy. So some forces do not do any work. This section explores forces that act at angles to the motion and forces that do no work.

Exploration 10.2

Forces which do work – and some which don't

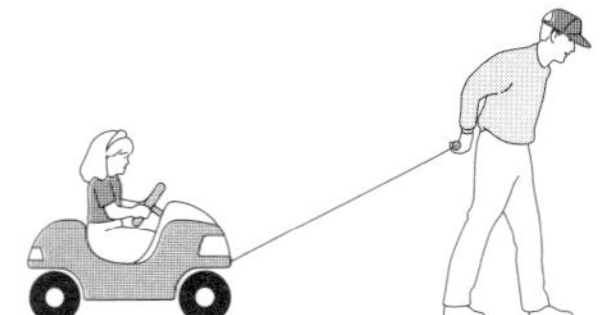

This child and her go-cart are being pulled along by a friend. Try to draw a diagram to show all of the forces acting on the go-cart. Which ones actually do the work? Which, if any, do *no* work at all? How could the forces be applied so that they did more work?

Forces doing work

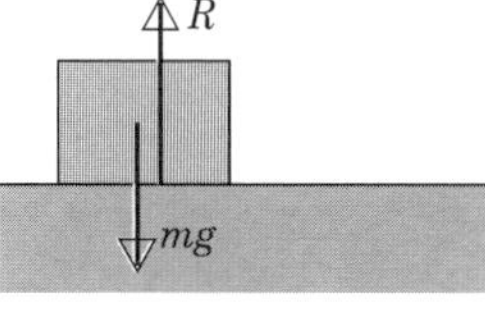

It is easiest to start by thinking of a simple object at rest. Its kinetic energy is not changing and so the forces are not doing any work.

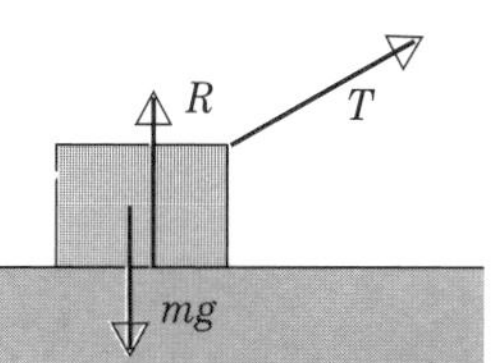

If a force, T, is applied as shown, this will do work, causing the object to gain kinetic energy; but if the surface is horizontal, then forces R and mg do no work. If T is applied horizontally rather than at an angle, then it will do even more work.

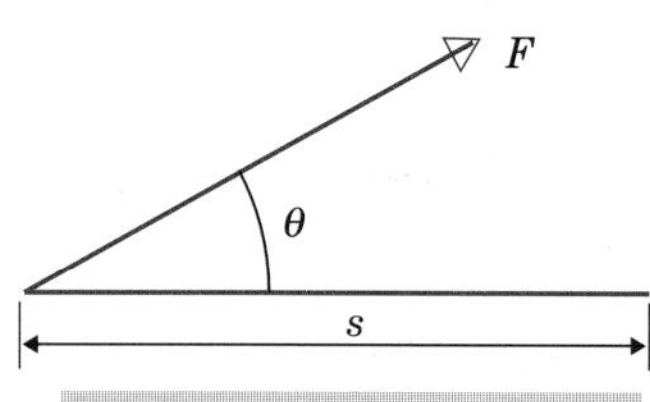
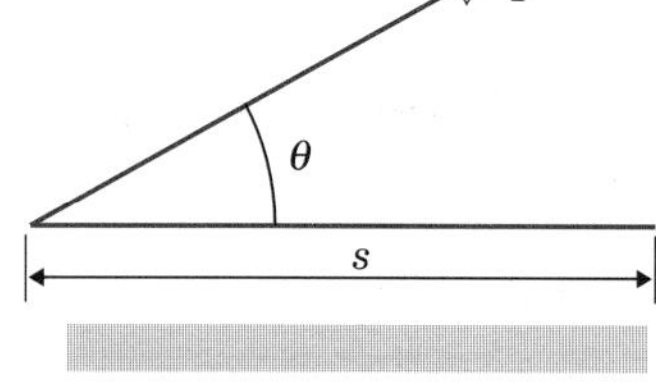

Forces that act at right angles to the direction of motion do no work because they do not cause the kinetic energy of the object to change. When a force acts at an angle to the direction of motion, then only the component in the direction of motion contributes to the work done. We need to be able to calculate this component. If a force of magnitude F acts at an angle θ to the direction of motion, then:

work done = $Fs\cos\theta$

This is the component of the force **in the direction of motion**, $R\cos\theta$ multiplied by s, the distance moved.

Example 10.7

At a swimming pool there is a slide that can be modelled as an incline of length 5 m at 40° to the horizontal.

a) *Find the work done by gravity as a swimmer of mass 45 kg uses the slide.*

b) *The swimmer was given a push and an initial speed of 1 m s^{-1} and enters the water moving at 5 m s^{-1}. Find her gain in kinetic energy.*

c) *Calculate the work done against friction.*

Solution

a) *The diagram shows the force of gravity which acts at 50° to the slope. To calculate the work done use $Fs\cos\theta$, with $F = 441$, $s = 5$ and $\theta = 50°$, so:*

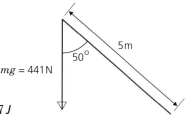

work done = $441 \times 5 \times \cos 50° = 1417\,J$

b) *The gain in kinetic energy of the swimmer is:*

$E_k = \frac{1}{2}mv^2 - \frac{1}{2}mu^2 = \frac{1}{2}\times 45\times 5^2 - \frac{1}{2}\times 45 \times 1^2 = 540\,J$

c) *The difference between the work done by gravity and the work done against friction will be equal to the change in kinetic energy.*

work done against friction $= 1417 - 540 = 877\,J$

Example 10.8

A button lift is used at a ski-slope. The skier is attached to a cable that makes an angle of 40° with the slope. The skier has mass 72 kg, and travels 50 m up the slope.

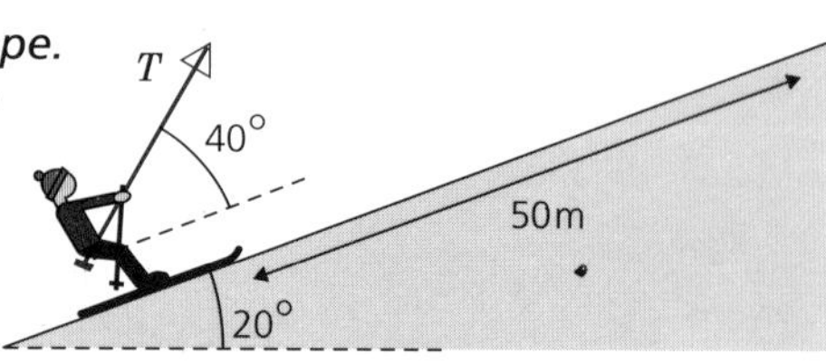

a) *Find the work done against friction (if this is assumed to be a constant force of 70 N) and the work done against gravity.*

b) *If the skier has a speed of 2 m s^{-1} at the top of the slope, find the work done by the tension force in the rope.*

c) *Find the magnitude of the tension force if it is assumed to be constant.*

d) *At the top of the slope the skier lets go of the lift and moves on a horizontal surface. How far does the skier move before stopping?*

Solution

a) *Work done against friction* $= 70 \times 50 = 3500\ J$
Work done against gravity $= 72 \times 9.8 \times 50 \times \cos 70° = 12\,066\,J$

b) *The kinetic energy of the skier at the top of the slope is:*

$$E_k = \tfrac{1}{2}mv^2 = \tfrac{1}{2} \times 72 \times 2^2 = 144\,J$$

The work done by the tension, less the work done against gravity and friction, gives the final kinetic energy of the skier.
Work done by tension $- (3500 + 12\,066) = 144$
Work done by tension $= 144 + 3500 + 12\,066 = 15\,710\,J$

c) *The work done by the force T is given by:*

$$T \times 50 \cos 40° = 15\,710$$

$$T = \frac{36\,796}{50 \cos 40°} = 410\ N$$

EXERCISES

10.4 A

1 In each case find the work done by the force illustrated as the object moves the distance given in the diagram.

a)

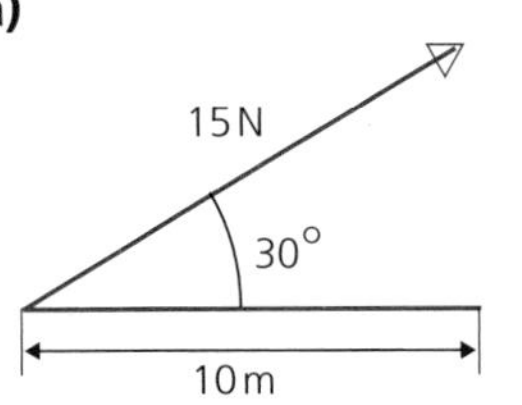

b) 1.1 m, 70°, 60 N

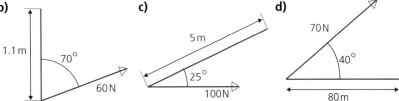

c) 5 m, 25°, 100 N

d) 70 N, 40°, 80 m

2 As a car of mass 1000 kg drives up a slope at 5° to the horizontal, it is acted on by a forward force of 9000 N. At the bottom of the slope it is travelling at $3\,\text{m s}^{-1}$ and it travels 300 m to the top of the slope. Assume there are no resistance forces.

a) Find the work done against gravity.

b) Find the work done by the 9000 N force and hence the change in the kinetic energy of the car.

c) Find the speed of the car at the top of the slope.

d) If there were in fact a resistance force of 700 N, what would be the speed of the car at the top of the slope?

3 A sledge of mass 5 kg slides on its own down a slope of length 10 m at 10° to the horizontal. It attains a speed of $3\,\text{m s}^{-1}$.

a) Find the work done by gravity.

b) Find the final kinetic energy of the sledge.

c) Find the work done against friction.

4 A 'death slide' is set up which has a length of 30 m and a rope that is assumed to be straight and at 20° to the horizontal. A resistive force that is assumed to be a constant 80 N acts on the pulley. A student of mass 69 kg tries the 'death slide'.

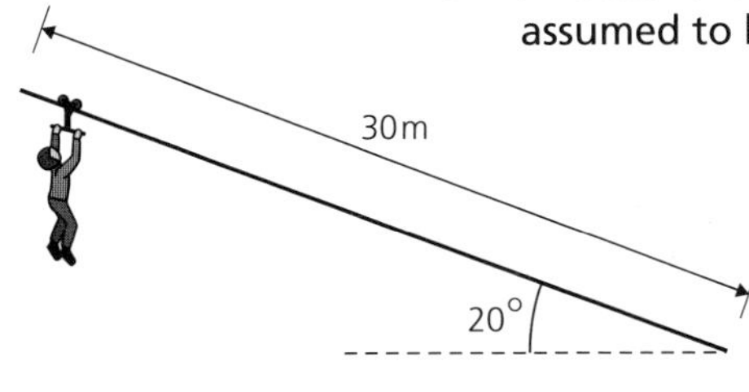

a) Find the work done by gravity and the work done against the resistance force.

b) Find the speed of the student at the bottom of the slide.

5 A tow truck is towing a broken down car as shown in the diagram. As the truck pulls out of a junction, it travels 80 m increasing speed from $5\,\text{m}\,\text{s}^{-1}$ to $13\,\text{m}\,\text{s}^{-1}$. Assume that the work done against resistance forces is 30% of the work done by the tow rope. The car has mass 950 kg.

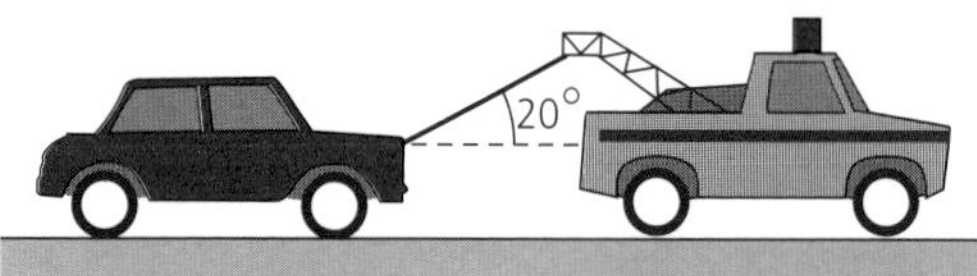

a) Find the change in kinetic energy of the car.
b) Find the work done by the tow rope and the work done against the resistance forces.
c) Find the tension in the tow rope and the magnitude of the resistance forces.
d) What assumption have you made about the road?

6 A cyclist of mass 64 kg free-wheels down a hill on a bicycle of mass 8 kg. The hill can be modelled as a slope at 6° to the horizontal.
a) Find the work done by gravity as the cyclist and cycle travel 50 m.
b) Find the speed reached by the cyclist.
c) In reality, the cyclist's speed is only $5\,\text{m}\,\text{s}^{-1}$. Find the work done against resistance forces.
d) After the 50 m slope, the cyclist free-wheels on a horizontal surface. Estimate how far the cyclist travels before stopping. What assumptions have you made?

7 A tug is attached by a tow rope to a ship of mass 300 tonnes. The tow rope makes an angle of 30° with the horizontal. The speed of the ship increases from $2\,\text{m}\,\text{s}^{-1}$ to $3.5\,\text{m}\,\text{s}^{-1}$ as the tug pulls the ship 500 m.

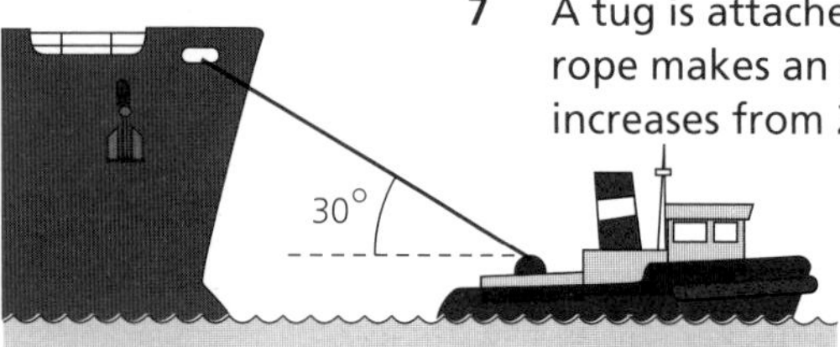

a) Find the gain in kinetic energy of the ship.
b) Express the work done by the tug in terms of T, the tension in the rope.
c) Find the tension in the tow rope.

8 A boat of mass 300 kg is winched up a slipway on a trolley. The slipway is at an angle of 20° to the horizontal and the winching rope is at 10° to the slipway. The boat is initially at rest and after moving 2 m, has reached a speed of $0.5\,\text{m}\,\text{s}^{-1}$. Assume that there is a constant resistance force of 50 N on the boat.

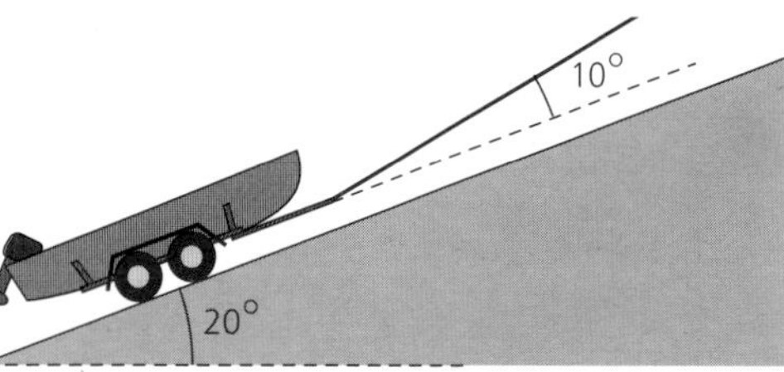

a) Find the gain in kinetic energy for the boat.
b) Find the work done against friction and the work done against gravity.
c) Find the work done by the winch.
d) Find the tension in the winch rope.

9 Three husky dogs are harnessed to a sledge of mass 500 kg as shown in the diagram. Assume that each dog exerts a force of 150 N.

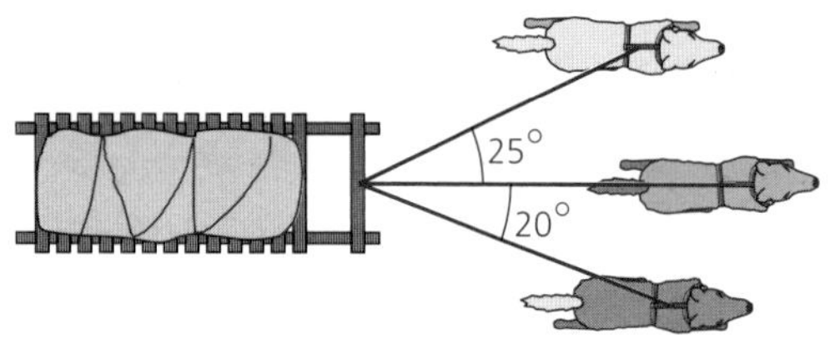

a) Find the work done on the sledge as it moves 100 m.
b) Find the speed of the sledge after 100 m.
c) If the speed is in fact 10% less than your answer to **b)**, find the work done against friction.

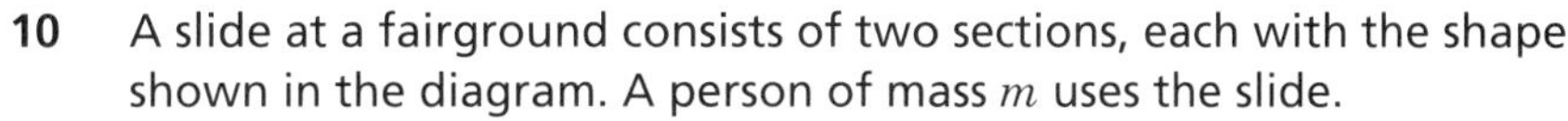

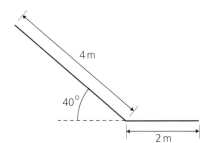

10 A slide at a fairground consists of two sections, each with the shape shown in the diagram. A person of mass m uses the slide.

a) Explain why the magnitude of the friction force is greater on the horizontal sections. Assume that the coefficient of friction between the slide and the user is 0.4.
b) Find the work done by gravity and the work done against friction on each section.
c) Find the total gain in kinetic energy of a user and their final speed.
d) What would be the effect of removing the horizontal sections of the slide?

EXERCISES

10.4 B

1 In each case find the work done as the force shown in the diagram acts and the object moves the distance illustrated.

a)
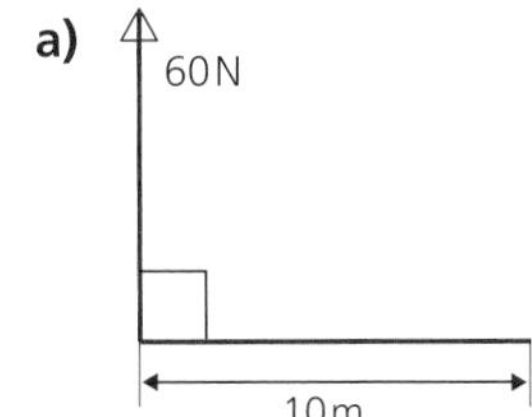

b)
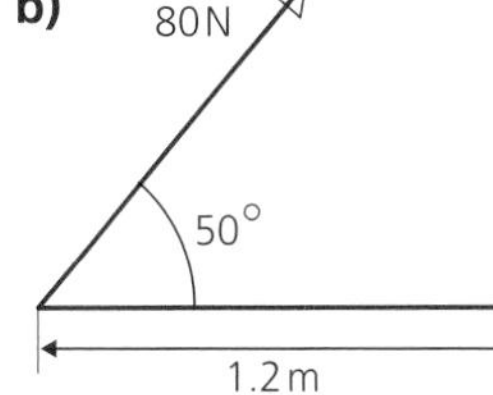

c)
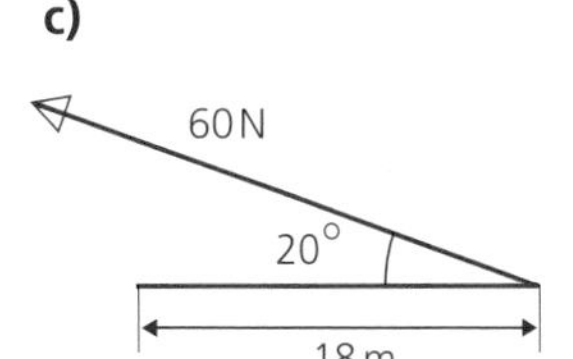

d)
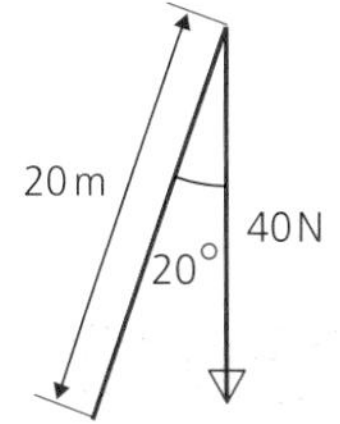

2 Alex, Bobbie and Chris are climbing up a slope and stop for a rest. Alex's mass is 74 kg, Bobbie's mass is 69 kg and Chris's mass is 62 kg. When they stop for a rest they have all done the same amount of work against gravity.
a) Find the work done against gravity by Chris.
b) Find the distances Alex and Bobbie have moved.

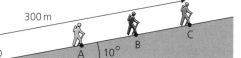

3 A slide at a leisure pool is 50 m long and is at a constant 3° to the horizontal. A child of mass 52 kg travels down the slide, reaching a speed of 2.4 m s^{-1} at the bottom of the slide.
a) Find the final kinetic energy of the child.
b) Find the work done by gravity.
c) Find the work done against the resistance forces.

4 As a car of mass 1110 kg travels 100 m down a hill its speed increases from 20 m s^{-1} to 22 m s^{-1}. The hill can be modelled as an incline at a constant angle of 4° to the horizontal.
a) Find the gain in kinetic energy of the car.
b) Find the work done by gravity.
c) What is the magnitude of the average resistance forces acting on the car?
d) What will be the speed of the car when it has travelled a total of 200 m down the hill? Describe the assumptions that you make to solve this problem.

5 A gardener pulls a roller of mass 80 kg across a lawn. A resistance force of 60 N acts as the roller is pulled. When the roller has been pulled 6 m it attains a speed of 0.8 m s^{-1}.

a) Find the kinetic energy of the roller when it is moving at 0.8 m s^{-1}.

b) Find the work done against the resistance forces acting.

c) What is the magnitude of the force exerted by the gardener on the roller?

d) What aspect of the motion of the roller has been ignored? Why do you think this may be significant in this case?

6 A loaded supermarket trolley has a mass of 75 kg and is at the top of a slope in a multi-storey car park. The trolley begins to roll down the slope, reaching a speed of $5\,\text{m}\,\text{s}^{-1}$ at the bottom. After it reaches the bottom and rolls on the level it slows down.

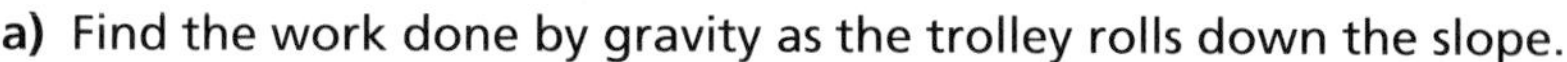

a) Find the work done by gravity as the trolley rolls down the slope.

b) Find the work done against any resistance forces.

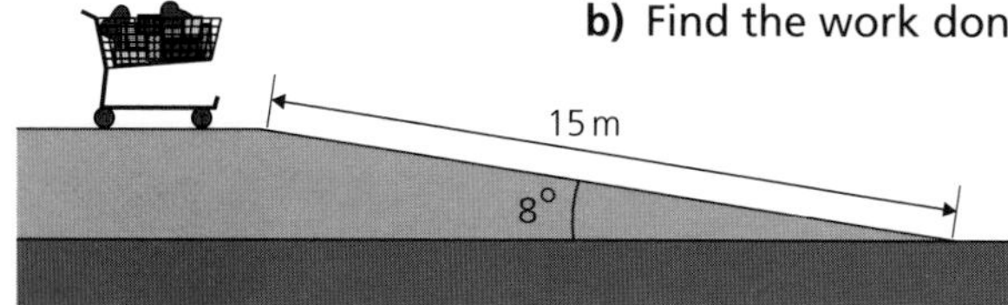

c) Assume that the resistance forces are constant and determine their magnitude.

d) How far does the trolley travel before it comes to rest on the horizontal car parking area?

7 A car of mass 1050 kg is travelling up a slope at 3° to the horizontal. Over a 100 m stretch its speed drops from $25\,\text{m}\,\text{s}^{-1}$ to $24\,\text{m}\,\text{s}^{-1}$.

a) Find the work done by the car as it travels the 100 m, assuming that there are no resistance forces.

b) How far will the car travel before it comes to rest?

c) How would your solutions to **a)** and **b)** change if there were a constant resistance force of 400 N?

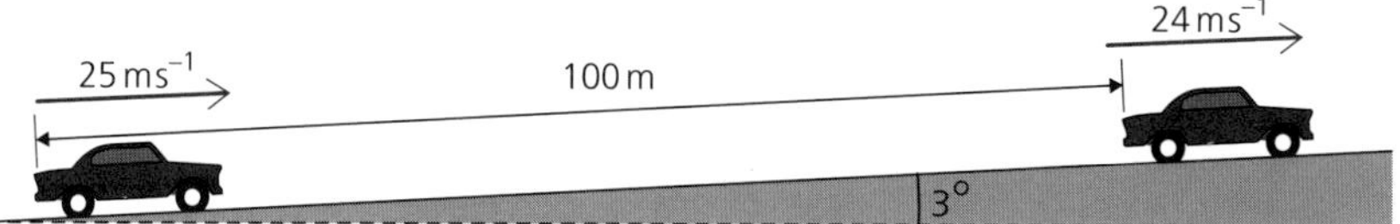

8 The company that manufactures children's toys claim that they have produced a new friction-free material that can be used for the manufacture of children's slides. The material is produced in 5 m lengths.

a) Traditionally slides are positioned at about 45° to the horizontal. Find the speed that a child would reach if a 5 m slide were replaced with the new material.

b) Find the angle, to the horizontal, at which a slide made of the new material should be positioned if a user should reach a speed of $3\,\text{m}\,\text{s}^{-1}$ at the bottom of the slide.

9 A woman is cutting the grass with a lawnmower that hovers. Her lawn is on a slope, as shown in the diagram. The lawnmower begins at rest at the bottom of the slope, and when 5 m up the slope is travelling at $2\,\text{m}\,\text{s}^{-1}$. The mass of the lawnmower is 18 kg.

a) Find the work done by the woman if there is a constant resistance force of 30 N parallel to the slope.

b) Find the magnitude of the force exerted on the lawnmower by its handle.

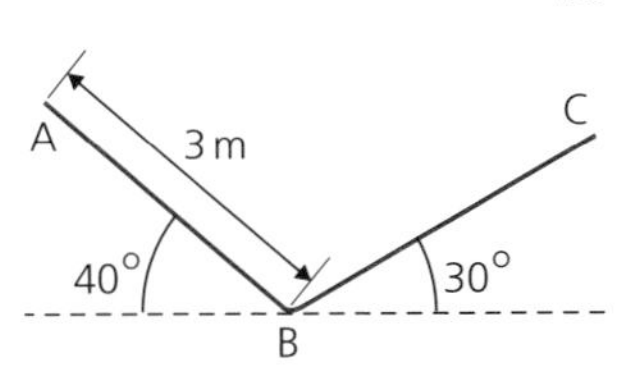

10 At a skateboard park there are two slopes that are to be modelled as shown in the diagram. The skateboarder travels down from A to B reaching a speed of $5\,\mathrm{m\,s^{-1}}$. Assume that skateboarder negotiates the turn at B without losing any energy. Formulate a simple model for the resistance force on the skateboarder and find how far she travels up the slope BC.

GRAVITATIONAL POTENTIAL ENERGY

When any object is positioned at a point above, and then released, it will gain kinetic energy as it falls. The amount of kinetic energy that it gains depends on how far it has to fall. When something is in a position where it can gain kinetic energy, then it is described as having **potential energy** (PE or E_P).

Exploration 10.3

Slopes and speeds

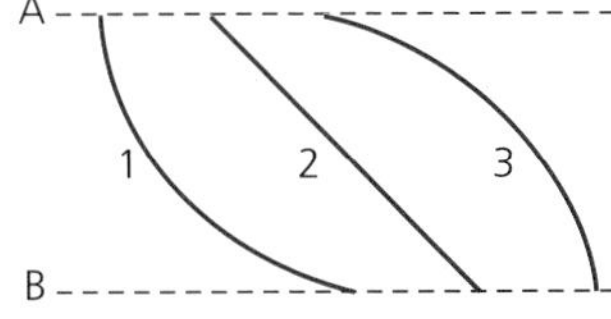

The diagram shows three curves: 1, 2 and 3. A small object is released at level A and allowed to slide to level B. Assume that no work is done against any resistance forces. How will the time of descent and the maximum speed reached compare for each slope? What about an object that is allowed to fall from level A straight to level B?

Conservative systems

Taking two of the above cases as examples, there is little doubt that the object will hit the ground most quickly when dropped rather than being constrained to slide down a curve.

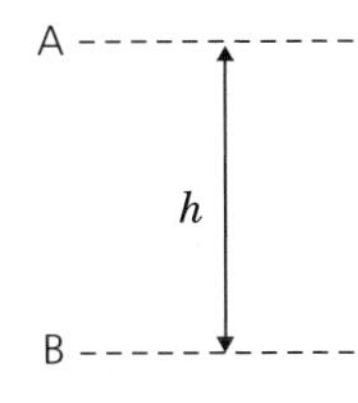

Now consider the ball that falls straight down, through a distance h. Then the work done by gravity is simply mgh, and this will be equal to the final kinetic energy, so:

$$mgh = \tfrac{1}{2}mv^2$$

$$gh = \tfrac{1}{2}v^2$$

$$v = \sqrt{2gh}$$

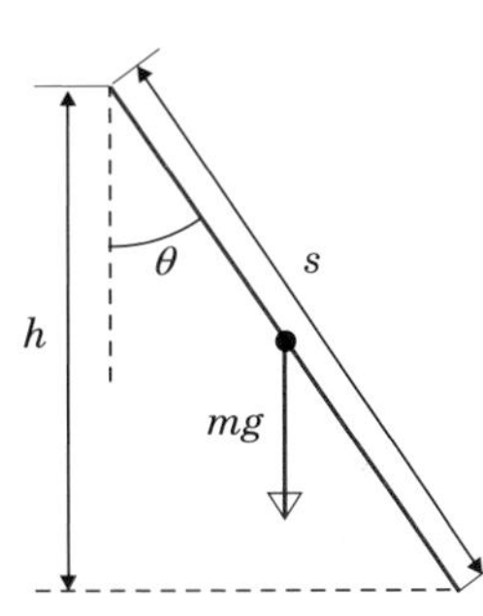

Now consider the straight slope, numbered 2. We can use trigonometry to calculate the length, s, as:

$$s = \frac{h}{\cos\theta}$$

So the work done by gravity will be:

$$mgs\cos\theta = mg \times \frac{h}{\cos\theta} \times \cos\theta = mgh$$

The object that slides down the slope will have the same final kinetic energy, and therefore the same final speed, as the object that fell straight down.

When an object has been raised through a certain height we say that it has **gravitational potential energy**. The amount of energy is simply mgh where h is the distance above some reference point. When an object loses height, this potential energy is converted to kinetic energy.

If, as an object moves, all the potential energy is converted to kinetic energy, then the situation is called a **conservative system**. If some work has to be done against resistance forces, so that not all the potential energy is converted to kinetic energy, this is a **non-conservative system**. In a conservative system, the path taken by an object does not affect the final speed attained.

Example 10.9

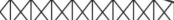

A carriage on a roller coaster ride has mass 150 kg and is at a height of 10 m above the lowest point of the ride.

a) *Find the potential energy of the carriage.*

b) *If it is moving at 1.2 m s⁻¹, find the speed it reaches at the lowest point.*

Solution

a) *The potential energy is given by mgh, so in this case:*

$$PE = 150 \times 9.8 \times 10 = 14700 \text{ J}$$

b) *The final kinetic energy will be the combined total of the initial kinetic energy and the initial potential energy.*

Final KE $= 108 + 14700 = 14808$ *J*

So now the final speed, v, can be found.

$$\tfrac{1}{2}mv^2 = 14\,808$$

$$\tfrac{1}{2} \times 150v^2 = 14\,808$$

$$v^2 = \frac{14\,808}{75}$$

$$v = 14.05 \text{ m s}^{-1}$$

Example 10.10

A bungee jumper, of mass 80 kg falls a distance of 24 m before reaching the lowest point during the jump. How much energy is stored in the stretched 'bungee' at this instant?

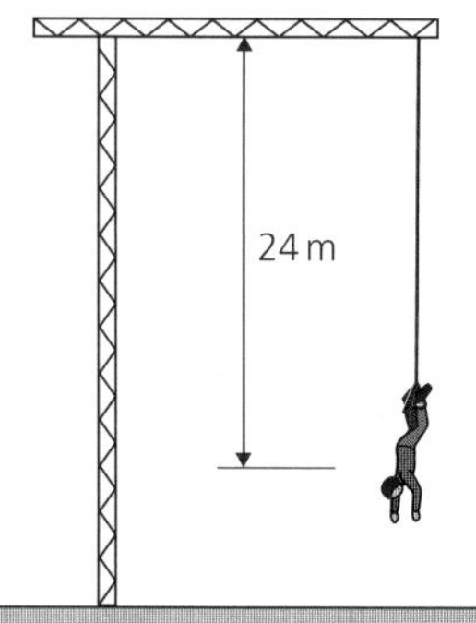

Solution

Assuming that there has been no loss of energy all the initial potential energy of the jumper will now be stored in the stretched elastic. Taking the lowest point reached as a reference point gives the initial potential energy:

$$mgh = 80 \times 9.8 \times 24 = 18\,816 \text{ J}$$

So the energy stored in the bungee is 18 816 J.

EXERCISES

10.5 A

1 Find the gravitational potential energy of each mass shown below, relative to the reference level O.

2 Find the speed at which each mass in question **1** would pass the reference level, O, if it is assumed that there is no loss of energy.

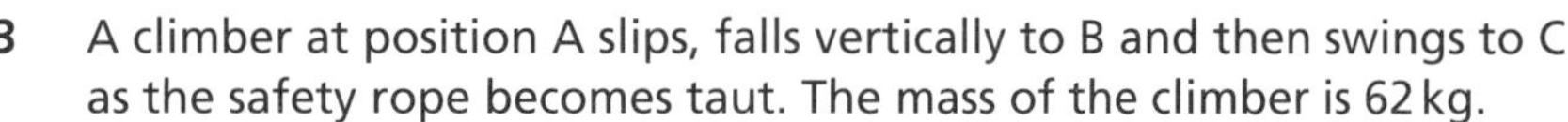

3 A climber at position A slips, falls vertically to B and then swings to C as the safety rope becomes taut. The mass of the climber is 62 kg.

a) Find the potential energy of the climber at A with reference to the point C, the lowest point reached.

b) Find the speed of the climber at point B.

c) Find the speed of the climber at point C.

d) What assumptions have you made?

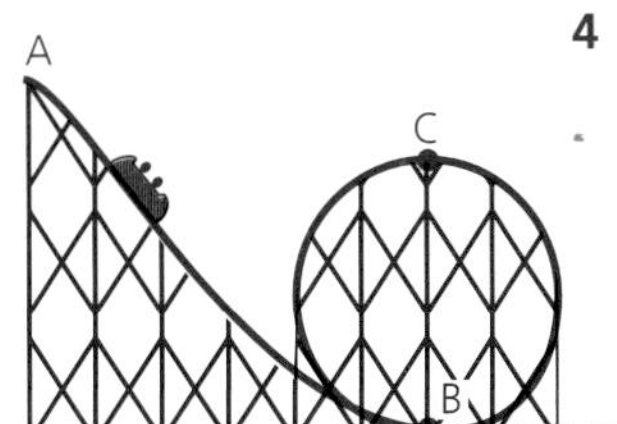

4 The diagram shows a roller coaster ride with a double loop-the-loop. The diameter of the first loop is 5 m and of the second loop is 4 m. The speed of the carriage which has mass 200 kg is $6\,\text{m}\,\text{s}^{-1}$ at C.

a) Find the speed of the carriage at point D.

b) Find the potential energy of the carriage at A, if it can be assumed to be at rest there.

c) Find the height of A above B.

5 A catapult can propel a stone of mass 50 grams vertically upwards to a height of 20 m.

a) Find the energy stored in the stretched elastic of the catapult.

b) Find the speed of the stone when it leaves the catapult.

c) What height would be reached when a 60 gram stone is launched?

6 A parachutist of mass 71 kg jumps from a height of 1000 m and hits the ground moving at $5\,\text{m}\,\text{s}^{-1}$.

a) Find the potential energy lost.

b) Find the work done against resistance forces during the jump.

7 A golf ball of mass 40 grams is struck so that it has an initial speed of $20\,\text{m}\,\text{s}^{-1}$. If the speed of the ball never drops below $11\,\text{m}\,\text{s}^{-1}$, find the maximum amount of potential energy that the ball has during the flight. What is the maximum height reached by the ball?

8 A rocket used at a fireworks display initially has a mass of 300 grams. At the point that all its fuel is used up it has a mass of 200 grams and is moving at $5\,\text{m}\,\text{s}^{-1}$. The rocket reaches a maximum height of 50 m.

a) Find the potential energy of the rocket at its highest point.

b) Find the height of the rocket when its fuel is used up.

c) Find the energy given to the rocket by its fuel.

9 A ball of mass 200 grams moves so that the position vector is given by:

$$\mathbf{r} = 4t\mathbf{i} + \left(8t - 5t^2\right)\mathbf{j}$$

where $\mathbf{i}$ and $\mathbf{j}$ are horizontal and vertical unit vectors respectively.

a) Find an expression for the potential energy of the ball.

b) Find an expression for the kinetic energy of the ball.

c) Is the total energy of the ball constant? If it is, prove the result or else give a counter example.

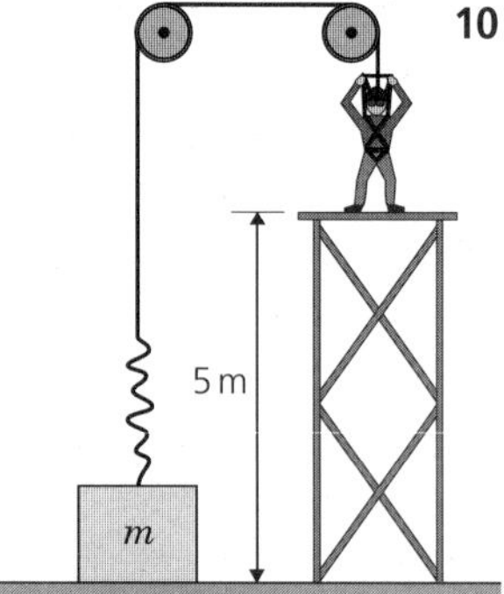

10 A simulator used at a parachuting centre consists of a harness attached to a mass, m, by a rope that passes over two pulleys. The person of mass M jumps from a platform at a height of 5 m above ground level.

a) Find the initial potential energy of the person.

b) After the person has fallen a distance x, the rope becomes taut and the mass, m, begins to rise. Express the speed of the person at this point in terms of x.

c) If the person just comes to rest at ground level, find x, in terms of m and M.

EXERCISES

10.5 B

1 A ball of mass 2 kg is dropped from rest at a height of 10 m.

a) What is the initial gravitational potential energy of the ball?

b) Calculate the speed at which the ball hits the ground.

2 A ball of mass 3 kg initially moves downwards with a speed of 8 m s^{-1} at a height of 12 m.

a) Find the total initial energy of the ball.

b) Calculate the speed at which the ball hits the ground.

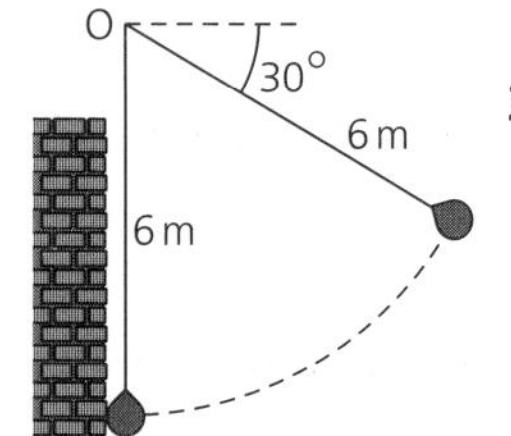

3 A large metal ball of mass 100 kg, attached to the end of a light chain of length 6 m, swings in the vertical plane in order to strike and hence knock over a brick wall. It is released from rest with the chain 30° from the horizontal.

a) Find how much gravitational potential energy is lost as the metal ball swings to its lowest position.

b) Calculate the speed of the ball just before impact with the wall.

4 The diagram shows a roller coaster ride with a single loop the loop. The diameter of the loop, CB is 5 m and the arc AB is a quarter of a circle of radius 10 m. The carriage which has mass 250 kg is at rest at A.

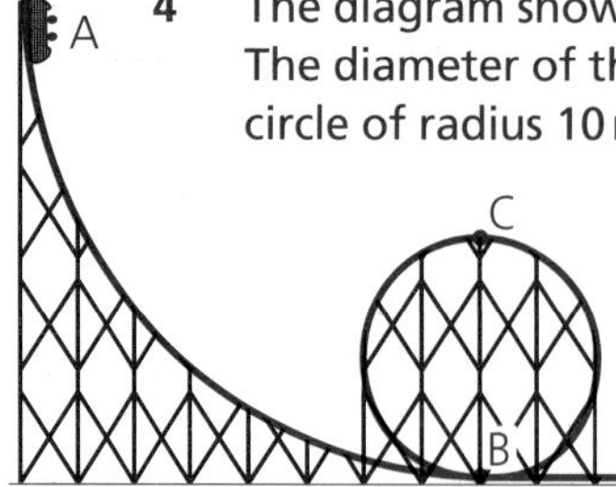

a) Find the kinetic energy and the speed of the carriage when it first reaches B.

b) Find speed of the carriage at C.

c) How would the results compare for a carriage of mass 300 kg?

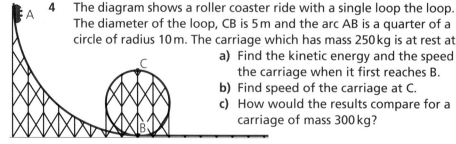

5 A catapult can propel a stone of mass 1 kg vertically upwards to a height of 8 m.

a) Find the energy stored in the stretched elastic of the catapult.

b) Find the speed of the stone when it leaves the catapult.
c) The catapult now propels the same stone horizontally as shown in the diagram. By using energy considerations, find the speed of the stone as it hits the ground.

6 It is estimated that a rugby ball of mass 800 grams reached a maximum height of 5 m and was then travelling at $2\,\mathrm{m\,s^{-1}}$.
a) Find the total energy of the rugby ball at this point.
b) Find the initial kinetic energy and hence the initial speed of the ball.

7 The terminal speed of a skydiver of mass 64 kg is $70\,\mathrm{m\,s^{-1}}$.
a) What is the kinetic energy of the skydiver at his terminal speed?
b) How far does the skydiver have to fall to reach this speed, assuming there is no resistance to his motion?
c) If in fact the skydiver has to fall 700 m before reaching terminal speed, find the work done against air resistance.

8 Tarzan, of mass 90 kg grabs hold of a rope that was initially 60° from the vertical. He swings through an angle of 90° and lets go of the rope. The length of the rope is 6 m. Calculate the speed of Tarzan when he lets go of the rope.
What effect does:
a) doubling the length of the rope,
b) Tarzan swinging with Jane (mass 45 kg),
have on Tarzan's speed when he lets go of the rope?

9 A ball of mass 100 grams moves on a parabolic path so that its kinetic energy varies between 20 J and 50 J.
a) Find the maximum height of the ball and its speed at this time.

Assume that the path of the ball is given by:

$$\mathbf{r} = at\,\mathbf{i} + (bt - 4.9t^2)\,\mathbf{j}$$

where a and b are constants and $\mathbf{i}$ and $\mathbf{j}$ are horizontal and vertical unit vectors respectively.
b) Find the values of a and b.

10 A bungee jumper of mass m jumps from a bridge of height 100 m.
a) After the jumper has fallen a distance x metres, the rope becomes taut. Find an expression for the speed of the jumper at this point.
b) The jumper reaches the ground with zero velocity. Calculate the energy stored in the rope at this point, in terms of m.

POWER

In everyday language, one car may be described as being more powerful than another. Experience suggests that cars that can increase their speed very quickly are more powerful than cars that increase their speed at a slower rate. Very heavy vehicles are also sometimes described as being powerful, not because they can increase their speeds rapidly, but because of the huge mass that may be involved.

Exploration 10.4

The word 'power' is often misused. What do you understand it to mean? Consider the cases below and try to place them in order, the most powerful first.

- A 10 tonne lorry that accelerates from 0 to 27 m s^{-1} (60 mph) in 20 seconds.
- A 1.2 tonne car that accelerates from 0 to 27 m s^{-1} (60 mph) in 6 seconds.
- An athlete of mass 65 kg who can accelerate from rest to 12 m s^{-1} in 2.5 seconds.

Exploration 10.5

Estimating your power

1 You will need a long staircase, the longer the better, a stopwatch, a ruler, bathroom scales and about six fellow mathematicians.
   - Find the time it takes for each member of the group to run to the top of the stairs.
   - Estimate (or measure) the height of the stairs and hence the work done by each student in running up the stairs.
   - Find the average rate of doing work for each student, by dividing the work done by the time taken. This gives you the average power of each student.

2 You could conduct a similar experiment in the school gym using a climbing rope. Are there any differences between the results of this experiment and the staircase experiment? Try to explain.

Power and energy

The more powerful something is, the more quickly it can increase its kinetic energy. Power can be defined as the rate of change of kinetic energy, but it is usual to define power as the rate at which work is done.

Power is the rate at which work is being done, or:

$$\text{power} = \frac{\mathrm{d}}{\mathrm{d}t}(\text{work done})$$

As for a constant force, the work done is Fs:

$$\text{power} = \frac{\mathrm{d}}{\mathrm{d}t}(Fs)$$

Differentiating using the product rule gives:

$$\text{power} = F\frac{\mathrm{d}s}{\mathrm{d}t} + s\frac{\mathrm{d}F}{\mathrm{d}t}$$

As the force is constant, $\frac{\mathrm{d}F}{\mathrm{d}t} = 0$ and so:

$$\text{power} = F\frac{\mathrm{d}s}{\mathrm{d}t} = Fv \quad \left(\text{since } \frac{\mathrm{d}s}{\mathrm{d}t} = v\right)$$

This expression for power is very useful but is based on the assumption of a constant force. The units of power would be J s^{-1} but a special unit, the watt (W), is introduced for power.

1 watt = $1\,\text{J s}^{-1}$

Traditionally the term 'horsepower' has been used as a unit of power. One horsepower is equivalent to 746 watts.

Example 10.11

A motorcycle has a top speed of 45 m s^{-1} (100 mph) and a maximum power output of 30 kW. Find the resistance force acting on the motorcycle at top speed.

Solution

At top speed, the forward force on the bike, F, is balanced by the resistance force, R, so $F = R$. Using the power equation: $P = Fv$ gives $30\,000 = F \times 45$

so $F = \dfrac{30\,000}{45} = 667N$

So the resistance force has magnitude 667N.

Example 10.12

The resistance force on a car is proportional to the speed squared. The top speed of the car is 50 m s^{-1} and the maximum power output is 60 kW. Find value of the constant of proportionality for the resistance force.

Solution

At top speed, the forward force, F, has the same magnitude as the resistance force R.

$F = R$

The resistance force is proportional to the speed squared, so:

$R \propto v^2$ *or* $R = kv^2$

where k is the constant of proportionality. At top speed $F = R = kv^2$. Using the power equation $P = Fv$ gives:

$60\,000 = kv^2 \times v = kv^3 = k \times 50^3$

$$k = \frac{60\,000}{50^3} = 0.48$$

So the constant of proportionality is 0.48 and the resistance force is modelled by $0.48v^2$.

Example 10.13

A cyclist pedalling flat out on the level can work at a maximum rate of 100 watts. If the resistance force on the cyclist can be modelled as:

$R = (4 + 2v)$

find the top speed of the cyclist.

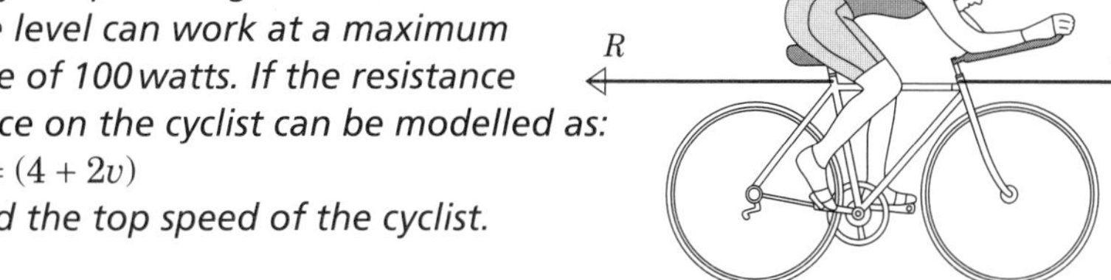

Solution

At top speed, the forward force, F, produced by the cyclist is equal to the resistance force R. So at top speed $F = (4 + 2v)$.
Using the power equation gives: $P = Fv$

$$100 = (4 + 2v)v = 4v + v^2 \Rightarrow v^2 + 2v - 50 = 0$$

Solving the quadratic equation gives:

$$v = \frac{-2 \pm \sqrt{4 + 200}}{2} = \frac{-2 \pm 14.3}{2} = 6.1 \text{ or } -8.1$$

so the top speed of the cyclist is 6 m s⁻¹ to one significant figure.

EXERCISES

10.6 A

1 Find the power required to maintain the motion described below.
a) A car travelling at 30 m s^{-1} experiencing a resistance force of 1700 N. What assumption have you made about the motion?
b) A cyclist travelling at 5 m s^{-1}, on a horizontal road, experiencing a resistance force of 50 N.
c) A man of mass 69 kg climbing a flight of stairs at a rate of four 20 cm steps per second.

2 A car with a maximum power output of 50 kW has a top speed of 40 m s^{-1}. Find the resistance force on the car at top speed.

3 A cyclist has a power output of 150 watts and experiences a resistance force of 45 N. Find the top speed of the cyclist.

4 The resistance force on a cyclist is assumed to be modelled by $R = \frac{v^2}{4}$. The cyclist can reach a maximum speed of 12 m s^{-1} on the level.
a) Find the forward force on the cycle at top speed.
b) Find the maximum power output of the cyclist.
c) The power output of another cyclist is 20% greater. Find the top speed for this cyclist.

5 A train of mass 150 tonnes can reach a speed of 10 m s^{-1} while travelling a distance of 400 m, from rest.
a) Find the acceleration of the train.
b) Estimate the power output of the train, at 10 m s^{-1}.
c) In what ways could your answers to **a)** and **b)** be criticised?

6 A crane can lift a load of mass 750 kg at speeds of up to 3 m s^{-1}.
a) Find the maximum power output of the crane.
b) If the crane is operating at its maximum power output, find the acceleration of the load when it is travelling at 1 m s^{-1}.
c) Express the acceleration of the mass in terms of its speed.

7 A car has a power output of 70 000 W. It experiences a resistance force of kv and has a top speed of 52 m s^{-1}.
a) Find the value of k.
The car is used to tow a caravan. The resistance force on the caravan is $\frac{3kv}{2}$.
b) Find the top speed of the car when towing the caravan.

8 A crane can lift a load of mass 500 kg at speeds of up to 2 m s^{-1}. What is the maximum speed at which it could lift a load of 1.8 tonnes?

9 A cyclist and bicycle have a combined mass of 75 kg and the cyclist is pedalling up a slope at 2° to the horizontal. The speed of the cyclist is a constant 3 m s^{-1}.

a) Find the component of gravity acting parallel to the slope.
b) Find the forward force exerted to maintain a constant speed.
c) Find the power output of the cyclist.
d) On a level road, the cyclist can reach a top speed of 10 m s^{-1}. Find the resistance force on the cyclist.
e) Is it reasonable to assume that there are no resistance forces on the cyclist when on the slope?

10 A car has a power output of 50 000 W and a mass of 1000 kg. On a horizontal road, it has a top speed of 50 m s^{-1}.

a) If the resistance forces on the car are proportional to the speed, i.e. $R = kv$, find the value of k.
b) A slope is at 5° to the horizontal. Find the component of gravity parallel to the slope.
c) Find the maximum speed of the car when going up the hill
d) Find the maximum speed of the car when going down the hill.

EXERCISES

10.6 B

1 Find the power required to maintain the motion described below.

a) A train that travels at 50 m s^{-1} against a resistance force of 5000 N.
b) A roller skater who travels at 4 m s^{-1} and who experiences a resistance force of 18 N.
c) A crane that lifts a load of 400 kg at a speed of 0.8 m s^{-1}.

2 A motorcycle with a top speed of 43 m s^{-1} on the level has a maximum power output of 22 kW.

a) Find the resistance force on the motorcycle at top speed.
b) When the motorcycle is carrying a pillion passenger the resistance force is increased by 5%. What is the top speed when carrying a pillion passenger?

3 As a supermarket trolley is pushed at 2.5 m s^{-1} a resistance force of 20 N acts.

a) Find the power output of the person pushing the trolley.
b) When the trolley is loaded the resistance force doubles. What happens to the power output of the person pushing?
c) What assumptions have you made to answer this question?

4 A simple model for the resistance forces experienced by a car is that they are proportional to the speed. A particular car has a maximum power output of 40 kW and a maximum speed of 32 m s^{-1}.

a) Find the constant of proportionality for the model described.
b) An identical car with a larger engine has a maximum power output that is 10% greater.

Find its maximum speed and compare with the speed of the original car.

5 The resistance force on a cyclist is modelled as a constant force of 90 N. At an instant the cyclist is moving forward on the level at $4\,\text{m s}^{-1}$ and accelerating at $0.3\,\text{m s}^{-2}$. The mass of the bicycle and cyclist is 78 kg.
a) Find the forward force acting on the cyclist.
b) What is the power output of the cyclist?

6 A child of mass 40 kg is running up a slope, inclined at 3° to the horizontal, and at a constant speed of $2\,\text{m s}^{-1}$.
a) Find the component of the force of gravity parallel to the slope.
b) Find the power output of the child.

7 A crane can lift a load of 600 kg at a maximum speed of $4\,\text{m s}^{-1}$.
a) Find the maximum power output of the crane.
b) If the mass is accelerating at $0.5\,\text{m s}^{-2}$, while the crane is operating at its maximum power output find the speed of the mass.

8 The power output of a car travelling at a constant $20\,\text{m s}^{-1}$ is 15 kW. Two possible models for the resistance forces are that they are proportional to the speed or to the square of the speed. The maximum power output of the car is 60 kW.
a) Find the constant of proportionality for each model.
b) Use each model to predict the top speed of the car.
c) Which model do you think is more realistic?

9 A car with a mass of 1000 kg can reach a top speed of $42\,\text{m s}^{-1}$ and has a power output of 54 kW.
a) Find the resistance force on the car at top speed.
b) Assume that the resistance force remains constant at the value you have calculated. Find the maximum speed of the car going up and down a slope inclined at 2° to the horizontal.
c) Is the assumption about the resistance force made in **b)** reasonable?

10 A lorry of mass 20 tonnes has a top speed, on the level, of $30\,\text{m s}^{-1}$ and a maximum power output of 100 kW.
a) Assume the resistance forces are proportional to the speed squared and find the constant of proportionality.
b) Find the maximum speed of the lorry going up a slope at 4° to the horizontal.

CONSOLIDATION EXERCISES FOR CHAPTER 10

1 A lorry of mass 6000 kg is moving along a horizontal road with its engine working at a constant rate of 42 kW. The total resistance to the motion of the lorry is 2800 N. At the instant when the lorry has speed $6\,\text{m s}^{-1}$, find:
a) the tractive force, **b)** the acceleration of the lorry.

(ULEAC Question 2, Specimen Paper M1, 1994)

2 A car of mass 852 kg is moving on a horizontal road. The resistance to motion is 95 N. Find the power of the engine at the instant when the speed is $15\,\text{m s}^{-1}$ and the acceleration is $1.2\,\text{m s}^{-2}$. State where the forces causing the motion of the car act and identify their nature.

(UCLES Question 4, Specimen Paper M2, 1994)

3 A car has an engine of maximum power 15 kW. Calculate the force resisting the motion of the car when it is travelling at its maximum speed of 120 km h^{-1} on a level road. Assuming an unchanged resistance and taking the mass of the car to be 800 kg, calculate, in m s^{-2}, the maximum acceleration of the car when travelling at 60 km h^{-1} on a level road.

(UCLES Question 11, Specimen Paper 3, 1994)

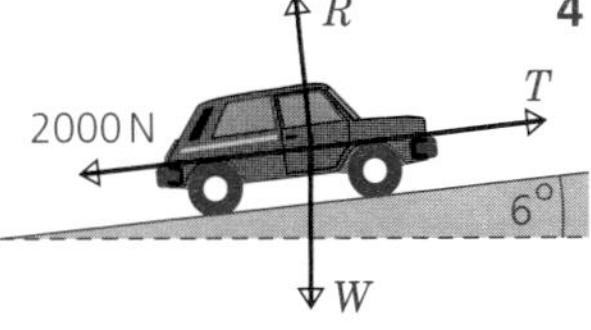

4 A car of mass 1 tonne is moving at a constant velocity of 60 km per hour up an inclined road which makes an angle of 6° with the horizontal.

a) Calculate the weight, W, of the car and the normal reaction, R, between the car and the road.

Given that the non-gravitational resistance down the slope is 2000 N, find:

b) the tractive force, T, which is propelling the car up the slope,

c) the rate at which T is doing work.

The engine has a maximum power output of 80 kW.

d) Assuming the resistances stay the same as before, calculate the maximum speed of the car up the same slope.

(MEI Question 3, Mechanics, January 1992)

5 The magnitude of the resistance to the motion of a motor coach is K newtons per tonne, where K is constant. The motor coach has mass 4.5 tonnes. When travelling on a straight horizontal road with the engine working at 39.6 kW, the coach maintains a steady speed of 40 m s^{-1}.

a) Show that $K = 220$.

The motor coach ascends a straight road, which is inclined at arcsin 0.3 to the horizontal, with the same power output and against the same constant resisting forces.

b) Find, in joules, to two significant figures, the kinetic energy of the motor coach when it is travelling at its maximum speed up the slope.

(ULEAC Question 4, Paper M1, January 1992)

6 Assume $g = 10$ m s^{-2} in this question. Define *power* and state whether it is a scalar or a vector.

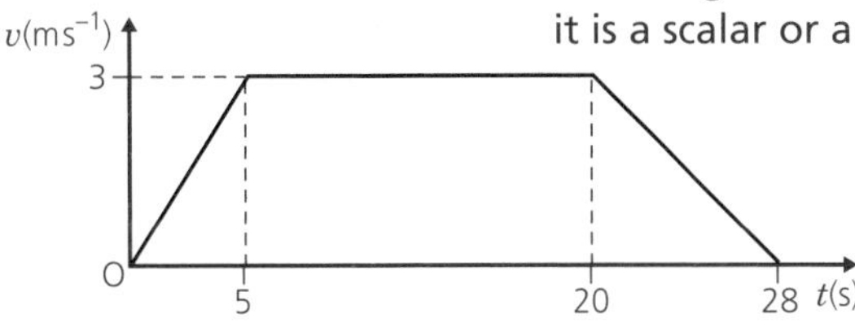

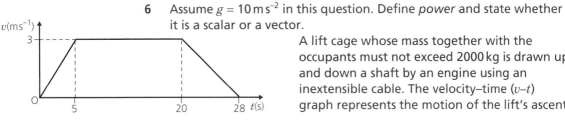

A lift cage whose mass together with the occupants must not exceed 2000 kg is drawn up and down a shaft by an engine using an inextensible cable. The velocity–time (v–t) graph represents the motion of the lift's ascent.

a) If the lift is fully loaded:

i) show that the power generated by the engine during the time when the lift has constant velocity is 60 kW,

ii) determine the maximum power generated during the lift's ascent.

b) Having reached the top of the shaft the lift descends, again fully loaded. Find the maximum power that the engine will need to develop in order that the lift's motion should have the same v–t graph when descending.

(Oxford Question 10, Paper 2, June 1990)

7 The diagram represents a path which consists of a slope AB, 90 m long, inclined at 25° to the horizontal, and a horizontal section BC. A boy on a skateboard starts from rest at A and glides down AB before coming to rest between B and C. The magnitude of the resistive forces opposing the motion are constant throughout the journey. The combined mass of the boy and skateboard is 40 kg and the boy reaches B with a speed of 14 m s^{-1}. By modelling the boy and his skateboard as a particle, find:

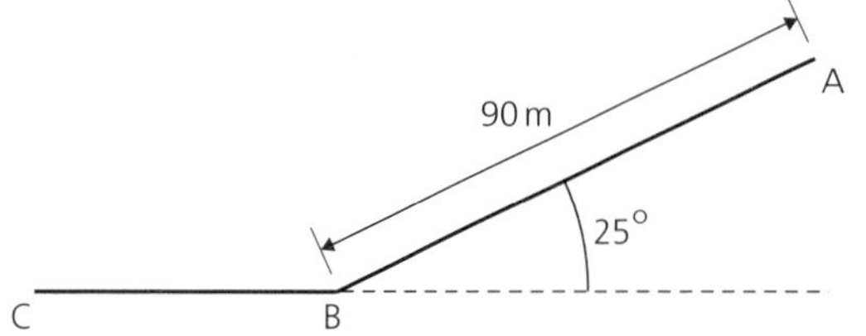

a) the energy lost, in J, by the boy and the skateboard in going from A to B,
b) the magnitude, in N, of the resistive forces,
c) the distance, in m, the boy travels along BC before coming to rest.

(ULEAC Question 4, Specimen Paper M1, 1994)

8 The total mass of a motorcyclist and his machine is 250 kg. The maximum power of the motorcycle is P kW. When the speed of the motorcycle is v m s^{-1}, the road resistance is kv newtons, where k is a constant. The motorcycle has a maximum steady speed of 25 m s^{-1} when ascending a hill inclined at an angle $\sin^{-1}\left(\frac{1}{20}\right)$ to the horizontal.

Taking $g = 10$ m s^{-2}, show that $8P = 5k + 25$

Given the maximum steady speed of the motorcycle down the same hill is 30 m s^{-1}, obtain another equation relating k and P. Deduce the values of k and P. Calculate, in m s^{-1}, the maximum steady speed of the motorcycle along a horizontal road.

(NEAB Question 10, Specimen Paper 6, 1994)

MATHEMATICAL MODELLING ACTIVITY

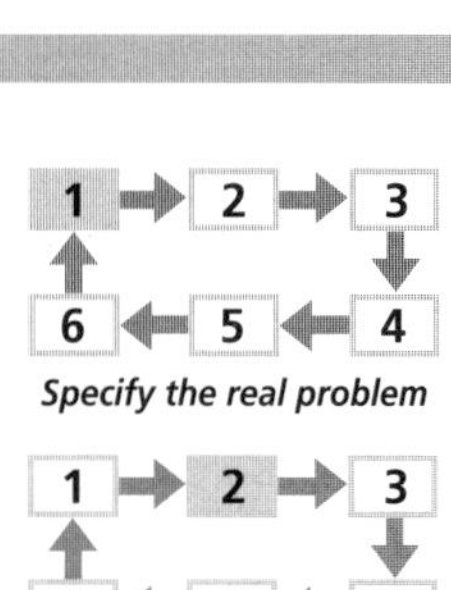

Problem statement

How much work do you do as you walk?

Set up a model

Make a list of the key features that you feel are important to consider when solving this problem.

The set of assumptions that have been set out below allow a first model to be formulated.

- The mass of the walker is m kg.
- The length of the walker's legs is l m and their pace has length p m.
- The walking takes place on level ground.
- As the person walks their body rises and falls slightly. The work done is simply that required to raise the body at each step.
- The walker's legs do not bend.

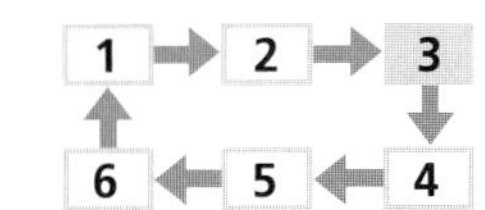

Formulate the mathematical problem

Formulate a mathematical problem

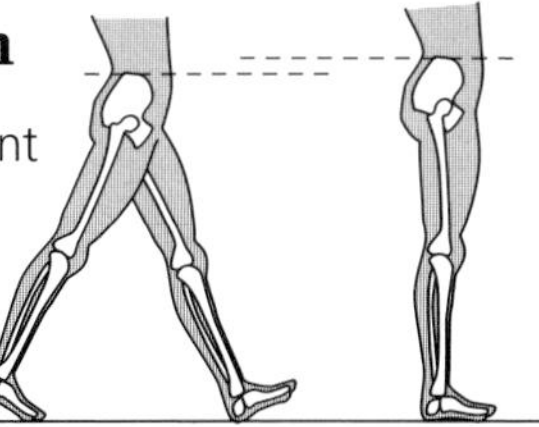

The diagram shows the walker at the start and mid-point of a step. Find the height gained during this phase of the step and the work done. Also find the work done as the walker travels 1 km.

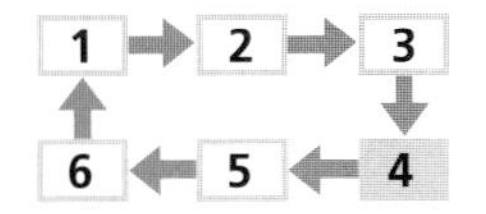

Solve the mathematical problem

Solve the mathematical problem

The diagrams show the legs of the walker at the start of the step and the mid-point.
The gain in height is:
$l - l\cos\alpha$
Show that:

$$\cos\alpha = \sqrt{1 - \frac{p^2}{4l^2}}$$

Now the gain in height can be expressed as:

$$l - l\cos\alpha = l\left(1 - \sqrt{1 - \frac{p^2}{4l^2}}\right)$$

and the work done per pace as:

$$mgl\left(1 - \sqrt{1 - \frac{p^2}{4l^2}}\right)$$

Show that the work done in travelling 1 km is:

$$\frac{1000\,mgl}{p}\left(1 - \sqrt{1 - \frac{p^2}{4l^2}}\right)$$

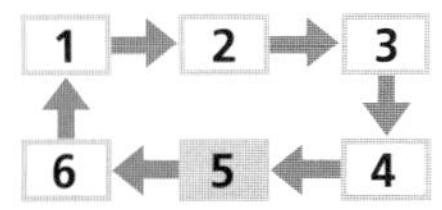

Interpret the solution

Interpretation

The result gives the work done as a person walks 1 km. For an adult typical values for the variables are:

$$m = 70\,\text{kg}$$
$$p = 0.8\,\text{m}$$
$$l = 0.9\,\text{m}$$

Using this value gives:

$$\text{work done} = \frac{1000 \times 70 \times 9.8 \times 0.9}{0.8} \times \left(1 - \sqrt{1 - \frac{0.8^2}{4 \times 0.9^2}}\right) = 80\,411\,\text{J}$$

Show that this is about the same as the work done when raising a 70 kg mass through 117 m.

Compare with reality

Compare with reality

Conduct the experiment below.

1 Find the power of a student who is climbing a flight of stairs as quickly as possible, assuming that work is only done against gravity.

2 Find the power of the same student who is walking as fast as possible on level ground, using the model above to find the work done.

3 Compare the results and comment.
Also comment on the model and suggest ways in which it could be improved.

POSSIBLE MODELLING TASKS

1 A runner maintains a constant speed while running and so would appear to have no change in kinetic energy and be doing no work. This is clearly not true. By considering the motion of the runner, develop a model that describes the work done by the runner.

2 On some steep hills, you will notice that there are lanes for slow-moving vehicles. Consider the effect of driving uphill on a lorry and recommend a policy for the provision of lorry lanes.

Summary

After working through this chapter you should:

- be aware of the different types of energy: potential and kinetic
- be able to calculate kinetic energies using $\frac{1}{2}mv^2$
- be able to calculate work done using Fs or $Fs\cos\theta$
- be able to use the result work done = change in kinetic energy
- be able to calculate gravitational potential energy using mgh
- be able to use power in the form $P = Fv$

Energy and work: Variable forces and scalar products

- *In Chapter 10 we covered the basic concepts of energy, work and power. In this chapter we extend the ideas developed for constant forces to variable forces.*
- *A definition of work done is introduced for use with vectors. This involves a new concept called the* ***scalar product***.

HOOKE'S LAW : A MODEL FOR THE TENSIONS IN SPRINGS

Springs and elastic strings exert forces that vary as they are stretched or compressed in the case of springs. A model to describe how the tension or compression varies was formulated by Hooke. The following exploration, which is a practical activity, explores this model.

Exploration 11.1

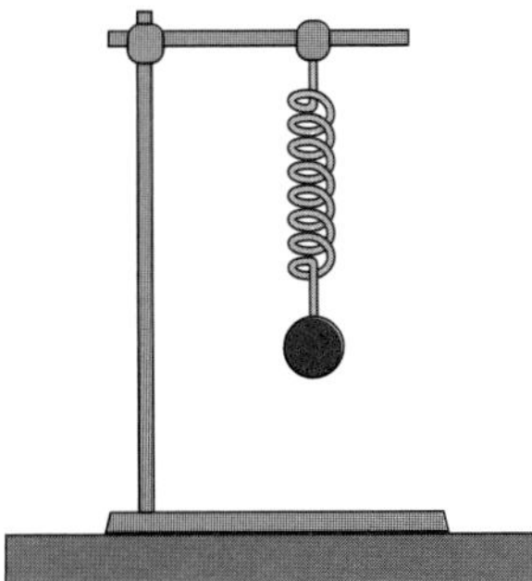

Forces in springs

- Set up the apparatus as shown in the diagram.
- What forces act on the mass on the end of the spring?
- The upward force exerted by the spring on the mass is the tension in the spring. When the mass is at rest this force is balanced by the force of gravity on the mass. Record the unstretched length of the spring. Add a small mass to the end of the spring and record how much it stretches. Repeat this process, recording the tension force and the extension of the spring in a table as shown in the example below. (Do not include the sample results shown.)

Mass (g)	Tension (N)	Extension (m)
50	0.49	0.02

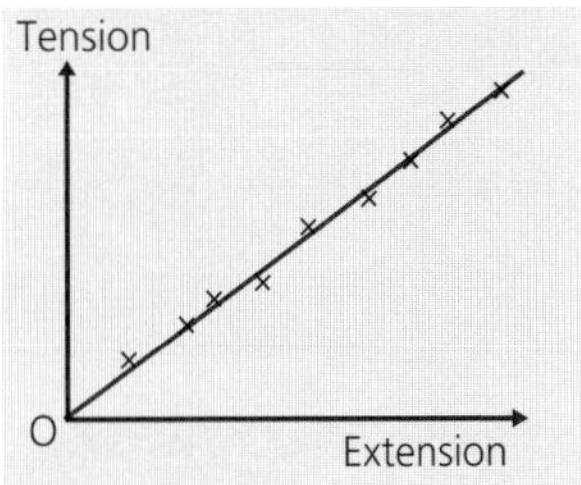

- Plot a graph of tension against extension, by drawing a line of best fit through your points. Comment on your results. It is reasonable to assume that there is a linear relationship. The gradient of the line is known as the **stiffness** of the spring.
- Join two identical springs together, to make a spring of twice the length, and repeat the experiment. How does the stiffness of the spring compare with that of the original?

Tension in springs

A spring that is stretched exerts a tension force, the magnitude of this force is ke, where k is the **spring stiffness** and e is the **extension** of the spring. The result is known as **Hooke's law**.

For springs of different lengths the spring stiffness is given by:

$$k = \frac{\lambda}{l}$$

where λ is the **modulus of elasticity** and l is the natural (or unstretched) length of the spring.

A compressed spring exerts a thrust force of ke, where e is the **compression**.

Elastic strings are similar to springs, but cannot exert thrust forces.

Example 11.1

A spring, of natural length 8 cm, stretches 4 cm when a 200 gram mass is attached to it and the whole system hangs vertically. Find:

a) *the tension in the spring,*
b) *the stiffness of the spring,*
c) *the modulus of elasticity.*

Solution

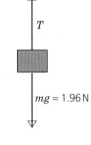

a) *The diagram shows the forces acting on the mass. When it is at rest the upward tension balances the downward force of gravity. As the mass is 0.2 kg the force of gravity is 1.96 N and so the tension is also 1.96 N.*

b) *Hooke's law states that:*

$$T = ke$$

Here $T = 1.96$ N and $e = 0.04$ m, giving:

$$1.96 = k \times 0.04$$

or:

$$k = \frac{1.96}{0.04} = 49 \ \mathrm{N\,m^{-1}}$$

c) *The stiffness and the modulus of elasticity are related by:*

$$k = \frac{\lambda}{l}$$

Here $k = 49$ and $l = 0.08$, so that substituting gives:

$$49 = \frac{\lambda}{0.08}$$

or

$$\lambda = 49 \times 0.08 = 3.92 \ \mathrm{N}$$

Note*: While k has units N m^{-1} the units of λ are N.*

Example 11.2

A 50 cm length of elastic extends 10 cm when a force of 10 N is applied to it.

a) *Find the modulus of elasticity of the elastic.*
b) *If a force is to double a length of the same type of elastic, what force must be applied?*

Solution

a) *Using Hooke's law in the form:*

$$T = \frac{\lambda}{l}e$$

with $T = 10$, $l = 0.5$ and $e = 0.1$ gives:

$$10 = \frac{\lambda}{0.5} \times 0.1$$

Solving for λ gives:

$$\lambda = \frac{10 \times 0.5}{0.1} = 50\,N$$

b) *If the elastic is to double in length then the extension will be equal to the natural length l, which in this case is 0.5 m. Using Hooke's law gives:*

$$T = \frac{50}{0.5} \times 0.5 = 50\,N$$

In fact the tension required to double the length of any spring or elastic string is equal to λ, the modulus of elasticity.

EXERCISES

11.1 A

1 When a 2 kg mass is hung from a spring the length of the spring increases from 20 cm to 32 cm.
  a) Find the stiffness of the spring.
  b) What is the modulus of elasticity?
  c) What would be the extension of a similar spring of natural length 10 cm supporting the same mass?

2 The pointer on a simple spring balance used by an angler moves 4.5 cm when it supports a fish of mass 800 grams.
  a) Find the stiffness of the spring.
  b) Find the extension for a 1.2 kg fish.

3 The spring in a set of bathroom scales is connected to the platform by a series of levers, so that the thrust in the spring is $\frac{1}{5}$ of the normal reaction acting on the platform. If the spring is compressed by 3 cm for a person of mass 60 kg, find:
  a) the stiffness of the spring,
  b) the compression for an 18 kg child.

4 A clock is controlled by a 40 gram mass that vibrates on the end of a spring.
  a) When the clock stops the mass remains at rest and the spring has an extension of 3 cm. Find the spring stiffness.
  b) The mass moves up and down 2 cm from its control position. What range of values does the tension in the spring cover as it moves up and down?

5 A spring has a natural length of 5 cm. When it supports a 100 gram mass it stretches to 7 cm.
  a) Determine the value of λ, the modulus of elasticity.

b) How far would a similar spring of length 8 cm stretch when supporting the 100 gram mass?

c) How far would a similar spring of length 3 cm stretch when supporting a 200 gram mass?

6 The diagram shows a block of mass 3 kg attached to a spring of natural length 10 cm and modulus of elasticity 40 N. The coefficient of friction between the block and the plane is 0.7 and the block is at rest.

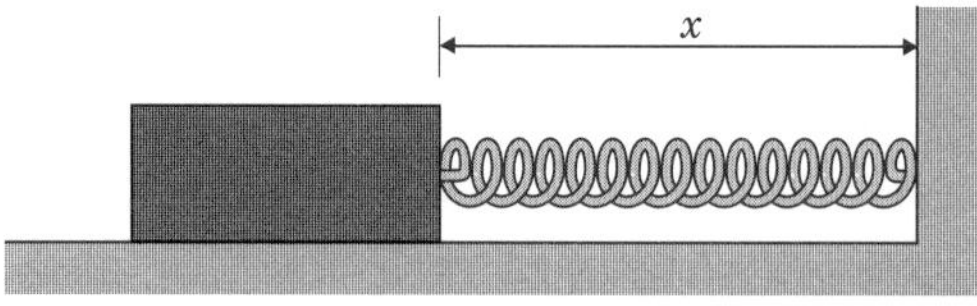

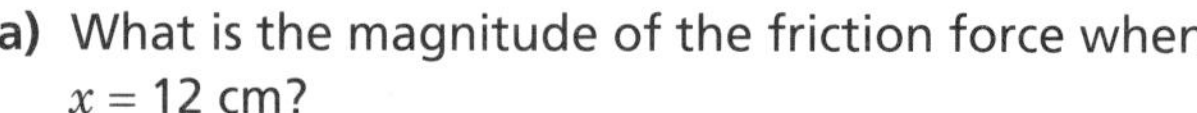

a) What is the magnitude of the friction force when $x = 12$ cm?

b) What are the maximum and minimum values of x?

7 The diagram shows two springs attached to an object of mass 250 grams, resting on a smooth surface. Both springs have natural lengths 40 cm and stiffness as illustrated.

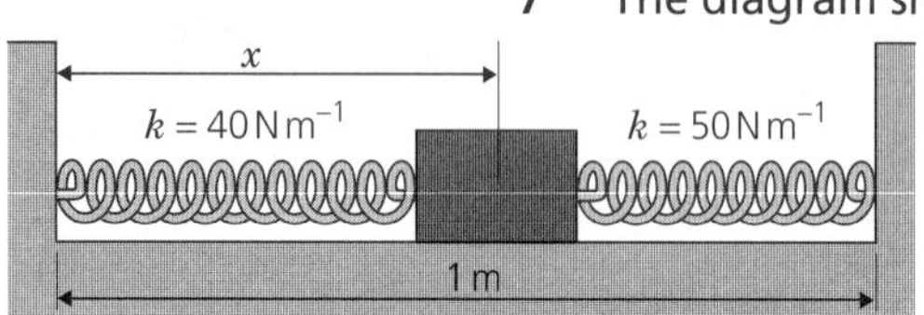

a) Express the extension of each spring in terms of x.

b) Find the value of x when the mass is at rest.

c) Find the acceleration of the mass when $x = 0.2$ m.

8 The diagram shows a spring of natural length d fixed to a 2 kg mass that slides freely on a vertical wire. When the mass is at rest the angle θ is 25°.

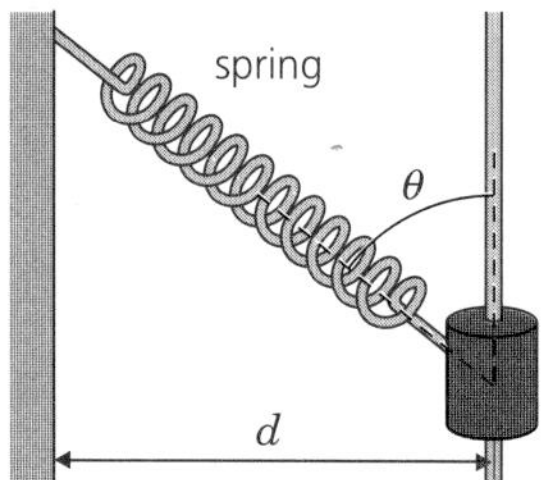

a) Find the tension in the spring.

b) If the distance d is 10 cm find the stiffness of the spring.

EXERCISES

11.1 B

1 When a mass of 3 kg is hung from an elastic thread of natural length 20 cm it increases in length to 25 cm.

a) Find the stiffness of the thread.

b) What is its modulus of elasticity?

c) What would be the extension of a thread of the same material but of natural length 15 cm supporting the same mass?

2 The pointer on a set of scales moves 3 cm along a straight scale, when it supports a bag of rice of mass 1 kg.

a) Find the stiffness of the spring.

b) Find the mass of a bag of rice which causes a movement of 2 cm.

3 The mass of a car is determined by parking the car on a weigh-bridge platform. The spring in the weigh-bridge is connected to the platform by a set of levers so that the thrust in the spring is $\frac{1}{300}$ of the normal reaction acting on the platform. If the spring is compressed by 15 cm for a car of mass 1.5 tonnes find:

a) the stiffness of the spring,

b) the compression for a motorcycle of mass 240 kg.

4 A particle of mass 100 grams oscillates vertically on the end of a spring. When hanging in equilibrium the spring is extended by 5 cm.

a) Find the stiffness of the spring.
b) If the oscillatory motion has an amplitude of 3 cm, what are the greatest and least tensions in the spring?

5 A spring has a natural length of 8 cm. When it supports a 200 gram mass it stretches to a length of 10 cm.
a) Determine the value of λ, the modulus of elasticity.
b) How far would a similar spring of length 10 cm stretch when supporting the 200 gram mass?
c) How far would a similar spring of length 4 cm stretch when supporting a 400 gram mass?

6 The diagram shows a block of mass m on a rough horizontal table.

The coefficient of friction between the block and the table is μ. The block is connected to a fixed wall by a spring of natural length l and modulus of elasticity λ. If the block is initially released from a position $x = a$ $(a < l)$ show that the acceleration of the mass is zero when:

$$x = l \pm \frac{\mu m g l}{\lambda}.$$

7 A particle of mass m is attached to the ceiling by two identical springs, each of natural length l and modulus of elasticity λ, as shown in the diagram.

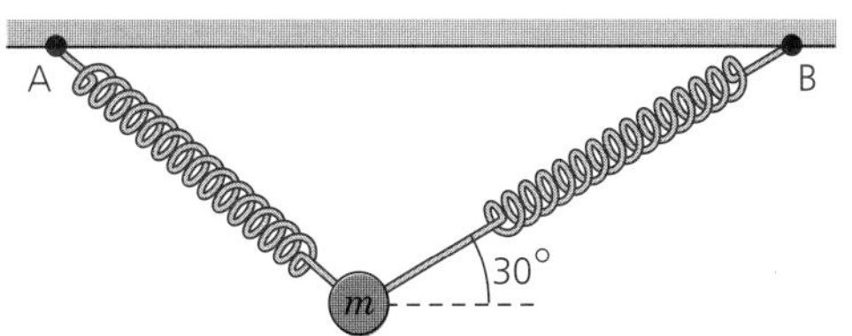

Obtain an expression for the depth of the mass m below the ceiling when the system is in equilibrium.

8 A simple model of a bow and arrow is to model the arrow as a particle and the bow as a horizontally mounted spring. The drawing of the bow is then modelled by the compression of the spring. The diagram shows the force F (newtons) which has to be applied to the spring to compress it a distance x (m).

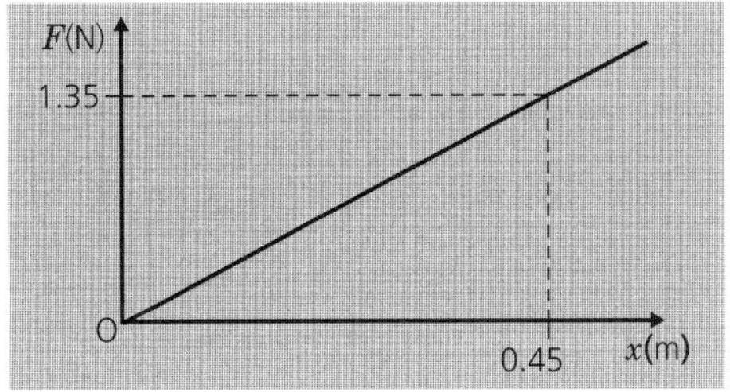

a) Determine the stiffness of the spring.
b) Assuming that an arrow has a mass of 28 grams determine the initial acceleration of the arrow as it leaves the bow.

THE WORK DONE BY A VARIABLE FORCE

The work done by a constant force has been established as Fs or $Fs \cos \theta$, but this cannot be applied, for example, to a spring because the applied force must vary as the tension on the spring varies. There are many other situations where it is useful to be able to calculate the work done by a variable force. For example as a space capsule containing astronauts returns to Earth the gravitational attraction increases as it gets closer to the Earth.

Exploration 11.2

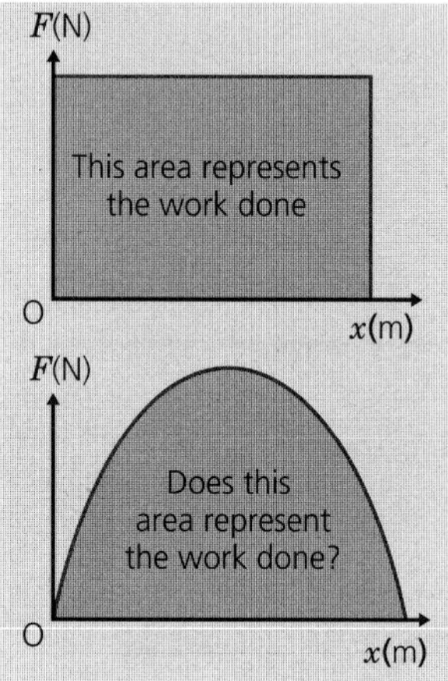

Forces on a bouncing ball

Consider a bouncing ball. How is the deformation of the ball linked to the force that is exerted on the ball? Try to illustrate your response with a graph.

Work done

So far, our results indicate that the greater the deformation, the greater the force. Here is a situation where the force acting depends on the deformation or the distance moved by the ball while in contact with the surface. In reality there are many other situations where force varies with distance. In this section we set out to find the work done by a variable force.

The work done by a constant force is Fd. This can be represented by the area under the graph as shown. Can the work done by a variable force be represented in the same way?

Example 11.3

Show that

$$\int_a^b F\,\mathrm{d}x = \tfrac{1}{2}mv^2 - \tfrac{1}{2}mu^2$$

where u is the speed when $x = a$ and v is the speed when $x = b$ and hence deduce that the work done by a variable force is $\int F\,\mathrm{d}x$.

Solution

First note that Newton's second law can be expressed in the form:

$$F(x) = m\frac{\mathrm{d}v}{\mathrm{d}t}$$

where the force is assumed to be a function of x.
Now using the chain rule:

$$\frac{\mathrm{d}v}{\mathrm{d}t} = \frac{\mathrm{d}v}{\mathrm{d}x} \times \frac{\mathrm{d}x}{\mathrm{d}t}$$

we can say that $F(x) = m\dfrac{\mathrm{d}x}{\mathrm{d}t} \times \dfrac{\mathrm{d}v}{\mathrm{d}x}$

or $F(x) = mv\dfrac{\mathrm{d}v}{\mathrm{d}x}$

Using the method of separation of variables gives:

$$\int_a^b F(x)\mathrm{d}x = m\int_u^v v\,\mathrm{d}v$$

where u is the speed when $x = a$ and v the speed when $x = b$.

Integrating the right hand side gives:

$$\int_a^b F(x)\mathrm{d}x = m\left[\frac{v^2}{2}\right]_u^v = \tfrac{1}{2}mv^2 - \tfrac{1}{2}mu^2$$

Now the right-hand side is the change in kinetic energy which is equal to the work done, so the work done by a variable force must be given by:

$$\int_a^b F(x)\mathrm{d}x$$

Example 11.4

A compressed spring is used to fire a ball of mass 50 grams in a pinball machine. The stiffness of the spring is 50 N m^{-1} and the spring is initially compressed by 0.1 cm.

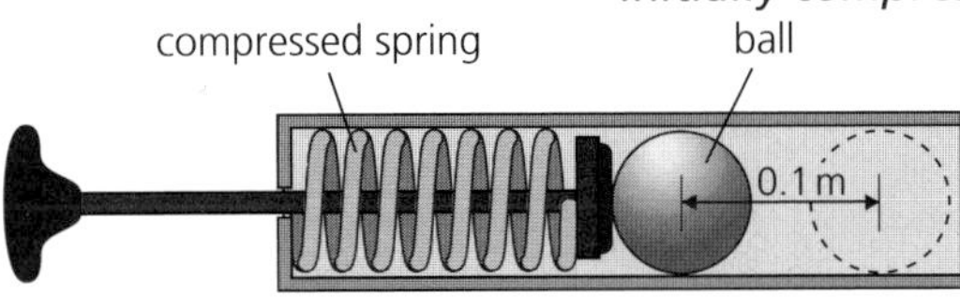

The ball launching mechanism from a pinball machine

a) *Sketch a graph to show how the tension, T, in the spring varies. Express T in terms of x, the distance of the ball from the position shown.*

b) *Find the work done by the tension force as the ball is launched.*

c) *Find the speed reached by the ball.*

Solution

a) *When the spring is compressed as shown, the tension in the spring is:*

$50 \times 0.1 = 5\,\mathrm{N}$

When the spring returns to its natural length the tension is 0 N and it varies linearly. So the tension varies as shown.

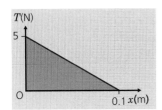

b) *The work done is given by the area under the line.*
Using the formula for the area of a triangle, this gives:
work done $= \frac{1}{2} \times 0.1 \times 5 = 0.25$ *J*

c) *Using work done = change in KE gives:*

$$0.25 = \tfrac{1}{2} \times 0.05v^2 - \tfrac{1}{2} \times 0.05u^2$$

As $u = 0$ *this simplifies to:*

$$0.25 = 0.025v^2$$

$$v^2 = \frac{0.25}{0.025} = 10$$

$$v = 3.2 \text{ m s}^{-1}$$

Example 11.5

As a car of mass 1.05 tonnes moves, the resultant force on it can be modelled as:

$$F(x) = 2000 - \frac{x^2}{5}$$

a) *Find the work done as the car travels 100 m.*

b) *Find the speed at the end of the distance.*

Solution

a) *Work done* $= \int_0^{100} 2000 - \frac{x^2}{5} \mathrm{d}x = \left[2000x - \frac{x^3}{15}\right]_0^{100}$

$$= 2000 \times 100 - \frac{100^3}{15} = 133\,333 \text{ J}$$

b) *Using work done = change in KE gives:*

$$133\,333 = \tfrac{1}{2} \times 1050 \times v^2 - \tfrac{1}{2} \times 1050 \times 0^2$$

so that $v^2 = 254$
and $v = 15.9 \text{ m s}^{-1}$

Example 11.6

A model for the force exerted on a ball bouncing against a vertical wall is $F(x) = kx(2a - x)$ for $x \le 2a$ where k is a constant and a is the maximum deformation of the ball.
For a particular bounce, $a = 0.005$ m and $k = 20\,000$.

a) *Sketch a graph of F against x and explain why the restriction $x \le 2a$ is needed.*

b) *Find the work done as the ball is brought to the point where it is instantaneously at rest.*

c) *If the mass of the ball is 50 grams, find the speed at which it hit the wall.*

Solution

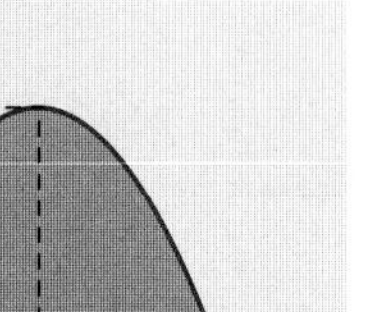

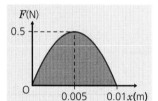

a) *Using $a = 0.005$ and $k = 20\,000$, $F(x) = 20\,000x\,(0.01 - x)$.*
$F(x)$ would be zero when $x = 0$ and when $x = 0.01$.
It would have its maximum value when $x = 0.005$, giving:
$20\,000 \times 0.005 \times 0.005 = 0.5\,N$
The model can only be valid for values of x up to a because the force must increase as x increases.

b) *The work done as the ball is brought to the point where it is instantaneously at rest is:*

$$\text{work done} = \int_0^{0.005} 20\,000x(0.01 - x)\mathrm{d}x = \int_0^{0.005} \left(200x - 20\,000x^2\right)\mathrm{d}x$$

$$= \left[100x^2 - \frac{20\,000x^3}{3}\right]_0^{0.005} = 1.7 \times 10^{-3}\ J$$

c) *As work done = change in KE*

$$1.7 \times 10^{-3} = \tfrac{1}{2} \times 0.05 \times v^2 - \tfrac{1}{2} \times 0.05 \times 0^2$$

$$v^2 = 6.8 \times 10^{-3}$$

$$v = 0.26\,m\,s^{-1}.$$

EXERCISES

11.2 A

1 As a car moves it experiences a resultant force modelled by:
$F(x) = 2000 - 5x$

a) Find the work done as the car moves 400 m.

b) If the car has mass 1000 kg and was initially at rest, find the speed reached by the car.

2 A driver applies the brakes so that the resultant force on the car is:
$F(x) = -(20 + 10x)$

a) Find the work done as the car travels 20 m.

b) The car has mass 1200 kg and had an initial speed of 15 m s^{-1}. What is the speed of the car when it has travelled the 20 m?

3 A chain of length 3 m and mass 60 kg is resting on the floor. A light rope that passes over a pulley is attached to the chain and used to raise the chain off the floor.

a) Show that the mass of the chain lifted off the floor is $20x$, when x is the length of the chain that has been lifted.

b) Find the work that must be done to lift the whole chain.

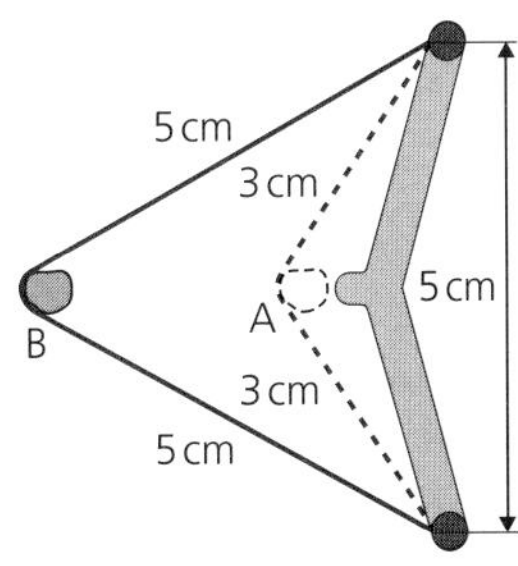

4 A catapult is made of elastic that has stiffness 100 N m^{-1} and natural length 6 cm. A stone of mass 20 grams is placed in the catapult and pulled back from position A to position B.

a) Find the work done in stretching the elastic.

b) Find the speed of the stone when it leaves the catapult at A.

5 When a snooker ball bounces against a cushion, the cushion is deformed. A snooker ball of mass 200 grams, travelling at 1.2 m s^{-1}, hits a cushion. The cushion is compressed 0.004 m before the ball stops and begins to rebound.

a) Find the kinetic energy of the ball when it hits the cushion.

b) Assume that the force exerted by the cushion is proportional to the compression and find the constant of proportionality.

6 The graph shows how the forward force on a train of total mass 30 tonnes varies as it leaves a station.

a) Find the work done in the first 300 m.

b) Find the speed at the end of the first 300 m.

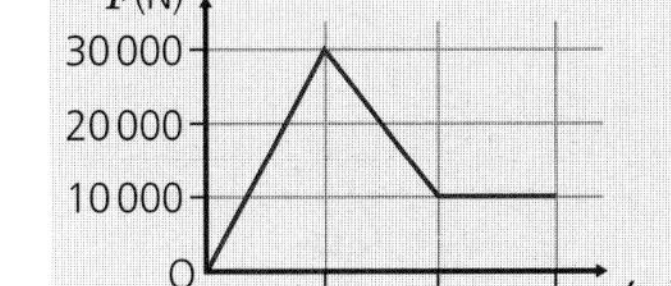

7 A builder is hoisting a leaky drum of water to the top of a scaffold. The mass of the drum is 20 kg and it initially contains 30 litres of water. (1 litre of water has a mass of 1 kg.) Assume that the water leaks out at a constant rate and that only 10 litres remain at the top of the scaffold, 5 m above the ground.

a) If the drum is raised at a constant speed show that its mass at height x is modelled by $m = 50 - 4x$.

b) Find the work done by the builder.

8 Forensic scientists assume that when a bullet enters a certain type of timber the retarding force is $100x^2$, where x is the distance penetrated, in cm. A bullet of mass 30 grams penetrates 8 cm into the timber.

a) Find the work done to stop the bullet moving.

b) Find the speed of the bullet when it enters the timber.

9 The magnitude of the force exerted by a magnet is modelled as $F(x) = \dfrac{4}{x^2}$ where x is the distance from the centre of the magnet in metres.

a) A small ball of mass 5 grams is initially 20 cm from the magnet. Find the work done as it moves to a position 10 cm from the magnet.

b) If the ball was initially moving at 7 m s^{-1}, find the speed of the ball when it is 10 cm from the magnet.

10 A space capsule of mass 500 kg is moving at a speed of 10 m s^{-1} when 1000 km above the surface of the Earth.

a) Find the work done by gravity on the space capsule as it returns to Earth, by using Newton's universal law of gravitation. (The radius of the Earth is 6.4×10^6 m.)

b) Find the speed of the capsule at the surface of the Earth.

c) Assume the capsule experiences a constant resistance force. The speed reached is in fact 30% less than that predicted in **b)**. Find the magnitude of the resistance force.

EXERCISES

11.2 B

1 A body of mass 50 kg is pulled along a straight line AB on a smooth horizontal surface by a force

$$F(x) = 2 - \frac{x}{100}$$

where x m represents the distance moved from A.

a) Find the work done as the body moves from A to B if AB = 5 m.

b) If the body was initially at rest, find the speed it has attained when it reaches B.

2 When the brakes are applied to a certain vehicle the resultant retarding force, F, may be expressed in terms of x, the distance travelled since the brakes were applied, by the equation:

$$F = -(6 + 20x)$$

a) Determine the work done as the vehicle travels 20 m.

b) If the vehicle has a mass of 824 kg and an initial speed of 20 m s^{-1} what is the speed of the vehicle after it has travelled the 20 m?

3 Construction sites often require the excavation of large holes for the foundations of buildings. One such hole has a rectangular cross section 2 m × 3 m and a depth of 25 m. The average density of the material removed from the hole is 900 kg m^{-3}. Determine the work done against gravity in excavating the hole, assuming that the material removed is spread thinly over the ground.

4 Two particles, each of mass m, lie on a smooth horizontal table and are connected by a spring of natural length l and stiffness $\frac{mg}{l}$.

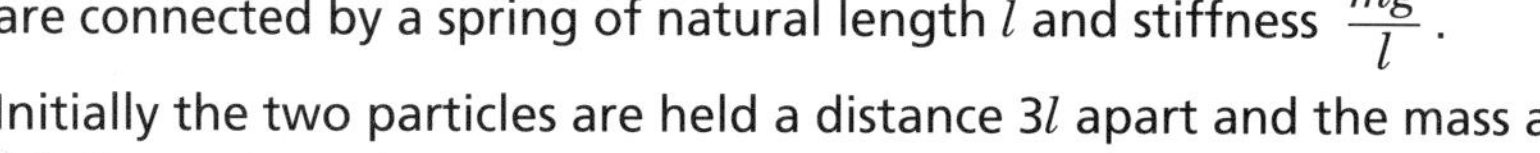

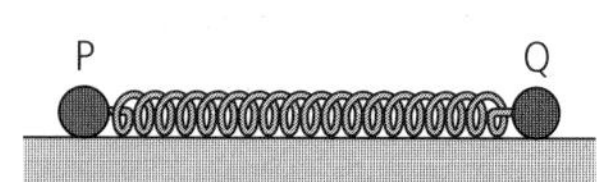

Initially the two particles are held a distance $3l$ apart and the mass at P is then released.

a) What is the speed of particle P when it reaches a position a distance l from Q?

b) If both P and Q were released simultaneously what would be the speed of each, when they are a distance l apart?

5 A cricket ball of mass 156 g is dropped vertically onto the ground and hits it at a speed of 6 m s^{-1}. Subsequent measurements reveal that the ground compressed a distance 0.005 m before the ball stopped and began to rebound.

a) Find the kinetic energy of the ball when it hits the ground.

b) Assume that the force exacted by the ground is proportional to the compression and determine the constant of proportionality.

6 The graph shows how the force on a car of mass 1 tonne varies as it leaves a set of traffic lights.

a) Find the work done in the first 150 m.

b) Determine the speed of the car after 150 m.

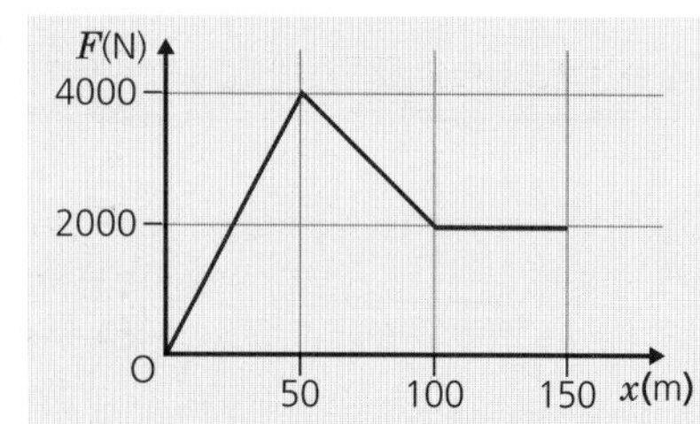

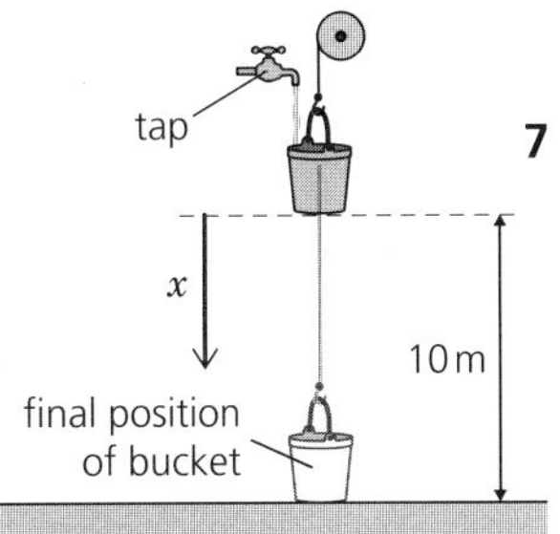

7 A bucket of mass 2 kg which is initially empty is being lowered 10 m from P to the ground while it is being filled with water at a constant rate. It is full when it reaches the ground and contains 18 litres of water. (1 litre of water has a mass of 1 kg.)
  a) If the bucket is lowered at a constant speed show that its mass at a distance x below P is modelled by $m = 2 + 1.8x$.
  b) Determine the work done in lowering the bucket to the ground.

8 Many car racing circuits have sand traps provided on corners so that if a car spins off the track if can be brought to rest quickly by the sand without causing injury to the driver or too much damage to the car. A racing car of mass 500 kg enters such a sand trap at a speed of 50 m s^{-1} and is brought to rest in a distance of 75 m. Assume the car travels along a straight line.
  a) Determine the work done in bringing the car to rest.
  b) Assuming that the resistive force provided by the sand is modelled as $F = kx^2$ determine the value of the constant k.

9 An object of mass m is attracted towards a fixed point by a force which is inversely proportional to the distance from the point and is given by

$$F = \frac{c}{x}$$

If the object has velocity V_0 when at a distance X from the fixed point, determine its velocity when it has moved to a point $\frac{1}{2}X$ from the fixed point.

10 A meteoroid initially at rest a large distance from the sun falls towards the sun under the influence of gravity.
  a) Determine the work done in falling to the surface of the sun assuming that 'a large distance from the sun' is represented mathematically as an infinite distance.
  b) Determine the velocity with which the meteoroid hits the sun. You may assume that the mass of the sun is 1.99×10^{30} kg and the radius of the sun 6.96×10^{8} m.

ELASTIC POTENTIAL ENERGY

A stretched or compressed spring has the ability to make an object move, gaining speed as it does move. Thus a stretched or compressed spring has the potential to create kinetic energy. The anount of energy that a spring can create as it is allowed to expand or contract is known as the **elastic potential energy** (EPE).

Exploration 11.3

Work in stretching a spring

- How much work do you need to do to stretch a spring?
- Do you need to do the same amount of work to increase the length from 5 cm to 10 cm as to increase it from 10 cm to 15 cm?
- Does it get harder to stretch a spring, the longer it gets?

The work done in stretching a spring

The work done in stretching a spring is $\frac{1}{2}ke^2$. This result can be derived theoretically, by considering the work done when a spring is stretched.
The work done by a variable force is given by $\int F(x)\mathrm{d}x$.

If we assume that $F(x)$, the force applied to stretch the string, is kx, where x is the extension of the spring, then if x increases from 0 to e, we obtain:

$$\text{work done} = \int_0^e kx\,\mathrm{d}x = \left[\tfrac{1}{2}kx^2\right]_0^e = \tfrac{1}{2}ke^2$$

so the elastic potential energy stored in the spring is $\frac{1}{2}ke^2$.

As the tension in the spring can also be expressed as:

$$T = \frac{\lambda}{l}x$$

the elastic potential energy can also be expressed as:

$$\frac{\lambda e^2}{2l}$$

Example 11.7

A ball of mass 100 grams is placed in a tube on top of a spring of stiffness 160 N m^{-1}. The spring is compressed by 2 cm.

a) *Find the work done to compress the spring.*

b) *Find the kinetic energy of the ball when it leaves contact with the spring.*

c) *Find the height reached by the ball above its point of release.*

Solution

a) *Using the result for the work done as* $\frac{1}{2}ke^2$ *gives:*

work done $= \frac{1}{2} \times 160 \times 0.02^2 = 0.032$ *J*

b) *When the spring is released, the energy given to the ball is the same as the work that was done in compressing the spring, namely 0.032 J.*

c) *At the highest point reached by the ball, its potential energy is equal to its initial potential energy, so:*

$mgh = 0.032$

$0.1 \times 9.8 \times h = 0.032$

$h = 0.33$ *m*

So the ball rises to a height of 33 cm.

Example 11.8

A spring of stiffness 100 N m^{-1} is used to project a ball of mass 50 grams in a pin-ball machine. The spring is compressed 20 cm and released to set the ball into motion. Find the speed of the ball if:

a) *the table is horizontal,*

b) *the table is at 10° to the horizontal.*

Assume all the energy from the spring is transferred to the ball.

Solution

When the spring is compressed, it has elastic potential energy.
This is given by:

$EPE = \frac{1}{2}ke^2 = \frac{1}{2}\times 100\times 0.2^2 = 2\,J$

a) *When the ball leaves the spring, all this energy is converted to the kinetic energy of the ball, so:*

$2 = \frac{1}{2}mv^2$

$v^2 = 80$

$v = 8.94\,m\,s^{-1}$

b) *On the sloping table, the ball gains gravitational potential energy as well as kinetic energy, so:*

$2 = \frac{1}{2}mv^2 + mgh$

The gain in height of the ball is:

$h = 20\sin 10° = 3.5\,cm$

20 cm, 10°, $h = 20\sin 10°$

So considering the energy gives:

$2 = \frac{1}{2}\times 0.05v^2 + 0.05\times 9.8\times 0.035$

$2 = 0.025v^2 + 0.017$

$v^2 = \frac{2 - 0.017}{0.025} = 79.32$

$v = 8.92\,m\,s^{-1}$

So the ball is only marginally slower on the sloping table.

EXERCISES

11.3A

1 A pop-up toy frog of mass 30 grams jumps up in the air when the spring suddenly expands. It jumps up 30 cm.

a) Find the potential energy of the frog when it reaches a height of 30 cm.

b) Find the elastic potential energy stored in the spring when it is compressed.

c) If the spring was compressed by 1 cm before the frog jumped, find the stiffness of the spring.

2 A train of mass 30 tonnes is travelling at 0.05 m s^{-1} when it hits a buffer containing a spring. The buffer brings the train to rest. Assume that the train runs on horizontal tracks.

a) Calculate the initial kinetic energy of the train.

b) What is the elastic potential energy stored in the spring, when the train stops?

c) If the stiffness of the spring is 2000 N m^{-1}, calculate how far it was compressed.

d) Explain why your answer to **b)** is likely to be incorrect.

3 A spring of stiffness 60 N m^{-1} is used to launch a ball of mass 50 grams up a sloping table, as shown in the diagram on the next page. The spring is compressed 5 cm and released to fire the ball.

a) Find the elastic potential energy stored in the spring when it is compressed.

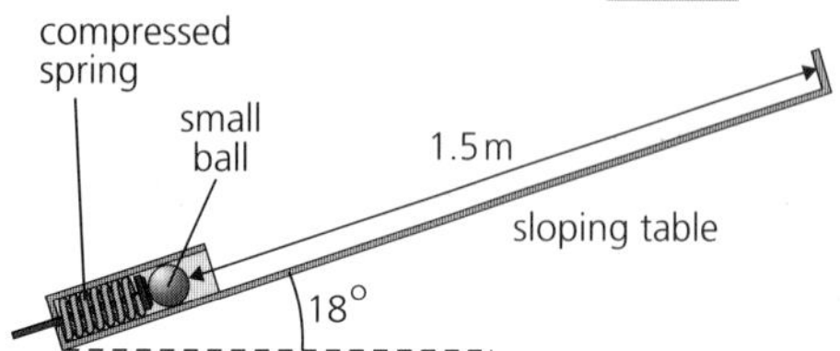

b) Assuming that all this energy is transferred to the ball, find its speed when it loses contact with the spring.

c) How far does the ball travel up the table?

d) Criticise the assumption made in part **b)**.

4 A bungee jumper of mass 65 kg falls through 20 m before the elastic rope becomes taut and then a further 20 m before stopping and starting to move upwards.

a) Find the potential energy lost during the jump.

b) Hence find the energy stored in the elastic rope when the jumper is at the lowest point.

c) Find the stiffness of the rope.

d) By considering the tension in the rope, find how far the jumper falls before they begin to slow down.

e) How far would the rope stretch for a jumper of mass 70 kg?

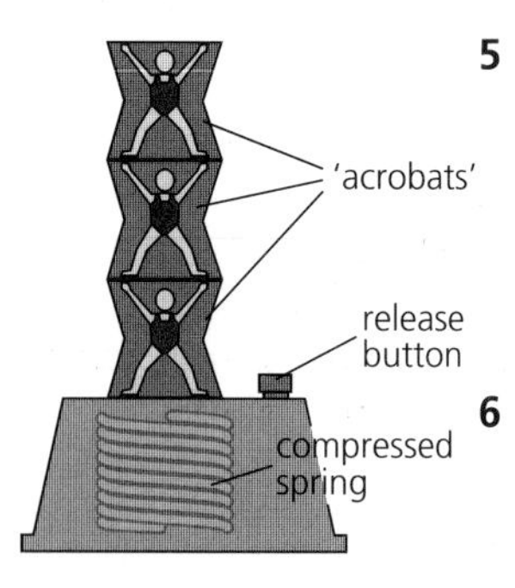

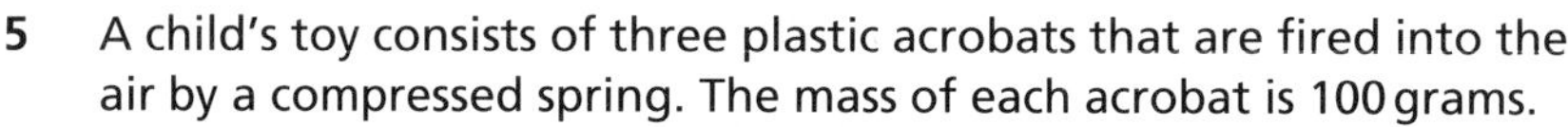

5 A child's toy consists of three plastic acrobats that are fired into the air by a compressed spring. The mass of each acrobat is 100 grams.

a) When used properly, each acrobat rises 20 cm. Find the elastic potential energy stored in the spring.

b) A child fixes a small object of mass 50 grams on the toy. Find the maximum speed of the object and the height it reaches.

6 **a)** The elastic potential energy of a spring is 100 J. Find the modulus of elasticity of the spring if its natural length has been doubled to 0.4 m.

b) If the natural length of the spring were decreased by 75%, what extension would be required to give an EPE of 100 J?

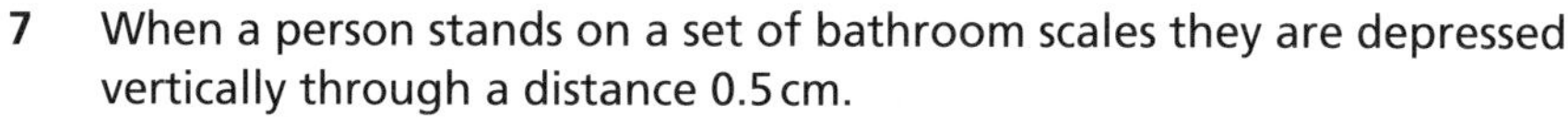

7 When a person stands on a set of bathroom scales they are depressed vertically through a distance 0.5 cm.

a) Find the loss in potential energy for a person of mass 62 kg.

b) If the spring inside the scales moves 2 cm, find its stiffness.

8 Two identical springs are arranged as shown in the diagrams. Both systems are stretched by the same amount. In which system is there the greater elastic potential energy?

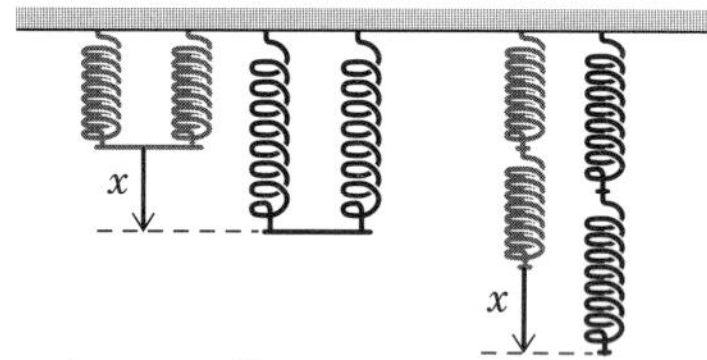

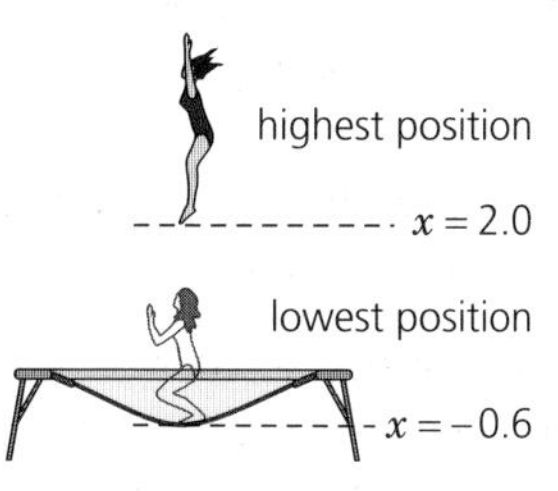

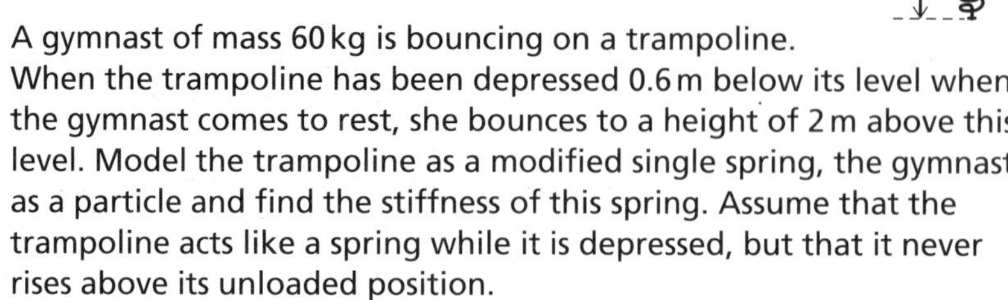

9 A gymnast of mass 60 kg is bouncing on a trampoline.
When the trampoline has been depressed 0.6 m below its level when the gymnast comes to rest, she bounces to a height of 2 m above this level. Model the trampoline as a modified single spring, the gymnast as a particle and find the stiffness of this spring. Assume that the trampoline acts like a spring while it is depressed, but that it never rises above its unloaded position.

a) Express the kinetic energy in terms of x, the displacement of the gymnast above the level of the unloaded trampoline, while she is in the air.

b) Repeat **a)** for the time that the gymnast is in contact with the trampoline.

c) Find the maximum speed of the gymnast.

10 An object of mass m rests on a spring of stiffness k, and natural length l. In equilibrium, the compression of the spring is $\frac{1}{5}l$. It is then depressed a further distance $\frac{1}{2}l$ and released. What is the maximum height reached by the mass?

EXERCISES

11.3 B

1 The spring of a vehicle suspension system has a stiffness of $4 \times 10^4\,\mathrm{N\,m^{-1}}$. How much work is done in compressing this spring from its natural length of 0.3 m to 0.25 m?

2 A toy gun shoots pellets horizontally by means of a compressed spring. To load the gun the pellet is placed against the horizontal spring which is then compressed and released. Determine the speed of the pellet when the spring reaches its natural length, if the mass of the pellet is 5 g, the stiffness of the spring is $200\,\mathrm{N\,m^{-1}}$ and the spring compression when fully loaded is 5 cm.

3 A spring of stiffness $50\,\mathrm{N\,m^{-1}}$ is used to launch a ball of mass 5 g up a slope inclined at an angle 30° to the horizontal.
The spring was initially compressed 5 cm and then released.

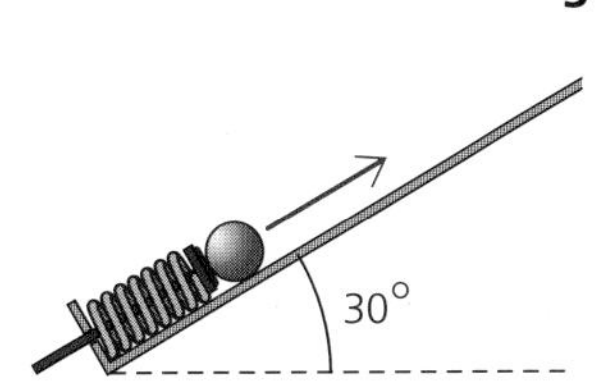

a) Find the elastic potential energy stored in the spring.
b) The speed of the ball when it loses contact with the spring (assume all the elastic potential energy is transferred to the ball).
c) How far up the slope does the ball travel?

4 Mountaineers use nylon rope to cushion themselves from the jolt experienced when their fall is suddenly arrested. A climber of mass 80 kg attached to a 12 m rope falls from a height of 6 m above the point at which the rope is anchored to the rock face. Model the rope as a spring of stiffness $4900\,\mathrm{N\,m^{-1}}$ and determine the extension of the rope in first bringing the climber to rest.

5 A spring of stiffness $k\,\mathrm{N\,m^{-1}}$ is used to launch a ball of mass m kg vertically. If the spring is initially compressed a distance x m determine the velocity of the ball when it loses contact with the spring.

a) If the ball is changed for another one of double the mass would the release velocity of this new ball be different and if so what would be its value?
b) Would changing the spring for one of half the original stiffness have any effect on the release velocity of the original ball? If so, what?
c) Suppose the stiffness of the spring and the mass of the ball are each doubled. What can you say about the release velocity this time?

6 A particle of mass 2 kg is attached to the origin by a spring of stiffness $40\,\mathrm{N\,m^{-1}}$. The particle is released from rest when the extension of the spring is 0.2 m. What is the velocity of the particle when the spring attains its natural length?

7 A particle of mass 1 kg moves on a smooth horizontal surface and is attached to a fixed point O by a spring of stiffness $2\,\mathrm{N\,m^{-1}}$. Initially the particle has an extension of 0.5 m and a velocity of $-2\,\mathrm{m\,s^{-1}}$.

a) What is the particle's maximum speed during the motion?
b) At what position is the particle momentarily at rest?

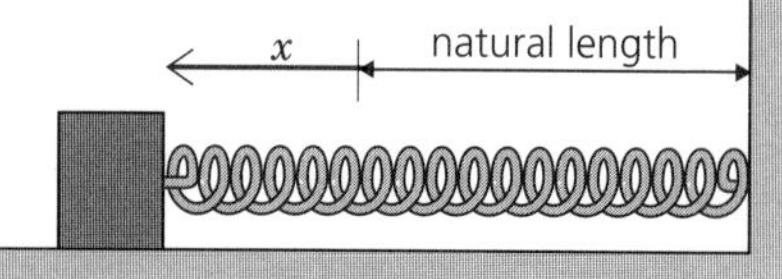

8 When rally cars are travelling at speed on rough roads they sometimes leave the ground on humps in the road. A certain car, of mass 1200 kg has suspension springs each of stiffness $7.0 \times 10^4\,\text{N m}^{-1}$. Assuming that the weight of the car is uniformly distributed over the four wheels and that the maximum height reached by the car above the horizontal surface on which it lands is 0.75 m, what is the instantaneous compression of each spring on landing?

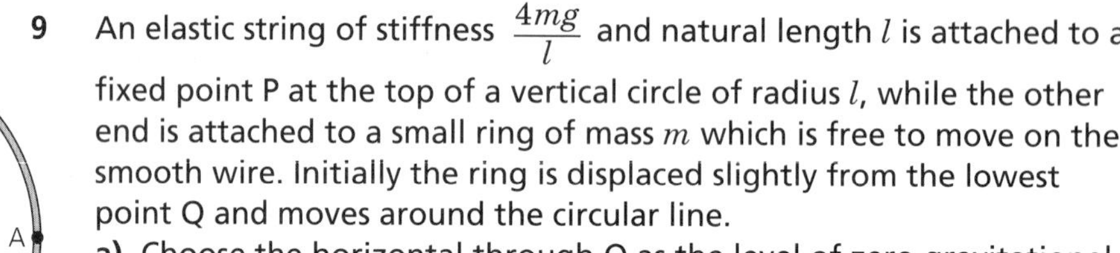

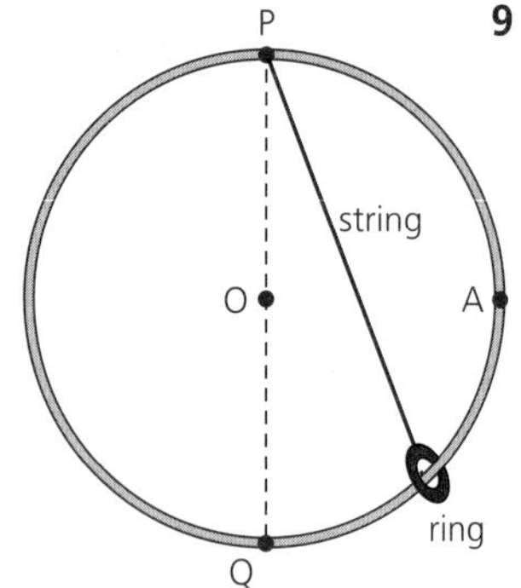

9 An elastic string of stiffness $\frac{4mg}{l}$ and natural length l is attached to a fixed point P at the top of a vertical circle of radius l, while the other end is attached to a small ring of mass m which is free to move on the smooth wire. Initially the ring is displaced slightly from the lowest point Q and moves around the circular line.

a) Choose the horizontal through Q as the level of zero gravitational potential energy of the system and hence write down the total energy of the system when the ring is at Q.

b) If the velocity of the ring when it reaches the point A (at the same horizontal level as O) is v write down the total energy of the system when the ring is at P. Apply the conservation of energy principle to obtain an expression for v.

c) Use an approach similar to **b)** to obtain an expression for the velocity of the ring when it reaches the position where the string becomes slack.

10 A block of mass 1 kg moves on a horizontal track and is accelerated by a compressed spring. While in contact with the spring the block moves on a smooth surface but when the spring attains its natural length the surface becomes rough (with a coefficient of friction of 0.2).

The spring stiffness is $100\,\text{N m}^{-1}$. Determine the initial compression of the spring if the block travels a distance 0.5 m on the rough surface before coming to rest.

WORK DONE AND THE SCALAR PRODUCT

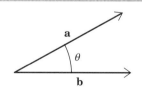

The scalar product is a product of two vectors that gives a result that is a scalar quantity. The scalar product of two vectors **a** and **b** is defined as:

$$\mathbf{a}.\mathbf{b} = ab\cos\theta$$

where a and b are the magnitudes of **a** and **b** and θ is the angle between them. The scalar product can be very useful for finding the work done by a force, when the forces and displacements are expressed as vectors.

Exploration 11.4

Investigating scalar products

- With the unit vectors $\mathbf{i}$, $\mathbf{j}$ and $\mathbf{k}$ defined in the normal way, consider the scalar product of various combinations of other unit vectors.
- What is the scalar product of a vector with itself?

Scalar products

You will probably have decided that:

$\mathbf{i}.\mathbf{i} = 1 \qquad \mathbf{j}.\mathbf{j} = 1 \qquad \mathbf{k}.\mathbf{k} = 1$

and that:

$\mathbf{i}.\mathbf{j} = 0 \qquad \mathbf{i}.\mathbf{k} = 0 \qquad \mathbf{j}.\mathbf{k} = 0$

The properties defined above are very useful as they make it very easy to find the scalar product of vectors expressed in the form:

$x\mathbf{i} + y\mathbf{j} + z\mathbf{k}$

You will recall that the definition of work done is:

work done $= Fs\cos\theta$

F θ s

If instead of simply allowing s to be a distance, we define it as a **displacement vector**, the result above is simply:

work done $= \mathbf{F}.\mathbf{s}$

Example 11.9

The force $\mathbf{F} = 3\mathbf{i} + 2\mathbf{j}$ acts while a particle is displaced by the vector $\mathbf{s} = 4\mathbf{i} + \mathbf{j}$. Find the work done by the force.

Solution

Work done $= \mathbf{F}.\mathbf{s} = (3\mathbf{i} + 2\mathbf{j}).(4\mathbf{i} + \mathbf{j}) = 3\mathbf{i}.4\mathbf{i} + 3\mathbf{i}.\mathbf{j} + 2\mathbf{j}.4\mathbf{i} + 2\mathbf{j}.\mathbf{j}$

$= 12\mathbf{i}.\mathbf{i} + 3\mathbf{i}.\mathbf{j} + 8\mathbf{j}.\mathbf{i} + 2\mathbf{j}.\mathbf{j}$

Since $\mathbf{i}.\mathbf{j} = 0$, $\mathbf{j}.\mathbf{i} = 0$ and $\mathbf{i}.\mathbf{i} = 1$, $\mathbf{j}.\mathbf{j} = 1$, this reduces to:

work done $= \mathbf{F}.\mathbf{s} = 12 + 2 = 14J$

In practice we would not write down the $\mathbf{i}.\mathbf{j}$ and $\mathbf{j}.\mathbf{i}$ terms as they are zero.

Example 11.10

Sometimes the vector $a\mathbf{i} + b\mathbf{j}$ is written as $\begin{pmatrix} a \\ b \end{pmatrix}$.

a) *Show that* $\begin{pmatrix} a \\ b \end{pmatrix}.\begin{pmatrix} c \\ d \end{pmatrix} = ac + bd$

b) *Find* $\begin{pmatrix} 4 \\ 7 \end{pmatrix}.\begin{pmatrix} 8 \\ -2 \end{pmatrix}$

Solution

a) *First convert to the usual notation, to give:*

$$\begin{pmatrix} a \\ b \end{pmatrix}.\begin{pmatrix} c \\ d \end{pmatrix} = (a\mathbf{i} + b\mathbf{j}).(c\mathbf{i} + d\mathbf{j}) = a\mathbf{i}.c\mathbf{i} + b\mathbf{j}.d\mathbf{j}$$

since $\mathbf{i}.\mathbf{j}$ and $\mathbf{j}.\mathbf{i} = 0$.

Then:

$$\begin{pmatrix} a \\ b \end{pmatrix}.\begin{pmatrix} c \\ d \end{pmatrix} = (a\mathbf{i} + b\mathbf{j}).(c\mathbf{i} + d\mathbf{j}) = ac\,\mathbf{i}.\mathbf{i} + bd\,\mathbf{j}.\mathbf{j} = ac + bd$$

Note that the top numbers are simply multiplied together, then the bottom numbers are multiplied and the two results added.

b) *Using part* **a)**:

$$\begin{pmatrix}4\\7\end{pmatrix}\cdot\begin{pmatrix}8\\-2\end{pmatrix}=4\times8+7\times(-2)=32-14=18$$

Example 11.11

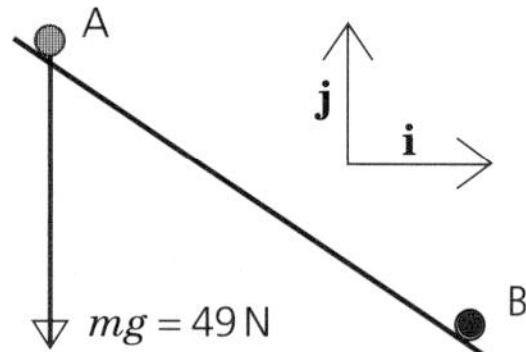

A particle of mass 5 kg moves from A to B. The position vector of A is $3\mathbf{i}+10\mathbf{j}$ *and the position of B is* $8\mathbf{i}+5\mathbf{j}$.

a) *Find the work done as the particle moves from A to B.*

b) *If the speed of the particle at A was 2 m s*$^{-1}$*, find the speed at B.*

Solution

a) *The force can be expressed as* $\mathbf{F}=0\mathbf{i}-49\mathbf{j}$.
The displacement is:
$\mathbf{s}=(8\mathbf{i}+5\mathbf{j})-(3\mathbf{i}+10\mathbf{j})=5\mathbf{i}-5\mathbf{j}$
The work done can now be found using the scalar product.
Work done $=\mathbf{F}.\mathbf{s}=(0\mathbf{i}-49\mathbf{j}).(5\mathbf{i}-5\mathbf{j})=245$ *J*

b) *Using:*
work done = change in KE
gives:

$$245=\tfrac{1}{2}\times5v^2-\tfrac{1}{2}\times5\times2^2=2.5v^2-10$$

$$v^2=\frac{255}{2.5}$$

$$v=10.1\ m\ s^{-1}$$

EXERCISES

11.4A

1 In each case below, find the work done by using the scalar product. All forces are in newtons and displacements in metres.

a) $\mathbf{F}=2\mathbf{i}+3\mathbf{j}$ $\quad\mathbf{s}=4\mathbf{i}-7\mathbf{j}$

b) $\mathbf{F}=10\mathbf{i}+6\mathbf{j}$ $\quad\mathbf{s}=50\mathbf{i}+2\mathbf{j}$

c) $\mathbf{F}=6\mathbf{i}+2\mathbf{j}$ $\quad\mathbf{s}=3\mathbf{i}-72\mathbf{j}$

d) $\mathbf{F}=6\mathbf{i}+6\mathbf{j}+10\mathbf{k}$ $\quad\mathbf{s}=3\mathbf{i}+2\mathbf{j}+4\mathbf{k}$

2 In each case find v given the force $\mathbf{F}$, the displacement $\mathbf{s}$ and the mass, m.

a) $\mathbf{F}=3\mathbf{i}+10\mathbf{j}$ $\quad\mathbf{s}=2\mathbf{i}+4\mathbf{j}$ $\quad u=2\,\mathrm{m\,s^{-1}}$ $\quad m=4\,\mathrm{kg}$

b) $\mathbf{F}=\mathbf{i}+\mathbf{j}$ $\quad\mathbf{s}=10\mathbf{j}$ $\quad u=1\,\mathrm{m\,s^{-1}}$ $\quad m=1\,\mathrm{kg}$

c) $\mathbf{F}=-4\mathbf{i}+10\mathbf{j}$ $\quad\mathbf{s}=3\mathbf{i}+2\mathbf{j}$ $\quad u=0.5\,\mathrm{m\,s^{-1}}$ $\quad m=2\,\mathrm{kg}$

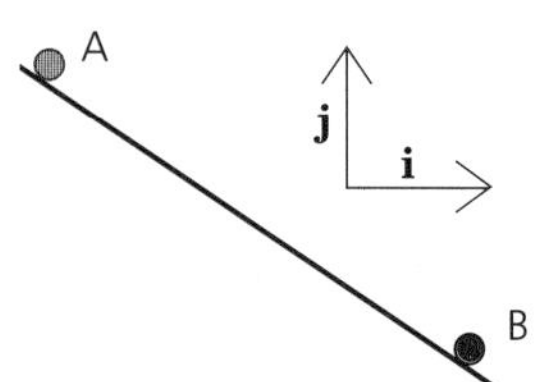

3 The position vector of point A is $10\mathbf{j}$ and the position vector of point B is $8\mathbf{i}+2\mathbf{j}$. A particle of mass 10 kg moves from A to B under the action of gravity alone.

a) Find the work done by gravity, assuming that there are no resistance forces present.

b) If the particle is moving at 3 m s^{-1} at A, find its speed at B.

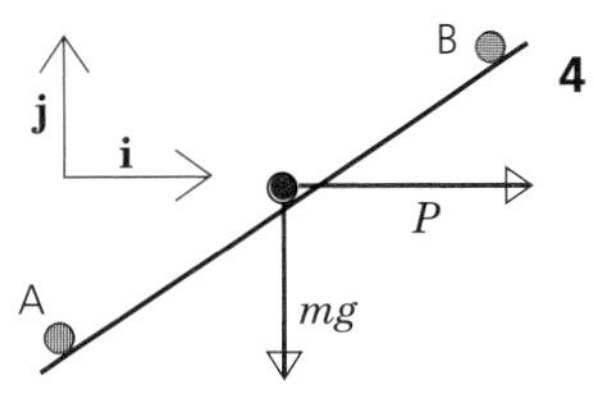

4 The forces shown act on a particle of mass 2 kg as it moves from A which has position $2\mathbf{i} + \mathbf{j}$ to B which has position $10\mathbf{i} + 5\mathbf{j}$.

a) Find the total work done in terms of P.

b) Find P if the speed of the particle does not change.

c) Find P if the speed of the particle increases from $2\,\text{m}\,\text{s}^{-1}$ to $3\,\text{m}\,\text{s}^{-1}$.

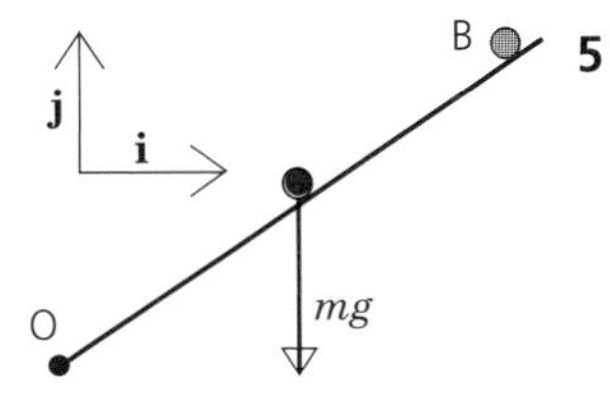

5 A particle of mass 3 kg has speed $10\,\text{m}\,\text{s}^{-1}$ when at the origin O and comes to rest at B. Assume that the slope is smooth and the particle travels along the slope. The position vector of B is $a\mathbf{i} + b\mathbf{j}$.

a) Find the work done in terms of a and b.

b) By considering the change in kinetic energy, find b.

c) Explain why there is no restriction on the value of a.

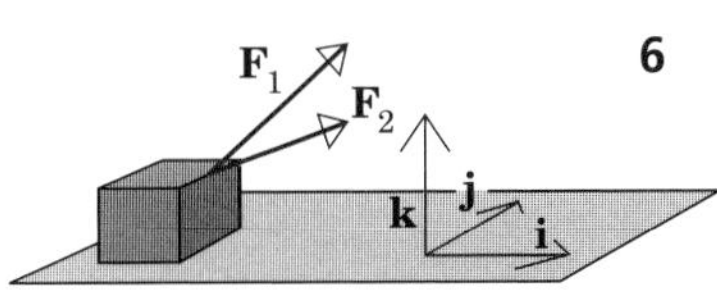

6 A box of mass 5 kg is acted on by two forces $\mathbf{F}_1 = 4\mathbf{i} + 6\mathbf{j} + 6\mathbf{k}$ and $\mathbf{F}_2 = 3\mathbf{i} + 2\mathbf{j} + 2\mathbf{k}$. The box slides on the smooth horizontal surface.

a) Find the magnitude of the normal reaction force on the box.

b) Explain why the force of gravity and the normal reaction do no work.

c) The displacement of the box is later $14\mathbf{i} + 16\mathbf{j}$ from its original position. Find the work done by $\mathbf{F}_1$ and $\mathbf{F}_2$.

d) What is the speed of the box at this point, if it was initially at rest.

7 Two forces, $4\mathbf{i} + 4\mathbf{j}$ and $6\mathbf{i} - 2\mathbf{j}$ act on a particle. Find the displacement if the particle moves in the direction of the resultant force and the work done by the forces is 156 J.

8 A particle moves so that its position vector is of the form $a\mathbf{i} + 2a\mathbf{j}$. The force $30\mathbf{i} + 20\mathbf{j}$ acts on the particle and does 350 J of work. Find the displacement of the particle.

9 When the force $5\mathbf{i} + 10\mathbf{j}$ acts on an object of mass 0.5 kg, its speed increases from $2\,\text{m s}^{-1}$ to $5\,\text{m s}^{-1}$ as it moves 2 m.

a) Find the gain in kinetic energy.

b) Find the magnitude of the force.

c) Find the angle between the force and the direction of motion.

10 The work done by the force $15\mathbf{i} - 10\mathbf{j}$ is 90 J, as an object moves 10 m. Find the displacement of the object in terms of $\mathbf{i}$ and $\mathbf{j}$.

EXERCISES

11.4 B

1 In each case below find the work done by using the scalar product.

a) $\mathbf{F} = \mathbf{i} + \mathbf{j} + \mathbf{k}$ $\quad \mathbf{s} = 2\mathbf{i} + 3\mathbf{j}$

b) $\mathbf{F} = -2\mathbf{i} + 3\mathbf{j} + 6\mathbf{k}$ $\quad \mathbf{s} = -\mathbf{i} + \mathbf{j}$

c) $\mathbf{F} = \mathbf{i} - 3\mathbf{j}$ $\quad \mathbf{s} = 3\mathbf{i} - 4\mathbf{j} + 12\mathbf{k}$

d) $\mathbf{F} = 3\mathbf{k}$ $\quad \mathbf{s} = 2\mathbf{i} - 3\mathbf{j}$

2 In each case find v given the force $\mathbf{F}$, the displacement $\mathbf{s}$, the initial speed v and the mass, m.

a) $\mathbf{F} = \mathbf{i} + 3\mathbf{j}$ $\quad \mathbf{s} = 5\mathbf{j}$ $\quad u = 2\,\text{m}\,\text{s}^{-1}$ $\quad m = 2\,\text{kg}$

b) $\mathbf{F} = 2\mathbf{i} + 3\mathbf{j} + 6\mathbf{k}$ $\quad \mathbf{s} = \mathbf{i} + 2\mathbf{j} + 3\mathbf{k}$ $\quad u = 3\,\text{m}\,\text{s}^{-1}$ $\quad m = 4\,\text{kg}$

c) $\mathbf{F} = \mathbf{i} + 2\mathbf{k}$ $\quad \mathbf{s} = \mathbf{j}$ $\quad u = 4\,\text{m}\,\text{s}^{-1}$ $\quad m = 1\,\text{kg}$

3 The position vector of the point A is $2\mathbf{i} + 2\mathbf{j} + 4\mathbf{k}$ and the position vector of the point B is $2\mathbf{i} + \mathbf{j} + 2\mathbf{k}$. A particle of mass 5 kg moves from A to B under the action of gravity alone (assume the positive axis is directed vertically upwards).

a) Determine the work done by gravity in the absence of any other forces.

b) If the speed of the particle at A is $2\,\text{m}\,\text{s}^{-1}$, find its speed at B.

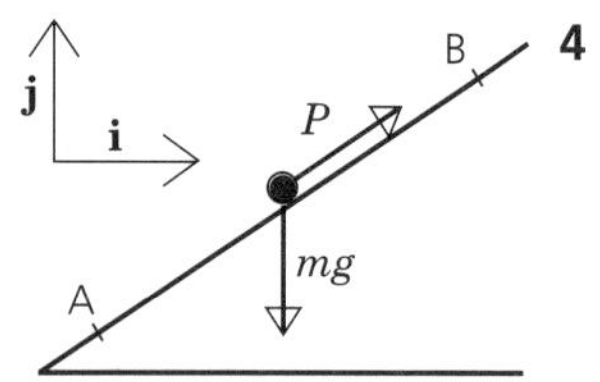

4 The forces shown act on a particle of mass 2 kg as it moves from A, which has position vector $2\mathbf{i} + 3\mathbf{j}$, to B which has position vector $3\mathbf{i} + 4\mathbf{j}$.

a) Find the total work done in terms of P.

b) If the speed of the particle at A is $2\,\text{m}\,\text{s}^{-1}$ and at B is $4\,\text{m}\,\text{s}^{-1}$ determine the magnitude of P.

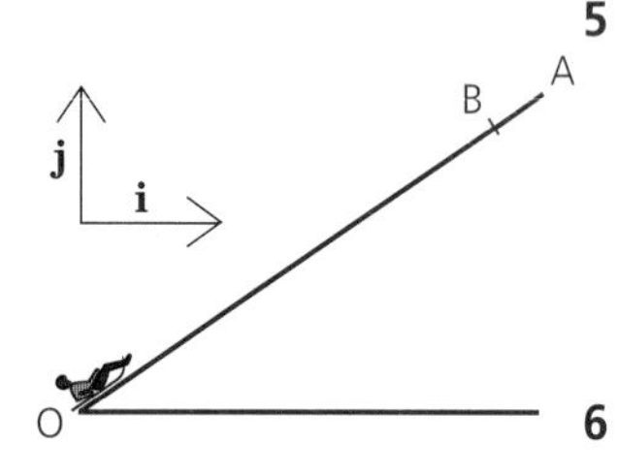

5 A toboggan and rider of total mass 80 kg have a speed of $10\,\text{m}\,\text{s}^{-1}$ as they embark on the slope OA at O. They come to rest at B.

a) Find the work done in terms of the coordinates (a, b) of B assuming that the motion takes place under gravity alone.

b) By considering the change in kinetic energy of the toboggan determine b.

6 A box of mass 3 kg slides on a smooth horizontal surface under the action of two forces $\mathbf{F}_1 = 2\mathbf{i} + 3\mathbf{j} + 6\mathbf{k}$ and $\mathbf{F}_2 = \mathbf{i} + \mathbf{j} + \mathbf{k}$. (You may assume that the horizontal surface is represented by the x-y plane and that the positive z-axis is directed vertically upwards from this.)

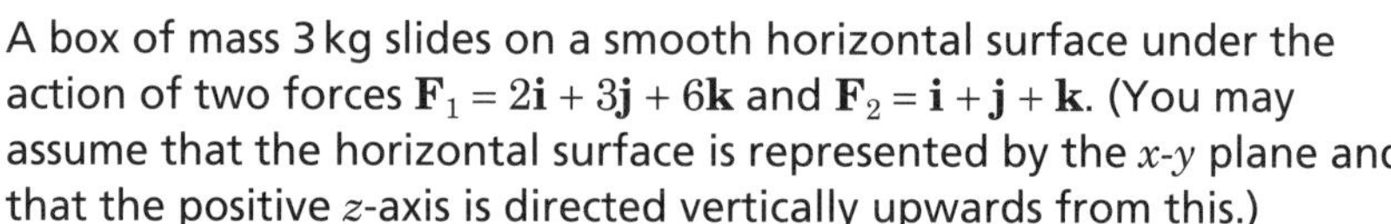

a) Find the magnitude of the normal reaction force on the box.

b) Why do neither the gravitational force nor the normal reaction do any work as the box moves?

c) Assuming that the box started from rest at the origin and that at some later time it is at the point with position vector $3\mathbf{i} + 4\mathbf{j}$ determine the work done by $\mathbf{F}_1$ and $\mathbf{F}_2$ and hence calculate the speed of the box at this point.

7 A particle located at the point A $(3\mathbf{i} + 2\mathbf{j} - \mathbf{k})$ is acted upon by a force $4\mathbf{i} - 3\mathbf{j} + 2\mathbf{k}$ which does 15 J of work in moving it, in the direction of the force, to B. Determine the coordinates of B.

8 When a force $30\mathbf{i} + 40\mathbf{j}$ acts on a body of mass 5 kg its speed increases from $2\,\text{m}\,\text{s}^{-1}$ to $4\,\text{m}\,\text{s}^{-1}$ while it moves a distance of 3 m.

a) Find the increase in kinetic energy of the body.

b) Find the magnitude of the force.

c) Find the angle between the force and the direction of motion.

9 A particle moves under the action of a force $\mathbf{F} = 3\mathbf{i} + 6\mathbf{k}$ such that its position vector is of the form $2a\mathbf{i} + a\mathbf{k}$.
Determine the displacement of the particle such that 54 J of work is done in the displacement.

10 The work done by the force $\mathbf{F} = 30\mathbf{i} + 40\mathbf{j}$ in moving an object 10 m is 300 J. Determine the angle between the force and the displacement.

CONSOLIDATION EXERCISES FOR CHAPTER 11

1 A spring has spring constant of 500 N m^{-1}.

a) What is the work done if it is stretched from its natural length by 60 cm?

b) What is the work done in stretching it a further 60 cm?

(SMP Question B2, Modelling with Circular Motion, Specimen Paper, 1994)

2 A ring of mass 1 kg is moved along a smooth horizontal wire from A to B, under the action of a force $\mathbf{F}$.
The vector from A to B, $\mathbf{AB}$, has magnitude 10 metres and direction $\begin{pmatrix}3\\4\end{pmatrix}$ and $\mathbf{F} = \begin{pmatrix}2\\3\end{pmatrix}$ newtons.

a) Find the work done by the force in moving the ring from A to B.

b) Given that the speed of the ring at A is 5 m s^{-1}, find its speed when it reaches B.

c) What effect would it have on your answer to **b)** if the ring were rough?

(SMP Question A1, Modelling with Circular Motion Specimen Paper, 1994)

3 A steel ball B, of mass 0.125 kg, is attached to one end of a light elastic string OB, the end O being attached to the ceiling. The modulus of elasticity of the string is 52.5 N, and the string has natural length 1.5 m. In equilibrium the ball is at E. Show that the depth of E below O is 1.54 m, correct to three significant figures. The ball is released from rest at O, and does not hit the floor. State an assumption necessary for conservation of energy to apply. Hence find:

a) the speed as B passes through E,

b) the maximum depth of B below O.

(UCLES Question 7, M2 Specimen Paper, 1994)

4 A mountain rescue team is investigating whether or not to use a new type of flexible rope. One end is attached to the top, S, of a fixed crane and a harness is attached to the other end. Kirsty, a member of the team, whose mass together with that of the harness is 60 kg, is lowered gently until she hangs at rest. The stretched rope is then 21 m long.

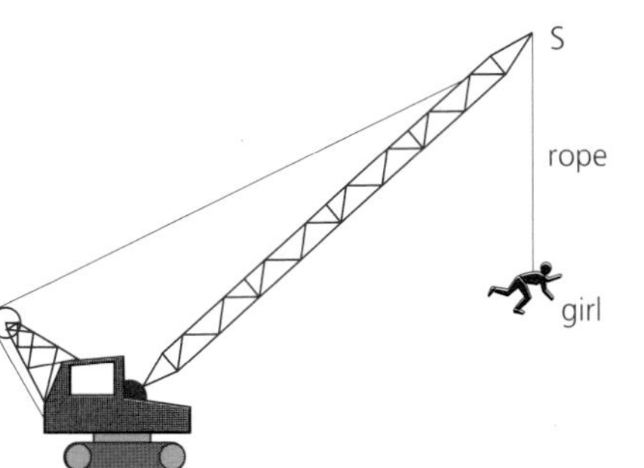

a) By modelling the rope as a light elastic string and Kirsty as a particle, estimate the modulus of elasticity of the rope.

Kirsty climbs back to S and releases herself to fall vertically, strapped in the harness attached to the rope. She comes to instantaneous rest at the point C at the end of her descent.

b) Estimate the length SC of the stretched rope,

c) Estimate the greatest speed that Kirsty achieves during her descent.

State any further assumptions you have made in modelling Kirsty's descent from S to C.

(ULEAC Question 9, Specimen Paper M2, 1994)

5 Show that the energy stored in a string of natural length a and modulus of elasticity λ, when stretched by a length x, is

$$\left(\frac{\lambda}{2a}\right)x^2$$

A string of natural length $2a$ and modulus of elasticity λ has its ends fixed to two points, A and B, which are at the same horizontal level and a distance $2a$ apart. The centre of the string is pulled back to a point C in the same horizontal plane as A and B such that ABC forms an equilateral triangle.
Find the tension in the stretched string and the energy stored in it.
A small mass, m, is placed inside the stretched string at the point C and the string released. The mass is catapulted through the mid-point of AB. Neglecting the effects of gravity find the speed of the mass as it leaves the string.

(MEI Question 2, Paper M3, January 1992)

MATHEMATICAL MODELLING ACTIVITY

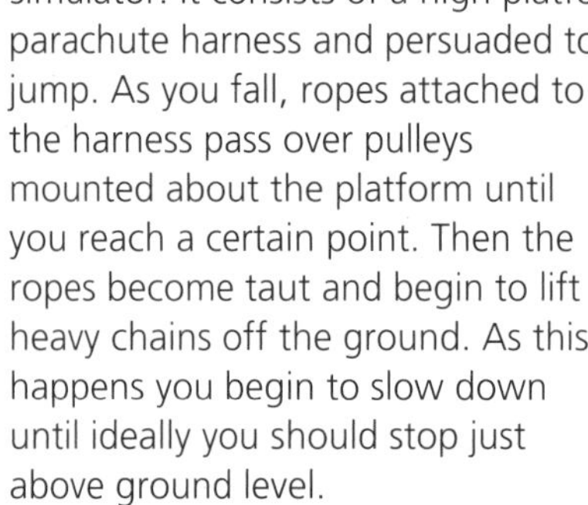

Specify the real problem

Problem statement

At the Eskdale Outward Bound Centre in the Lake District, there is a free-fall simulator. It consists of a high platform, where you are strapped into a parachute harness and persuaded to jump. As you fall, ropes attached to the harness pass over pulleys mounted about the platform until you reach a certain point. Then the ropes become taut and begin to lift heavy chains off the ground. As this happens you begin to slow down until ideally you should stop just above ground level.

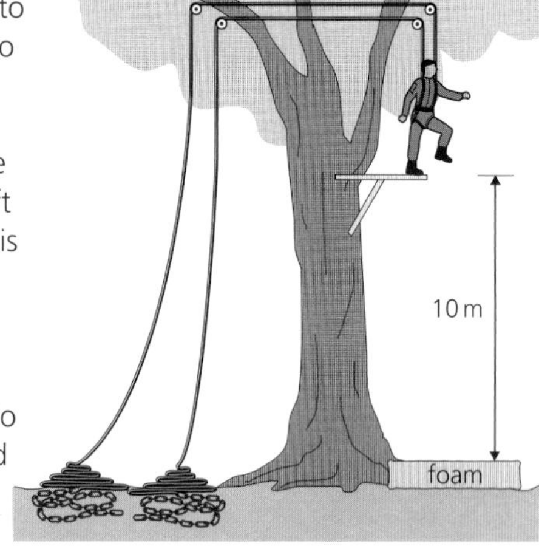

How far can a person be allowed to free-fall before the chains are lifted off the ground?

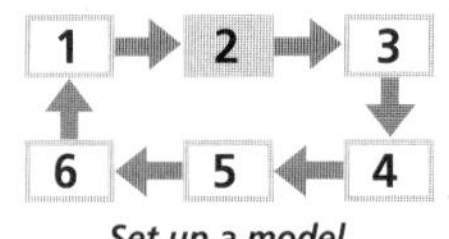

Set up a model

Set up a model

Identify the important features that should be considered and set out a list of assumptions that will allow a simple model to be formulated.
The list below gives a set of assumptions that could be used to solve this problem.

- There is no air resistance present.
- There is no friction in the system.
- The masses of the ropes are negligible.
- The person is modelled as a particle of mass m.
- The chains are modelled as a particle of mass M.
- The person falls a distance x before the ropes become taut and start to lift the mass.
- The speed of the person reaches zero as they reach the ground.

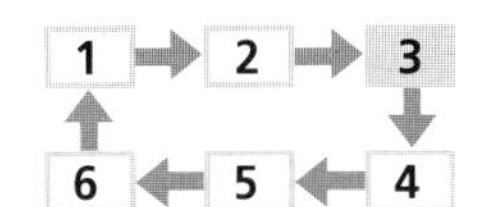

Formulate the mathematical problem

Formulate a mathematical problem

The total energy of the system should remain constant, so the initial potential energy of the person will be transformed to the final potential energy of the mass. Express x in terms of m and M.

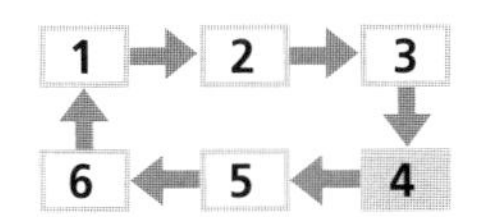

Solve the mathematical problem

Solve the mathematical problem

The initial potential energy of the person is simply:
$mgh = m \times 9.8 \times 10 = 98m$
If the person falls a distance x before the ropes become taut, then the mass will rise a distance $10 - x$, to that its final potential energy is:
$Mgh = M \times 9.8 \times (10 - x) = 98M - 9.8Mx$
Equating the two energies gives:
$98m = 98M - 9.8Mx$
Show that:

$$x = \frac{10(M - m)}{M}$$

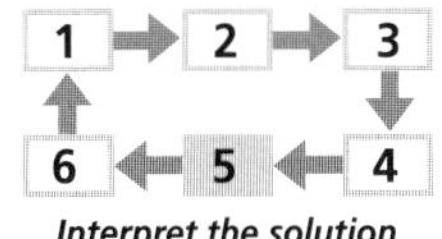

Interpret the solution

Interpreting the solution

Consider a person of mass 70 kg, so that:

$$x = \frac{10(M - 70)}{M}$$

The graph shows x against M.

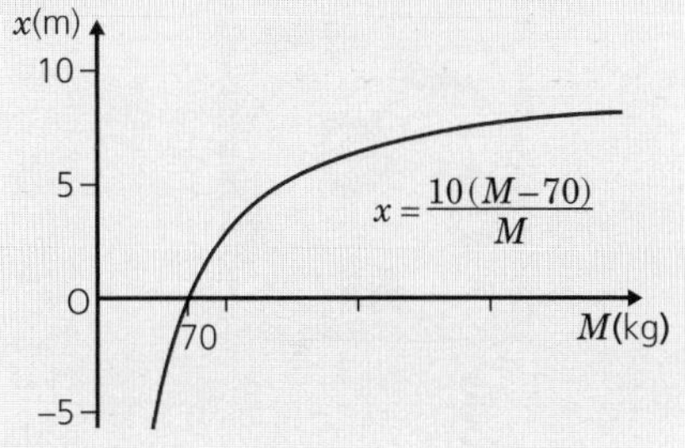

From the graph it is possible to see that the mass M must be at least 70 kg and that this will not allow any free-fall. To fall for a distance of 5 m requires a mass of 140 kg.

Note also how the gradient of the graph decreases. This shows that for a small increase in the free-fall distance, increasingly greater masses are needed. The result of this will be larger and larger impulses on the person as the rope becomes taut.

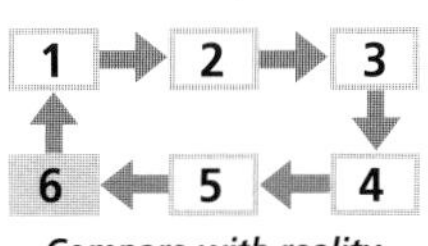

Compare with reality

Compare with reality

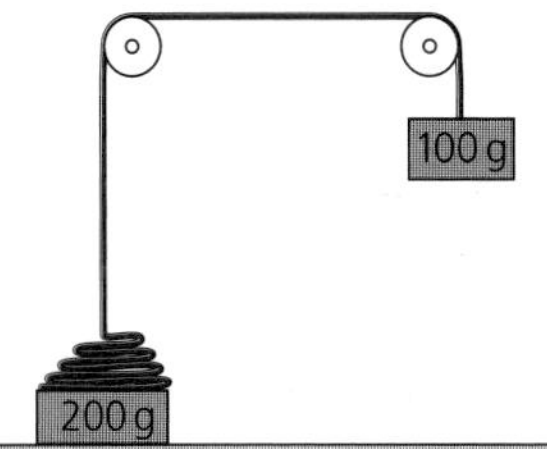

It will be difficult to compare with the real simulator, but you can set up a simple experiment as shown right, predicting the amount of slack needed for the 100 gram mass *just* to touch the floor.
The obvious criticism that must be made of this simple model is that there will be a very substantial jerk when rope becomes taut. A refined model using a chain should be considered.
Other factors that will affect the solution are air resistance and friction. As energy will be lost by both of these, they will not cause a danger to the person on the simulator. If they are a problem the mass could be gradually reduced until a person does just reach the ground.

Refining the model

In reality, a chain is used instead of a single mass. Assume that each metre of the chain has a mass of 40 kg.
Show that when h m of chain have been lifted off the ground the work done against gravity is $98h^2$.
As for the simple model the potential energy lost by the person will be $98m$.
If the person free-falls a distance x then the length of chain lifted is $10 - x$, so the potential energy gained by the chain is $98(10 - x)^2$.
Equating these energies gives:

$$98(10 - x)^2 = 98m$$

or:

$$(10 - x)^2 = m$$

Rearranging for m gives:

$$x = 10 - \sqrt{m}$$

So for a person of mass 70 kg the free-fall distance is $x = 10 - \sqrt{70} = 1.6$ m

POSSIBLE MODELLING TASKS

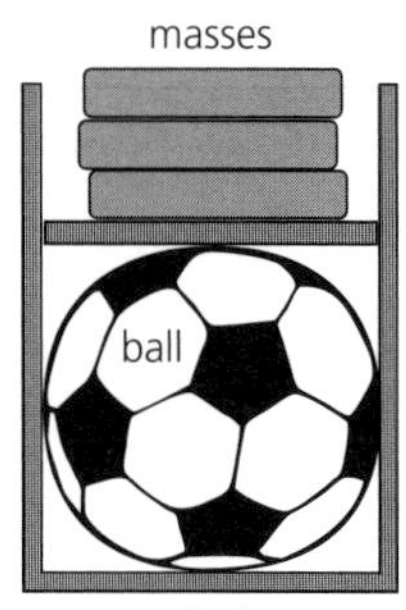

1 Place a ball in a cylinder and load it with various masses. Try to develop a model of how the compression is linked to the forces applied. Estimate how much compression the ball would experience when dropped from different heights.
2 Drop a ball into a tank of water from different heights. Do you think that the upward force on the ball will vary with the depth? Try to formulate a suitable model, to test your ideas.
3 The latest attraction at the Lightwater Valley Theme Park in North Yorkshire is known as the 'Ultimate roller coaster' and is claimed to be the longest ride in the world. It is 2.27 km long, contains two 30 m drops and it is predicted will reach speeds of 96 km h^{-1}. Verify these predictions.
4 The picture shows a fire escape that was installed at a 26-storey hotel in Benidorm. The chute is a big elasticated tube that has been coated with a fire resistant material. Investigate how objects would move in such a chute.

Instructions
1. Take your shoes off. Sit with your legs hanging into the chute. Hold the upper chute firmly. Slide slowly into the chute.
2. If you fall too fast you can slow down by moving your arms and legs outwards.
3. Handicapped and unconscious people can be taken down on the shoulders of other people.
4. Hold small children in your arms to protect them from the sides.

Summary

After working through this chapter you should:

- know and be able to use Hooke's law in the form:

  $T = ke$ or $T = \dfrac{\lambda e}{l}$

- know that the work done by a variable force is:

  $\int F \, \mathrm{d}x$

- know that the elastic potential energy in a spring is:

  $\frac{1}{2}ke^2 \quad \text{or} \quad \dfrac{\lambda e^2}{2l}$

- know that the work done can be found using the scalar product **F.r**.

Circular motion at constant speed

- *There are many examples of circular motion that can be modelled in a simple way using circular motion.*
- *In this chapter we explore the specific requirements that the forces on an object must satisfy if it is to move in a circle.*

CIRCULAR PATHS

Exploration 12.1

These objects follow circular paths.

- a conker on a string,
- a car navigating a sharp icy corner,
- a fairground 'chair-o-plane' ride.

There are many questions we could ask about motions of this type.

- What happens as a fairground ride slows down?
- What happens if a car hits black ice?
- What happens if the conker string breaks?

Discuss the following questions for each example.

- What forces are acting on the object?
- How would you describe circular motion?
- What makes objects move in circles?
- What specific problems might you meet in each case?

Horizontal circular motion

These are examples of **horizontal circular motion** in which each object moves in a horizontal plane. In this chapter we shall explore motion of this type, in which the object moves with constant speed. However, although the speed is constant, since the direction of the object keeps changing its direction of motion the velocity is not constant. Consequently the object is accelerating so that there is a net force acting on it.

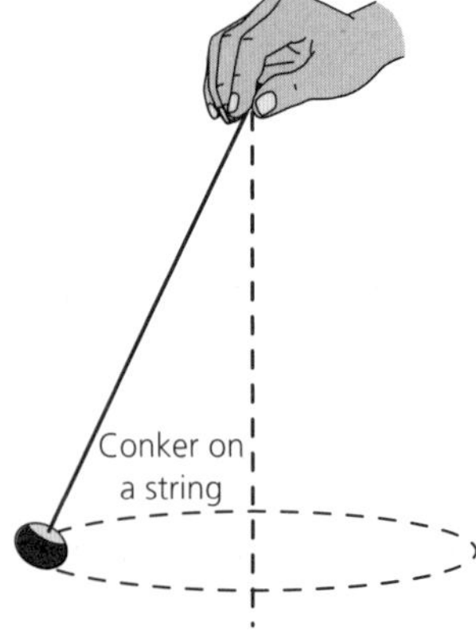

Your own experience of the three examples may have led you to say that:

- as the fairground ride slows down, the radius of the rider's motion decreases,
- if the car hits black ice the car will tend to skid,
- if the conker string breaks the conker will fly off initially along the tangent to the circular path.

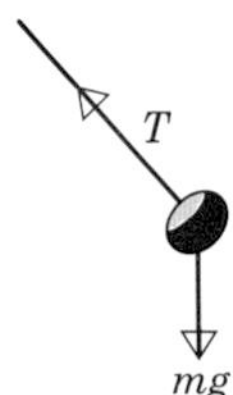

Consider the motion of the conker in more detail. There are two forces acting on the conker – the force of **gravity** and the **tension** in the conker string.

The features of the motion that we would want to describe might be:

- the speed of the conker,
- the angle that the string makes with the vertical,
- how large the tension in the string needs to be to maintain the motion.

Clearly these features are likely to be related. In this chapter, we shall be investigating problems associated with the size of these features and relationships between them. One important point to observe from each of the three examples of circular motion is that to maintain circular motion there needs to be a force towards the centre of the circle. If we cut the conker string, circular motion is no longer possible and the conker becomes a projectile moving freely under gravity.

Exploration 12.2

Practical activity – Penny on a turntable

This practical activity uses equipment from the Unilab Mechanics Kit.

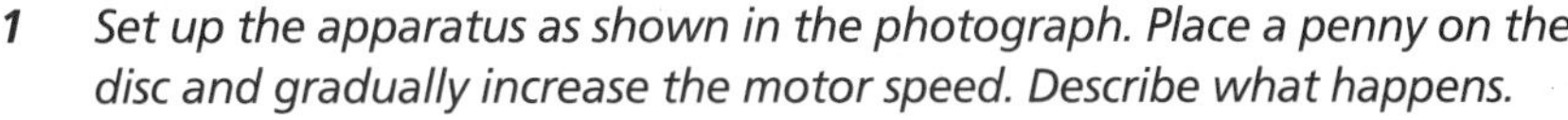

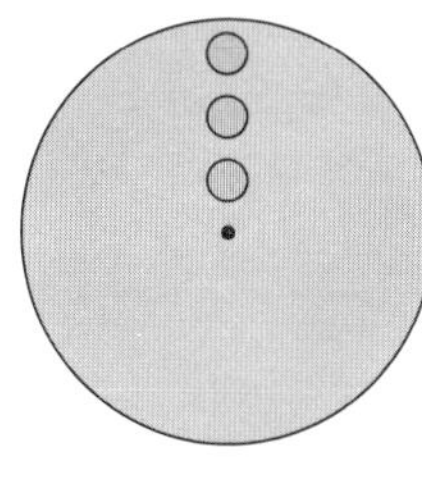

1. *Set up the apparatus as shown in the photograph. Place a penny on the disc and gradually increase the motor speed. Describe what happens.*
2. *Describe the path of the penny when it is not slipping. Explain why the penny does not have a constant velocity.*
3. *Explain why the penny is accelerating.*
4. *What forces act on the penny while it is moving in a circle? Draw diagrams to show these forces.*
5. *Which will slip first, a penny near the centre or a penny near the edge of the disc? Place three pennies on the disc, as shown, and gradually increase the motor speed.*
6. *Will a heavy coin slip at the same motor speed as a lighter coin? Try out a simple experiment a few times to see if you get consistent results.*
7. *Place a coin close to the edge of the disc. How does it move when it leaves the disc?*

Results

This physical situation is modelled fully in Example 12.7, when you can compare your answers with the mathematical solution.

CALCULATOR ACTIVITY 12.1

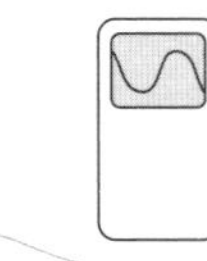

Firstly scale your axes using the range settings:

$x_{min} = -6$; $x_{max} = 6$
$y_{min} = -4$; $y_{max} = 4$
$T_{min} = 0$; $T_{max} = 2\pi$; Step/Ptch = 0.1

1 With your calculator in parametric plotting mode and radian mode, plot the path of an object that moves with position vector $\mathbf{r} = \cos t\,\mathbf{i} + \sin t\,\mathbf{j}$.
Describe the path that is produced.

2 Now try these.
a) $\mathbf{r} = 2\cos t\,\mathbf{i} + 2\sin t\,\mathbf{j}$
b) $\mathbf{r} = 4\cos t\,\mathbf{i} + 4\sin t\,\mathbf{j}$

3 Clear your screen. Now plot the path of an object that moves with position vector $\mathbf{r} = \cos(2t)\,\mathbf{i} + \sin(2t)\,\mathbf{j}$.
What differences do you notice when compared with $\mathbf{r} = \cos t\,\mathbf{i} + \sin t\,\mathbf{j}$?

Results

In each case the position vector is in the form $\mathbf{r} = r\cos(\omega t)\,\mathbf{i} + r\sin(\omega t)\,\mathbf{j}$ and the path of the object is a circle. The greater the value of r the larger is the radius of the circle and the greater the value of ω the more quickly the circle is completed. These features will be important as we model motion in a circle with constant speed.

Position vector and angular speed

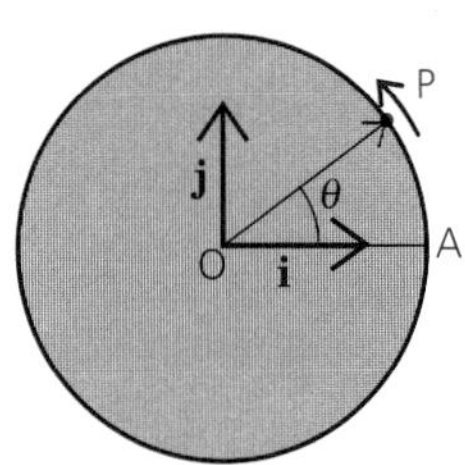

Consider an object that moves with constant speed around a circular path with centre O and radius r. If the object starts at point A, then suppose that after t seconds the position vector **OP** will have turned through an angle θ. So at time t the position vector of the object will be:

$$\mathbf{r} = r\cos\theta\,\mathbf{i} + r\sin\theta\,\mathbf{j}.$$

As time increases so θ increases. The rate of increase of θ is called the **angular speed** and is usually denoted by the Greek letter omega, ω. If the speed of the object is constant then ω is also constant.

Hence $\dfrac{d\theta}{dt} = \omega$ and integrating this we find $\theta = \omega t$.

In this integration we have set the constant of integration to zero assuming that $t = 0$ when the object is at the point A.
We can now write the position vector of an object moving in a circle with constant speed as a function of time t. This gives:

$$\mathbf{r} = r\cos(\omega t)\,\mathbf{i} + r\sin(\omega t)\,\mathbf{j}$$

for appropriate perpendicular unit vectors **i** and **j**.

In circular motion angles are measured in radians, so the angular speed ω is usually measured in radians per second ($rad\,s^{-1}$). Sometimes the angular speed may be measured in other units, for example in revolutions per minute (rpm).

Example 12.1

The small hand of a clock completes one revolution every 12 hours and the big hand one revolution every hour. Find the angular speed of each hand in rad s⁻¹ and rpm.

Solution

The small hand completes one revolution or 2 radians in 12 hours. or 43 200 s. So the angular speed in radians per second is:

$$\omega = \frac{2}{43200} = 1.45 \times 10^{-4}\,rad\,s^{-1}$$

The big hand completes one revolution or 2π radians in 1 hour, or 3600 seconds. So the angular speed in radians per second is:

$$\omega = \frac{2}{3600} = 1.75 \times 10^{-3}\,rad\,s^{-1}$$

To express this in revolutions per minute the small hand completes one revolution in $12 \times 60 = 720$ minutes so we have:

$$\omega = \frac{1}{720} = 1.39 \times 10^{-3}\,rpm$$

and for the big hand

$$\omega = \frac{1}{60} = 1.67 \times 10^{-2}\,rpm$$

Example 12.2

The flywheel in a motorcycle engine is rotating at 5000 rpm. Find its angular speed in $rad\,s^{-1}$.

Solution

The flywheel completes 5000 revolutions in one minute or 60 seconds. So it turns through $2\pi \times 5000 = 10\,000\pi$ radians in 60 seconds.

So the angular speed is given by $\omega = \frac{10000}{60} = 524\,rad\,s^{-1}$

Velocity and angular velocity

The **speed** of a car travelling round a roundabout is shown on the car's speedometer but its **velocity** is a **vector** with **direction**.

When specifying the velocity of an object moving with a circular path, we need to identify its direction as well as its magnitude. We will show that for circular motion at constant speed, at any instant the direction of the velocity is along the tangent to the circular path. For **angular velocity** the magnitude is the angular speed, discussed above, but its direction is a bit more tricky. In fact it is defined to be perpendicular to the plane of the motion, although the explanation for this will have to wait. For the moment it is convenient to take the angular velocity as positive for an anticlockwise rotation and negative for a clockwise rotation.

In order to obtain the velocity vector for an object moving round a circle, it is simply necessary to differentiate the position vector with respect to time. Starting with

$$\mathbf{r} = r\cos(\omega t)\,\mathbf{i} + r\sin(\omega t)\,\mathbf{j}$$

and differentiating gives

$$\mathbf{v} = -r\omega\sin(\omega t)\,\mathbf{i} + r\omega\cos(\omega t)\,\mathbf{j}$$

The magnitude of this vector is the actual speed of the object describing circular motion.

$$\begin{aligned} v = |\mathbf{v}| &= \sqrt{(-r\omega\sin(\omega t))^2 + (r\omega\cos(\omega t))^2} \\ &= \sqrt{r^2\omega^2\sin^2(\omega t) + r^2\omega^2\cos^2(\omega t)} \\ &= \sqrt{r^2\omega^2(\sin^2(\omega t) + \cos^2(\omega t))} \\ &= \sqrt{r^2\omega^2} \quad (\text{since } \sin^2(\omega t) + \cos^2(\omega t) = 1) \\ &= r\omega \end{aligned}$$

So the speed of the object is given by the product of the radius and the angular speed. It is sometimes useful to express this in the form

$$\omega = \frac{v}{r}$$

The velocity vector can be expressed as $\mathbf{v} = -v\sin(\omega t)\,\mathbf{i} + v\cos(\omega t)\,\mathbf{j}$.

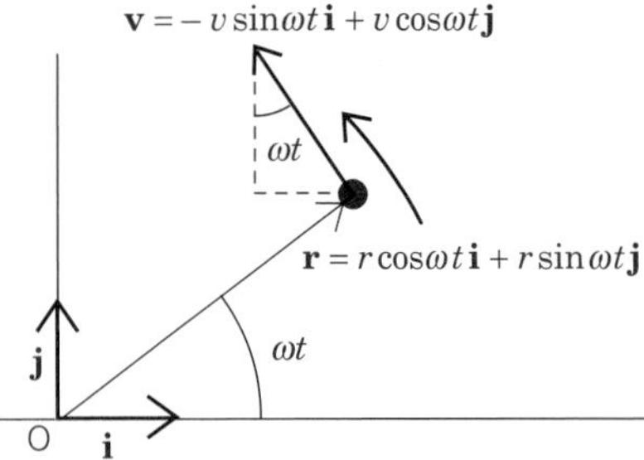

At each instant of time, the velocity vector of an object in circular motion is at right angles to its position vector. This is illustrated in the diagram.

The position vector points outwards along a radius from the centre of motion. This is called the **radial direction**. The velocity vector points at right angles to this radial direction and this is called the **transverse direction**. Notice that the transverse direction is that of ωt increasing in an anticlockwise sense.

Example 12.3

A fairground roundabout takes three seconds to complete one revolution. Rachel and Ben sit on seats that are 5 m and 3 m from the centre of rotation respectively.

For each child find:

a) *their position vector and velocity vector,*

b) *their speed.*

Solution

First we need to find the angular speed. The ride completes one revolution or 2π radians in three seconds so

$$\omega = \frac{2}{3} \text{ rad s}^{-1}$$

a) *Choosing unit vectors* $\mathbf{i}$ *and* $\mathbf{j}$ *in the horizontal plane, the position vectors will be given in the form:*

$\mathbf{r} = r\cos(\omega t)\,\mathbf{i} + r\sin(\omega t)\,\mathbf{j}$

So for Ben:

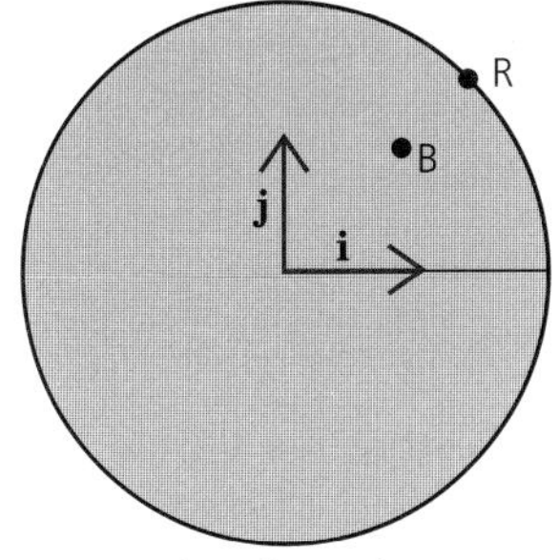

Looking from above

$$\text{r} = 3\cos\left(\frac{2\ \ t}{3}\right)\mathbf{i} + 3\sin\left(\frac{2\ \ t}{3}\right)\mathbf{j}$$

and differentiating gives:

$$\mathbf{v} = -2\ \sin\left(\frac{2\ \ t}{3}\right)\mathbf{i} + 2\ \cos\left(\frac{2\ \ t}{3}\right)\mathbf{j}$$

For Rachel:

$$\mathbf{r} = 5\cos\left(\frac{2\ \ t}{3}\right)\mathbf{i} + 5\sin\left(\frac{2\ \ t}{3}\right)\mathbf{j}$$

and differentiating gives:

$$\mathbf{v} = -\frac{10}{3}\ \sin\left(\frac{2\ \ t}{3}\right)\mathbf{i} + \frac{10}{3}\ \cos\left(\frac{2\ \ t}{3}\right)\mathbf{j}$$

b) *Their speeds are given by $v = r\omega$, so for Ben:*

$$v = 3 \times \frac{2}{3} = 2\pi = 6.28\, m\, s^{-1}$$

and for Rachel:

$$v = 5 \times \frac{2}{3} = \frac{10}{3} = 10.47\, m\, s^{-1}$$

Example 12.4

A motorcyclist describes part of a circle as she goes round a roundabout. The radius of the circle is 20 m and the speed of the motorcyclist is 30 mph. Find the angular speed (in rad s^{-1}) and the position vector of the motorcyclist in terms of suitable unit vectors.

Solution

Converting 30 mph to m s^{-1} gives

$$v = \frac{30 \times 1609}{3600} = 13.4\, m\, s^{-1}.$$

The angular speed ω can be obtained using

$$\omega = \frac{v}{\text{r}} = \frac{13.4}{20} = 0.67\, rad\, s^{-1}.$$

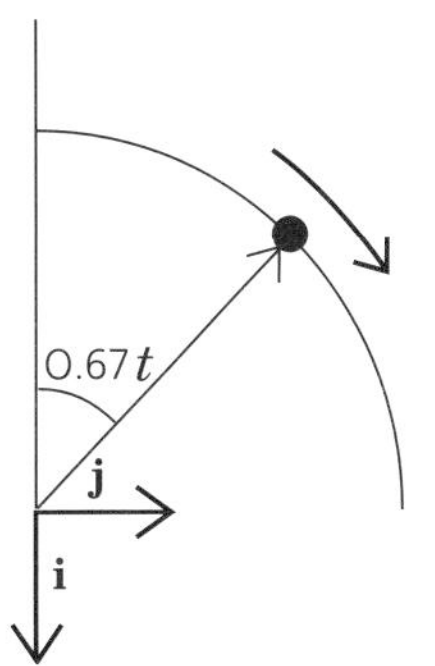

So the position vector will be given by: $\mathbf{r} = -20\cos(0.67t)\mathbf{i} + 20\sin(0.67t)\mathbf{j}$ *where* $\mathbf{i}$ *acts along the direction of approach to the roundabout and* $\mathbf{j}$ *is perpendicular to* $\mathbf{i}$ *as shown.*

EXERCISES

12.1 A

1 The Earth rotates about its axis once every 24 hours.
   a) What is the angular speed of the Earth about its axis?
   b) The radius of the Earth is 6.4×10^6 m. What is the speed of a person on the surface of the Earth at the equator?

2 A car is moving at $10\,\text{m s}^{-1}$ and has wheels of radius 25 cm.
   a) How long does it take for the wheel to complete one revolution?
   b) What is the angular speed of the wheel?

3 Complete the table below.

Object	Angular speed (rad s^{-1})	Time for one revolution	Revolutions per minute
Engine flywheel			500
Record			45
Fairground ride	$\pi/3$		
Planet		20 hours	

4 A car with wheels of diameter 50 cm tows a trailer that has wheels of diameter 40 cm. Find the angular speed of both wheels when the car and trailer are travelling at $20\,\text{m s}^{-1}$.

5 An object moves with position vector

$$\mathbf{r} = 4\cos(3t)\mathbf{i} + 4\sin(3t)\mathbf{j}$$

   a) What is the speed of the object?
   b) What is the angular speed of the object?
   c) How long would it take to complete one revolution?
   d) How many revolutions would it complete in one minute?

6 A chain passes over a circular sprocket of radius 10 cm. If the chain is moving at $0.8\,\text{m s}^{-1}$, find the angular speed of the sprocket.

7 Emma gets into a big wheel at the fairground and sits on a seat that is at its lowest point. The wheel has radius 5 m and takes 10 seconds to complete one revolution. Using unit vectors that are horizontal and vertical, find position and velocity vectors for Emma at time t after the ride starts if:
   a) the origin is at the centre of the wheel,
   b) the origin is at the lowest point of the ride.

8 The moon completes one orbit around the Earth every 28 days. Assume that the moon describes a circular path of radius 3.84×10^8 m. Calculate the speed at which the moon is moving.

9 A satellite is in geostationary orbit, which means it always stays above the same point on the Earth's surface. It is at a height of 4.24×10^7 m above the centre of the Earth, and it always stays above the same point on the equator. Find the speed of this satellite.

10 A spin drier rotates at 1100 rpm and has a drum of radius of 22 cm. Find the speed of the clothes in the drier.

EXERCISES

12.1 B

1 A fairground ride rotates once every five seconds.
a) What is the angular speed of the ride?
b) If the radius of the ride is 5 m, what is the speed of a person at the edge of the ride?

2 A bicycle has wheels of radius 40 cm. The bicycle is moving at $5\,\mathrm{m\,s^{-1}}$.
a) Find the circumference of the wheel in metres.
b) How long does it take for the wheel to complete one revolution?
c) Find the angular speed of the wheel.

3 For each situation below, convert the angular speed from rpm to $\mathrm{rad\,s^{-1}}$ and find the time for one revolution.
a) A motorcycle engine's flywheel rotating at 5000 rpm.
b) A record rotating at 33 rpm.
c) The PTO shaft of a tractor rotating at 200 rpm.
d) A clockwork mobile that it is rotating at 10 rpm.

4 An object moves so that its position is given by:

$$\mathbf{r} = 3\cos(2t)\mathbf{i} + 3\sin(2t)\mathbf{j}$$

where $\mathbf{i}$ and $\mathbf{j}$ are perpendicular unit vectors.
a) Find an expression for the velocity of the object.
b) What is the speed of the object?
c) How long does it take for the object to complete one revolution?

5 A drive belt is moving at $20\,\mathrm{m\,s^{-1}}$ as it passes over a pulley, of radius 0.15 m.
a) Assume that the belt does not slip on the pulley and find the angular speed of the pulley in $\mathrm{rad\,s^{-1}}$.
b) Convert the angular speed to rpm.

6 A car is moving at $20\,\mathrm{m\,s^{-1}}$. The wheels of the car have radius 25 cm. A small mass is attached to the wheel 15 cm from the centre.
a) Find the angular speed of the wheel.
b) What is the speed of the mass?

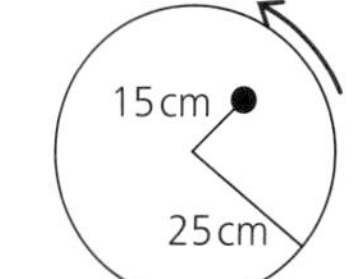

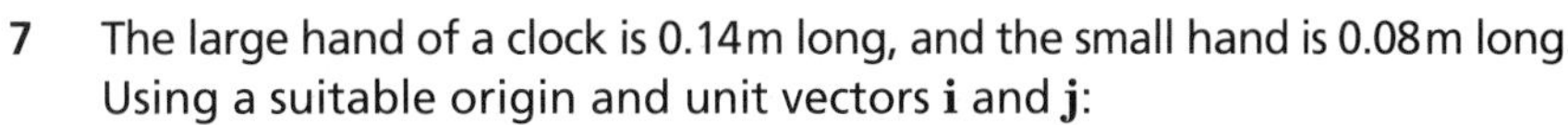

7 The large hand of a clock is 0.14 m long, and the small hand is 0.08 m long. Using a suitable origin and unit vectors $\mathbf{i}$ and $\mathbf{j}$:
a) find an expression for the position vectors of the tip of each hand,
b) find the velocity of the tip of each hand.

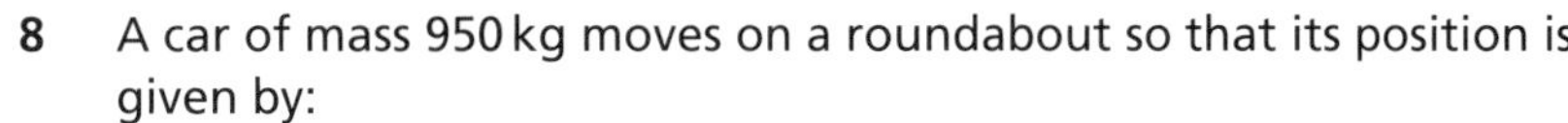

8 A car of mass 950 kg moves on a roundabout so that its position is given by:

$$\mathbf{r} = 50\cos\left(\frac{\pi}{10}t\right)\mathbf{i} + 50\sin\left(\frac{\pi}{10}t\right)\mathbf{j}$$

where $\mathbf{i}$ and $\mathbf{j}$ are perpendicular unit vectors.
a) Find the velocity of the car.
b) Show that the kinetic energy of the car is constant, and find its value.

9 A particle is constrained to move in a circle by a length of string so that its position vector is:

$$\mathbf{r} = 0.2\cos(20t)\mathbf{i} + 0.2\sin(20t)\mathbf{j}$$

a) Find the velocity of the particle.
b) When $t = 3$ the string breaks, find the velocity of the particle at this time.
c) Find the position of the particle when $t = 5$, stating any assumptions you make.

10 A child sits on the edge of a roundabout of radius 2.5 m holding a ball.
a) Find the speed of the ball if the roundabout rotates once every two seconds.
b) Sketch the path of the ball if the child lets go of it. Assuming that the ball is 1.5 m above ground level when released, how long does it take to reach the ground?
c) If the child throws the ball horizontally towards the centre of the roundabout at $3\,\text{m s}^{-1}$, find its actual velocity. Sketch the path of the ball as viewed from above.

THE ACCELERATION VECTOR

The acceleration vector is one of the most important vectors in mechanics because it acts in the same direction as the resultant force acting on the object. If you know the acceleration you can find the force and vice-versa. We have seen above that although the speed of circular motion might be constant the velocity is changing because the direction of motion is continually changing. Where there is a change of velocity an object has an acceleration, but how are its magnitude and direction related to radius r and angular speed ω?

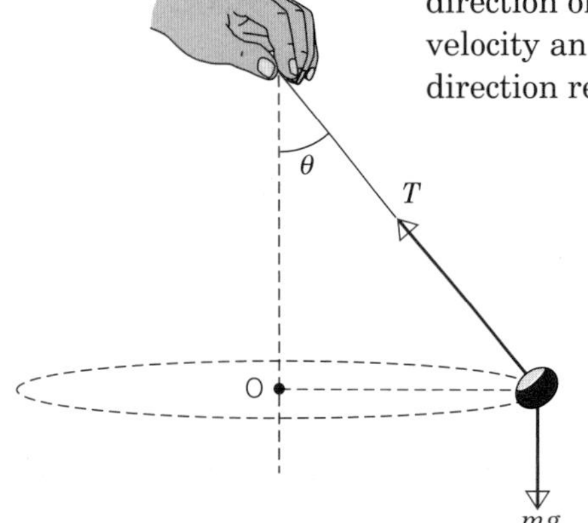

To get some insight into the answer, consider the motion of a conker on a string moving in a horizontal circle with constant speed. We can identify two forces acting on the conker: the **tension** force along the string and the force of **gravity**. The vertical component of the tension will balance the force of gravity ($T\cos\theta = mg$) and the horizontal component of the tension ($T\sin\theta$) is the resultant force. This points towards the centre of the circle, so we might deduce that the acceleration vector also points towards the centre of the circle (using Newton's second law).

The magnitude is perhaps more difficult to deduce. Intuitively, we could argue that as the conker is swung harder the radius of the circle and the angular speed of the conker both increase so that the angle θ increases. Then mathematically we can say that as the angle θ increases the tension increases (since $T\cos\theta = mg$) and so the acceleration will increase.

So from this physical situation we can suggest that for circular motion with constant speed, the acceleration vector:

- points inwards along the radius
- has magnitude which increases when r and ω increase.

Now we shall see if we can confirm these results using calculus. The acceleration vector can be obtained by differentiating the velocity vector with respect to time. Starting with:

$$\mathbf{v} = -r\omega\sin(\omega t)\mathbf{i} + r\omega\cos(\omega t)\mathbf{j}$$

and differentiating gives:

$$\begin{aligned}\mathbf{a} &= -r\omega^2\cos(\omega t)\mathbf{i} - r\omega^2\sin(\omega t)\mathbf{j}\\ &= -\omega^2(r\cos(\omega t)\mathbf{i} + r\sin(\omega t)\mathbf{j})\\ &= -\omega^2\mathbf{r}\end{aligned}$$

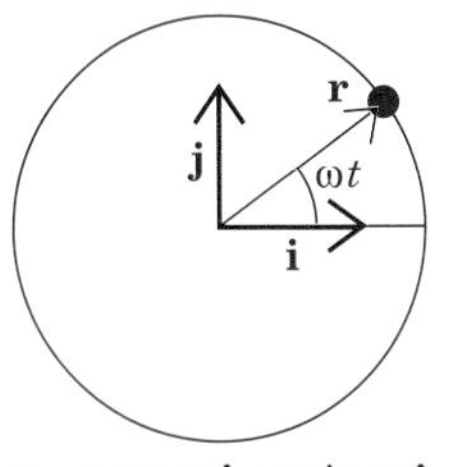

$\mathbf{r} = r\cos\omega t\,\mathbf{i} + r\sin\omega t\,\mathbf{j}$
$\mathbf{v} = -r\omega\sin\omega t\,\mathbf{i} + r\omega\cos\omega t\,\mathbf{j}$

The acceleration vector is related to the position vector. The negative sign indicates that the acceleration vector is directed towards the centre of the circle. The magnitude of the acceleration is $r\omega^2$ where r is the radius of the circle. This confirms the experimental evidence for the conker on a string.

It is sometimes useful to be able to express the acceleration of an object describing a circle in terms of its speed rather than its angular speed. Substituting for $\omega = \dfrac{v}{r}$ in $a = r\omega^2$ we get:

$$a = \frac{v^2}{r}$$

We now have the main results needed to model any circular motion with constant speed. These results are summarised in the following figures.

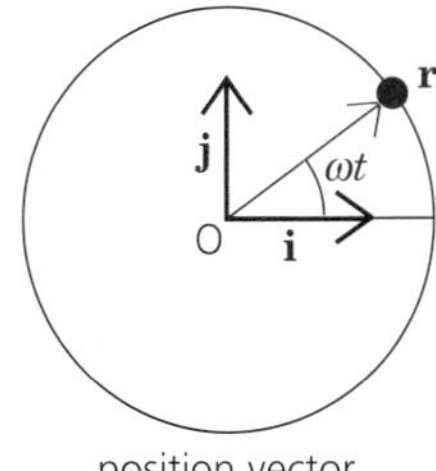

position vector
$\mathbf{r} = -r\cos\omega t\,\mathbf{i} + r\sin\omega t\,\mathbf{j}$

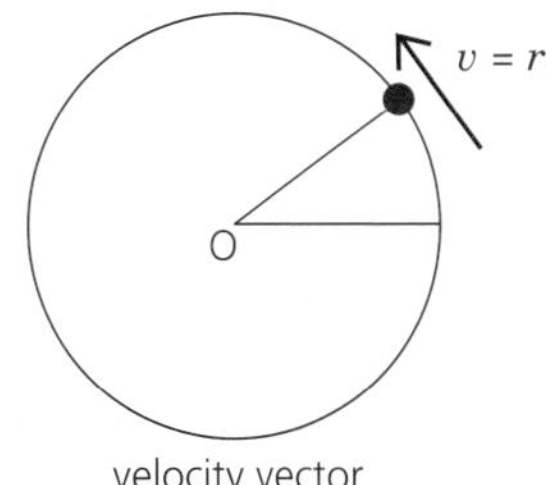

velocity vector
$\mathbf{v} = -v\sin\omega t\,\mathbf{i} + v\cos\omega t\,\mathbf{j}$

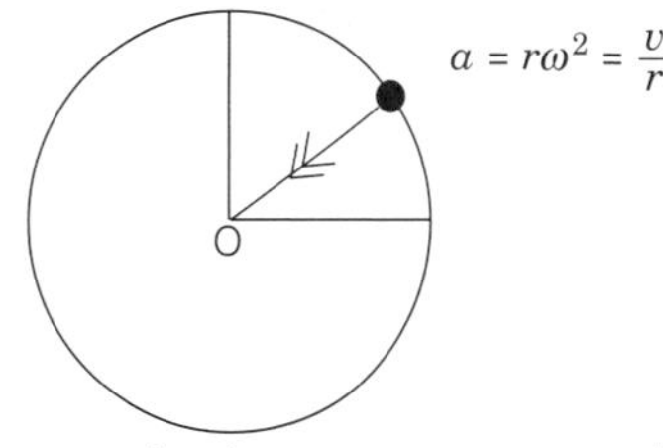

acceleration vector
$\mathbf{a} = -a\cos\omega t\,\mathbf{i} - a\sin\omega t\,\mathbf{j}$

Newton's second law tells us that the resultant force required for any motion acts in the same direction as the acceleration. Hence the resultant force required for circular motion at constant speed to be sustained acts *towards* the centre of the circle. This may be intuitively obvious for the chairoplanes ride and the conker on a string. For the car going round the roundabout the frictional force directed towards the centre of the circle is essential to maintain the circular path.

Example 12.5

A car of mass 1100 kg is travelling at 12 m s^{-1} round a roundabout of radius 50 m. Find the acceleration of the car and the force needed to keep the car moving in a circle.

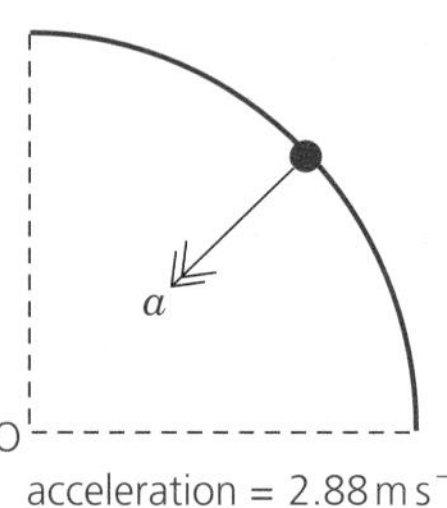

acceleration = 2.88 m s^{-2}
radially inwards

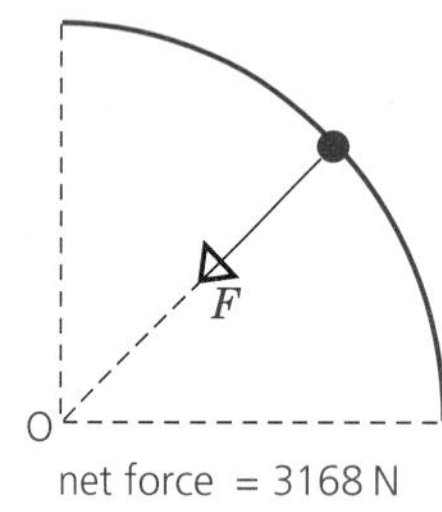

net force = 3168 N
radially inwards

Solution

The magnitude of the acceleration is given by:

$$a = \frac{v^2}{r} = \frac{12^2}{50} = 2.88\,m\,s^{-1}$$

As the car is describing the circle at constant speed, the acceleration must act towards the centre of the circle. The resultant force must also have the same direction, towards the centre of the circle. The magnitude of this resultant force is given by $F = ma = 1100 \times 2.88 = 3168$ N.
This force is supplied by friction.

Example 12.6

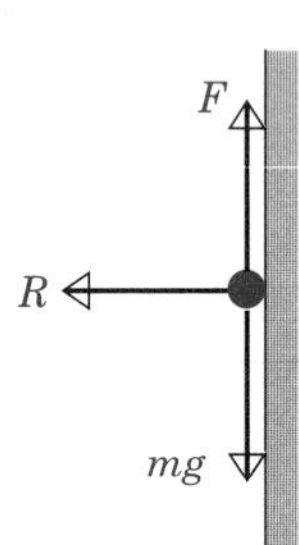

A fairground ride – 'wall of death' – consists of a cylindrical drum that rotates. Riders stand inside the drum and lean against the drum wall. When the drum rotates the floor is lowered, but the riders appear to stick to the sides of the drum i.e. they do not slide down the wall. The radius of the drum is 3 m and the coefficient of friction between the riders and the drum is 0.7. Find the minimum speed at which the drum must rotate if the riders are to stay in position as the floor drops.

Solution

Modelling the rider as a particle and identifying the forces acting on the rider gives the diagram shown. Gravity acts downwards, friction acts upwards (opposing the tendency of the rider to slide down) and a normal reaction acts towards the centre of the drum.
The friction force must balance the force of the gravity so:

$F = mg$ (1)

The normal reaction must provide an acceleration of $r\omega^2$ toward the centre so that:

$R = mr\omega^2 = 3m\omega^2$ (2)

As the coefficient of friction is 0.7, in the limiting case when the rider is about to slide:

$F = \mu R = 0.7R$ (3)

Using equations (1) and (3) gives:

$$0.7R = mg \Rightarrow R = \frac{mg}{0.7}$$

Now substituting into equation (2) gives:

$$\frac{mg}{0.7} = 3m\omega^2 \Rightarrow \omega^2 = \frac{g}{2.1} = 4.67$$

$\omega = 2.16$ *rad s*$^{-1}$

Example 12.7

Penny on a turntable

This example models the practical activity of Exploration 12.2.
The photograph shows a horizontal flat disc which rotates about an axis through its centre. A penny is placed on the disc and the disc rotates so that the penny remains at rest relative to the disc. Investigate the motion of the penny as the angular speed of the disc and the initial position of the penny changes.

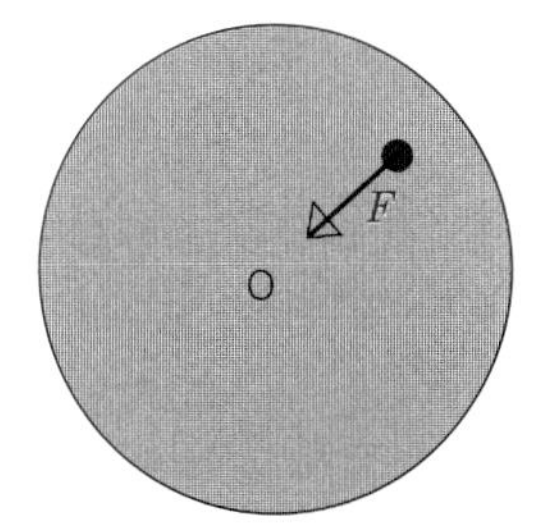

view from above

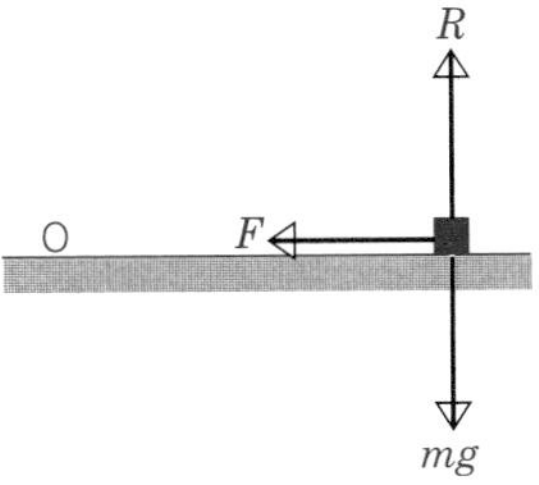

view from side

Solution

While the penny is at rest relative to the disc it describes a horizontal circle with constant speed, the same speed as the disc. The penny is accelerating because its direction of motion is changing. The figure shows the forces on the penny which is modelled as a particle.

*There are three forces acting on the penny: the force of **gravity**, the **normal reaction** and **friction** towards the centre. Remember that it is essential that the friction is towards the centre because the acceleration is towards the centre.*

The vertical forces are in balance so that $R = mg$

Applying Newton's second law to the particle gives $F = mr\omega^2$.

If the coefficient of friction is denoted by μ *then* $F \le \mu R$. *Substituting for* R *and* F *gives* $mr\omega^2 \le \mu R = \mu mg$

Dividing by m *we get* $r\omega^2 \le \mu g$.

For a given angular speed ω *the penny must be within a circle of radius* $\frac{\mu g}{\omega^2}$; *if the penny is placed on the disc outside this circle, it cannot remain at rest as the disc rotates. The motion of the penny relative to the disc is a complicated path as the photograph below shows. (The mathematical description of this motion is beyond the scope of this book.) Similarly, for a given initial position of the penny on the disc, the angular speed must be less than* $\sqrt{\frac{\mu g}{r}}$ *for the penny to remain at rest relative to the disc.*

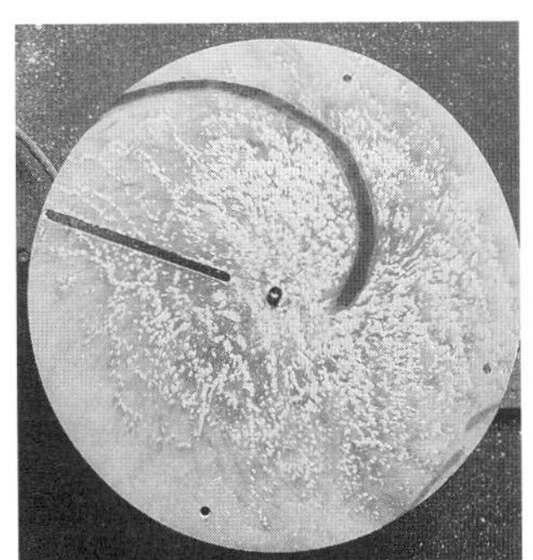

As the angular speed ω *increases the friction force* F *increases until limiting friction is reached. Beyond this critical value of* ω, *the friction is no longer sufficient to maintain the relative rest between the penny and disc so that the penny slides, following the path shown in the photograph.*

An important thing to notice about the solution is that it is independent of the mass of the penny. According to the theory, heavy masses will behave in the same way as light masses. However, once the penny begins to slide the mass does become important in the subsequent motion.

Look back at Exploration 12.2 and compare your answers with the mathematical description in this example.

Banked tracks

In a cycle velodrome, the track is banked so that the outside of the circuit is higher than the inside. This feature is also found on bends on some roads and on railway tracks. The photograph shows an example of banking.

The banking is intended to reduce the tendency for the vehicle to slide outwards as it travels round the circular path at high speeds. The next example shows how to model problems involving banked tracks.

Example 12.8

A cyclist is describing part of a circle of radius 30 m at a constant speed on a track banked at an angle of 30° to the horizontal. Find the speed of the cyclist for a model in which friction is assumed to be negligible.

Solution

Assume that the cyclist is a particle and neglect friction forces. In the plane perpendicular to the motion there are two forces: the force of gravity and the normal reaction.

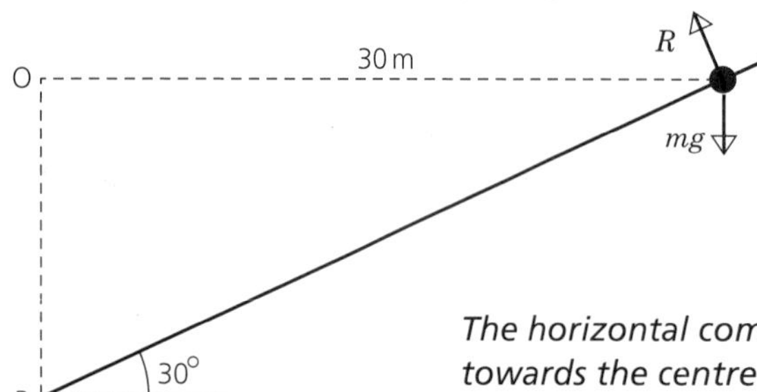

The particle is moving with constant speed in a circular path with centre O. (Note that the centre of the circle is in a horizontal plane not at the bottom of the track B.) The force of gravity must be balanced by the vertical component of R, so:

$$mg = R\cos 30° \qquad (1)$$

The horizontal component of R must provide the acceleration towards the centre of the circle otherwise circular motion would not be possible, so:

$$\mathrm{m}\frac{v^2}{30} = \mathrm{R}\cos 60° \qquad (2)$$

Eliminating R from equations (1) and (2) gives:

$$\mathrm{m}\frac{v^2}{30} = \left(\frac{\mathrm{mg}}{\cos 30°}\right)\cos 60°$$

Solving for v^2:

$$u^2 = \frac{30 \times \mathrm{g} \times \cos 60°}{\cos 30°}$$

and then $v = 13\,m s^{-1}$.

Without the banking, circular motion would not be possible, because without friction there would be no force towards the centre of the circle to provide the acceleration.

EXERCISES

12.2 A

The following data are used in some of these exercises.
The radius of the Earth is 6.37×10^6 m and the mass of the Earth is 5.98×10^{24} kg.
A geostationary orbit is one where the orbiting object remains above the same point on the Earth.

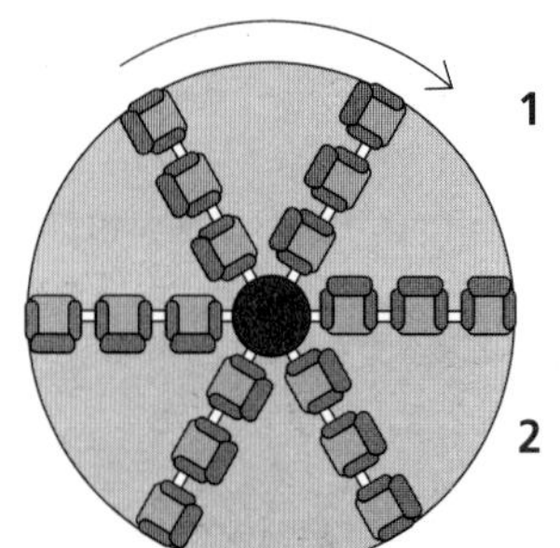

1 Calculate the acceleration experienced in each situation described below.

a) A car travelling at a constant 12 m s^{-1} on a bend of radius 80 m.

b) A child on a roundabout of radius 2 m in a park, that rotates at 1 rad s^{-1}.

c) A fairground big wheel, with radius 5 m, that completes one revolution every 10 seconds.

2 Where should you sit on the fairground ride shown if you want to experience the greatest acceleration? Explain why.

3 A car of mass 1200 kg is trying to negotiate a roundabout of radius 50 m. The coefficient of friction between the tyres and the road is 0.9.

a) Draw a diagram showing the forces acting on the car in the plane perpendicular to the direction of motion of the car.

b) Find the magnitude of the maximum friction force that can act on the car.

c) Calculate the maximum magnitude of the acceleration of the car.

d) Calculate the maximum speed that the car can travel round the roundabout.

e) Would a lighter car be able to get round the roundabout at a speed of $25\,\mathrm{m\,s^{-1}}$?

4 A satellite is to be placed in a circular orbit around the Earth.

a) Draw a diagram to show the forces acting on the satellite while it is in orbit.

b) The satellite is to orbit the Earth at a height of 450 000 m above the Earth's surface. Find the magnitude of the acceleration of the satellite. (**Hint:** At this distance from the Earth you will need to use Newton's law of universal gravitation.)

c) Find the angular speed of the satellite required for this orbit.

d) How long does it take for the satellite to orbit the Earth once?

5 An aeroplane of mass 20 000 kg is flying with constant velocity in a circular path of radius 500 m while waiting to land. Assume that the lift force on the aeroplane is perpendicular to the aeroplane.

a) Draw a diagram to show the forces acting on the aeroplane if it is banked at an angle of 15° to the horizontal.

b) By considering the vertical components of the forces acting, calculate the magnitude of the lift force.

c) Use your result from **b)** to find the resultant force on the aeroplane, and hence its acceleration.

d) Find the speed of the aeroplane.

e) The air traffic controllers want to instruct a lighter aeroplane to circle in an identical way. What speed should they specify?

6 **a)** Modelling the Earth as a sphere rotating about an axis through its centre, calculate the angular speed of the Earth in $\mathrm{rad\,s^{-1}}$. Hence deduce the angular speed of a satellite in geostationary orbit.

b) Find an expression for the resultant force on a satellite in a geostationary orbit above the equator in terms of its mass, m, and the radius, r, of the orbit.

c) Find an expression for the magnitude of the gravitational force acting on the satellite in terms of G, m and M, the mass of the Earth.

d) Use your answers to **b)** and **c)** to find the radius of the geostationary orbit.

7 A crate is placed on the back of a lorry but not secured in any way. The coefficient of friction between the crate and the back of the lorry is 0.6.

a) A bend has a radius of 20 m and the speed of the lorry is $15\,\mathrm{m\,s^{-1}}$. Does the crate slip?

b) What is the maximum speed that the lorry can travel round the bend, if the crate is not to slip?

8 A cycle track is circular and banked at 30° to the horizontal, with a radius of 60 m.

a) In a first model, ignoring the friction force, find the maximum speed of the cyclist.

b) If the coefficient of friction between the tyres and the track is 0.7, find the maximum speed of the cyclist on the track if friction is included.

9 **a)** Draw a diagram to show the forces acting on a person standing on the surface of the Earth.

b) Calculate the acceleration of the person due to the rotation of the Earth.

c) Use your answer to calculate the difference between the normal reaction and the force of gravity on a person of mass m.

d) Is it reasonable to ignore the effects of the Earth's rotation when considering objects at rest on the surface of the Earth?

10 On the spin cycle the drum of a washing machine rotates at 1100 rpm about a horizontal axis so that clothes in the drum move in a vertical circle with constant speed.

a) Draw diagrams to show the forces acting on an item of clothing, modelled as a particle of mass M, at the top and bottom of the drum.

b) Find the acceleration of the clothes in the drum, which has radius 45 cm.

c) Express the magnitude of the normal reaction in terms of M, at the top and bottom of the drum.

11 A motorcyclist performs a loop the loop stunt by riding inside a cylinder, which has a horizontal axis of symmetry and a radius of 4 m. The rider ensures that the motorbike maintains a constant speed at all times.

a) Describe what happens to the normal reaction force on the motorcycle as it completes the loop.

b) Explain why at the slowest safe speed the normal reaction can drop to zero at the top of the cylinder.

c) Calculate the slowest safe speed for the motorcycle.

EXERCISES

12.2 B

The following data are used in some of these exercises.
The radius of the Earth is 6.37×10^6 m and the mass of the Earth is 5.98×10^{24} kg.
A geostationary orbit is one where the orbiting object remains above the same point on the Earth.

1 Calculate the acceleration experienced in each situation described below.

a) A particle in a centrifuge that describes a circle of radius 8 cm, and rotates at 400 rpm.

b) An aeroplane travelling at 200 m s^{-1} in an arc of a circle of radius 500 m.

c) A jetski travelling at 5 m s^{-1} in a circle of radius 20 m.

2 Find the magnitude of the force required for each object to move as described.

a) A car of mass 1000 kg, travelling at 10 m s^{-1} on a roundabout of radius 60 m.

b) A children's roundabout in a playground that rotates at 3 rad s^{-1} (The child has mass 40 kg and sits 1.2 m from the centre of the roundabout.)

c) A satellite of mass 300 kg in a circular orbit of radius 7×10^8 m, which completes one revolution every 8 hours.

3 A shirt in a spin drier that rotates about a vertical axis can be modelled as a particle. The drum of the drier rotates at 1000 rpm and has radius 30 cm.

a) Find the acceleration of the shirt.

b) If the shirt has a mass of 1 kg when wet, find the reaction force exerted by the drum.

4 A particle moves so that its position vector is

$$\mathbf{r} = 6\cos(2\ t)\mathbf{i} + 6\sin(2\ t)\mathbf{j}$$

where $\mathbf{i}$ and $\mathbf{j}$ are perpendicular unit vectors.

a) Find the acceleration vector.

b) Find the magnitude of the acceleration.

5 A child of mass 42 kg is sitting on the edge of a roundabout of radius 1.4 m in a playground.

a) Find the normal reaction force acting on the child.

b) Find the angular speed of the roundabout when the child would slip off, if the coefficient of friction between the child and the roundabout is 0.6.

c) How would your answer change for a heavier or lighter child?

d) Do you think that there is likely to be an accident, while the roundabout is in normal use?

6 A car of mass 1200 kg travels on a road banked at 5° to the horizontal. Assume that there is no friction present.

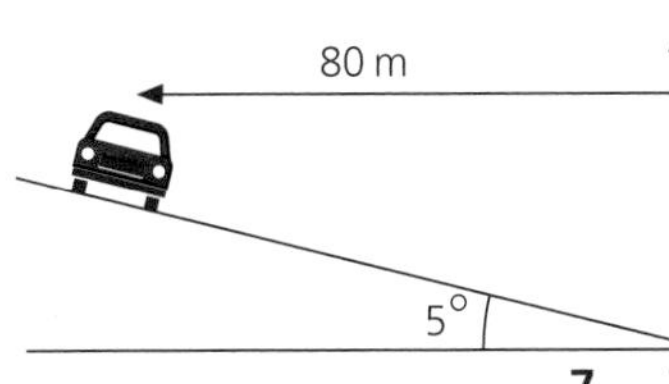

a) Find the normal reaction force on the car.

b) Find the speed of the car if it describes a circle of 80 m.

c) Comment on how helpful it would be to have a 5° banked road, as part of an approach to a motorway.

7 The coefficient of friction between the tyres of a racing motorcycle and the surface of the race track is 1.0. Find the greatest speed at which a bend of radius 150 m can be taken.

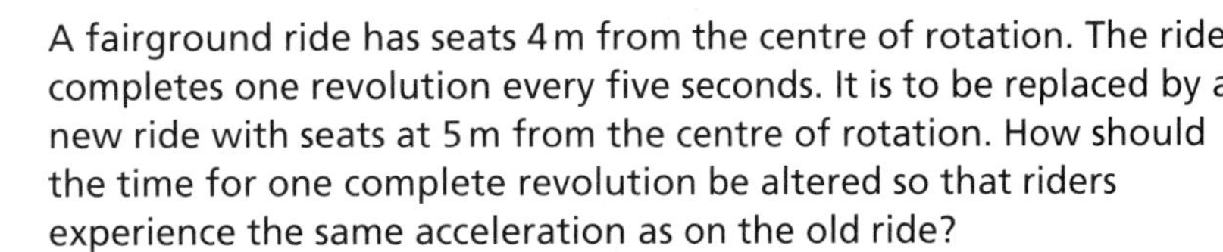

8 A fairground ride has seats 4 m from the centre of rotation. The ride completes one revolution every five seconds. It is to be replaced by a new ride with seats at 5 m from the centre of rotation. How should the time for one complete revolution be altered so that riders experience the same acceleration as on the old ride?

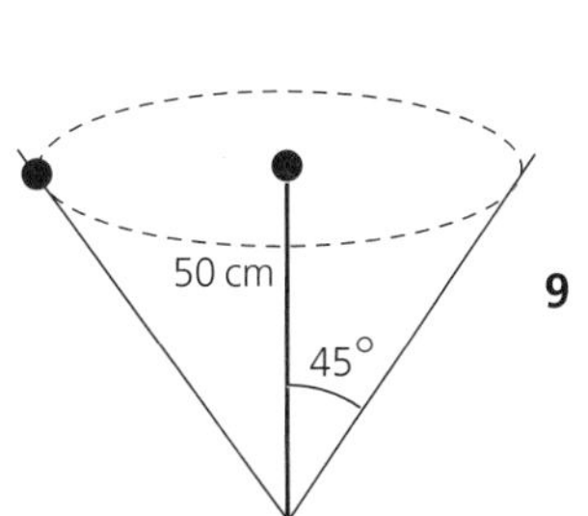

9 A marble of mass 20 grams describes a circle inside the cone illustrated. The centre of the circle is 50 cm above the bottom of the cone.

a) Find the speed of marble and the size of the normal reaction force.

b) What role, if any, does friction have in this situation?

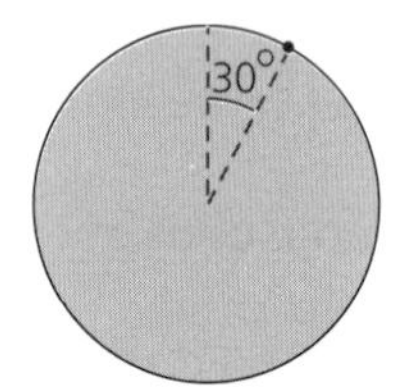

10 To be most effective the clothes in a tumble drier should not always stick to the side of the drier. A designer considers that ideally the clothes should fall away at the position shown in the diagram. Calculate the rate of rotation for a drum of radius 40 cm.

THE CONICAL PENDULUM

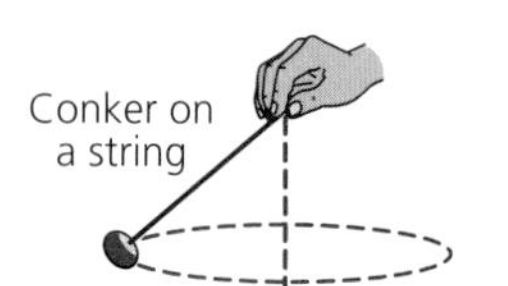

The chairoplanes theme park ride and the conker on the string shown in the figure are both examples of **conical pendulums**. Such a system consists of an object, attached to a fixed point by an inextensible string, made to rotate in a horizontal circle. We begin this section with a practical activity to explore the properties of a conical pendulum.

Exploration 12.3

Practical activity

For this practical activity you will need the conical pendulum components from the Unilab Mechanics Kit.

1 *Set up the apparatus as shown in the photograph. You may find it helpful to use a slide projector or an OHP to shine a bright light on the apparatus, so that it casts a shadow onto a wall or screen.*

2 *What forces act on the object at the end of the string while it is rotating? Illustrate these forces on a diagram.*

Without further experimentation, answer the following questions about the motion of the object.

3 *What happens when the angular speed of the system is increased? Is there a maximum angle that can be obtained between the string and the post?*

4 *What happens if you change the length of the string?*

5 *What happens if you change the mass at the end of the string?*

Now check your answers by carrying out simple experiments.

6 *Describe what happens to the velocity and acceleration of the mass as it rotates.*

Before modelling the conical pendulum, here are two examples which will give some insight as to how to proceed in the general case.

Example 12.9

A child of mass 40 kg swings on the end of a rope of length 4 m, describing a horizontal circle of radius 1.5 m. Find the angular speed of the child and the tension in the rope.

Solution

Modelling the child as a particle, the diagram below shows the circular motion with centre O and the forces acting.

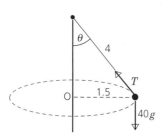

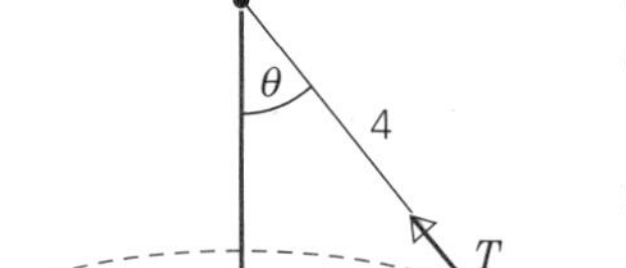

The vertical component of the tension must balance the force of gravity, so using $g = 9.8\,\text{m s}^{-2}$ gives:

$T\cos\theta = 40 \times 9.8 = 392$ *(1)*

As the child describes a circle, the acceleration is $r\omega^2$ towards the centre of the circle. The horizontal component of the tension provides this acceleration, so:

$T\sin\theta = 40 \times 1.5\omega^2$ *(2)*

Eliminating T from equations (1) and (2) gives:

$$\frac{392}{\cos\theta} \times \sin\theta = 60\omega^2$$

so:

$$\omega^2 = \frac{392}{60}\tan\theta \qquad (3)$$

From the diagram and using simple trigonometry:

$$\tan\theta = \frac{1.5}{\sqrt{4^2 - 1.5^2}} = 0.405$$

Now the value for ω is found by substituting for $\tan\theta$ in equation (3).

$$\omega^2 = \frac{392}{60} \times 0.405 = 2.65$$

$\omega = 1.63$ *rad* s^{-1}

So the angular speed is 1.63 rads^{-1}*. The tension is found using equation (1).*

$T\cos\theta = 392$

and $\cos\theta = \dfrac{\sqrt{4^2 - 1.5^2}}{4} = 0.927$

Solving for T:

$$T = \frac{392}{\cos\imath} = \frac{392}{0.927} = 422.9 \text{ N.}$$

Example 12.10

A toy aeroplane is suspended from a ceiling by a piece of string of length 50 cm. When the aeroplane is wound up it describes a horizontal circle and the string makes an angle of 30° to the vertical. Find the speed of the aeroplane.

Solution

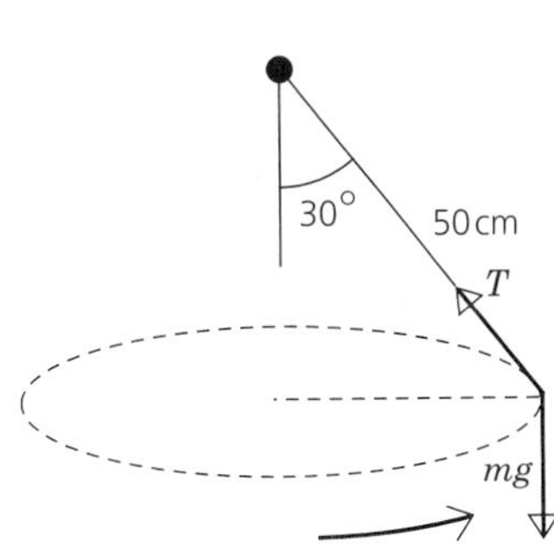

The forces acting on the aeroplane, modelled as a particle, are the tension and the force of gravity.

The vertical component of the tension must balance the force of gravity, so:

$T\cos 30° = mg$ *(1)*

The aeroplane must have an acceleration of $\frac{v^2}{r}$ towards the centre of the circle, and the horizontal component of the tension must cause this acceleration, so Newton's second law gives:

$$T\sin 30° = \frac{mv^2}{r} \qquad (2)$$

Dividing equation (1) by equation (2) eliminates T to give:

$$\tan 30° = \frac{v^2}{rg}$$

Solving for v^2 gives:

$v^2 = rg\tan 30°$

The radius of the circle is given by:
$r = 0.5\sin 30° = 0.25$
Now v can be found.
$v^2 = rg\tan 30° = 0.25 \times 9.8 \times \tan 30° = 1.415$
$v = 1.19\,m\,s^{-1}$

Angles and speed

These two examples show us how to proceed for the general conical pendulum with an object of mass m, attached to an inextensible string of length l, describing circular horizontal motion with constant speed. Suppose that the radius of the circular motion is r and the angle of the string to the vertical is θ. The figure shows the object modelled as a particle, and the forces which are acting. These are the tension in the string and the force of gravity.

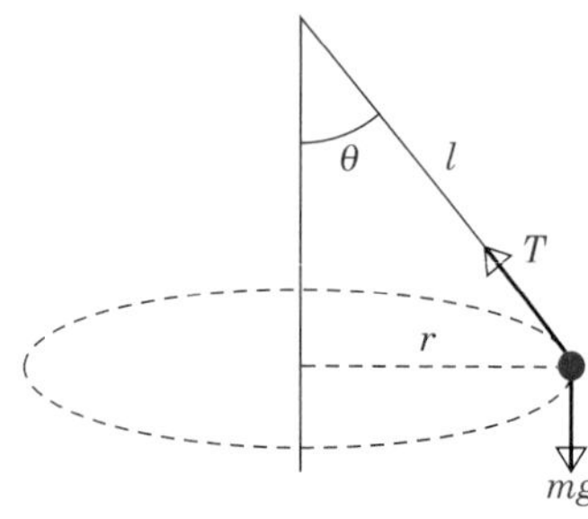

The vertical component of the tangent balances the force of gravity so

$$T\cos\theta = mg \qquad (1)$$

For circular motion to be possible, there must be a force towards the centre of the circle O. This force is the horizontal component of the tension. Applying Newton's second law gives:

$$T\sin\theta = mr\omega^2$$

where ω is the constant angular speed.
From the figure:

$$r = l\sin\theta$$

giving

$$T\sin\theta = ml\omega^2\sin\theta$$

and dividing both sides by $\sin\theta$ gives:

$$T = ml\omega^2$$

Substituting for T from equation (1) gives:

$$\frac{mg}{\cos\theta} = ml\omega^2$$

and solving for $\cos\theta$ gives:

$$\cos\theta = \frac{g}{l\omega^2}$$

This is a general formula relating the angle between the conical pendulum and the vertical to the length of the string and the angular speed. From this result we can make three mathematical deductions:

- the motion of the object is independent of its mass,
- as the length l increases, $\cos\theta$ decreases and so the angle θ increases,
- as the angular speed ω increases, $\cos\theta$ decreases and so the angle θ increases.

The consequences of these results are important in many fairground rides which can be modelled by the conical pendulum. For example, on a chairoplanes ride, the rider will travel at the same angle to the vertical, whatever their mass. This is rather fortunate! Imagine two adjacent seats, one containing a heavyweight boxer and the other containing a light ballet dancer. If the motion depended on the mass of the rider it could be unfortunate for the ballet dancer on the inside!

EXERCISES

12.3 A

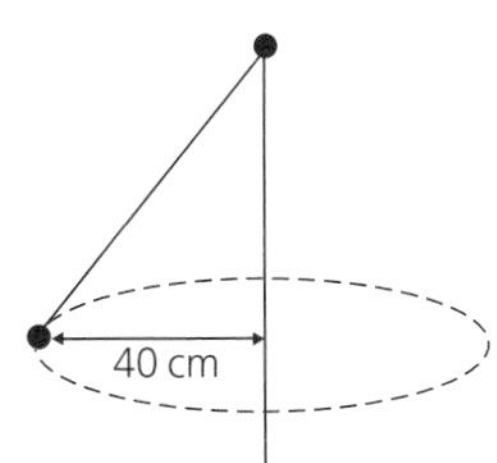

1 A conker on the end of a length of string is set in horizontal motion with a speed of $4\,\text{m s}^{-1}$, so that it describes a circle.
   a) Draw a diagram to show the forces acting on the conker.
   b) Write down two equations by considering first the vertical component of the tension and secondly the horizontal component of the tension and the acceleration of the conker.
   c) Eliminate the tension from the two equations above and find the angle that the string makes with the vertical.

2 A child of mass 50 kg swings on the end of a rope of length 6 m and describes a horizontal circle of radius 2 m.
   a) Find the tension in the rope.
   b) Find the angular speed of the child and the time taken to complete one circle.
   c) What assumptions have been made about the rope?
   d) Two children, of masses 40 kg and 45 kg, swing together on the rope in the same way. What happens to the tension and the time to complete one circle?

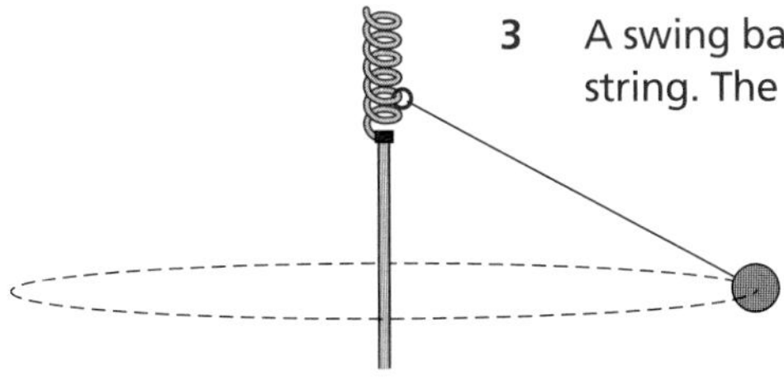

3 A swing ball game consists of a ball of mass 100 grams attached to a string. The ball is hit so that it moves at $2\,\text{m s}^{-1}$, in a circle of radius 1.2 m.
   a) Find the angular speed of the ball.
   b) Find the angle that the string makes with the vertical post.
   c) Find the tension of the string.

4 At a School Fayre a ball on the end of a piece of string has to be swung to hit a skittle. The diagram shows the set up, and the position of the skittle.
Find the speed at which the ball should be thrown if it is to hit the skittle.

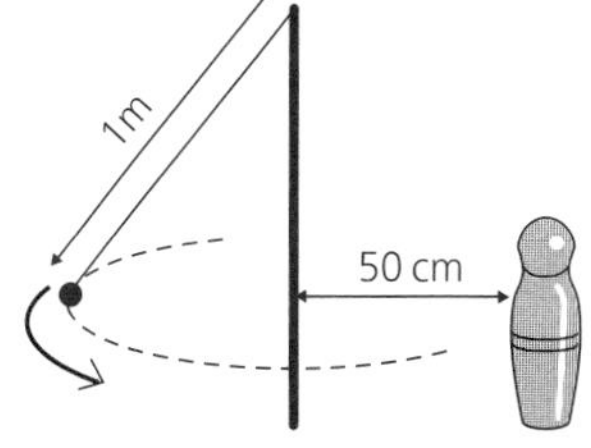

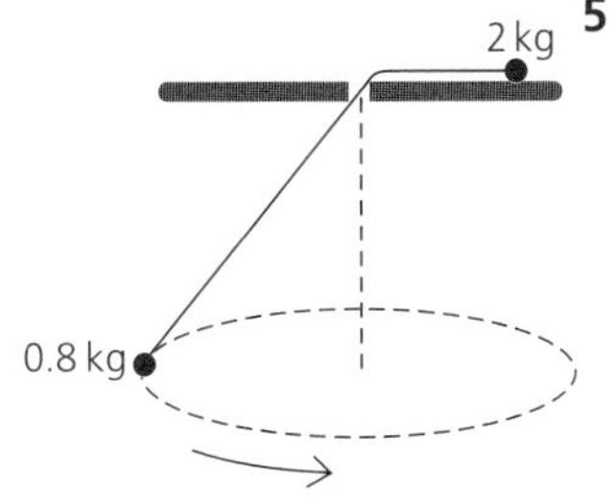

5 An object of mass 2 kg rests on a rough horizontal table with coefficient of friction 0.5. It is attached to another object of mass 0.8 kg by a light inelastic string which passes through a small hole in a table. This object describes a horizontal circle with constant speed $2\,\text{m s}^{-1}$ and the object on the table is on the point of slipping.
   a) Draw a diagram showing the forces acting on each object.
   b) Find the tension in the string.
   c) Calculate the radius of the circle and the length of the string below the table.

6 The diagram shows an object attached by two strings. The object has mass 200 grams and completes three revolutions per second.
   a) Find the acceleration of the object.
   b) Find the tension in each string.

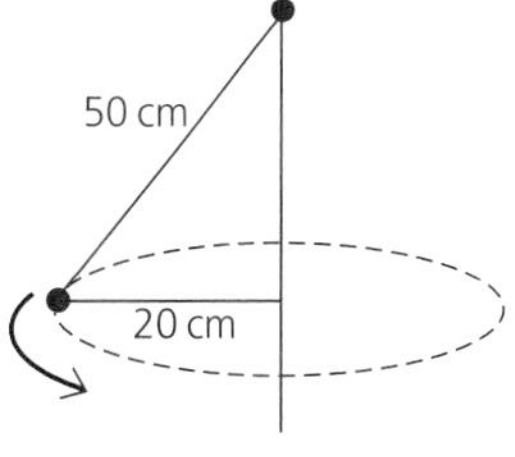

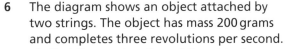

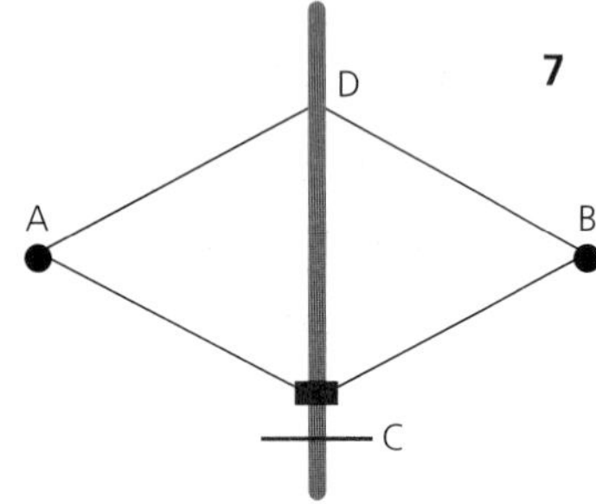

7 The mechanism shown in the diagram is designed to regulate the flow of steam from a boiler, to control the speed of an engine. It is called an 'engine governor'. It consists of four light rods together with two metal spheres A and B. The rods AD and BD are smoothly jointed to a smooth vertical spindle at D. A small ring is smoothly jointed to AC and BC at C and can slide on the spindle below D. When a valve is open the ring rests on a smooth horizontal ledge fixed to the spindle. The system rotates with the engine drive shaft about the axis CD.
As the angular speed increases so the ring slides up the spindle, controlling the flow of steam from the boiler and the pressure needed to drive the engine.

a) What assumptions do you think would be appropriate when modelling this system?

b) Draw a diagram to show the forces on one sphere.

c) For a particular system, spheres of mass 250 grams are used on arms of length 15 cm. The mass of the ring is 100 grams. At the optimum operating speed the governors rotate at 120 rpm. What angle do the arms make with the vertical?

d) What rate of rotation is needed before the system can be effective if the radius of each sphere is 2.4 cm?

e) What would happen if you assume the mass of the ring is zero?

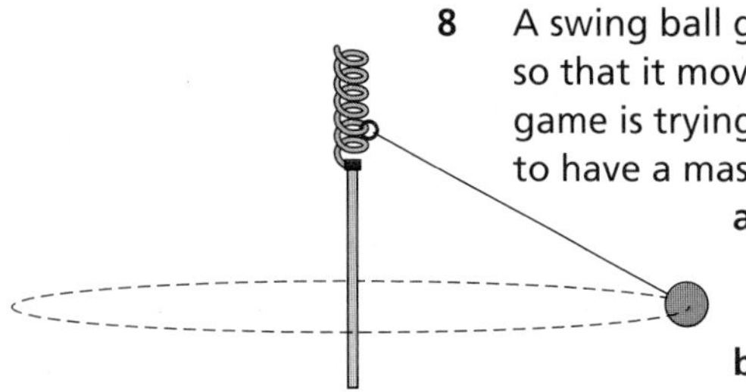

8 A swing ball game consists of a ball attached to a string. The ball is hit so that it moves in a horizontal circle. The designer of a swingball game is trying to decide how strong the string needs to be. The ball is to have a mass of 120 grams and the string has a length of 1.2 m.

a) Estimate the greatest speed at which a swing ball is likely to move and then determine the greatest tension that the string must be able to withstand.

b) Find a general expression for tension, T, in terms of the speed of the ball, v, and its mass, m.

EXERCISES

12.3 B

1 A conical pendulum consists of a string of length 1.2 m and an object of mass 3 kg. The string makes an angle of 20° to the vertical as the object describes a circle.

a) Find the radius of the circle.

b) Use the vertical components of the force to find the tension in the string.

c) Find the angular speed of the object, by considering the horizontal components of the forces acting on it.

2 A heavy ball, of mass 200 kg, that is used for demolition work, is set into motion so that it moves in a horizontal circle at a constant speed. The cable supporting the ball is 6 m long and makes an angle of 5° with the vertical.

a) Find the tension in the cable, by considering the vertical components of the forces acting.

b) Find the speed of the ball.

3 Two children play a game, using conkers on the ends of strings of length 40 cm. They take turns to swing the conker round in a circle and try to get their hand as close as possible to the surface of a table. One child's hand is 12 cm above the table. Find the speed of the conker.

4 Find the speed of each particle and the tension in each string shown below.

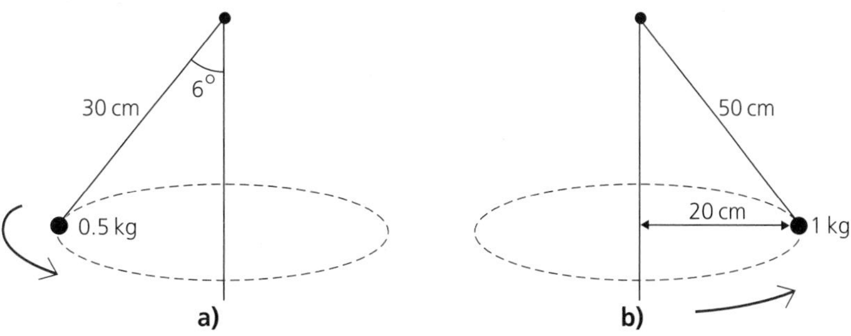

5 The diagram shows a length of string that has been passed through a hole in a table. Masses are attached to the ends of each string as shown in the diagram.
The mass on the table top is on the point of slipping.

a) Draw two diagrams to show the forces acting on each mass.
b) Find the tension in the string.
c) Find the acceleration of the lower mass.
d) What is the coefficient of friction between the 100 gram mass and the surface?
e) Is the value that you have obtained reasonable?

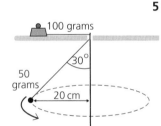

6 A conical pendulum consists of a string of length l, and a mass, m. The string makes an angle α with the vertical. The mass moves with speed v.

a) Find an expression for the tension, T, in the string in terms of m, v, l and α.
b) Find an expression for $\tan\alpha$ in terms of g, v and r.

7 A mass, M, is at rest on the outer surface of a smooth cone. It is attached by a length of string to another mass, m, which describes a circle of radius r.

a) Find the tension in the string.
b) Find the angular speed of the moving mass in terms of g, r, m and M.
c) Express M in terms of m.
d) If the centre of the circle is 50 cm below the vertex of the cone find ω.

8 An object, of mass 2 kg, is attached by two ropes to a spindle that rotates at a constant angular speed of 10 rad s^{-1}. Find the tension in each string, given that the object describes a circle of radius 1 m.

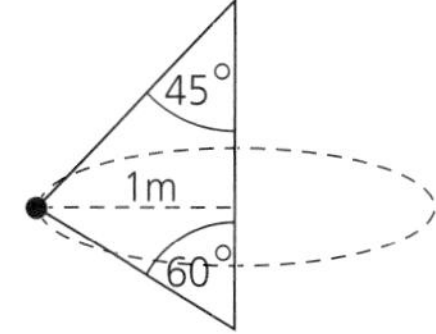

CONSOLIDATION EXERCISES FOR CHAPTER 12

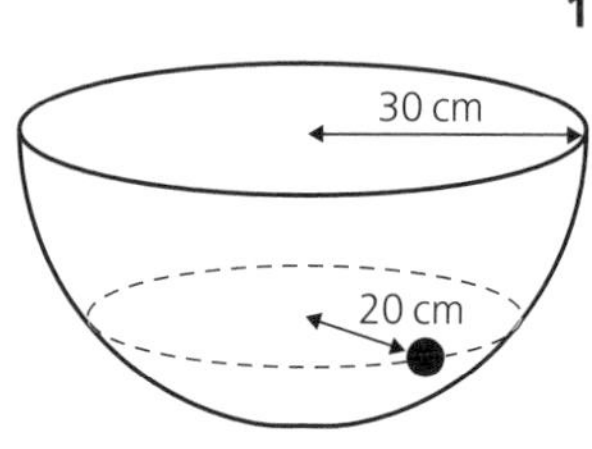

1 A small marble of mass 5 grams is moving in a horizontal circle of radius 20 cm on the smooth inner surface of a hemispherical bowl of radius 30 cm (see diagram).

a) Find, in newtons, the magnitude of the force exerted on the marble by the bowl.

b) Find, in m s^{-1}, the speed of the marble.

(UCLES Specimen Paper, 1994)

2 A particle is attached to the end of a light inextensible string OA of length 1 m. The end O is fixed at a distance of 1 m above the horizontal ground. The particle describes a horizontal circle at a height of 0.5 m above the ground.
Calculate the steady speed of the particle.

(Oxford Specimen Paper, 1994)

3 A particle P of mass 3 kg moves under the action of a single force $\mathbf{F}$ so that its position vector at time t is given by:

$$\mathbf{r} = \cos 2t\,\mathbf{i} + \sin 2t\,\mathbf{j} - 6t\,\mathbf{k}$$

a) Find the velocity and the momentum of P at time t.

b) Express $\mathbf{F}$ in terms of t.

c) Show that the direction of $\mathbf{F}$ is perpendicular to the velocity for all values of t.

Give a brief description of the motion of P given that the direction of $\mathbf{k}$ is vertically upwards.

(NEAB Specimen Paper, 1994)

4 A child of mass 30 kg amuses herself by swinging on a 5 m rope attached to an overhanging tree. She is hanging on the lower end of the rope and 'swinging' in a horizontal circle of radius 3 m.

a) Draw a diagram to show the forces acting on the girl.

b) Find the tension on the rope.

c) Show that the time she takes to complete one circle is approximately 4 seconds.

d) State any assumptions you have made about the rope.

e) The girl's older brother then swings on his own on the rope, in a horizontal circle of the same radius. Show that the tension in the rope is now $\dfrac{5mg}{4}$ where m is his mass.

Find the time that it takes for him to complete one circle.

(AEB Specimen Paper, 1994)

5 The position vector (in metres) at time t seconds, relative to a fixed origin O, of a particle P moving in a horizontal plane is given by:

$$\mathbf{r} = \frac{2}{\ }\left(\mathbf{i}\cos\ \ t + \mathbf{j}\sin\ \ t\right).$$

Describe as clearly as you can the path of P and determine the velocity of P at time t seconds. Given that P is of mass 0.2 kg find the horizontal force acting on it at time t seconds.

(WJEC Specimen Paper, 1994)

6 A space-ship S, of mass M, near the moon, experiences a gravitational force, of magnitude F, which is directed towards O, the centre of the moon. It is known that

$$F = \frac{Mk}{r^2} \text{ where OS} = r \text{ and } k \text{ is constant.}$$

a) By modelling the moon as a sphere of radius 2×10^6 m and by taking the acceleration due to gravity at the moon's surface to be of magnitude 1.6 m s^{-2}, find the value, in m^3 s^{-2}, of k.

The space-ship S moves around the moon in a circular orbit, centre O and radius 3×10^6 m.

b) Estimate the speed of S, giving your answer in m s^{-1}.

(ULEAC Specimen Paper, 1994)

MATHEMATICAL MODELLING ACTIVITY

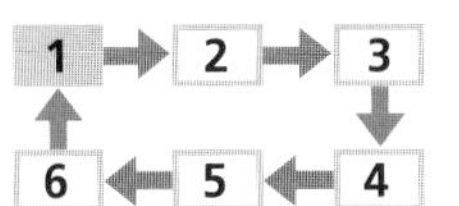

Specify the real problem

Problem statement

Many people, when they first see a chairoplanes ride at a theme park or fun fair, are amazed that the angle of inclination of the support chain for each chair appears to be independent of the size of the passenger – child or adult – or even whether the chair is occupied at all. Develop a model which could be used to explain this phenomenon. Use it to predict how a passenger's ride will be affected by other design features of the ride.

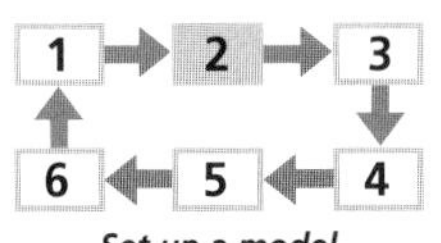

Set up a model

Set up a model

The first stage in setting up a model is to identify the important features:

- the length of the chains,
- the angular speed of rotation,
- the mass of a passenger.

Are there any other important factors that you feel should be included?

Having considered these features, we can list a set of assumptions:

- the chairoplane can be modelled as a conical pendulum,
- the mass of each chair is negligible,
- the chains are all of length l (m),
- the angular speed of rotation is constant at ω (rad s^{-1}),
- the path of each passenger of mass m (kg) is a horizontal circle when the chairoplane is rotating with constant angular speed,
- the support chains are all attached at the same point, vertically above the centre of rotation of the passengers,
- air resistance effects can be ignored.

These assumptions allow a simple first model to be formulated.

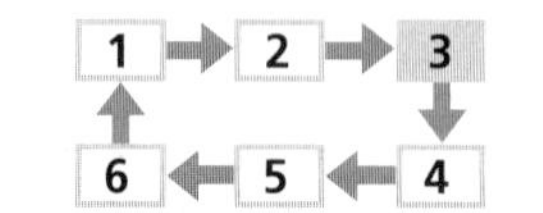

Formulate the mathematical problem

Formulate a mathematical problem

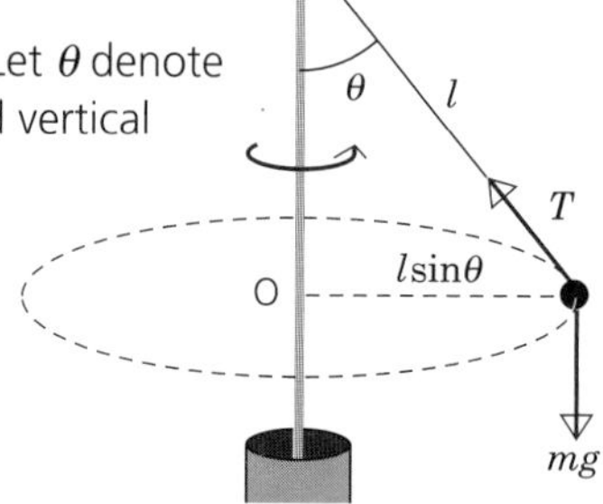

Consider one of the passengers of mass m. Let θ denote the inclination of her chain to the downward vertical and T be the tension in the chain.

Explain why the radial component of acceleration is $l\omega^2\sin\theta$ directed towards the centre of the circular path.

1 2 3 6 5 4

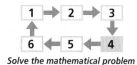

Solve the mathematical problem

Mathematical solution

Now applying Newton's second law of motion radially gives:

$$m\omega^2 l\sin\theta = T\sin\theta \qquad (1)$$

Furthermore, since the path of the passenger is assumed to be a horizontal circle then the net vertical force must be zero and so:

$$mg = T\cos\theta \qquad (2)$$

and from equation (1):

$$T = ml\omega^2 \qquad (3)$$

then from (2):

$$\cos\theta = \frac{g}{l\omega^2} \qquad (4)$$

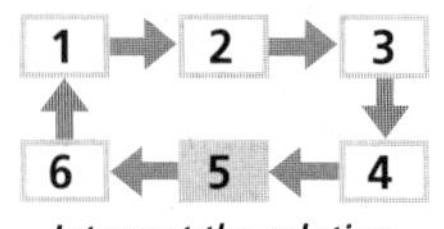

Interpret the solution

Interpretation

Equation (4) is independent of m which means that the angle of inclination of each chain does not depend on the mass of the passenger or even on whether the chair is occupied at all. This agrees with the observed behaviour. *What else does equation (4) tell us? Devise a simple experiment to confirm your conclusions.*

Equation (3) shows that the tension in each chain is increased if the mass of the passenger is increased and/or the chain is lengthened and/or the angular speed is increased. Equation (4) shows that the angular inclination would be greater in an amusement park on the moon for the same values of l and ω.

1 2 3 6 5 4

Compare with reality

Compare with reality

The fact that the angle of inclination of the chains does not depend on the attached mass is confirmed by our observations. The major criticism of the model developed so far relates to the point of attachment of the support chains. In fact they are not all attached to the same point on the axis of rotation but are attached to points on the circumference of a circular disc with its centre on the axis of rotation, as shown in the diagram.

Refining the model

Develop a revised model, based upon the above criticism, and solve the resulting revised problem to show that the angle of inclination θ now satisfies the equation:

$$R\cot\theta + l\cos\theta = \frac{g}{\omega^2} \qquad (5)$$

where R is the radius of the circular disc.

Interpretation of the refined model

a) The absence of the mass, m, from equation (5) means that the angle θ remains independent of the mass of the passenger, in agreement with our observations.

b) While solution of equation (5) will give a more realistic solution to the problem than that of equation (4), it is evident that equation (5) is considerably more complicated. It is not possible to obtain an analytical solution. Instead we could adopt a graphical approach.

Use a graphics package or graphics calculator to plot graphs of ω against θ for a range of different values of R (e.g. $R = 1.0, 1.5, 2.0, 2.5$ m) and $l = 2$ m.

Is it possible for the chains to become parallel to the ground? Explain your answer. How do the radius of the disc and the speed of rotation affect the angle of inclination?

POSSIBLE MODELLING TASKS

1 How fast should the drum of a tumble drier rotate?

2 When 'preparing' the bride and groom's car at weddings, a favourite trick is to place small stones inside the hub caps. What happens when the car is driven away?

3 A pendulum fixed inside a car can be used to estimate the radius of curvature of bends when the car is driven round them at constant speed. Use this experiment to estimate the radius of a bend.

4 A car negotiating a bend on a banked road can slide in two ways: if the speed of the car is too slow it could tend to slide inwards down the road; however, if it is too fast the car could tend to slide outwards. Investigate the motion of cars on banked racing tracks.

Summary

After working through this chapter you should:

- be able to identify and model circular motion with constant speed
- know that for circular motion with constant angular speed ω and radius r that the velocity has magnitude $v = r\omega$ and is directed in the transverse direction and the acceleration has magnitude $a = \dfrac{v^2}{r} = r\omega^2$ and is directed radially inwards towards the centre
- know that for circular motion with constant speed the required force is directed radially inwards towards the centre
- be able to apply the principles of circular motion to the conical pendulum.

Circular motion with variable speed

- *Objects that change their speed as they move often describe circular paths.*
- *There will be two components of acceleration: one towards the centre and one along the tangent.*

CIRCULAR MOTION WITH VARIABLE SPEED

Very few examples of circular motion actually involve constant speed. Nevertheless, the principles developed in Chapter 12 can be an ideal starting point when modelling situations where circular motion involves variable speed. Almost all of the examples considered in Chapter 12 would have passed through a phase where the speed of the objects considered was increasing. For example, a car that drives on a roundabout will almost certainly vary its speed. A child on a fairground ride will gain speed as the ride is set into motion. In this chapter we consider how the acceleration varies when circular motion takes place with variable speeds.

Exploration 13.1

Loop the loop

This practical activity uses equipment from the Unilab Mechanics Kit. Alternatively, similar equipment can be obtained from a good toy shop!

1 Set up the apparatus as shown in the photograph.

2 Release the car from different heights, to find the minimum height from which it will *just* loop the loop. You have probably modelled the car as a particle. List some other factors which would make the model more realistic but have so far been ignored.

3 Repeat step **2** with a marble or small ball. What deficiencies would there be if the marble was modelled as a particle in an analysis of the problem?

4 On which part of the track is the car most likely to leave the track?

5 What forces act on the car while it is in contact with the track?

6 What forces act on the car when it has left the track and is falling?

7 State and explain the condition for the car to remain in contact with the track.

Forces and circular motion

You might have expected the car to loop the loop if it were released from a height level with the top of the loop, but should have discovered that this is not the case. At the top of the loop the car must still be moving, or else it will fall off, and so it must still retain some kinetic energy.

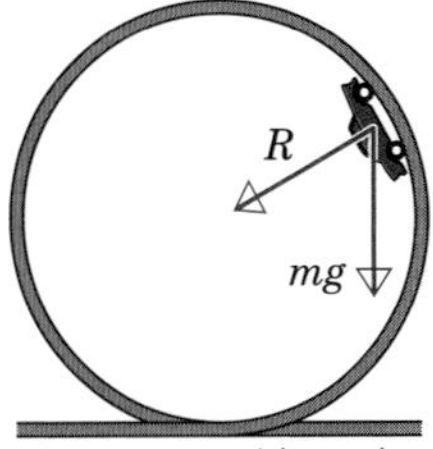

In contact with track

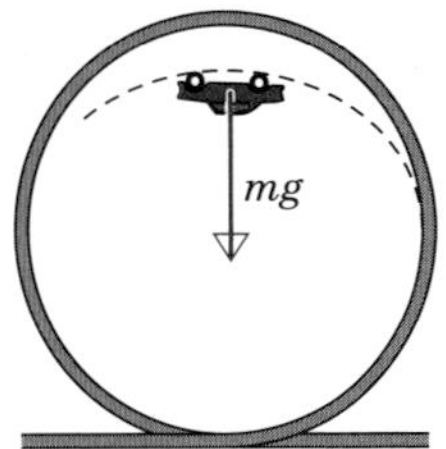

Lost contact with track

As the car moves on the track there are two forces acting, the force of **gravity** and the **normal reaction** R, if we assume that resistance forces have a negligible affect on the motion of the car. When the car leaves the track we can assume that the only force acting is gravity. The point on the track at which the normal reaction, R, drops to zero is the critical point, because it is at this point that the car will leave the track.

Exploration 13.2

Acceleration in circles

Imagine that you are sitting on a fairground roundabout that is at rest. What acceleration would you experience as the speed of the roundabout increases to its normal operating speed?

Circular motion and acceleration

As you would be describing a circle there must be a component of acceleration **towards the centre** of the circle. In addition there would be a component of acceleration **along a tangent** to the circle as the speed increases. It is the combination of these two components that will give the actual acceleration. The diagram shows two examples. In the first, the ride is just starting, so the component towards the centre is small because the speed is small, and so the resultant is close to the tangent. In the second, the ride is close to its operating speed. The component towards the centre is now much larger and the tangential component is smaller as the speed increases more slowly. The resultant is now directed more towards the centre of the circle.

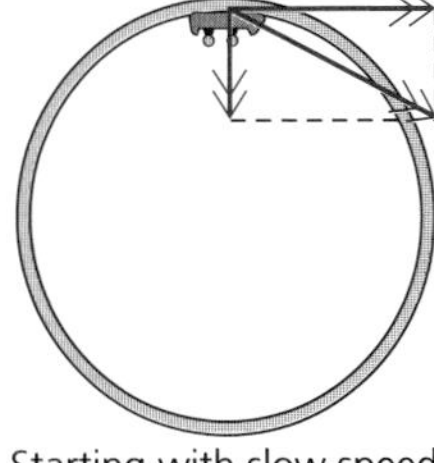
Starting with slow speed

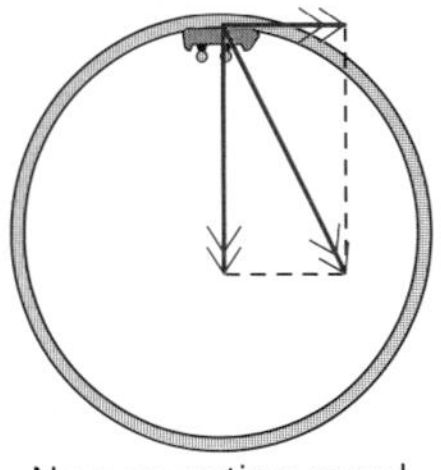
Near operating speed

To derive expressions for the velocity and acceleration of an object describing circular motion it is necessary to differentiate the position vector twice. In order to help interpret the results it is useful to define unit vectors $\mathbf{e}_r$ and $\mathbf{e}_t$ as well as $\mathbf{i}$ and $\mathbf{j}$.

BRIDGWATER COLLEGE LIBRARY

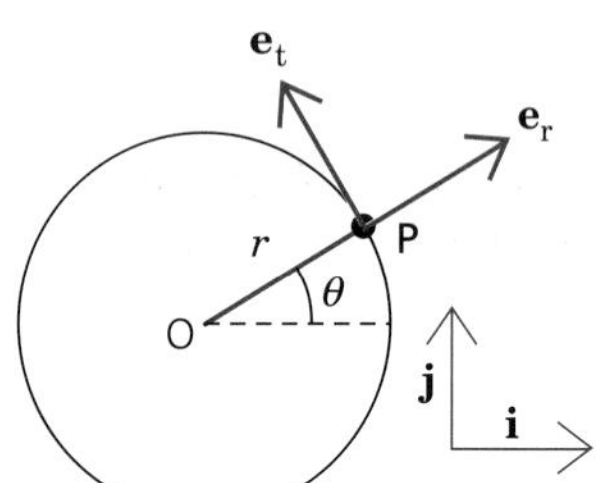

The unit vectors $\mathbf{e}_r$ and $\mathbf{e}_t$ are directed outwards along the radius and along the tangent respectively as shown. These new unit vectors can be expressed in terms of **i** and **j** as:

$$\mathbf{e}_r = \cos\theta\mathbf{i} + \sin\theta\mathbf{j}$$

and

$$\mathbf{e}_t = -\sin\theta\mathbf{i} + \cos\theta\mathbf{j}$$

The position vector can now be written as $\mathbf{r} = r\mathbf{e}_r$, where r is the radius of the circle. However to begin with, **r** is expressed as:

$$\mathbf{r} = r\cos\theta\mathbf{i} + r\sin\theta\mathbf{j}$$

Differentiating gives:

$$\mathbf{v} = \frac{\mathrm{d}\mathbf{r}}{\mathrm{d}t} = -r\sin\theta\frac{\mathrm{d}\theta}{\mathrm{d}t}\mathbf{i} + r\cos\theta\frac{\mathrm{d}\theta}{\mathrm{d}t}\mathbf{j}$$

The term $\frac{\mathrm{d}\theta}{\mathrm{d}t}$ has to be introduced because it is no longer reasonable to assume that the angular velocity is constant and we must therefore consider the case where θ is any function of t.

Note that **v** can be expressed as:

$$\mathbf{v} = r\frac{\mathrm{d}\theta}{\mathrm{d}t}(-\sin\theta\mathbf{i} + \cos\theta\mathbf{j}) = r\frac{\mathrm{d}\theta}{\mathrm{d}t}\mathbf{e}_t$$

So the velocity has magnitude $r\frac{\mathrm{d}\theta}{\mathrm{d}t}$ and is directed along the tangent.

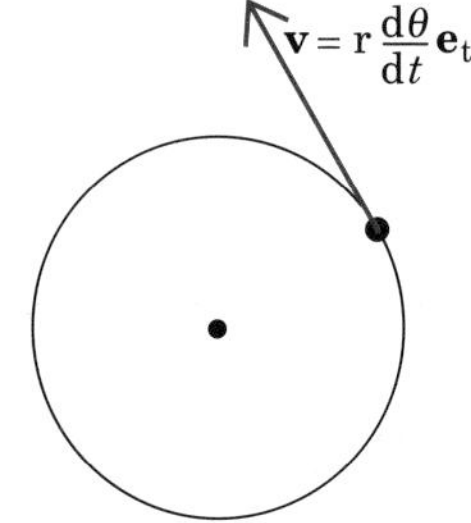

The velocity is differentiated using the product rule to give the acceleration:

$$\mathbf{a} = \frac{\mathrm{d}\mathbf{v}}{\mathrm{d}t} = \left(-r\sin\theta\frac{\mathrm{d}^2\theta}{\mathrm{d}t^2} - r\cos\theta\left(\frac{\mathrm{d}\theta}{\mathrm{d}t}\right)^2\right)\mathbf{i} + \left(r\cos\theta\frac{\mathrm{d}^2\theta}{\mathrm{d}t^2} - r\sin\theta\left(\frac{\mathrm{d}\theta}{\mathrm{d}t}\right)^2\right)\mathbf{j}$$

$$= r\frac{\mathrm{d}^2\theta}{\mathrm{d}t^2}(-\sin\theta\mathbf{i} + \cos\theta\mathbf{j}) - r\left(\frac{\mathrm{d}\theta}{\mathrm{d}t}\right)^2(\cos\theta\mathbf{i} + \sin\theta\mathbf{j})$$

$$= r\frac{\mathrm{d}^2\theta}{\mathrm{d}t^2}\mathbf{e}_t - r\left(\frac{\mathrm{d}\theta}{\mathrm{d}t}\right)^2\mathbf{e}_r$$

This expression can be simplified by noting that:

$$v = r\frac{\mathrm{d}\theta}{\mathrm{d}t}$$

and that this can be differentiated to give:

$$\frac{\mathrm{d}v}{\mathrm{d}t} = r\frac{\mathrm{d}^2\theta}{\mathrm{d}t^2}$$

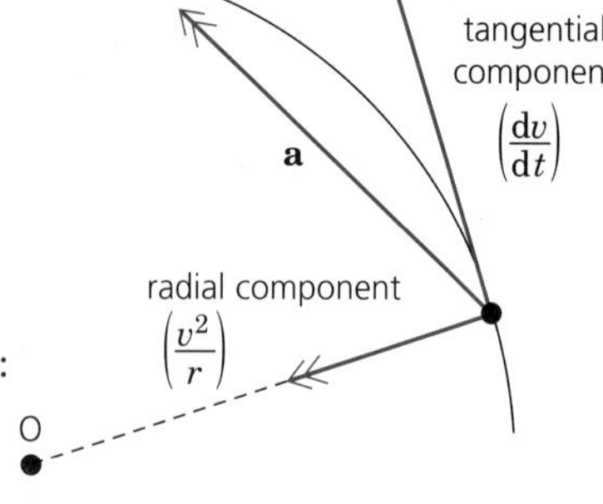

Substituting these into the expression for the acceleration gives:

$$\mathbf{a} = \frac{\mathrm{d}v}{\mathrm{d}t}\mathbf{e}_t - \frac{v^2}{r}\mathbf{e}_r$$

So the acceleration has two components. One acts towards the centre with magnitude $\frac{v^2}{r}$ which corresponds to the acceleration for the constant speed case. The second component is along the tangent and has magnitude $\frac{\mathrm{d}v}{\mathrm{d}t}$, the rate of change of the speed.

Example 13.1

A car drives round a roundabout of radius 35 m. Its speed increases uniformly from 4 m s^{-1} to 12 m s^{-1} in a four-second period. Find an expression for the acceleration of the car.

Solution

The acceleration of the car will be given by:

$$\mathbf{a} = \frac{dv}{dt}\mathbf{e}_t - \frac{v^2}{r}\mathbf{e}_r$$

The rate of change of speed can be used to find an expression for v. As the speed increases uniformly from 4 m s^{-1} to 12 m s^{-1} over a four-second period:

$$\frac{dv}{dt} = \frac{12-4}{4} = 2\,\text{m s}^{-2}$$

This can be integrated to give:

$$v = \int 2\,dt = 2t + c$$

As the car has an initial speed of 4 m s^{-1}, the value of c is 4, so:

$$v = 2t + 4$$

Now the acceleration can be expressed as:

$$\mathbf{a} = \frac{dv}{dt}\mathbf{e}_t - \frac{v^2}{r}\mathbf{e}_r = 2\mathbf{e}_t - \frac{(2t+4)^2}{35}\mathbf{e}_r$$

Example 13.2

A loop the loop track has a loop of radius 20 cm.

a) *Find the speed of the car at the top of the loop if it is just to loop the loop.*

b) *Find the height of release needed to reach this speed.*

Solution

a) *If the car loops the loop it will remain in contact with the track and so the normal reaction force R will always act. In the case of the car that just loops the loop the reaction force will be zero at the top of the loop, i.e. $R = 0$.*

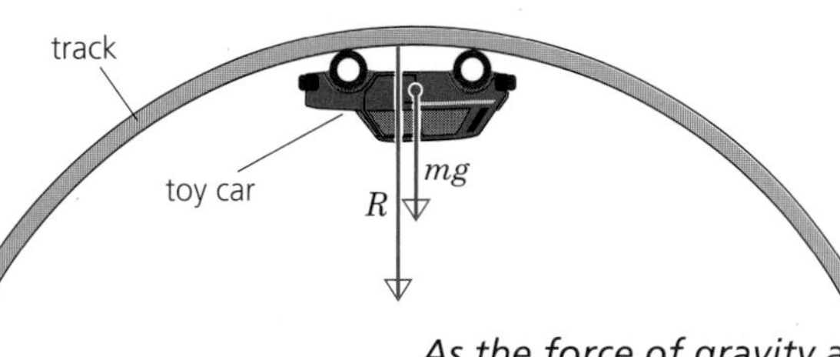

As the force of gravity acts straight down the resultant force is simply mg towards the centre of the circle. The car must have an acceleration of $\frac{v^2}{r}$ towards the centre, so applying Newton's second law gives:

$$mg = m\frac{v^2}{r}$$

or

$$v^2 = rg = 0.2 \times 9.8 = 1.96$$

and so $v = 1.4\,\text{m s}^{-1}$.

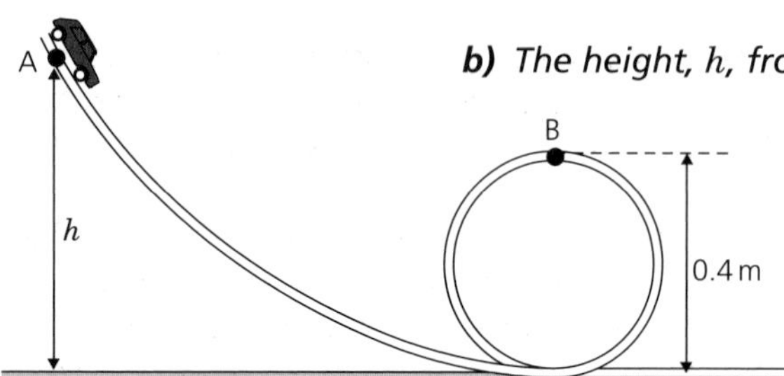

b) *The height, h, from which to release the car can be determined by conservation of energy.*

PE of car at A = PE of car at B + KE of car at B

$$mgh = mg \times 0.4 + \tfrac{1}{2} m \times 1.4^2$$
$$9.8h = 3.92 + 0.98$$
$$h = \frac{4.9}{9.8} = 0.5\ m$$

Example 13.3

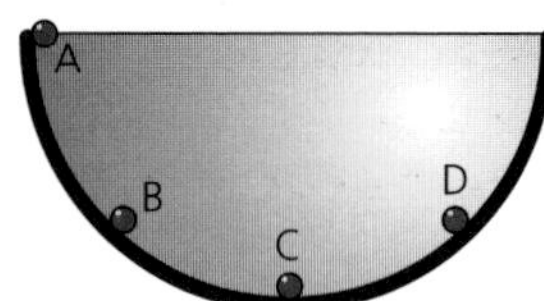

A marble is released at the edge of a hemispherical bowl. Sketch the approximate direction of the acceleration of the marble at each position shown.

Solution

Assume that A is the initial position. Here the speed of the marble is zero, so the component of acceleration towards the centre is zero. Thus the acceleration is along the tangent as shown.

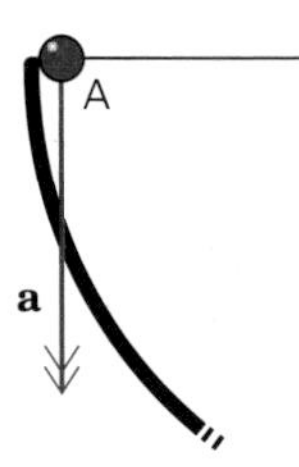

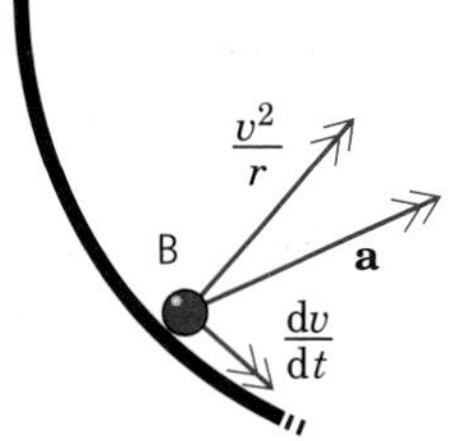

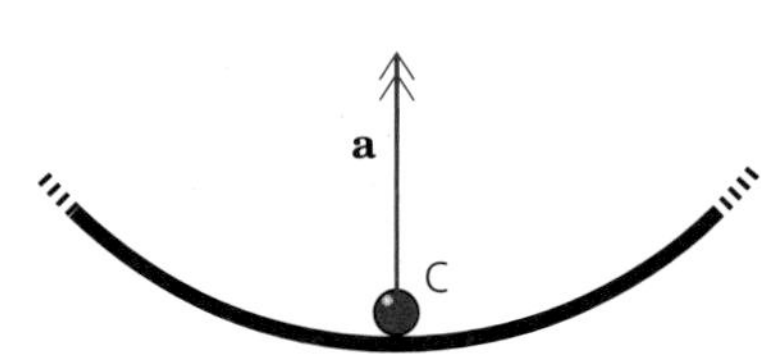

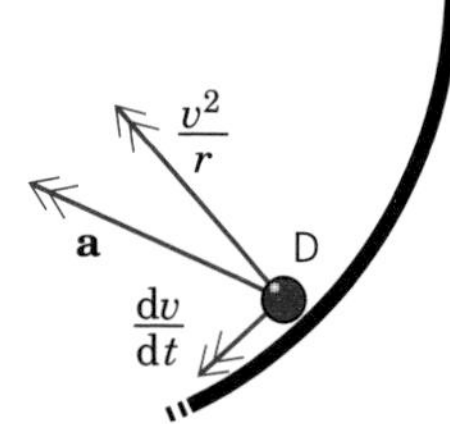

At B the marble is gaining speed so $\frac{dv}{dt}$ is positive and as $v \neq 0$ there is also a component of acceleration towards the centre of the circle of magnitude $\frac{v^2}{r}$.

At C the speed of the marble has reached a maximum value so $\frac{dv}{dt}$ is zero. So at this position there is simply a component of acceleration towards the centre of the circle.

At D the marble is losing speed so $\frac{dv}{dt}$ is negative. This means that the component of the acceleration along the tangent is in the opposite direction to the motion. There is still a component of acceleration towards the centre of the circle.

EXERCISES

13.1 A

1 A car is driving round a roundabout. Draw diagrams to indicate the direction of its acceleration if:

a) it has a constant speed, **b)** it is slowing down,

c) it is gaining speed.

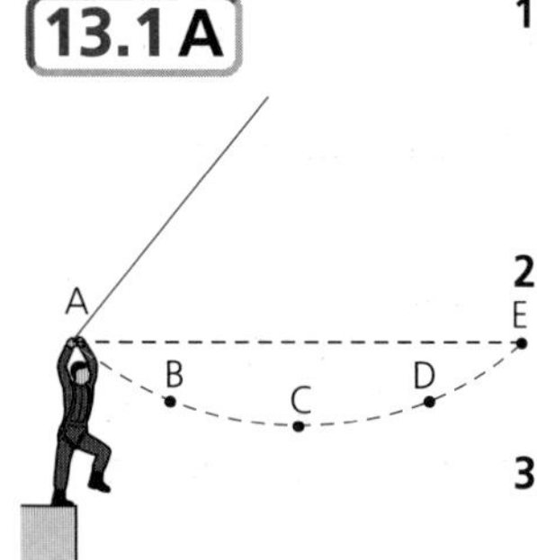

2 A marine on an assault course swings from A to E on a rope. Draw diagrams to show the acceleration of the marine at each of the positions A, B, C, D and E.

3 The speed of a motorcycle on a roundabout of radius 50 m is increasing from $4\,m\,s^{-1}$ at a rate of $5\,m\,s^{-2}$. Find the magnitude of the acceleration of the motorcycle after three seconds.

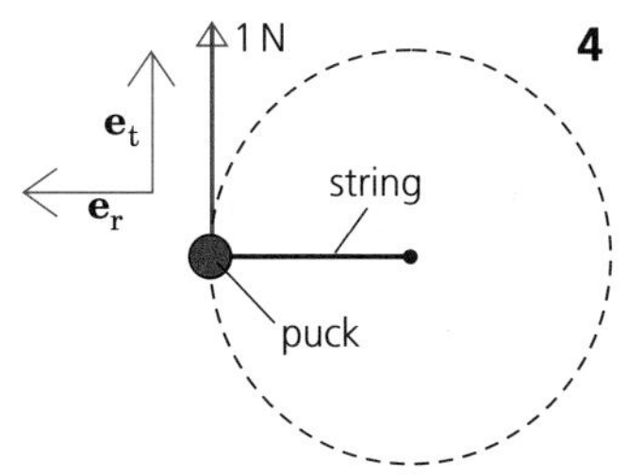

4 A puck of mass 2 kg hovers on a smooth glass surface and is attached by a string of length 50 cm to a fixed point. A force of magnitude 1 N acts at right angles to the string as shown. The puck moves in a circle with the string taut.

a) Find the resultant force on the puck using the unit vectors $\mathbf{e}_r$ and $\mathbf{e}_t$.

b) Show that $\frac{dv}{dt}$ is a constant and hence find an expression for v.

c) How long does it take for the puck to complete one revolution? Find the speed and the magnitude of the acceleration at this time.

d) Find an expression for the tension in the string.

5 A new type of slide has been introduced at an amusement park. It consists of a smooth hemisphere. Children climb up inside and slide off it as shown. Safety experts recommend that to be safe the children should always remain in contact with the slide.

a) Use conservation of energy to show that the speed of a child is given by:

$$v = \sqrt{2gr(1 - \cos\theta)}$$

where r is the radius (in metres) of the hemisphere and θ is the angle between the child and the upward vertical through C.

b) Explain what happens to the forces on a child if they lose contact with the slide.

c) Find the value of θ when the children will leave the slide.

b) Recommend a revised design for the slide that would be considered safe.

6 A car is on a roundabout of radius 50 m. If the coefficient of friction between the tyres and the road is 0.8, what is the maximum rate at which the car can increase its speed when it is travelling at 7 m s^{-1}?

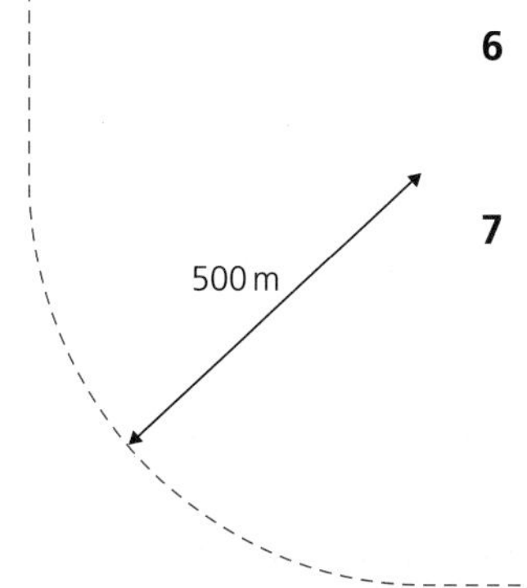

7 An aeroplane that was travelling vertically downwards levels out by moving through a quarter of a circle as shown. The circle has radius 500 m and its speed reduces uniformly from 100 m s^{-1} to 70 m s^{-1} during the ten seconds it takes to turn. Find expressions for both the radial and tangential components of the acceleration of the plane during the turn in terms of time, t.

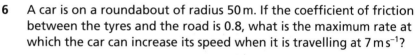

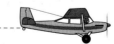

8 A cyclist is resting at the top of a humped-back bridge. The cyclist starts to free-wheel from the top of the bridge.

a) Find an expression in terms of θ for the speed of the cyclist as he moves on the circular part of the bridge.

b) Use the answer to **a)** to find an expression for the magnitude of the acceleration in terms of θ.

9 A gymnast is swinging in vertical circles about a horizontal bar. Model the gymnast as a particle of mass 50 kg at a distance of 0.95 m from the bar. The gymnast just completes the circle. Find the speed of the gymnast at the bottom of the circle and the tension in her arms at this position.

10 Part of a skateboard track in a park has a cross-section as shown in the figure. While in this section the maximum speed of one child is measured as $4\,\text{m}\,\text{s}^{-1}$.

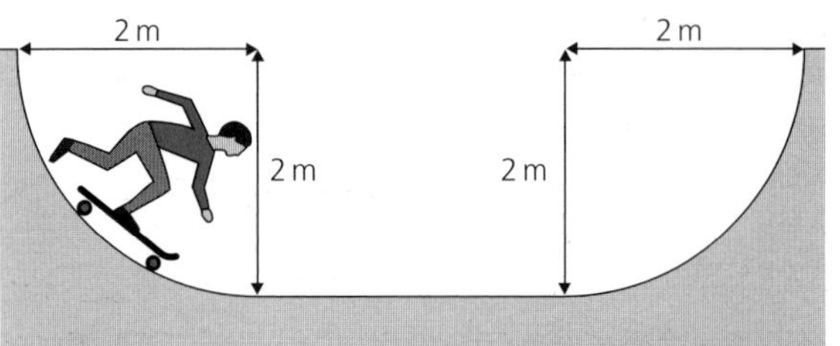

a) What is the maximum height reached by the child?

b) What is the reaction between the skateboard and the track when on the horizontal section if the combined mass of the child and skateboard is 45 kg?

c) What is the reaction between the child and the skateboard when at the highest point?

d) What physical principle did you use to answer part **a)**? Do you think it reasonable in this situation?

EXERCISES

13.1 B

1 A child's electric train is moving around a circular track. Draw diagrams to indicate the direction of its acceleration if:

a) it has a constant speed,

b) it is slowing down,

c) it is gaining speed.

2 An electric light suspended from a ceiling is swinging to and fro in the wind between the extreme points A and E, as shown in the diagram. Draw diagrams to show the acceleration of the light at each of the positions indicated.

3 A racing car, initially travelling at $30\,\text{m}\,\text{s}^{-1}$, is accelerating around a corner (assumed to be a circular arc of radius 75 m). If its speed is increasing at a rate of $5\,\text{m}\,\text{s}^{-2}$, find the magnitude of the acceleration of the car after 0.5 seconds.

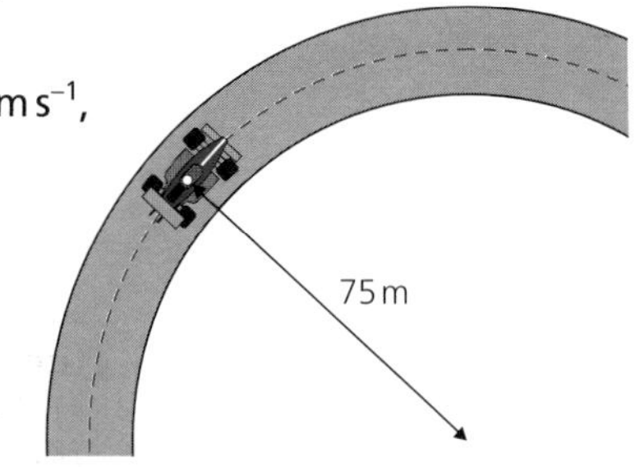

4 A child is sitting on a sledge which is tied to a fixed post in a snow-covered horizontal field by a rope of length 5 m. The combined mass of the child and sledge is 30 kg, and the sledge is pulled by another child who exerts a constant force of magnitude 10 N at right angles to the rope as shown. (You may assume that the rope is taut throughout and that the force always acts along a tangent to the circle.)

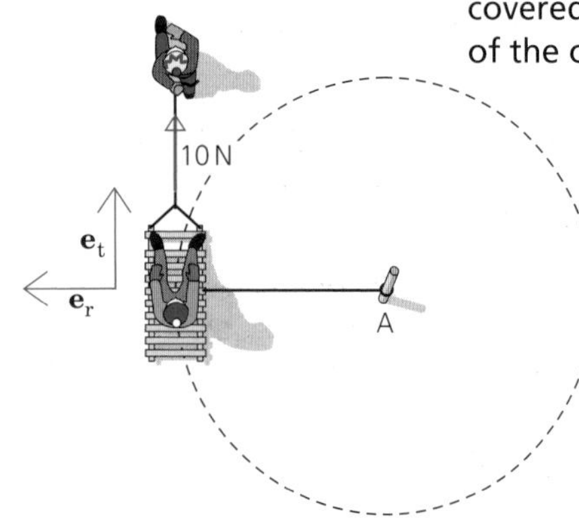

a) Find the resultant force on the sledge using the unit vectors $\mathbf{e}_r$ and $\mathbf{e}_t$.

b) Show that $\frac{dv}{dt}$ is constant and find an expression for v.

c) How long does it take for the sledge to complete one revolution? Find the speed and the magnitude of the acceleration at this time.

d) Find an expression for the tension in the rope?

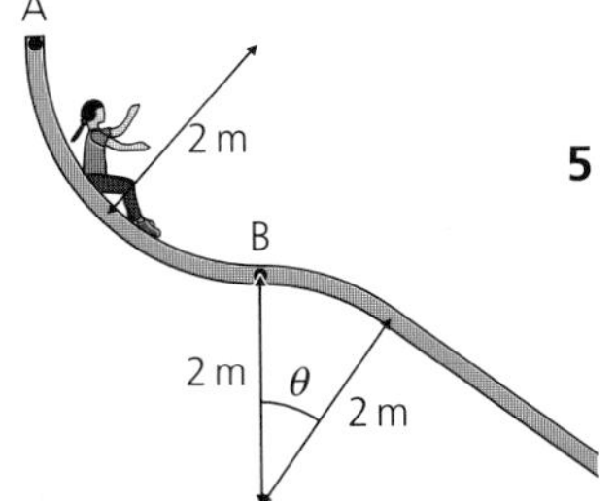

5 The diagram shows a design of a slide for a children's adventure park. The children should always remain in contact with the slide. The children begin at rest at A. Assume there is no friction on the slide.

a) Find the speed of a child at B.

b) Find the maximum value of θ if the child is not to lose contact with the slide.

6 A car is on a roundabout of radius 50 m. The coefficient of friction between the tyres and the road is 0.6. What is the greatest speed at which the car can safely negotiate the roundabout if the road surface is horizontal? What is the maximum possible rate of increase of speed when $v = 10\ \text{m s}^{-1}$?

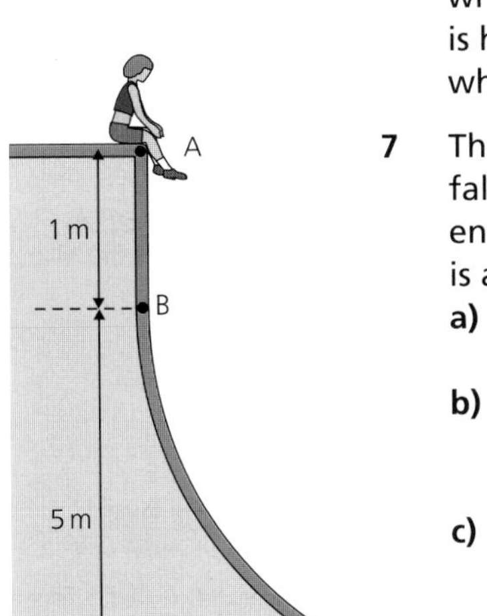

7 The figure shows a vertical drop attraction in a theme park. After falling, under gravity alone, down the smooth wall AB, the child enters the smooth circular quadrant segment BC. During this phase he is accelerated due to the circular motion.

a) Use conservation of energy to determine the speed of the child at the two points B and C.

b) Find an expression for the tangential acceleration of the child, in terms of θ (the angle between the child and the horizontal radius through B) as he travels from B to C.

c) Obtain an expression for the velocity $v\ \text{m s}^{-1}$ of the child while on the segment BC and hence write down an expression for the radial acceleration of the child.

8 A race-walker is to can be modelled simply as a particle situated at the top of two light rigid rods representing her legs. Race-walkers are disqualified from races if they are found to be 'lifting', that is if both feet are off the ground simultaneously.

a) By modelling the path of the walker's centre of mass as an arc of a circle, show that the above model leads to the conclusion that the maximum legal race-walking speed is less than about $3.1\ \text{m s}^{-1}$, assuming that the leg length is 1 m.

b) Given that the maximum speeds actually observed in competition are of the order of $4\ \text{m s}^{-1}$, in what ways could the above model be criticised?

9 A particle of mass m is attached to a fixed point O by a string of length L. It is initially held, with the string taut, at the same horizontal level as O and then released from rest.

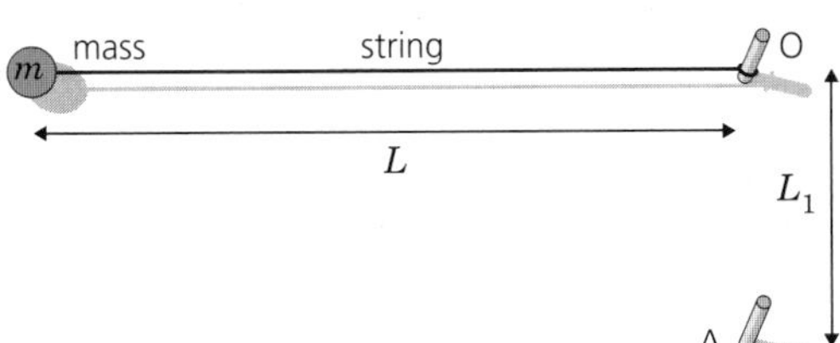

When the particle is vertically below O the string catches on a peg A (at a distance L_1 below O). The particle subsequently moves on a circular path centred on A. Show that complete revolutions about A are not possible unless $L_1 > \frac{3}{5}L$.

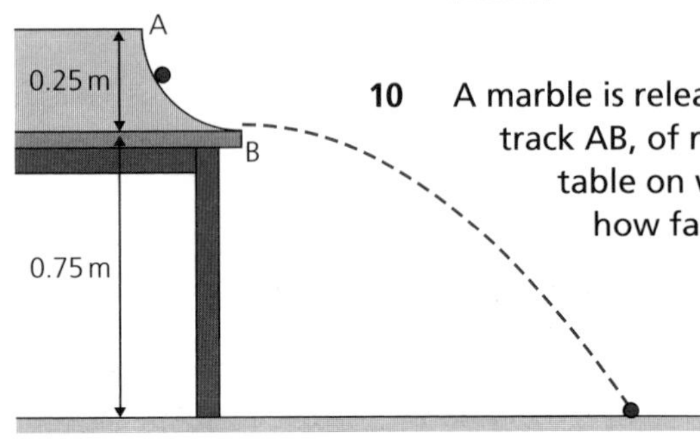

10 A marble is released from the point A and rolls down the circular track AB, of radius 0.25 m, to B where it becomes a projectile. If the table on which the track is mounted has a height of 0.75 m, how far from the table does the marble land?

Sketch a graph to show how the magnitude of the acceleration of the marble varies. What simplifying assumptions did you make in obtaining your solution? How might they be criticised?

CONSOLIDATION EXERCISES FOR CHAPTER 13

1 A particle P of mass m lies inside a fixed smooth hollow sphere of internal radius a and centre O. When P is at rest at the lowest point, A, of the sphere, it is given a horizontal impulse of magnitude mu. The particles loses contact with the inner surface of the sphere at the point B, where ∠AOB 120°.

a) Show that $u^2 = \frac{7}{2}ga$.
b) Find the greatest height above B reached by P.

(ULEAC Question 7, Paper M2, January 1993)

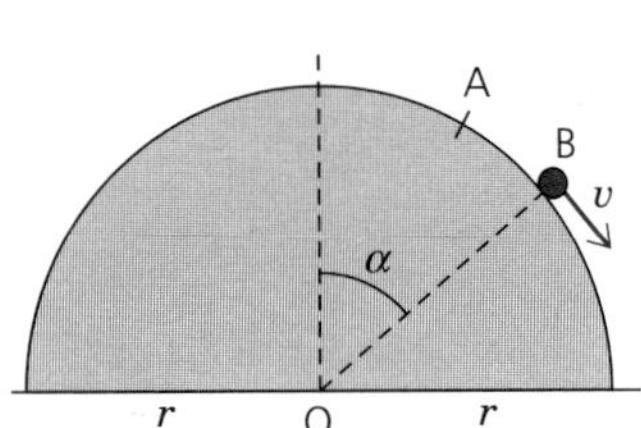

2 The diagram shows a smooth solid hemisphere H of radius r fixed with its plane surface, centre O, in contact with horizontal ground. A particle is released from rest on the surface of H at a point A such that OA makes an angle $\arccos\frac{7}{8}$ with the upward vertical. The particle slides freely until it leaves the surface of H at the point B with speed v. Given that OB makes an angle α with the upward vertical:
a) show that $\cos\alpha = \frac{7}{12}$,
b) find v^2 in terms of g and r.

Given that the particle strikes the ground with speed u:
c) find u^2 in terms of g and r.

(ULEAC Question 8, Paper M2, January 1992)

3 A ball B, of mass 20 grams, is attached to one end of a light inextensible string of length 80 cm. The other end of the string is attached to a fixed point O, and B hangs in equilibrium. A horizontal impulse of magnitude 0.1 N s is applied to B.

Show that, provided the string has not become slack, the speed v m s^{-1} of B, when OB makes an angle θ with the downward vertical at O, is given by:

$$v^2 = 9.30 + 15.70\cos\theta$$

where the coefficients are correct to two decimal places.
Find a similar expression for the tension in terms of θ, and find also the value of θ at the point where the string becomes slack.
Briefly describe the motion B after the string becomes slack.

(UCLES Specimen Paper M3, 1994)

MATHEMATICAL MODELLING ACTIVITY

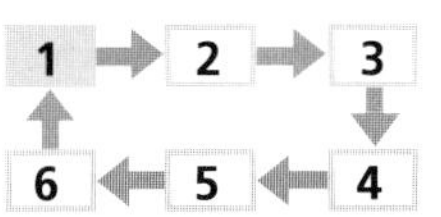

Specify the real problem

Problem statement

At a theme park there are plans to build a new slide. It will have a vertical section and then a section in the shape of a quarter of a circle, with a final horizontal section. The owners of the park are concerned that the riders should experience the maximum possible safe acceleration to gain the greatest excitement. The space that is available for the slide is limited and so the horizontal section should not exceed 20 m. The diagram shows the shape of the slide.

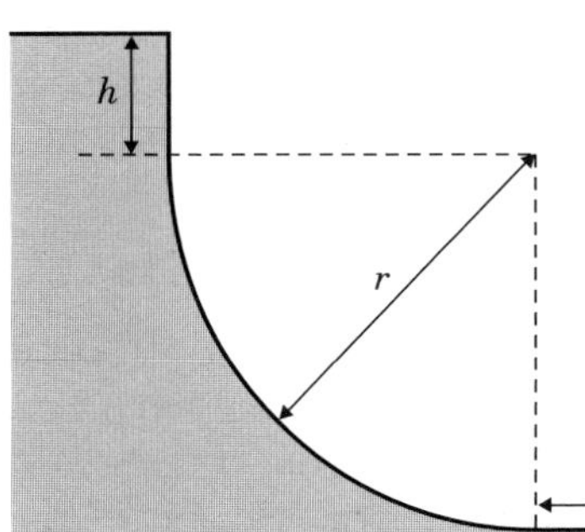

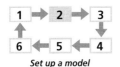

Set up a model

Set up a model

Make a list of the key features that should be considered when solving this problem.

The following list of assumptions will be used to provide a starting point for the solution of the problem.

- The riders on the slide can be modelled as particles.
- There is no air resistance.
- There is only friction on the horizontal section and here $\mu = 0.4$.
- The people fall a distance h.
- The radius of the circle is r.
- Riders should not experience an acceleration of magnitude greater than $3g$.

Comment on these, in the light of your list of key features, adding further assumptions if necessary.

Formulate the mathematical problem

Formulate a mathematical problem

Find the maximum speed of the riders when they reach the horizontal section if they are to stop at the end of it. Use this to find a relationship between h and r. Then find expressions for the magnitude of the acceleration experienced by riders on each stage.

Solve the mathematical problem

Mathematical solution

While on the horizontal section the riders will experience a force of $-\mu mg$ and an acceleration of $-\mu g$, or in this case $-0.4 \times 9.8 = -3.92\,\text{m}\,\text{s}^{-2}$.

Using $v^2 = u^2 + 2as$ with $v = 0$, $a = -3.92\,\text{m}\,\text{s}^{-2}$ and $s = 20\,\text{m}$ gives:

$$0 = u^2 + 2 \times (-3.92) \times 20 \Rightarrow u^2 = 156.8$$

so that:

$$u = 12.5\,\text{m}\,\text{s}^{-1}$$

As there is assumed to be no friction or air resistance on the first two stages of motion, conservation of energy can be applied. The potential energy lost while using the ride is $h + r$. So using conservation of energy gives:

$mg(h+r) = \frac{1}{2}mv^2 \Rightarrow g(h+r) = \frac{1}{2} \times 156.8 \Rightarrow h + r = 8 \text{ m}$

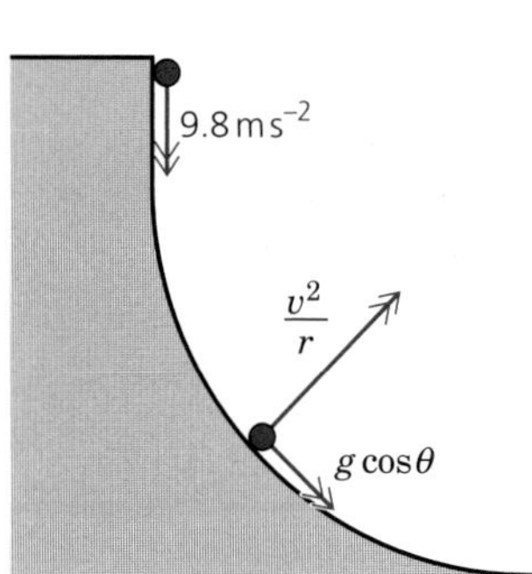

The acceleration of the rider is simply 9.8 m s^{-2} during the first stage.

During the second stage the acceleration has two components, $\frac{v^2}{r}$ towards the centre of the circle and $\frac{dv}{dt}$ along the tangent. In this case $\frac{dv}{dt}$ is given by the component of gravity parallel to the tangent. This can be expressed as $g\cos\theta$, where θ is as defined in the diagram. The acceleration on the final stage is –3.92 m s^{-2} as calculated previously.

To calculate the speed of the riders on the circular section, it is necessary to use conservation of energy. The potential energy lost will be given by:

$mg(h + r\sin\theta)$

as $(h + r\sin\theta)$ is the total distance fallen.

So $\frac{1}{2}mv^2 = mg(h + r\sin\theta) \Rightarrow v^2 = 2g(h + r\sin\theta)$.

Now the component of the acceleration towards the centre can be expressed in terms of θ:

$$\frac{v^2}{r} = \frac{2g}{r}(h + r\sin\theta)$$

As $h + r = 8$ the h can be replaced by $8 - r$ to give:

$$\frac{2g}{r}(8 - r + r\sin\theta)$$

Now the magnitude of the acceleration can be calculated as:

$$a = \sqrt{\left(\frac{2r}{g}(8 - r + r\sin\theta)\right)^2 + (g\cos\theta)^2}$$

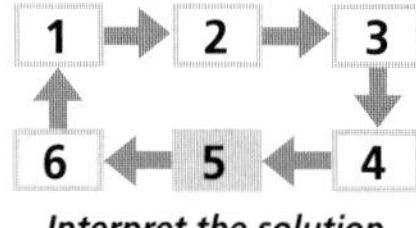

Interpret the solution

Interpretation

The expression for the magnitude of the acceleration appears to be complicated and difficult to deal with. A graphical approach helps to show what is happening to the acceleration as a rider descends the slide. The graphs below show how the magnitude of the acceleration varies with θ, for different values of r.

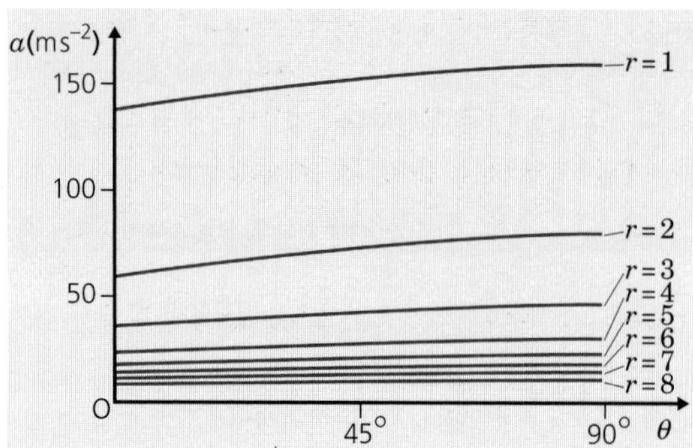

Note that the magnitude increases to a maximum when $\theta = 90°$ and that for the smaller values of r the magnitude of the acceleration can be excessive, for example with $r = 1$ it reaches 157 m s^{-2}.

As the maximum acceleration is reached when $\theta = 90°$, it is important to consider this value.

Show that the maximum magnitude of the acceleration experienced is $\frac{16g}{r}$.

If the maximum acceleration that is to be experience is $3g$ then:

$$3g = \frac{16g}{r} \Rightarrow r = \frac{16}{3} = 5\tfrac{1}{3}\text{ m}$$

This radius will produce a design that satisfies the requirements for the design, but there are some factors that could affect the model. One is the effect of air resistance and friction on the circular part of the slide. These would slow riders down to some extent, and therefore would reduce the maximum acceleration that could be experienced, which could vary considerably due to the different clothes that people wear. It may be possible that some riders would not have stopped moving at the end of the 20 m horizontal section.

List any further criticisms that could be made of the model used.

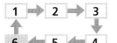

Compare with reality

Compare with reality

It is difficult to compare the results of this model with reality. However the next time you visit a theme park with such a slide you might try to estimate the dimensions of a slide and compare with the results here.

POSSIBLE MODELLING TASKS

1 Investigate the acceleration experienced as cars travel around a roundabout. You could consider cars that start at rest, or that move on to the roundabout at speed, or the effects of using different exits.
Alternatively you may wish to consider the acceleration of a person on a fairground ride as it starts up and increases its angular speed to its normal operating value, or as its angular speed decreases.

2 Investigate the effects of resistance forces on the car used in Exploration 13.1, the 'Loop the loop' activity. Try to develop a model for the resistance forces and use this to make revised predictions about the height of release needed.

Summary

After working through this chapter you should know that:

- when objects follow circular paths with variable speed they experience one component of acceleration towards the centre of the circle and another along a tangent to the circle
- the radial component of the acceleration has magnitude
$$\frac{v^2}{r} \text{ or } r\left(\frac{d\theta}{dt}\right)^2$$
- the tangential component of the acceleration has magnitude
$$\frac{dv}{dt} \text{ or } r\frac{d^2\theta}{dt^2}$$

Simple harmonic motion

- *Simple harmonic motion is the motion performed by objects such as pendulums and springs.*
- *Simple harmonic motion can be used to model the motion of many vibrating or oscillating systems.*
- *In this chapter we assume that there are no resistance forces in the models and systems we consider.*

SPRINGS AND SIMPLE HARMONIC MOTION (SHM)

There are many situations in which springs are involved in vibrating systems. One of the most familiar is probably the use of springs in the suspension systems of motor vehicles. Bungee jumpers also experience vibration, although they are attached to an elastic rope rather than a spring! In both of these examples the vibrations **decay** very quickly, and it is important that they do. In this chapter ideal systems where the vibrations do not decay will be considered and this first section deals specifically with vibrating spring systems.

Exploration 14.1

Practical activity

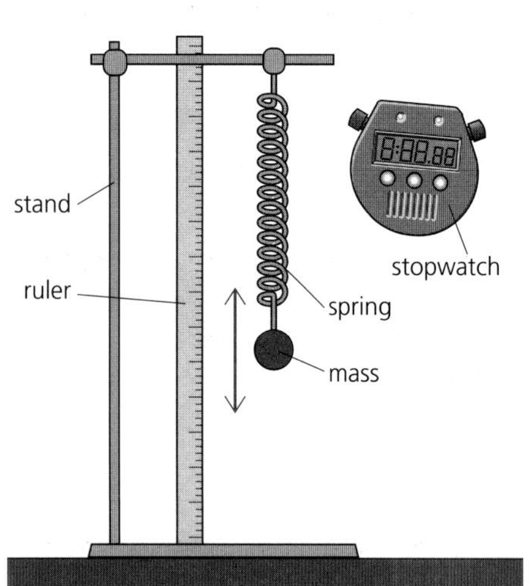

1. Hang a spring on a stand and attach a mass to the bottom end of it. Pull the mass down a small distance from the equilibrium position and record the time taken for, say, ten oscillations.
   Calculate the **period of oscillation**, the time taken for one complete oscillation.
2. Look at each of the factors suggested below, and decide if it will affect the period of the oscillations and then try an experiment to check:

   a) the mass of the spring,
   b) the initial displacement,
   c) the stiffness of the spring.

   Hint: Two springs joined end to end and will have a different stiffness.

CALCULATOR ACTIVITY 14.1

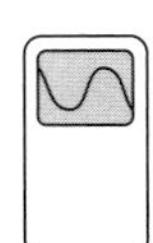

In this activity you explore graphs of some oscillations that could be used to model the motion of a spring mass system.

1 Make sure that your calculator is in radian mode, and set the Range or Window settings as below.
$x_{min} = 0$; $x_{max} = 10$; $x_{scl} = 1$
$y_{min} = -5$; $y_{max} = 5$; $y_{scl} = 1$

2 Draw the graph of $y = \sin x$. What is the period of this oscillation? What is the amplitude of the oscillation? (The amplitude is the distance that it moves from its equilibrium position.)

3 Draw the graph of each example given below. In each case state the period and amplitude.
a) $y = 3\sin x$ **b)** $y = 4\sin 2x$ **c)** $y = 4\sin 3x$
d) $y = 2\cos x$ **e)** $y = 3\sin 2x$

4 Using your results verify that the period of an oscillation of the form $\sin(\omega t)$ or $\cos(\omega t)$ has period $\frac{2\pi}{\omega}$.

5 Plot the graphs, $\sin x$, $\cos x$ and $-\cos x$. Which best models the motion of a mass that is:
a) pulled down and released,
b) pushed up and released,
c) given an impulse while at its equilibrium position?

6 Try to produce a graph to show how a mass oscillates if:
a) it has amplitude 5, period π and is started by being pulled down and released,
b) it is raised 3 units and released, and has period $\frac{\pi}{2}$,
c) it is given an initial upward velocity and has amplitude 4 and period 0.5.

Example 14.1

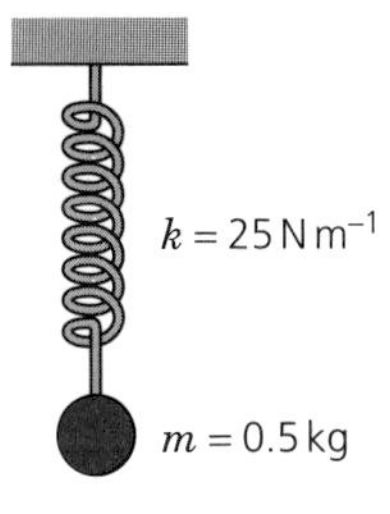

The diagram shows a 0.5 kg mass hanging on a spring that has stiffness 25 N m^{-1}.

a) *Find the resultant force on the mass in terms of x, its displacement from the equilibrium position.*
b) *The mass is pulled down and released. Show that its acceleration can be modelled by the equation:*

$$\frac{d^2x}{dt^2} = -50x$$

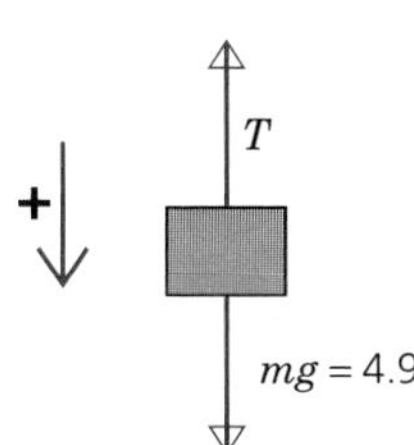

Solution

a) *The diagram shows the forces acting, the tension in the spring and the force of gravity. When the spring is at rest in the equilibrium position the tension is 4.9 N. When it is moved down a distance x, from the equilibrium position, the tension increases by $25x$ N, to give a total of $4.9 + 25x$.*

Defining the downward direction as positive gives a resultant force:

$mg - T = 4.9 - (4.9 + 25x) = -25x$

b) *As the displacement of the mass is x, its acceleration will be $\dfrac{d^2x}{dt^2}$. So Newton's second law gives:*

$$0.5\frac{d^2x}{dt^2} = -25x$$

$$\Rightarrow \frac{d^2x}{dt^2} = -50x$$

Example 14.2

An object of mass m is suspended from a spring of stiffness k. Find an expression for the acceleration of the mass if it is allowed to oscillate freely, in terms of x, its displacement from its equilibrium position.

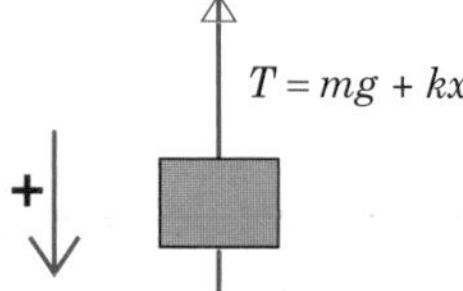

Solution

When in its equilibrium position the tension in the spring is mg. When the mass is moved a distance x from the equilibrium position the tension increases by kx, so the tension is $mg + kx$.

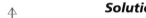

The resultant force is:

$mg - T = mg - (mg + kx) = -kx$

Applying Newton's second law allows us to find the acceleration, $\dfrac{d^2x}{dt^2}$.

$$m\frac{d^2x}{dt^2} = -kx$$

$$\Rightarrow \frac{d^2x}{dt^2} = \frac{-k}{m}x$$

The SHM equation

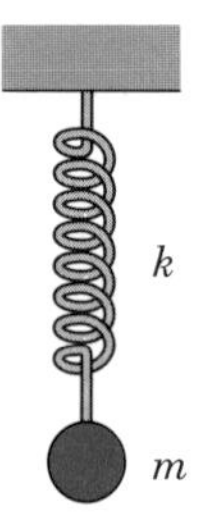

The acceleration of a mass–spring system is given by:

$$\frac{d^2x}{dt^2} = \frac{-k}{m}x.$$

An equation like this cannot be solved by direct integration, because the right-hand side is in terms of x. The following examples show that the solution of this equation is the form:

$$x = A\cos(\omega t + \alpha)$$

When the acceleration of an object and its position are of the form given above then the motion is described as **simple harmonic motion** or SHM.

Example 14.3

The acceleration of an oscillating mass–spring system is given by:

$$\frac{d^2x}{dt^2} = -9x$$

a) *Show that if the displacement of the mass is $x = A\cos 3t$, this satisfies the equation above.*

b) *If the amplitude of the oscillations is 0.06 state the value of A.*

c) *Sketch the graph of displacement against time, indicating on the graph the period. Also describe how the motion begins if the downward direction is positive.*

Solution

a) *If* $x = A\cos 3t$

then $\dfrac{dx}{dt} = -3A\sin 3t$

and $\dfrac{d^2x}{dt^2} = -9A\cos 3t = -9x$

since $x = A\cos 3t$

So $x = A\cos 3t$ *satisfies the initial differential equation.*

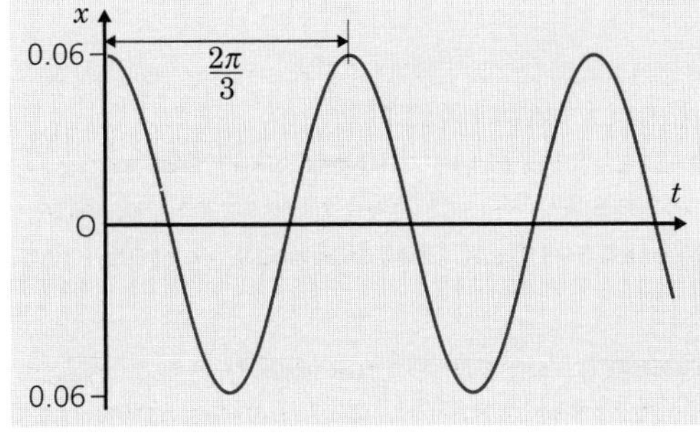

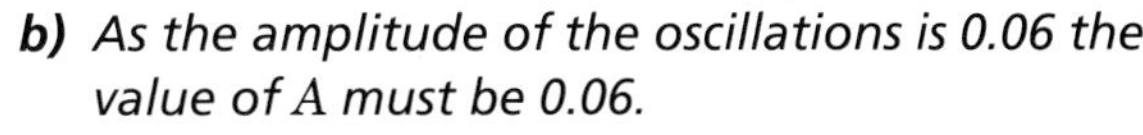

b) *As the amplitude of the oscillations is 0.06 the value of* A *must be 0.06.*

c) *The graph is shown.*

Note that the one complete oscillation is completed when:

$3t = 2$

or $t = \dfrac{2\pi}{3}$

So the period of the oscillation is $\dfrac{2\pi}{3}$.

Initially the mass is displaced in the positive direction and released. As the position direction has been defined as downward, this means that the mass was pulled down and released.

Example 14.4

Show that:

$$x = A\cos(\omega t + \alpha)$$

is a solution of the differential equation:

$$\frac{d^2x}{dt^2} = \frac{-k}{m}x$$

that arises from a vibrating mass–spring system, and find ω *in terms of* m *and* k. *Find the period of the oscillations.*

Solution

Starting with $x = A\cos(\omega t + \alpha)$

and differentiating gives:

$$\frac{dx}{dt} = -A\omega\sin(\omega t + \alpha)$$

and:

$$\frac{d^2x}{dt^2} = -A\omega^2\cos(\omega t + \alpha)$$

As $x = A\cos(\omega t + \alpha)$ *this is equivalent to:*

$$\frac{d^2x}{dt^2} = -\omega^2 x$$

This will satisfy the differential equation:

$$\frac{d^2x}{dt^2} = \frac{-k}{m}x$$

if $\omega^2 = \dfrac{k}{m}$

One oscillation will be completed when:

$\omega t = 2\pi$ or $t = \dfrac{2\pi}{\omega} = 2\pi\sqrt{\dfrac{m}{k}}$

So the period will be $2\pi\sqrt{\dfrac{m}{k}}$

Example 14.5

A clock is regulated by a 10 gram mass hanging on a spring of stiffness 10 N m^{-1}.

a) *Find an expression for the acceleration,* $\frac{d^2x}{dt^2}$*, of the mass, where* x *is the displacement from the equilibrium position.*

b) *Find an expression for* x*, if the mass is set in motion by being pulled down 1.5 cm and released.*

c) *State the period of the oscillations that take place.*

Solution

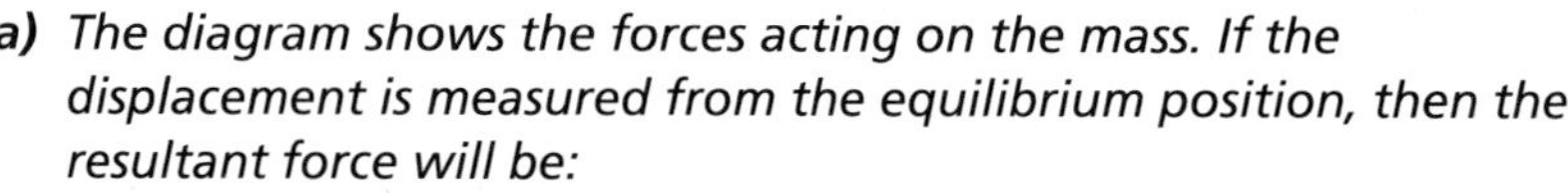

a) *The diagram shows the forces acting on the mass. If the displacement is measured from the equilibrium position, then the resultant force will be:*

$mg - T = mg - (mg + kx) = -kx$

Using Newton's second law gives:

$$m\frac{d^2x}{dt^2} = -kx$$

Substituting for m *and* k *gives:*

$$0.01\frac{d^2x}{dt^2} = -10x$$

or $$\frac{d^2x}{dt^2} = -1000x$$

b) *The displacement* x *will be of the form:*

$x = A\cos(\omega t + \alpha)$

where $\omega^2 = 1000$ *or* $\omega = 31.6$*. As the mass is initially displaced 1.5 cm or 0.015 m, the amplitude* A *will be 0.015, giving:*

$x = 0.015\cos(31.6t + \alpha)$

When $t = 0$*,* $x = 0.015$*, giving:*

$0.0015 = 0.015\cos\alpha$

so $\alpha = 0$ *and*

$x = 0.015\cos 31.6t$

c) *One oscillation will be completed when:*

$31.6t = 2\pi$

or $t = \dfrac{2\pi}{31.6} = 0.205$ *seconds*

Now back to the bungee jumper who has been hanging around since the beginning of this chapter.

Example 14.6

The bungee jumper has mass 60 kg and is attached to an elastic string of stiffness 50 N m^{-1}. The bungee jumper is moving at 10 m s^{-1} when the elastic becomes taut. Assume there is no air resistance. Find the period of the oscillations of the jumper.

Solution

As the stiffness of the elastic is 50 N m^{-1}, the resultant force will be –50x. Newton's second law can be applied to give:

$$60\frac{d^2x}{dt^2} = -50x$$

or $$\frac{d^2x}{dt^2} = -\frac{50}{60}x$$

The solution of this equation will of the form
$x = A\cos(\omega t + \alpha)$, *where* $\omega = \sqrt{\frac{50}{60}} = 0.913$.

To find the period consider the time taken for the complete oscillation. When $\omega t = 2\pi$ *one oscillation will have been completed giving:*

$$t = \frac{2\pi}{\omega} = \frac{2\pi}{0.913} = 6.88 \text{ seconds}$$

EXERCISES

14.1 A

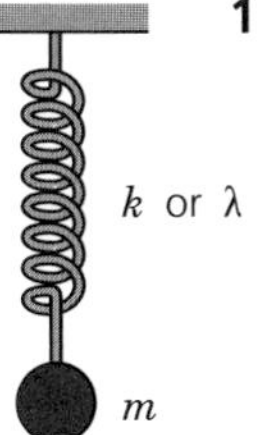

1 For each mass–spring system shown below, find an expression for $\frac{d^2x}{dt^2}$ the acceleration of the mass in terms of x its displacement from the equilibrium position.

a) $k = 20$ N m^{-1} $m = 1$ kg
b) $k = 20$ N m^{-1} $m = 20$ kg
c) $\lambda = 60$ $l = 0.04$ m $m = 4$ kg
d) $\lambda = 20$ N $l = 0.2$ m $m = 2$ kg

2 The diagram shows an object of mass 0.1 kg and two springs of stiffness 30 N m^{-1}.

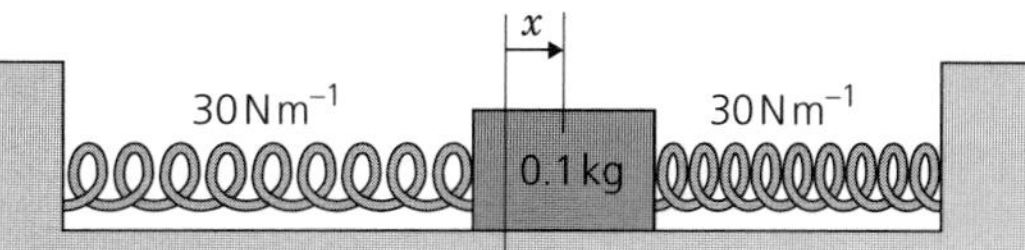

a) Find the resultant force on the mass in terms of x, the displacement of the mass from its equilibrium position.
b) Hence find the acceleration of the mass in terms of x.
c) Repeat **a)** and **b)** if the right-hand spring is replaced with one of stiffness 60 N m^{-1}.

3 Show that each differential equation below is satisfied by the expression, $x = A\cos(\omega t + \alpha)$ and find the value of ω:

a) $\frac{d^2x}{dt^2} = -4x$ **b)** $\frac{d^2x}{dt^2} = -100x$

4 Find the period for each mass–spring system defined below, if a single mass is suspended from a spring with the characteristics given.

a) $k = 24$ N m^{-1}, $m = 60$ grams
b) $k = 40$ N m^{-1}, $m = 1$ kg
c) $m = 100$ grams, $\lambda = 4$ N, $l = 0.2$ m
d) $m = 200$ grams, $\lambda = 50$ N, $l = 0.1$ m

In which cases do you think it would be feasible to carry out an experiment to determine the period by observing the number of oscillations in a given time interval?

5 Two springs with stiffness $40\,\text{N m}^{-1}$ are attached to a 0.5 kg mass as shown in the diagram. It is pulled 3 cm to the right and released.

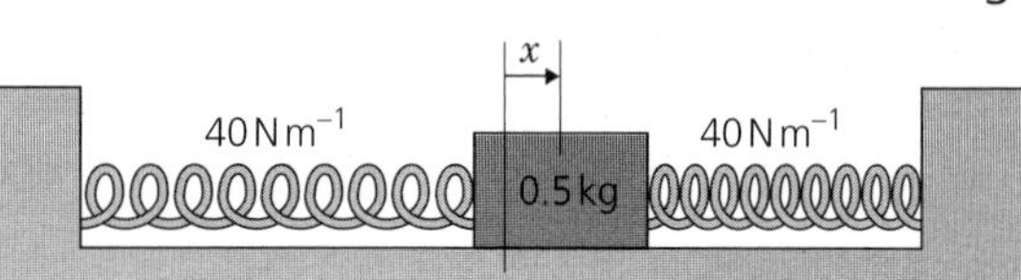

a) Find the acceleration of the mass, in terms of x, its displacement to the right of its equilibrium position.
b) Find the period of the oscillations.
c) Find an expression for x.
d) If the mass were initially moved to the left instead of the right, how would your answer change?

6 Two identical springs and masses are arranged as shown. How will the period of oscillation of the systems compare?

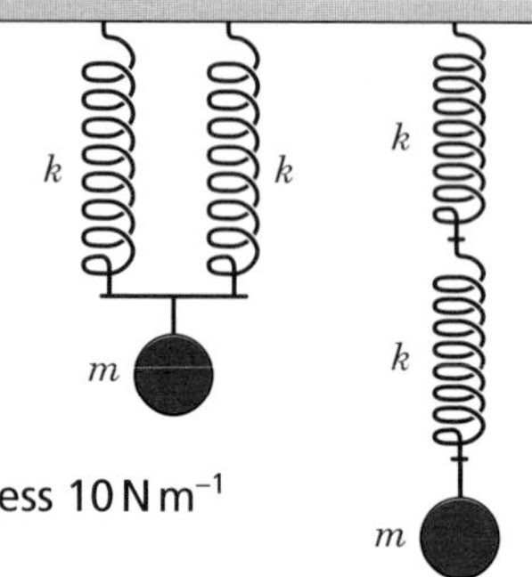

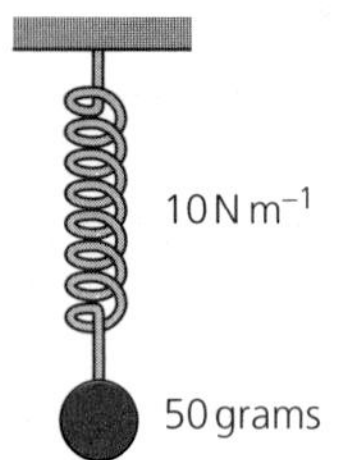

7 A clock manufacturer uses a spring of stiffness $10\,\text{N m}^{-1}$ and a mass of 50 grams, to regulate a clock.
a) Find the period of oscillation.
A clock is taken to the moon, where $g = 1.6\,\text{m s}^{-2}$.
b) Find the resultant force on the mass and hence its acceleration.
c) Find the period of the oscillations when the clock is on the moon.
d) What difference is there between the oscillations of two identical clocks, one on Earth and the other on the moon?

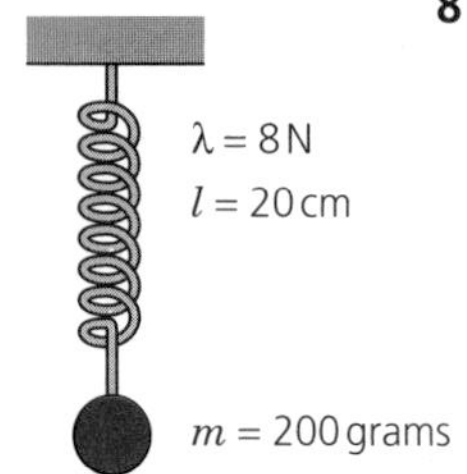

8 The mass–spring system shown is at rest in its equilibrium position. The mass is set into motion by being given an initial upward velocity of $2\,\text{m s}^{-1}$.
a) Show that the displacement of the mass from its equilibrium position is given by:
$$x = A\cos(\omega t + \alpha)$$
and find the value of ω.
b) Use the fact that when $t = 0$, $x = 0$, to find the value of α.
c) Find an expression for the velocity of the mass and hence find the value of A.

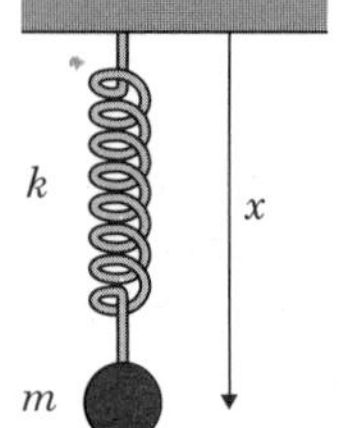

9 The diagram shows a spring of stiffness k, and natural length l, attached to a mass m. The displacement, x, of the mass is measured from the point of suspension.
a) Show that the acceleration of the mass is:
$$\frac{d^2x}{dt^2} = -\frac{k}{m}x + g + \frac{kl}{m}$$
b) If $k = 6.4\,\text{N m}^{-1}$, $l = 0.4\,\text{m}$, $m = 0.4\,\text{kg}$ and $g = 9.8\,\text{m s}^{-2}$, show that:
$$x = 0.02\cos 4t + 1.0215$$
models the motion of the mass if it is pulled down 0.02 m from the equilibrium position and released.
c) Find the period of the motion.
d) How close does the mass get to the point of suspension?

10 The stiffness of a spring k is given by $k = \frac{\lambda}{l}$ where λ is the modulus of elasticity and l is the natural length of spring.

a) Find an expression, in terms of m, λ and l, for the acceleration of a mass attached to the spring when it is a distance x from its equilibrium position.

b) What is the period of the motion of the mass?

EXERCISES

14.1 B

1 For each mass–spring system below, find an expression for $\frac{d^2x}{dt^2}$, the acceleration of the mass in terms of x its displacement from the equilibrium position.

a) $k = 30\,\text{N m}^{-1}$ $\quad m = 1\,\text{kg}$

b) $k = 30\,\text{N m}^{-1}$ $\quad m = 15\,\text{kg}$

c) $\lambda = 40\,\text{N}$ $\quad l = 0.5\,\text{m}$ $\quad m = 2\,\text{kg}$

d) $\lambda = 30\,\text{N}$ $\quad l = 60\,\text{cm}$ $\quad m = 5\,\text{kg}$

k or λ, l

m

2 The diagram shows an object of mass 0.2 kg and two springs of stiffness $40\,\text{N m}^{-1}$.

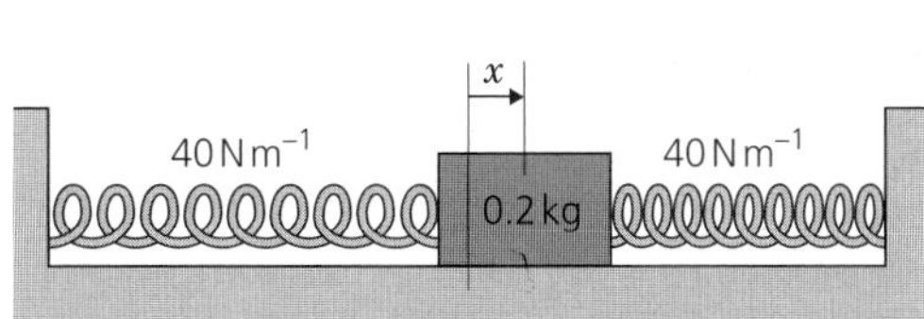

a) Find the resultant force on the mass in terms of x, the displacement of the mass from its equilibrium position.

b) Hence find the acceleration of the mass in terms of x.

c) Repeat **a)** and **b)** if the right-hand spring is replaced with one of stiffness $80\,\text{N m}^{-1}$.

3 Show that each differential equation below is satisfied by the expression $x = A\cos(\omega t + \alpha)$ and find the value of ω.

a) $\frac{d^2x}{dt^2} = -25x$ $\qquad$ **b)** $\frac{d^2x}{dt^2} = -81x$

4 Find the period for each mass–spring system below, if a single mass is suspended from a spring with the characteristics given.

a) $k = 36\,\text{N m}^{-1}$, $\quad m = 90$ grams

b) $k = 20\,\text{N m}^{-1}$, $\quad m = 100$ grams

c) $m = 200$ grams, $\quad \lambda = 5\,\text{N}$, $\quad l = 0.4\,\text{m}$

d) $m = 1\,\text{kg}$, $\quad \lambda = 40\,\text{N}$, $\quad l = 0.8\,\text{m}$

5 Two springs with stiffness $60\,\text{N m}^{-1}$ are attached to a 0.4 kg mass as shown in the diagram. It is pulled 5 cm to the right and released.

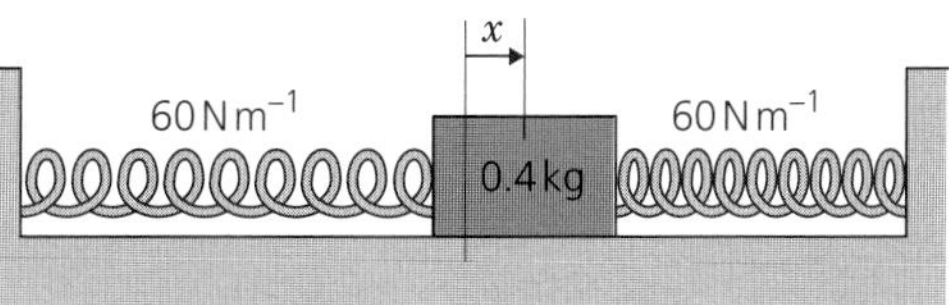

a) Find the acceleration of the mass, in terms of x its displacement to the right of its equilibrium position.

b) Find the period of the oscillations.

c) Find an expression for x.

d) If the mass were initially moved to the left instead of the right, how would your answer change?

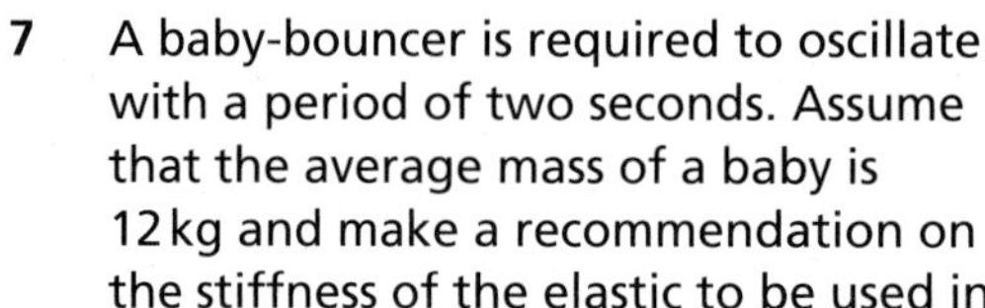

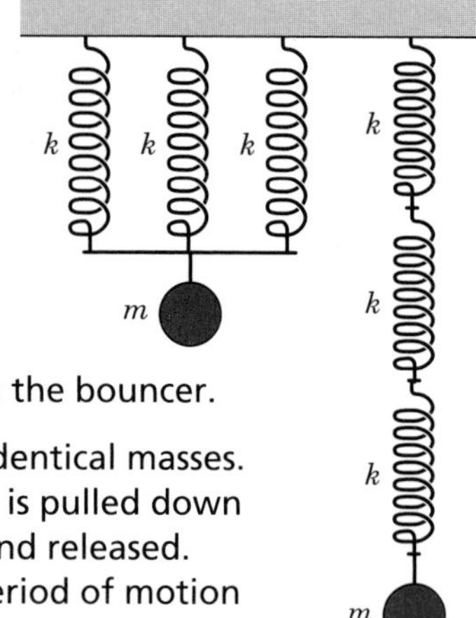

6 Three identical springs and masses are arranged as shown. How will the period of oscillation of each system compare?

7 A baby-bouncer is required to oscillate with a period of two seconds. Assume that the average mass of a baby is 12 kg and make a recommendation on the stiffness of the elastic to be used in the bouncer.

8 Two identical springs are attached to identical masses. One is pulled down 3 cm and the other is pulled down 4 cm from their equilibrium positions and released.

a) How would the amplitude of the period of motion for each system compare?

b) If the stiffness of the springs is 40 N m^{-1} and the mass is 100 grams, find the period.

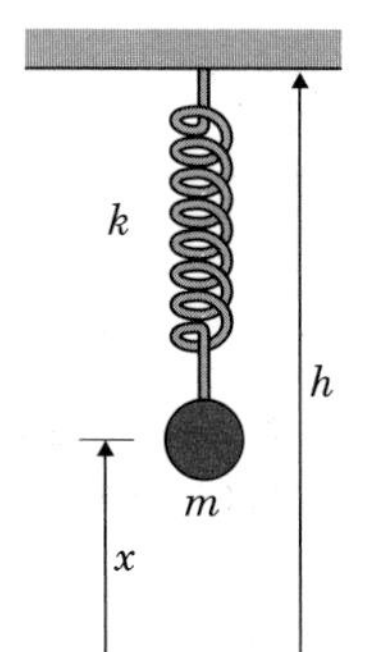

9 The diagram shows a spring of stiffness k, and natural length l, attached to a mass m. The displacement, x, of the mass is measured from a point a distance h below the point of suspension.

a) Show that the acceleration of the mass is:

$$\frac{d^2x}{dt^2} = -g - \frac{k}{m}(l + x - h)$$

b) If $k = 23.52\,\text{N m}^{-1}$, $l = 0.6\,\text{m}$, $m = 1.2\,\text{kg}$, $g = 9.8\,\text{m s}^{-2}$ and $h = 2\,\text{m}$ show that $x = 0.9 - 0.3\cos\sqrt{19.6}t$ models the motion of the mass if it is pulled down a distance of 0.03 m from the equilibrium position and released.

c) Find the period of the motion.

d) How close does the mass get to the point of suspension?

10 The natural length of a spring is l and its modulus of elasticity is λ. When a mass m is attached to the spring and allowed to oscillate about the equilibrium position, the displacement from this position is labelled x.

a) Find an expression for the tension in the spring in terms of m, g, l, λ and x.

b) Find an expression for the resultant force on the mass.

c) Find the period of the motion in terms of m, l and λ.

AN ENERGY APPROACH TO SHM

As we have seen, applying Newton's second law leads to a detailed description of the motion of a spring system describing simple harmonic motion. Often, though, problems can be solved using a simpler approach based on energy considerations. We start by considering the transfer of potential and kinetic energy as the system vibrates.

Exploration 14.2

The graph shows displacement from the equilibrium position, plotted against time, for a mass that is oscillating on the end of a spring. At

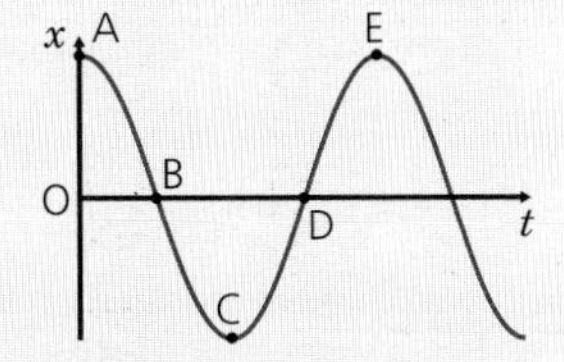

each point indicated consider how much potential and kinetic energy the mass has and what happens to this energy.

Energy considerations

At some positions the gravitational potential energy of the mass is at its lowest value and the kinetic energy is zero. This is a situation where much of the energy of the system is stored in the spring. Energy stored in an elastic string or spring is known as **elastic potential energy**.

We studied elastic potential energy in Chapter 11. The work done in stretching a spring of stiffness k a distance e from its natural length, where $T(x)$ is the tension in the spring, is:

$$\int_0^e T(x)\mathrm{d}x = \int_0^e kx\,\mathrm{d}x = \left[\tfrac{1}{2}kx^2\right]_0^e = \tfrac{1}{2}ke^2$$

so the elastic potential energy, stored in the spring, is:

$$\tfrac{1}{2}ke^2 \text{ or } \frac{\lambda e^2}{2l}$$

where λ is the modulus of elasticity and l the natural length of the spring.

Example 14.7

The diagram shows a spring of stiffness, k, attached to a mass m.

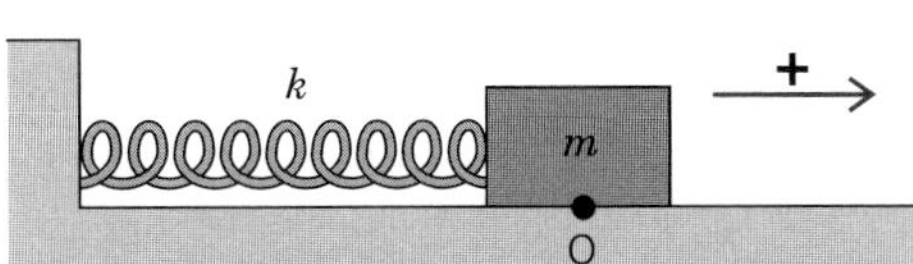

The mass is pulled a distance a to the left and released. Show that velocity of the mass can be modelled by:

$$v = \omega\sqrt{a^2 - x^2}$$

where x is the extension of the spring. What important assumption has to be made about the system?

Solution

When the mass is pulled a distance a to the right the elastic potential energy (EPE) will be:

EPE $= \tfrac{1}{2}ke^2 = \tfrac{1}{2}ka^2$

This is the total energy of the system. As the mass moves this will converted into kinetic energy (KE) and back to elastic potential energy. At any time the kinetic energy is $\tfrac{1}{2}mv^2$ and the elastic potential energy is $\tfrac{1}{2}kx^2$ so by conservation of energy:

total energy = KE + EPE $\Rightarrow \tfrac{1}{2}ka^2 = \tfrac{1}{2}mv^2 + \tfrac{1}{2}kx^2$

Rearranging for v gives:

$$ka^2 = mv^2 + kx^2$$
$$mv^2 = ka^2 - kx^2$$
$$v^2 = \frac{k}{m}\left(a^2 - x^2\right)$$

Earlier in this chapter ω was introduced and defined as $\omega^2 = \frac{k}{m}$, so the equation becomes:

$$v^2 = \omega^2 (a^2 - x^2)$$

or $v = \omega\sqrt{a^2 - x^2}$

It is important to realise that this result is only true if not energy is lost from the system, for example by doing work against air resistance.

The result demonstrated above holds true for all cases of simple harmonic motion.

Example 14.8

A 2 kg mass suspended on a spring of stiffness 200 N m⁻¹ is at rest in its equilibrium position. The mass is struck so that it leaves the equilibrium position moving upwards at 0.4 m s⁻¹.

a) *Find the period of the motion.*

b) *Find the amplitude.*

Solution

a) The period for simple harmonic motion, as described by the mass, is $T = \frac{2\pi}{\omega}$, and for the mass–spring system $\omega^2 = \frac{k}{m}$. So in this case:

$$\omega^2 = \frac{200}{2} = 100$$
$$\omega = 10$$

So the period is:

$$T = \frac{2\pi}{\omega} = \frac{2\pi}{10} = \frac{\pi}{5} = 0.63 \text{ seconds.}$$

b) The amplitude can be found using:

$$v = \omega\sqrt{a^2 - x^2}$$

Initially $v = 0.4$ and $x = 0$, so using $\omega = 10$, gives:

$$0.4 = 10\sqrt{a^2 - 0}$$

$$a = \frac{0.4}{10} = 0.04\,m$$

Example 14.9

A thin steel rod is displaced and released. Assume that the tip of the rod describes simple harmonic motion with period 0.01 seconds. If the tip of the rod moves between extremities 0.4 m apart find the maximum speed of the tip of the rod.

Solution

Fist the value of must be found. Since the period is $\frac{2\pi}{\omega}$, this gives:

$$0.01 = \frac{2\pi}{\omega}$$

so

$$\omega = \frac{2\pi}{0.1} = 200\pi$$

0.4 m

Now using

$$v = \omega\sqrt{a^2 - x^2}$$

and noting that $a = 0.2$ and $x = 0$ when the tip has its maximum speed gives:

$$v = 200\pi\sqrt{0.2^2 - 0} = 40\pi = 126\,m\,s^{-1}$$

EXERCISES

14.2 A

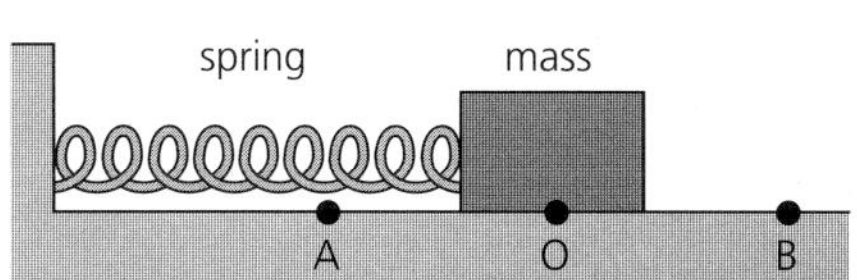

1 A spring has stiffness $50\,\text{N m}^{-1}$ and one end is attached to a 100 gram mass, which is free to move on a smooth horizontal plane. The other end of spring is fixed as shown in the diagram.

a) Find the energy stored in the spring when it is moved to B, 5 cm from O, the position of the mass, when the spring is at its natural length.

b) State the KE of the mass as it passes through O, and hence find the speed of the mass at this point.

c) Repeat **a)** and **b)** for an initial displacement of 10 cm.

2 A mass–spring system performs simple harmonic motion with amplitude 5 cm and period 2 seconds. Find the maximum speed attained by the mass.

3 A bottle bobbing up and down in a tank of water oscillates with simple harmonic motion. The amplitude of the motion is 6 cm and the period is 1.2 seconds. Find the maximum speed reached by the bottle.

4 The manufacturer of an electric saw claims that the saw completes two cycles per second and that the blade reaches a maximum speed of $0.75\ \text{m s}^{-1}$. Find the amplitude of the motion.

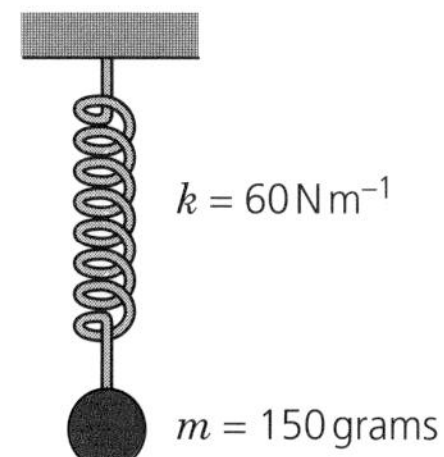

5 A mass–spring system consists of a spring and mass as shown. When set into motion the system describes simple harmonic motion, with an initial speed of $0.8\,\text{m s}^{-1}$ at its equilibrium position.

a) Find the period of the motion.

b) What is the greatest speed attained by the mass during the motion?

6 The piston in a car engine describes what is approximately simple harmonic motion.

a) If the engine is operating at 5000 rpm find ω.

b) The piston moves up and down between points 12 cm apart during each cycle. Find the maximum speed of the piston.

c) Find the speed of the piston when it is 1 cm from its lowest position.

7 The motion of the needle of a sewing machine can be modelled as simple harmonic motion. The top of the needle moves between points 1.2 cm above and 0.6 cm below the flat surface of the sewing machine. The needle reaches a maximum speed of $5\,\text{cm s}^{-1}$.

a) What is the amplitude of the motion?

b) What is the speed of the needle when it pierces the material resting on the flat surface?

c) An improved model reduces the period of the motion by 10%. What happens to the speed of the needle when it enters the cloth?

8 The displacement of an object from its equilibrium position is given by:

$$x = a\cos\omega t$$

a) Find an expression for its speed in terms of a, ω and t by differentiating the expression for x.

b) Substitute the expression for x into equation $v = \omega\sqrt{a^2 - x^2}$ and show that you get the same result in part **a)**.

9 An object describes simple harmonic motion. When 1 cm from the centre of its oscillations it moves at $2\,\text{m}\,\text{s}^{-1}$ and when 2 cm from its centre it moves at $0.5\,\text{m}\,\text{s}^{-1}$. Find the amplitude and period of the motion.

10 An electric jig saw has a blade that describes simple harmonic motion. It is essential that the speed of the blade halves from its maximum of $20\,\text{cm}\,\text{s}^{-1}$ by the time the blade has moved 0.3 cm from its mid-point.

a) Find the amplitude of the motion.

b) Find the period of the motion.

EXERCISES

14.2 B

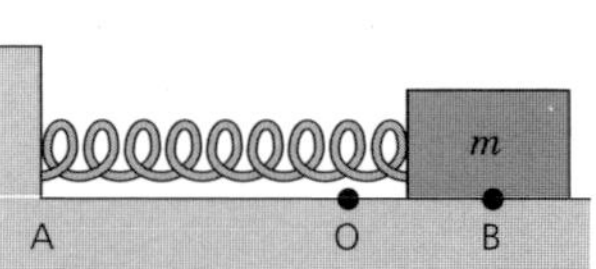

1 The diagram shows a spring lying along a smooth, horizontal surface with one end fixed at A and the other attached to a 250 gram mass held stationary at B. The natural length of the spring is AO and the stiffness of the spring is $60\,\text{N}\,\text{m}^{-1}$.

a) If the distance OB is 15 cm find the speed with which the mass passes through O after being released from B.

b) Repeat **a)** with the distance OB equal to 20 cm.

2 When a mass is suspended from a spring it oscillates with simple harmonic motion. If the amplitude is 9 cm and the period is 0.36 seconds, find the maximum speed attained by the mass.

3 The end of a metal beam is oscillating with simple harmonic motion. The amplitude of the motion is 1.5 cm and the period is 0.3 seconds. Find the maximum speed reached by the end of the beam.

4 When a certain guitar string is plucked it oscillates with simple harmonic motion. If the amplitude is 4 mm and it vibrates 60 times per second, find the maximum speed of the string.

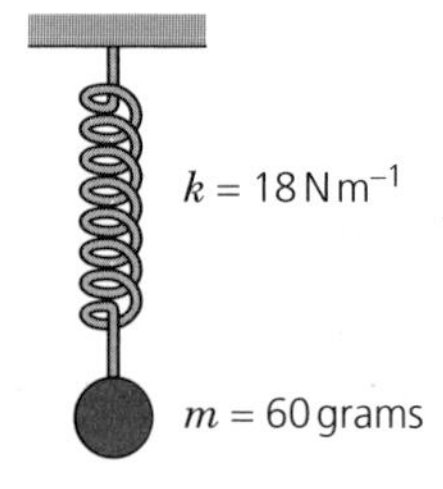

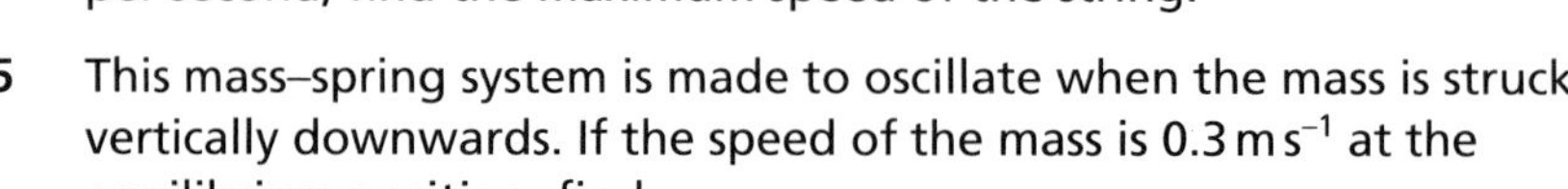

5 This mass–spring system is made to oscillate when the mass is struck vertically downwards. If the speed of the mass is $0.3\,\text{m}\,\text{s}^{-1}$ at the equilibrium position, find:

a) the period and amplitude of the motion,

b) the greatest speed attained by the mass during the motion.

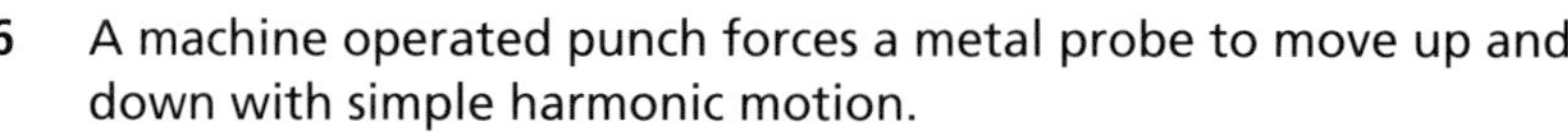

6 A machine operated punch forces a metal probe to move up and down with simple harmonic motion.

a) If the probe completes 450 oscillations every minute, find ω.

b) The probe moves between two points which are 15 cm apart. Find the maximum speed of the probe.

c) Find the speed of the probe when it is 2.5 cm from its lowest position.

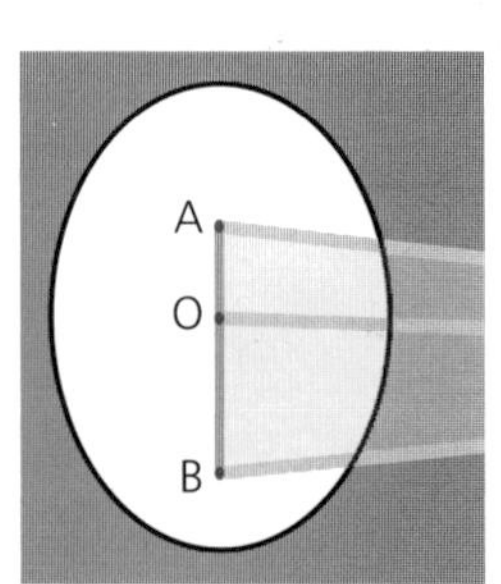

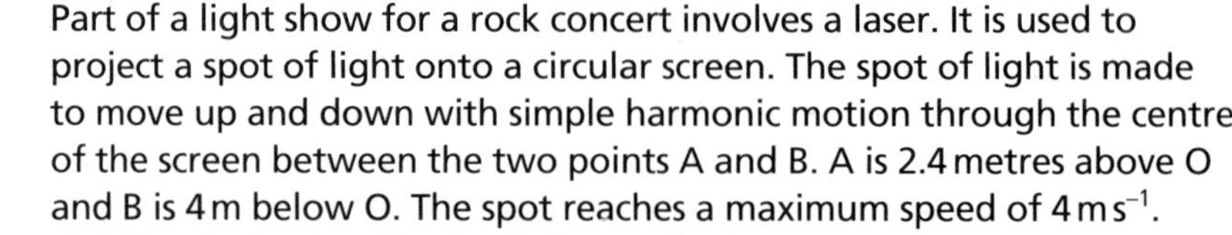

7 Part of a light show for a rock concert involves a laser. It is used to project a spot of light onto a circular screen. The spot of light is made to move up and down with simple harmonic motion through the centre of the screen between the two points A and B. A is 2.4 metres above O and B is 4 m below O. The spot reaches a maximum speed of $4\,\text{m}\,\text{s}^{-1}$.

a) What is the amplitude of the motion?

b) What is the speed of the spot as it passes through O?

c) The period is then changed, increasing it by 20%. What effect does this have on the speed of the spot as it passes through the point O?

8 The displacement of an object from its equilibrium position is given by $x = a\sin\omega t$.

a) Find an expression for its speed in terms of a, ω and t by differentiating the expression for x.

b) By substituting the expression for x into the equation $v = \omega\sqrt{a^2 - x^2}$ show that you get the same results as in part **a)**.

9 An object is moving with simple harmonic motion about an equilibrium position O. When 3 cm from O it moves at $4\,\text{m s}^{-1}$ and when 6 cm from O it moves at $1\,\text{m s}^{-1}$. Find the amplitude and the period of the motion.

10 The suspension system in a car allows a wheel to oscillate vertically with simple harmonic motion. The designers have been instructed to ensure that at a displacement of 8 cm from the equilibrium position the vertical speed of the wheel is one-quarter of the maximum speed. If the maximum vertical speed is $2\,\text{m s}^{-1}$, find:

a) the amplitude of the motion,

b) the period of the motion.

THE SIMPLE PENDULUM

It is not only spring systems that vibrate with simple harmonic motion. The previous section has included references to other situations where simple harmonic motion takes place, for example in a sewing machine. A pendulum is an example of motion that is almost simple harmonic. In this section we look at the motion of a pendulum and the assumptions that must be made in order to model the vibrations as simple harmonic motion.

Exploration 14.3 *A pendulum*

1 Many clocks use pendulums to regulate their time-keeping. What factors do you think affect the motion of the pendulum.

2 The pendulum bob describes an arc of a circle. How does the acceleration of the pendulum change as the bob moves along the arc?.

Exploration 14.4 *Practical activity*

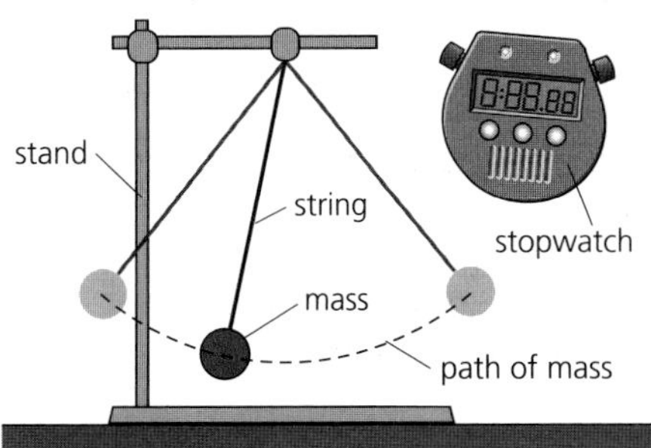

1 Set up a simple pendulum as shown in the diagram.

2 Decide which of the features listed are likely to affect the period of the oscillations of the pendulum.

a) The length of the string.

b) The mass on the end of the string.

c) The release position.

3 Devise some simple experiments to verify your answers to **2**.

Example 14.10

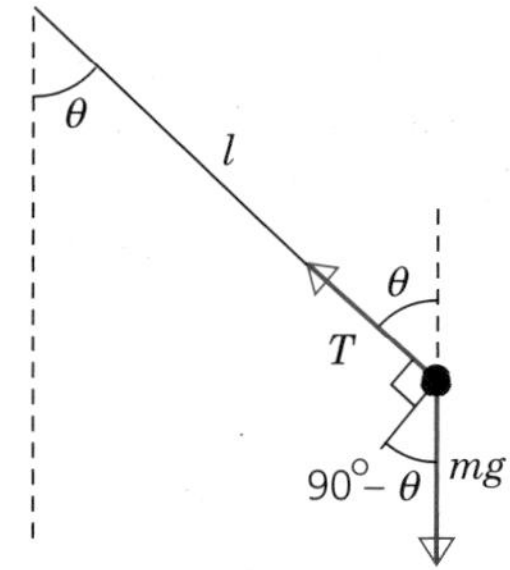

A pendulum consists of a mass on the end of a string of length l.

a) *Find the component of the resultant force on the mass, in terms of θ, the angle that the pendulum makes with the vertical, that is directed along the tangent.*

b) *Explain why the distance moved by the pendulum is $l\theta$.*

c) *By differentiating the result in **b)** find the tangential component of acceleration.*

d) *Apply Newton's second law to express $\frac{d^2\theta}{dt^2}$ in terms of a.*

Solution

a) *The tangential component of the resultant force is* $-mg\cos(90° - \theta) = -mg\sin\theta$*, since* $\cos(90° - \theta) = \sin\theta$.

b) *If θ is measured in radians, the distance moved is given by the arc OA. The length of an arc is $r\theta$, so in this case, the distance is $l\theta$ as the radius of the circle is equal to the length of the string.*

c) *The acceleration can be found by differentiating the distance travelled twice. The first derivative gives the speed:*

$$v = l\frac{d\theta}{dt}$$

and the second the acceleration:

$$a = l\frac{d^2\theta}{dt^2}$$

d) *Applying Newton's second law, $F = ma$, gives:*

$$-mg\sin\theta = ml\frac{d^2\theta}{dt^2}$$

so that:

$$\frac{d^2\theta}{dt^2} = -\frac{g\sin\theta}{l}$$

So we can calculate the acceleration of the mass at any point along the arc described by the pendulum.

Exploration 14.4

A graphical investigation

1 Plot graphs of $f(\theta) = \theta$ and $g(\theta) = \sin\theta$ (a computer package or a graphics calculator may be useful here), using radians.

2 By comparing the two graphs decide if it is reasonable to approximate $\sin\theta$ to θ, and for what range of values this is acceptable.

You will find that for small values of θ it would appear to be reasonable to approximate $\sin\theta$ for θ, and so the expression for the acceleration that we have just derived:

$$\frac{d^2\theta}{dt^2} = -\frac{g\sin\theta}{l}$$

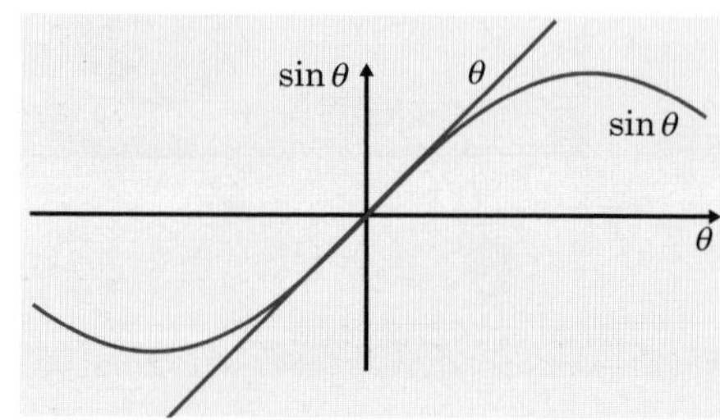

can reasonably be approximated to:

$$\frac{d^2\theta}{dt^2} = -\frac{g\theta}{l}$$

The advantage of this approximation is that the differential equation can be recognised as describing simple harmonic motion. The motion of a simple pendulum can be modelled as simple harmonic provided that the angle between the string and the vertical remains small. The graph shows how θ and $\sin\theta$ compare.

Acceleration of the pendulum

The acceleration of an object describing circular motion will always have two components, $\frac{v^2}{r}$ towards its centre and $\frac{dv}{dt}$ along a tangent. These expressions can also be expressed in terms of θ, the angle turned, as

$$r\left(\frac{d\theta}{dt}\right)^2 \text{ and } r\frac{d^2\theta}{dt^2}$$

This second form is very useful when dealing with the simple pendulum.

Example 14.11

The motion of a simple pendulum is modelled by the differential equation:

$$\frac{d^2\theta}{dt^2} = -\frac{g\sin\theta}{l}$$

a) *Use the approximation, $\theta = \sin\theta$, to simplify this equation stating when the approximation is valid.*

b) *Show that $\theta = A\cos(\omega t + \alpha)$ satisfies this differential equation and express ω in terms of g and l.*

c) *Find the period of the motion of the pendulum.*

Solution

a) *The approximation, $\theta = \sin\theta$, is valid for small angles and so is reasonable to use with a simple pendulum. This gives the differential equation:*

$$\frac{d^2\theta}{dt^2} = -\frac{g\theta}{l}$$

b) *Differentiating $\theta = A\cos(\omega t + \alpha)$ twice gives:*

$$\frac{d\theta}{dt} = -A\omega\sin(\omega t + \alpha)$$

and

$$\frac{d^2\theta}{dt^2} = -A\omega^2\cos(\omega t + \alpha) = -\omega^2\theta$$

So the expression for θ satisfies the differential equation provided

$$\omega^2 = \frac{g}{l}$$

c) *One oscillation will have been completed when $\omega t = 2\pi$, which in this case gives:*

$$\sqrt{\frac{g}{l}}t = 2\pi \text{ or } t = 2\pi\sqrt{\frac{l}{g}}$$

So the period of a simple pendulum is $2\pi\sqrt{\frac{l}{g}}$.

Example 14.12

A pendulum consists of a string of length 0.9 m and a mass.

a) *Find the period of its motion.*

b) *The pendulum is set into motion, from its lowest point with a speed of 0.5 m s^{-1}. Find an expression for θ, if it initially increases.*

Solution

a) *The period is given by:* $T = 2\pi\sqrt{\frac{l}{g}}$

so in this case: $2\pi\sqrt{\frac{0.9}{9.8}} = 1.9$ seconds

b) *An expression for θ will be of the form* $\theta = A\cos(\omega t + \alpha)$

where $\omega^2 = \frac{g}{l}$. *In this case:*

$\omega = \sqrt{\frac{9.8}{0.8}} = 3.3 \Rightarrow \theta = A\cos(3.3t + \alpha)$

Initially $v = 0.5$, *but since* $v = r\frac{d\theta}{dt}$ *we can say that:*

$\frac{d\theta}{dt} = \frac{v}{r} = \frac{0.5}{0.9} = 0.56 \, rad\, s^{-1}$

Differentiating the expression for θ gives:

$\frac{d\theta}{dt} = -A\omega\sin(\omega t + \alpha)$

As initially $\frac{d\theta}{dt} = 0.56$ *and* $\theta = 0$, *we have:*

$0 = A\cos\alpha$ *(1)*

and:

$0.56 = -3.3A\sin\alpha$

or: $-0.17 = A\sin\alpha$ *(2)*

From equation (1) $\cos\alpha = 0$, *so* $\alpha = \frac{\pi}{2}$ *or* $-\frac{\pi}{2}$.

Considering equation (2), if $\alpha = \frac{\pi}{2}$, $A\sin\alpha$ *is positive and so does not satisfy the equation.*

Thus $\alpha = -\frac{\pi}{2}$

Now to find A, using $\alpha = -\frac{\pi}{2}$ *and equation (2) gives:*

$-0.17 = A\sin\left(-\frac{\pi}{2}\right) \Rightarrow -0.17 = -A \Rightarrow A = 0.17$

So θ can be modelled by $\theta = 0.17\cos\left(3.3t - \frac{\pi}{2}\right)$

EXERCISES

14.3 A

1 A clock has a pendulum of length 8 cm. Find the time taken for one complete oscillation.

2 A clock manufacturer requires a pendulum for a clock with a period of 0.5 seconds. Find the length of the pendulum required for this clock.

3 A clock uses a pendulum of length 10 cm.
a) Find the period of the clock on Earth.
b) Find the period of the clock on the moon, when $g = 1.6\,\text{m}\,\text{s}^{-2}$.

4 A new design of clock has been launched for space travellers. It contains a clock with a set of detachable pendulums for use on different planets. A pendulum of length 4 cm is supplied for the moon.
a) Find the period of this pendulum.
b) What length pendulum would be required for use on Earth?
c) A space traveller lands on an unknown planet and notices that the clock runs fast with either pendulum. What can you deduce about g on this planet?

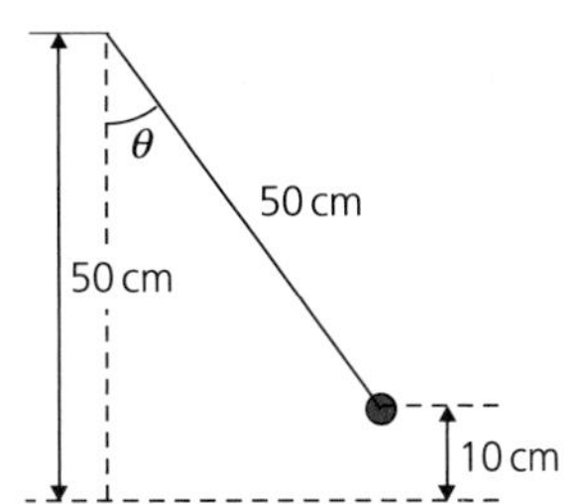

5 A simple pendulum has a string length of 50 cm.
a) Find the period of the pendulum.
b) The pendulum bob is released from a point 10 cm above its lowest point. Find the angle, θ, between the string and the vertical in this position, giving your answer in radians.
c) The angle can be expressed as $\theta = A\cos\omega t$. Find the values of A and ω.

6 The diagrams each show a simple pendulum that is released from rest in the position shown. In each case find the values of A and ω in radians if $\theta = A\cos\omega t$.

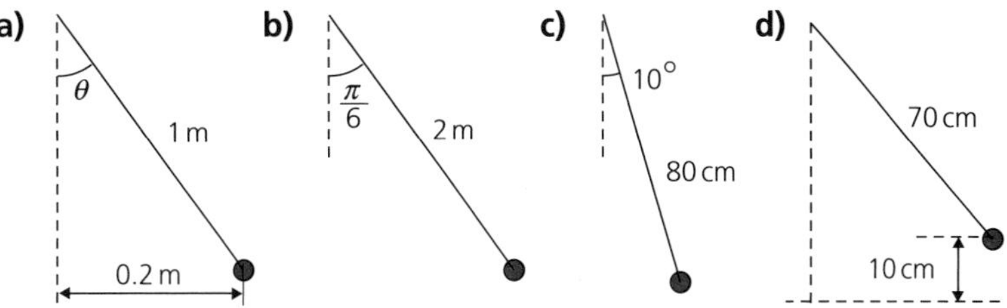

In which case it is most appropriate to assume that simple harmonic motion takes place and in which case is it least appropriate?

7 An antiques enthusiast is trying to restore a grandfather clock. It is now working, but the pendulum completes 60 swings in 62 seconds, instead of one minute. Two suggestions are made to him. The first is to attach a small mass to the back of the pendulum. The second is to adjust the length of the pendulum by moving the point where the mass is attached. Model the clock pendulum as a simple pendulum.
a) Explain whether or not you would recommend the first idea and why.
b) Find the length of the pendulum.
c) What adjustment should be made to give the required period? Is this feasible?
d) Explain why it may not be reasonable to model a clock pendulum by a simple pendulum.

8 A monkey jumps so that he is travelling horizontally when he grabs a tyre hanging on a rope. The tyre and the monkey have an initial speed of $0.7\,\text{m}\,\text{s}^{-1}$. The monkey then swings backwards and forwards on the rope. The motion of the monkey can be modelled by:

$$\theta = A\cos(\omega t + \alpha)$$

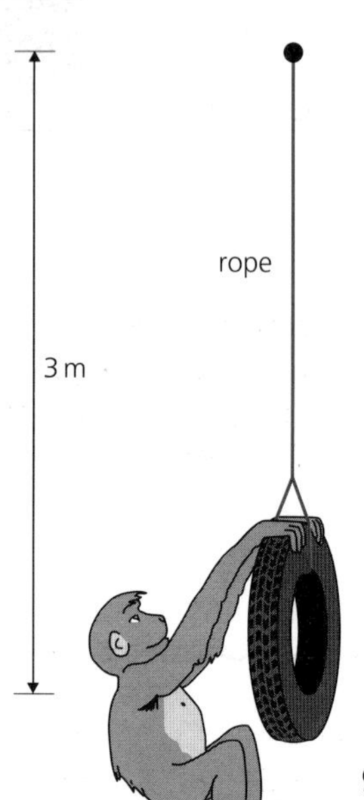

a) Find the value of ω.
b) Find the initial kinetic energy of the monkey in terms of m, the total mass of the monkey and the tyre. Use this to find the maximum gain in height made by the monkey, since he grabbed the tyre.
c) Calculate the value of A, the amplitude of the oscillations.
d) Use the fact that $v = r\frac{d\theta}{dt}$ to find the initial value of $\frac{d\theta}{dt}$ and then use this to determine α.
e) If the monkey had been travelling at 1.0 m s^{-1} when he grabbed the tyre, how would this have affected:
  i) its initial motion,
  ii) the amplitude of the subsequent motion,
  iii) the period of the subsequent motion?

9 Each pendulum shown below is set into motion so that the bob moves with the speed shown.

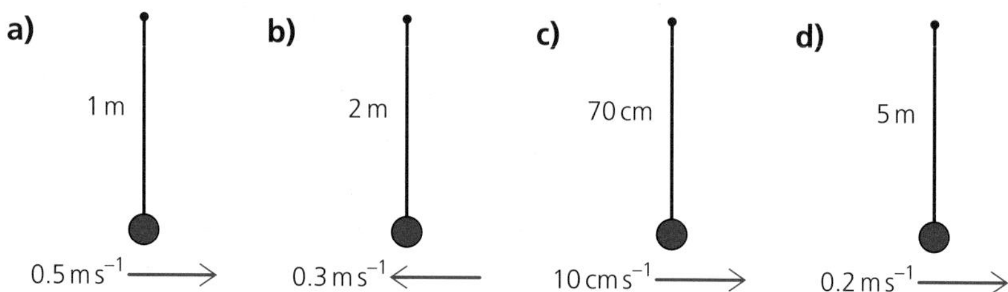

In each case:
  i) Find the greatest height reached.
  ii) If the motion can be modelled by $\theta = A\cos(\omega t + \alpha)$ find the values of A, ω and α.

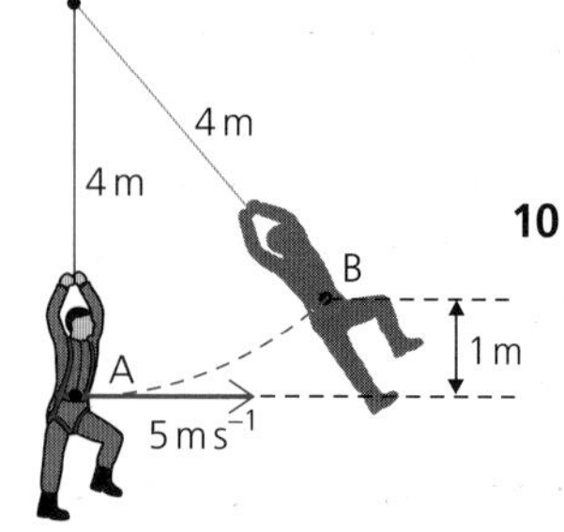

10 A marine of mass 70 kg swings on a rope of length 4 m in a circular arc from point A toward B which is 1 m higher. His initial speed is 5 m s^{-1}.
a) Find the initial kinetic energy of the marine and show that it is possible for him to get to B.
b) Find the time it takes to get to B, and his speed at B.

EXERCISES

14.3 B

1 A clock uses a pendulum of length 12 cm. Find the time taken for one complete oscillation.

2 Find the length of a pendulum if it is to have period of 0.75 seconds, on Earth.

3 A clock uses a pendulum of length 18 cm.
a) Find the period of the pendulum on Earth.
b) Find the period of the pendulum on the moon where $g = 1.6$ m s^{-2}.

4 **a)** For a clock to have a pendulum which makes one complete oscillation every two seconds, what length must the pendulum be?
b) **i)** If the clock were taken to a planet which had a gravitational pull eleven times that of Earth what would the period of the pendulum be?
  ii) If the period was required to be the same as that on Earth, what length must the pendulum be?

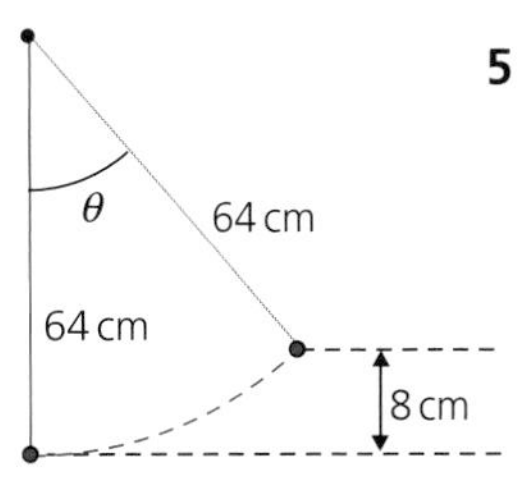

5 A simple pendulum has a string length of 64 cm.

a) Find the period of the pendulum.

b) The pendulum bob is held with the string taut and a point which is 8 cm vertically above the lowest point of the swing. Find the angle, θ, between the string and the vertical, giving your answer in radians.

c) The angle can be expressed as $\theta = A\cos(\omega t + \alpha)$. Find the values of A and ω.

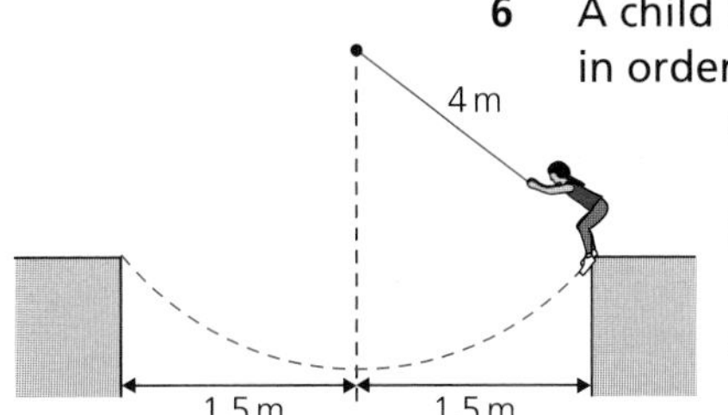

6 A child in an adventure playground swings on a rope of length 4 m, in order to cross a 3 m gap.

a) Find the angle in radians that the rope initially makes with the vertical.

b) Find the period of the oscillation that would take place if the child were to swing backwards and forwards on the rope.

c) The angle θ can be modelled as $\theta = A\cos(\omega t + \alpha)$. Find the values of A and ω.

d) If the child crosses the gap in one swing, how long does it take?

e) Comment on the validity of the model introduced in **c)**.

7 A clock is examined and its pendulum is found to complete 60 swings in 55 seconds, instead of one minute.

a) If the clock can be modelled as a simple pendulum, explain what effect changing the mass of the bob weight will have upon the period.

b) Find the length of the pendulum.

c) How should the length of the pendulum be adjusted so that it completes 60 swings in 1 minute?

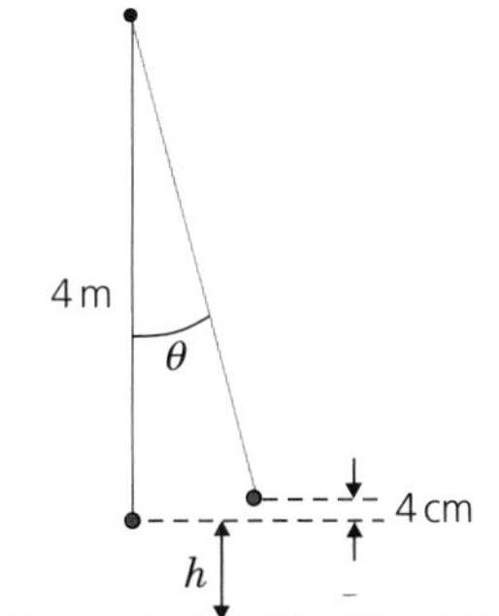

8 A child runs towards a swing which is at rest a distance h above the ground. After jumping on the swing he is carried to a point such that his height above the ground is 4 cm greater than h.
The motion of the child is modelled by $\theta = A\cos(\omega t + \alpha)$.

a) Find the value of ω.

b) Find the speed of the boy the instant after he made contact with the swing.

c) Calculate the value of A, the amplitude of the oscillations that follow.

d) Find the initial value of $\frac{d\theta}{dt}$ and then use this to find α.

9 A child is swinging in a swing at the park which has ropes of length 2.5 m. The child is pushed by her father, every time she comes to rest in front of him. How often does he have to push the swing?

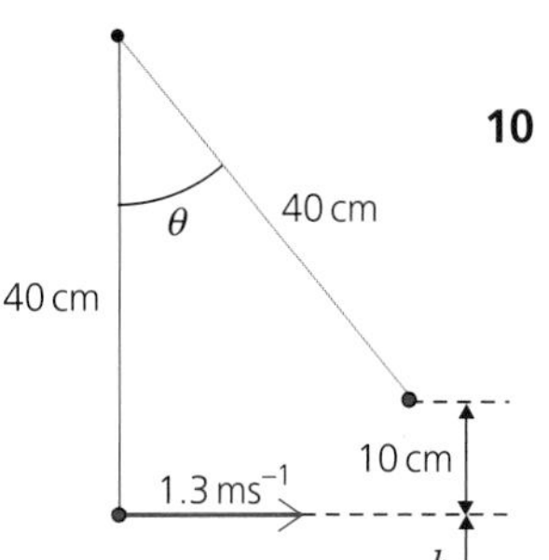

10 A simple pendulum consists of a string of length 40 cm hanging vertically with a 500 gram mass attached at the end. The mass is at rest at a distance h above the surface of a table when it is given a horizontal speed of $1.3\,\text{m s}^{-1}$.

a) By conservation of energy show that the mass cannot reach a point which is 10 cm further above the surface of the table.

b) What would the minimum initial speed have to be in order for the mass to reach this point?

c) List any assumptions you have made in arriving at your answer for **b)**.

CONSOLIDATION EXERCISES FOR CHAPTER 14

1 A body P is moving in simple harmonic motion, in a straight line, with centre O and amplitude 3 m. Given that the speed of P when OP = 2 m is 1 m s^{-1}, find the period of the motion.

(UCLES Question 3, M3 Specimen Paper, 1994)

2 A simple pendulum has a period of 1 second. It is tested in two towns, A and B. It has 3601 oscillations in an hour in town A and 3599 oscillations in an hour at town B.

a) Compare the values of g for each of the two towns.

b) Give a possible explanation for the difference in g between the two towns.

(SMP Question A1, Modelling with Differential Equations Specimen Paper 1994)

3 A light elastic string, of natural length l and modulus $2mg$, has one end attached to a fixed point A. A particle P, of mass m, is attached to the other end of the string and hangs freely in equilibrium at the point O, vertically below A.

a) Find, in terms of L, the length OA.

The particle P is pulled down a vertical distance h below O, and released from rest at time $t = 0$. At time t, the displacement of P from O is x.

b) Show that, while the string is taut:

$$\frac{\mathrm{d}^2 x}{\mathrm{d}t^2} = -\frac{2gx}{L}$$

c) State the set of values of h for which P performs complete simple harmonic oscillations.

You are now given that $h = \frac{1}{3}L$.

d) Find the time at which P first comes instantaneously to rest.

e) Find the greatest speed of P.

(ULEAC Question 9, M2, January 1993)

4 The three points O, B, C lie, in order, on a straight line l on a smooth horizontal plane with OB = 0.3 m, OC = 0.4 m.

A particle P describes simple harmonic motion with centre O along the line l. At B the speed of the particle is 12 m s^{-1} and at C its speed is 9 m s^{-1}. Find:

a) the amplitude of the motion,

b) the period of the motion:

c) the maximum speed of P,

d) the time to travel from O to C.

This simple harmonic motion is caused by a light elastic spring attached to P, the other end of the spring is fixed at a point A on l where A is on the opposite side of O to B and C and AO = 2 m. Given that P has mass 0.2 kg, find the modules of the spring and the energy stored in it when AP = 2.4 m.

(AED Question 11, Paper 3, June 1991)

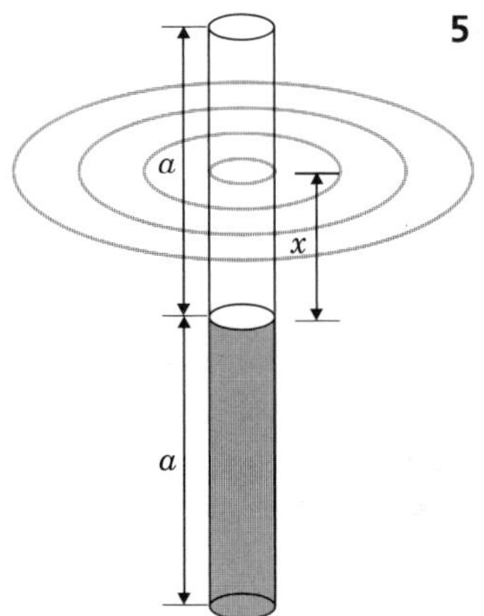

5 A thin uniform rod of mass m and length $2a$ has a small mass m fixed to one end. The weighted rod floats vertically in a liquid and when in equilibrium half of its length is submerged. The top of the rod is depressed until three-quarters of its length is submerged and is then released. At any time the upward force on the rod due to submersion is proportional to the length of the rod which is submerged.

The diagram shows the rod when its centre is a distance x below the surface. Find the equation of motion of the rod.

By noting the form of this equation, and assuming an appropriate general solution, find the displacement x of the centre of the rod as a function of time. What is the principal feature which is likely to invalidate this model?

(MEI Question 4, Mechanics 3 Paper 9, January 1992)

MATHEMATICAL MODELLING ACTIVITY

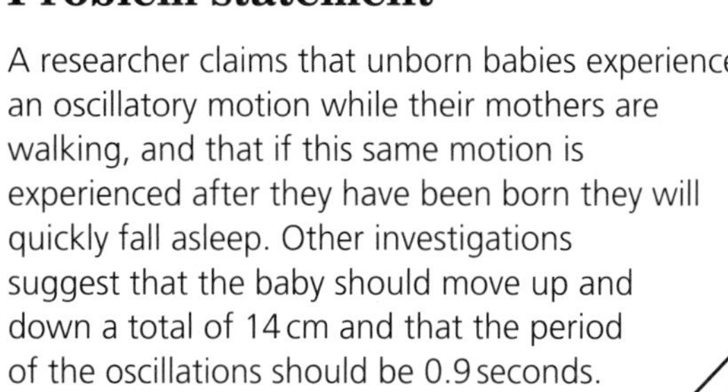

Specify the real problem

Problem statement

A researcher claims that unborn babies experience an oscillatory motion while their mothers are walking, and that if this same motion is experienced after they have been born they will quickly fall asleep. Other investigations suggest that the baby should move up and down a total of 14 cm and that the period of the oscillations should be 0.9 seconds.

The researcher aims to produce a cradle suspended by ropes that will swing to reproduce the motion of the walking expectant mother.

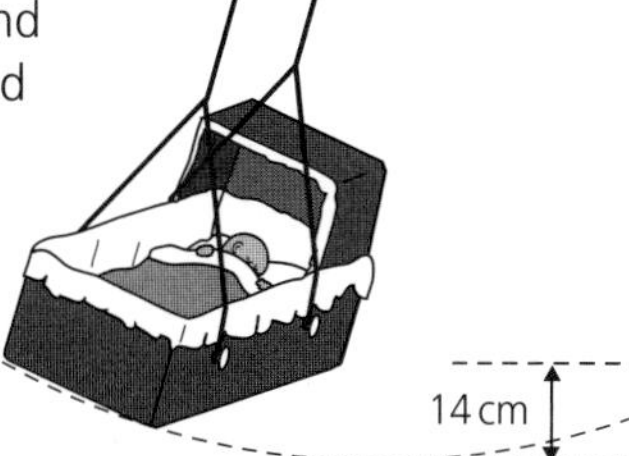

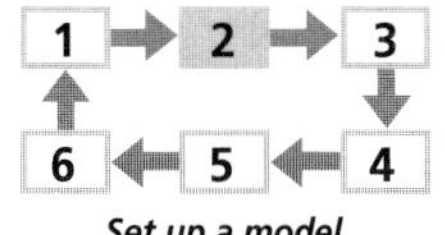

Set up a model

Set up a model

The list below contains some assumptions based on the key features of the problem that should be taken into account.

- The string does not stretch and has length l.
- The baby lies still and does not move or roll inside the cradle.
- The motion is simple harmonic.
- The mass of the baby and cradle is m.

Add ***at least two*** *further assumptions to the list.*

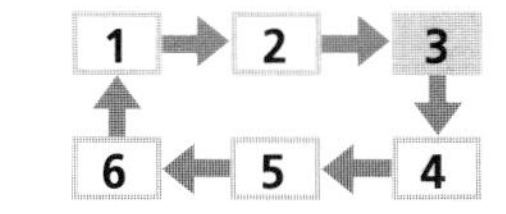

Formulate the mathematical problem

Formulate a mathematical problem

Use the standard results for a simple pendulum to find the length of the string to give a period of 0.9 seconds, and the initial angle that the string should make with the vertical.

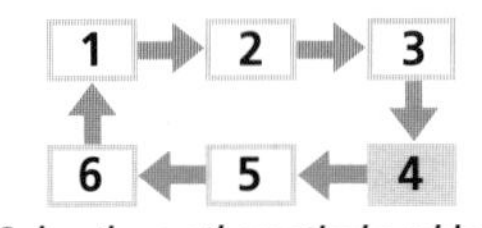

Solve the mathematical problem

Mathematical solution

The period, T, of a simple pendulum is given by:

$$T = 2\pi\sqrt{\frac{l}{g}}$$

where l is the length of the string.
Show that to give a period of 0.9 seconds a string of length 0.2 m is required.
In order that the total movement of the cradle is 14 cm the initial position illustrated is required.
Show that the angle θ that the string initially makes with the vertical is 72.5°.

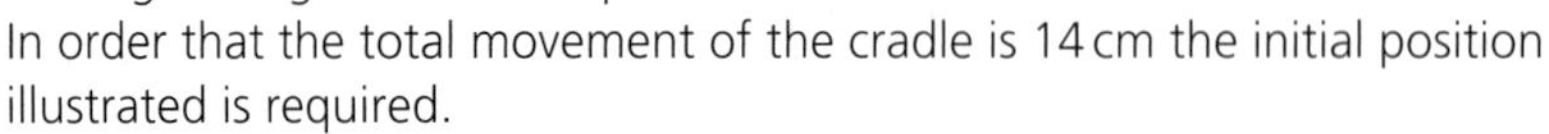

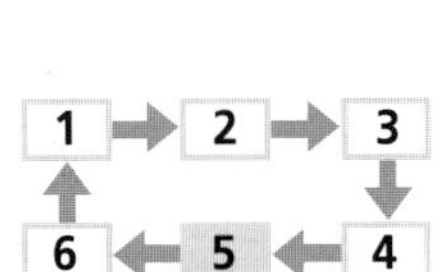

Interpret the solution

Interpretation

The solution shows that it is possible to obtain the desired period by using a string of length 20 cm. In practice this will be difficult to produce because of the size of the cradle. The diagram shows three ways of attaching the string, by assuming that the radius of the swing is measured from different points on the cradle.

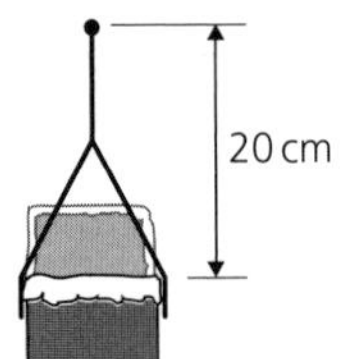

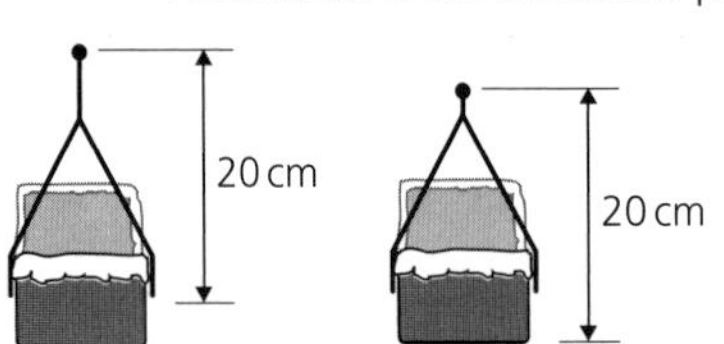

The centre diagram probably shows the most sensible approach, but as the cradle will be close to its supports it may be difficult to get the baby in and out.

The required starting position creates a difficulty because the angle between the string and the vertical is large. This invalidates the assumption that the motion is simple harmonic and means that the formula used to calculate the length is not valid (see Exploration 14.6).
So it is impossible to fulfil the criteria of the designer with simple harmonic motion; we have to consider alternative approaches.

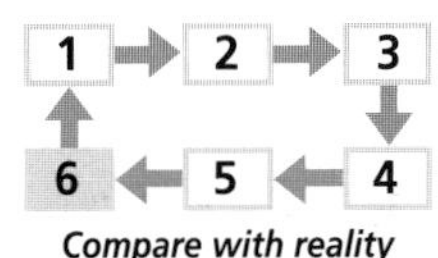

Compare with reality

Compare with reality

Set up an experimental model of the cradle using for example a cardboard box containing a couple of textbooks.

Verify that if the centre of the box describes an arc of radius 20 cm the motion has a period of 0.9 seconds, for small swings. Does the same result hold true for swings of a large amplitude?

Refinement of the model

Two strategies are open to the designer, one is to keep the period as a priority and use a small amplitude, the other is to keep the amplitude of the motion as a priority.

Find the length of the string required if the motion is to be close to simple harmonic and the amplitude of the motion is to satisfy the design activities. Calculate the period and compare with the design specifications.

POSSIBLE MODELLING TASKS

1 Investigate whether it would be feasible to use a simple pendulum to determine the height of a mountain.

2 Investigate what would happen to a mass–spring system if it were placed in a lift.

3 Consider other oscillating systems and try to model their motion, to decide whether or not it is simple harmonic. For example:
a) water in a U-tube, **b)** a marble in a bowl.

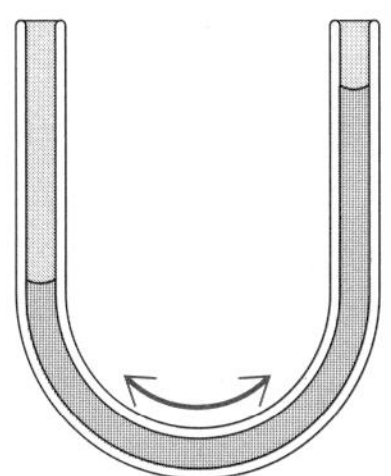

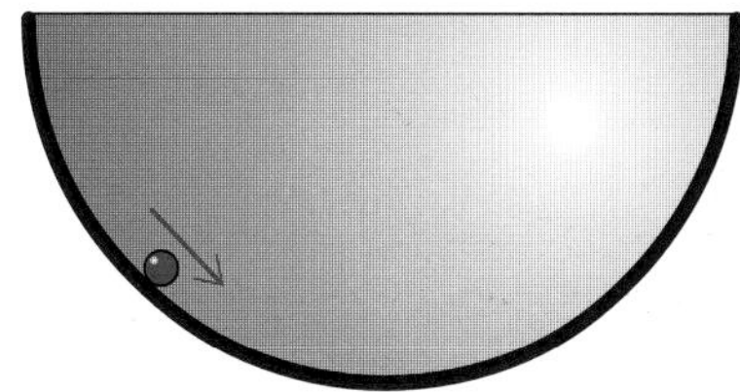

Summary

After working through this chapter you should:

- be aware that if:

$$\frac{d^2x}{dt^2} = -\omega^2 x$$

then simple harmonic motion will take place with:

$$x = A\cos(\omega t + \alpha)$$

- know that the period of oscillation for a mass spring system is:

$$2\pi\sqrt{\frac{m}{k}} \text{ or } 2\pi\sqrt{\frac{ml}{\lambda}}$$

- know that the velocity of an object describing simple harmonic motion can be found using:

$$v = \omega\sqrt{a^2 - x^2}$$

- know that the period of a simple pendulum is:

$$2\pi\sqrt{\frac{l}{g}}$$

Dimensions and dimensional analysis

- *All quantities in mechanics can be expressed in terms of the three fundamental quantities: mass, length and time.*
- *Models for the relationships between different quantities can be determined by using these fundamental quantities.*

DIMENSIONS

As an example, the units of momentum can be expressed as either N s or kg m s^{-1}. The second set of units is expressed in terms of the fundamental quantities of **mass**, **length** and **time**, rather than other units that depend on these. In this section we shall explore the relationships between different quantities and these three fundamental quantities.

Exploration 15.1

Comparing units

1 Explain why kg m^2s^{-2} is an alternative unit for energy, first by considering kinetic energy and then by considering potential energy.

2 Find alternative units in terms of kilograms (kg), metres (m) and seconds (s) for:

a) watts **b)** newtons.

Expressing units in terms of dimensions

We have seen how various units can be expressed in terms of the fundamental quantities. But because there are so many different units that can be used for the same quantities, for example, speed can be measured in miles per hour (mph), metres per second (m s^{-1}), kilometres per hour (km h^{-1}) etc., it is more convenient to introduce a general approach.

The dimensions of a quantity show how it is related to the three fundamental quantities of mass, length and time. The notation used is M for mass, L for length and T for time. The dimensions are **independent of any units**. A square bracket notation is used, for example $[v]$ means the dimensions of v, the velocity:

$$[v] = LT^{-1}$$

and the dimensions of acceleration are written as:

$[a] = LT^{-2}$

In problems that also involve heat, the fourth fundamental quantity of **temperature** is also needed.

Example 15.1

Find the dimension of:
a) *force,* **b)** *area,* **c)** *kinetic energy.*

Solution

a) *Using Newton's second law, $F = ma$, gives:*

$[F] = [ma] = [m] \times [a] = MLT^{-2}$

Note that the dimensions of a product such as ma are equal to the ***product*** *of the dimensions of each term. In general $[xy] = [x] \times [y]$.*

b) *Area is the product of two lengths, so:*

$[A] = L^2$

c) *Kinetic energy is given by $\frac{1}{2}mv^2$*

$[KE] = [m] \times [v^2] = M(LT^{-1})^2 = ML^2T^{-2}$

Note that the dimensions of v^2 are $[v]^2$. In general $[x^n] = [x]^n$.

Note that $\frac{1}{2}$ is simply a number and has no dimensions. There are many dimensionless quantities. The next example shows two examples of dimensionless constants, but not all constants are dimensionless.

Example 15.2

Show that π and μ, the coefficient of friction, are dimensionless.

Solution

First using $A = \pi r^2$, the formula for the area of a circle:

$[A] = [\pi] \times [r]^2$

$L^2 = [\pi] \times L^2$

$[\pi] = \dfrac{L^2}{L^2} = 1$

So π is dimensionless.

Now consider μ. In limiting friction $F = \mu R$,

$[F] = [\mu] \times [R]$

The dimensions of force are MLT^{-2}, so:

$MLT^{-2} = [\mu]MLT^{-2}$

$[\mu] = \dfrac{MLT^{-2}}{MLT^{-2}} = 1$

So μ is dimensionless.

Example 15.3

a) *Hooke's law states that $T = ke$. Find the dimensions of k.*

b) *A model for air resistance is $R = kv^2$. Find the dimensions of k.*

Solution

a) *Note T is a force and so has dimensions MLT^{-2} and e is the extension which has dimension L.*

$[T] = [k] \times [e]$

$\mathrm{MLT}^{-2} = [k]\,\mathrm{L}$

$$[k] = \frac{\mathrm{MLT}^{-2}}{\mathrm{L}} = \mathrm{MT}^{-2}$$

b) *Here R has dimensions MLT^{-2} and v has dimensions LT^{-1}.*

$[R] = [k] \times [v]^2$

$\mathrm{MLT}^{-2} = [k]\mathrm{L}^2\mathrm{T}^{-2}$

$$[k] = \frac{\mathrm{MLT}^{-2}}{\mathrm{L}^2\mathrm{T}^{-2}} = \frac{\mathrm{M}}{\mathrm{L}} = \mathrm{ML}^{-1}$$

Example 15.4

The motion of a free-wheeling cyclist is modelled by:

$$v\frac{\mathrm{d}v}{\mathrm{d}x} = kv^2$$

Find the dimensions of k.

Solution

This expression includes a derivative, so its dimensions are obtained as follows using the property that the derivative is related to the ratio of two small changes in the quantities. These small changes have the same dimension as the quantities themselves. So to find the dimensions of $\frac{\mathrm{d}v}{\mathrm{d}x}$:

$$\left[\frac{\mathrm{d}v}{\mathrm{d}x}\right] = \frac{[\mathrm{d}v]}{[\mathrm{d}x]} = \frac{\mathrm{LT}^{-1}}{\mathrm{L}} = \mathrm{T}^{-1}$$

Now:

$$[v]\left[\frac{\mathrm{d}v}{\mathrm{d}x}\right] = [k][v]^2 \Rightarrow \left[\frac{\mathrm{d}v}{\mathrm{d}x}\right] = [k][v]$$

$$\mathrm{T}^{-1} = [k]\mathrm{LT}^{-1}$$

$$[k] = \frac{\mathrm{T}^{-1}}{\mathrm{LT}^{-1}} = \mathrm{L}^{-1}$$

Dimensional consistency

Any equation that models a situation in the real world must be **dimensionally consistent**. This means that both sides of an equation must have the same dimensions, and that quantities that are added together must have the same dimensions. This idea of dimensional consistency is very useful, both for checking that models or formulae are correct, or at least plausible, and – even more important – for assisting in the formulation of models for different situations.

Example 15.5

Show that each of the following equations are dimensionally consistent.

a) $s = ut + \frac{1}{2}at^2$ *(the formula for an object moving with constant acceleration)*

b) $a = \frac{v^2}{r}$ *(the magnitude of the acceleration of an object in circular motion at a constant speed)*

Solution

a) *The dimensions of* s *are simply* L, *so the dimensions of both other terms must also be* L.

Their dimensions are given by:

$[ut] = [u] \times [t] = \mathrm{LT}^{-1} \times \mathrm{T} = \mathrm{L}$

and $\left[\frac{1}{2}at^2\right] = \left[\frac{1}{2}\right] \times [a] \times [t]^2 = 1 \times \mathrm{LT}^{-2} \times \mathrm{T}^2 = \mathrm{L}$

so the equation is dimensionally consistent.

b) *Both sides of the expression must have the same dimensions.*

$[a] = \mathrm{LT}^{-2}$

and $\left[\frac{v^2}{r}\right] = \frac{[v]^2}{[r]} = \frac{\mathrm{L}^2\mathrm{T}^{-2}}{\mathrm{L}} = \mathrm{LT}^{-2}$

so the equation is dimensionally consistent.

Example 15.6

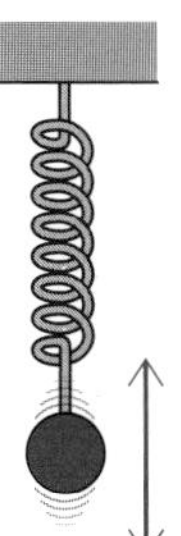

While a student is attempting to formulate a model for the motion of a simple mass–spring system, as illustrated in the diagram, she assumes that its period depends on the mass, m, *spring stiffness,* k *and natural length,* l.

If the period, P, *is assumed to be of the form:*

$P = \alpha m^a k^b l^c$

where α *is a dimensionless constant, find the values of* a, b *and* c.

Solution

The dimensions of P *are simply* T.

Noting that the dimension of k *are* MT^{-2} *and simplifying gives:*

$[m]^a[k]^b[l]^c = \mathrm{T}$

$\mathrm{M}^a\,(\mathrm{MT}^{-2})^b\,\mathrm{L}^c = \mathrm{T}$

$\mathrm{M}^a\,\mathrm{M}^b\,\mathrm{T}^{-2b}\,\mathrm{L}^c = \mathrm{T}$

$\mathrm{M}^{a+b}\,\mathrm{L}^c\,\mathrm{T}^{-2b} = \mathrm{TM}^0\mathrm{L}^0$

For dimensional consistency:

$a + b = 0$ *because the power of* M *is 0,*

$c = 0$ *because the power of* L *is 0,*

$-2b = 1$ *because the power of* T *is 1.*

Solving these three simultaneous equations gives $c=0$, $b=-\frac{1}{2}$ *and* $a=\frac{1}{2}$.

So $P = \alpha m^{\frac{1}{2}} k^{-\frac{1}{2}} l^0 = \alpha\sqrt{\frac{m}{k}}$

Example 15.7

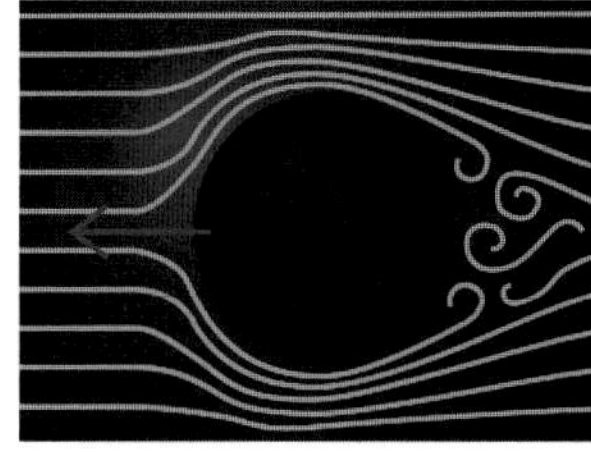

A model for the air resistance force, R, *on a sphere assumes:*

$R = \alpha m^a r^b v^c$

where α *is a dimensionless constant,* m *is the mass of the sphere,* r *is the radius of the sphere and* v *its speed. Find* a, b *and* c.

Solution

The dimensions of R *are* MLT^{-2} *so:*

$\mathrm{MLT}^{-2} = [m]^a\,[r]^b\,[v]^c = \mathrm{M}^a\,\mathrm{L}^b\,(\mathrm{LT}^{-1})^c = \mathrm{M}^a\,\mathrm{L}^b\,\mathrm{L}^c\,\mathrm{T}^{-c} = \mathrm{M}^a\,\mathrm{L}^{b+c}\,\mathrm{T}^{-c}$

For dimensional consistency:

$a = 1$

$b + c = 1$

$c = 2$

So $a = 1$, $b = -1$ *and* $c = 2$, *which gives:*

$$R = mr^{-1}v^2 = \frac{mv^2}{r}$$

EXERCISES

15.1 A

1 By considering the formula:

arc length $= r\theta$

show that angles are dimensionless.

2 Hooke's law can be expressed as:

$$T = \frac{\lambda x}{l}$$

Find the dimensions of λ.

3 Newton's law of gravitation states that the attraction between two bodies is given by:

$$\frac{Gm_1m_2}{d^2}$$

Find the dimensions of G.

4 For circular motion at constant speed the acceleration, a, is given by:

$$a = r\omega^2$$

where r is the radius of the circle. Find the dimensions of ω.

5 Determine whether each equation below is dimensionally consistent.

a) $v = u + at$ (the constant acceleration equation)

b) $Fd = \frac{1}{2}mv^2 - \frac{1}{2}mu^2$ (the equation for conservation of energy)

c) $P = 2\pi\sqrt{\frac{m\lambda}{l}}$ (the period of a spring mass system)

6 Pressure, P, is given by the formula:

$$P = \frac{F}{A}$$

where F is the applied force and A is the area on which it acts. Find the dimensions of P.

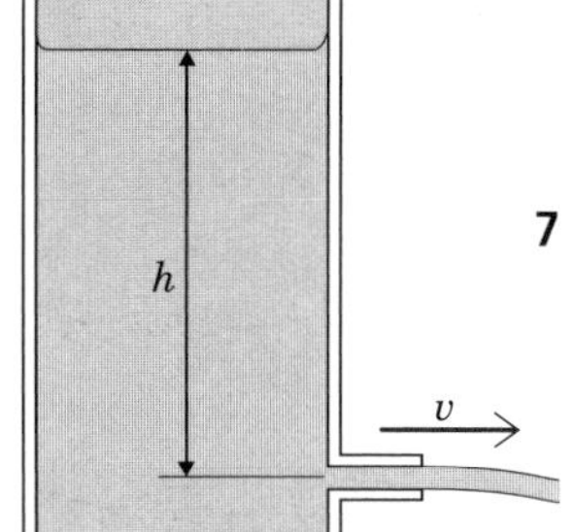

7 The speed, v, at which water leaves a tank is modelled by:

$$v = \alpha h^a g^b A^c$$

where α is a dimensionless constant, h is the height of water in the tank, g the acceleration due to gravity and A the area of the opening. Determine the value of b and state the relationship between a and c.

8 **a)** Find the dimensions of power.

b) The power, P, of a water pump is assumed to be of the form:

$$P = \alpha m^a h^b t^c$$

where m is the mass of water raised through a height h in time t and α is a dimensionless constant. Find the values of a, b and c.

9 When modelling the motion of a simple pendulum it is assumed that for small angles:

$$\sin\theta = \theta$$

Is this assumption dimensionally consistent?

10 **a)** Find the dimensions of frequency.
b) A student assumes that the frequency of a note on a stringed instrument depends on the tension in the string, T, and the length of the string, l. What other factor does the student need to include?
c) Find a possible expression for the frequency.

EXERCISES

15.1 B

1 Show that the coefficient of restitution is dimensionless.

2 **a)** Find the dimensions of energy.
b) The energy stored in a stretched spring is given by $\frac{\lambda x^2}{2l}$.
Use this to find the dimensions of λ.

3 A particle falling in a resistive medium experiences a resultant force given by:

$$F = mg - kv^n$$

Find the dimensions of k if:
a) $n = 1$, **b)** $n = \frac{1}{2}$.

4 For an object describing simple harmonic motion, the displacement, x, measured from an equilibrium position can be modelled by:

$$x = A\cos\omega t$$

where A is the amplitude of the motion. Find the dimensions of ω.

5 **a)** Find the dimensions of the coefficient of friction.
b) Explain why this suggests that the coefficient of friction does not depend on the areas of the surfaces in contact.

6 Determine whether each equation below is dimensionally consistent.
a) $v^2 = u^2 + 2as$ (the constant acceleration equation)
b) A model for air resistance states that the resistance force, R, on a sphere moving with speed v is given by:

$$R = 0.6mrv$$

where m is the mass of the sphere and r the radius.
c) The maximum speed, v, at which a car of mass, m, can go round a bend of radius, r, is given in terms of the coefficient of friction as:

$$\frac{mv^2}{r} = \mu R$$

where R is the normal reaction on the car, and μ is the coefficient of friction between the car and the road.

7 Power can be defined as either Fv or as $\frac{\mathrm{d}}{\mathrm{d}t}(Fx)$. Find the dimensions of power using both definitions.

8 When the resultant force on a flywheel is zero, but the total moment is not zero, the flywheel will rotate with an increasing angular speed. The angular speed, ω, is related to the total moment X.

a) State the dimensions of ω.

b) By considering the dimensions of each expression below determine which could be correct. The mass of the flywheel is m and its radius is a.

$$X = \tfrac{1}{2}ma^2\frac{d\omega}{dt} \qquad X = \frac{ma^2\omega}{t} \qquad X = ma\frac{d\omega}{dt} \qquad X = \frac{ma\omega}{4t}$$

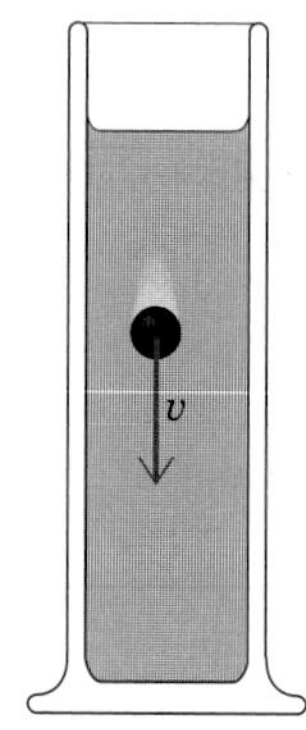

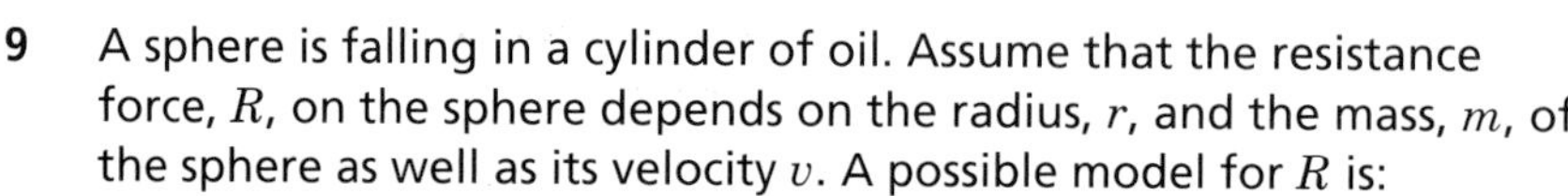

9 A sphere is falling in a cylinder of oil. Assume that the resistance force, R, on the sphere depends on the radius, r, and the mass, m, of the sphere as well as its velocity v. A possible model for R is:

$$R = k\,m^a\,r^b\,v^c$$

where k is a dimensionless constant.

a) Explain why it is possible for c to be equal to 2.

b) Determine the values of a and b.

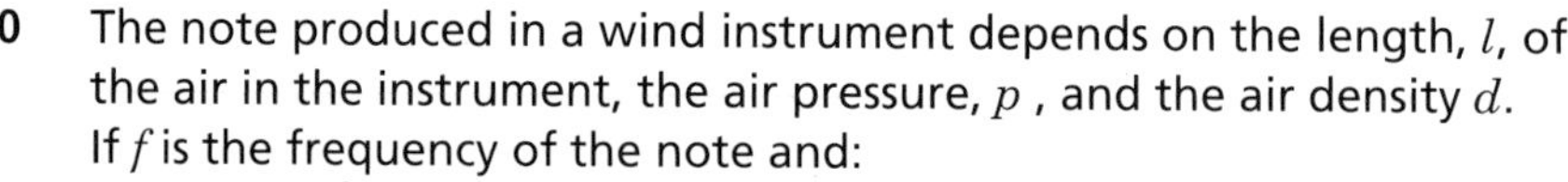

10 The note produced in a wind instrument depends on the length, l, of the air in the instrument, the air pressure, p , and the air density d. If f is the frequency of the note and:

$$l = k f^a d^b p^c$$

determine the values of a, b and c. What assumptions have you made about the constant k?

CONSOLIDATION EXERCISES FOR CHAPTER 15

1 Stuart and Michael are planning to drop a water bomb on their sister Karen from the bedroom window. Whilst they are waiting they decide to investigate the retarding force on the water bomb as it falls. They assume that the magnitude of the retarding force, R, depends upon the cross-sectional area, A, of the bomb, the air density, ρ, and the velocity, v, to some power c.

$$R = \alpha A \rho v^c$$

Find the value of c, if α is dimensionless.
Modelling the water bomb, of mass 10 grams, as a sphere of diameter 180 mm, moving through the air of density 1.2 kg m^{-3} and travelling at a constant velocity of 8 m s^{-1}. Find α.

2 The period of vibration, T, of a liquid drop is given by:

$$T = k\,r^a\,\rho^b\,\tau^c$$

where r is the radius of the drop ρ is the density, τ is the surface tension and k is a constant. Find the values of a, b and c.

3 The string on a violin is plucked and a wave is set up. The velocity, c, of the wave depends upon the mass of the string, m, its length, l, and the tension, T.

$$c = k m^a l^b T^d$$

Find a, b and d, if k is a dimensionless constant.

4 When two particles rotate about each other under the influence of their mutual attraction, the period of their motion is given by:

$$P = \frac{2\pi d^a}{G^b(m+M)^c}$$

where d is the distance between them, m and M are their masses and G is the constant of universal gravitation. The force that the two particles exert on each other is:

$$F = \frac{GmM}{d^2}$$

a) Find the dimensions of G.
b) Find the values of a, b and d.

5 The volume rate of flow R $\mathrm{m^3\,s^{-1}}$ of a liquid with a viscosity η $\mathrm{kg\,m^{-1}\,s^{-1}}$ through a cylindrical pipe of length a m and an internal radius r m is believed to be of the form:

$$R = k\eta^w a^x r^y P^z$$

where k is a non-dimensional constant and P $\mathrm{N\,m^{-2}}$ is the pressure difference between the ends of the pipe. Using dimensional considerations:

a) show that $w = -1$ and $z = 1$,
b) find the relationship between x and y.

Observations of a water supply provided at constant pressure and viscosity produced the following recordings:

Pipe length a (m)	100	200
Internal radius r (m)	0.050	0.050
Flow rate R ($\mathrm{m^3\,s^{-1}}$)	0.420	0.210

c) Derive a formula for R in terms of a and r for such a supply.
d) Find the rate of flow of water through 1000 m of pipe with an internal radius of 0.075 m under the same pressure and with the same viscosity.

(MEI Question 1, M4 Paper 10, June 1992)

MATHEMATICAL MODELLING ACTIVITY

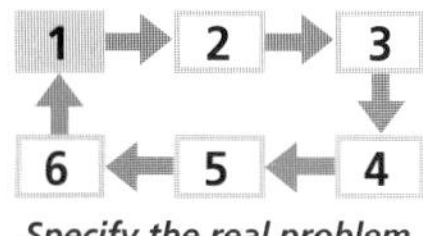

Specify the real problem

Problem statement

A simple pendulum consists of a mass that is allowed to swing on the end of a length of string.

On what factors do you consider the period of a simple pendulum might depend? Formulate an expression for the period of a simple pendulum.

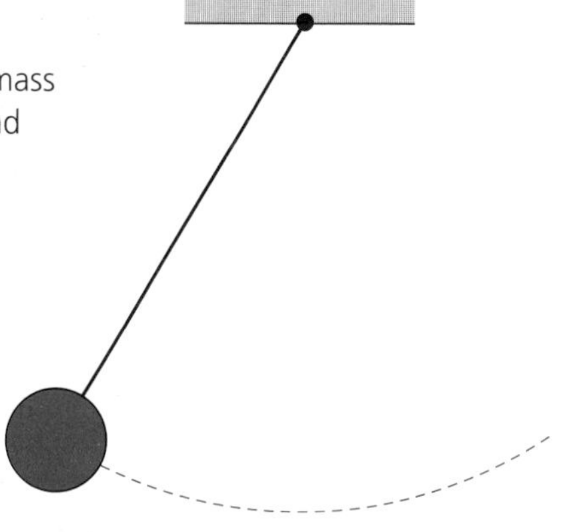

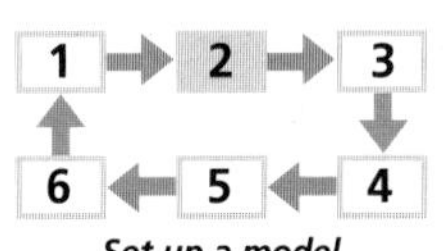
Set up a model

Formulation

Assume that the pendulum is set into motion in its lowest position by a sharp blow. The factors on which the period could depend are:

- length of string l
- mass of pendulum bob m
- acceleration due to gravity g
- initial speed u

List any other factors that you feel have been omitted from this list.

Formulate the mathematical problem

Formulate a mathematical problem

Assume a proposed model for the period, P, is:
$P = kl^a m^b g^c u^d$
where a, b, c and d are constants and k is dimensionless.

Find the values of a, b, c and d.

Solve the mathematical problem

Solution of mathematical problem

The dimensions of each quantity involved in the model are:
$[P] = \mathrm{T}$ $[g] = \mathrm{LT}^{-2}$ $[l] = \mathrm{L}$ $[u] = \mathrm{LT}^{-1}$ $[m] = \mathrm{T}$
Considering the dimensions of the model:
$P = kl^a m^b g^c u^d$
gives: $\mathrm{T}^1 = \mathrm{L}^a\,\mathrm{M}^b\,(\mathrm{LT}^{-2})^c\,(\mathrm{LT}^{-1})^d = \mathrm{L}^{a+c+d}\,\mathrm{M}^b\,\mathrm{T}^{-2c-d}$

For dimensional consistency the equations below must be satisfied.
$a + c + d = 0$
$b = 0$
$-2c - d = 1$

By considering these equations it can be shown that:
$a = c + 1$
and $d = -2c - 1$
as well as $b = 0$

The original expression can now be given as:

$$P = kl^{c+1}m^0 g^c u^{-2c-1} = k\left(\frac{lg}{u^2}\right)^c \times \frac{l}{u}$$

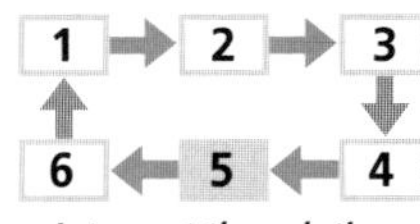
Interpret the solution

Interpretation

No clear result has been obtained as the value of c could not be determined. The model does suggest that increasing the initial speed will decrease the period, or increasing the length of the string will increase the period.

A single pendulum can be easily set up to test these predictions.

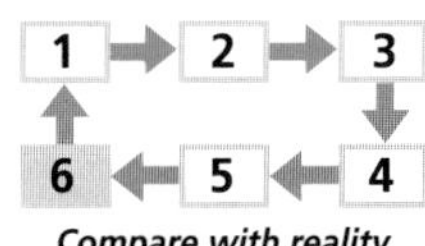
Compare with reality

Compare with reality

Set up an experiment to test the predictions described above. Comment on the results that you obtain.

You will have probably found that one of the factors listed in the initial formulation stage does not affect the period. Removing this factor from the model will provide an opportunity to refine the model.

Refining the model

Assume that the period does not depend on the initial speed. A revised model would be of the form:

$P = k l^a m^b g^c$

For this refined model determine the values of a, b and c, comment on the results you obtain and if possible determine the value of k by an experiment.

Summary

After working through this chapter you should be able to:

- determine the dimensions of a quantity in terms of the three fundamental quantities of M, L and T
- check for dimensional consistency and use this approach to predict or formulate models.

MECHANICS

16

Moments, couples and equilibrium

- *When forces are applied to objects, they can cause a change in motion in a straight line and/or make an object begin to rotate.*
- *For equilibrium the turning effects of forces must be considered, as well as the resultant force.*

MOMENTS

In earlier chapters we have considered the effects of applying forces in different ways, but so far all the objects to which the forces have been applied have been modelled as particles. This is often quite a reasonable assumption to make, but in some cases it is important to take account of *where* the force is applied – its **point of application**. For example, it is easier to open a door if you push the door at the edge furthest from the hinges, rather than in the middle.

In this chapter we shall explore how the point at which a force is applied can be important in mechanics, and develop the idea of the moment of a force.

Exploration 16.1 *A car park barrier*

The diagram shows a barrier situated at the entrance to a car park. It always rises to its open position when the bar is released, unless it is held down by a clip. Where should you pull on the barrier to make closing it as easy as possible? Explain why.

Applying forces

Experience and intuition suggest that we should pull at the end of the barrier that is furthest from the pivot, provided that we can reach it. When forces are applied to the barrier they cause it to rotate about its pivot. The further away from the pivot the force is applied, the greater the turning effect. This turning effect depends on the force, the direction in which it acts and the distance from the pivot. Exploration 16.2 examines the relationship between these factors.

Exploration 16.2 *Practical activity*

1 Try to gain access to a barrier similar to the one in Exploration 16.1. The barrier should rise to its open position when it is released. Attach a spring balance and rope or string to the barrier at different positions, as shown in the diagram. Use the spring balance to hold the barrier in a horizontal position. Record your results in a table like the one below.

Distance from pivot (m)	Force required (N)

Using your data, deduce a relationship between the force and distance.

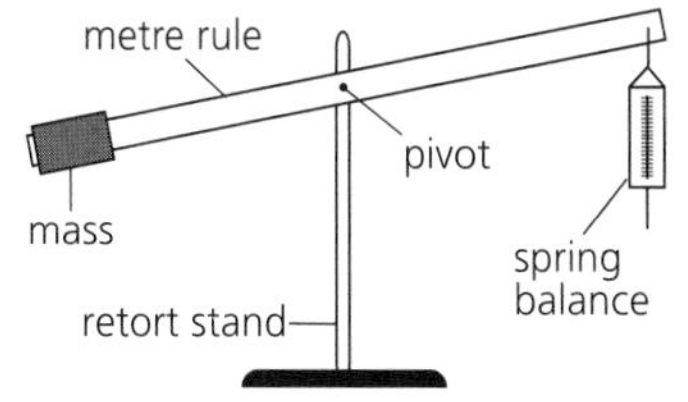

If you can't find a suitable barrier, you can improvise one from a metre rule pivoted on a rod, by drilling a hole in it and attaching a mass to one end.

2 Repeat experiment 1, but instead of letting it rest in a horizontal position, try holding the barrier or metre rule at the particular angle, e.g. 45° to the horizontal. How does this affect the results you obtain?

3 Now return to the apparatus and pull down the barrier so that it is horizontal. This time do not hold the spring balance vertical, but at an angle as shown in the diagram. Describe how the force required varies with the angle.

Forces acting

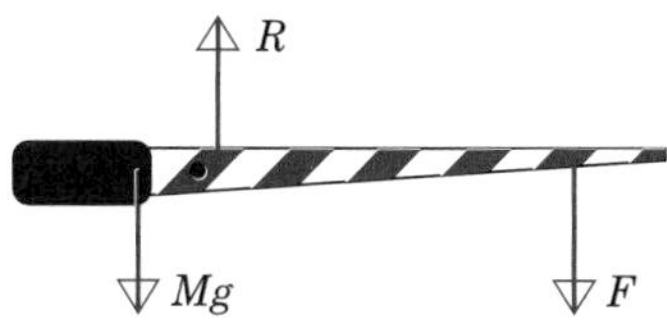

The diagram shows the forces acting on the barrier considered in the explorations. The three forces are Mg, the force of gravity which can be assumed to act at a point to the left of the pivot, R, the reaction force exerted by the pivot and F, the force exerted by the string.

The two forces Mg and F each have a turning effect on the barrier, but in opposite directions, so their turning effects are balanced when they both act. If one force were removed, the barrier would begin to turn. For example if F were removed, the barrier would start to rise. The force, R, at the pivot does not have a turning effect. No matter how large a force is applied at the pivot, the barrier will not begin to turn.

The turning effect of a force is the **moment** of the force. The moment depends on the force and where it is applied.

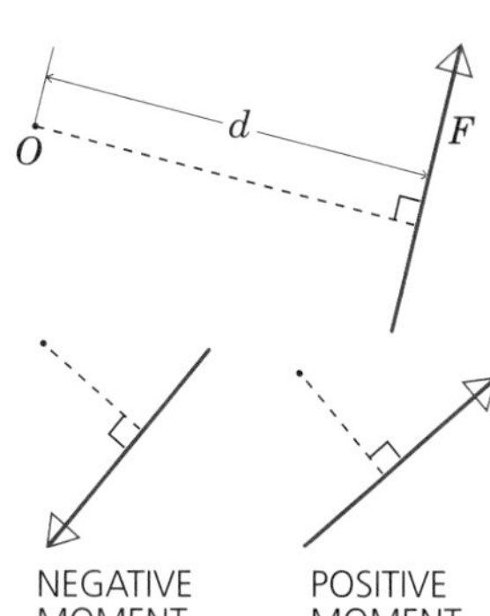

The moment of the force F about the point O is Fd, where d is the perpendicular distance from O to the force.

Moments can cause either clockwise or anti-clockwise rotation. It is conventional to define clockwise moments as negative and anti-clockwise moments as positive. The units for moments are newton-metres (Nm).

Example 16.1

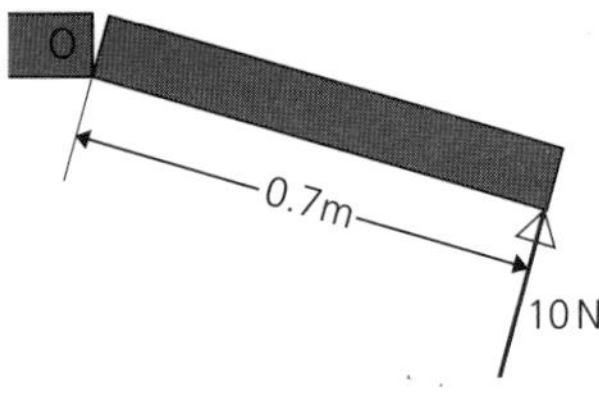

The diagram shows a door, hinged at O. A force of 10 N acts at right angles to the door, at a distance of 0.7 m from O. Find the moment of the force about O.

Solution

The moment will be anti-clockwise and so positive. Here $F = 10\,\mathrm{N}$ *and* $d = 0.7\,\mathrm{m}$, *so:*
moment $Fd = 10 \times 0.7 = 7\,\mathrm{Nm}$

Example 16.2

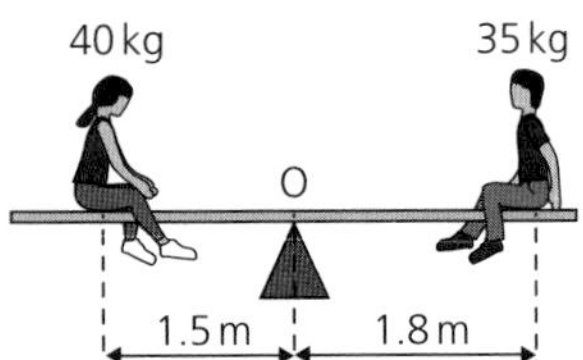

A simple seesaw consists of a plank pivoted at its centre, as shown in the diagram. Two children sit on it in the positions shown, and then the plank is allowed to move.

a) *Draw a diagram to show the forces acting on the plank.*
b) *Find the moment of each force about the pivot.*
c) *Find the total moment about O and state what happens when the plank is allowed to move.*

Solution

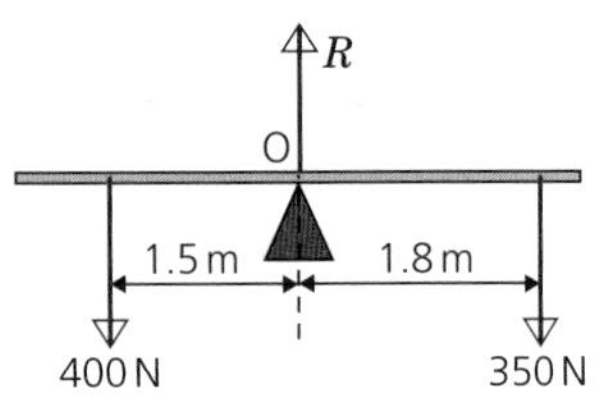

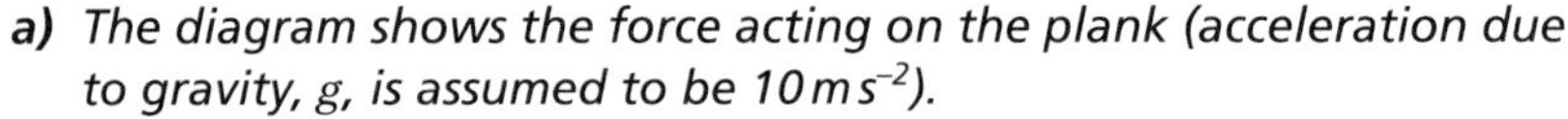

a) *The diagram shows the force acting on the plank (acceleration due to gravity, g, is assumed to be 10 m s*$^{-2}$*).*

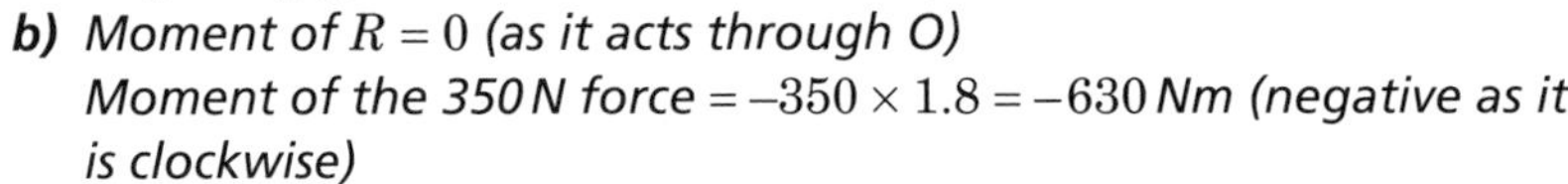

b) *Moment of* $R = 0$ *(as it acts through O)*
Moment of the 350 N force $= -350 \times 1.8 = -630$ *Nm (negative as it is clockwise)*
Moment of the 400 N force $= 400 \times 1.5 = 600$ *Nm*

c) *The total moment is simply found by adding the individual moments.*
Total moment $= 0 - 630 + 600 = -30$ *Nm*
As this is negative the plank would begin to rotate clockwise.

Resolving to find moments

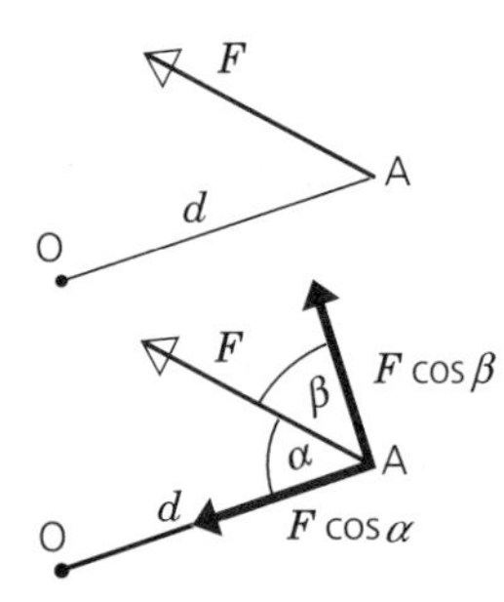

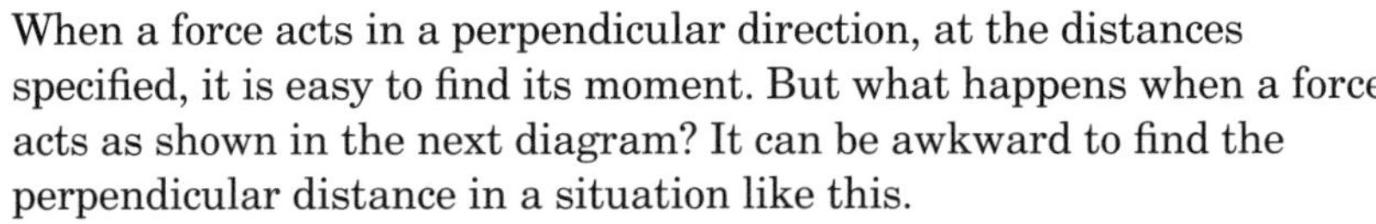

When a force acts in a perpendicular direction, at the distances specified, it is easy to find its moment. But what happens when a force acts as shown in the next diagram? It can be awkward to find the perpendicular distance in a situation like this.

To solve problems like this, we need to split the force, F, into two components, one **parallel** to the line OA and the other **perpendicular** to OA. The moment of the component $F\cos\alpha$ is zero because it is along the line through O. The moment of the component $F\cos\beta$ is simply $Fd\cos\beta$.

Example 16.3

The diagram shows a lever OA.
A force of 100 N acts on the lever as shown.

a) *Find the component of the force perpendicular to OA.*
b) *Find the moment of the force acting.*
c) *Explain why the moment is negative.*

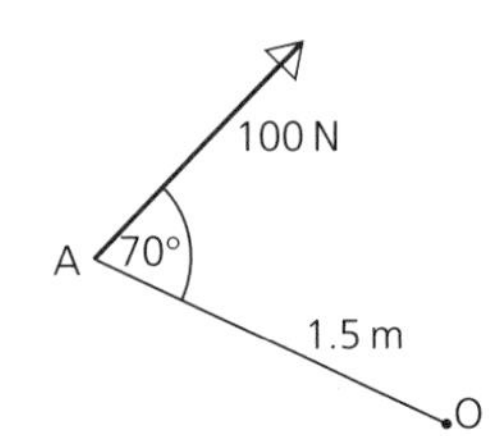

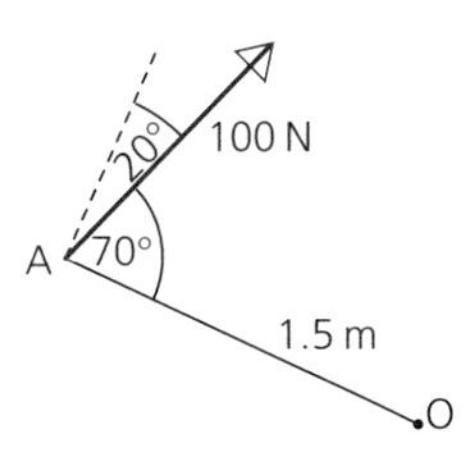

Solution

a) *The component of the force perpendicular OA is:*
$100 \cos 20° = 94$ *N*

b) *The moment of the force about O is then:*
$-1.5 \times 94 = -141$ *N*

c) *The moment is negative because the force would cause a clockwise rotation.*

Example 16.4

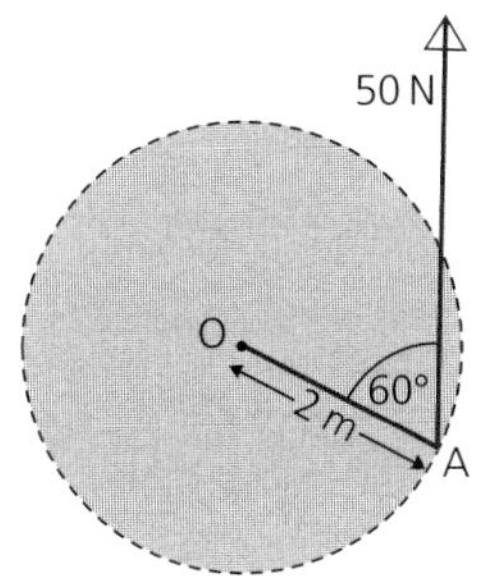

A force is applied to a playground roundabout as shown in the diagram.

a) *Show that the component of the force perpendicular to OA is 43.3 N.*

b) *Find the moment of the force about O.*

c) *Describe how a greater moment could be obtained with the same force.*

Solution

a) *The component of the force perpendicular to the radius is OA is given by:*
$50 \cos 30° = 43.3$ *N*

b) *Moment about O* $= 2 \times 43.3 = 86.6$ *Nm*

c) *If the force is applied along a tangent, the component of the force perpendicular to OA is 50 N, which gives a moment of 100 Nm.*

EXERCISES

16.1 A

1 Find the moment of each force about the point O.

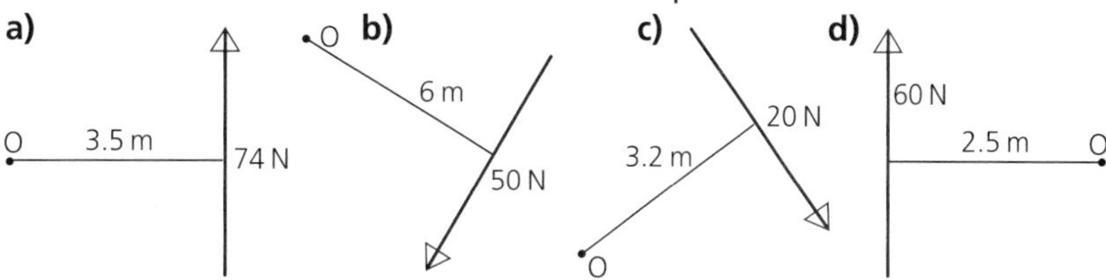

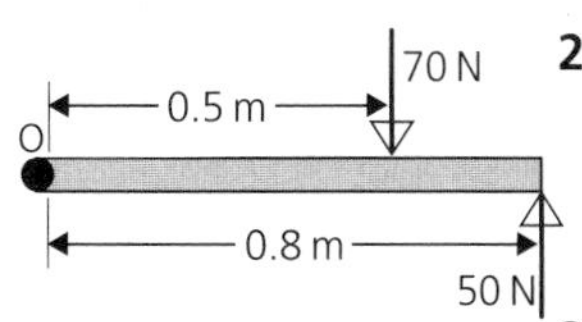

2 Two people are trying to open a door by pushing on opposite sides. They exert forces as shown.

a) Find the moment of both forces about O.

b) What is the total moment? What happens to the door?

3 Find the total moment about O for each object shown.

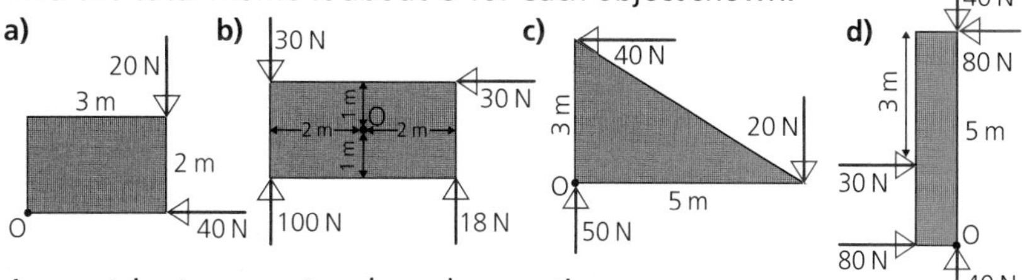

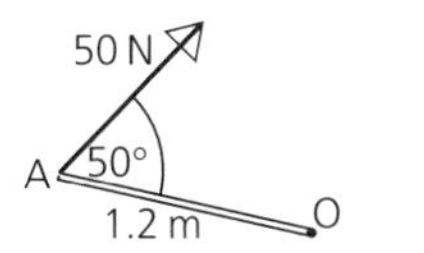

4 A man tries to operate a lever by exerting a force of 50 N as shown in the diagram.

a) Show that the component of the force perpendicular to OA is 38.3 N.

b) Find the moment of the force about O.

c) If a moment of 70 Nm would just turn the lever, is it possible for the man to move the lever with the 50 N force?

5 For each of these diagrams, find the moment of each force about the point O.

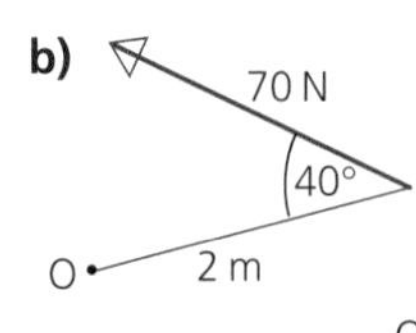

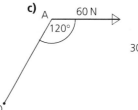

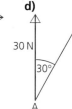

6 Three children exert forces as shown on a playground roundabout. Find the total moment on the roundabout.

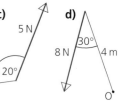

7 Find the distance OA if the magnitude of the moment of each force illustrated below is 30 Nm.

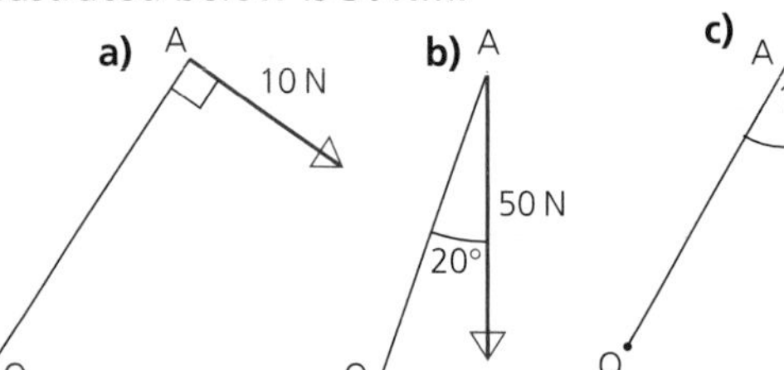

8 The diagram shows a spanner of length 30 cm being used to undo a nut.

A constant vertical force of 100 N acts as shown. Find θ if the moment of this force is:
a) 15 Nm, **b)** 10 Nm.

EXERCISES

16.1 B

1 Find the moment of each force about the point O, stating if your answer is a clockwise or anticlockwise moment.

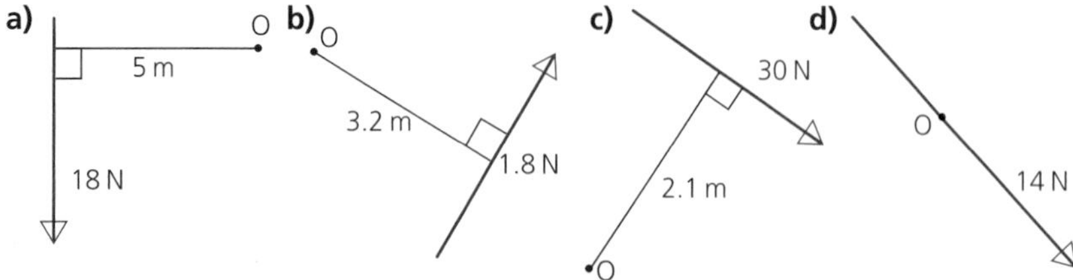

2 Ann and Bob have masses of 20 kg and 25 kg respectively. They sit on a seesaw with Ann 1.5 m from the centre.

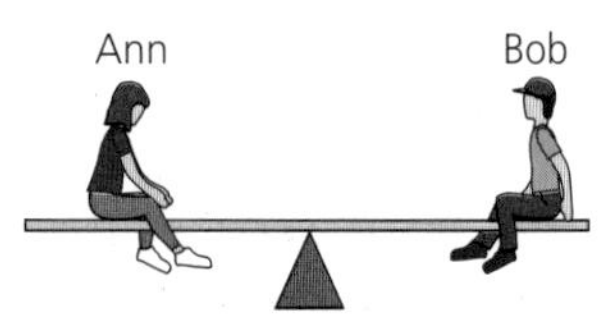

a) How far from the centre is Bob, if the total moment about its centre is zero?

b) Both children move 40 cm further away from the centre. Find the total moment on the seesaw. What happens to the seesaw now?

3 Find the resultant moment about O for each object shown.

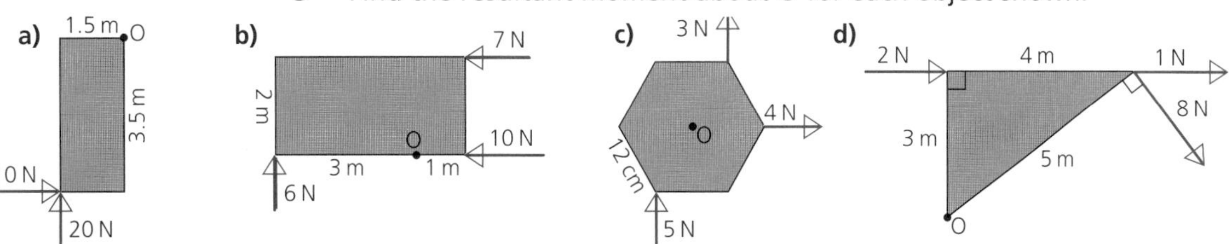

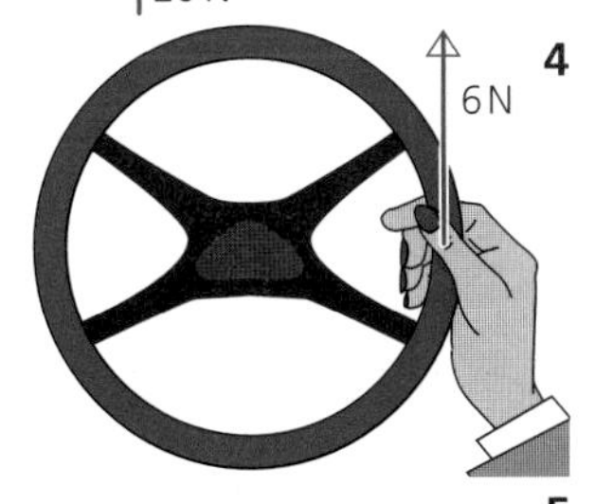

4 A driver steers a car by pushing upward on the steering wheel of radius 20 cm with a force of 6 N.

a) If the driver's hand is in the position shown in the diagram, calculate the moment about the centre of the wheel.

b) As the wheel turns does the moment remain constant? Give reasons for your answer.

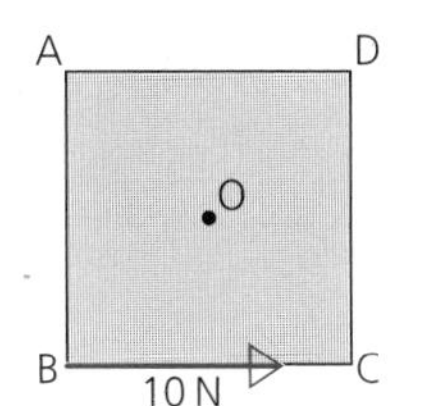

5 In the diagram, O is at the centre of a square ABCD. If the moment produced by a force of 10 N acting along the line BC is 4 Nm, calculate the length of a side of the square.

6 For each diagram, find the moment of the given force about the point O.

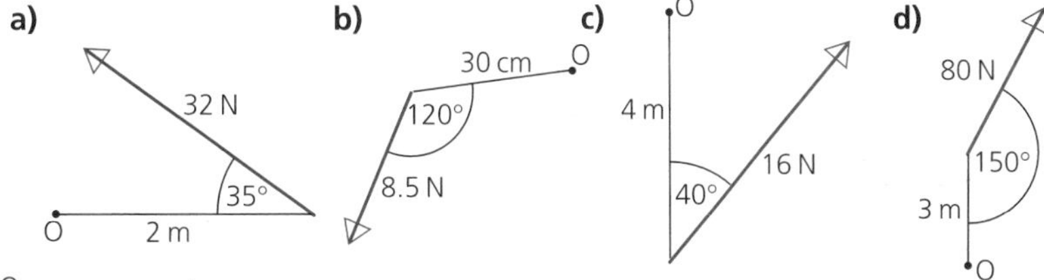

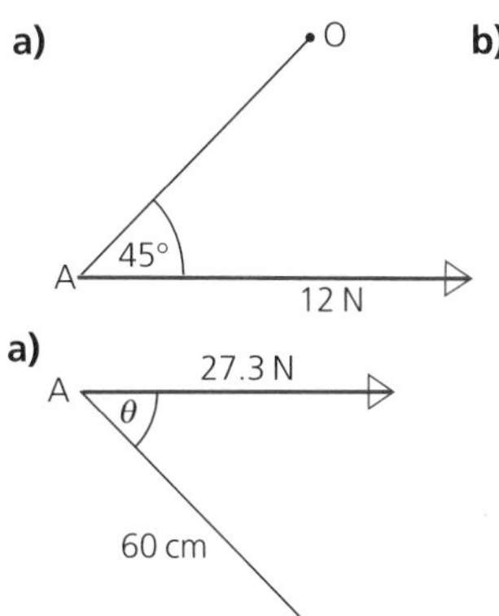

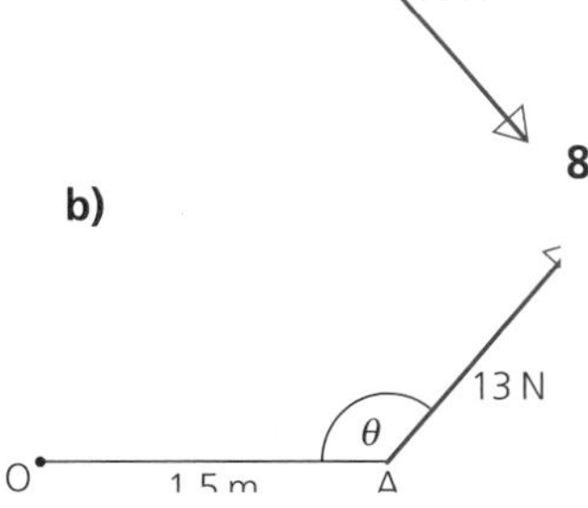

7 Find the distance OA if the moment about O which is produced by the given force is 24 Nm in each case.

8 Find the angle θ if the moment about O which is produced by the given force is 9 Nm in each case.

Explain why there are two possible values for θ in each of the above.

MOMENTS AND EQUILIBRIUM

When forces act on a particle, if the **resultant** force is zero, its motion will not change. However for a larger body the forces may be applied in such a way that although the resultant is zero, there is a rotation due to the turning effects of the forces. For example think about a Catherine wheel firework. The forces acting on it will have a resultant of zero, because it remains in the same place, but the forces will have a turning effect which causes it to start to spin. The ideas of equilibrium are extended in this section to take account of the turning effects of the forces acting.

Take a few minutes to think about the following questions.

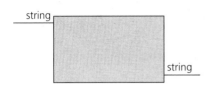

1 Two strings are attached to a sheet of material as shown in the diagram. What happens if the strings are pulled?
2 Is it possible to position the strings so that there is no rotation?
3 In Chapter 3, we found that if the resultant force on an object is zero it will not accelerate. What additional requirement is necessary for there to be no movement at all?

Requirements for equilibrium

For equilibrium there are two requirements that must both be satisfied.

a) The resultant force must be zero.
b) The total moment must be zero.

Example 16.5

A book of mass 1.5 kg stands on a bookshelf in the position shown in the diagram. If the mass of the shelf is negligible, find the reaction forces exerted at each end of the shelf.

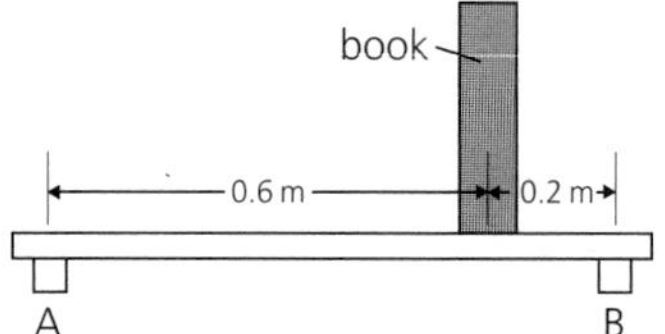

Solution

The diagram shows the forces acting on the shelf, which are the reactions R and S at the ends plus a 14.7 N force where the book rests.

As the upward forces must balance the downward forces:
$R + S = 14.7$
Taking moments about A gives:
moment of $R = 0$
moment of force of 14.7 N $= -0.6 \times 14.7 = -8.82$
moment of $S = 0.8S$
Total moment $= 0.8S - 8.82$
As the total moment must be zero:
$0.8S - 8.82 = 0$

$$S = \frac{8.82}{0.8} = 11.0\ N$$

Now using $R + S = 14.7$ *with* $S = 11.0$ *gives:*
$R + 11.0 = 14.7 \Rightarrow R = 14.7 - 11.0 = 3.7\ N$

Example 16.6

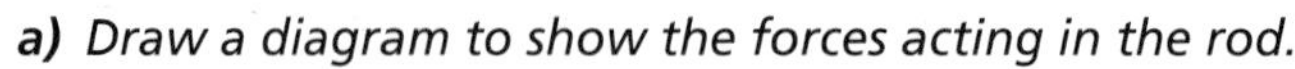

A metal rod of mass 2 kg is supported as shown in the diagram.

a) *Draw a diagram to show the forces acting in the rod.*
b) *By taking moments about O, find the reaction force exerted by the other support.*
c) *By considering the resultant force, find the reaction force exerted on the rod at O.*

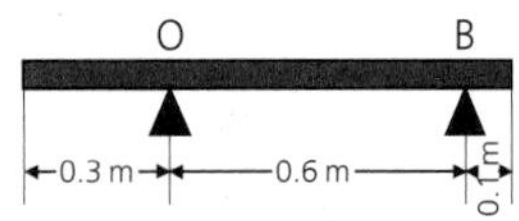

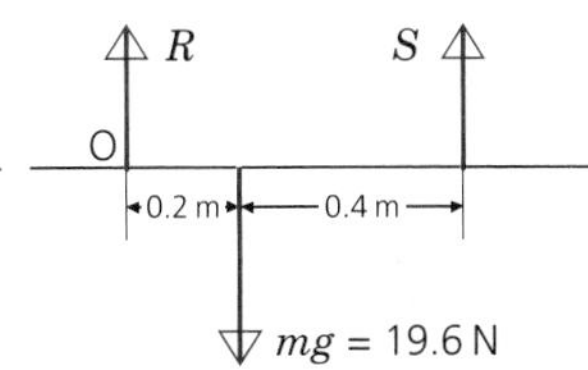

Solution

a) The diagram shows the forces acting, reaction forces R and S at the two supports and the force of gravity, 19.6 N acting down. Assume that the force of gravity acts at the centre of the rod.

b) Taking moments about O for each force:
moment of $R = 0$
moment of $S = 0.6S$
moment of $mg = -0.2 \times 19.6 = -3.92$
As the rod is at rest the total moment must be zero, so:
$0.6S - 3.92 = 0$
$\Rightarrow \; S = \dfrac{3.92}{0.6} = 6.53$ N

c) The resultant force on the rod must be zero and so:
$R + S = 19.6$
As $S = 6.53$ this becomes:
$R + 6.53 = 19.6 \;\Rightarrow\; R = 13.07$ N

Example 16.7

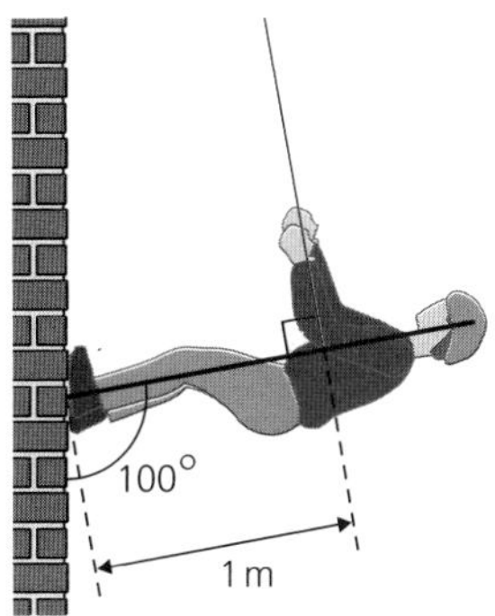

A climber of mass 60 kg has his feet resting against a vertical wall and a rope attached to him to support him. Assume that the force of gravity acts at the same point that the rope is attached. Find the magnitude of the forces acting on the climber.

Solution

The forces acting on the climber are shown in the diagram.
The component of mg perpendicular to OA is:
$60g\cos 10° = 588\cos 10° = 579$ N
Taking moments about O gives:
$T \times 1 - 579 \times 1 = 0$
and rearranging gives:
$T = 579\text{N}$
Now the resultant force on the climber is:
$(R - T\cos 80°)\mathbf{i} + (F + T\cos 10° - 588)\mathbf{j}$
As the forces are in equilibrium:
$R - T\cos 80° = 0$
$\Rightarrow \; R = 579\cos 80° = 100$ N
and $F + T\cos 10° - 588 = 0$
$\Rightarrow \; F = 588 - 579\cos 10° = 18$ N

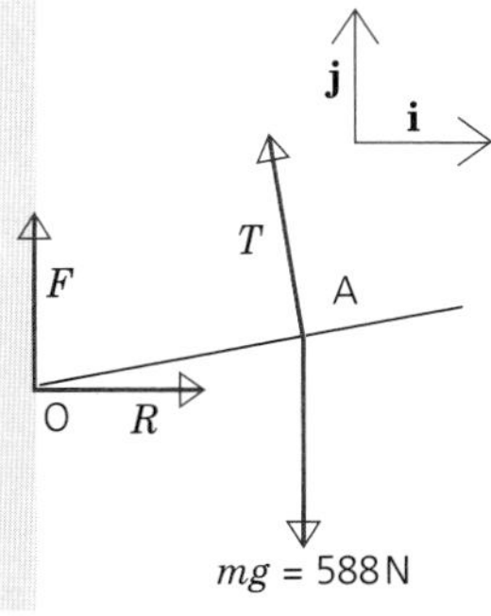

Example 16.8

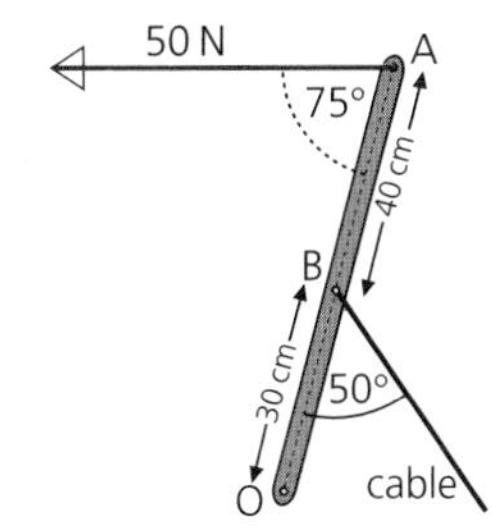

The lever shown is at rest and is pivoted at O and a force of 50 N is applied at A. A cable is attached to the lever at B.

a) By taking moments about O, find the tension in the cable.

b) Find the magnitude of the force exerted on the lever by the pivot at O.

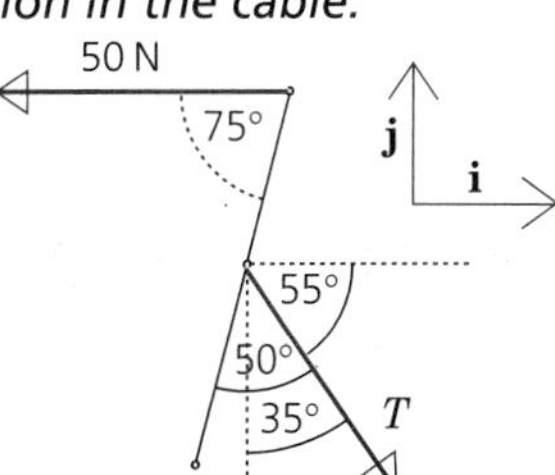

Solution

a) The component of the 50 N force perpendicular to OA is $50\cos 15° = 48.3$ N and the moment of this force about O is:
$0.7 \times 48.3 = 33.8$ Nm

The component of T perpendicular to OA is:
$T \cos 40° = 0.766\,T$
and the moment of T about O is then:
$-0.3 \times 0.766T = 0.230T$
As the lever is in equilibrium the total moment must be zero, so:
$33.8 - 0.230T = 0$
$T = \dfrac{33.8}{0.230} = 147\text{ N}$

b) *The force exerted by the pivot on the lever can be found since the resultant force must be zero. The two other forces can be expressed in terms of* $\mathbf{i}$ *and* $\mathbf{j}$ *as* $-50\mathbf{i}$ *and* $147 \cos 55°\mathbf{i} - 147 \cos 35°\mathbf{j}$.
If the reaction force exerted by the pivot is $a\mathbf{i} + b\mathbf{j}$*, then the resultant is:*
$(147 \cos 55° - 50 + a)\mathbf{i} + (b - 147 \cos 35°)\mathbf{j}$
So for equilibriumN:
$147 \cos 55° - 50 + a = 0$
$a = 50 - 147 \cos 55° = -35\,N$
and: $b - 147 \cos 35° = 0$
$b = 147 \cos 35° = 120\,N$
So the force is $-35\mathbf{i} + 120\mathbf{j}$*, which has a magnitude:*
$\sqrt{35^2 + 120^2} = 125\,N$

EXERCISES

16.2 A

1 The diagram shows a metal bar of mass 20 kg resting on two supports and the reaction forces that they exert.

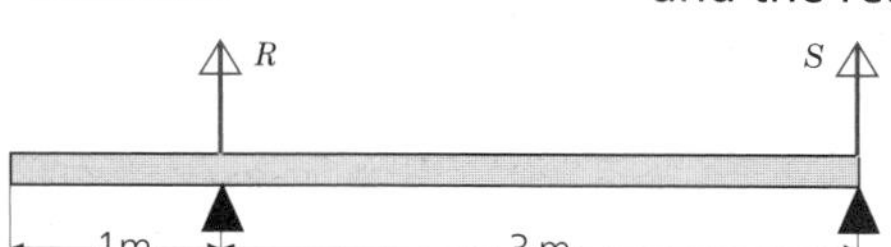

a) Draw a diagram to show where the force of gravity acting on the bar acts.
b) By taking moments show that $S = 65\frac{1}{3}$ N.
c) Find R.

2 The diagram shows a beam that rests on two supports and carries a heavy load as shown. Which of the following statements is true?

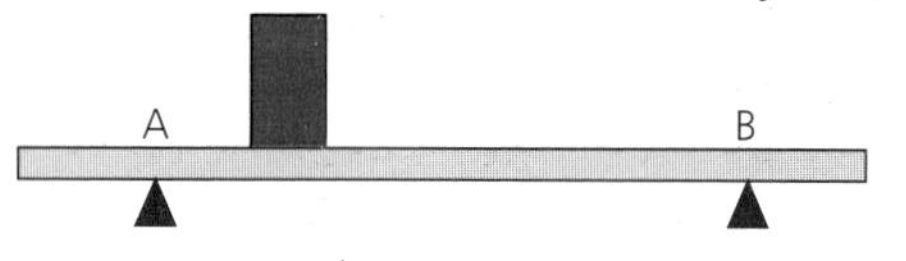

a) The reaction force on the plank at A is equal to the reaction force at B.
b) The reaction force on the plank is greater at A than B.
c) The reaction force on the plank is greater at B than A.
d) The total of the reaction forces on the plank at A and B is zero.
e) There are only two reaction forces acting on the plank.

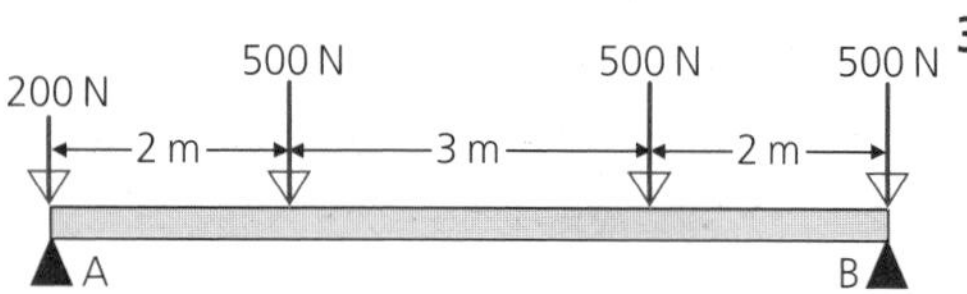

3 A beam supports the loads shown. Find the reaction forces acting on the beam if:
a) its mass is negligible,
b) its mass is 100 kg.

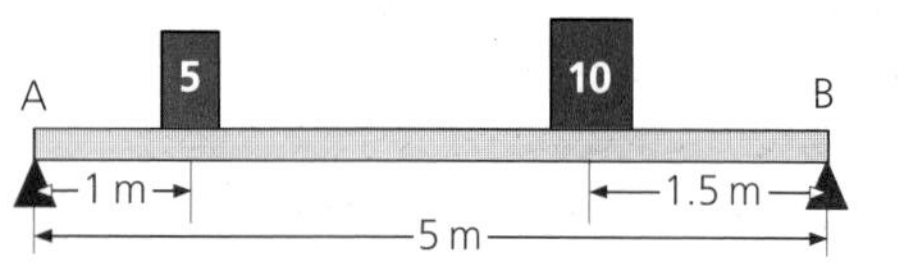

4 A light shelf is supported by brackets at A and B. Objects of masses 5 kg and 10 kg are placed in the positions shown. Find the reaction forces exerted by the two brackets.

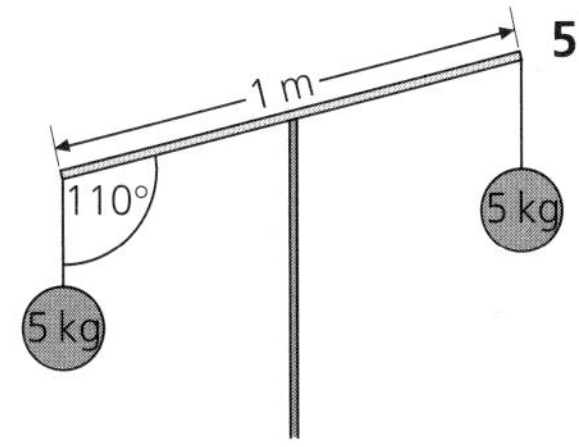

5 The diagram shows a ruler smoothly pivoted at its centre with equal masses suspended from it at each end. It is held in the position shown and then released.

a) What do you think will happen when the ruler is released?

b) Find the moment, about the centre of the ruler, of each force acting on the ruler and then the total moment. Does this confirm the prediction you made in part **a)**?

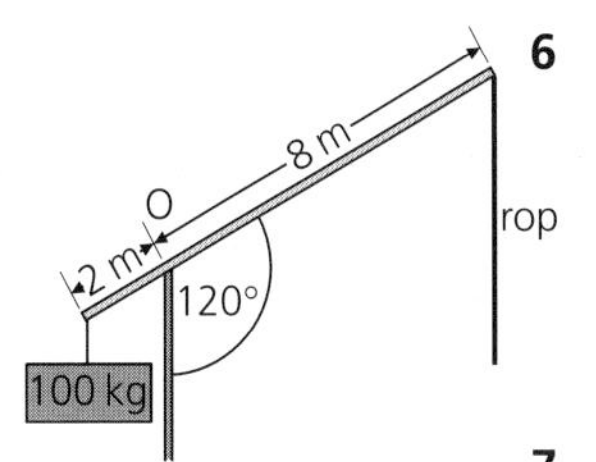

6 The diagram shows a light beam and a rope.

a) If the beam makes the angle shown with the vertical, find the tension in the rope when the system is at rest, by taking moments about O.

b) If the beam is horizontal and the rope is still vertical, what happens to the tension in the rope?

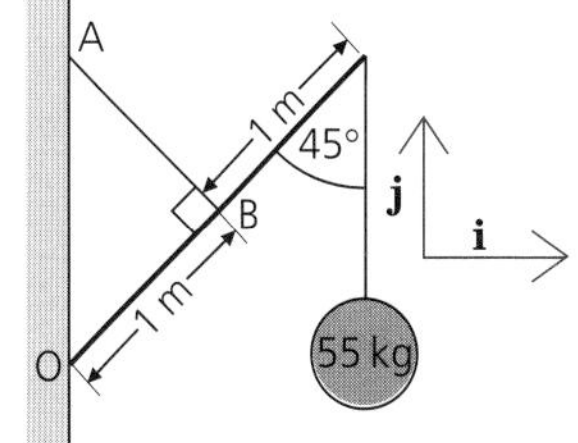

7 The diagram shows a simple crane, which is supporting a load of mass 55 kg.

a) Find the tension in each cable.

b) Find the force exerted on the rod by the pivot at O, and the tension in the cable AB in terms of the unit vectors $\mathbf{i}$ and $\mathbf{j}$.

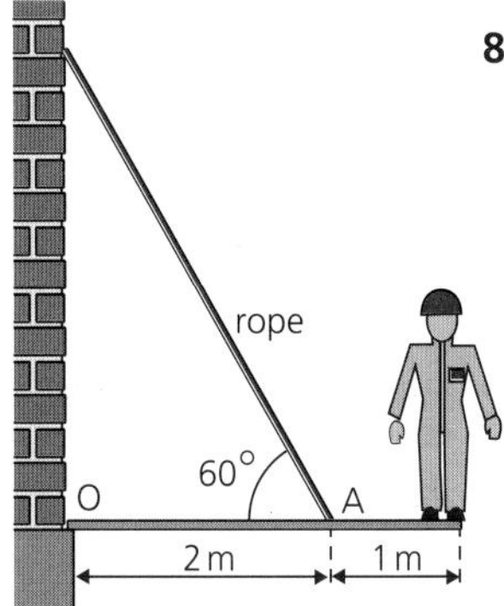

8 A plank has length 3 m and mass 20 kg. One end rests on a narrow ledge at O and a rope is fixed to the point A as shown. A man of mass 75 kg starts at the far end of the plank. Assume the plank remains at rest, but is on the point of slipping.

a) Draw a diagram to show the forces acting on the plank.

b) By taking moments about the point A, where the rope is attached, find the friction force acting on the end of the plank.

c) By taking moments about O, find the tension in the rope.

d) Find the reaction force exerted by the wall on the end of the plank.

e) Find the coefficient of friction between the wall and the plank. Is this value realistic?

9 A builder places a plank of length 4 m and mass 20 kg so that it rests on the back of a lorry, as shown in the diagram. Assume that there is no friction between the plank and the lorry.

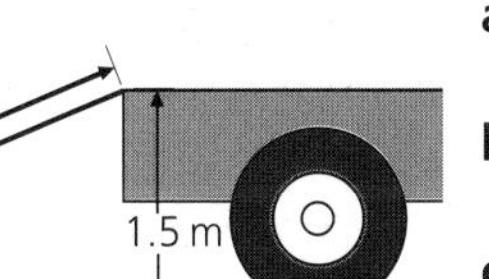

a) By taking moments about the lower end of the plank find the force the lorry exerts on the plank.

b) Find the normal reaction force and the friction force acting at the bottom of the plank.

c) Find the minimum value of the coefficient of friction between the plank and the ground.

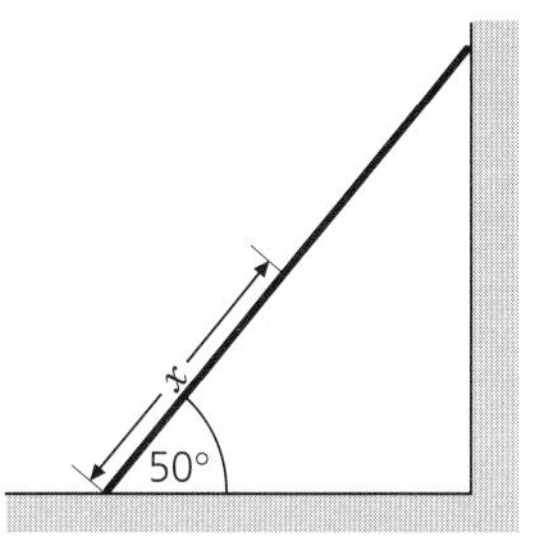

10 A ladder of length 2 m and mass 10 kg rests against a smooth wall. A firewoman of mass 60 kg climbs the ladder. The coefficient of friction between the ground and the ladder is 0.45.

a) Find the reaction and friction forces on the bottom of the ladder, when the firewoman is in the position shown.

b) When the firewoman climbs along the ladder it begins to slip. How far up the ladder can she get before it begins to slip?

EXERCISES

16.2 B

1 A metal bar of mass 100 kg rests on two supports. Find the magnitude of M and N as shown on the diagram.

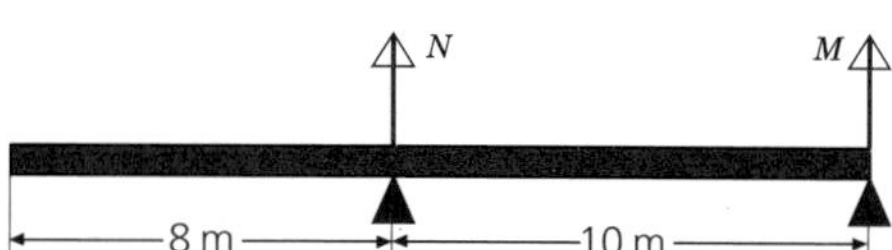

2 The diagram shows a beam that rests on two supports and carries a heavy load as shown.

a) How many reaction forces act on the plank?

b) How does the reaction at A compare with the reaction at B?

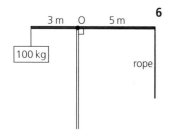

3 The beam in the diagram supports the loads shown. Find the reaction forces acting on the beam if:

a) its mass is negligible,

b) its mass is 50 kg.

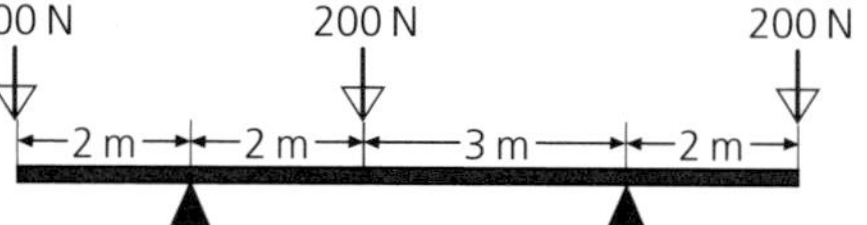

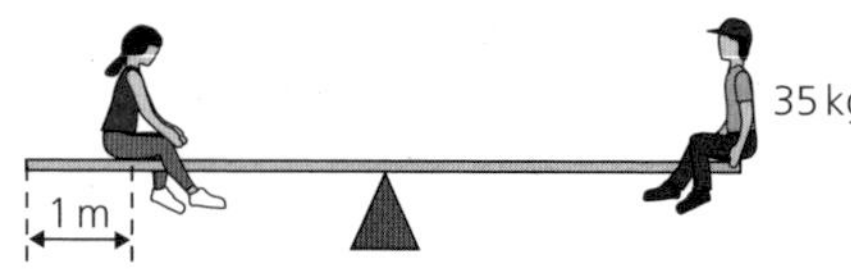

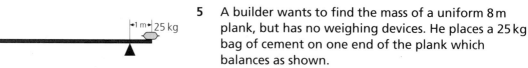

4 A boy of mass 35 kg sits on the end of a light 6 m plank which pivots at its centre. When a girl sits 1 m from the other end the plank balances. Find the mass of the girl.

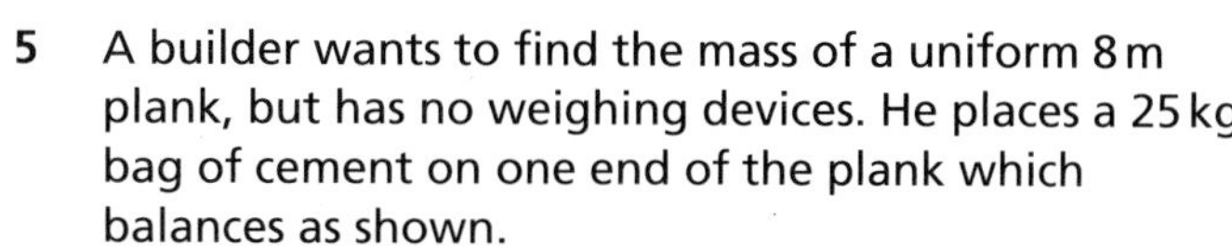

5 A builder wants to find the mass of a uniform 8 m plank, but has no weighing devices. He places a 25 kg bag of cement on one end of the plank which balances as shown.

a) Draw a diagram to show where the forces on the plank act.

b) By taking moments about the pivot point, find the mass of the plank.

6 The diagram shows a beam of mass 20 kg with a mass of 100 kg attached to one end and a rope attached to the other end.

a) By taking moments, find the tension on the rope.

b) Find the reaction at O.

The same beam is then allowed to tilt at 30° to the horizontal, and the rope is held so that it forms a right angle with the beam.

c) By taking moments about O find the tension in the rope.

d) Find the magnitude and direction of the reaction at O.

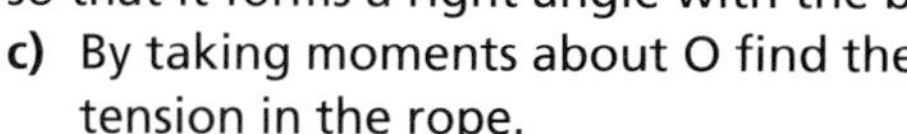

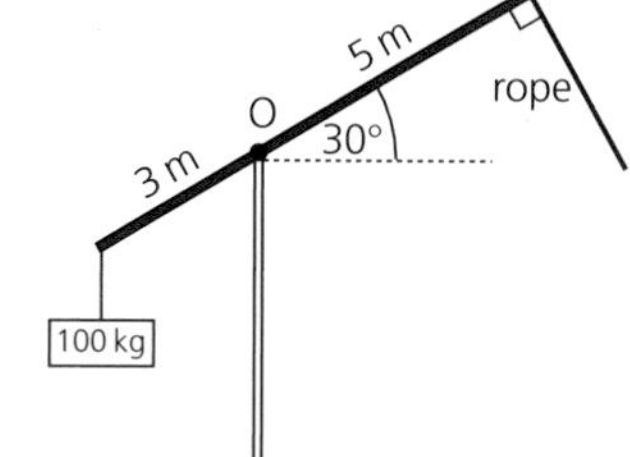

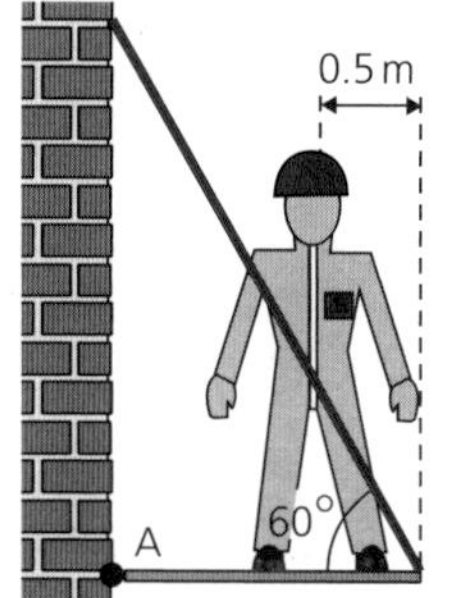

7 A man of mass 70 kg stands on a uniform platform of mass 30 kg and width 1.5 m. The platform is smoothly hinged to the wall at A and is kept horizontal by two parallel ropes attached to the two outer corners of the platform, inclined at 60° to the horizontal.

a) Draw a diagram showing all the forces acting on the platform (represent the reaction at the hinge as a horizontal and a vertical component).

b) By taking moments about A, find the tension in each rope.

c) Find the horizontal and vertical components of the reaction force at A.

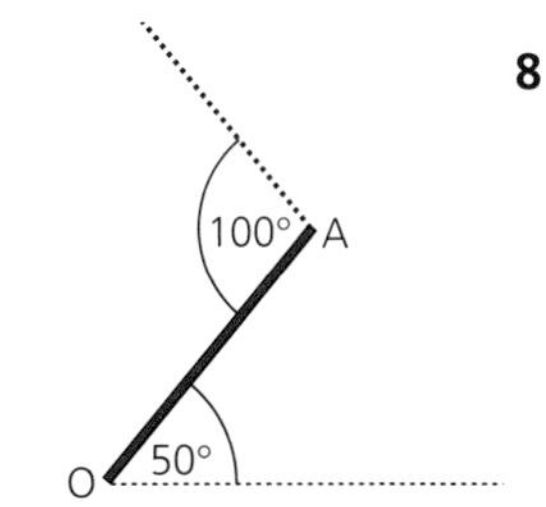

8 A uniform hatch door of a yacht, of mass 15 kg, is held open at 50° to the horizontal by a rope as shown. The door is hinged at O, and OA is 1.5 m.

a) Find the tension in the rope.

b) Find the vertical and horizontal components of the reaction force at the hinge.

c) Find the magnitude and direction of the reaction force at the hinge.

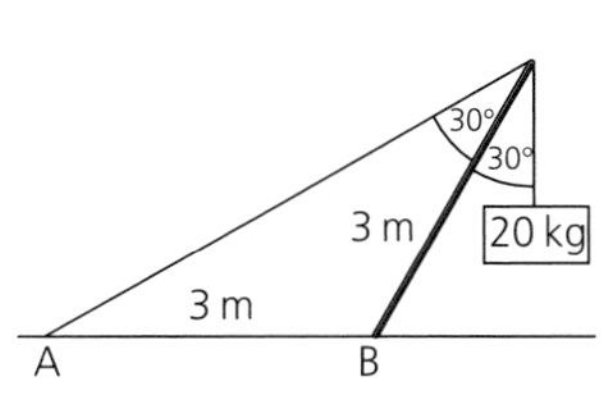

9 Some children construct a simple crane to support a load of 20 kg as shown in the diagram. The rope is fixed to a peg in the ground at A and the rod rests on the ground at B.

a) By taking moments about B, find the tension in the rope.

b) Find the magnitude of the force in the rod.

c) If the bottom of the rod is about to slip, find the value of the coefficient of friction, μ.

10 A uniform ladder of mass m, inclined at 45° to the vertical, is at rest with one end on rough horizontal ground and the other end against a smooth vertical wall. The ladder is on the point of slipping. If the coefficient of friction between the ladder and the ground is μ, find the value of μ.

COUPLES

It is possible for two or more forces to act so that their resultant is zero, whilst at the same time their resultant moment is not zero. In this section we explore the special properties of pairs of forces that act in this way.

Exploration 16.3

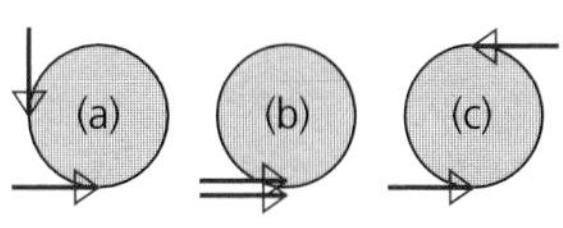

Pushing a roundabout

The diagrams show different ways of pushing a roundabout in a children's playground.

In each case, consider the force exerted on the roundabout at its centre and the total moment on the roundabout. Does any one way of pushing offer any advantage over the others?

Parallel forces

Two forces that have the same magnitude and are parallel, but are not collinear, i.e. that do not lie along the same line, will have a zero resultant but cause a **turning effect**. Such a pair of forces is called a **couple**. In part **c)** of the diagram above the forces form a couple, while in the other cases they do not, but could be expressed as a combination of a force and a couple.

Example 16.9

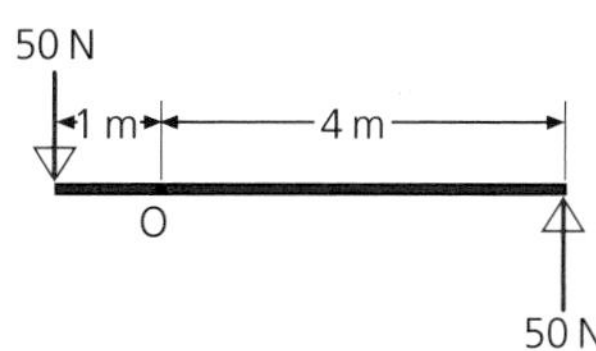

A light rod is pivoted about a point, as shown in the diagram. The forces acting form a couple.

a) *Find the moment of the forces about the pivot.*

b) *If the pivot is moved to the centre of the rod, find the moment of the forces about this new pivot point.*

Solution

a) *The moment is given by:*
$50 \times 1 + 50 \times 4 = 250$ *Nm*

b) *With the pivot in the centre, both forces act at a distance of 2.5 m from the pivot, so that the moment is:*
$2 \times 2.5 \times 50 = 250$ *Nm*
Note that the moment is the same about any point. This is true for any couple.

Example 16.10

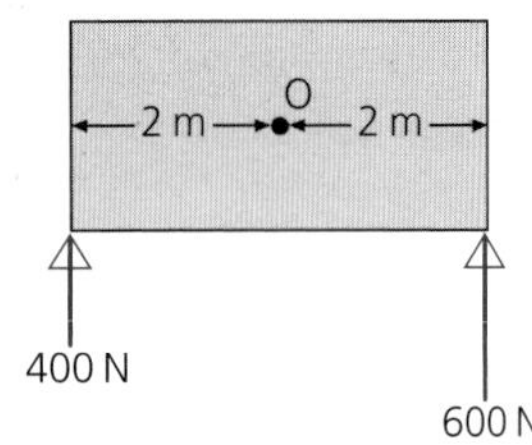

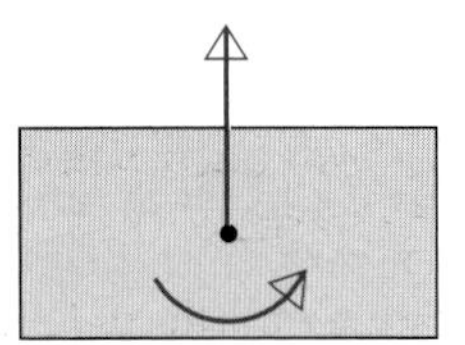

Two forces act on a crate as shown in the diagram.

a) *Find the resultant force on the crate and the total moment, about the point O, of the forces acting.*

b) *The forces acting are equivalent to a single force that acts at O and a couple. State the magnitude of the force and the couple.*

c) *Describe what happens to the crate as it moves.*

Solution

a) *The resultant force is simply 1000 N, in the same direction as the original forces. The moments are calculated about the point O.*
The moment of the 600 N force is $2 \times 600 = 1200$ *Nm*
The moment of the 400 N force is $-2 \times 400 = -800$ *Nm*
So the total moment is 400 *Nm.*

b) *The system is equivalent to a 1000 N force at O plus a couple of 400 Nm.*

c) *As the forces are applied the crate will accelerate forwards and begin to turn anticlockwise.*

Example 16.11

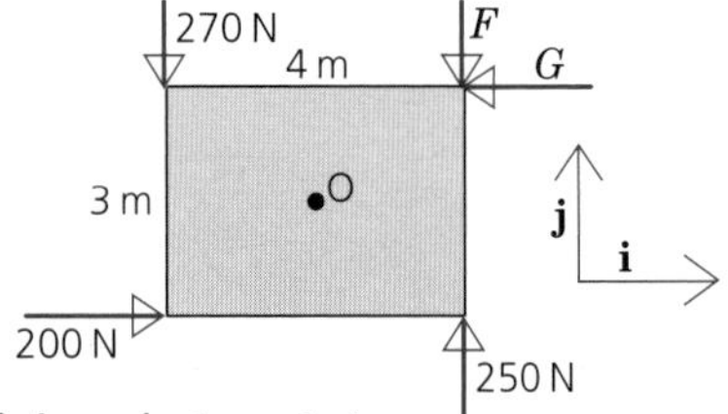

Five builders are trying to move a large tank into its correct position. They exert forces as shown in the diagram. The point O is at the centre of the tank.

a) *Find the forces F and G, if the tank is only to rotate.*

b) *Find the moment of each force about the point O, and state the magnitude of the couple that is equivalent to the system.*

c) *What important assumptions have you made about the tank?*

Solution

a) *If the tank is only to rotate, then the resultant force must be zero.*
The resultant is: $(200 - G)\mathbf{i} + (250 - 270 - F)\mathbf{j}$
and so: $G = 200$ N *and* $F = -20$ *N*

b) *Calculating moments about the point O gives:*
moment of $F = 2 \times 20 = 40$ *Nm*
moment of $G = 1.5 \times 200 = 300$ *Nm*
moment of the 250 N force $= 2 \times 250 = 500$ *Nm*
moment of the 200 N force $= 1.5 \times 200 = 300$ *Nm*
moment of the 270 N force $= 2 \times 270 = 540$ *Nm*
So the total is 1680 Nm. The system is equivalent to a couple of 1680 Nm anticlockwise.

c) *We have assumed that there is no friction between the ground and the bottom of the tank.*

EXERCISES

16.3 A

1 For each part of the diagram, state if the forces acting are equivalent to a couple and if so find the magnitude of the couple about the point marked.

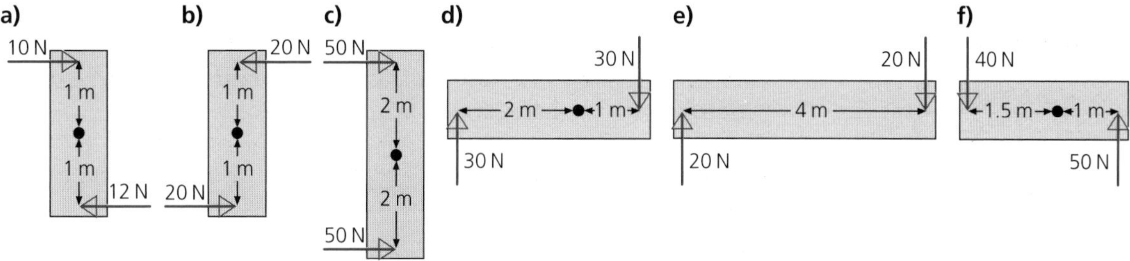

2 The diagram shows a light rod with points A, B, C, D and E all 1 m apart. A 20 N force acts straight up at B and a 20 N force acts straight down at D.

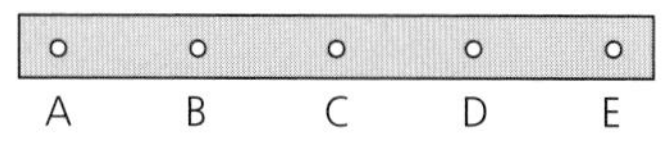

a) Find the combined moment of the two forces about each point.

b) What can you deduce about the magnitude of the couple?

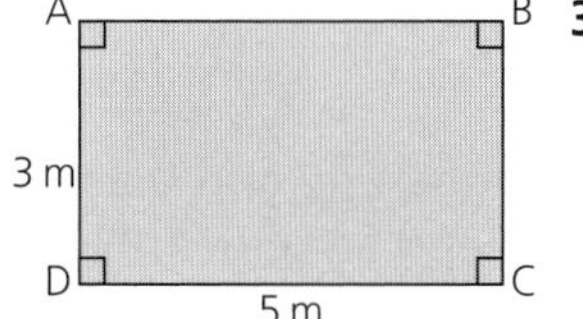

3 The diagram shows a lamina. Where should the forces stated below act in order to produce a couple of the required magnitude?

a) Two forces of 20 N to give a couple of 100 Nm.

b) Two forces of 600 N to give a couple of 800 Nm.

c) Two forces of 80 N to give a couple of 0 Nm.

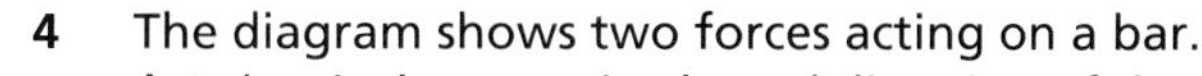

4 The diagram shows two forces acting on a bar.

a) What is the magnitude and direction of the extra force required if the system is to be equivalent to a couple?

b) Where should this force act if the moment of the couple is to be 100 Nm?

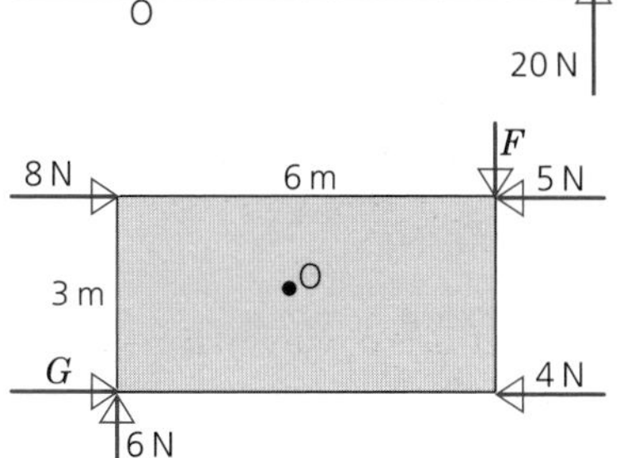

5 The diagram shows the forces that act on a toy box as some children all try to get hold of it. The box rotates, but does not move.

a) Find the magnitude of the forces F and G.

b) By taking moments about the centre of the box, show that the system is equivalent to a couple. State its magnitude.

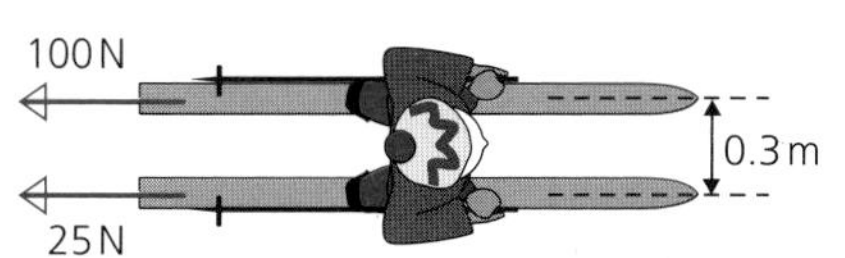

6 A skier reaches a horizontal surface at the bottom of a ski slope. One ski moves off the snow and onto a different surface, so that the forces shown act.

a) What is the resultant force on the skier?

b) The system of forces is equivalent to a single force at the centre of the skier and a couple. State the magnitude of this force and find the magnitude of the couple.

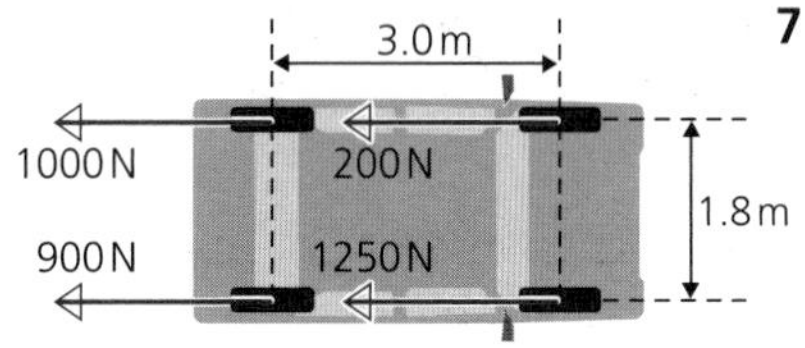

7 A car driver, travelling on a surface that has some oily patches, applies the brakes, which lock so that the car skids. The forces shown in the diagram act on each wheel of the car. These forces are equivalent to a single force and a couple at the centre of the car. Find the size of the force and the couple.

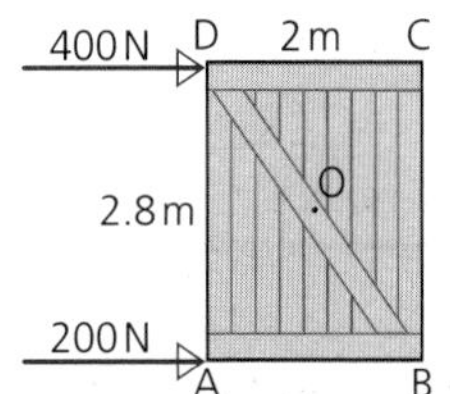

8 **a)** The two forces shown in the diagram act on a heavy crate. A single force and a couple at O, the centre of the crate, form an equivalent system. State the magnitude of the force and the couple.

b) An additional 800 N force is applied at right angles to the side AB, so that the three forces are equivalent to a single force acting at O. Where should the 800 N force be applied?

EXERCISES

16.3 B

1 In each case state if the forces acting are equivalent to a couple and, if so, find the magnitude of the couple about the point marked.

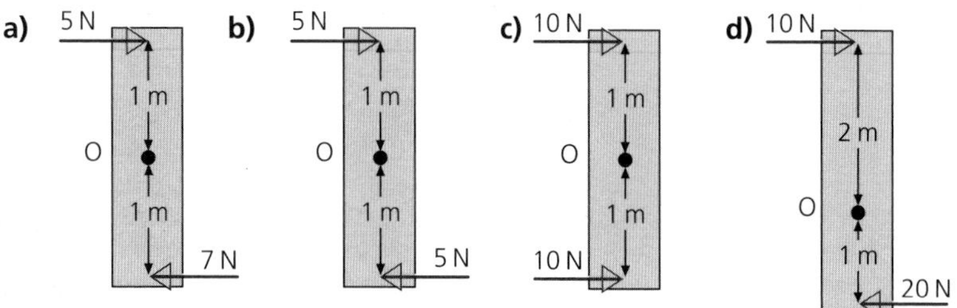

2 The diagram shows a light rod on which the points A, B, C and D are all 1 m apart. A force of 10 N acts straight up at B and force of 10 N acts straight down at D.

A B C D

a) Find the magnitude of the combined moment of the two forces about each point.

b) What can you deduce about the magnitude of the couple?

A B 3 m D 4 m C

3 The figure shows a rectangular lamina. For each of the force and couple combinations stated below, indicate two possible positions for the lines of action of the forces which will produce the desired couple.

a) Two forces of 50 N to give a couple of 150 Nm.

b) Two forces of 400 N to give a couple of 1600 Nm.

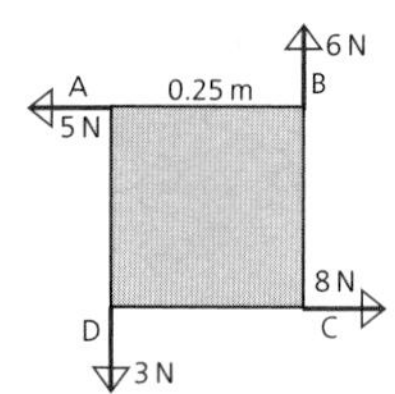

4 The diagram shows a square lamina ABCD of side 0.25 m, which is subject to the force system shown. If the system is reduced to a force acting through D, together with a couple, find the magnitude and direction of the force and the moment of the couple.

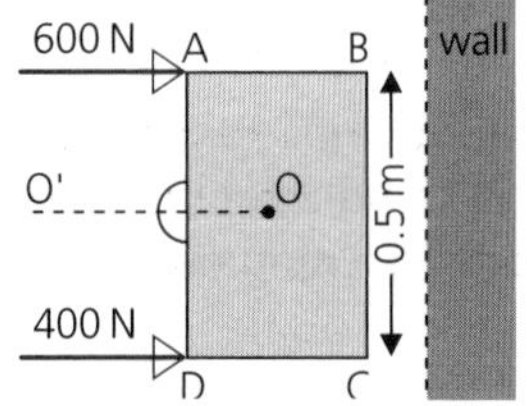

5 Two removal men are pushing a full filing cabinet into position against a wall by exerting forces as shown in the figure.

a) Where should one removal man, working on his own, push, and with what force, in order to move the filing cabinet in the same way?

b) Reduce the original two man force system to a single force, F, acting along O′O and a couple, G.

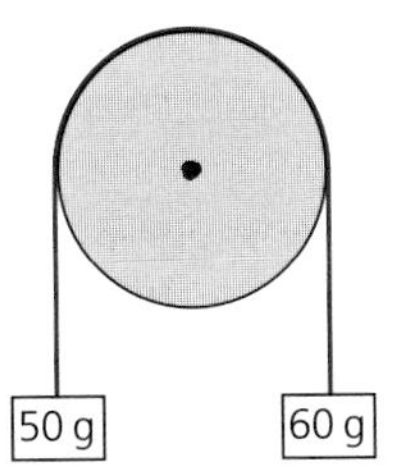

6 Although engineers strive to produce frictionless bearings there is always some friction present. This manifests itself as a frictional couple opposing the rotation of the bearing.

A pulley of radius 1 cm has two masses of 60 grams and 50 grams suspended over it by a light string, as shown in the figure. Due to the roughness in the bearing the system remains stationary. Determine the magnitude of the frictional couple acting on the pulley.

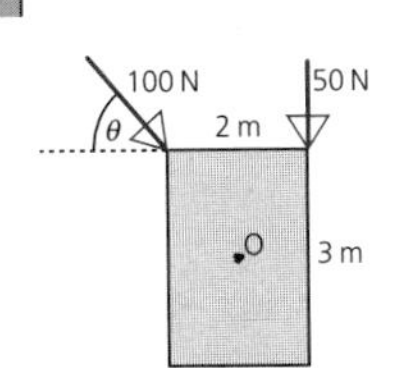

7 The figure shows a plan view of a lock gate and the bar used to push it open or closed. The diagrams A and B show two possible methods which two people could adopt in order to operate the lock gate.

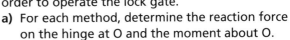

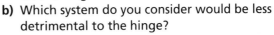

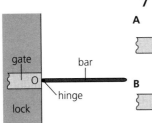

a) For each method, determine the reaction force on the hinge at O and the moment about O.

b) Which system do you consider would be less detrimental to the hinge?

8 The diagram shows two forces that act on a crate. The two forces are equivalent to a single force acting at O, the centre of the crate.

a) Show that: $1 + 3\cos\theta = 2\sin\theta$
and that: $\theta \approx 72°$

b) Find the magnitude of the single force.

CONSOLIDATION EXERCISES FOR CHAPTER 16

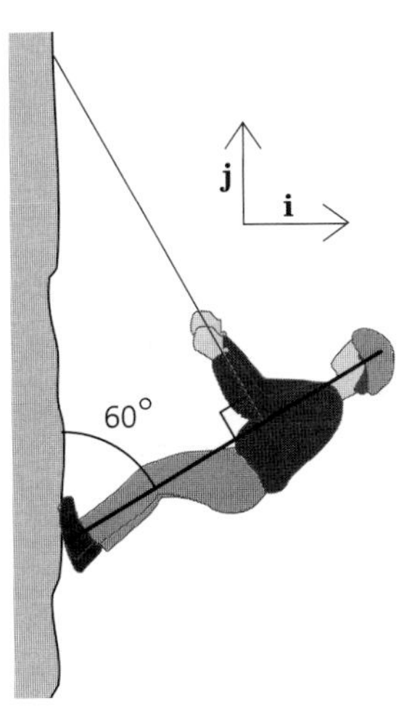

1 (In this question take the acceleration due to gravity, g, as $9.8\,\text{m s}^{-2}$.)
The diagram shows a man of mass 70 kg at rest while abseiling down a vertical cliff. Assume that the rope is attached to the man at his centre of mass. You should model the man as a rod and assume he is not holding the rope.

a) Draw a diagram to show the forces acting on the man.

b) The angle between the legs and the cliff is 60°. By taking moments show that the man can only remain in position if the coefficient of friction between his feet and the wall is greater than or equal to $\frac{1}{\sqrt{3}}$.

c) In the position shown the rope is at 90° to the man's body. Express the resultant force on the man in terms of the unit vectors **i** and **j** which act horizontally and vertically, respectively.

d) Show that the tension in the rope is 594 N correct to three significant figures.

e) When the man has descended a further distance he again stops and remains at rest. How would the magnitude of the normal reaction now compare with its value in the position illustrated above?

(AEB Question 5, Specimen Paper)

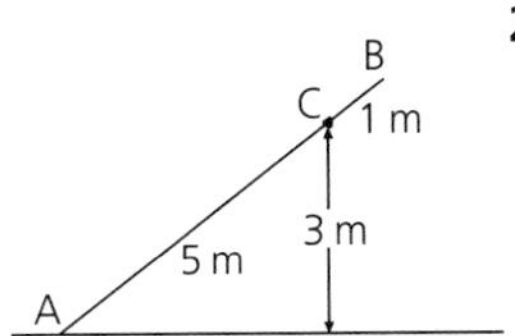

2 A smooth horizontal rail is fixed at a height of 3 m above a horizontal playground with a rough surface. A straight uniform pole AB, of mass 20 kg and length 6 m, is placed to rest at a point C on the rail with the end A on the playground. The vertical plane containing the pole is at right angles to the rail. The distance AC is 5 m and the pole rests in limiting equilibrium, as shown in the diagram.

Calculate:

a) the magnitude of the force exerted by the rail on the pole, giving your answer to the nearest N,

b) the coefficient of friction between the pole and the playground, giving your answer to two decimal places,

c) the magnitude of the force exerted by the playground on the pole, giving your answer to the nearest N.

(ULEAC Question 9, M2, June 1992)

3 **a)** **i)** Draw a diagram showing the forces acting on an inclined ladder which is standing on a horizontal floor and leaning against a vertical wall.

ii) Explain why the ladder cannot be in equilibrium if the floor is frictionless, even if the wall is rough.

b) A uniform ladder of length 8 m and mass 20 kg is inclined at 60° to the horizontal, against a smooth vertical wall. A 60 kg man is standing on the ladder x m from its lower end. The horizontal floor has coefficient of friction 0.4 with the base of the ladder. The ladder is about to slip.

i) Show that the frictional force on the ladder is $32g$ N.

ii) Find the reaction of the wall on the ladder.

iii) By taking moments about the base of the ladder, or otherwise, show that x is about 6.06 m.

(MEI Question 1, M2, January 1992)

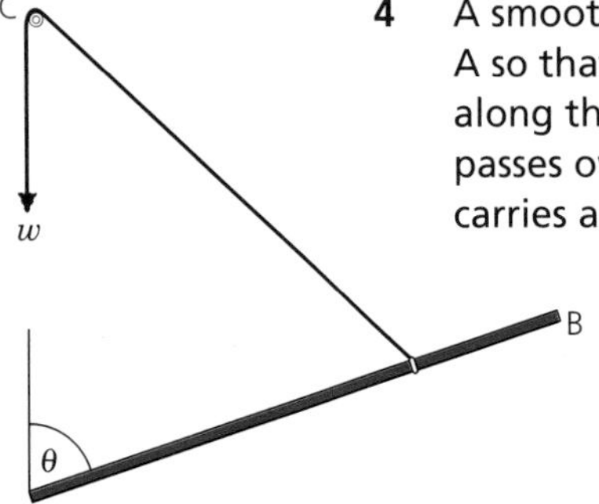

4 A smooth uniform rod AB, of length $3a$ and weight $2w$, is pivoted at A so that it can rotate in a vertical plane. A light ring is free to slide along the rod. A light inextensible string is attached to the ring and passes over a fixed smooth peg at a point C, a height $4a$ above A, and carries a particle of weight w hanging freely, as illustrated.

a) Give reasons why in equilibrium, as shown, the string will be at right angles to the rod.

b) Show the angle θ that the rod makes to the vertical in equilibrium is given by $\tan\theta = \frac{4}{3}$.

c) Find the magnitude of the force of the pivot on the rod A in terms of w.

(Oxford Question 8, Specimen Paper M6)

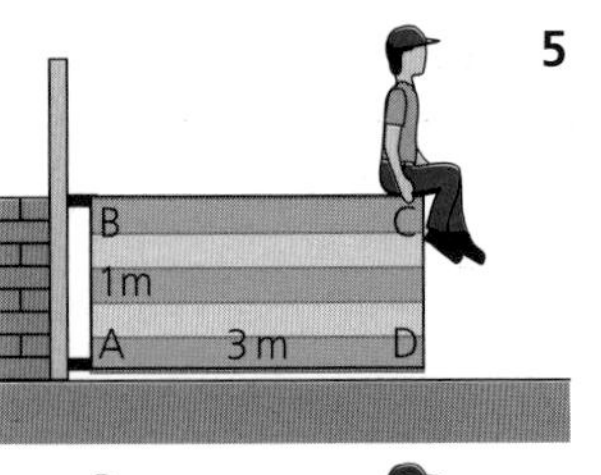

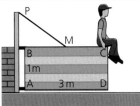

5 A rectangular gate ABCD, where AB = 1 m and AD = 3 m, is supported by smooth pins at A and B, where B is vertically above A. The pins are located in such a way that the force at B is always horizontal. The gate has mass 120 kg and it can be modelled as a uniform rectangular lamina. A boy, of mass 45 kg, sits on the gate with his centre of mass vertically above C, as in the first diagram. Find the magnitudes of the forces on the gate at B and at A.

In order to support the gate, the owner fits a cable attaching the midpoint M of BC to a point P, vertically above B and such that BP is 1.5 m. The boy once again sits on the gate at C, as in the second diagram, and it is given that there is now no force acting at B. Find the tension in the cable and the magnitude of the force now acting at A.

(UCLES Question 8, Specimen Paper M2)

MATHEMATICAL MODELLING ACTIVITY

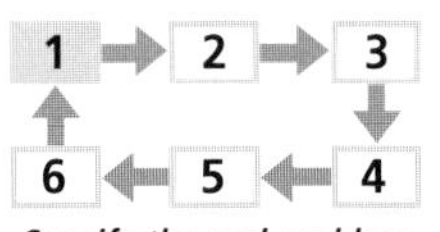

Specify the real problem

Problem statement

The diagram shows a design of temporary road signs. A mass is placed at A to stop the sign blowing over in high winds. How should the mass relate to the size of the sign?

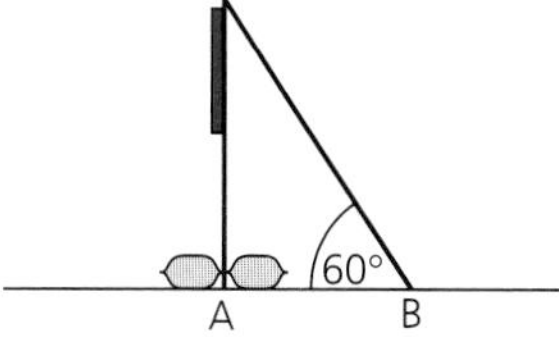

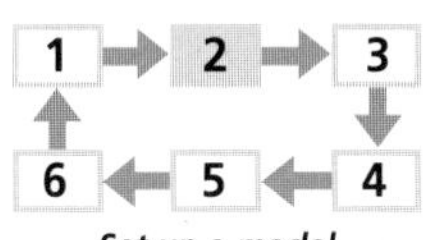

Set up a model

Set up a model

Make a list of the important features that you feel should be included in a solution to this problem. The assumptions set out below allow a first attempt to be made to solve the problem.

- The wind strikes the sign at right angles and is brought to rest.

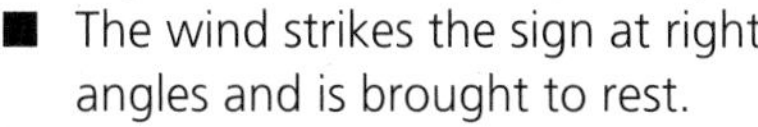

- The maximum wind speed is $30\,\mathrm{m\,s^{-1}}$.
- The density of air is $1.29\,\mathrm{kg\,m^{-3}}$.
- The effect of the wind can be considered as a single force acting at the centre of the sign.

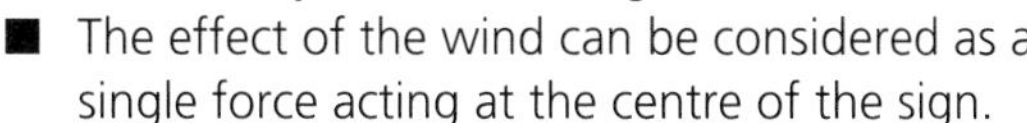

- The mass of the frame supporting the sign is negligible.

- The sign has width a, height b and its base is h m above the ground.

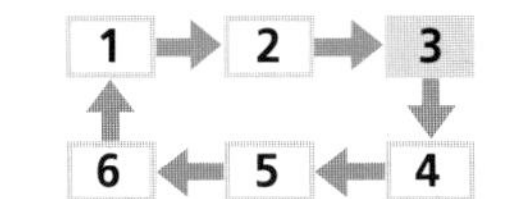

Formulate the mathematical problem

Formulate a mathematical problem

Find the force exerted on the sign by the wind.
Find the moment of this force about B.
Find the mass that must be placed at A if the sign is to remain at rest.

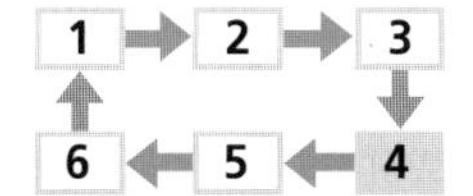

Solve the mathematical problem

Mathematical solution

First, find the force exerted by the wind. The volume of air hitting the sign each second is:
$V = 30ab\,\mathrm{m}^3$
The mass of this air is found by multiplying the volume by its density to give:
$\mathrm{M} = 30ab \times 1.29 = 38.7ab\,\mathrm{kg}$

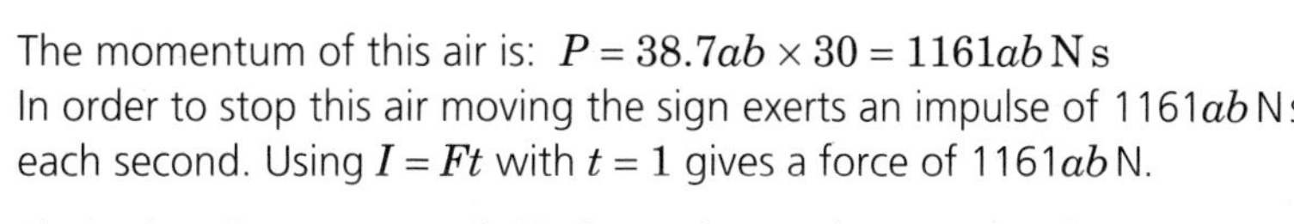

The momentum of this air is: $P = 38.7ab \times 30 = 1161ab\,\mathrm{N\,s}$
In order to stop this air moving the sign exerts an impulse of $1161ab\,\mathrm{N\,s}$ each second. Using $I = Ft$ with $t = 1$ gives a force of $1161ab\,\mathrm{N}$.

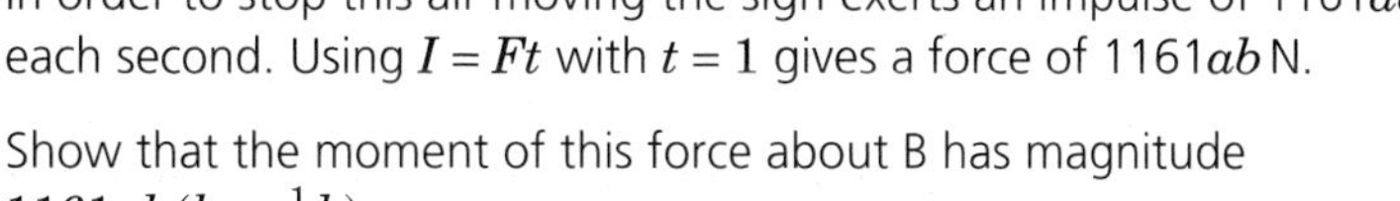

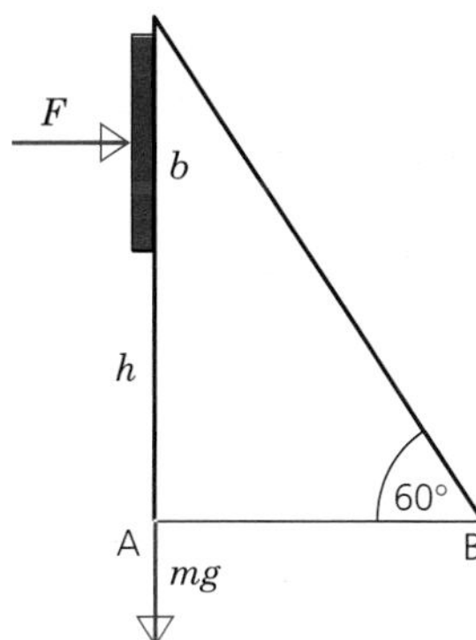

Show that the moment of this force about B has magnitude
$1161ab(h + \tfrac{1}{2}b)$
and that the moment of the force of gravity on the mass at A has magnitude

$$\frac{mg(h+b)}{\tan 60^\circ}$$

Verify that: $$m = \frac{1161ab\left(h+\frac{1}{2}b\right)\tan 60^\circ}{g(h+b)}$$

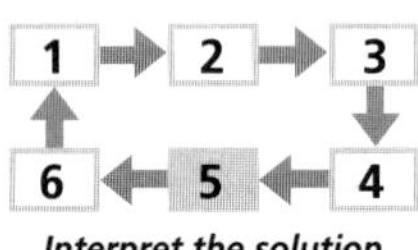

Interpret the solution

Interpretation

Now the required mass can be calculated for any rectangular sign. As an example assume a sign has $a = b = 2\,\mathrm{m}$ and $h = 1.5\,\mathrm{m}$. Then:

$$m = \frac{1161 \times 2 \times 2(1.5+1)\tan 60°}{9.8(1.5+2)} = 586 \text{ kg}$$

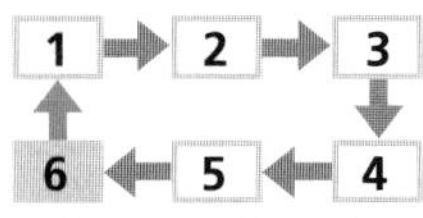

Compare with reality

Compare with reality

Often temporary road side signs are weighted with sand bags. Estimate the mass of a sand bag and comment on how this compares with the solution obtained above. (The density of sand is $2000\,\mathrm{kg\,m^{-3}}$.)

Refinement of the model

Clearly the sign could also blow over if the wind was in the opposite direction. Suggest how this could be prevented.

POSSIBLE MODELLING TASKS

1 Is it possible to design a road barrier that stays up when it is up and down when it is down? Investigate designs that might have this property.

2 Consider the ramp that is used for loading vehicles into a ferry. Investigate how the tension in the cables changes as the ramp is lifted.

3 An empty tipper lorry raises its back. Investigate how the forces on the back of the lorry vary as it is tipped. Consider also the forces acting on the lorry.

4 If a ladder is placed at a low angle against a wall, as shown, it will almost certainly slip. Investigate to see if there is a minimum safe angle that you could recommend to those who use ladders regularly.

Summary

After working through this chapter you should:

- be able to calculate the magnitude of the moment of a force using
  a) the magnitude force multiplied by the perpendicular distance, Fd
  b) the magnitude of the component of the force perpendicular to the defined distance, multiplied by the distance
- know that a positive moment indicates an anticlockwise turning effect and a negative moment a clockwise turning effect
- know that for equilibrium both the resultant force and the total moment must be zero
- know that a couple consists of two forces of which the resultant force is zero, but the combined moment is not zero.

MECHANICS

17 Centre of mass

- *The force of gravity on a body can be assumed to act at a point known as the centre of mass.*
- *The centre of mass of a moving object will follow the path that would be taken by a particle of the same mass.*

GRAVITY ACTING ON A BODY

In many of the situations that have been discussed so far in this book the force of gravity has been assumed to act at the centre of an object. If the object is to be modelled as a particle then this is very reasonable. In other cases it is reasonable to assume that the force of gravity acts at the centre of the object, for example a plank or a metal rod. But in yet other cases it is difficult to know where to consider the force of gravity to be acting, for example a tennis racquet, a model or toy bird (as shown in the diagram) or even a human body. In this chapter we explore this issue and meet the term **centre of mass**, which is used to describe the point in an object where the force of gravity can be assumed to act.

Exploration 17.1

Centre of mass of a rocking bird

Consider the rocking bird toy, shown in the diagram above. Assume that the only two forces acting on the bird are the force of **gravity** downwards and an upward **normal** reaction.

- Consider the forces on the bird when it is at rest. What information does this give you about the position of the centre of mass?
- If the centre of mass were above the level of the shelf, what would happen when the bird rocked?
- Explain why the centre of mass *must* be below the level of the shelf.

Forces in equilibrium

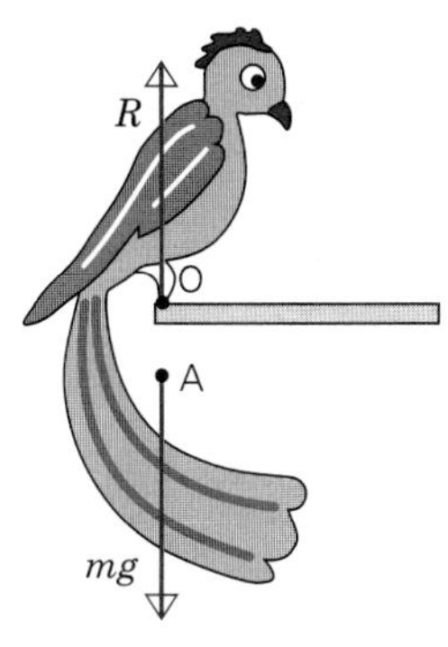

When the bird is at rest, both the resultant force and the total moment of the forces acting must be zero. For this condition to be satisfied the normal reaction and the force of gravity must act along the same straight line. As the normal reaction must act on the pin the force of gravity must act along a vertical line through the pin.

The diagram shows one possible position for the force of gravity to act. If it were to act as shown, the centre of mass would be at the point A. You can see that the centre of mass may not actually be on the bird itself.

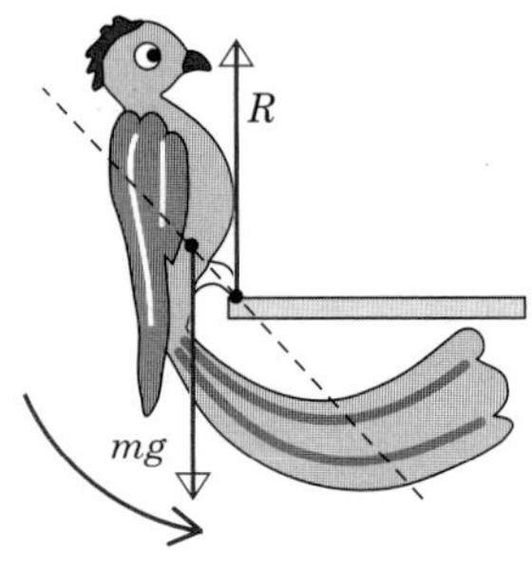

So the centre of mass must lie on the line shown, but whereabouts on this line? If the centre of mass is above the level of the shelf the forces will act as shown when the top of the bird is pushed to the left.

The effect of these forces would be to cause the bird to continue to turn anticlockwise until it falls off the shelf.

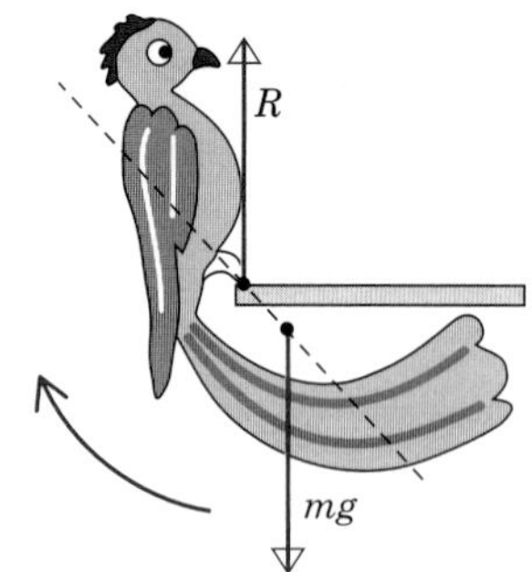

If the centre of mass is below the level of the shelf, then the forces would act as shown and produce a clockwise turning effect, so that the bird turns back towards its equilibrium position.

Exploration 17.2

The centre of mass of a triangle

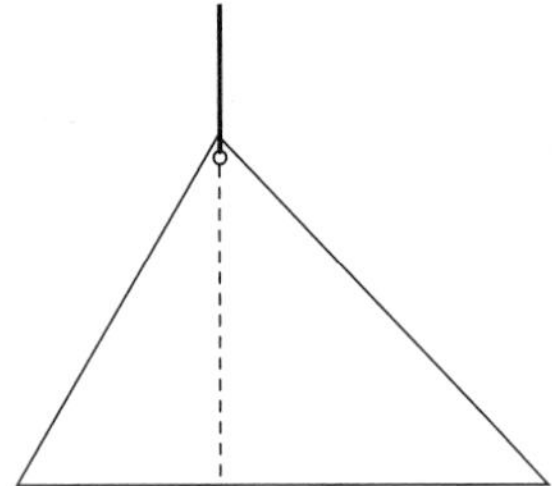

The aim of this activity is to find a general result that can be used to find the position of the centre of mass of any triangle.

- Cut out a triangle from a piece of stiff card. Make a hole in one corner and hang the triangle from a length of string. When the triangle is at rest draw a line that is an extension of the string, as shown in the diagram. (It may help to use a length of string and a mass to form a plumb line.)
- Why must the centre of mass of the triangle lie on the line that you have drawn?
- Suspend the card from each of the other two corners. By drawing vertical lines, find the position of the centre of mass.
- Repeat for some different triangles. Try to find a general rule to find the centre of mass of any triangle. (**Hint:** Measure the **height** of the triangle and the distance of the centre of mass from the **base**.)

The position of the centre of mass

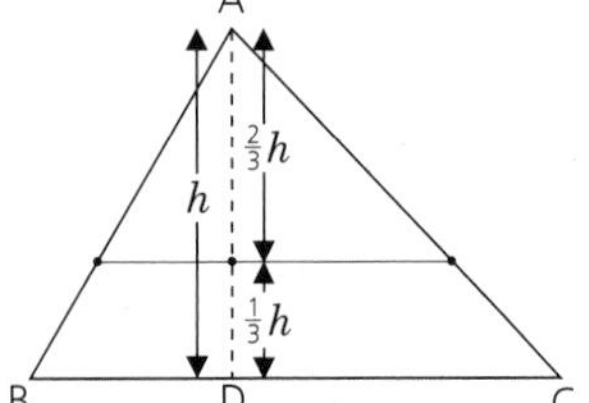

The centre of mass of a triangle is always at a distance $\frac{1}{3}$ of its height from its base. For the triangle ABC shown the height is h and so the centre of mass is a distance $\frac{1}{3}h$ from the base BC, so it lies on the line EF.

The process can be repeated, using each side as the base to find the position of the centre of mass.

In this diagram, if AB is taken as the base the height is 9 cm and so the centre of mass is 3 cm from AB. If BC is taken as the base the height is 15 cm and so the centre of mass is 5 cm from BC. This allows the centre of mass to be pin-pointed, as shown in the diagram.

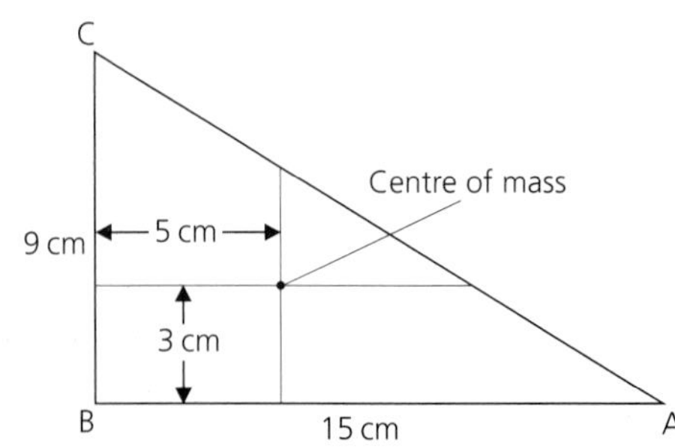

Systems of particles

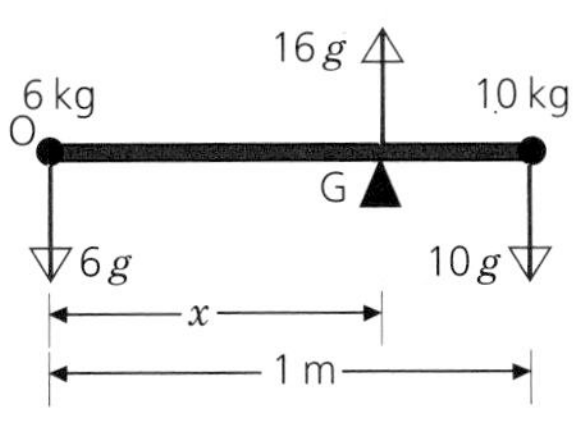

Imagine a light rod with masses fixed at each end. Somewhere along the length of that rod there is a point where the rod would balance as shown in the diagram. This point where the rod balances is its centre of mass.

An upward force of 16g N must act at G. Assuming that the centre of mass is at distance x from O, and taking moments about O gives:
$16gx = 6g \times 0 - 10g \times 1$

This simplifies to $16x = 10 \Rightarrow x = \frac{5}{8} = 0.625$ m

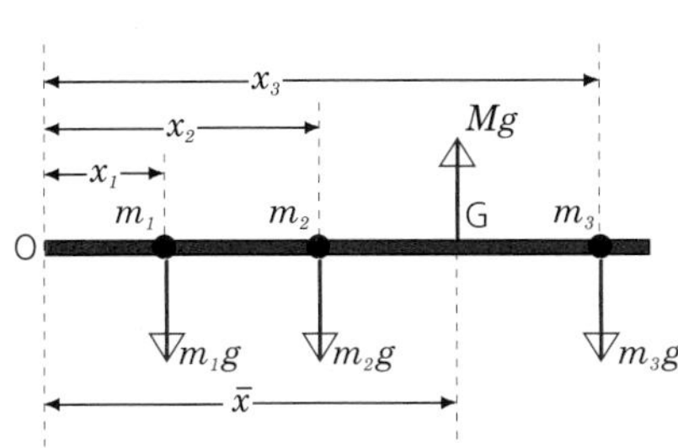

This approach can be extended to give a general formula for the centre of mass.

The rod shown has masses m_1, m_2, ... at distances x_1, x_2, ... from O. If gravity acts then the forces shown act on each mass and they are balanced by an upward force at G of magnitude:

$$m_1g + m_2g + \ldots = Mg$$

acting at the centre of mass, where M is the total mass. If the distance of the centre of mass from O is $\bar{x}$, then taking moments about O gives $Mg\bar{x} = m_1gx_1 + m_2gx_2 + \ldots$

which simplifies to $M\bar{x} = m_1x_1 + m_2x_2 + \ldots$

$$\Rightarrow \bar{x} = \frac{1}{M}\sum_{i=1}^{n} m_ix_i$$

We shall use this result in the following examples.

Example 17.1

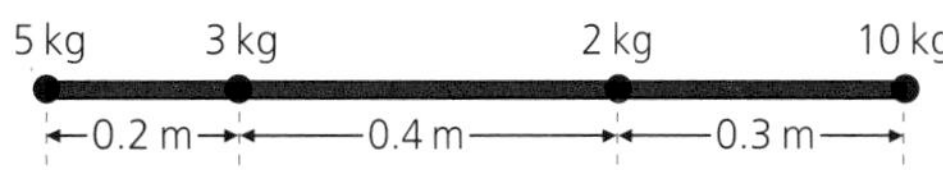

Four masses are attached to a rod as shown. Find the centre of mass of the system.

Solution

By measuring distances from the left-hand end of the rod, and using the result $\bar{x} = \frac{1}{M}\sum_{i=1}^{n} m_ix_i$ *gives:*

$$\bar{x} = \frac{1}{20}(5\times0 + 3\times0.2 + 2\times0.6 + 10\times0.9) = \frac{1}{20}\times(0.6 + 1.2 + 9) = \frac{10.8}{20} = 0.54m$$

So the centre of mass is 0.54 m from the left-hand end of the rod.

Frameworks

The diagram shows a light square framework with masses fixed to each corner.

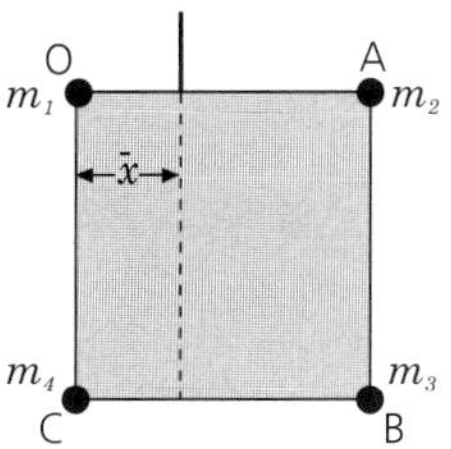

It is possible to attach a string to a point on the side OA so that the square hangs vertically. The centre of mass then lies on the dotted line shown at a distance $\bar{x}$ from OC.

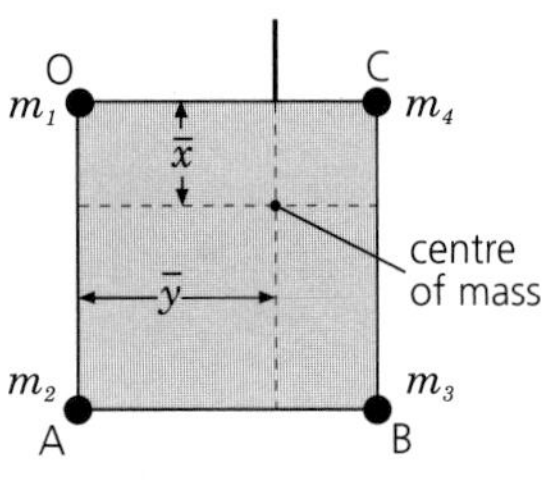

The value of $\bar{x}$ can be found by using the general formula:

$$\bar{x} = \frac{1}{M}\sum_{i=1}^{n} m_i x_i$$

where x_i is the distance of each particle from OC.

The framework could be hung from OC as shown, with the centre of mass at the point of intersection of the two dotted lines. If the distance of each mass from OC is y_i then:

$$\bar{y} = \frac{1}{M}\sum_{i=1}^{n} m_i y_i$$

If unit vectors are introduced, so that each particle has its position given by $\mathbf{r}_i = x_i\mathbf{i} + y_i\mathbf{j}$ then the position of the centre of mass is:

$$\bar{\mathbf{r}} = \bar{x}\mathbf{i} + \bar{y}\mathbf{j} = \left(\frac{1}{M}\sum_{i=1}^{n} m_i x_i\right)\mathbf{i} + \left(\frac{1}{M}\sum_{i=1}^{n} m_i y_i\right)\mathbf{j}$$

Example 17.2

The diagram shows a light rectangular framework with masses fixed at each corner.

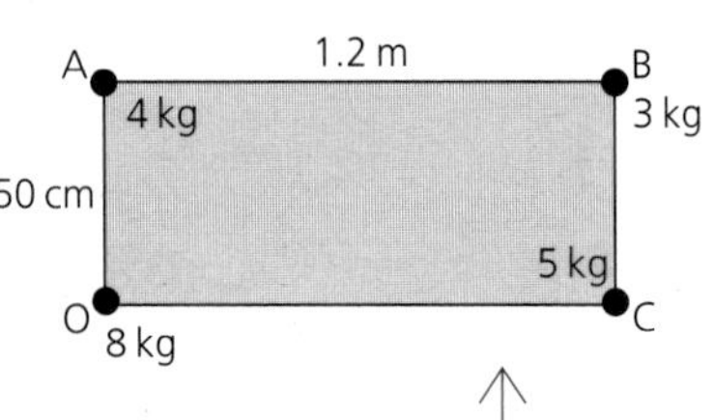

*a) Find the centre of mass of the system in terms of the unit vectors **i** and **j**.*

b) If the framework is suspended from A, find the angle the side OA makes with the vertical.

Solution

a) First note the position vector of each mass:

$\mathbf{r}_O = 0\mathbf{i} + 0\mathbf{j}$ $\quad m_O = 8$

$\mathbf{r}_A = 0\mathbf{i} + 0.5\mathbf{j}$ $\quad m_A = 4$

$\mathbf{r}_B = 1.2\mathbf{i} + 0.5\mathbf{j}$ $\quad m_B = 3$

$\mathbf{r}_C = 1.2\mathbf{i} + 0\mathbf{j}$ $\quad m_C = 5$

Using the result: $\bar{\mathbf{r}} = \left(\frac{1}{M}\sum_{i=1}^{n} m_i x_i\right)\mathbf{i} + \left(\frac{1}{M}\sum_{i=1}^{n} m_i y_i\right)\mathbf{j}$

and noting that $M = 20$ *gives:*

$$\mathbf{r} = \tfrac{1}{20}(8 \times 0 + 4 \times 0 + 3 \times 1.2 + 5 \times 1.2)\mathbf{i}$$
$$+ \tfrac{1}{20}(8 \times 0 + 4 \times 0.5 + 3 \times 0.5 + 5 \times 0)\mathbf{j}$$
$$= \tfrac{9.6}{20}\mathbf{i} + \tfrac{3.5}{20}\mathbf{j} = 0.48\mathbf{i} + 0.175\mathbf{j}$$

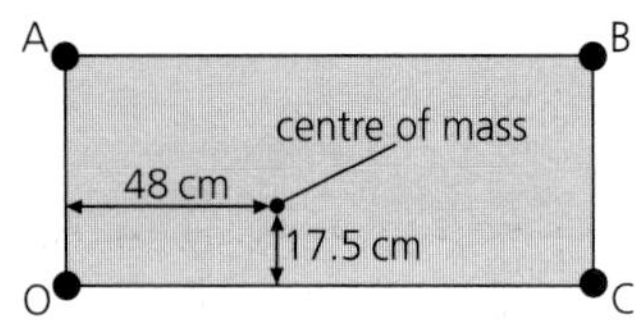

So the position of the centre of mass is as shown in the diagram on the left.

b) When the framework is suspended from A it will hang so that the centre of mass is directly below A as shown in the diagram to the right.

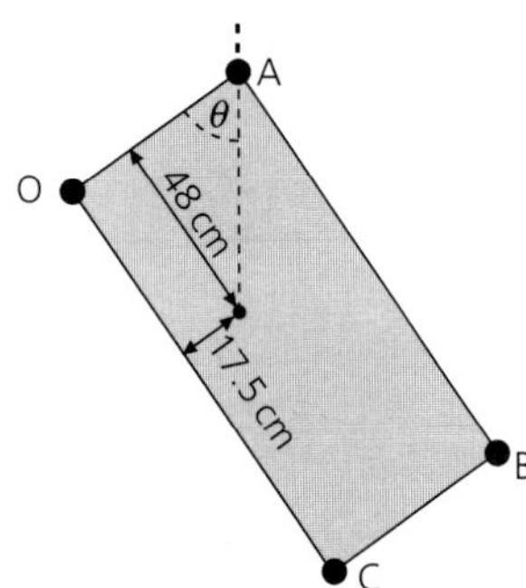

We need to find the angle marked θ. Using trigonometry gives:

$$\tan\theta = \frac{48}{50-17.5} = 1.477 \Rightarrow \theta = 56°$$

Example 17.3

The diagram shows a light triangular plate with bolts of mass 100 grams fixed at two corners. Find the mass of the third bolt if the centre of mass is on the dotted line, then find the position vector of the centre of mass.

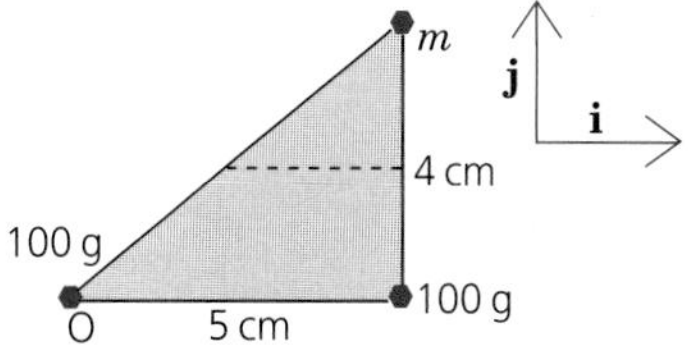

Solution

Taking O as the origin and using the unit vectors $\mathbf{i}$ *and* $\mathbf{j}$ *as defined in the diagram, the centre of mass has position vector* $\bar{\mathbf{r}} = \bar{x}\mathbf{i} + 0.02\mathbf{j}$

Now using $\bar{y} = \frac{1}{M}\sum_{i=1}^{n} m_i y_i$ *gives:*

$$0.02 = \frac{1}{0.2+m} \times 0.04m \Rightarrow 0.004 + 0.02m = 0.04m \Rightarrow 0.004 = 0.02m \Rightarrow m = 0.2\,kg$$

So the mass is 0.2 kg or 200 grams.
The value of $\bar{x}$ *can now be calculated using:*

$$\bar{x} = \frac{1}{M}\sum_{i=1}^{n} m_i x_i = \frac{1}{0.4}(0.1\times 0 + 0.1\times 0.05 + 0.2\times 0.05) = \frac{0.015}{0.4} = 0.0375$$

So the position vector of the centre of mass is $\bar{\mathbf{r}} = 0.0375\mathbf{i} + 0.02\mathbf{j}$.

EXERCISES

17.1A

1 Light rectangular frameworks, as in the diagram, have masses fixed at each corner. Find the position of the centre of mass, for each example below, in terms of the unit vectors $\mathbf{i}$ and $\mathbf{j}$ with reference to the corner marked O.

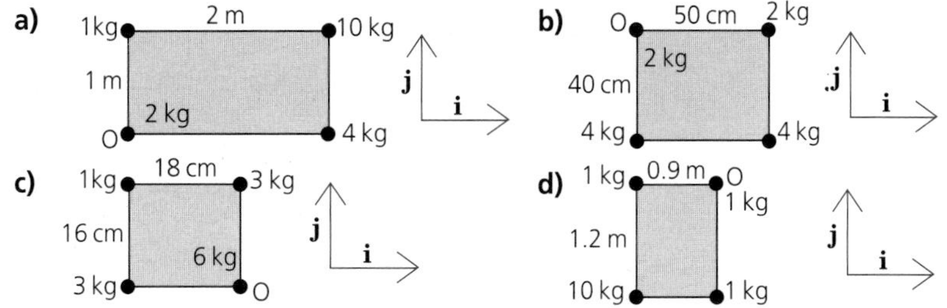

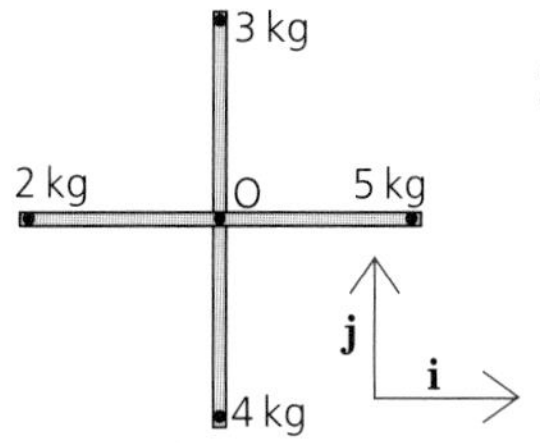

2 The diagram shows a light cross-shaped frame that is free to rotate about its centre, O. Four masses are fixed to the framework as shown, each 30 cm from O.

a) Find the position of the centre of mass when the framework is in the position shown.

b) Describe the path of the centre of mass when the framework rotates.

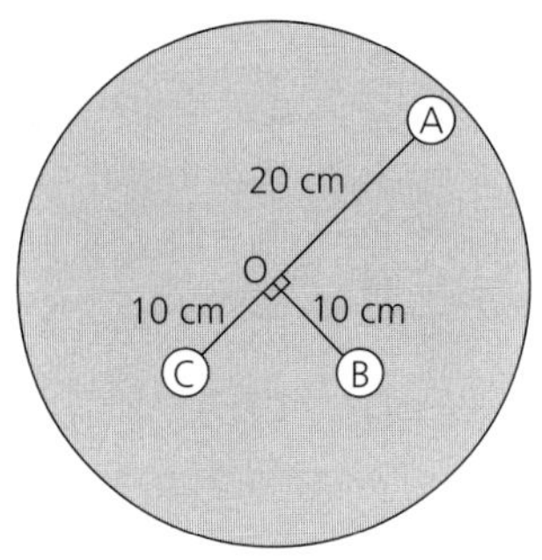

3 A light disc has three masses fixed to it as shown in the diagram. The mass A is 50 grams, B is 100 grams and C is 80 grams.

a) Find the distance of the centre of mass from O.

b) If a 100 gram mass is attached to the disc at O, what happens to the centre of mass?

4 The diagram shows a light rod to which two masses have been attached.

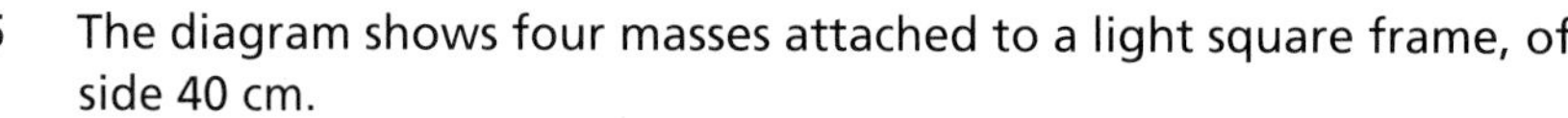

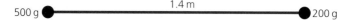

a) Find the position of the centre of mass of the rod and masses.

b) The 500 gram mass is replaced by a 300 gram mass. What mass must be attached to the other end if the centre of mass is to remain in the same place?

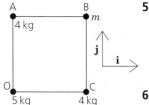

5 The diagram shows four masses attached to a light square frame, of side 40 cm.

a) Find the position of the centre of mass of the frame in terms of m, $\mathbf{i}$ and $\mathbf{j}$.

b) Find m if the centre of mass is to be at a distance of 20 cm from O.

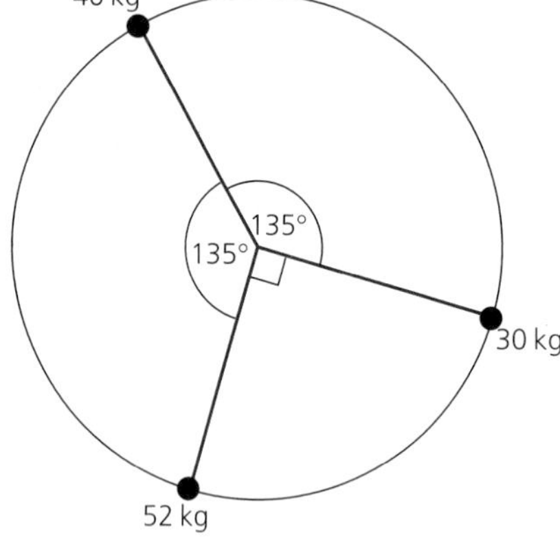

6 A roundabout in a children's playground has radius 1.6 m. Three children sit as shown in the diagram.

a) Find the distance of the centre of mass from the centre, assuming that the mass of the roundabout is zero.

b) Describe how the position of the centre of mass would change if the mass of the roundabout were to be considered.

c) Suggest how the mass of the roundabout could be considered in this problem.

d) If the roundabout is represented as a point of mass 200 kg at its centre, find the position of the centre of mass of the roundabout and the children.

7 A man of mass 68 kg climbs a ladder of mass 20 kg, inclined at 60° to the horizontal, and of length 8 m. Model the man as a point of mass and the ladder as a light rod with a point mass at its centre.

a) Find the highest and lowest positions of the centre of mass while the man is on the ladder.

b) Find the position of the man, when the centre of mass is $\frac{1}{4}$ of the way up the ladder.

8 A light triangular frame has masses attached to it as shown in the diagram.
 a) Find M and m, if the centre of mass is at G in the position shown.
 b) Is it possible for the centre of mass to lie on the hypotenuse of the triangle? Explain why.

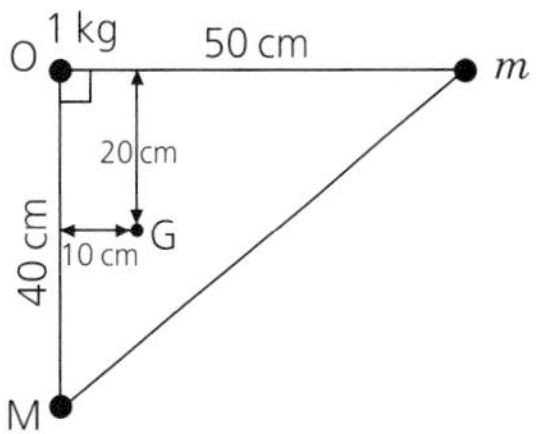

EXERCISES

17.1 B

1 For each of the following light frameworks, find the position of the centre of mass in terms of the unit vectors **i** and **j** with reference to the corner marked O.

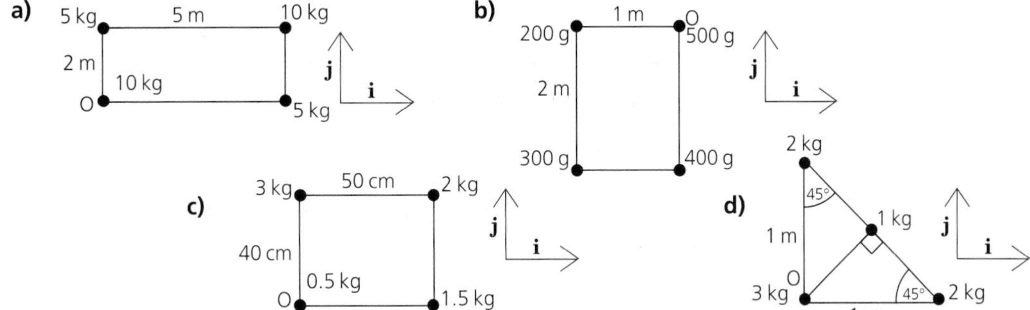

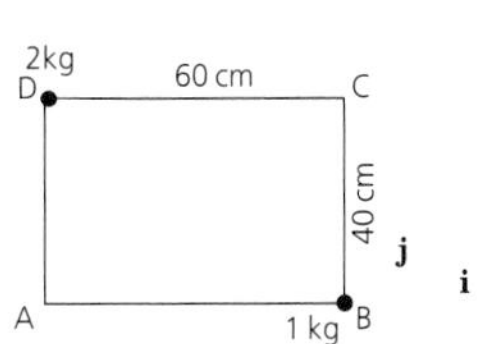

2 A light sheet of metal ABCD has masses attached to it as shown in the diagram.
 a) Find the position of the centre of mass relative to A.
 b) The sheet is suspended from A. What angle does the side AB make with the vertical?

3 A disc of mass 100 grams has three masses fixed to it as shown.
 a) Find the distance of the centre of mass from O.
 b) Find the distance of the centre of mass from O if the mass of the disc is negligible.

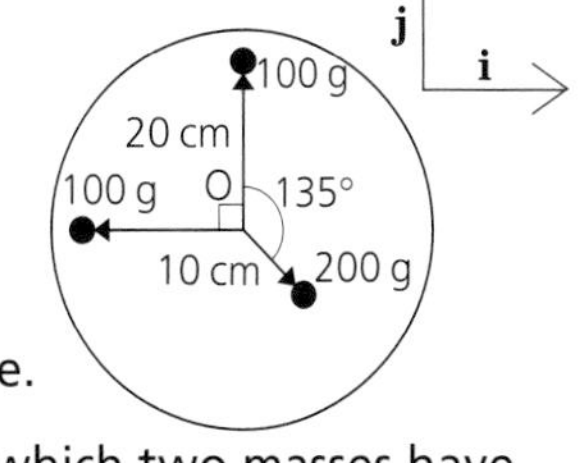

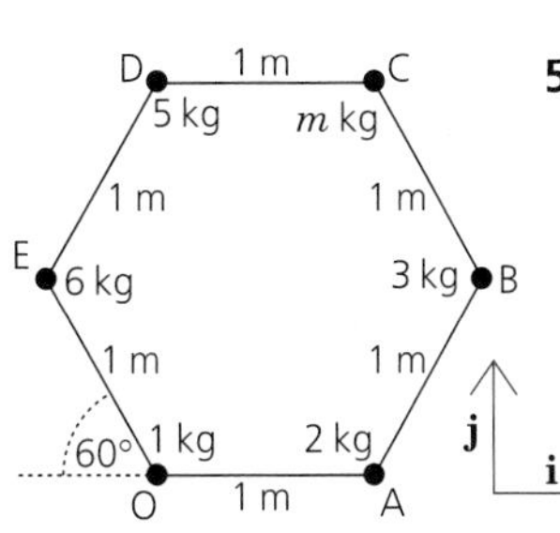

4 The diagram shows a light, L-shaped rod to which two masses have been attached.
 a) Find the position of the centre of mass of the rod and masses.
 b) The 3 kg mass is replaced by a 2 kg mass. What mass should be attached to the other end of the rod if the centre of mass is to stay in the same place?

O 0.2 m 3 kg; 0.2 m; 2 kg; i; j

5 Six masses are attached to a light regular hexagonal plate, as shown in the diagram.
 a) Find the position of the centre of mass of the plate and the masses, in terms of m, **i** and **j** relative to O.
 b) Find m if the position of the centre of mass is $0.4\mathbf{i} + \alpha\mathbf{j}$.
 c) For this value of m, find the value of α.

6 A roundabout in a children's playground has radius 2 m. Four children sit on the roundabout as shown in the diagram.

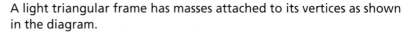

a) Find the distance of the centre of mass from the centre of the roundabout, assuming that the mass of the roundabout is zero.

b) Describe how the position of the centre of mass would change if the mass of the roundabout were to be considered.

c) Is it reasonable to assume that the mass of the roundabout is zero in this case?

7 In order to find the mass of a plank AB of length 4 m, a builder attaches a 10 kg mass to the end, A, of the plank. The plank balances when it is horizontal and resting on a metal bar that is 1 m from A. Find the mass of the plank.

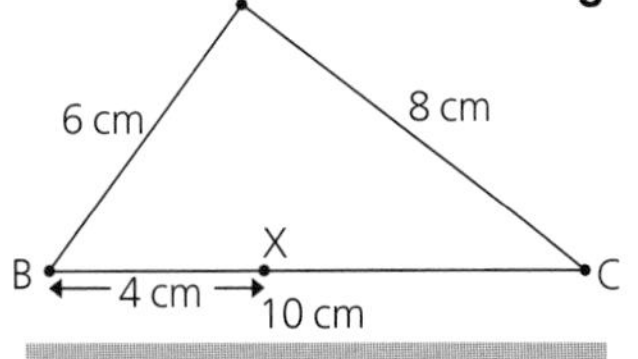

8 A light triangular frame has masses attached to its vertices as shown in the diagram.

a) What must be the mass at A if the centre of mass is to be at X?

b) If the masses at each vertices A, B and C are m, find the centre of mass of the framework.

CENTRES OF MASS FOR COMPOSITE BODIES

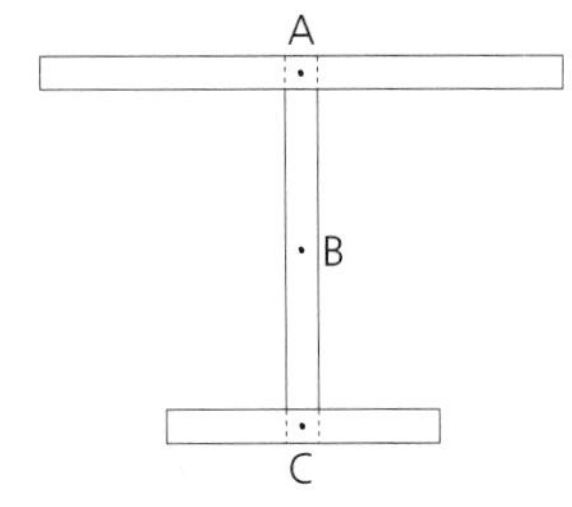

In reality very few things actually consist of systems of particles, but many can be **modelled** as systems of particles, so we can find their centre of mass. For example a table that consists of a top, a base and a pole can be modelled as three particles as shown in the diagram. Thus a composite body can be treated as a collection of particles which can then be considered individually.

The centre of mass can be found by considering particles at A, B and C. The particle at A has the same mass as the table top. The particle at B has the same mass as the pole and the particle at C has the same mass as the base. We assume that each particle is placed at the centre of mass of the component of the body that it represents.

Example 17.4

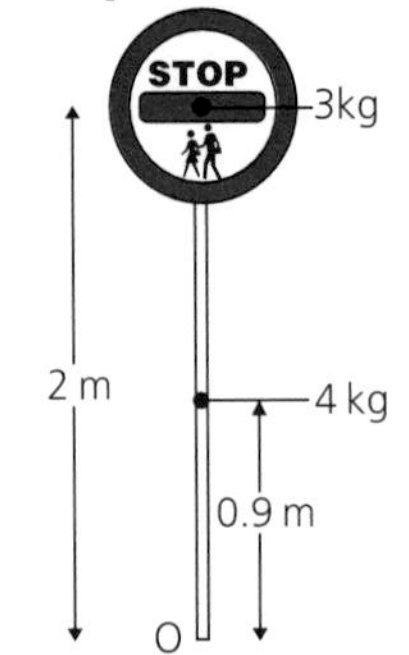

A 'lollipop stick' used by a person on a school crossing is made up of a pole of mass 4 kg and length 1.8 m, and a disc of mass 3 kg and radius 0.2 m. Find the height from the base of the stick to the centre of mass of the lollipop stick.

Solution

Assume that there is no overlap of the pole at the disc. The diagram shows the two components and their centres of mass. If the distance of the centre of mass from O is $\bar{x}$, then:

$$\bar{x} = \frac{2\times3+0.9\times4}{3+4} = 1.37\ m$$

Example 17.5

A parasol has three components and these can be modelled as a circular disc of mass 3 kg, fixed to a pole of mass 2 kg and length 2 m and inclined at 60° to the horizontal. The pole is fixed to a base of mass 12 kg. Find the position of the centre of mass of the complete parasol.

Solution

The diagram shows the parasol's three components, the origin O at the centre of the base and the unit vectors $\mathbf{i}$ *and* $\mathbf{j}$*, and three particles at A, B and C that can be used to represent each component.*

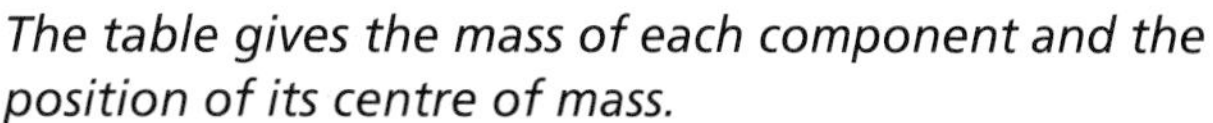

The table gives the mass of each component and the position of its centre of mass.

	Mass (kg)	Position
A	12	$0.05\mathbf{j}$
B	2	$\cos 60°\mathbf{i} + \cos 30°\mathbf{j}$
C	3	$2\cos 60°\mathbf{i} + 2\cos 30°\mathbf{j}$

So the position of the centre of mass, $\bar{\mathbf{r}}$*, is given by:*

$$\bar{\mathbf{r}} = \frac{12 \times 0.05\mathbf{j} + 2 \times (\cos 60°\mathbf{i} + \cos 30°\mathbf{j}) + 3(2\cos 60°\mathbf{i} + 2\cos 30°\mathbf{j})}{12 + 2 + 3}$$

$$= \frac{4\mathbf{i} + 7.53\mathbf{j}}{17} = 0.235\mathbf{i} + 0.443\mathbf{j}$$

Example 17.6

The diagram shows a hoist used for lifting engines out of cars. It is made from three metal bars with masses and lengths as shown. Assume that the width of the bars is negligible and find the position of the centre of mass relative to O in terms of the unit vectors $\mathbf{i}$ *and* $\mathbf{j}$*.*

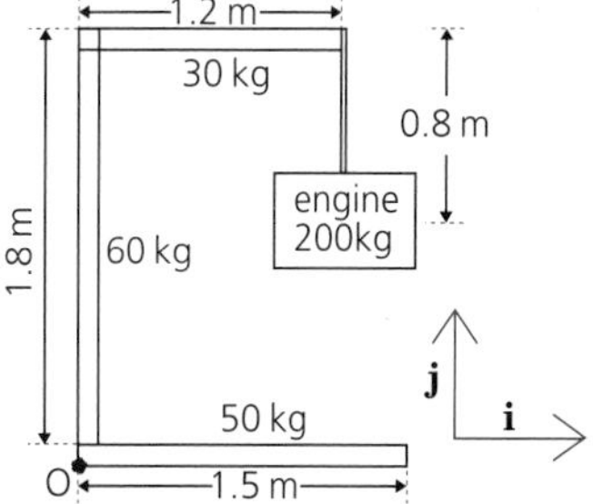

Solution

The table below gives the mass of each component and its centre of mass.

	Mass (kg)	Position of centre of mass
Bottom bar	50	0.75i
Upright bar	60	0.9j
Top bar	30	0.6i + 1.8j
Engine	200	1.2i + 1.0j

The position of the centre of mass, $\bar{\mathbf{r}}$*, is then given by:*

$$\bar{\mathbf{r}} = \frac{50 \times 0.75\mathbf{i} + 60 \times 0.9\mathbf{j} + 30(0.6\mathbf{i} + 1.8\mathbf{j}) + 200(1.2\mathbf{i} + 1.0\mathbf{j})}{50 + 60 + 30 + 200}$$

$$= \frac{295.5\mathbf{i} + 308\mathbf{j}}{340} = 0.87\mathbf{i} + 0.91\mathbf{j}$$

EXERCISES

17.2 A

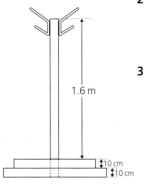

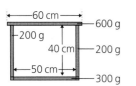

1 A wheel has mass 0.7 kg and radius 25 cm. A 50 gram mass is attached to the edge of the wheel. How far does the centre of mass move from the centre of the wheel when the mass is added?

2 The diagram on the left shows a coat stand. The base is made of two sections, the lower one with a mass of 5 kg and the upper one with a mass of 2 kg. The mass of the upright is 4 kg. Find the height of the centre of mass of the coat stand.

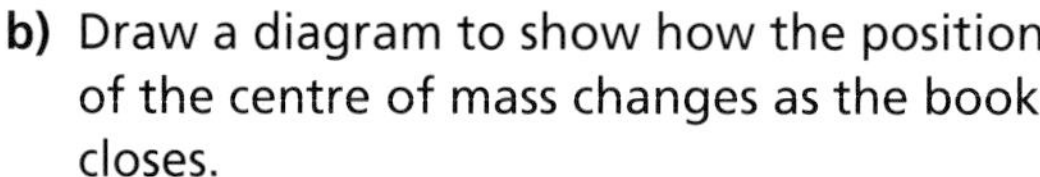

3 The diagrams on the right show a book modelled as two hinged sections, each of mass 200 grams, as it is being closed.

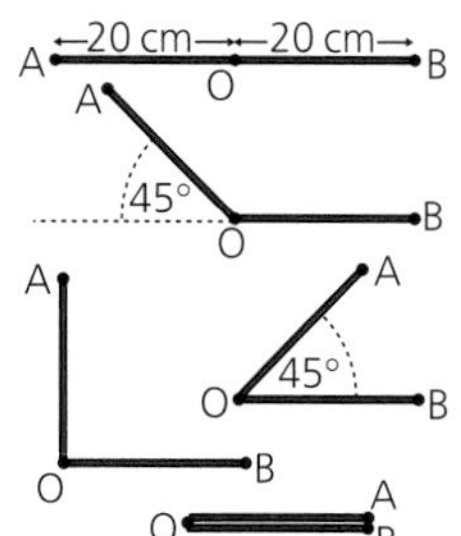

a) Find the position of the centre of mass for each case illustrated.

b) Draw a diagram to show how the position of the centre of mass changes as the book closes.

4 The diagram on the left shows a wooden drawer. Assuming the mass of the base is negligible, find the position of the centre of mass of the drawer.

5 A children's slide is made up of a metal ladder, of length 1.2 m and mass 8 kg, and a reinforced plastic slide of length 1.8 m and mass 6 kg. The two sections are joined as shown in the diagram.

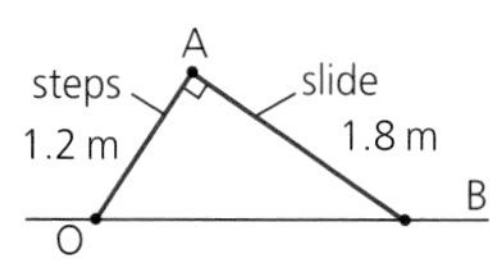

Find the position of the centre of mass of the slide if:

a) no one is using the slide,

b) a child of mass 20 kg sits at A.

State clearly any assumptions that you make.

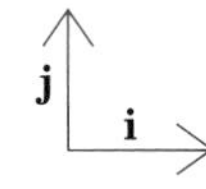

6 The diagram shows the cross-section of a table made up of four sections with dimensions and masses shown in the diagram.

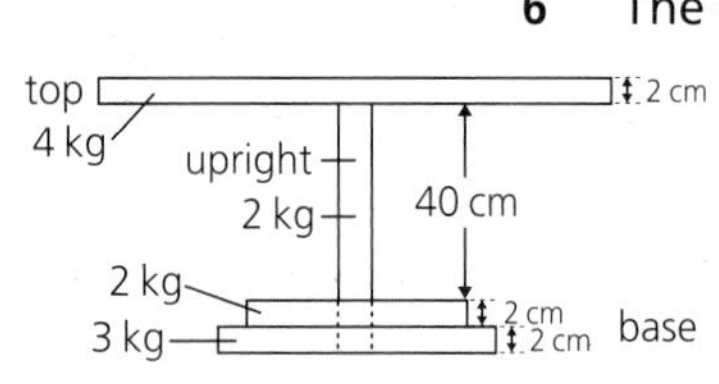

a) Find the height of the centre of mass.

b) A jug of water of mass 3 kg is paced on the table 30 cm from its centre. Find the new height of the centre of mass and its distance from the centre of the table.

7 The diagrams show the highest and lowest positions of a person doing press-ups. The person is modelled as a straight rod of length 1.5 m and mass 60 kg (labelled A), a sphere of mass 6 kg and radius 12 cm (labelled B) and a rod of mass 6 kg that varies in length between 20 and 45 cm (labelled C).

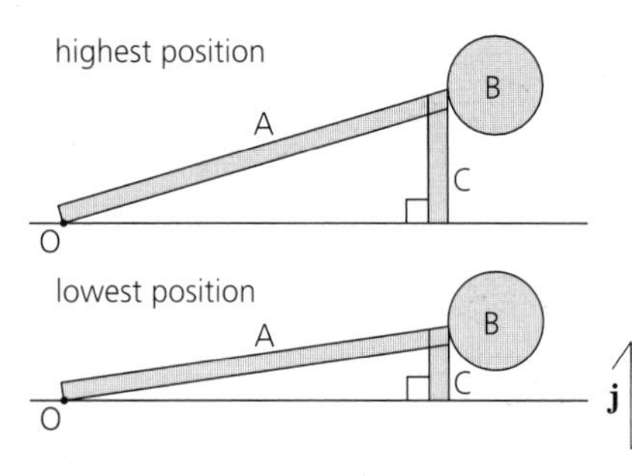

a) Find the position of the centre of mass in each position.

b) Find the height gained by the centre of mass.

c) Find the work done during the lifting stage of each press up.

d) Comment on the assumptions that the arms can be modelled as straight rods that are perpendicular to the ground.

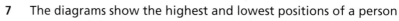

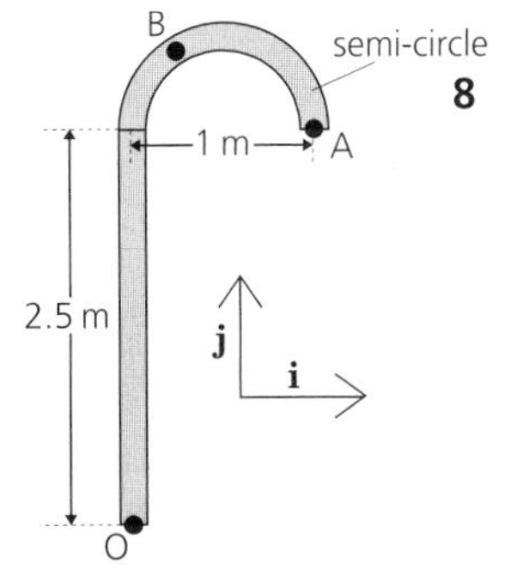

8 The diagram shows a design for a stand used in a city centre. Baskets are hung from A and a hook is fixed to the stands at B so they can be lifted by a small crane.

a) Find the distance of the centre of mass of the stand from a vertical line through O. Assume the stands are made from uniform tubes.

b) The local council wish to be able to remove and replace the stand easily so that the long straight section hangs vertically when the stands are lifted by a rope attached to the hook at B. Find the position of B relative to O in terms of **i** and **j**.

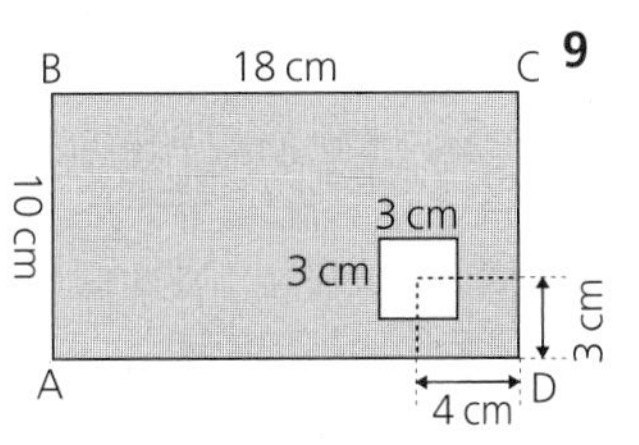

9 A steel plate has a square hole of side 3 cm cut in it.

a) If the mass of the original plate was 3.6 kg, find the mass lost when the hole was drilled.

b) The centre of mass is at a distance x from AB. By taking moments about AB show that the equation $1620 = 171x + 126$ holds and find the distance of the centre of mass from AB.

c) Find the distance of the centre of mass from BC.

d) The plate is suspended from the corner A by a rope. Find the angle between AB and the rope.

10 A drinks can of radius r and height H of mass 50 grams contains 250 grams of drink when full.

a) Find an expression for the position of the centre of mass of the can and drink, when the height of the drink is h.

b) Find the lowest possible position for the centre of mass.

EXERCISES

17.2 B

1 A wheel has mass 1.1 kg and radius 50 cm. A 100 gram mass is attached to one edge of the wheel and a 250 gram mass is attached diagonally opposite the first mass, at the edge of the wheel. Find the distance of the centre of mass from the centre of the wheel.

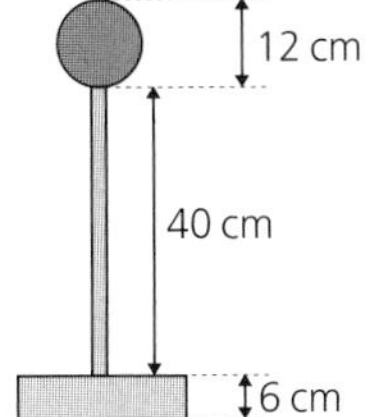

2 A lamp can be modelled as a sphere of mass 100 grams, a metal base of mass 2 kg and a vertical rod of mass 500 grams. Find the height of the centre of mass of the lamp if the dimensions are as shown in the diagram.

3 The diagram on the right shows a simple model of a crane.

a) Find the centre of mass relative to the point O, if the angle θ is:

i) 80° **ii)** 45° **iii)** 15°.

b) Draw a diagram to show how the position of the centre of mass changes as the jib of the crane is raised or lowered.

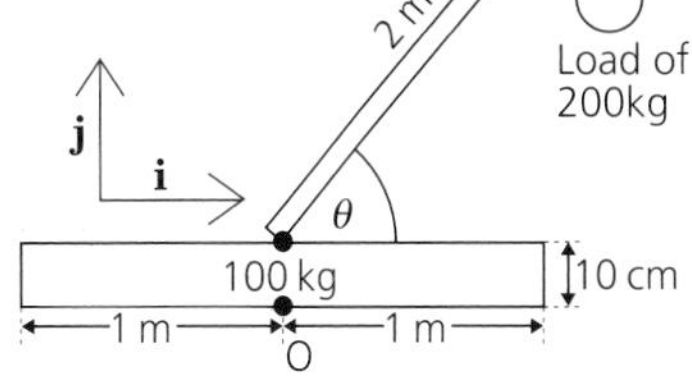

4 **a)** Find the position of the centre of mass of the uniform lamina shown in the diagram.

b) The lamina is suspended from the point A. Find the angle that the side AB makes with the vertical.

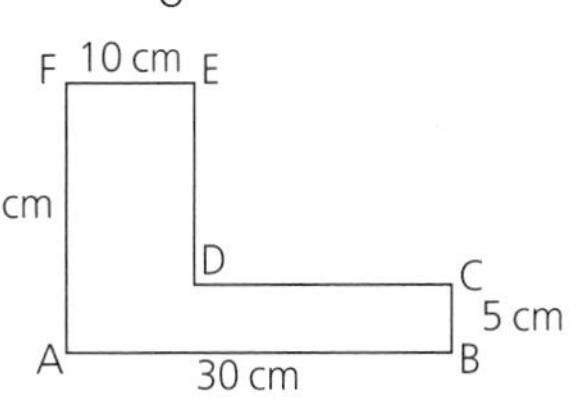

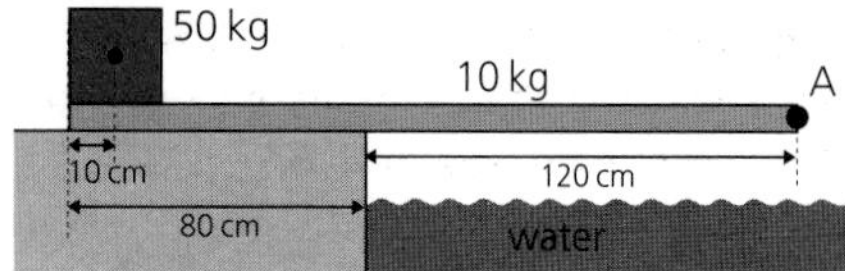

5 A makeshift diving board is made by fixing a concrete mass to the end of a plank, as shown in the diagram.
   a) Find the distance of the centre of mass from A.
   b) What is the largest vertical force that could be exerted on the diving board at A, if it does not tip?

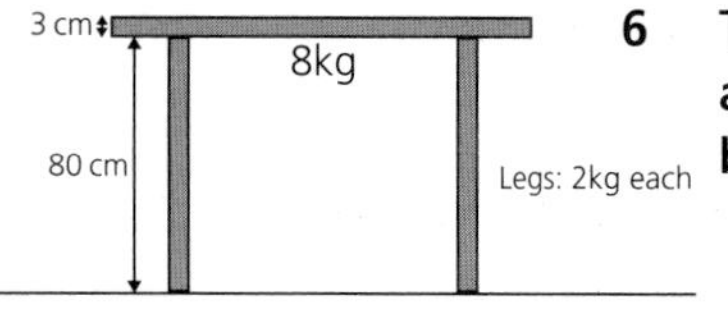

6 The diagram shows a side view of a table that has four legs and a top.
   a) Find the height of the centre of mass of the table.
   b) A small uniform box of mass 4 kg is placed on the table. Find the new height of the centre of mass of the table if the height of the box is: **i)** negligible, **ii)** 20 cm.

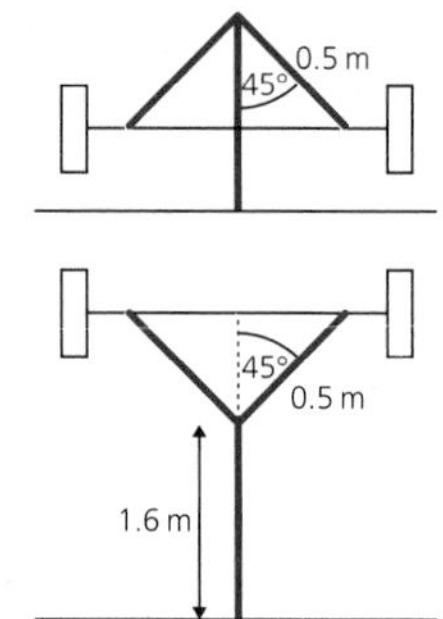

7 The diagrams show the highest and lowest positions of a weightlifter lifting a set of weights. The weightlifter is modelled as a straight rod of length 1.6 m and mass 80 kg. Assume the weightlifter's arms are light rods of length 0.5 m. The bar is assumed to be light and the masses at each end are 100 kg each. The diagrams show two positions of the weightlifter and weights.
   a) Find the height of the centre of mass for each position.
   b) Find the height gained by the centre of mass as it is lifted from the lower to the higher position.
   c) Find the work done during the lifting stage.

8 A circular metal plate has a radius of 20 cm. A hole of radius 10 cm is drilled in it as shown in the diagram. Find the angle that the diagonal AB makes with the vertical when the plate is suspended from A.

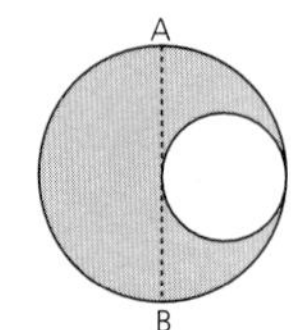

9 A tank has a mass of 1 kg and contains 9 kg of water. The tank is tipped until the water is on the point of flowing over the edge AE. Find the position of the centre of mass when the tank is in this position.

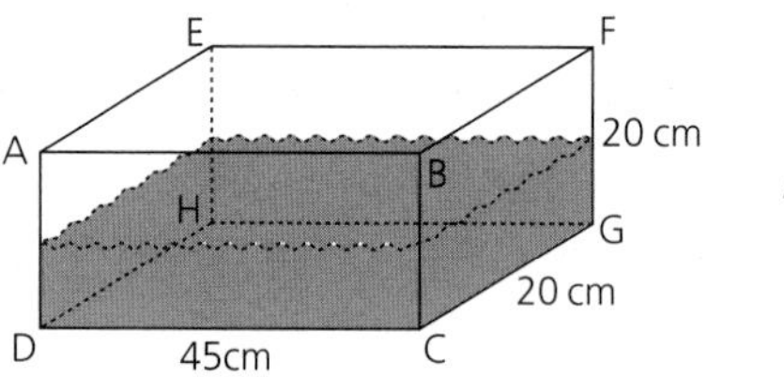

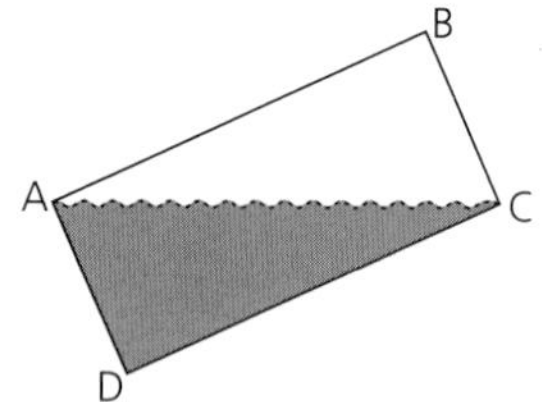

10 A cylindrical tank of mass 4kg, radius 40 cm and height 1.2 m is slowly filled with water at a rate of 250 cm^3 every second. The mass of 1000 cm^3 of water is 1 kg. Find the height of the centre of mass at time t seconds and sketch a graph to show how the height of the centre of mass varies with time.

FINDING THE POSITION OF THE CENTRE OF MASS BY INTEGRATION

In many cases the position of the centre of mass of a body can be found by modelling it as if it were made up of simpler shapes and using the approach for composite bodies. In many other cases, though, this is impossible. For example, where is the centre of mass of a semi-circular plate? In this section we shall develop an approach that can be used for objects that have some symmetry. First we shall consider the plate shown in the diagram, which has the x-axis as a line of symmetry.

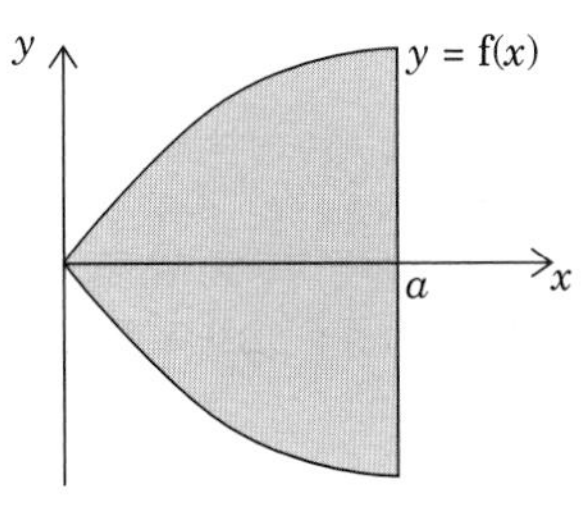

Assume that the plate is made from a material which has uniform density, giving it a mass of ρ per unit area, so that its mass is simply ρ multiplied by its area. Then the mass of the plate is given by:

$$m = 2\rho\int_0^a y\,\mathrm{d}x$$

Now imagine that the shape is split into a number of strips of width δx. One is shown in the diagram.

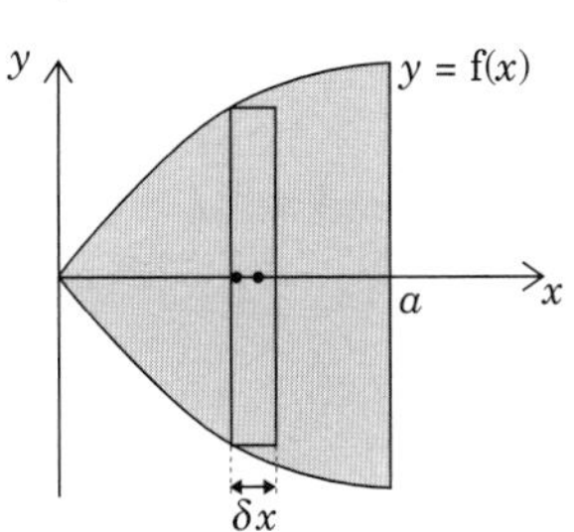

The mass of each strip is $2\rho y\delta x$, and its distance from O is simply x. If the whole shape were split into similar strips then the distance of the centre of mass from O would be given by:

$$\bar{x} = \frac{\sum(2\rho y\delta x)x}{m}$$

If the width of the strips, δx, is allowed to tend to zero, then the expression for $\bar{x}$ becomes: $\bar{x}\ \dfrac{\int_0^a 2\rho yx\,\mathrm{d}x}{m}$

and using the result obtained for m this gives: $\bar{x} = \dfrac{\int_0^a 2\rho yx\,\mathrm{d}x}{\int_0^a 2\rho y\,\mathrm{d}x} = \dfrac{\int_0^a yx\,\mathrm{d}x}{\int_0^a y\,\mathrm{d}x}$

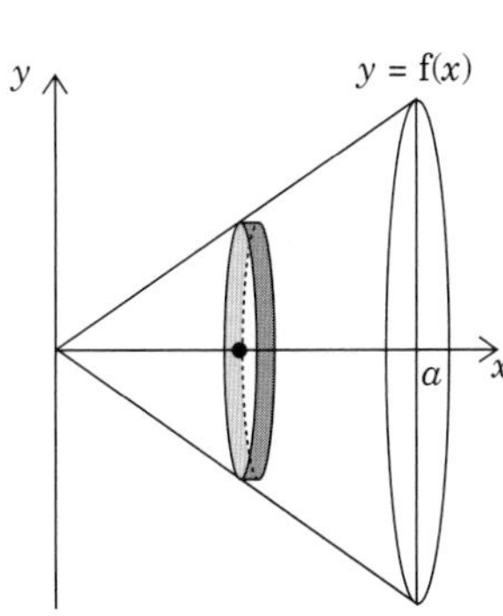

This result is very important. It can be used to find the centre of mass of any lamina that has a line of symmetry. A similar result can be obtained for solid objects that can be formed by rotating a curve around the x-axis, for example for a cone or a hemisphere.

The diagram shows how, when working with solid shapes, we must consider discs instead of the strips we used for flat plates or laminas. If the density of the material is ρ then the volume of each disc is $\pi\rho y^2\delta x$.

The mass of the solid is given by: $m = \int_0^a \rho\pi y^2\,\mathrm{d}x$

and the centre of mass is given by: $\bar{x} = \dfrac{\int_0^a \rho\pi y^2 x\,\mathrm{d}x}{\int_0^a \rho\pi y^2\,\mathrm{d}x} = \dfrac{\int_0^a y^2 x\,\mathrm{d}x}{\int_0^a y^2\,\mathrm{d}x}$

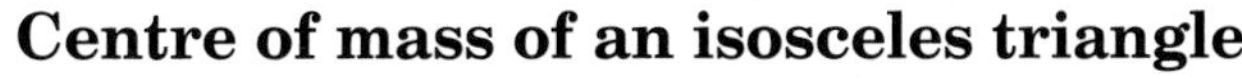

Centre of mass of an isosceles triangle

Example 17.7

Show that the distance of the centre of mass of an isosceles triangles of height h is $\frac{2}{3}h$ from its top.

Solution

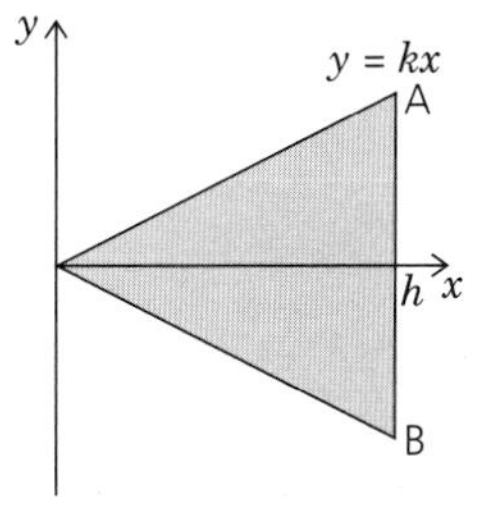

The diagram shows an isosceles triangle.
The sloping sides of the triangle are straight lines, so the line OA will be given by $f(x) = kx$. *Now the centre of mass can be found.*

$$\bar{x} = \frac{\int_0^h xy\,\mathrm{d}x}{\int_0^h y\,\mathrm{d}x} = \frac{\int_0^h kx^2\,\mathrm{d}x}{\int_0^h kx\,\mathrm{d}x} = \frac{\left[\frac{1}{3}kx^3\right]_0^h}{\left[\frac{1}{2}kx^2\right]_0^h} = \frac{\frac{1}{3}kh^3}{\frac{1}{2}kh^2} = \frac{2}{3}h$$

So the centre of mass is at a distance $\frac{2}{3}h$ from the top of the triangle.

Centre of mass of a cone

Example 17.8

The diagram shows a truncated cone of height 5 cm. Find the distance of the centre of mass from O.

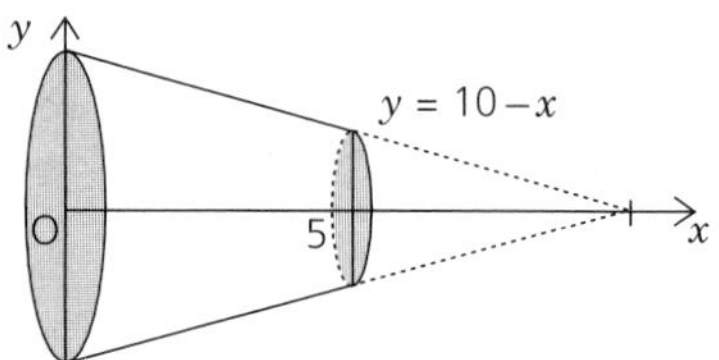

Solution

The centre of mass can be found using: $\bar{x} = \dfrac{\int_0^a y^2 x\,dx}{\int_0^a y^2\,dx}$

In this case $y = 10 - x$ *and* $a = 5$ *which gives:*

$$\bar{x} = \frac{\int_0^5 (10-x)^2 x\,dx}{\int_0^5 (10-x)^2\,dx} = \frac{\int_0^5 \left(100x - 20x^2 + x^3\right)dx}{\int_0^5 \left(100 - 20x + x^2\right)dx}$$

$$= \frac{\left[50x^2 - \frac{20}{3}x^3 + \frac{1}{4}x^4\right]_0^5}{\left[100x - 10x^2 + \frac{1}{3}x^3\right]_0^5} = \frac{50\times 5^2 - \frac{20}{3}\times 5^3 + \frac{1}{4}\times 5^4}{100\times 5 - 10\times 5^2 + \frac{1}{3}\times 5^3} = \frac{6875}{12}\times\frac{3}{875} = \frac{55}{28}$$

So the centre of mass is 1.96 cm from the base.

Centre of mass of a hemisphere

Example 17.9

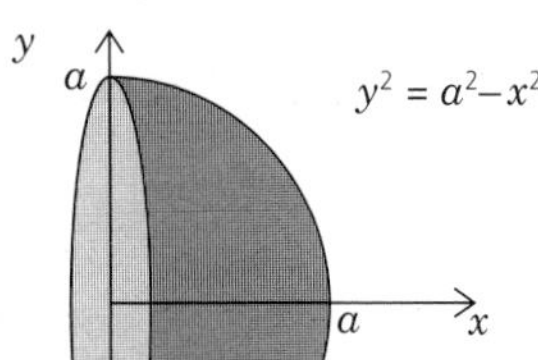

Find the position of the centre of mass of a solid hemisphere of radius a.

Solution

The diagram shows the hemisphere.

The equation of the circle of radius a *is* $x^2 + y^2 = a^2$ *or* $y^2 = a^2 - x^2$.

The mass of the hemisphere can be calculated without integration as:

$m = \rho \times \frac{1}{2} \times \frac{4}{3}\pi a^3 = \frac{2}{3}\rho\pi a^2$

The centre of mass can now be found using:

$$\bar{x} = \frac{\int_0^a \pi\rho y^2 x\,dx}{m} = \frac{\int_0^a \pi\rho\left(a^2 - x^2\right)^2 x\,dx}{\frac{2}{3}\pi\rho a^3} = \frac{3}{2a^3}\int_0^a \left(xa^2 - x^3\right)dx$$

$$= \frac{3}{2a^3}\left[\frac{a^2x^2}{2} - \frac{x^4}{4}\right]_0^a = \frac{3}{2a^3}\left(\frac{a^4}{2} - \frac{a^4}{4}\right) = \frac{3}{2a^3}\times\frac{a^4}{4} = \frac{3a}{8}$$

So the centre of mass is at a distance $\frac{3}{8}a$ *from the base of the hemisphere.*

EXERCISES

17.3A

1 Find the centre of mass of the trapezium shown in the diagram.

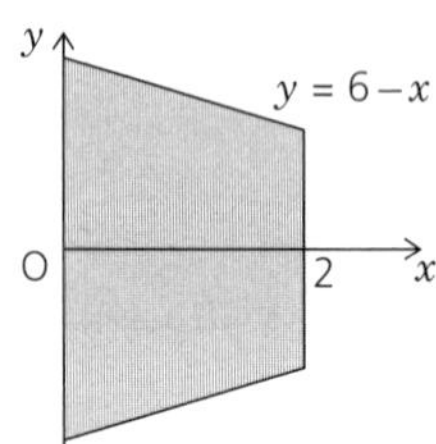

2 A region is bounded by the curves $y = x^2$, $y = -x^2$ and $x = 1$. A sheet of material has the same shape as this region. Find the position of the centre of mass of the shape.

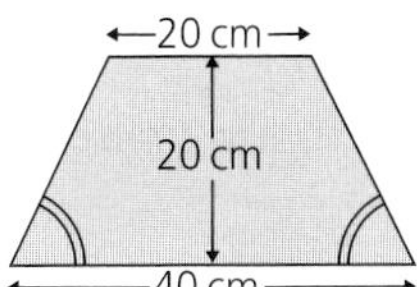

3 A steel plate is in the shape of a trapezium as shown in the diagram. Find the position of the centre of mass of the plate.

4 A lamina has the same shape as a region bounded by the curves $y = \sqrt{x}$, $y = -\sqrt{x}$ and $x = 4$. Find the position of the centre of mass of this shape.

5 **a)** Show that the centre of mass of a semicircular lamina of radius a is at a distance $\frac{4a}{3\pi}$ from the straight edge.

b) Find the centre of mass of the lamina shown in the diagram.

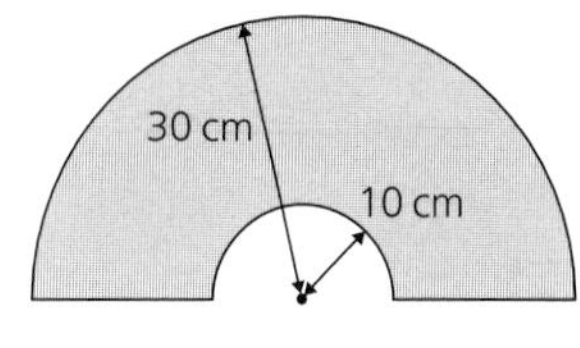

6 The diagram shows a cone that has been split to form a smaller cone and a solid known as a frustrum.

a) Find the distance of the centre of mass of the smaller cone from O.

b) Show that the centre of mass of the frustrum is $\frac{11}{56}h$ from the base of the frustrum.

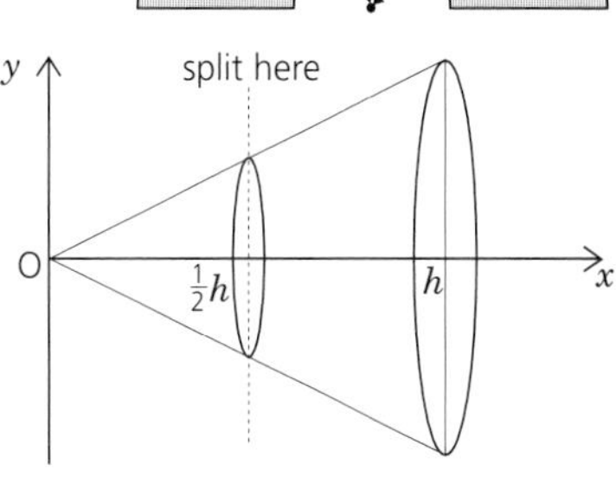

7 A solid object is formed by rotating the area under the curve $y = x^2$ around the x-axis. Find the position of the centre of mass of the solid if the curve is rotated for values of x:

a) between 0 and 2, **b)** between 1 and 2, **c)** between 0 and a.

8 A component for a pulley is formed by the solid enclosed when the curve $y = x^2 - 2x + 2$ and lines $x = 0$ and $x = 2$ are rotated about the x-axis to give the outline of the solid. Find the distance of the centre of mass from its base.

9 The diagram shows a hemisphere that has been cut into two parts. By integrating between the limits $\frac{1}{2}a$ and a find the centre of mass of the smaller part of the hemisphere.

10 The diagram shows part of cone that has had a hole drilled through its centre.

a) Find the centre of mass of the object before the hole is drilled.

b) State the position of the centre of mass of the cylinder of material that is removed during drilling.

c) Find the centre of mass of the drilled core.

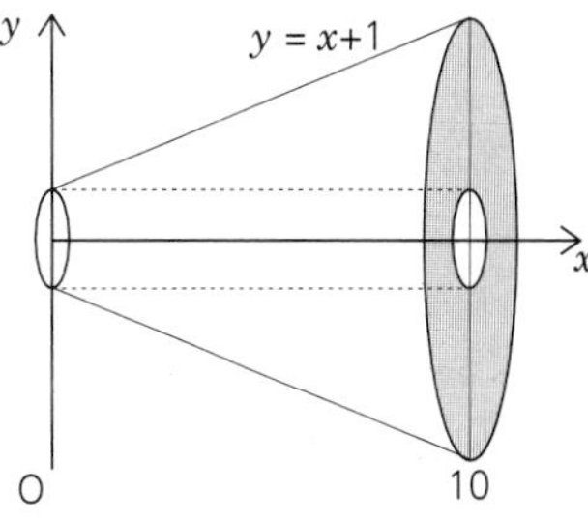

EXERCISES

17.3 B

1 Find the distance from O of the centre of mass of the trapezium shown in the diagram.

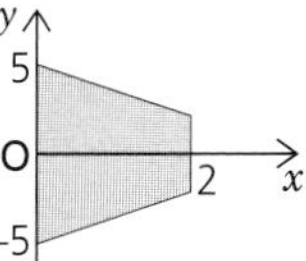

2 A region is bounded by curves $y = x^3$, $y = -x^3$ and the line $x = 1$. A sheet of material has the same shape as the region. Find the position of the centre of mass of the sheet of material.

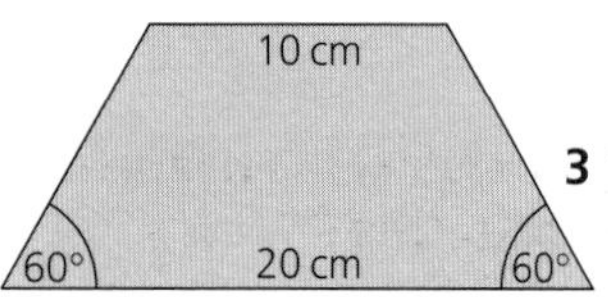

3 A steel plate is in the shape of a trapezium as shown in the diagram. Find the position of the centre of mass of the plate.

4 A lamina has the same shape as a region bounded by the curves $y = x + x^2$, $y = -(x + x^2)$ and $x = 2$.
 a) Find the distance of the centre of mass from the line $x = 0$.
 b) A similar shape is bounded by $y = x + x^2 + 2$, $y = 2 - (x + x^2)$ and $x = 2$. Find the coordinates of its centre of mass.

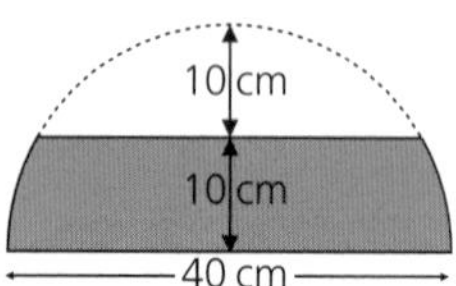

5 Find the centre of mass of the lamina shaded in the diagram. The lamina was formed by cutting off the top half of a semi-circle.

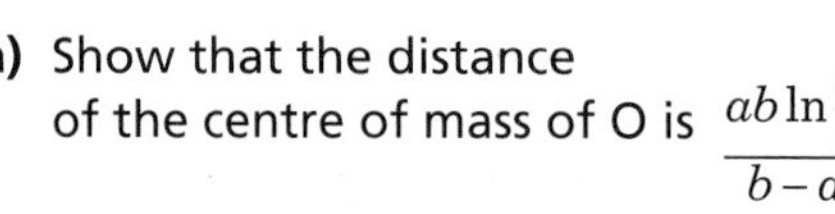

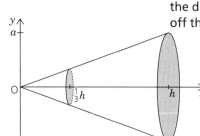

6 The diagram shows a cone that is to be split into two parts.
 a) Show that the centre of mass of the cone is $\frac{3}{4}h$ from O.
 b) Find the distance of the centre of mass of the smaller cone from O.
 c) Find the distance of the centre of mass of the frustrum from its base.

7 A solid object is found by rotating the area under the curve $y = \frac{1}{x}$ around the x-axis between the lines $x = a$ and $x = b$.

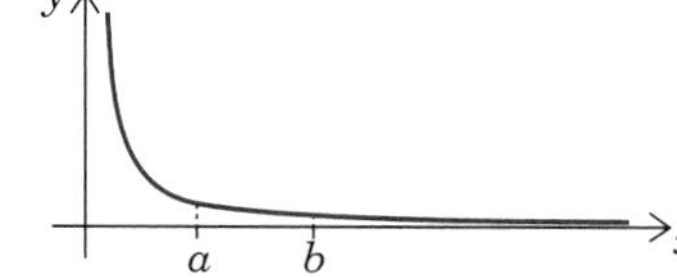

 a) Show that the distance of the centre of mass of O is $\dfrac{ab\ln\frac{a}{b}}{b-a}$
 b) Find the distance of the centre of mass from the line $x = a$.

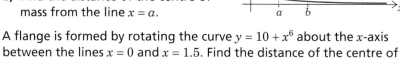

8 A flange is formed by rotating the curve $y = 10 + x^6$ about the x-axis between the lines $x = 0$ and $x = 1.5$. Find the distance of the centre of mass from the end of the flange with the smallest diameter.

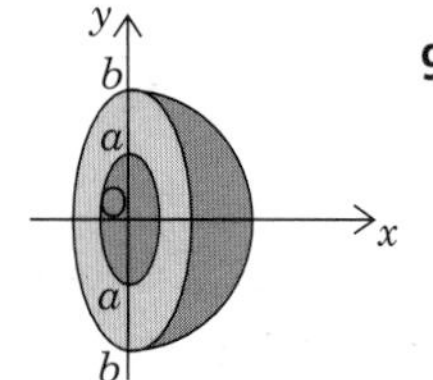

9 a) A hollow hemisphere is formed as shown in the diagram. Show that the centre of mass is at a distance $\bar{x}$ from O, where $\bar{x} = \dfrac{3(b^4 - a^4)}{8(b^3 - a^3)}$.
 b) Find the position of the centre of mass of a hemispherical shell of outer radius 10 cm and thickness 2 mm.

10 A hollow cone is formed as shown in the diagram. Find the distance of its centre of mass from O.

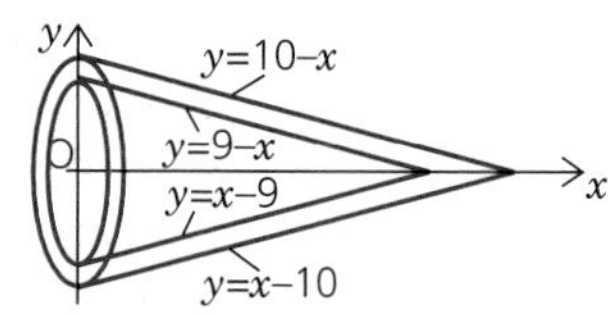

CONSOLIDATION EXERCISES FOR CHAPTER 17

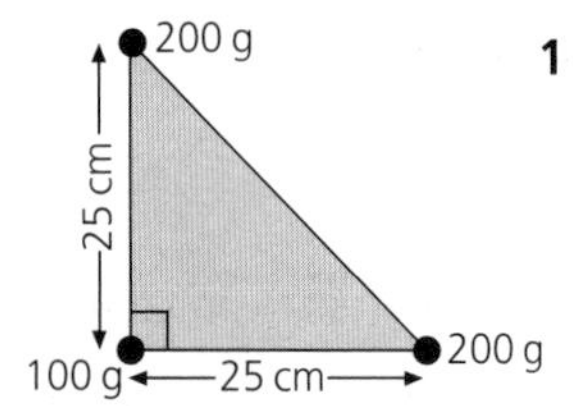

1 Three particles of masses 200 grams, 200 grams and 100 grams are connected by light rods. The rods and masses form a right-angled isosceles triangle with the 100 grams mass at the right angle. The rods of equal length are 25 cm. Find the position of the centre of mass of the system relative to the 100 grams mass.

(Nuffield Question 4, Mechanics 2 Specimen Paper, 1994)

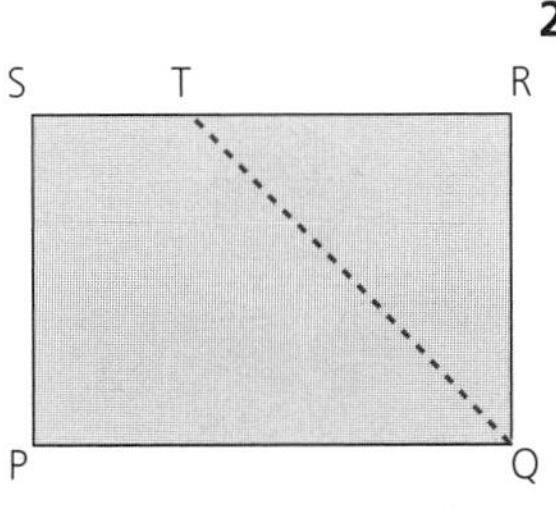

2 The diagram shows a rectangular sheet PQRS of uniform thin metal with PQ = 4 m and QR = 3 m. T is a point on RS such that RT = 3 m. The sheet is folded about the line QT until R lies on PQ.

a) Find the distances from PQ and PS of the centre of mass of the folded sheet.

The folded sheet is freely suspended from the point S and hangs in equilibrium.

b) Calculate the angle of inclination of the edge PS to the vertical.

(AEB Question 6, June 1994)

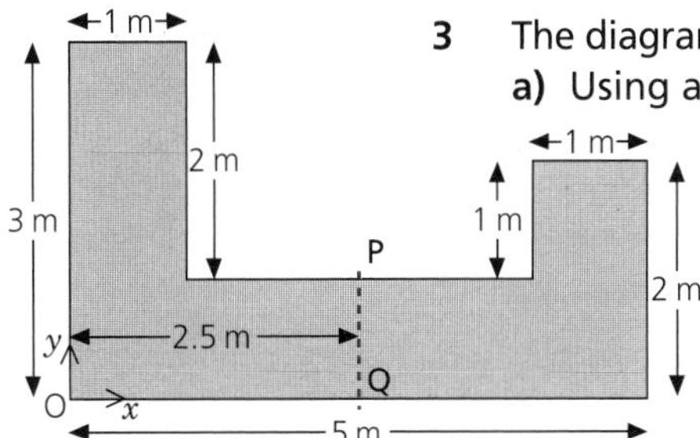

3 The diagram gives the dimensions of the design of a uniform metal plate.

a) Using a coordinate system with O as origin, the x- and y-axes as shown and 1 metre as 1 unit, show that the centre of mass has y-coordinate 1 and find its x-coordinate.

The design requires the plate to have its centre of mass halfway across (i.e. on the line PQ in the diagram), and in order to achieve this a circular hole centred on $(\frac{1}{2}, \frac{1}{2})$ is considered.

b) Find the appropriate radius for such a hole and hence explain why this idea is not feasible.

It is then decided to cut two circular holes of each radius r, both centred on the line $x = \frac{1}{2}$. The first hole is centred at $(\frac{1}{2}, \frac{1}{2})$ and the centre of mass of the plate is to be at P.

c) Find the value of r and the coordinates of the centre of the second hole.

(MEI Question 2, M2 Paper 8, June 1992)

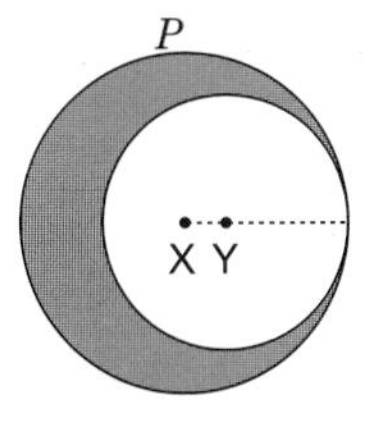

4 The diagram shows a lamina which consists of a heavy uniform circular disc, centre X and radius R, from which a circular hole, centre Y and radius r, has been cut, where $r < R$. The centre of mass of the lamina is at a distance $\frac{4}{9}r$ from X and $XY = R - r$. By taking moments about X, or otherwise, show that $R = \frac{5}{4}r$.

(ULEAC Question 4, M1 Specimen Paper, 1994)

5 A right circular cylindrical container has a horizontal base but no top. The container is made from uniform thin sheet metal. Each square centimetre (cm^2) of this metal has mass 0.002 kg. The base radius of the container is 10 cm and the height is 20 cm.

a) Show that the mass of the container is π kg.

b) Find, in cm, the distance of the centre of mass of the container from the base.

Molten metal is poured into the container to a depth of y cm, measured from the base. Each cubic centimetre (cm^3) of this molten metal has mass 0.003 kg.

c) Show that the centre of mass of the container and the molten metal is at a distance S cm from the base, where:

$$S = \frac{3y^2 + 160}{6y + 20}$$

d) Find the value of y for which S is stationary, giving your answer to one decimal place.

(ULEAC Question 2, Paper 3, January 1991)

MATHEMATICAL MODELLING ACTIVITY

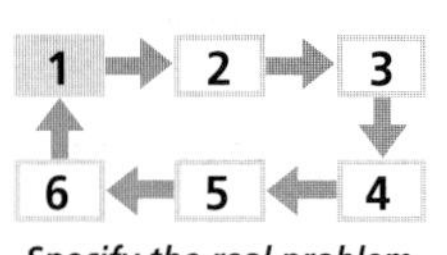

Specify the real problem

Problem statement

Athletes in events such as the pole vault and high jump often adopt positions so that their centre of mass is as far below their bodies as possible as they go over the bar. How far below their body is it reasonable to assume that their centre of mass can be?

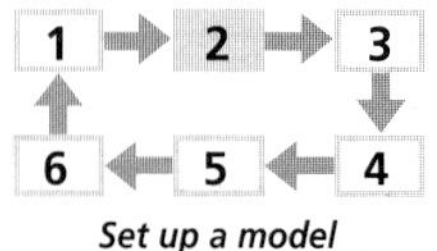

Set up a model

Set up a model

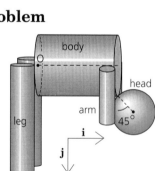

The solution to this problem will be based on a model of a human body as five cylinders and a sphere. The sphere represents the head and the body, the arms and legs are all modelled as cylinders.

The table below gives some data that suggest how the mass of the body is divided among these five components and the approximate dimensions of each component.

Component	Mass (%)	Length (m)	Radius (m)
Body	52	0.65	0.18
Arms	5	0.75	0.04
Legs	15	0.95	0.09
Head	8	–	0.10

In addition it will be assumed that the body joints are completely flexible.

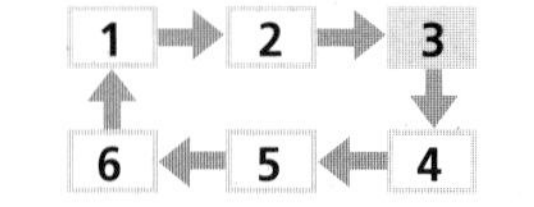

Formulate the mathematical problem

Formulate a mathematical problem

Assume that the athlete can take up the position shown in the diagram.
Find the position of the centre of mass of the athlete with reference to O and in terms of **i** and **j**.

1 2 3
6 5 4
Solve the mathematical problem

Mathematical solution

As the body is made up of several components, each can be considered as if it were a particle located at its centre of mass. The table below gives the mass of each component and the position of its centre of mass.

	Mass	Position of centre of mass
Body	$0.52m$	$0.325\mathbf{i}$
Arms (both)	$0.10m$	$0.63\mathbf{i} + 0.375\mathbf{j}$
Legs (both)	$0.30m$	$-0.045\mathbf{i} + 0.475\mathbf{j}$
Head	$0.08m$	$0.721\mathbf{i} + 0.251\mathbf{j}$

The position of the centre of mass can now be found.

$$\bar{\mathbf{r}} = \frac{0.52m \times 0.325\mathbf{i} + 0.1m(0.63\mathbf{i} + 0.375\mathbf{j}) + 0.3m(-0.045\mathbf{i} + 0.475\mathbf{j}) + 0.08m(0.721\mathbf{i} + 0.251\mathbf{j})}{m}$$

$$= 0.276\mathbf{i} + 0.200\mathbf{j}$$

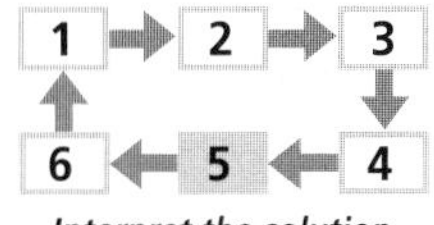

Interpret the solution

Interpretation

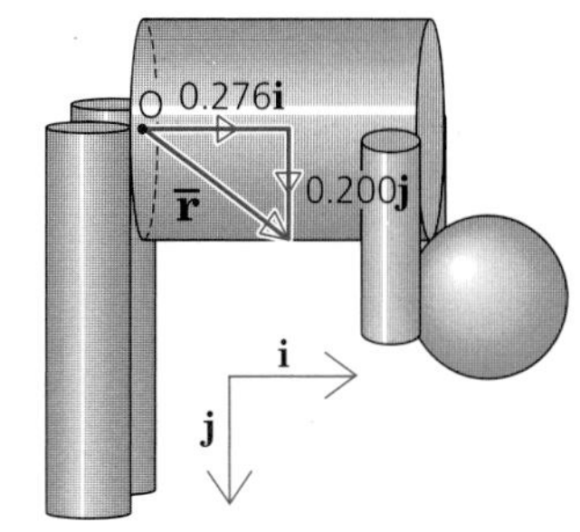

The diagram shows the position of the centre of mass.

Note that it has just moved below the body of the athlete. The advantage gained is quite considerable compared to an athlete who crosses the bar in a horizontal position, as the athlete can clear a bar 20 cm higher. It is interesting to note that the centre of mass may pass under the bar, while the athlete goes over it.

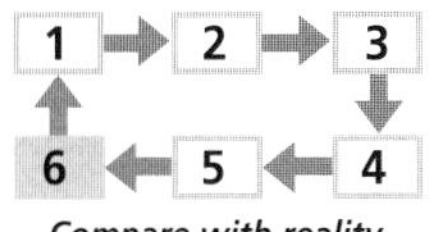

Compare with reality

Compare with reality

The position that the athlete is assumed to have adopted is rather extreme. In fact it is probably unobtainable in real life. It may also be difficult for an athlete who adopts this position to get their arms and legs over the bar.

Refining the model

By examining photographs, try to establish the type of positions that athletes use in events like the high jump. Assume a more realistic position for an athlete and explore how this affects the position of their centre of mass.

POSSIBLE MODELLING TASKS

1 A road barrier should *stay* down when it is down and up when it is up. Design a barrier that is made from metal tubes of mass 10 kg per metre and a single mass, so that when the barrier is down the centre of mass is to the right of the pivot, and when the barrier is up the centre of mass is to the left of the pivot.

2 Not all objects have uniform density or uniform mass per unit area. You can create such objects by cutting a shape from a piece of card and sticking smaller pieces of card to it. Create or find such an object. Find its centre of mass and confirm your prediction experimentally.

3 Investigate how the centre of mass of the human form moves during participation in various sports. You could use the data given in the mathematical modelling activity.

4 Find the centre of mass of an object, for example a tennis racquet. Mark this point clearly on the racquet. Then throw the object so that it spins and rides its motion. Replay in slow motion or a frame at a time. Describe the path of the centre of mass.

Summary

After working through this chapter you should be able to:

- find the centre of mass of a system of particles using:

$$\bar{\mathbf{r}} = \bar{x}\,\mathbf{i} + \bar{y}\,\mathbf{j} = \left(\frac{1}{M}\sum_{i=1}^{n} m_i x_i\right)\mathbf{i} + \left(\frac{1}{M}\sum_{i=1}^{n} m_i y_i\right)\mathbf{j} = \frac{\sum m_i \mathbf{r}_i}{\sum m_i}$$

- find the centre of mass of a composite body by modelling it as a system of particles
- find the centre of mass of a lamina that has the x-axis as a line of symmetry using:

$$\bar{x} = \frac{\int_0^a \rho xy\,\mathrm{d}x}{m} = \frac{\int_0^a xy\,\mathrm{d}x}{\int_0^a y\,\mathrm{d}x}$$

- find the centre of mass of a solid object that has been formed by rotating a region around the x-axis using:

$$\bar{x} = \frac{\int_0^a \rho\pi xy^2\,\mathrm{d}x}{m} = \frac{\int_0^a xy^2\,\mathrm{d}x}{\int_0^a y^2\,\mathrm{d}x}$$

MECHANICS

18 Conditions for sliding and toppling

- *When a force is applied to a body that is at rest it may remain at rest, slide or topple over.*
- *The effect of the force depends on a number of factors including the point of application, the coefficient of friction and the centre of mass of the body.*

SLIDING OR TOPPLING

A force can have different affects on an object depending on where and how it is applied. Applying a horizontal force to the top of a cereal packet may cause it to topple over, while applying the same force at the base may cause the packet to slide. In this chapter we explore the ideas of sliding and toppling and try to predict which will happen in a given situation.

Exploration 18.1

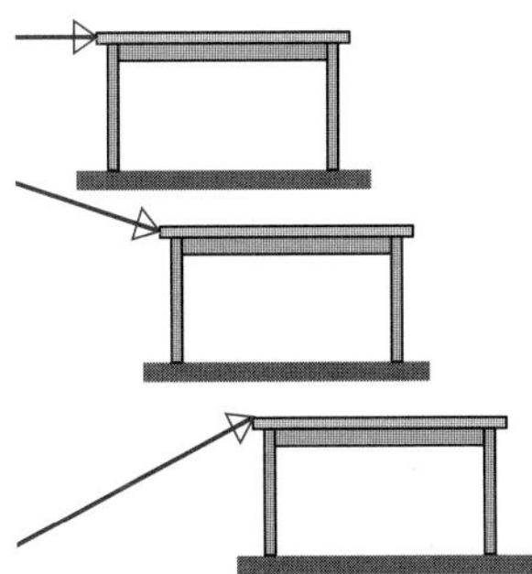

Practical activity

- A force is applied to a table as shown in the diagram. What could happen? What is most likely to happen? Try it out and see. You could experiment with tables of different dimensions, if they are available, or place them on different surfaces.
- What difference does it make if the force is applied at an angle, as shown in the diagrams?
- What difference does it make if a heavy object is placed on the table, or if someone sits on it? Try out an experiment with the heavy object in a variety of positions.

On the point of sliding or toppling

Exploration 18.1 showed that objects can either slide or topple. To decide if an object will slide or topple we need to examine two cases, when the object is *on the point of* sliding and when it is *on the point of* toppling.

- When an object is on the point of sliding the magnitude of friction force is equal to μR.
- When the object is about to topple the reaction force acts at the corner about which the object would topple.

This is illustrated in the following examples.

Example 18.1

A box of mass 5 kg has the dimensions shown in the diagram. The coefficient of friction between the box and the floor is 0.5. A horizontal force of magnitude Q is applied to the box as shown.

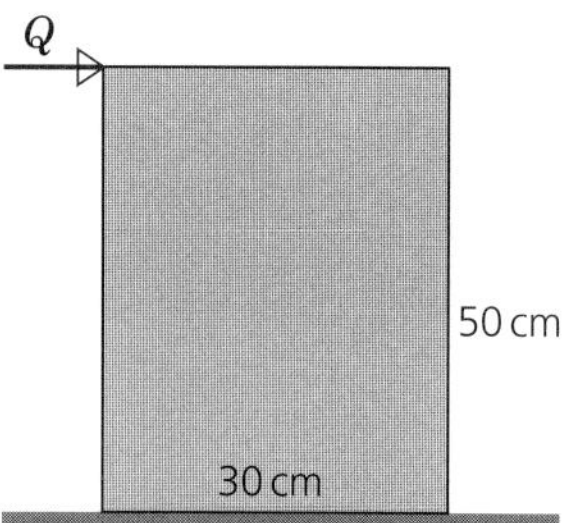

a) *Find Q if the box is on the point of sliding.*
b) *Find Q if the box is about to topple.*
c) *Does the box slide or topple?*

Solution

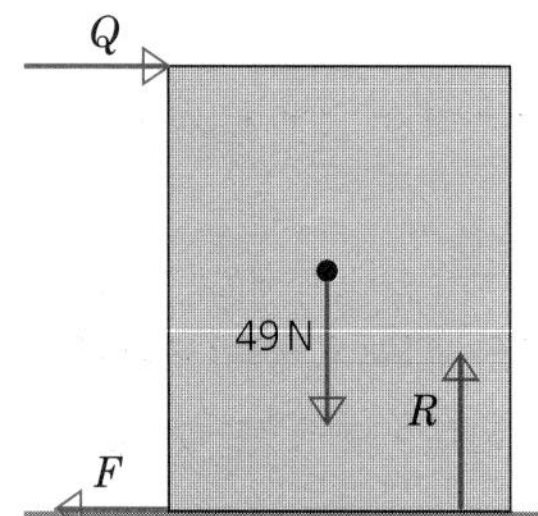

The diagram on the left shows the forces acting on the box.

a) *If the box is about to slide then:*
$Q = F = \mu R$
As the vertical forces must be in equilibrium $R = 49$ N and so:
$Q = \mu R = 0.5 \times 49 = 24.5$ *N*

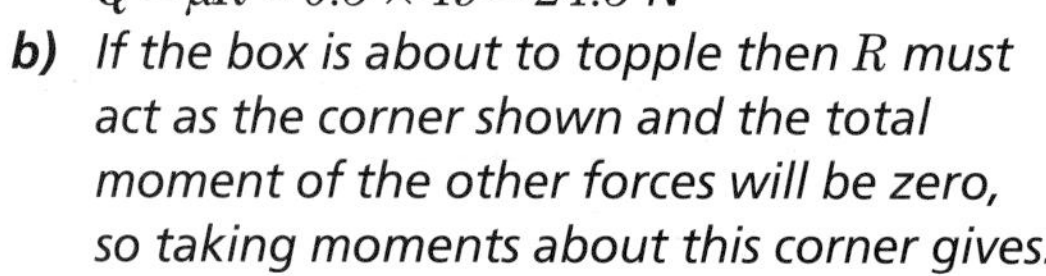

b) *If the box is about to topple then R must act as the corner shown and the total moment of the other forces will be zero, so taking moments about this corner gives:*

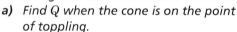

$$0.15 \times 49 - 0.5Q = 0 \Rightarrow Q = \frac{0.15 \times 49}{0.5} = 14.7$$

c) *A force of 24.5 N is required for the box to slide, but only 14.7 N for the box to topple. So the box will topple before it slides.*

Example 18.2

A force of magnitude Q is applied horizontally to the top of a cone of mass 1 kg. The centre of mass of the cone is 10 cm from the base.

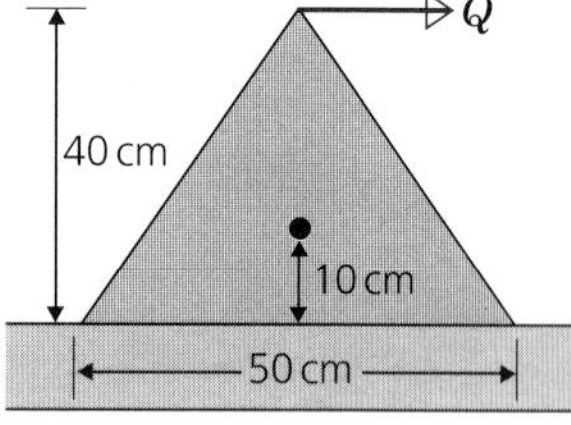

a) *Find Q when the cone is on the point of toppling.*
b) *Find Q when the cone is on the point of sliding, in terms of μ the coefficient of friction.*
c) *What can be deduced about μ if the cone slides before it topples?*

Solution

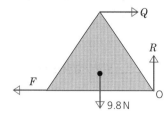

a) *When the cone is on the point of toppling the forces will act as shown in the diagram, and the total moment about O will be zero. So taking moment about O gives:*

$$9.8 \times 0.25 - 0.40 \times Q = 0$$

$$\Rightarrow \quad Q = \frac{9.8 \times 0.25}{0.40} = 6.125 \text{ N}$$

b) *If the cone is on the point of sliding $Q = F = \mu R$*
As the vertical forces are in equilibrium:
$R = 9.8$ *N* $\Rightarrow Q = 9.8\mu$

c) If the cone slides before it topples, then the force for sliding must be less than the force for toppling.

$$9.8\mu < 6.125 \Rightarrow \mu < \frac{6.125}{9.8} = 0.625$$

Sliding and toppling on slopes

When an object is placed on a slope, it may remain at rest. Alternatively, it could slide down the slope or topple over. We now explore the ideas of sliding and toppling on slopes.

Exploration 18.2

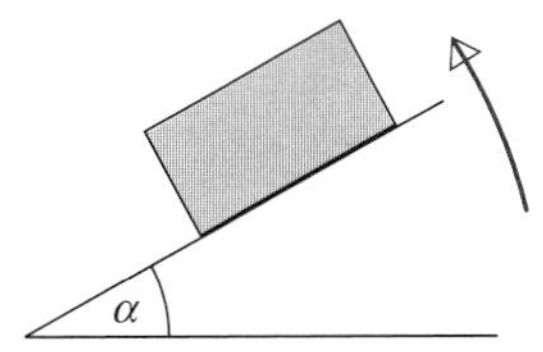

Practical activity

A box is placed on a board and one end of the board is lifted. What could happen to the box?

If the box slides down the slope, what is the relationship between the angle, α, and the coefficient of friction between the box and the board?

If the box topples, what is the relationship between the angle, α, and the dimensions of the box?

The angle of friction

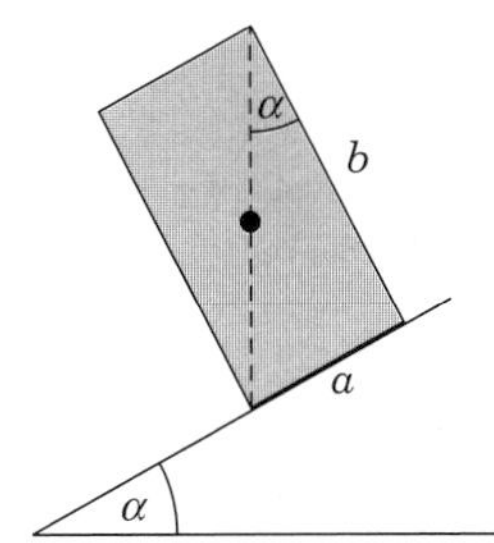

If the box is on the point of sliding down the slope, then the angle between the slope and the horizontal will be the **angle of friction**. $\tan \alpha = \mu$

If the box is *on the point of toppling*, the centre of mass of the box will be directly above the corner about which it will topple. So for a uniform box as illustrated:

$$\tan \alpha = \frac{a}{b}$$

Example 18.3

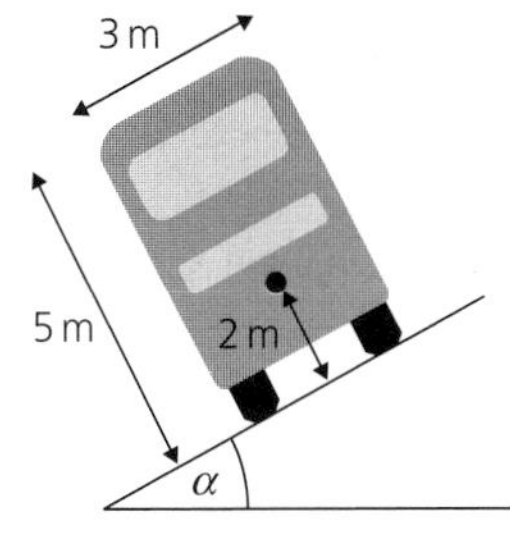

A bus has dimensions as shown and its centre of mass is 2 m above the road. It is placed on a test ramp and tilted at an angle α to the horizontal.

a) *At what angle would the bus topple?*

b) *At what angle would the bus slide on the ramp if the coefficient of friction between its tyres and the ramp is 0.8?*

c) *Does the bus slide or topple first?*

Solution

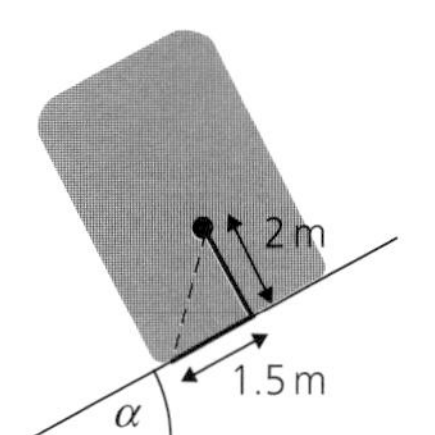

a) *The bus topples when the centre of mass is above the bottom corner as shown. In this position:*

$$\tan \alpha = \frac{1.5}{2} \Rightarrow \alpha = 36.9°$$

b) *The bus slides, at the angle of friction, when $\tan\alpha = \mu$. So in this case $\tan\alpha = 0.8 \Rightarrow \alpha = 38.7°$.*

c) *As the angle for toppling is less than the angle for sliding the bus will topple.*

EXERCISES

18.1 A

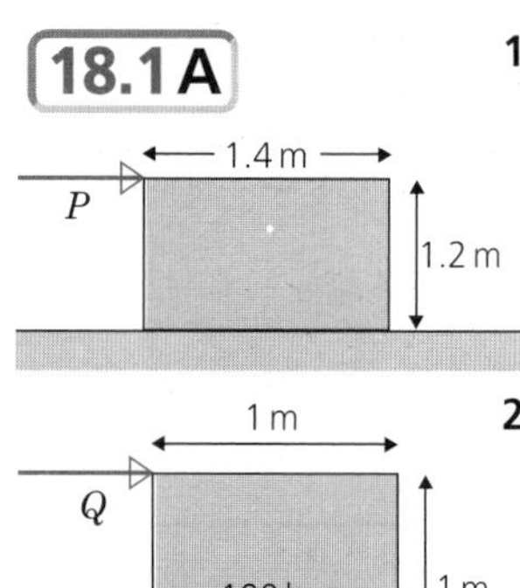

1 The diagram shows a crate of mass 50 kg. A horizontal force of magnitude P acts at the top of the crate as shown. The coefficient of friction between the crate and the ground is 0.6.

a) What is P when the box is about to slide?

b) What is P when the box is about to topple?

c) Does the box slide or topple?

2 The diagram shows a force of magnitude Q that is applied as shown to a cylinder of mass 100 kg. The coefficient of friction between the cylinder and the ground is 0.707.

a) Show that $Q = 693$ N when the cylinder is about to slide.

b) Find Q when the cylinder is about to topple.

c) Explain why the cylinder topples before it slides.

d) For what range of values of μ would the cylinder slide before it topples?

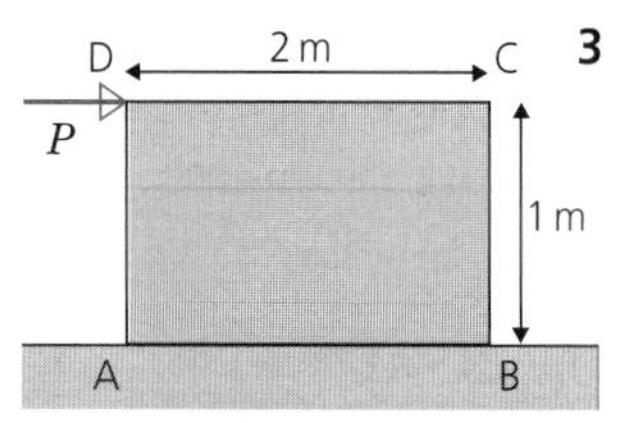

3 The diagram shows a crate at rest on a rough surface. The coefficient of friction between the crate and the ground is μ. A horizontal force of magnitude P is applied at D. P increases gradually.

a) For what range of values of μ will the crate topple before it slides?

b) What difference does it make if the mass of the crate is doubled?

c) What difference does it make if the direction of P is reversed?

4 In each example below find the values of α at which the box shown would topple or slide and state which happens in each case.

a)

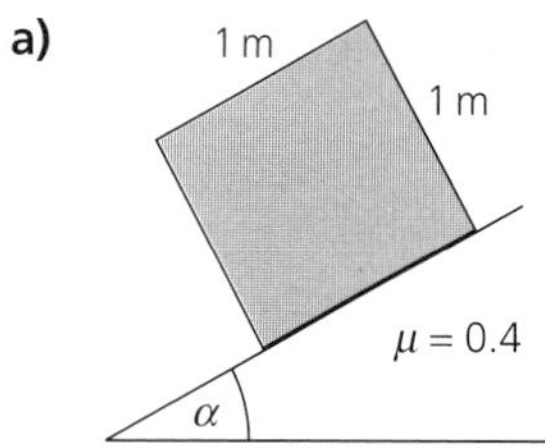

b)

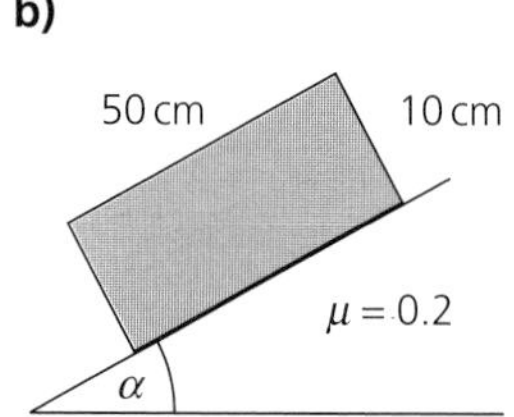

c)

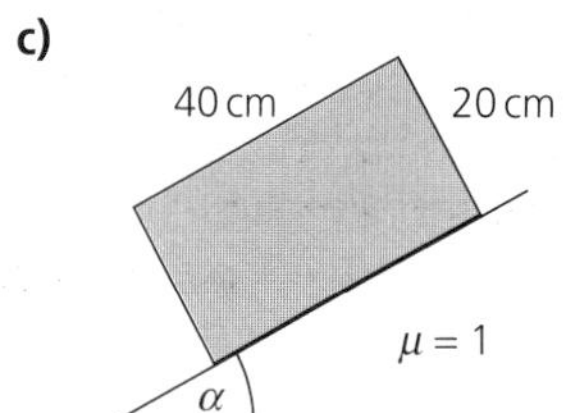

d)

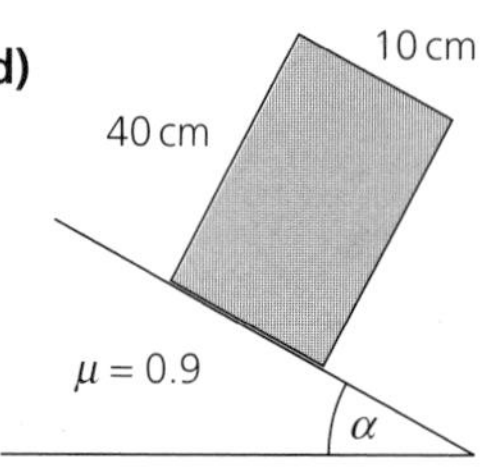

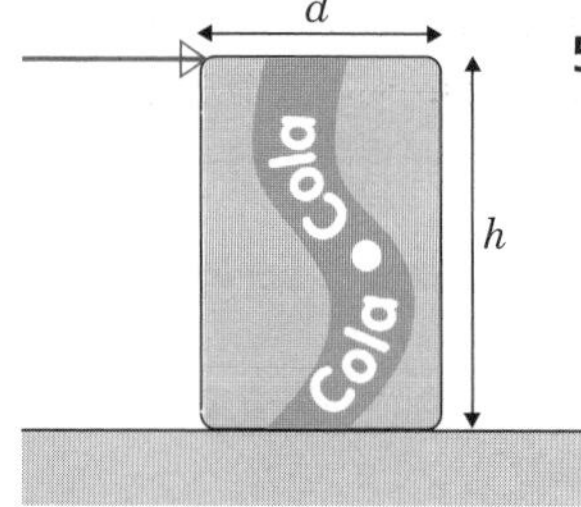

5 At a school fete there is a stall where cans of cola have to be knocked over by competitors. Two students suggest two different strategies. One suggests filling the empty can completely with sand. The other suggests filling the can one-third full with sand. Assume the mass of the can is negligible in this problem.

a) Show that for the full can to topple a horizontal force of $\dfrac{mgd}{2h}$ must be applied at the top of the can, as shown in the diagram.

b) What force must be applied for the one-third full can to topple?

c) Show that the full can will slide before it topples if $\mu < \frac{d}{2h}$.

d) Find a similar result for the can that is one-third full.

e) Comment on your results, recommend how far to fill a can and state any important features of the problem that have been overlooked.

6 A force of magnitude P acts horizontally on a lamina which is in the shape of a parallelogram. The mass of the lamina is 2 kg. The coefficient of friction between the lamina and the ground is 0.4.

a) Find the centre of mass of the lamina, in terms of $\mathbf{i}$ and $\mathbf{j}$ relative to the origin O.

b) If P is gradually increased, what happens to the lamina?

c) What happens if the force is applied at A in the opposite direction?

7 The diagram shows a car parked on a steep slope.
The centre of mass is at A. If the angle α were increased would the car slide or topple first and at what angle? Assume the coefficient of friction between the tyres and the surface is 0.8.

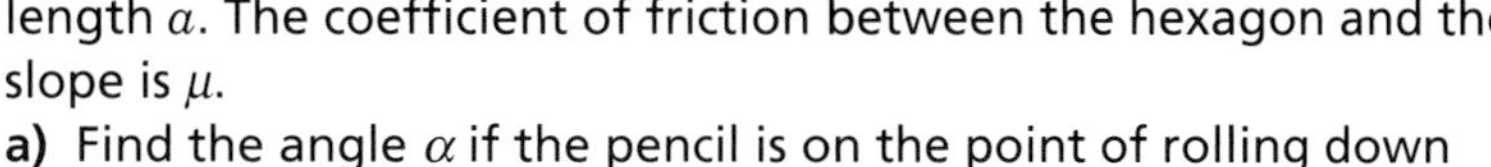

8 The diagram shows a pencil with hexagonal cross-section of side length a. The coefficient of friction between the hexagon and the slope is μ.

a) Find the angle α if the pencil is on the point of rolling down the slope.

b) Find an inequality involving μ if the pencil rolls before it slides.

c) Repeat (a) and (b) for a pencil with a regular octagonal cross-section.

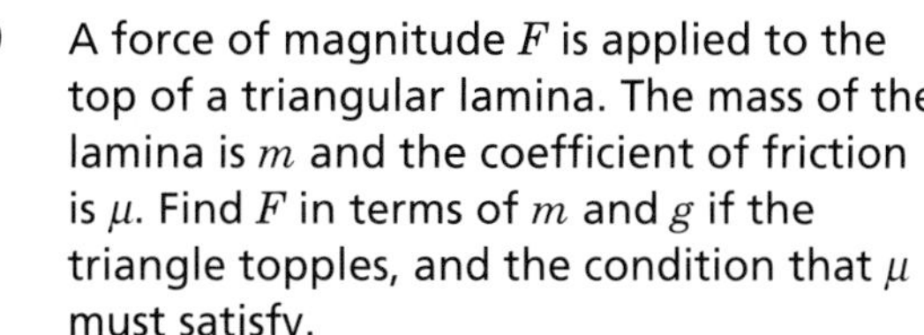

9 A force of magnitude F is applied to the top of a triangular lamina. The mass of the lamina is m and the coefficient of friction is μ. Find F in terms of m and g if the triangle topples, and the condition that μ must satisfy.

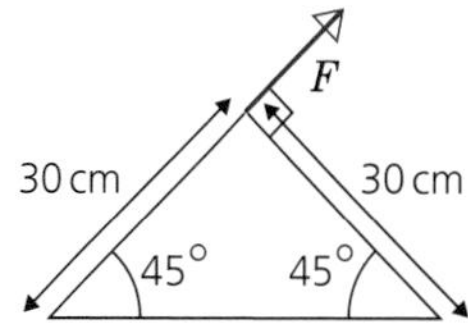

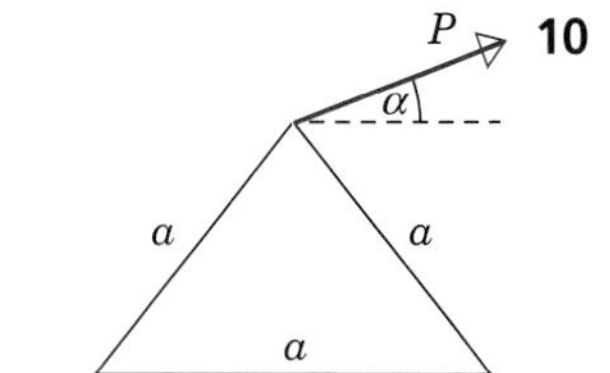

10 The diagram shows a force of magnitude P that acts at an angle α to the horizontal on an equilateral triangular lamina. The coefficient of friction between the lamina and the surface is μ.

a) Find P if the lamina is on the point of sliding.

b) Find P if the lamina is about to topple.

c) Find a condition for μ if the lamina slides before it topples.

EXERCISES

18.1 B

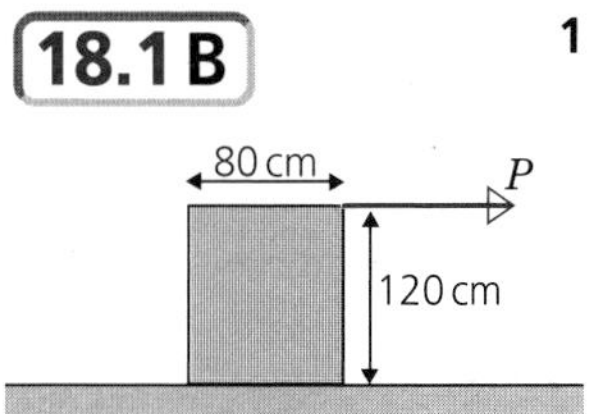

1 A horizontal force of magnitude P acts at the top of a box of mass 40 kg as shown in the diagram. The coefficient of friction between the box and the ground is 0.7.

a) What is P when the box is about to slide?

b) What is P when the box is about to topple?

c) Does the box slide or topple first?

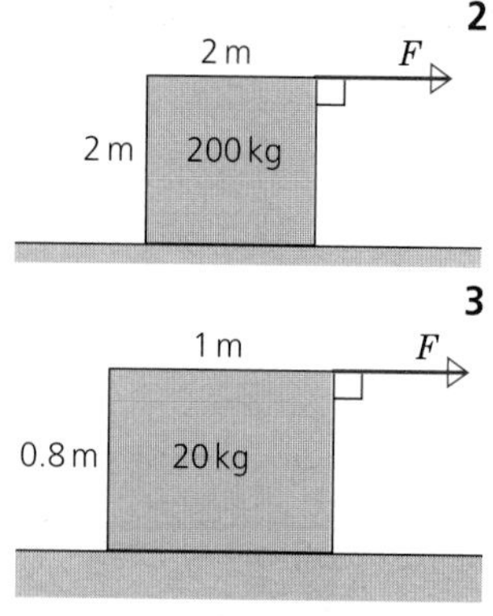

2 A force of magnitude F acts as shown on a crate of mass 200 kg. The coefficient of friction between the crate and the ground is 0.6.

a) Show that when $F = 980$ N the crate is about to topple.

b) Find F if the crate is about to slide.

c) Does the crate slide or topple first?

3 The diagram shows a box at rest and a horizontal force acting on it. The coefficient of friction between the box and the ground is μ.

a) Find the magnitude of the force if the box is on the point of toppling.

b) Find the magnitude of the force, in terms of μ, if the box is about to slide.

c) What can be deduced about μ if the box slides before it topples?

4 In each case below, what happens to the box illustrated as the angle α is increased?

a)

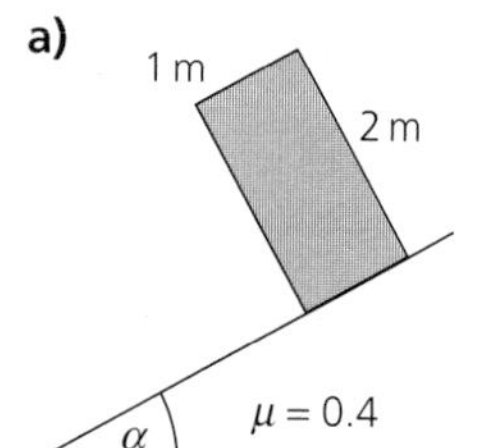

b)

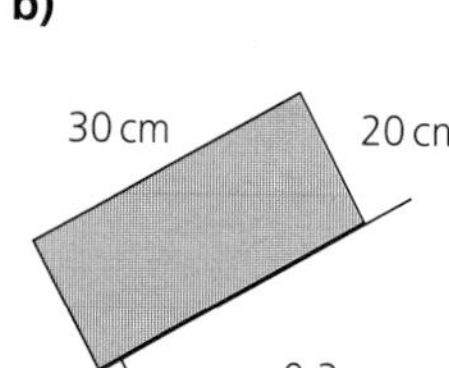

c)

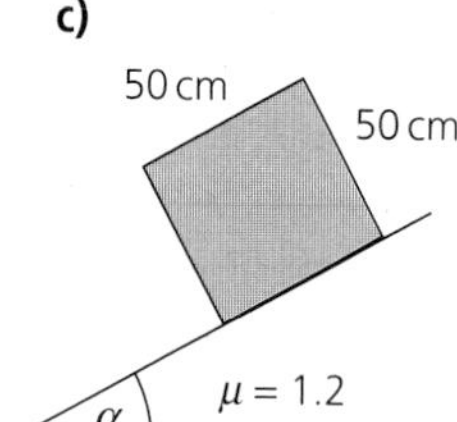

d)

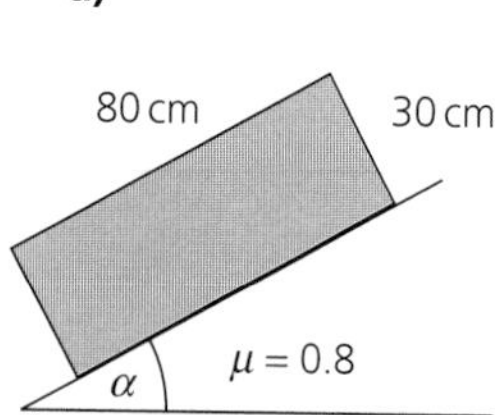

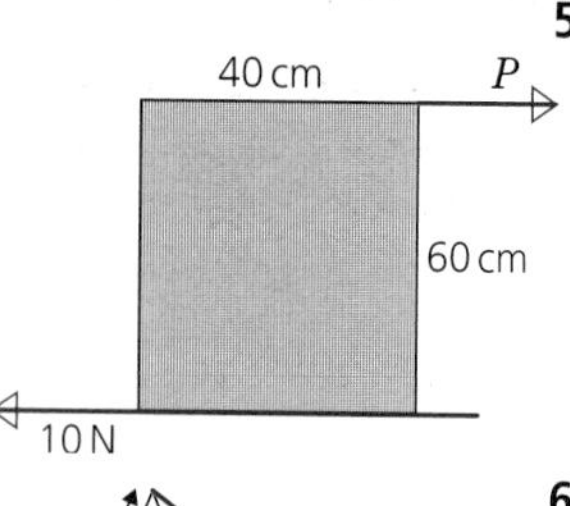

5 The diagram shows two forces that are applied to a crate of mass 4 kg. The coefficient of friction between the crate and the horizontal surface is 0.4.

a) Find P if the box is about to slide.

b) Find P if the box is about to topple.

c) What happens to the box?

d) Would you answer to **c)** change if the 10 N force were removed?

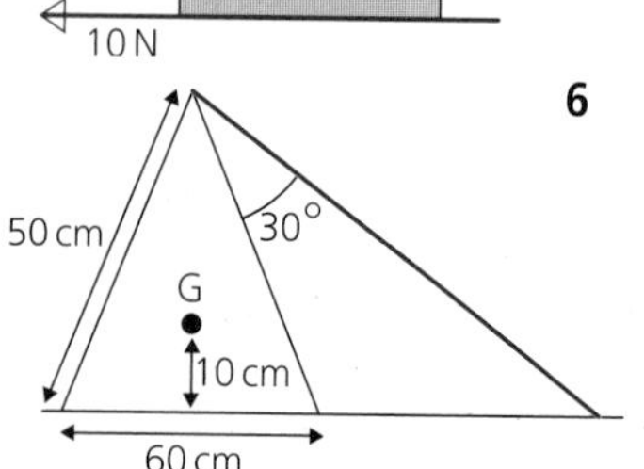

6 A cable is attached to the top of a cone of mass 40 kg and a winch on the ground. The winch gradually increases the tension in the cable. The centre of mass of the cone is at G and the coefficient of friction between cone and the ground is μ.

a) Find the tension in the cable if the cone is on the point of toppling.

b) Express the tension in the cable if the cone slides in terms of μ.

c) For what range of values of μ will the cone slide before it topples, as the tension in the cable is increased?

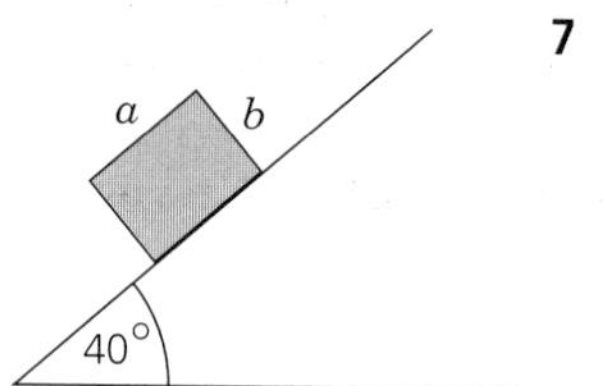

7 At a factory boxes are placed on a slope at 40° to the horizontal and allowed to slide down ready to be loaded onto lorries. The coefficient of friction between the boxes and the slope is 0.2.

a) Explain why the boxes can slide down the slope.

b) If the base of a box is of length a and the height of the box is b, find the value of b, in terms of a, for which the box is about to topple.

8 For each lamina shown below, state which forces could cause toppling about the point X.

a)

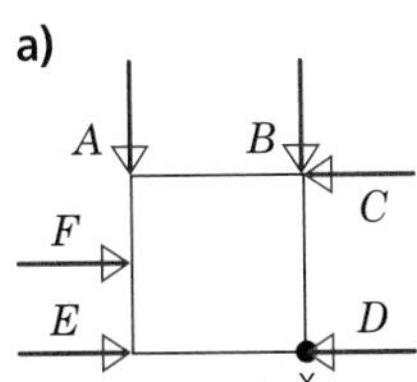

b)

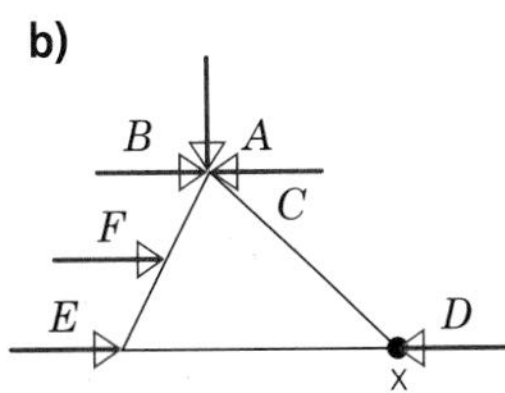

9 A force of magnitude P is applied as shown to the lamina, of mass m, illustrated. The coefficient of friction between the lamina and the ground is μ. Find P if the lamina topples, and the condition for μ if it topples before it slides.

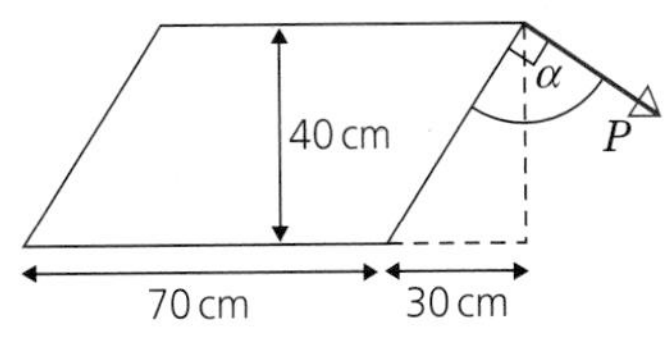

10 The diagram shows a lamina of mass m. The force of magnitude P is shown and the coefficient of friction between the lamina and the surface is μ. Find a condition for μ if the lamina slides before it topples.

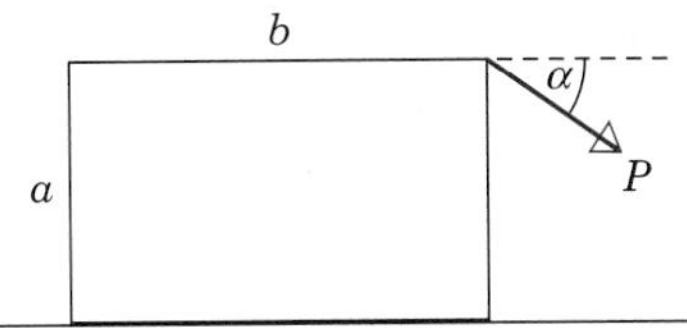

CONSOLIDATION EXERCISES FOR CHAPTER 18

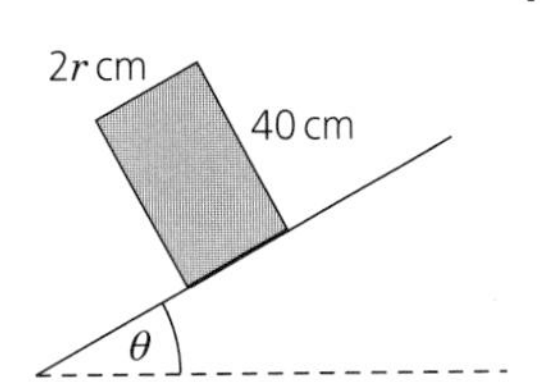

1 A uniform right cylinder has height 40 cm and base radius r cm. It is placed with its axis vertical on a rough horizontal plane. The plane is slowly tilted, and the cylinder topples when the angle of inclination (see diagram) is 20°. Find r.

What can be said about the coefficient of friction between the cylinder and the plane?

(UCLES Question 2, M2 Specimen Paper, 1994)

2 A bird table is made from a uniform square base of side 0.3 m with mass 5 kg, a uniform square top of side 0.5 m and mass 2 kg. A uniform thin rod of length 1.6 m and mass 1 kg connects the centres of the top and base. The top and base have negligible thickness.

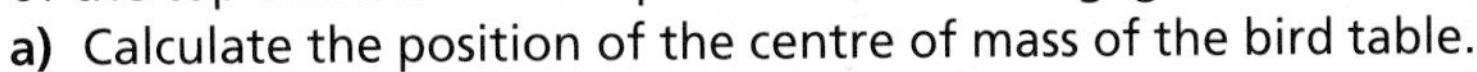

a) Calculate the position of the centre of mass of the bird table.

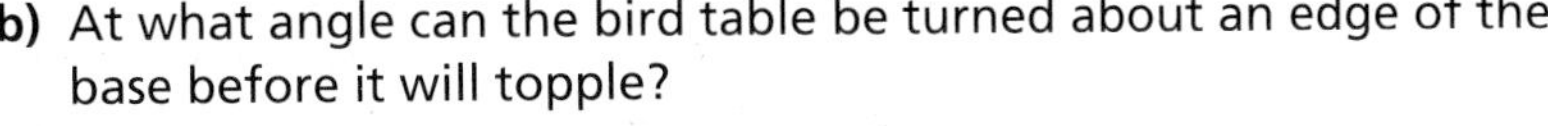

b) At what angle can the bird table be turned about an edge of the base before it will topple?

It is decided to make the base heavier so that the bird table can be tipped at 40° to the horizontal before it topples. The base still has negligible thickness.

c) Show that the centre of mass must now be about 0.18 m above the base.

d) What is the new mass of the base?

(MEI Question 2, M2, January 1992)

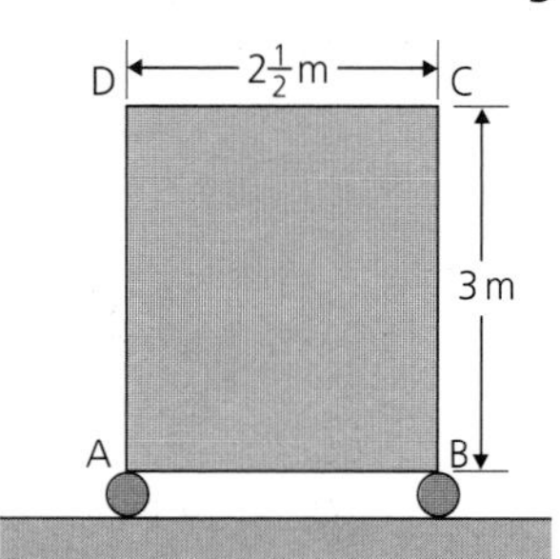

3 The diagram shows a uniform rectangular sliding door ABCD of weight W N. The door is $2\frac{1}{2}$ m wide and 3 m high. The door rests in a vertical plane and has, at corners A and B, small wheels which rest upon a fixed horizontal rail. The coefficient of friction between the wheel at B and the rail is μ and there is no friction between the wheel at A and the rail. It is required to open the door by applying a horizontal force in the vertical plane ABCD to a vertical edge of the door at a point $1\frac{1}{2}$ m above the rail.

a) Show that if $\mu = \frac{5}{6}$ the magnitude of the force required to slide the door in the direction BA must be greater than $\frac{5}{24}W$ N.

b) Show that if $\mu = \frac{5}{9}$ the magnitude of the force required to slide the door in the direction AB must be greater than $\frac{5}{12}W$ N.

c) Given that $\mu < \frac{5}{6}$ and that the horizontal force is applied to the edge AD in the direction AB, show that the door will tend to turn about the wheel at B before it will slide in the direction AB.

(ULEAC Question 6, Mathematics Paper 3, January 1990)

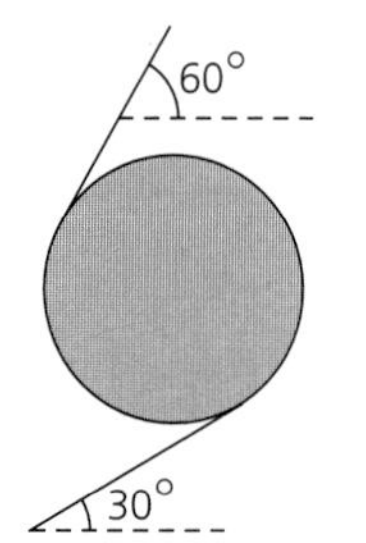

4 A uniform cylindrical oil drum is held in equilibrium, with its curved surface on a rough slope, of inclination 30° to the horizontal, by a light rope attached to a point on the circumference of the oil drum. The rope is pulled upwards at an angle of 60° to the horizontal and is tangential to the drum (see diagram). The axis of the drum is horizontal and the rope is perpendicular to the axis of the drum. Find the least possible value for the coefficient of friction between the oil drum and the slope.

(UCLES Question 5, Linear Further Maths Paper 2 Specimen Paper, 1994)

MATHEMATICAL MODELLING ACTIVITY

1 → 2 → 3 → 4 → 5 → 6

Specify the real problem

Problem statement

If a square table were designed to have legs at the mid-point of each side, instead of each corner, would it be very stable?

1 → 2 → 3 → 4 → 5 → 6

Set up a model

Set up a model

Some important features that should be considered are:

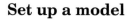

- size of the table
- mass of the table
- whether the legs will be included in the model
- the mass to be supported
- position of mass to be supported.

Are there other important features that you feel should be included?

Having considered these features, a set of assumptions can now be set up.

- The table is 1 m square.
- We shall consider only the table top and ignore the mass of the legs. Assume the table has a mass of 12 kg so that the force of gravity is $12g$ N.
- We shall want to vary the mass, m, which needs to be supported.

- The position of the load which will cause the greatest tendency to overbalance the table is at a corner, so we shall assume the mass is concentrated at one corner.

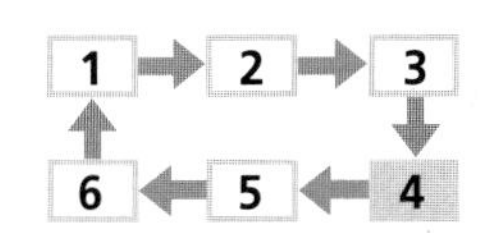

Formulate the mathematical problem

Formulate a mathematical problem

The diagram shows the forces and the positions where they act. Find when the table topples.

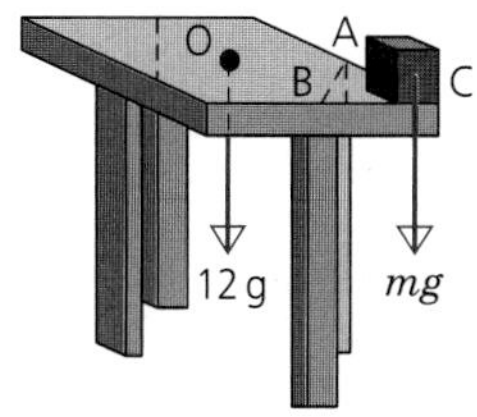

Mathematical solution

Solve the mathematical problem

Clearly the principle of moments will be important. We can see from the above diagram that the table will topple about the line AB if we are considering the table top only. To calculate the moments we need to find the perpendicular distances from the line AB to both O and C.

It can quickly be seen that symmetry requires these distances to be equal.

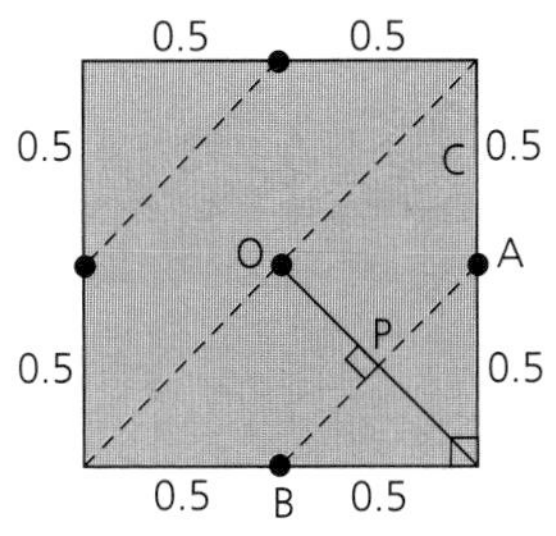

First calculate AB.

$$AB = \sqrt{0.5^2 + 0.5^2} = 0.7071\,\text{m}$$

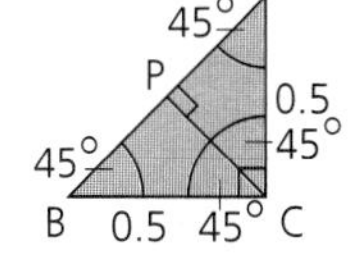

Then as AP = $\frac{1}{2}$AB, the length AP = 0.3536 m

and also PC = 0.3536 m.

For the table to overturn, clockwise moment ≥ anticlockwise moment

$\Rightarrow mg \times 0.3536 \geq 12g \times 0.3536 \Rightarrow m \geq 12$.

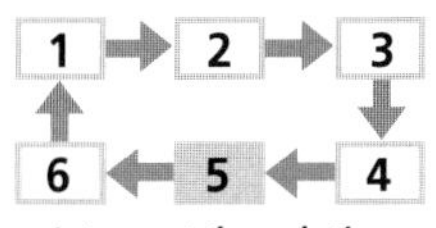

Interpret the solution

Interpretation

The table top will become unstable when a mass of a 12 kg is placed at the corner. Two questions immediately come to mind.

1. Is it a coincidence that the maximum mass is equal to the mass of the table top?
2. Is it a reasonable answer?

To answer question 1, start with a square table of side P metres and let the mass of the table top be M kg.

Work through the mathematical solution again, and check if $m \geq M$.
To answer question 2, you could compare with reality.

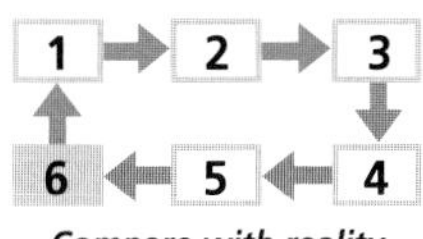

Compare with reality

Compare with reality (practical activity)

Try to find a wooden chessboard (or any square piece of rigid material) to represent the table top. Find the mass of your square.

Support your square with four 'legs' (white-board pens are quite good). Place masses on the corner of your table until it collapses and note the weight of the masses applied.

You may have already found reasons for criticising the model and the result you now have may confirm that some improvements are necessary.

Refining the model

There are two questions which can be asked now.

1 Why did your chessboard/pen model support slightly more weight than was predicted?
2 How would the model change if we considered a real table with 'heavy' legs joined to the table top?

To answer question 1, produce an improved mathematical solution. (**Hint:** Consider a better position for the vector representing the supported weight.) Continue to compare with reality.

To answer question 2, again produce an improved mathematical solution. (**Hint:** Consider the line about which the table would topple.)

At this stage you may wish to explore some other ideas.

POSSIBLE MODELLING TASKS

1 Consider rectangular (or other shaped) tables as an extension of the modelling activity.

2 Consider the problem of the modelling activity. Could the legs be positioned in such a way that they join the table top at a point three-quarters of way along each side, or another fraction?

3 Use a toppling experiment to determine the position of the centre of mass for an unusual or irregular solid.

Summary

After working through this chapter you should be able to:

- determine whether an object will slide or topple
- describe the conditions when this will happen.

Frameworks

- *There are very many examples of frameworks in real life, including some bridges, the steel structures that support some roofs and electricity pylons.*
- *The forces acting on a framework can be very complex, but simplifying assumptions can be used to allow the forces in each member of a framework to be modelled.*

FRAMEWORKS AROUND US

The photograph shows part of a framework in a roof structure. This type of design is very common and gives much more strength than a single beam or girder. It is important for designers to be able to predict the forces that will act in each **member** of such a framework. In this chapter we develop an approach that allows the forces to be estimated subject to some simple, but important assumptions.

Exploration 19.1

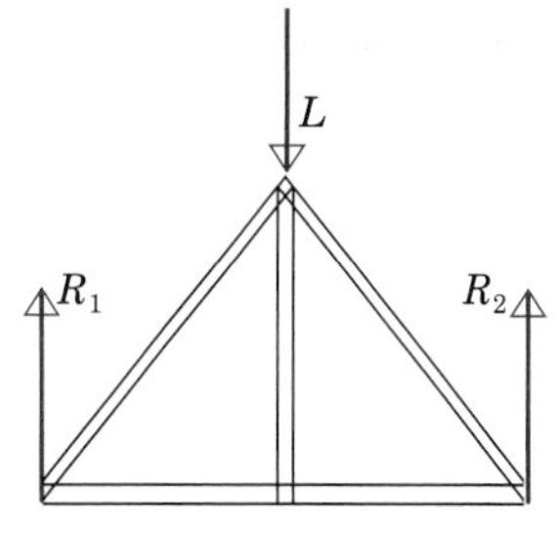

A simple framework

The diagram shows a simple framework that might be found in the roof of an older house or other building. Assume that the framework supports a load, L, at the top and itself is supported by the forces R_1 and R_2.

- What are the magnitudes of R_1 and R_2?
- Consider one member of the framework, describe the forces acting on it.

Give examples of frameworks that you have seen.

Modelling a framework

There may be many factors to consider at the joints of a framework, for example when rods are welded together or fixed by a number of bolts or nails as shown in the diagram.

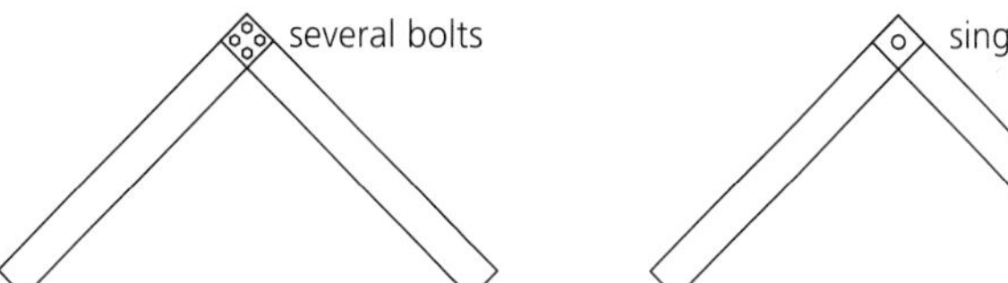

The force of gravity on each member may also complicate the situation. A first model of the forces acting in a framework is based on the following two assumptions.

- The members of a framework are light rods and their mass is assumed to be zero.
- All the joints are smooth pin-joints. This means that all the members at any joint are fixed by a pin, about which they can freely pivot. Imagine two strips of card with holes punched in them, joined by a simple clip.

Example 19.1

The diagram shows a roof truss. The roof exerts three forces on the truss as illustrated.

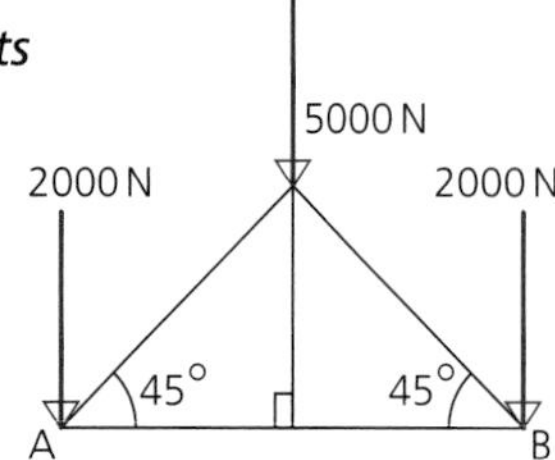

By modelling the truss as a light pin-jointed framework find:

a) *the external forces acting on the truss at the supports A and B,*

b) *the force in each member.*
Comment on the design of the truss.

Solution

a) *First consider the truss as a single object. The diagram shows the external forces acting on the truss. As all these forces must be in equilibrium* $R + S = 9000$

Due to the symmetry of the truss:

$R = S \Rightarrow R = S = 4500$ *N*

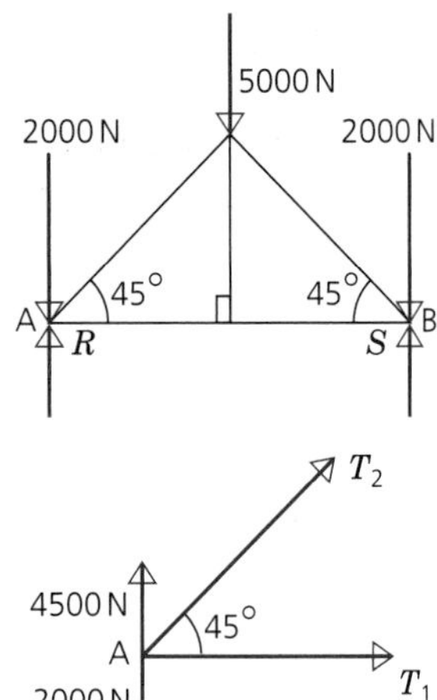

b) *To find the forces in each member, start by considering the point A. Assume that all the rods are in tension and so they exert forces that act toward their centres. The diagram shows the forces at A. The resultant force at A must be zero. Considering the vertical components of the forces gives:*

$4500 - 2000 + T_2 \cos 45° = 0$

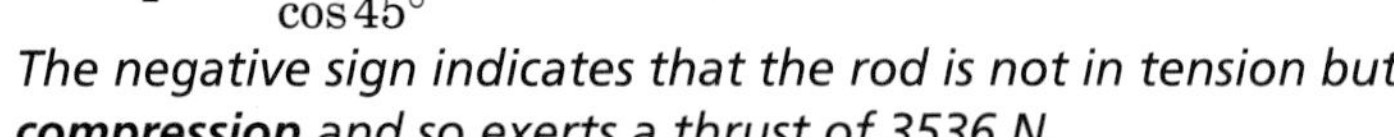

$\Rightarrow \quad T_2 = \dfrac{2000 - 4500}{\cos 45°} = -3536$ *N*

The negative sign indicates that the rod is not in tension but ***compression*** *and so exerts a thrust of 3536 N.*

Now consider the horizontal component of the forces, which gives:

$T_2 \cos 45° + T_1 = 0$

$\Rightarrow -3536 \cos 45° + T_1 = 0$

$T_1 = 3536 \cos 45° = 2500$ *N*

As this is positive this member is in tension.

Now consider the peak of the truss. The diagram shows the forces acting at this point. Note that due to the symmetry of the truss the force in the other sloping member will also be 3536 N. As there is only one unknown force here, only the vertical components need be considered to give:

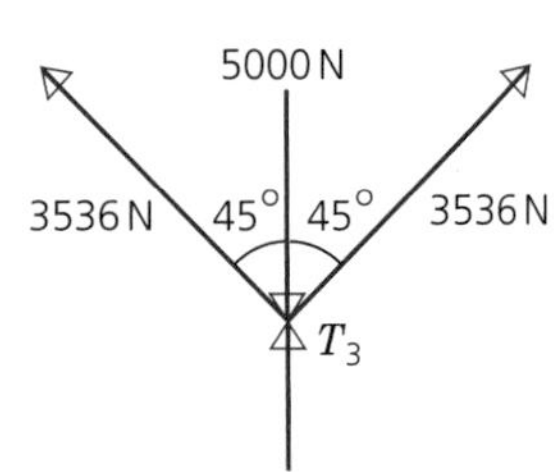

$2 \times 3536 \cos 45° + T_3 - 5000 = 0$

$\Rightarrow T_3 = 5000 - 2 \times 3536 \cos 45° = 0$

As there is no force in the vertical member it appears to serve no useful purpose in the truss.

Example 19.2

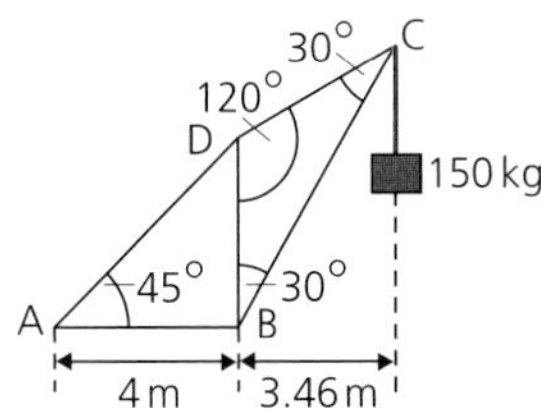

The diagram shows a simple crane that is supporting a load of 150 kg.

a) *Find the external forces that act on the crane where it is fixed to the ground at A and B.*

b) *Find the force in each member, stating whether it is in tension or compression.*

c) *Which member is most likely to fracture during a lift?*

Solution

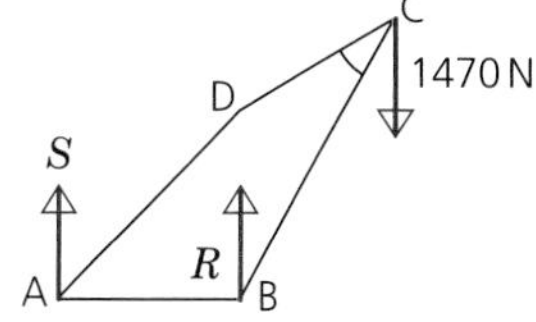

a) *Considering the framework as a single body and taking moments about A gives:* $4R = 7.46 \times 1470$

$$\Rightarrow R = \frac{7.46 \times 1470}{4} = 2742$$

Also by considering the resultant force:

$R + S = 1470$

$\Rightarrow 2742 + S = 1470$

$S = 1470 - 2742 = -1272\ N$

So an upward force of 2742 N acts at B and a downward force of 1272 N acts at A.

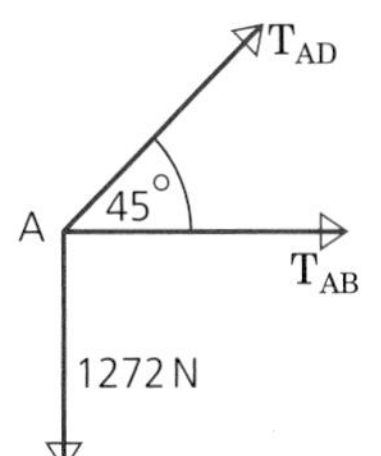

b) *Begin by considering the point A. The diagram shows the forces acting.*

All the rods have been assumed to be in tension and T_{AB} *denotes the tension in the rod AB, etc. All the forces at each joint must be in equilibrium.*

Considering the vertical components of the forces gives:

$T_{AD} \cos 45° - 1272 = 0$

$$\Rightarrow T_{AD} = \frac{1272}{\cos 45°} = 1799\ N$$

The rod AD is in tension as T_{AD} *is positive. Considering the horizontal components gives:*

$T_{AD} \cos 45° + T_{AB} = 0$

$\Rightarrow 1799 \cos 45° + T_{AB} = 0$

$T_{AB} = -1779 \cos 45° = -1272\ N$

The rod AB is in compression as T_{AB} *is negative.*

Now consider the forces at B, which are shown on the diagram.

Considering the horizontal components of the forces gives:

$T_{BC} \cos 60° + 1272 = 0$

$$\Rightarrow T_{BC} = -\frac{1272}{\cos 60°} = -2544N$$

As T_{BC} *is negative the rod BC is in compression.*

Now considering the vertical components gives:

$T_{BD} + T_{BC} \cos 30° + R = 0$

$\Rightarrow T_{BD} - 2544 \cos 30° + 2742 = 0$

$T_{BD} = 2544 \cos 30° - 2742 = -539\ N$

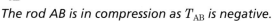

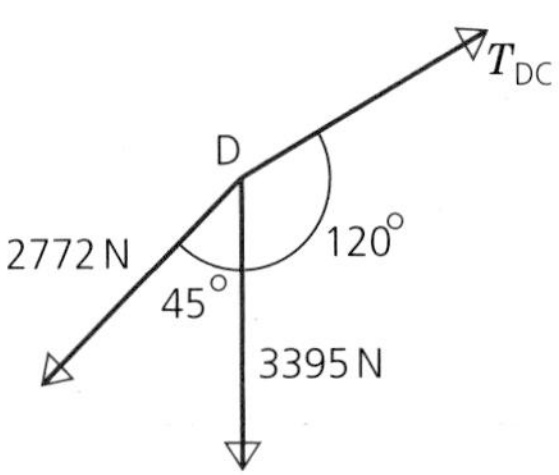

As T_{BD} is negative, there is a thrust in BD.
Now consider the point D. The diagram shows the forces acting at D. Considering the horizontal components of the forces gives:

$T_{DC} \cos 30° - 1799 \cos 45° = 0$

$T_{DC} = \dfrac{1799 \cos 45°}{\cos 30°} = 1469\ N$

As T_{DC} is positive it is in tension.
The diagram summarises the forces present in the framework.

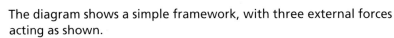

Tension (1469 N), Thrust (539 N), Tension (1799 N), Thrust (2544 N), Thrust (1272 N)

c) *The member BC has to exert the greatest force and so is the member that is most likely to fracture.*

EXERCISES

19.1 A

1 The diagram shows a light pin-jointed framework, with three external forces acting as shown.

5000 N, C, R, S, A, 30°, D, 60°, B, 10 m

a) Which rods do you think will be in tension?
b) Calculate the length BD.
c) Take moments about A to find S.
d) Find R.
e) Consider the forces acting at A and find the tension or thrust in AB and AC.
f) What forces act in BC?
g) Compare your answers in **e)** and **f)** with your predictions in **a)**.

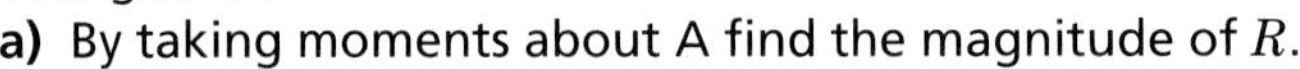

2 The diagram shows a simple framework, with three external forces acting as shown.

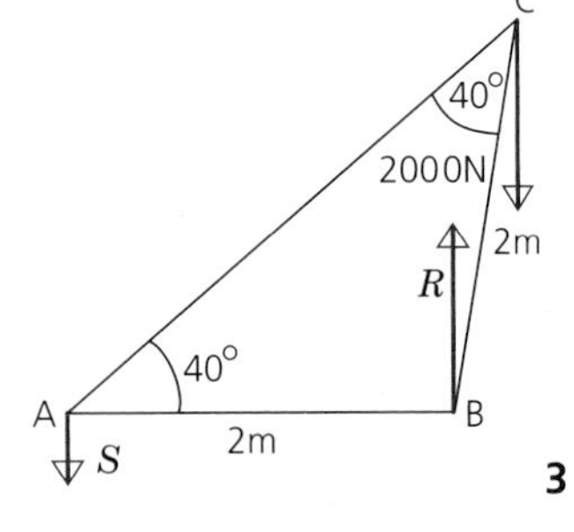

a) By taking moments about A find the magnitude of R.
b) By considering the resultant force on the framework find S.
c) Consider the forces acting at A to find the force in AC and then the force in AB.
d) Find the force in BC.
e) For each member of the framework, state whether it is in tension or compression.

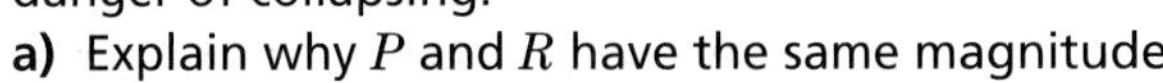

3 Three scaffolding poles joined at A, B and C form a framework that is used to support a wall that is in danger of collapsing.

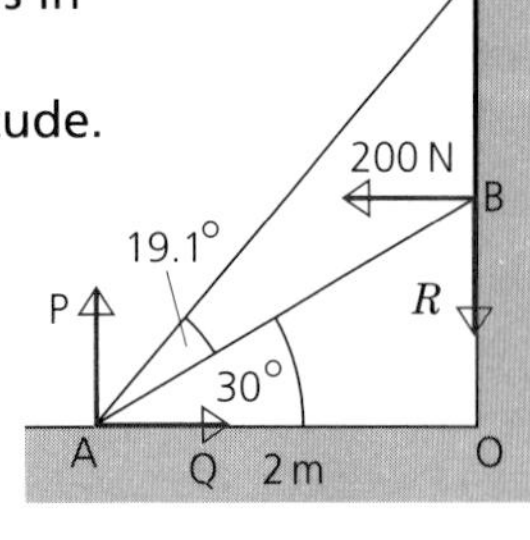

a) Explain why P and R have the same magnitude.
b) State the magnitude of the force Q.
c) Show that OB = BC and that $P = 348$ N.
d) Find the force in each pole, stating if it is in tension or compression.
e) What assumption have you made about the poles?

4 The diagram shows a car jack which supports a load of 1800 N.

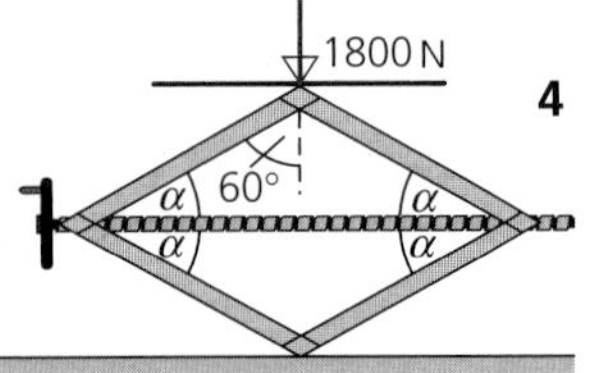

a) Which components of the jack would you expect to exert tensions and which thrusts?
b) Confirm your predictions by finding the force in each component when $\alpha = 30°$.

5 The diagram shows a simple framework, with three external forces P, Q and R acting on it.

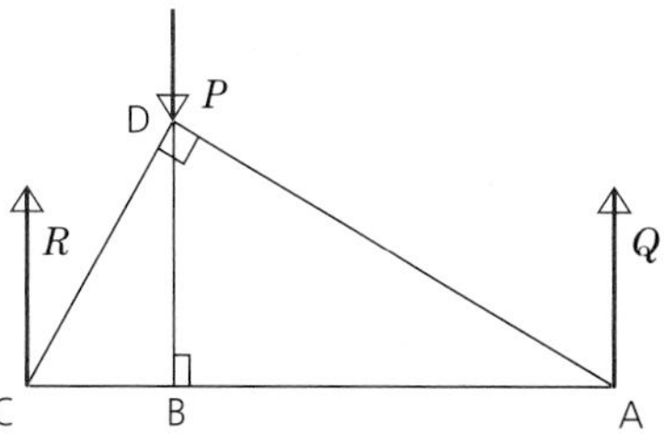

The forces on the framework are in equilibrium. By considering the forces acting at B, explain why the force in BD is zero.

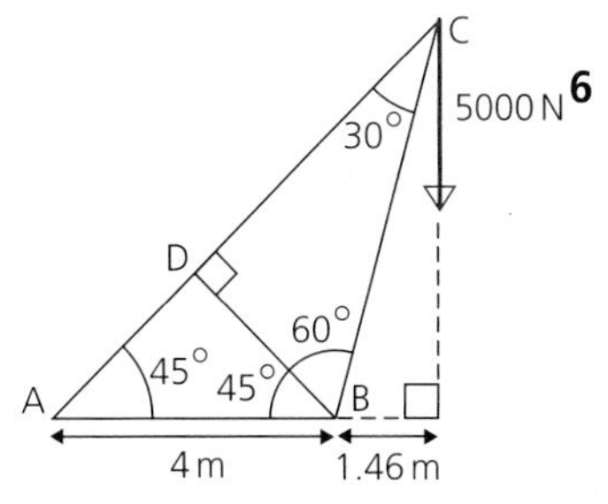

6 The diagram shows a design for a simple crane that supports a load of 5000 N at C.

By making suitable assumptions and giving your answers correct to two significant figure, find:

a) the vertical forces that act at A and B,

b) the force in each member, stating whether this force is a tension or a thrust.

7 The diagram shows a light pin-jointed framework, made up of equilateral triangles.

a) Find the magnitude of the forces R and S.

b) Describe how the symmetry of the framework can help you to find the forces in each rod.

c) Find the force in each rod.

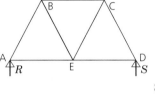

8 The framework shown in the diagram consists of light pin-jointed rods, joined to cables at A and D, and a smooth hinge at B.

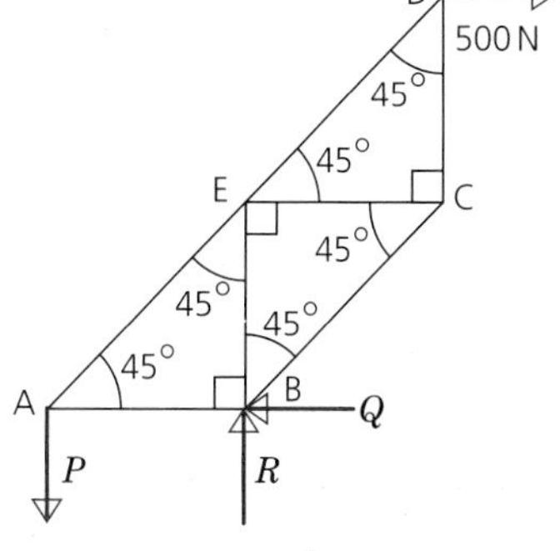

a) Find the magnitude of the force P that acts at A.

b) Find the horizontal component, Q, and the vertical component, R, of the force exerted by the hinge on the framework.

c) Find the force in each rod.

d) Which rod is most likely to break?

9 For the framework shown find the force in each member and state whether it is in tension or compression. All the triangles of the framework are equilateral.

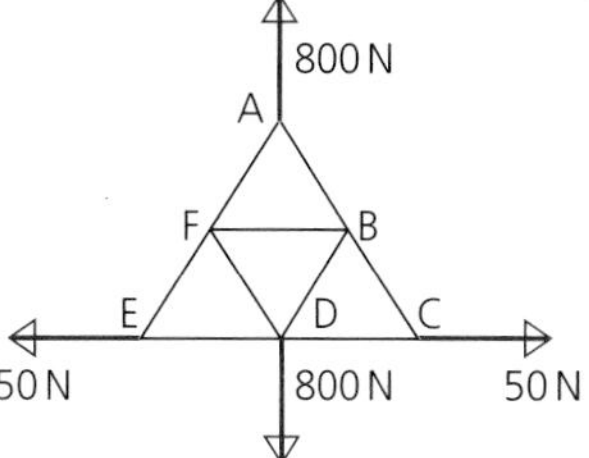

10 The diagram shows a framework that is part of a crane.

The framework is pivoted at E and a force Q acts at F. By making suitable assumptions, find the external forces on the framework and the force in each member.

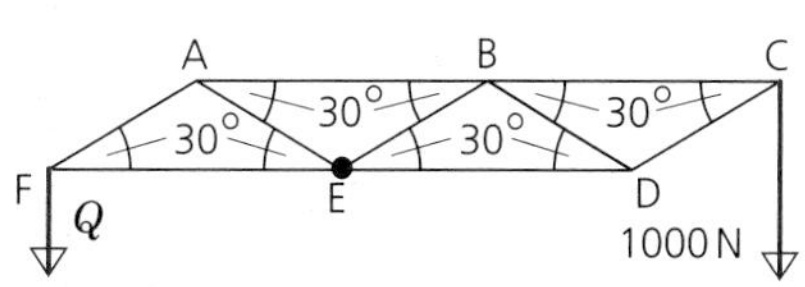

EXERCISES

19.1 B

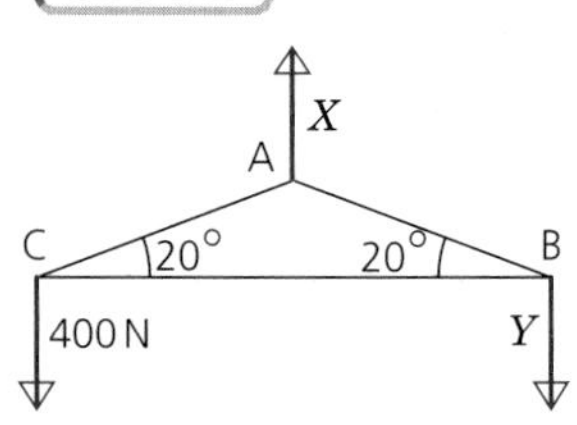

1 The diagram shows a simple light pin-jointed framework, with three external forces acting as shown.

a) By taking moments about the point B, find the magnitude of the force X.

b) Find Y by considering the resultant force on the framework.

c) By considering the forces at C, find the force in CA and then the force in CB. Which rod is in tension and which is in compression?

d) What is the force in AB?

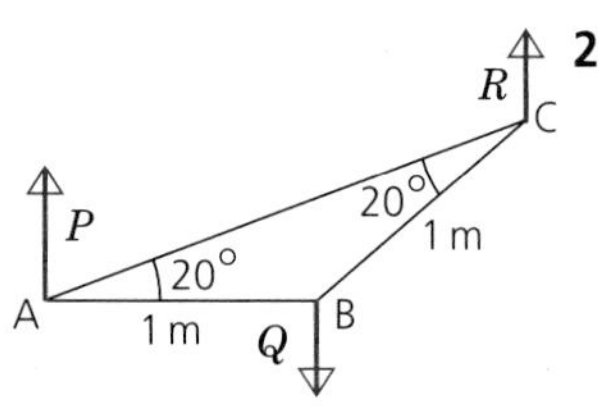

2 The diagram shows a framework with three external forces acting on it.

a) What assumptions can be made to allow you to estimate the forces in each rod?

b) If the force Q has magnitude 500 N, find P and R.

c) Find the force in each member of the framework and state if it is in tension or compression.

3 The diagram shows part of a linkage used to connect a trailer to a tractor. The tractor exerts a 5000 N force at C and the linkage is connected to the trailer at A and B, where forces of magnitude P act.

a) Find P.

b) Find the force in each member if they are assumed to be light and pin-jointed. Also state whether the members are in tension or compression.

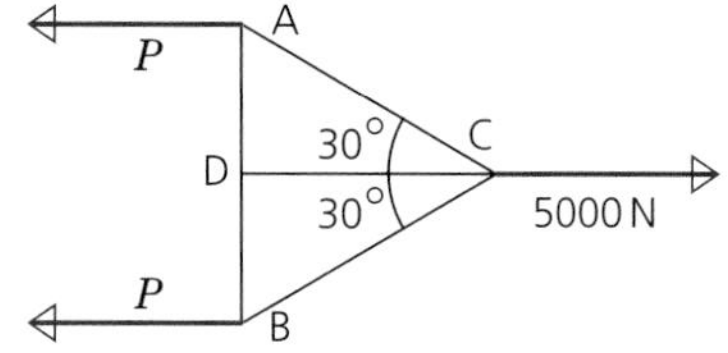

c) Comment on the design of the linkage.

4 The diagram shows a simple crane that is attached to the ground at A and B and supports a load of 6000 N at C as shown.

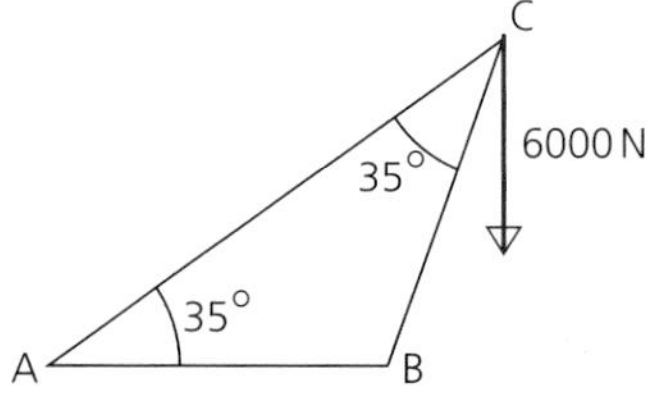

a) Which members do you think will be in tension and which in compression?

b) Find the forces acting at A and B.

c) By modelling the crane as a light pin-jointed framework, find the force in each member and compare with your predictions made in **a)**.

5 The diagram shows a light pin-jointed framework that is hinged at D.

a) By taking moments, find the magnitude of R.

b) By considering the resultant force on the framework, find P and Q the horizontal and vertical components of the force exerted by the hinge.

c) Find the force in each member of the framework.

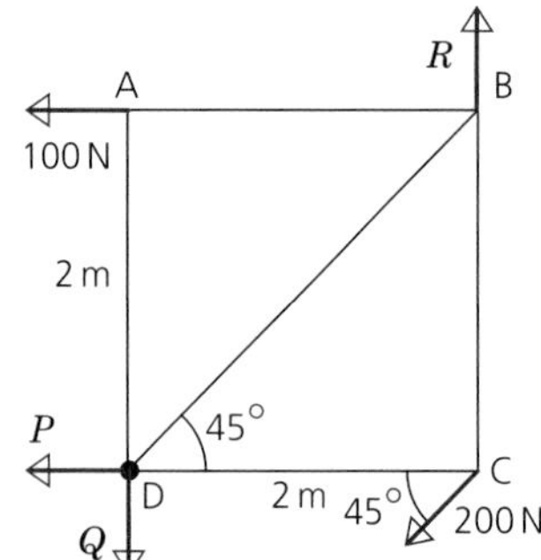

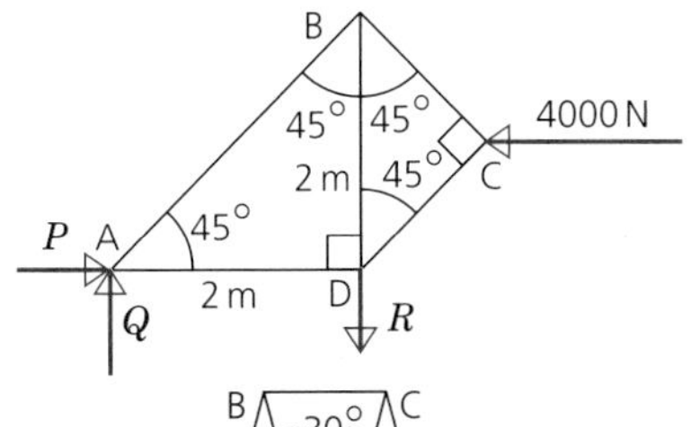

6 The diagram shows the design of a new type of railway buffer. A train exerts a force of 4000 N on the buffer. Forces P, Q and R act to maintain equilibrium.
   a) Find the magnitude of the forces P, Q and R.
   b) Find the force in each member of the buffer by modelling it as a light pin-jointed framework.

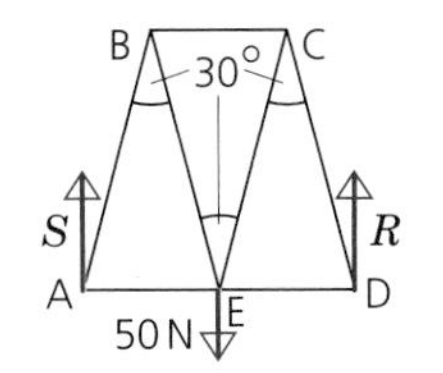

7 The diagram shows a pin-jointed framework that consists of three isosceles triangles.

Find the force in each rod. Clearly illustrate your results on a diagram that indicates whether forces are tensions or thrusts.

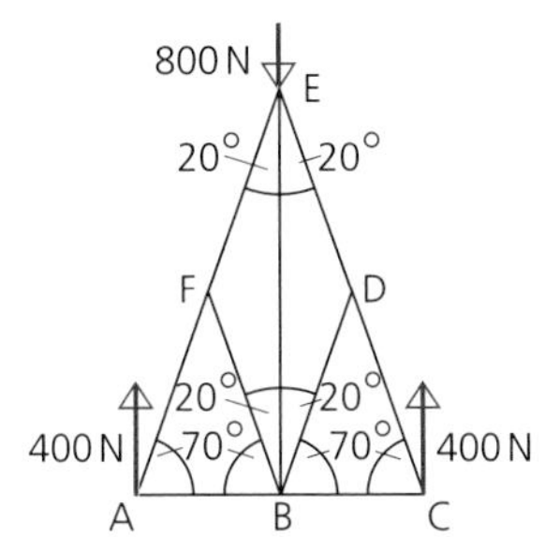

8 Find the force in each rod of the pin-jointed structure shown in the diagram on the left.

9 The diagram below shows a light pin-jointed framework that is freely pivoted on a hinge at D. A cable attached at C exerts a force of 160 N as shown. Another force acts at A in the direction shown. The framework remains in equilibrium. Find the magnitude of the force acting at A, the force in each member and the horizontal and vertical components of the forces acting on the framework at D.

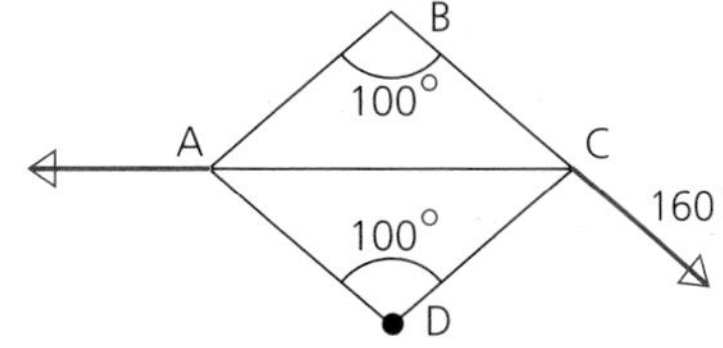

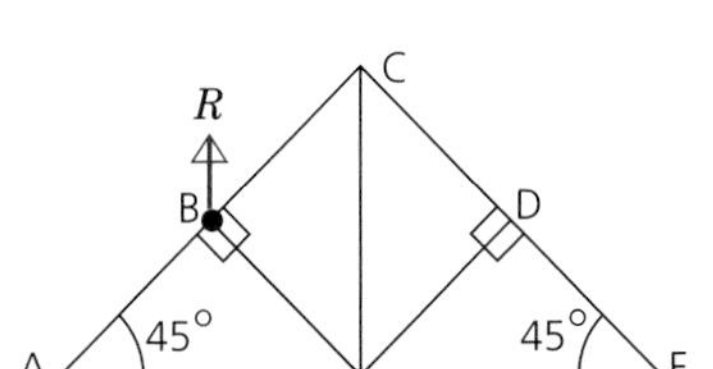

10 The diagram shows a light pin-jointed structure that is acted on by forces P, Q and R.
   a) Explain why $P = 3Q$, and state R in terms of P.
   b) If P is 90 N find the force in each rod.
   c) Use a diagram to illustrate the forces in each rod, showing if they are in tension or compression.

CONSOLIDATION EXERCISES FOR CHAPTER 19

1 The diagram shows a framework that is hinged at A. A 100 kg mass hangs from B and a cable attached at C maintains equilibrium.
   a) State any simplifying assumptions that would help you to proceed with analysis of the forces acting in this situation.
   b) Find the tension in the cable attached to C if it is horizontal.
   c) Which rods would you expect to be in tension and which in compression?
   d) Find the force in each of the rods.
   e) Find the magnitude and direction of the reaction force at A.

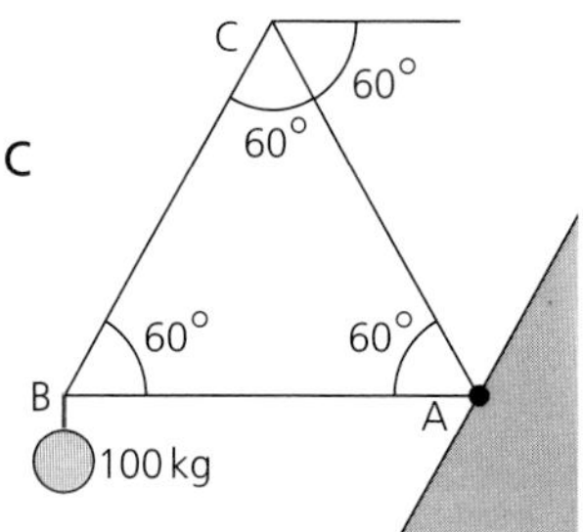

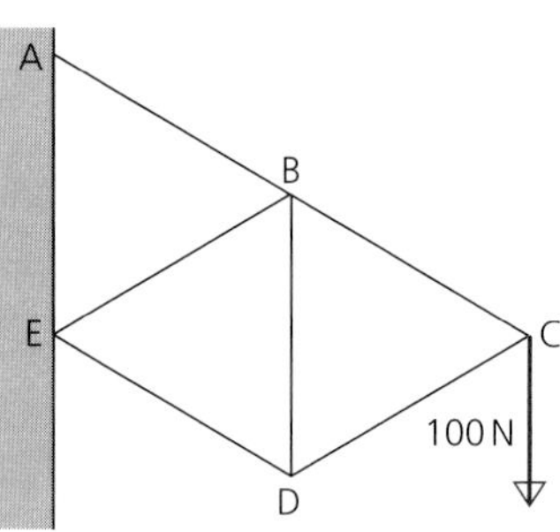

2 The diagram shows a light pin-jointed framework, made up of equilateral triangles.

a) Which rods do you expect to be in compression and which in tension?

b) Show that the forces in each rod have the same magnitude and confirm your predictions made in **a)**.

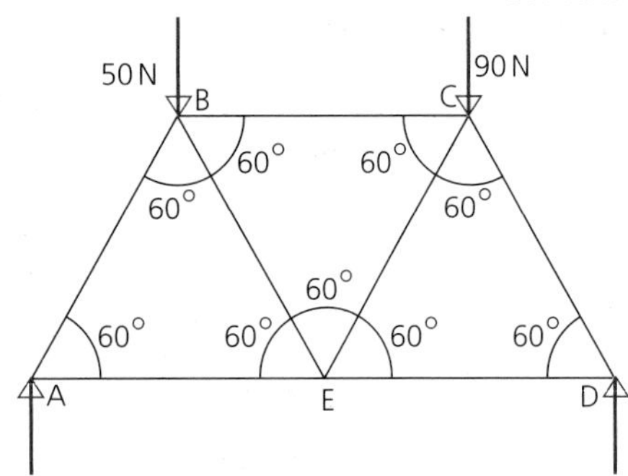

3 The figure shows a framework consisting of seven equal smoothly-jointed light rods AB, BC, DC, DE, AE, EB and EC. The framework is in a vertical plane with AE, ED and BC horizontal and is simply supported at A and D. It carries vertical loads of 50 N and 90 N at B and C respectively.

Find:

a) the reactions at A and D,

b) the magnitudes of the forces in AB, AE and BC.

(AEB Question 7, Paper 3, June 1991)

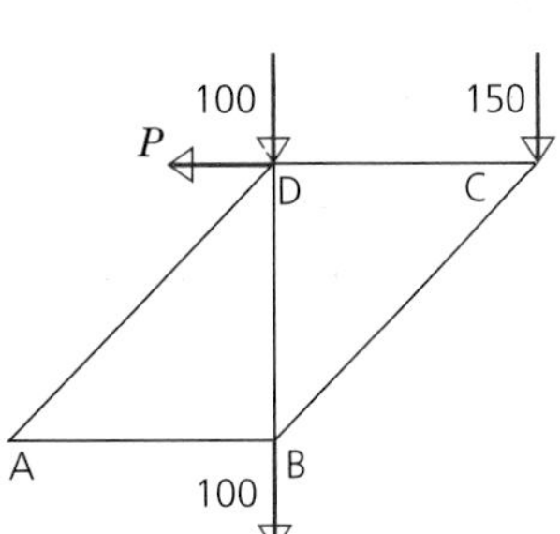

4 The diagram shows a framework A, B, C, D of freely-hinged light rods loaded at B, C and D with weights of magnitude 10 N, 150 N and 100 N respectively. The framework is in a vertical plane, smoothly hinged at the fixed point A and equilibrium is maintained by a horizontal force of magnitude P newtons at D. The rods AD and BC are inclined at $\frac{1}{4}\pi$ to the horizontal, the rods AB and CD are horizontal and rod BD is vertical. Find:

a) the value of P,

b) the magnitude and direction of the reaction at the hinge A,

c) the magnitude of the forces in each of the rods, stating in each case whether the rod is in tension or compression.

(AEB Question 10, Paper 3, June 1994)

MATHEMATICAL MODELLING ACTIVITY

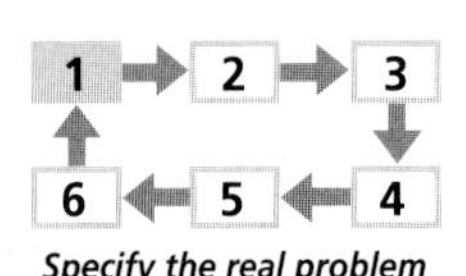

Specify the real problem

Problem statement

The diagram shows a crane of a fairly common design.

What loads can such a crane be expected to lift and what forces are exerted by the main components?

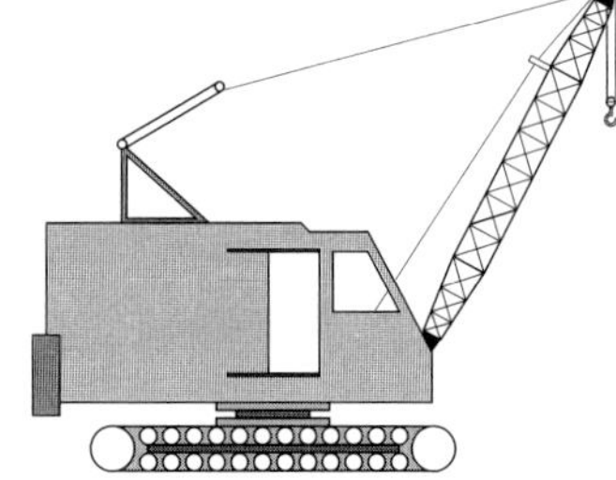

1 → 2 → 3
6 ← 5 ← 4

Set up a model

Set up a model

The assumptions below allow a first model to be formulated.

- The jib of the crane can be modelled as a pin-jointed light rod of length 60 m.
- The mass of the crane is 130 tonnes, and its body can be modelled as a rectangular block with its centre of mass at the centre.

■ The dimensions of this model of the crane are as illustrated in the diagram.

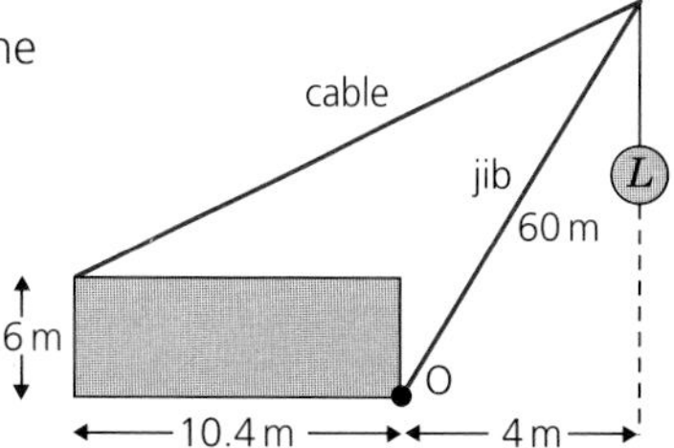

When the crane is lifting its maximum load the reaction force acts at O.

What other factors, if any, do you think should be considered in this problem?

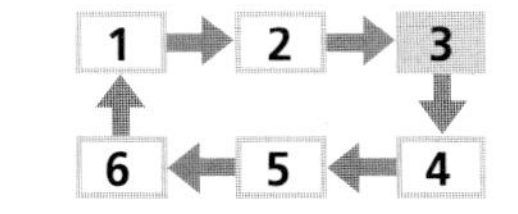

Formulate the mathematical problem

Formulate a mathematical problem

Find the maximum load that the crane can lift, if the crane is on the point of toppling. Also find the tension in the cable and the thrust in the jib.

Solve the mathematical problem

Solution

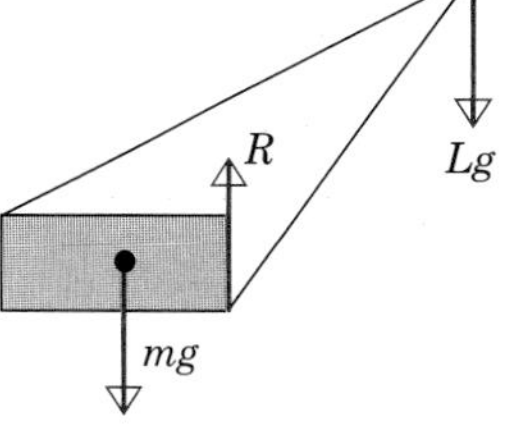

First consider the crane as a whole.

The diagram shows the three external forces, Lg, mg and R. Taking moments about O gives:

$$4Lg = 5.2 \times 1.3 \times 10^5 g$$

$$\Rightarrow L = \frac{5.2 \times 1.3 \times 10^5}{4} = 169\,000 \text{ kg} = 169 \text{ tonnes}$$

So the maximum load that can be supported is 169 tonnes.

Now consider the forces acting at the top of the jib, which are shown in the diagram.

T_1 β α $1.96 \times 10^5 gN$ T_2

Show that $\alpha = 3.8°$ and $\beta = 10.6°$.

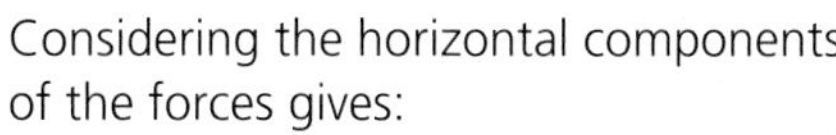

Considering the horizontal components of the forces gives:

$$T_1 \cos 75.6° + T_2 \cos 86.2° = 0$$

Considering the vertical components of three forces gives:

$$1.96 \times 10^5 \times 9.8 + T_2 \cos 3.8° + T_1 \cos 14.40° = 0$$

Show that these equations can be solved to give:

$T_1 = 6.92 \times 10^5$ *N* and $T_2 = -2.60 \times 10^6$ *N*

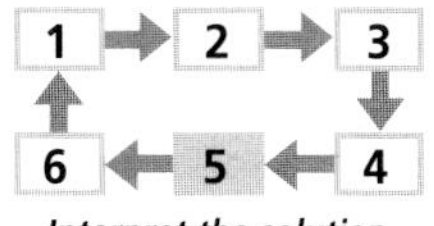

Interpret the solution

Interpretation

The calculations suggest that a maximum load of 169 tonnes could be lifted. As this load would leave the crane on the point of toppling this would not in reality be safe and a smaller load would have to be stipulated as a maximum.

The tension T_1 in the cable can be assumed to have a maximum value of 6.92×10^5 N and the thrust in the jib a maximum of 2.60×10^6 N. The fact that T_2 is negative confirms that a thrust is present.

Compare with reality

Compare with reality

The dimensions given were based on a Demag cc 600 crane which can lift a maximum load of 140 tonnes. The model, then, compares favourably with this result, as it would be risky to try to lift 169 tonnes.

Two major factors that have been ignored are the triangular framework at the rear of the crane and the fact that the mass would accelerate as it is initially lifted.

Refining the model

Consider the effect of adding a triangular framework at the rear of the crane.

POSSIBLE MODELLING TASKS

1 Construct a framework using the UniLab Structures kit. Predict the forces in each member when the framework is supporting a load and confirm your results with a forcemeter.

2 Identify some examples of frameworks that you have seen in your locality. Model them as light, pin-jointed structures and predict the forces that they must exert.

Summary

After having worked through this chapter you should be able to:

- model a framework as a system of light rods that are smoothly pin-jointed
- find external force on a framework by considering it as a whole
- find the forces in individual members by considering the forces at each joint to be in equilibrium.

MECHANICS

20 Differential equations in mechanics

- *Some resistance forces depend on the speed of an object. When they occur, they often lead to differential equations that cannot be solved by direct integration.*
- *In this chapter we use the method of separation of variables to solve these differential equations.*

FORMING EQUATIONS

When solving dynamics problems we almost always need to apply Newton's second law to produce an expression for the acceleration of the object under consideration. If this expression is a function of time, t, then the acceleration can be equated to $\frac{d^2x}{dt^2}$ (rate of change of displacement or distance per second per second) or $\frac{dv}{dx}$ (rate of change of velocity per second) and then integrated. However if the acceleration is a function of x or v, we have to use the method of **separation of variables**.

If the acceleration is a function of x it is useful to use an alternative expression for $\frac{dv}{dt}$, that involves dx not dt. This is obtained using the chain rule:

$$\frac{dv}{dt} = \frac{dx}{dt} \times \frac{dv}{dx} = v\frac{dv}{dx}$$

Example 20.1

A swimmer of mass 60 kg pushes off the side of a swimming pool and moves with an initial speed of 4 m s^{-1} under the water. There is a resistance force of 120v acting on the swimmer as she moves through the water. How far does the swimmer move before she comes to a halt?

Solution

The resultant force on the swimmer is simply –120v, so the acceleration is –2v. Using $\frac{dv}{dt}$ for the acceleration gives:

$$\frac{dv}{dt} = -2v$$

Separating the variables and integrating gives:

$$\frac{1}{v}\frac{dv}{dt} = -2 \Rightarrow \int \frac{1}{v}dv = \int -2dt \Rightarrow \ln v = -2t + c$$

Making v the subject of the expression gives:

$$v = e^{-2t+c} \Rightarrow v = Ae^{-2t}$$

where $A = e^c$. The value of A can now be determined using the initial conditions $v = 4$ when $t = 0$. This gives:

$$4 = Ae^0 \quad A = 4 \Rightarrow v = 4e^{-2t}$$

Now assume that $v = 0$ when $t = \infty$.
The distance that the swimmer travels can now be found by integrating. This distance is given by:

$$\int_0^\infty 4e^{-2t}\,dt = \left[-2e^{-2t}\right]_0^\infty = \left(-2e^{-2\infty}\right) - \left(-2e^0\right) = 0 + 2 = 2$$

The distance is 2 m.

Example 20.2

The terminal speed of a parachutist is estimated to be 5 m s^{-1}. Assume the air resistance is proportional to the mass and speed of the parachutist, and that the parachutist is moving at 20 m s^{-1} when the parachute opens and becomes effective.

a) *Formulate a simple model for the air resistance on the parachutist.*

b) *Find expressions for the speed of the parachutist and the distance fallen.*

Solution

a) *The air resistance is given by:* $R = mkv$

where k is an unknown constant.

*When the parachutist is at terminal speed the **resultant** force on the parachutist is zero.*

So $mg = mkv$

since $v = 5$ this gives $k = \dfrac{g}{5}$

So the resistance force, R, can be modelled as:

$$R = \frac{mgv}{5}$$

b) *With the downward direction defined as positive, the resultant force on the parachutist is:*

$$mg - R = mg - \frac{mgv}{5} = mg\left(1 - \frac{v}{5}\right)$$

Applying Newton's second law with $a = \dfrac{dv}{dt}$ *gives:*

$$m\frac{dv}{dt} = mg\left(1 - \frac{v}{5}\right) \Rightarrow \frac{dv}{dt} = g\left(1 - \frac{v}{5}\right)$$

Now this can be integrated using the method of separation of variables.

$$\frac{1}{1 - \frac{1}{5}v}\frac{dv}{dt} = g \Rightarrow \int \frac{1}{1 - \frac{1}{5}v}\,dv = \int g\,dt \Rightarrow -5\ln\left(1 - \tfrac{1}{5}v\right) = gt + c$$

This can be rearranged to make v the subject.

$$\ln\left(1-\tfrac{1}{5}v\right)=-\tfrac{1}{5}(gt+c)$$
$$1-\tfrac{1}{5}v=\mathrm{e}^{-\frac{1}{5}(gt+c)}$$
$$\tfrac{1}{5}v=1-\mathrm{e}^{-\frac{1}{5}(gt+c)}$$
$$v=5\left(1-\mathrm{e}^{-\frac{1}{5}(gt+c)}\right)$$
$$=5\left(1-A\mathrm{e}^{\frac{-gt}{5}}\right)$$

where $A=\mathrm{e}^{-\frac{c}{5}}$. Now the value of A can be determined using the initial conditions, $v = 20$ when $t = 0$. Substituting these values gives:

$$20=5\left(1-A\mathrm{e}^{0}\right)$$
$$4=1-A\Rightarrow A=-3$$

So the expression for v becomes:

$$v=5\left(1+3\mathrm{e}^{-\frac{gt}{5}}\right),$$

This can now be integrated to give the distance travelled as the parachute opens:

$$x=\int\left(5+15\mathrm{e}^{-\frac{gt}{5}}\right)\mathrm{d}t$$
$$=5t-\frac{5}{g}\times 15\mathrm{e}^{-\frac{gt}{5}}+c$$
$$=5t-\frac{75}{g}\mathrm{e}^{-\frac{gt}{5}}+c$$

If the value of $x = 0$ when $t = 0$, then c can be determined by substituting these values. This gives:

$$0=-\frac{75}{g}\mathrm{e}^{0}+c$$

$$\Rightarrow\quad c=\frac{75}{g}\mathrm{e}^{0}=\frac{75}{g}$$

Now the distance fallen can be expressed as:

$$x=5t-\frac{75}{g}\mathrm{e}^{-\frac{gt}{5}}+\frac{75}{g}=5t+\frac{75}{g}\left(1-\mathrm{e}^{-\frac{gt}{5}}\right)$$

Example 20.3

A bungee jumper of mass 75 kg jumps and reaches a speed of 15 m s^{-1} when the rope becomes taut. The stiffness of the rope is 100 N m^{-1}. Assume g = 10 m s^{-2}.

a) *Find an expression for the speed of the jumper in terms of the distance fallen after the rope becomes taut.*

b) *Find the maximum speed of the jumper.*

c) *Find the maximum extension of the rope.*

d) *Sketch a graph of v against the extension of the rope as the jumper descends.*

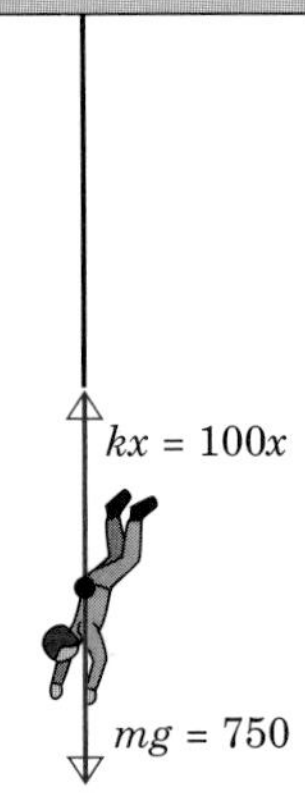

Solution

a) *The diagram shows the forces acting and the downward direction defined as positive.*
The resultant force is:
$mg - kx = 750 - 100x$
where x is the extension of the rope after it becomes taut.

So the acceleration is described by: $a = 10 - \frac{4x}{3}$

As the acceleration depends on x it is appropriate to use $v\frac{dv}{dx}$ for the acceleration, so:

$$v\frac{dv}{dx} = 10 - \frac{4x}{3}$$

Integrating gives:

$$\int v\,dv = \int\left(10 - \frac{4x}{3}\right)dx$$

$$\frac{v^2}{2} = 10x - \frac{4x^2}{6} + c$$

$$v^2 = 20x - \frac{4x^2}{3} + 2c$$

The initial conditions, $x = 0$ and $v = 15$ can be used to find the value of the constant c.
$225 = 2c \Rightarrow c = 112.5$
so $v^2 = 20x - \frac{4}{3}x^2 + 225$

b) *The speed is a maximum when $\frac{dv}{dx} = 0$. However, when v is maximum, v^2 will also be a maximum, so we can find when $\frac{d(v^2)}{dx} = 0$. This is easier to deal with.*

Differentiating the expression for v^2 gives:

$$\frac{d(v^2)}{dx} = 20 - \frac{8}{3}x$$

So for the maximum speed:

$$20 - \frac{8}{3}x = 0 \Rightarrow x = \frac{60}{8} = 7.5 \text{ m}$$

Substituting into the expression for v^2 gives:

$$v^2 = 20\times 7.5 - \frac{4\times 7.5^2}{3} + 225 = 300 \text{ and } v = 17.3 \text{ m s}^{-1}$$

c) *At the lowest point $v = 0$, so* $0 = 20x - \frac{4}{3}x^2 + 225$

Solving this quadratic equation gives:

$$x = \frac{-20 \pm \sqrt{20^2 - 4\times\left(-\frac{4}{3}\right)\times 225}}{2\times\left(-\frac{4}{3}\right)} = \frac{-20 \pm \sqrt{1600}}{-\frac{8}{3}} = 22.5 \text{ or } -7.5$$

d) *The sketch should include the following three points:*
i) *an initial speed of 15 m s^{-1} when $x = 0$,*
ii) *a maximum speed of 17.3 m s^{-1} when $x = 7.5$ m,*

iii) *a speed of 0 when $x = 22.5$ m.*

These are illustrated in the graph.

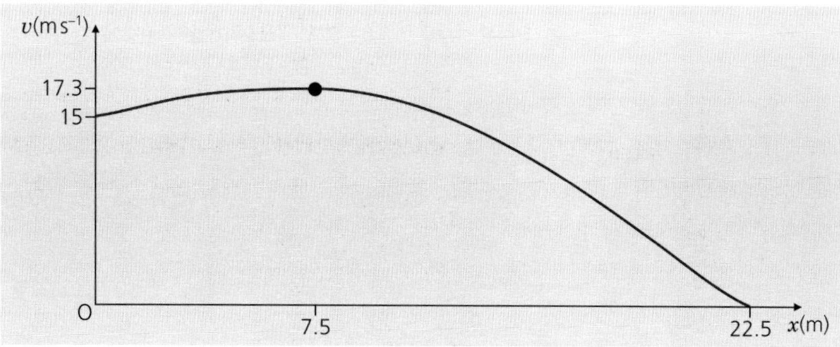

EXERCISES

20.1 A

1 As a cyclist free-wheels he experiences a resistive force of magnitude $40v$. Assume that the combined mass of the cycle and cyclist is 80 kg.

a) Find an expression for $\frac{dv}{dt}$, that models the motion of the cyclist freewheeling on a horizontal surface.

b) If the initial velocity of the cyclist is 10 m s^{-1} find an expression for the speed of the cyclist at time t.

c) Find an expression for the distance travelled from the initial position in terms of t. Use this to show that the cyclist travels a distance of 20 m before stopping.

2 A skydiver in free-fall has a terminal speed of 60 m s^{-1}. Assume that the air resistance is proportional to the speed of the skydiver.

a) Show that $\frac{dv}{dt} = g\left(1 - \frac{v}{60}\right)$.

b) Find an expression for v, if the skydiver is initially at rest.

c) How long does it take for the skydiver to reach a speed equal to half of his terminal speed?

3 A small boat of mass 250 kg experiences a resistance force that can be modelled as having magnitude mkv^2. The boat has a power output of 1000 watts and a top speed of 7.4 m s^{-1}.

a) By considering the top speed of the boat show that $k \approx 0.01$.

b) The boat cuts out its motors while travelling at its top speed. Find an expression for the acceleration of the boat at this time.

c) By using $v\frac{dv}{dx}$ find an expression for the velocity of the boat when it has travelled a distance x.

d) How far does the boat have to travel before its speed is reduced to 0.1 m s^{-1}?

4 The acceleration of a parachutist can be modelled by the expression

$$a = g\left(1 - \frac{v}{4}\right)$$

a) What is the terminal speed of the parachutist?

The parachutist was moving at 10 m s^{-1} when the parachute opened.

b) Find an expression for v in terms of t.

c) Find an expression for x in terms of t.

d) Use these results to find the speed and the distance travelled by the parachutist after ten seconds.

5 The differential equations below are possible models for different examples of motion.

A $\frac{dv}{dt} = g\sin 5° - \frac{v}{5}$ B $\frac{dv}{dt} = -\left(g\sin 5° + \frac{v}{5}\right)$

C $\frac{dv}{dt} = g - \frac{v^2}{20}$ D $v\frac{dv}{dt} = -g + \frac{v}{4}$

Which could model:

a) a car freewheeling downhill,

b) a cyclist freewheeling uphill,

c) a ball going upwards,

d) a ball going downwards?

6 A particle of mass 1 kg is attached to a fixed point by a length of elastic cord of stiffness 5 N m^{-1}, and natural length 1 m. The particle has an initial speed of 20 m s^{-1} and moves on a smooth horizontal surface. The particle is initially at the point where the elastic is attached. How far does the particle move before it comes to rest?

7 A bungee jumper of mass 60 kg uses an elastic rope of length 20 m of stiffness 100 N m^{-1}. Assume that there is no air resistance.

a) Find the speed of the bungee jumper when the rope becomes taut.

b) Find an expression for the speed of the bungee jumper in terms of the extension of the rope.

c) Find the maximum speed of the bungee jumper and the maximum extension of the rope.

8 The acceleration of a falling skydiver is modelled by $a = 10 - 0.3v$ where v is the velocity at time t.

a) What is the terminal speed? Is the skydiver's parachute open or is he in free-fall?

b) Find the speed of the skydiver after having fallen for ten seconds from rest.

9 The forces acting on an unpowered probe of mass m that is propelled from a submarine at a speed of 20 m s^{-1} are in equilibrium vertically, but include a horizontal resistance force of magnitude mkv^2.

a) Find expressions for v in terms of t and in terms of x.

b) After ten seconds the speed of the probe has dropped to 5 m s^{-1}. Find k and the distance travelled by the probe at this time.

10 Two identical balls of mass m are allowed to fall from rest. They are subject to a resistance force $-mv$. The first ball is released when $t = 0$ and the second is released when $t = T$. Find expressions for the distance fallen by each ball and describe what happens to the distance between them.

EXERCISES

20.1 B

1 A bullet of mass 40 grams enters a sand bank travelling at 80 m s^{-1}. It is subject to a resistance force of $20v$ N as it moves through the sand.

a) Find the acceleration of the bullet.

b) Find an expression for the speed of the bullet in terms of t.

c) Find an expression for the distance travelled by the bullet in terms of t.

d) Find how far the bullet travels before it stops.

2 A car with a power output of 40 kW has a top speed of 50 m s^{-1}.

a) Find a simple model for the resistance forces in the car by assuming that they are proportional to its speed.

b) The car, which has mass 800 kg, is travelling at 20 m s^{-1} when the driver allows it to free-wheel until it comes to rest. Find the distance that the car travels before it stops.

3 A boat with maximum power output of 30 kW can reach a top speed of 10 m s^{-1}. The mass of the boat is 1200 kg.

a) What can be deduced about the resultant force on the boat when it is travelling at its top speed?

b) By assuming the resistance forces on the boat are proportional to its speed find a simple model for the resistance force.

c) By assuming that the forward force on the boat is a constant and always takes its maximum value show that: $\dfrac{dv}{dt} = \dfrac{10 - v}{4}$

d) Find an expression for v in terms of t, if $v = 0$ when $t = 0$.

e) How long would it take the boat to reach a speed of 5 m s^{-1}?

4 A ship of mass m is subject to a resistance force of magnitude kv^2.

a) Suggest the factors on which k might depend.

b) The ship's motors stop, so that it is only acted upon by the resistance force. Show that when the ship has travelled a distance $\dfrac{m}{k}\ln 2$ its initial speed has been halved.

5 The four differential equations below could be used to model several different situations.

A $\dfrac{dv}{dt} = -g - kv$ B $x\dfrac{dv}{dx} = \dfrac{k}{x}$

C $\dfrac{dv}{dt} = -\dfrac{kv^2}{m}$ D $x\dfrac{dv}{dx} = g - kv^2$

Which of the equations could model:

a) a ball bearing moving towards a magnet,

b) a bullet fired directly up into the air,

c) an unpowered torpedo moving horizontally through the sea,

d) a sky diver in free-fall?

6 A magnet of radius 2 cm exerts a force of magnitude $\frac{10m}{x}$ N on a ball bearing, of mass m, when it is at a distance x m, from the centre of the magnet. Find the speed of the ball bearing when it hits the magnet, if it is initially at rest at a distance of 10 cm from the centre of the magnet.

7 A particle of mass, 100 grams, is attached to a 1m length of elastic of stiffness 20 N m^{-1}. The particle is projected with an initial speed of 10 m s^{-1} from the point where the other end of the elastic is fixed on a smooth table.

a) Find the maximum displacement of the particle.

b) State two factors that would in reality affect the motion of the particle. Revise the differential equation that you used to solve this problem, clearly defining any variables that you introduce.

8 A cycle and cyclist of total mass 75 kg experience a forward force of magnitude $\frac{150}{v}$ and a resistance force of magnitude $75kv$.

a) Find an expression for the acceleration of the cyclist and used it to obtain an expression for v, in terms of k, t and an unknown constant if the cyclist starts at rest.

b) The maximum speed of the cyclist is 12 m s^{-1}. Use this to find the value of k.

9 Emergency parcels are to be dropped from a helicopter to stranded hill-walkers. The parcels fall under gravity and are subject to an air resistance force of magnitude $\frac{1}{50}mv^2$. The parcels are designed to withstand an impact at a speed of 10 m s^{-1}. Find the maximum height from which the parcels can be dropped.

10 A torpedo, of mass m, is fired horizontally at 50 m s^{-1} from a stationary submarine. It continues to move horizontally and after three seconds hits a target at a speed of 20 m s^{-1}. Assume that the torpedo is subject to a resisting force of magnitude mkv^3, where v is the speed of the torpedo and k a constant.

By considering $\frac{dv}{dt}$ and $\frac{dv}{dx}$ find the distance of the target from the submarine.

CONSOLIDATION EXERCISES FOR CHAPTER 20

1 A small stone, S, of mass 0.005 kg, falls from rest in a tank of oil in which the resistance to motion is $0.2v$ N, where v m s^{-1} is the speed of S at time t seconds.

a) Show that the acceleration of S at time v is $(9.8 - 40v)$ m s^{-2}.

b) Find v in terms of t.

(ULEAC Question 3, Paper M2 January 1992)

2 A particle of mass 0.2 kg is moving in a straight line under the action of a resistive force which, when the particle has speed v m s^{-1}, is of magnitude kv^3 N, where k is a constant. Given that the force is of magnitude 80 N when $v = 2$, find k.

Show that, at time t seconds, v satisfies: $\frac{dv}{dt} = -50v^3$

Find the time taken for the speed to decrease from 2 m s^{-1} to 1 m s^{-1}.

(AEB Question 5, Paper 3, June 1991)

3 A skydiver, with her equipment, has a mass of 60 kg. She drops vertically from an aeroplane and falls freely from rest under her own weight. Air resistance exerts a retarding force of kv N where k kg s^{-1} is a constant and v m s^{-1} is her speed. Under these circumstances she will eventually reach a terminal speed of 50 m s^{-1}.

a) Explain how her motion is modelled by a differential equation of the form:

$$m\frac{dv}{dt} = mg - kv$$

b) Find the value of k.

c) Find her speed as a function of time.

d) Find the time taken for her to reach 95% of her terminal speed.

(MEI Question 3, M3 June 1993)

4 A particle P of mass m falls vertically into a container of liquid, entering the liquid with speed U. The particle continues to fall vertically, and when it has fallen a distance x through the liquid, its speed is v. The motion of P through the liquid is resisted by a force of magnitude mkv^2, where k is a constant. Show that:

$$x = \frac{1}{2k}\ln\frac{g - kU^2}{g - kv^2}.$$

(ULEAC Question 5, M2, January 1993)

5 An animal A runs in a straight line and has an acceleration of $0.05(20v - v^2)$ m s^{-2}, where v m s^{-1} is the speed of A.
Show that, at time t seconds:

$$\frac{dv}{dt}\left(\frac{1}{v} + \frac{1}{20 - v}\right) = 1$$

Given that at time $t = 0$, A passes through a point O with speed 4 m s^{-1}, show that at time t seconds the speed of A is given by:

$$v = \frac{20e^t}{4 + e^t}$$

Hence, or otherwise, find the distance of the animal from O when $t = 9$.

(AEB Question 11, Paper 3, June 1994)

MATHEMATICAL MODELLING ACTIVITY

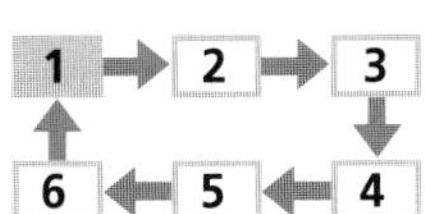

Specify the real problem

Problem statement

Experienced parachutists claim that when they jump from an aeroplane their speed decreases and then increases. Investigate the claim and suggest when a parachutist should open his parachute.

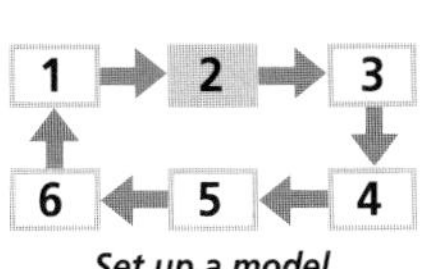

Set up a model

Set up a model

Make a list of the key features that you feel should be taken into account in solving this problem and suggest suitable values for any constants, such as terminal speed, that you introduce.

The assumptions given below represent one way of starting the problem.

- The aeroplane is travelling horizontally at 200 m s^{-1} when the parachutist jumps out.

- The terminal speed of parachutist in freefall is 70 m s^{-1}.
- The air resistance is proportional to the speed and mass of the parachutist.
- The acceleration due to gravity is 10 m s^{-2}.
- The parachutist is modelled as a particle.

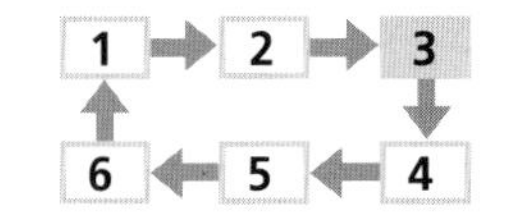

Formulate the mathematical problem

Formulate a mathematical problem

1 Find a simple model for the air resistance force.
2 Find the resultant force on the parachutist and derive an expression for their acceleration.
3 Find the velocity of the parachutist and determine when its magnitude is a minimum.

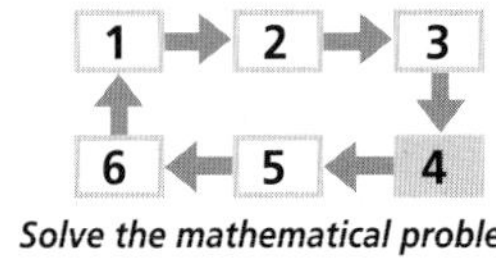

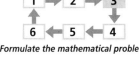

Solve the mathematical problem

Mathematical solution

1 Consider a parachutist falling vertically at the terminal speed of 70 m s^{-1}. If the air resistance is proportional to the speed and mass, it will have magnitude kmv or in this case $70km$ where k is an unknown constant. However as it balances the force of gravity $70km = mg$ so that $k = \frac{1}{7}$. The air resistance then has magnitude $\frac{1}{7}mv$.
2 The diagram shows the forces acting on the parachutist based on the assumption that the air resistance is make up of a horizontal and a vertical component.

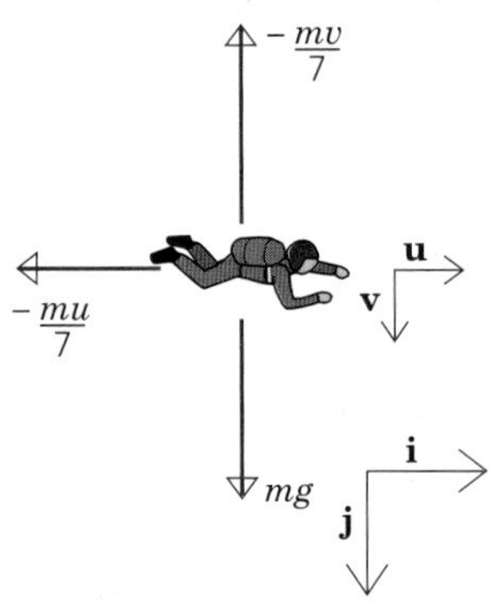

Note that v is the vertical component of the velocity and u the horizontal component of the velocity. So the resultant force, $\mathbf{F}$, is:

$$\mathbf{F} = -\tfrac{1}{7}mu\,\mathbf{i} + \left(mg - \tfrac{1}{7}mv\right)\mathbf{j}$$

The acceleration can now be found as:

$$\mathbf{a} = -\tfrac{1}{7}u\,\mathbf{i} + \left(g - \tfrac{1}{7}v\right)\mathbf{j} \;\Rightarrow\; \frac{\mathrm{d}u}{\mathrm{d}t}\mathbf{i} + \frac{\mathrm{d}v}{\mathrm{d}t}\mathbf{j} = -\tfrac{1}{7}u\,\mathbf{i} + \left(g - \tfrac{1}{7}v\right)\mathbf{j}$$

Considering the components separately gives:

$$\frac{\mathrm{d}u}{\mathrm{d}t} = -\tfrac{1}{7}u \quad \text{and} \quad \frac{\mathrm{d}v}{\mathrm{d}t} = g - \tfrac{1}{7}v$$

Show that using separation of variables and interpreting gives:

$$u = 200\mathrm{e}^{-\frac{1}{7}t} \quad \text{and} \quad v = 70\left(1 - \mathrm{e}^{-\frac{1}{7}t}\right)$$

(**Hint:** The initial conditions are $u = 200$ and $v = 0$ when $t = 0$.)

Now the actual speed, V, of the parachutist is given by:

$$V = \sqrt{u^2 + v^2} \quad \text{or} \quad V^2 = u^2 + v^2$$

Substituting the expressions for u and v gives:

$$V^2 = 40\,000\,\mathrm{e}^{-\frac{2}{7}t} + 4900\left(1 - 2\mathrm{e}^{-\frac{1}{7}t} + \mathrm{e}^{-\frac{2}{7}t}\right) = 4900 - 9800\,\mathrm{e}^{-\frac{1}{7}t} + 44\,900\,\mathrm{e}^{-\frac{2}{7}t}$$

3 For V to be a minimum $\frac{\mathrm{d}V}{\mathrm{d}t}$ must be zero.
Differentiating gives:

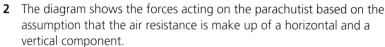

$$2V\frac{\mathrm{d}V}{\mathrm{d}t} = 1400\,\mathrm{e}^{-\frac{1}{7}t} - \frac{89\,800}{7}\mathrm{e}^{-\frac{2}{7}t}$$

so that when $\frac{dV}{dt} = 0$:

$$1400e^{-\frac{1}{7}t} - \frac{89\,800}{7}e^{-\frac{2}{7}t} = 0$$

Multiplying by $e^{\frac{1}{7}t}$ gives:

$$1400 - \frac{89\,800}{7}e^{-\frac{1}{7}t} = 0$$

and then solving for t gives:

$$t = -7\ln\frac{1400 \times 7}{89\,800} = 15.5$$

So $t = 15.5$ s.

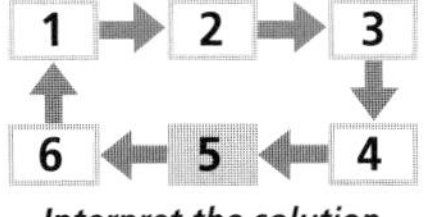

Interpret the solution

Interpretation

The parachutist reaches a minimum speed after 15.5 seconds. This is the best time to open the parachute. One danger is that the parachutist may have travelled a considerable distance in this time.

Find the position of the parachutist after 15.5 s, relative to the starting point and verify that he has fallen approximately 649 m.

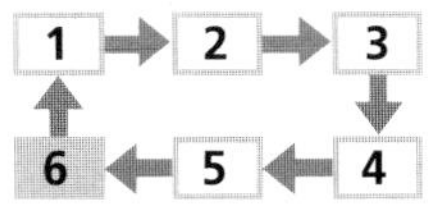

Compare with reality

Compare with reality

If possible, discuss your findings with an experienced parachutist!

POSSIBLE MODELLING TASKS

1 Find out about a ship or boat. Gather data on the top speed and power output of its engines. Use this to develop a model for the resistance forces acting on a moving boat. Then apply this model to produce a table of stopping distances for that boat similar to the stopping distances given for cars in the *Highway Code*. Sample data for a Type 22 Frigate: power output 40 MW, maximum speed 31 knots.

2 When parachutists jump from a balloon they fall vertically to the ground. What is the minimum safe height from which a parachutist should be able to jump?

Summary

After you have worked through the chapter you should be able to:

- use the different forms of acceleration: $\frac{d^2x}{dt^2}$, $\frac{dv}{dt}$ and $v\frac{dv}{dx}$
- set up differential equations to describe real situations
- solve problems using separation of variables.

Relative motion

- *When a ship sails in a current or tidal stream it is important to include the effects of the current on the motion of the ship.*
- *It is essential that aeroplanes and ships are not allowed to travel on courses that will cause them to collide. Their relative motion can be used to check that they will not collide.*

RELATIVE AND RESULTANT MOTION

In this chapter we are concerned with relative motion. A canoe paddled in a fast-moving river would travel relative to the water, but the speed and direction of the flow of the river would be important factors in determining the **resultant motion** of the canoe. Two boats could be travelling in fixed directions. To the crew on one boat, the direction of motion of other boat appears to be different from what it really is, because they are observing the motion of the second boat relative to their own motion.

Exploration 21.1

Paddling a canoe across a river

A canoeist tries to paddle straight across a river, at a constant speed, by paddling at right angles to the bank.

- What could happen to the canoeist as she tries to cross the river? Sketch some possible paths.
- If the canoeist wants to land at a point directly opposite her starting point, in which direction should she paddle?

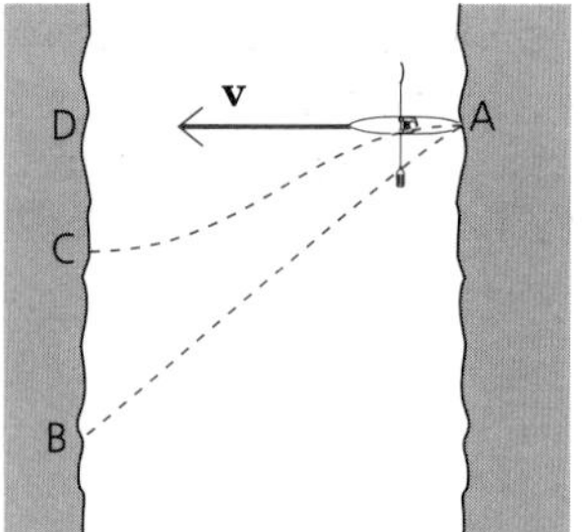

Resultant motion

The canoeist will clearly be swept downstream and could land at a point on the opposite bank below the starting point. The path AB assumes there is a strong current that is constant across the river. The path AC assumes there is a weaker current, but that it is stronger in the centre of the river.

If the canoeist wants to cross from A to D then the canoe must have a component of its velocity, equal to the speed of the current, but directed *up* the river. If the maximum speed of the canoeist is equal to the speed of the current, then it is impossible for her to cross the river without being swept downstream.

Example 21.1

A man tries to sail a dinghy north-east at 4 m s^{-1} in a current that is flowing due south at 2 m s^{-1}. Find the actual speed and direction of motion of the dinghy.

Solution

The diagram shows the velocity of the dinghy and the velocity of the current, as well as the unit vectors $\mathbf{i}$ *and* $\mathbf{j}$.

The velocity of the dinghy relative to the current is:

$\mathbf{V}_\mathrm{D} = 4\cos45°\mathbf{i} + 4\cos45°\mathbf{j}$

and the velocity of the current is:

$\mathbf{V}_\mathrm{C} = -2\mathbf{j}$

The actual velocity $\mathbf{V}$ *of the dinghy is:*

$\mathbf{V} = \mathbf{V}_\mathrm{D} + \mathbf{V}_\mathrm{C} = 4\cos45°\mathbf{i} + (4\cos45° - 2)\mathbf{j} = 2.83\mathbf{i} + 0.83\mathbf{j}$

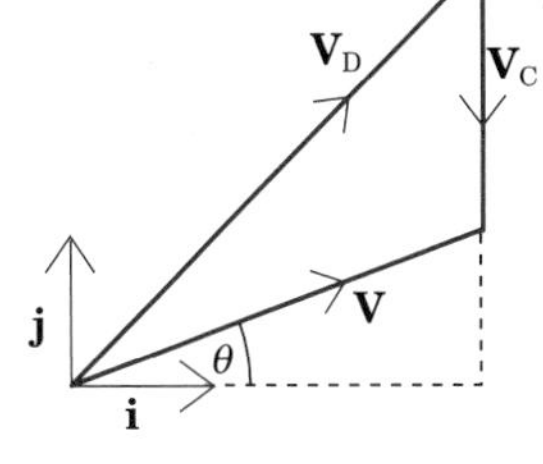

The actual velocity is illustrated in the diagram, which shows its relationship to $\mathbf{V}_\mathrm{D}$ *and* $\mathbf{V}_\mathrm{C}$.

The resultant speed is given by:

$$V = \sqrt{2.83^2 + 0.83^2} = 2.95 \text{ m s}^{-1}$$

The direction of motion is given by:

$$\theta = \tan^{-1}\frac{0.83}{2.83} = 16.3°$$

So the actual motion of the ship is 2.95 m s^{-1} on a bearing of 073.7°.

Example 21.2

A man can row at a speed of 1.2 m s^{-1} and is trying to cross a river 10 m wide, that has a current that moves at 0.4 m s^{-1}.

a) *The man wants to follow a path perpendicular to the bank. In what direction should he row the boat?*

b) *How long does it take the man to cross the river?*

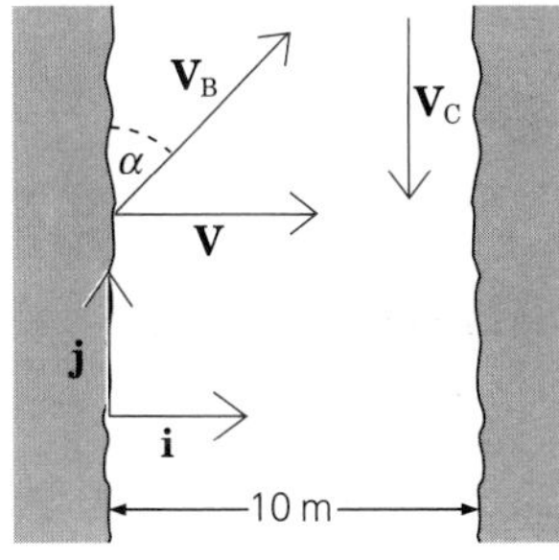

Solution

a) *The man must row so that there is a component of his velocity upstream to cancel out the effect of the current on the boat. Assume that the velocity of the boat relative to the current is at an angle* α *to the bank as shown in the diagram.*

The velocity of the boat relative the current is:

$\mathbf{V}_\mathrm{B} = 1.2\sin\alpha\mathbf{i} + 1.2\cos\alpha\mathbf{j}$

and the velocity of the current is:

$\mathbf{V}_\mathrm{C} = -0.4\mathbf{j}$

The resultant velocity $\mathbf{V}$ *is then:*

$\mathbf{V} = \mathbf{V}_\mathrm{B} + \mathbf{V}_\mathrm{C} = 1.2\sin\alpha\mathbf{i} + 1.2\cos\alpha\mathbf{j} - 0.4\mathbf{j} = 1.2\sin\alpha\mathbf{i} + (1.2\cos\alpha - 0.4)\mathbf{j}$

For the boat to cross at right angles to the bank the component of the velocity parallel to the banks must be zero.

$$1.2\cos\alpha - 0.4 = 0 \Rightarrow \cos\alpha = \frac{0.4}{1.2} = 0.333 \Rightarrow \alpha = 70.5°$$

So the man must row at an angle of 70.5° to the bank.

b) *The component of velocity perpendicular to the bank is given by:*
$1.2\sin\alpha = 1.2\sin 70.5° = 1.13 \text{ m s}^{-1}$
So the time taken to cross the river is $t = \frac{10}{1.13} = 8.8$ *seconds.*

Relative velocity

In the examples we have considered so far, where a current flows, the actual velocity $\mathbf{V}_B$ is given by:

$$\mathbf{V}_B = \mathbf{V}_C + \mathbf{V}_{BC}$$

where $\mathbf{V}_C$ is the velocity of the current and $\mathbf{V}_{BC}$ the velocity of the boat relative to the current. This can be rearranged to give:

$$\mathbf{V}_{BC} = \mathbf{V}_B - \mathbf{V}_C$$

The velocity of the boat relative to the current is equal to the velocity of the boat minus the velocity of the current.

This idea can be extended to find the relative velocity of one object relative to another. If A has velocity $\mathbf{V}_A$ and B has velocity $\mathbf{V}_B$, then the velocity of B relative to A is given by:

$$\mathbf{V}_{BA} = \mathbf{V}_B - \mathbf{V}_A$$

Example 21.3

A ship sails due east at 3 m s^{-1} *in thick fog. A tanker sails north east at a speed of 4 m s*$^{-1}$*. The tanker is initially 200 m due south of the ship.*

a) *Find the velocity of the ship relative to the tanker.*
b) *Find the position of the ship relative to the tanker.*
c) *Do the two vessels collide?*

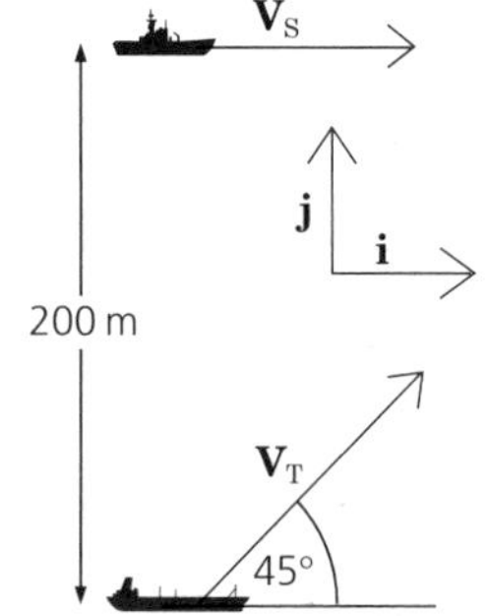

Solution

The diagram shows the initial positions of the ship and the tanker and their velocities.

a) *The velocity of the ship relative to the tanker,* $\mathbf{V}_{ST}$*, is given by:*
$\mathbf{V}_{ST} = \mathbf{V}_S - \mathbf{V}_T = 3\mathbf{i} - (4\cos 45°\mathbf{i} + 4\cos 45°\mathbf{j}) = 0.17\mathbf{i} - 2.83\mathbf{j}$
To the crew of the ship the tanker would appear to have an easterly and a southerly components of velocity.

b) *The displacement of the ship relative to the tanker is given by:*
$\mathbf{r}_{ST} = \mathbf{r}_0 + \mathbf{V}_{ST}t$
where $\mathbf{r}_0$ *is the initial position of the ship relative to the tanker. So:*
$\mathbf{r}_{ST} = 200\mathbf{j} + (0.17\mathbf{i} - 2.83\mathbf{j})t = 0.17t\ \mathbf{i} + (200 - 2.83t)\mathbf{j}$

c) *For the two vessels to collide* $\mathbf{r}_{ST}$ *must be* $0\mathbf{i} + 0\mathbf{j}$ *or which gives:*
$200 - 2.83t = 0$ *(1)*
$0.17t = 0$ *(2)*
Equation (1) gives $t \approx 71$ *seconds and equation (2) gives* $t = 0$*. So clearly the two vessels never collide.*

Example 21.4

A yacht gets into difficulties and drifts due south at 0.5 m s^{-1}*. A lifeboat is launched to search for the yacht. The lifeboat travels from a point 5000 m due west on a bearing of 100° at a speed of 8 m s*$^{-1}$*.*

a) *Find the velocity of the yacht relative to the lifeboat.*

b) *Find the position of the yacht relative to the lifeboat.*
c) *Find the distance between the two vessels.*
d) *What is the minimum distance between the two vessels?*

Solution

The diagram shows the initial positions and the velocities of the two vessels.

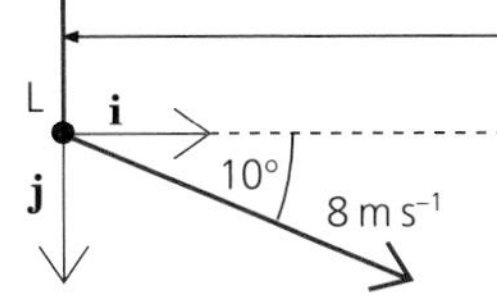

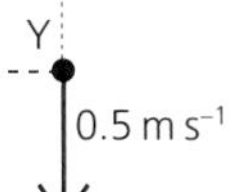

a) *The velocity of the yacht* $\mathbf{V}_\mathrm{Y}$ *is* $0.5\mathbf{j}$. *The velocity of the lifeboat* $\mathbf{V}_\mathrm{L}$ *is:*
$\mathbf{V}_\mathrm{L} = 8\cos 10°\mathbf{i} + 8\cos 80°\mathbf{j} = 7.88\mathbf{i} + 1.39\mathbf{j}$
The velocity of the yacht relative to the lifeboat is:
$\mathbf{V}_\mathrm{YL} = \mathbf{V}_\mathrm{Y} - \mathbf{V}_\mathrm{L} = 0.6\mathbf{j} - (7.88\mathbf{i} + 1.39\mathbf{j}) = -7.88\mathbf{i} - 0.89\mathbf{j}$

b) *The position of the yacht relative to the lifeboat is:*
$\mathbf{r}_\mathrm{YL} = \mathbf{r}_0 + \mathbf{V}_\mathrm{YL}t = 5000\mathbf{i} + (-7.88\mathbf{i} - 0.89\mathbf{j})t = (5000 - 7.88t)\mathbf{i} - 0.89t\mathbf{j}$

c) *The distance, x, between the two vessels is given by:*
$x^2 = (5000 - 7.88t)^2 + (-0.89t)^2 = 25\,000\,000 - 78\,800t + 62.89t^2$

To find the minimum distance between the vessels, the value t for which $\frac{\mathrm{d}x}{\mathrm{d}t} = 0$ *must be found. Differentiating the expression above gives:*

$$2x\frac{\mathrm{d}x}{\mathrm{d}t} = -78\,800 + 125.78t$$

When $\frac{\mathrm{d}x}{\mathrm{d}t} = 0$ *this becomes:*

$$-78\,800 + 125.78t = 0 \Rightarrow t = \frac{78\,800}{125.78} = 626.5 \text{ seconds}$$

The minimum distance between the vessels is now given by:

$$x = \sqrt{25\,000\,000 - 78\,800 \times 626.5 + 62.89 \times 626.5^2} = 561\,m$$

EXERCISES

1 A man rows a boat at 1.2 m s^{-1} at right angles to a river bank. The width of the river is 6 m and a current flows in the river at 0.3 m s^{-1}.

a) Write down the resultant velocity, in terms of the unit vectors $\mathbf{i}$ and $\mathbf{j}$ that are perpendicular and parallel to the bank respectively.
b) Find the time it takes for the man to cross the river.
c) How far downstream does the boat travel?

2 A canoeist can paddle at 2 m s^{-1}. A river is 10 m wide and has a current that flows at a speed of 1.2 m s^{-1}. The canoeist tries to cross the river.

a) If the canoeist paddles at right angles to the bank of the river, find the resultant velocity, in terms of appropriate unit vectors, and how far downstream the canoeist travels while crossing the river.

b) The canoeist paddles at an angle to the bank so as to follow a path that is at right angles to the bank. Find the direction in which the canoeist must paddle and the time it takes to cross the river.

3 Some students are trying to determine the wind speed by an experiment. They throw a ball up into the air so that its initial velocity is vertical. It is in the air for 2.4 s and travels a horizontal distance of 3.84 m.

a) Show that a possible estimate of the wind speed is 1.6 m s^{-1}.

b) Criticise this estimate.

4 An aeroplane flies at 100 m s^{-1} on bearing of 060°, in air that is moving due north at 20 m s^{-1}.

a) Find the resultant velocity of the aeroplane, in terms of the unit vectors $\mathbf{i}$ and $\mathbf{j}$ which are east and north respectively.

b) If the pilot is not aware of the wind, how far north would the aeroplane be from its proposed course after 1 hour?

5 A train is travelling horizontally at 20 m s^{-1} and a raindrop is falling vertically at 10 m s^{-1}. Find the velocity of the raindrop, relative to the train, in terms of the horizontal and vertical unit vectors $\mathbf{i}$ and $\mathbf{j}$.

6 Two ferries, A and B, leave a port at the same time. A sails due north at 6 m s^{-1} and B sails due east at 5 m s^{-1}.

a) Find the velocity of A relative to B, in terms of the unit vectors $\mathbf{i}$ and $\mathbf{j}$ which are directed east and north respectively.

b) Find the position of A relative to B.

c) Find an expression for the distance between the two ships.

7 A tanker is initially at A and heading east at 4 m s^{-1}, while a fishing boat is initially at B and heading south at 3 m s^{-1}.

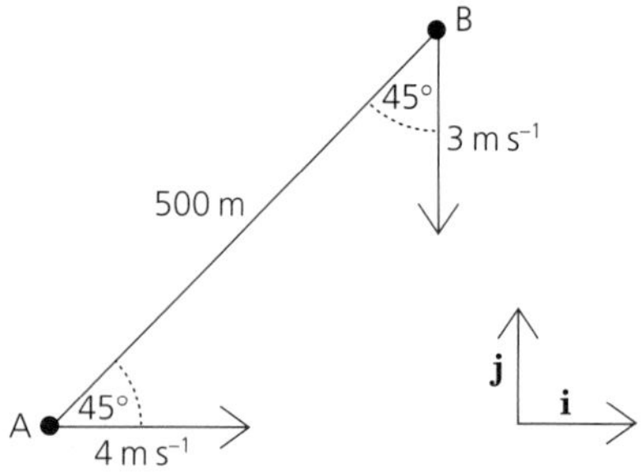

B
45°
3 m s⁻¹
500 m
j
i
A
45°
4 m s⁻¹

a) Find the velocity of the tanker relative to the fishing boat, in terms of the unit vectors $\mathbf{i}$ and $\mathbf{j}$ which are east and north respectively.

b) Show that the position of the tanker, relative to the fishing boat is: $(4t - 354)\mathbf{i} + (3t - 354)\mathbf{j}$.

c) Find the minimum distance between the two vessels.

8 An aeroplane is initially 5 km due north of a helicopter. The aeroplane flies east at 100 m s^{-1}. The helicopter flies north east at 50 m s^{-1}.

a) Find the velocity of the aeroplane relative to the helicopter, in terms of the unit vectors $\mathbf{i}$ and $\mathbf{j}$ which are east and north respectively.

b) Find the position of the aeroplane relative to the helicopter.

c) Find an expression for the distance between them and find its minimum value.

9 A yacht that is suspected of smuggling activities is sailing south west at 2 m s^{-1}. A motor launch is initially 2 km due south of the yacht, and travels at a speed of 5 m s^{-1} to intercept the yacht. In which direction should the launch travel to intercept the yacht in the shortest possible time? Find the time taken for the launch to reach the yacht.

10 A sailing boat is 3 km on a bearing of 300° from a port when it begins to sink. The crew radio for help before escaping in a life raft. A rescue boat immediately leaves the port and travels at 4 m s^{-1}, heading towards the position when the boat sank. The life raft drifts due south at 0.5 m s^{-1}. How close does the rescue boat get to the life raft? Do you think the crew are seen by the rescue boat?

EXERCISES

21.1B

1 A radio controlled aeroplane is flying at 12 m s^{-1} in a wind that is blowing at right angles to its path at 5 m s^{-1}. The unit vectors $\mathbf{i}$ and $\mathbf{j}$ are parallel and perpendicular to the wind respectively.

a) Find the resultant velocity of the aeroplane, in terms of $\mathbf{i}$ and $\mathbf{j}$.

b) Find the distance travelled by the aeroplane in 20 seconds.

2 A boy attempts to swim across a river in which there is a current that flows at 0.2 m s^{-1}. The boy can swim at 0.6 m s^{-1}. The width of the river is 7.8 m.

a) If the boy swims at right angles to the bank, how far downstream does he move?

b) How should he swim in order not to be swept downstream while crossing the river?

c) What is the quickest way for him to cross the river?

3 A canoeist paddles across a river at an angle of 45° to the bank, pointing upstream, at a speed of 1.4 m s^{-1}. The width of the river is 10 m. Find the speed of the current if the canoeist:

a) moves 2 m upstream,

b) moves 3 m downstream, while crossing the river.

4 A captain sails his boat at 4 m s^{-1} on a bearing of 120° unaware of a current that flows at 2 m s^{-1} due south.

a) Find the actual speed and direction of the boat.

b) How far south is the boat from its intended position after 20 minutes?

5 A raindrop falling vertically at 4 m s^{-1} when it hits a cyclist travelling horizontally at 8 m s^{-1}.

a) Find the velocity of the raindrop relative to the cyclist when it hits the cyclist in terms of the horizontal and vertical unit vectors $\mathbf{i}$ and $\mathbf{j}$.

b) Find the speed of the raindrop relative to the cyclist.

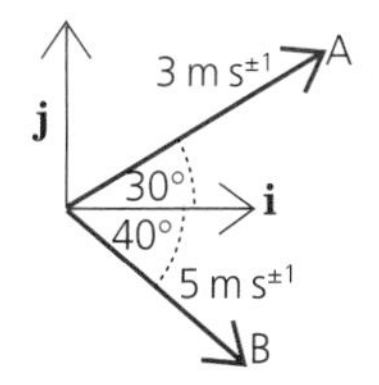

6 Two birds leave a chimney pot and fly in the directions shown in the diagram.

a) Find the velocity of B relative to A, in terms of the unit vectors $\mathbf{i}$ and $\mathbf{j}$.

b) Find the position of B relative to A.

c) Find an expression for the distance between the two birds, at time t.

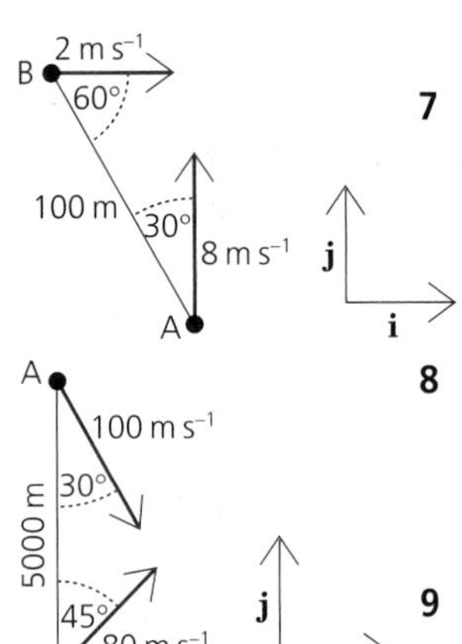

7 A speed boat is at A heading north at 8 m s^{-1} and dinghy is at B sailing east at 2 m s^{-1}.

a) Find the velocity and position of the dinghy relative to the speed boat.

b) Find an expression for the distance between them.

c) How close does the speed boat get to the dinghy?

8 Two aeroplanes, A and B, start from the positions in the diagram, with velocities as shown.

a) Find the velocity and position of A relative to B.

b) Find the minimum distance between the two aeroplanes.

9 A person walking with an umbrella at 4 m s^{-1} finds that rain hits her at a height of 60 cm above the ground. If the umbrella has the dimensions shown, find the speed at which the rain is falling, if it is assumed to be falling vertically at a constant speed.

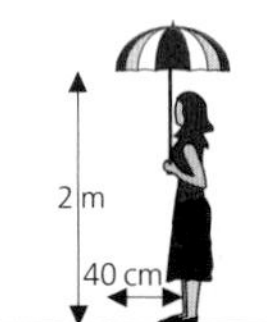

10 A ship is sailing on a bearing of 20° at a speed of 4 m s^{-1}. A helicopter is to travel to the ship. Initially the helicopter is 50 km east of the ship and can travel at a speed of 40 m s^{-1}. In what direction should the helicopter fly in order to meet the ship as soon as possible?

CONSOLIDATION EXERCISES FOR CHAPTER 21

1 At noon, two ships, A and B, have position vectors $(3\mathbf{i} + 2\mathbf{j})$ km and $(-\mathbf{i} + 4\mathbf{j})$ km respectively, referred to an origin O. The velocities of A and B are $(-\mathbf{i} + 5\mathbf{j})$ km h^{-1} and $(2\mathbf{i} + \mathbf{j})$ km h^{-1} respectively. The position vector of A relative to B, at t hours after noon is $\mathbf{r}$ km. Show that:

a) $\mathbf{r} = (4 - 3t)\mathbf{i} + (-2 + 4t)\mathbf{j}$

b) A and B are nearest to each other at 12.48 p.m.

(ULEAC Question 7, M1 Specimen Paper, 1994)

2 Two cyclists, C and D, are travelling with constant velocities $(5\mathbf{i} - 2\mathbf{j})$ m s^{-1} and $8\mathbf{j}$ m s^{-1} respectively, relative to a fixed origin O.

a) Find the velocity of C relative to D.

At noon, the position vectors of C and D are $(100\mathbf{i} + 300\mathbf{j})$ metres and $(150\mathbf{i} + 100\mathbf{j})$ metres respectively, referred to O. At t seconds after noon, the position vector of C relative to D is $\mathbf{s}$ metres.

b) Show that $\mathbf{s} = (-50 + 5t)\mathbf{i} + (200 - 10t)\mathbf{j}$.

c) By considering $|\mathbf{s}|^2$, or otherwise, find the value of t for which C and D are closest together.

(ULEAC Question 7, M1, January 1993)

3 A river with long straight banks is 500 m wide and flows with a constant speed of 3 m s^{-1}. A man rowing a boat at a steady speed of 5 m s^{-1}, relative to the river, sets off from a point A on one bank so as to arrive at the point B directly opposite A on the other bank. Find the time taken to cross the river.

A woman also sets off at A rowing at 5 m s^{-1} relative to the river and crosses in the shortest possible time. Find this time and the distance downstream of B of the point at which she lands.

(AEB Question 5, Paper 3, June 1991)

4 A model speed boat which travels at a speed of 5 m s^{-1} in still water is to cross a flowing river from one bank to a point directly opposite on the other bank. The river is 20 m wide and flows at a speed of 3 m s^{-1} parallel to the banks.

a) In what direction should the speed boat be launched in order to cross the river directly?

b) How long will it take the speed boat to make the crossing?

c) Give a reason why the speed boat would not be able to cross the river directly if the river flows at a speed of 5 m s^{-1} or more.

(Oxford Question 2, M6 Specimen Paper, 1994)

5 A canal has long straight parallel banks that ran north/south. The canal has width 25 m. There is a uniform current of 1.5 m s^{-1} towards the south. A girl wishes to row from a point O on the west bank to a point P on the east bank. The point on the east bank directly opposite O is E, and P is south of E with $\angle$OPE $= \theta$ (see diagram). Her speed in still water is 1 m s^{-1}. Show that she can only row directly to P if $\sin\theta \le \frac{2}{3}$.

Given that $\sin\theta = \frac{1}{3}$, find the direction in which she should point the boat in order to reach P as quickly as possible. Find the time taken in this case.

(UCLES Question 7, M3 Specimen Paper, 1994)

6 A car A is travelling with a constant velocity of 20 km h^{-1} due west and a cyclist B has a constant velocity of 16 km h^{-1} in the direction of the vector $(-4\mathbf{i} + 3\mathbf{j})$, where $\mathbf{i}$ and $\mathbf{j}$ are unit vectors due east and due north respectively. At noon, A is 1.2 km due north of B. Take the position of A at noon as the origin and obtain expressions for the position vectors of A and B at time t hours after noon, and hence show that the position vector of A relative to B is $\mathbf{r}$ km, where:

$$5\mathbf{r} = 6\{-6t\mathbf{i} + (1 - 8t)\mathbf{j}\}$$

Deduce that the distance between A and B is d km, where:

$$25d^2 = 36(100t^2 - 16t + 1)$$

Hence show that the minimum separation between A and B is 720 m and find the time at which this occurs.

(AEB Question 8, Paper 3, Winter 1994)

MATHEMATICAL MODELLING ACTIVITY

Problem statement

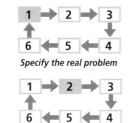

Specify the real problem

A rowing boat is rowed across a river so that its velocity relative to the current is at right angles to the bank. Find the path that the boat would follow.

Set up a model

Set up a model

A very simple model would be set up, assuming that the river flows at the same speed right across its width. In the solution suggested here, we shall consider alternative models for the speed of the current that vary across the river.

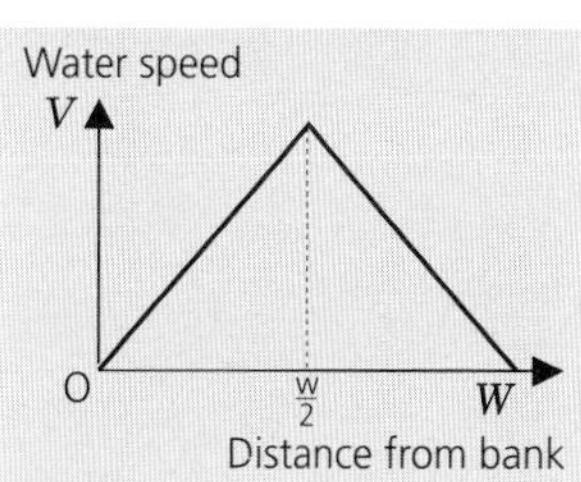

Suggest various ways in which the flow might vary across the river.

A first model assumes that the speed of the water varies as shown in the graph on the left.

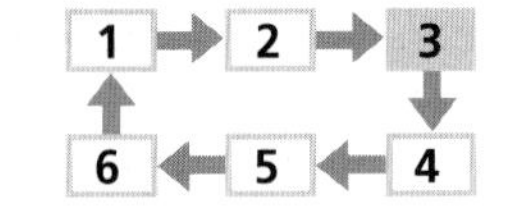

Formulate the mathematical problem

Formulate a mathematical problem

A boat has a constant velocity relative to the water of magnitude u and is perpendicular to the bank. Find the path of the boat if the speed of the water is given in the graph above.

Solve the mathematical problem

Mathematical solution

The speed of the water is given by: $v = \dfrac{2Vx}{w}$ for $0 \le x \le \frac{1}{2}w$

$$v = 2V\left(1 - \frac{x}{w}\right) \quad \text{for} \quad \tfrac{1}{2}w \le x \le w$$

The equations above are in terms of x, the displacement of the boat perpendicular to the bank. If the speed of the boat relative to the water is u, then $x = ut$, so the equations become:

$$v = \frac{2Vut}{w} \quad \text{for} \quad 0 \le t \le \frac{w}{2u}$$

$$v = 2V\left(1 - \frac{ut}{w}\right) \quad \text{for} \quad \frac{w}{2u} \le t \le \frac{w}{u}$$

These expressions can be integrated to give the displacement, y, of the boat perpendicular to the bank.

$$y = \frac{Vut^2}{w} \quad \text{for} \quad 0 \le t \le \frac{w}{2u}$$

$$y = 2V\left(t - \frac{ut^2}{2w}\right) \quad \text{for} \quad \frac{w}{2u} \le t \le \frac{w}{u}$$

Explain why the term $-\dfrac{Vw}{2u}$ appears in the second expression.

The path of the boat is then given by: $\mathbf{r} = x\mathbf{i} + y\mathbf{j}$

where x and y are defined as above, and $\mathbf{i}$ and $\mathbf{j}$ are unit vectors perpendicular and parallel to the respectively.

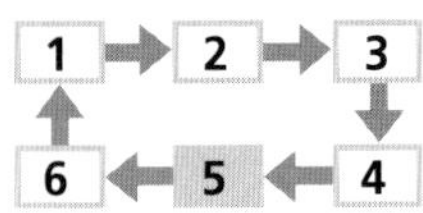

Interpret the solution

Interpretation

Consider a river of width 10 m and a current that reaches a maximum speed of $2\,\text{m s}^{-1}$. Also assume the rowing boat can be rowed at a speed of $1\,\text{m s}^{-1}$. The expressions for x and y then become:

$x = t$ and

$y = 0.2t^2$ for $0 \le t \le 5$

$y = 4(t - 0.05t^2) - 10$ for $0 \le t \le 10$

The path can then be plotted as shown.

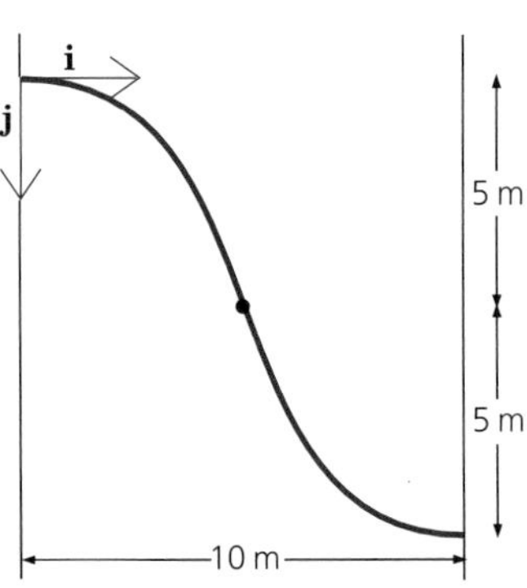

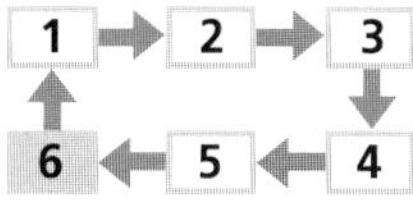

Compare with reality

Compare with reality

It may be possible for you to examine the path of a boat or canoe as it crosses a river. Alternatively you may wish to investigate the validity of the model that describes how the speed of the current varies across a river. A possible experiment could involve dropping corks or similar buoyant objects into a river and observing how long it takes them to travel specified distances at different distances from the bank. You may suggest alternative models for the way the current changes and revise the solution in the light of your alternative models.

POSSIBLE MODELLING TASKS

1 Investigate the affect of a wind in the path of a rugby ball or football.

2 How should someone paddle a canoe to cross a river in a straight line perpendicular to the bank if the speed of the water varies across the river?

Summary

After working through this chapter you should:

- be able to find the actual velocity when an object moves in a medium that is itself moving, for example a boat in a flowing river
- know that the velocity of object A relative to object B is: $\mathbf{V}_{AB} = \mathbf{V}_A - \mathbf{V}_B$
- be able to find relative positions.

BRIDGWATER COLLEGE LIBRARY

Answers

CHAPTER 1
Mechanics and the real world

Exercises 1.1A *(p.6)*

1 a) No **b)** Yes or No (students may feel length of car is significant) **c)** No **d)** Yes
3 a) No **b)** No **c)** Yes
4 a) Yes **b)** No **c)** Yes (or no if area reached by satellite to be considered)

Exercises 1.1B *(p.6)*

1 a) Yes **b)** Yes unless close to poles or bar **c)** No **d)** Yes
2 a) e.g. car wheel, football, rolling marble
b) e.g. golf ball, snooker or pool ball, boomerang
3 At the 'centre' of the child, probably a little above the swing seat.
4 a) No **b)** Yes **c)** No

CHAPTER 2
Force

Exercises 2.1A *(p.10)*

1 a) Yes **b)** No **c)** No **d)** No
e) Yes if the skateboard travels in a straight line. **f)** Yes **g)** No
3 a) The speed is likely to increase, until it reaches a maximum.
b) If friction is constant then the net force is constant however if friction depends on speed then the net force is likely to reduce in magnitude.

Exercises 2.1B *(p.10)*

1 a) Yes **b)** No **c)** Yes **d)** No
e) No **f)** Yes **g)** Yes
2 a) The lorry moves at constant speed in a straight line.
b) When they have come to rest at the end of the jump.
c) If the ice is so smooth that there is no friction.
d) If the buoyant force is equal to the force of gravity.
3 a) Increases to a maximum.
b) Decreases to zero.
4 a) Decreases to zero.
b) Decreases to zero.

Exercises 2.2A *(p.13)*

1 a) 245 N **b)** 0.0294 N
c) 17 640 N
2 900 kg
4 No, unless air resistance is significant.
5 104 N
6 2.8×10^{-7} N
7 15.13 m s^{-2}
8 290.2 N
9 672.4 N
10 3.46×10^{8} m

Exercises 2.2B *(p.13)*

1 a) 666.4 N **b)** 0.49 N
c) 1.96×10^{6} N
2 286 grams
4 Fall together in both cases, but slower on the moon.
5 384.8 N
6 15 m s^{-2}
7 2.67×10^{-7} N
8 1.76×10^{28} kg
9 37.43 N
10 6.05 m s^{-2}

Exercises 2.3A *(p.18)*

1 a) 735 N **b)** 5.88 N **c)** 6.53 N
d) 49 N
3 Book exerts a reaction force of 19.6 N. Ground exerts a reaction force of 78.4 N on each leg. Gravity exerts force of 294 N.
4 Reaction forces have magnitude 19.6 N, 68.6 N and 98 N.
5 5
6 a) 78.4 N to left **b)** 0 N
c) 29.4 N to left
7 a) 280 N **b)** 200 N
8 294 N
9 a) 10 143 N **b)** 7326 N
10 b) Between box and sledge 98 N. Between sledge and ground 205.8 N. **c)** No
11 a) 4900 N **b)** 3332 N **c)** 9
d) 1200 N

Exercises 2.3B *(p.20)*

1 a) 490 N **b)** 4.9 N **c)** 19 600 N
2 a) 49 N Equal force on each leg. **b)** 19.6 N Equal force on each leg. **c)** 12 250 N Equal force on each wheel.
4 Forces due to gravity: 1470 N on sledge and 245 N on box. Reaction forces: 245 N between box and sledge, and 1715 N between sledge and ground.
5 2352 N on bottom of box, 2058 N on top of box.
6 a) 8 N to left **b)** 2 N to right
c) 0 N **d)** 1 N to left
7 a) 0.98 N **b)** 0.98 N
c) $F > 0.98$ N **d)** $F < 0.98$ N
8 a) 695.8 N
b) Force greater than 146.1 N.
9 $0.581 \le \mu \le 0.867$.
10 a) 3920 N **b)** P
11 a) 2450 N **b)** 3920 N
c) Rear wheels 500 N, front wheels 4400 N.

Exercises 2.4A *(p.23)*

1 14.7 N
2 9.8 N
3 a) 0.49 N **b)** Decreases going up, increases going down.
4 a) 49 N, 39.2 N **c)** 9.8 N
5 a) 4.9 N **b)** 3.43 N **c)** 3.43 N
6 b) 19.6 N, 49 N **c)** 3.2 N, 8 N
7 a) 5684 N **b)** 5684 N
c) Greater than 5684 N
8 a) 122.5 N **b)** Equal tension in each rope, child sits in middle, seat has no mass.
9 2450 N
10 a) 29.4 N, 49 N
b) 29.4 N, 29.4 N
c) 29.4 N, 19.6 N, 9.8 N

Exercises 2.4B *(p.24)*

1 98 N
2 a) 88.2 N **b)** 10.20 kg
3 a) 10 N **b)** 1.02 kg
4 a) 98 N **b)** 80 N
5 2744 N
6 b) 98 N, 127.4 N
c) 98 N, 127.4 N
7 a) 58.8 N, 58.8 N.
b) 29.4 N, 39.2 N, 9.8 N
8 490 N, 735 N, Top two cables have equal tensions.
9 a) 58.8 N, 39.2 N, 9.8 N
b) 1.176 m, 0.784 m, 0.196 m

Consolidation Exercises for Chapter 2 *(p.26)*

2 a) Buoyant force = mg
c) Equal in magnitude as it goes through its equilibrium level.
4 b) Friction causes the marble to roll.

CHAPTER 3
Vectors and forces

Exercises 3.1A *(p.31)*

1 a) $28.19\mathbf{i} + 10.26\mathbf{j}$
b) $-45.89\mathbf{i} - 65.53\mathbf{j}$
c) $-21.13\mathbf{i} + 45.32\mathbf{j}$
d) $4.91\mathbf{i} - 3.44\mathbf{j}$
2 a) $\begin{pmatrix} 4 \\ 6.93 \end{pmatrix}$ **b)** $\begin{pmatrix} 9.40 \\ -3.42 \end{pmatrix}$
c) $\begin{pmatrix} -8.68 \\ 49.24 \end{pmatrix}$ **d)** $\begin{pmatrix} -20 \\ 34.64 \end{pmatrix}$
3 a) $5\mathbf{i} + 3\mathbf{j}$, 5.83 N, 31.0° to $\mathbf{i}$
b) $3.14\mathbf{i} + 2.13\mathbf{j}$, 3.79 N, 34.2° to $\mathbf{i}$ **c)** $4.52\mathbf{i} + 7.74\mathbf{j}$, 8.96 N, 59.7° to $\mathbf{i}$ **d)** $-6.69\mathbf{i} + 0.45\mathbf{j}$, 6.71 N, 176.2° to $\mathbf{i}$ **e)** $-1.5\mathbf{i}$ $-0.866\mathbf{j}$, 1.73 N, −150.0° to $\mathbf{i}$
f) $1.05\mathbf{j}$, 1.05, 90° to $\mathbf{i}$
4 a) 15.62 N, 50.2° **b)** 35.78 N, 63.4° **c)** 13.60 N, 36.0°
d) 102.4 N, −77.6° **e)** 20 N, −143.1° **f)** 5 N, 126.9°
5 a) 3.61 N, 33.7° to $\mathbf{i}$
b) 9.22 N, 130.6° to $\mathbf{i}$
c) 4.47 N, −153.4° to $\mathbf{i}$
d) 2.24 N, 153.4° to $\mathbf{i}$
6 a) $5.14\mathbf{j} + 6.13\mathbf{j}$ **b)** $5.19\mathbf{i} + 3\mathbf{j}$
c) $80\mathbf{i}$ **d)** $59.1\mathbf{i} + 10.4\mathbf{j}$
7 a) $-25\mathbf{i} - 20\mathbf{j}$ **b)** $-65\mathbf{i} - 13.3\mathbf{j}$

Exercises 3.1B *(p.32)*

1 a) $6\mathbf{j}$ **b)** $49.24\mathbf{i} + 8.68\mathbf{j}$
c) $-13.79\mathbf{i} + 2.43\mathbf{j}$
d) $18.79\mathbf{i} - 6.84\mathbf{j}$
e) $-3\mathbf{i} - 5.20\mathbf{j}$
f) $3.54\mathbf{i} + 3.54\mathbf{j}$
2 a) $\mathbf{i} - \mathbf{j}$ **b)** $2.36\mathbf{i} + 1.05\mathbf{j}$
c) $-15.32\mathbf{i} + 50.00\mathbf{j}$
d) $5.63\mathbf{i}$ **e)** $2.32\mathbf{i} + 25.16\mathbf{j}$
3 a) $\begin{pmatrix} 2.05 \\ 5.64 \end{pmatrix}$ **b)** $\begin{pmatrix} 28.19 \\ -10.26 \end{pmatrix}$
c) $\begin{pmatrix} -86.60 \\ 50 \end{pmatrix}$ **d)** $\begin{pmatrix} -20.52 \\ -56.38 \end{pmatrix}$
4 a) 10.44 N, 73.3° to $\mathbf{i}$
b) 125 N, −61.4° to $\mathbf{i}$
c) 89.2 N, −160.3° to $\mathbf{i}$
d) 44.72 N, 153° to $\mathbf{i}$
e) 10 N, −53° to $\mathbf{i}$
f) 62.2 N, 11° to $\mathbf{i}$
5 a) 8.06 N, 29.7° to $\mathbf{i}$
b) 6.32 N, −18.4° to $\mathbf{i}$
c) 10.2 N, −11.3° to $\mathbf{i}$
d) 10 N, 126.9° to $\mathbf{i}$
6 a) $40\mathbf{j}$ **b)** $-20\mathbf{i}$ **c)** $-34.64\mathbf{i} - 20\mathbf{j}$
d) $38.57\mathbf{i} + 45.96\mathbf{j}$
7 a) $-64.20\mathbf{i} - 13.97\mathbf{j}$
b) $10\mathbf{i} + 40.36\mathbf{j}$

Exercises 3.2A *(p.38)*

1 a) $-18.54\mathbf{i} + 97.42\mathbf{j}$
b) $18.54\mathbf{i} - 97.42\mathbf{j}$
c) 99.17 N, −79.2° to $\mathbf{i}$
2 a) 490 N **b)** Cable 1: $-T\cos30°\mathbf{i} + -T\cos60°\mathbf{j}$
Cable 2: $-490\mathbf{j}$ Rod: $-R\mathbf{i}$
c) $T = 980$ N, $R = 848.7$ N
d) 498 N, 980 N, 762 N
3 b) $-T_1\cos50°\mathbf{i} + T_1\cos40°\mathbf{j}$
$T_2\cos40°\mathbf{i} + T_2\cos50°\mathbf{j}$ $-mg\mathbf{j}$
c) 37.76 N **d)** 6 kg
4 b) $R\mathbf{j}$, $F\mathbf{i}$, $-392\cos80°\mathbf{i}$
$-392\cos10°\mathbf{j}$ **c)** $R = 386.04$ N, $F = 68.07$ N **d)** 196 N
5 b) Equal in magnitude but opposite in direction.
c) 45 m s^{-1} **d)** 20.9 m s^{-1}
6 a) $376.2\mathbf{i} + 36.26\mathbf{j}$
b) $-376.2\mathbf{i} - 36.26\mathbf{j}$
c) 377.94 N, −174.5° to $\mathbf{i}$
d) 25.86° **e)** No because the cables will sag.
7 a) 31.5 N **b)** 16.48 N
8 Large ball 2.14 N and 0.86 N
Small ball 2.94 N and 0.86 N
9 a) 346 N
b) 451.6 N and 399.5 N
10 b) $(1637 - F)\mathbf{i} + (11\,646 - R)\mathbf{j}$
c) 1637 N

Exercises 3.2B *(p.40)*

1 a) $-1121\mathbf{i} + 684\mathbf{j}$
b) $1121\mathbf{i} - 684\mathbf{j}$
c) 1313 N, 31.4°
2 a) $(T_2 - T_1\cos 50°)\mathbf{i} +$
$(T_1\cos 40° - 400)\mathbf{j}$
b) 522 N **c)** 336 N
3 b) $-648\mathbf{i} - 9287\mathbf{j}$
c) $F = 648$ N, $R = 9287$ N
4 b) 206 N **c)** 15.75 N
5 a) 21.5 N **b)** Rope probably not horizontal, there will be some friction. **c)** 21.8 N
6 a) 2.33 N
7 a) 158.7 N **b)** 0.231
c) No account has been taken of other forces such as air resistance on the skier. Estimate the magnitude of the air resistance.
8 $Q = 298$ N, $P = 51.8$ N
9 2521 N
10 a) 11.76 N, 14.7 N **b)** 1.68 kg

Consolidation Exercises for Chapter 3 *(p.41)*

1 a) 390 N, 39.07° **b)** 291 N
c) The cables will hang down.
2 173 N
3 a) $191\mathbf{i} + 19.4\mathbf{j}$
b) $-191\mathbf{i} - 19.4\mathbf{j}$
c) 192 N **e)** 16.3°
4 a) $p = 2$, $q = -6$ **b)** 6.32 N.
c) 18°
5 155.6 N, 185.4 N, 348.4 N, 70° to horizontal
6 $T = \dfrac{mg\sin\alpha}{\cos\theta}$

CHAPTER 4
Kinematics in one dimension

Exercises 4.1A *(p.47)*

1 a) 5 × 140 = 700 m **b)** $140T$
2 a) $\frac{1}{15}$ m s^{-2}, $v = \frac{1}{15}t$

b) i) $\frac{10}{3}$m ii) 53 $\frac{1}{3}$m v) $\frac{T^2}{30}$

3 b) $s = 4.9t^2$ c) 1.92 s

4 b) $v = 10 + 1.2\,T$

c) i) 65 m ii) $10T + 0.6T^2$

5 b) −1.5 m s^{-2}

c) $v = 18 - 1.5t$

d) i) 47.25 m ii) $18T - 0.75T^2$

e) 108 m

Exercises 4.1B *(p.48)*

1 a) 200 m b) $20t$ m

2 a) $\frac{1}{3}$, $\frac{1}{3}$m s^{-2}

b) $v = \frac{1}{3}t$ c) i) 4.17 m

ii) 37.5 m iii) $\frac{1}{6}t^2$

d) For example speed increases more rapidly at first, a greater distance would be travelled.

3 a) 0.8 m s^{-2} b) $0.4t^2$

c) 1.73 s, 1.39 m s^{-1}

4 a) 0.5 m s^{-2}, constant acceleration b) $70 + 0.5t$.

c) 3200 m d) $70t + 0.25t^2$

5 b) 15 s c) 225 m d) −2 m s^{-2}

Exercises 4.2A *(p.51)*

1 a) 1.43 s b) 14 m s^{-1}

2 a) 0.114 m s^{-2} b) 5 m s^{-2}

c) 1.219 m s^{-2}.

That the accelerations are constant.

3 a) 4.77 m b) 9.66 m s^{-1}

4 a) −3.6 m s^{-2} b) 1.67 s

5 a) −1.12 m s^{-2}

b) Yes, after 178 m. c) 10.7 s

d) 4.3 s e) 1.61 m s^{-2}

f) 15 s g) 13.3 m s^{-1}

h) Constant acceleration.

6 a) 18 m b) 24 s c) 40 s

d) 7 m s^{-1}

7 3840 m

8 65 m

9 a) $v = 2a$, $s = 2a$ b) $s = 20a$

c) 4.55 m s^{-2}

10 a) 4.43 m s^{-1} b) 3.70 m s^{-1}

c) 49 cm

Exercises 4.2B *(p.52)*

1 a) 0.2 m s^{-2} b) 2 m s^{-2}

c) 1.2 m s^{-2}

2 a) $t = 0.204$ s

b) 1.204 m c) 4.86 m s^{-1}

3 18 m s^{-2}

4 a) 1.955 m s^{-2}, 6.72 s

b) 501 m, 22.4 s

5 a) 5 s

b) 25 m

c) 10 m s^{-1}

d) In practice the ball is unlikely to resurface at 10 m s^{-1}.

6 a) 600 m b) 60 s

7 a) aT b) $20aT - aT^2$

c) $\frac{1}{15}$m s^{-1}, $\frac{1}{3}$m s^{-1}

8 a) 2 m s^{-2} b) 20 m s^{-1} c) 200 m

9 a) −5 m s^{-1} b) Yes

c) 97 m beyond traffic lights.

d) He is safe, but foolhardy!

10 a) 5.56 m s^{-1}

b) The cyclist will probably reach a constant speed and remain at that speed for most of the ride.

Consolidation Exercises for Chapter 4 *(p.54)*

1 a) 0.1 m s^{-2} b) 155 s

2 a) BC stopping

CD waiting

DE gaining speed

b) 0.5 m s^{-2}, 2500 m

c) 0.2 m s^{-2}, 6250 m

d) 325 s

3 a) 75 m s^{-1} b) 1.875 m s^{-2}

c) $\frac{1}{2}$

4 0.2 m s^{-1}, 0 m s^{-1}, 0.15 m s^{-1}

5 a) Maximum speed = 22.2 m s^{-1}, area = 26.64 km, so not possible.

b) i) $T = 60$ s

ii) 0.37 m s^{-2}

iii) Crossing time increases by approximately 1 minute.

CHAPTER 5

Motion and vectors

Exercises 5.1A *(p.61)*

1 a) $t = 0$ $\mathbf{r} = 0 + 0\mathbf{j}$

$t = 1$ $\mathbf{r} = 4\mathbf{i} + 0.1\mathbf{j}$

$t = 2$ $\mathbf{r} = 8\mathbf{i} + 0.4\mathbf{j}$

$t = 3$ $\mathbf{r} = 16\mathbf{i} + 1.6\mathbf{j}$

$t = 4$ $\mathbf{r} = 24\mathbf{i} + 3.6\mathbf{j}$

b) Slope parallel to $\mathbf{j}$ direction.

2 a) $\mathbf{r}_A = 10\mathbf{j}$ $\mathbf{r}_B = 8\mathbf{i} - 2\mathbf{j}$

b) $4\mathbf{i} + 4\mathbf{j}$, 045°, 5.66 km

c) Note $\mathbf{r} = 3\mathbf{i} + 5.5\mathbf{j} \Rightarrow t = 1.5$.

3 a) $t = 0$ $\mathbf{r} = 0\mathbf{i} + 1\mathbf{j}$

$t = 0.5$ $\mathbf{r} = 4\mathbf{i} + 7.25\mathbf{j}$

$t = 1$ $\mathbf{r} = 8\mathbf{i} + 11\mathbf{j}$

$t = 1.5$ $\mathbf{r} = 12\mathbf{i} + 12.25\mathbf{j}$

$t = 2$ $\mathbf{r} = 16\mathbf{i} + 11\mathbf{j}$

$t = 2.5$ $\mathbf{r} = 20\mathbf{i} + 7.25\mathbf{j}$

$t = 3$ $\mathbf{r} = 24\mathbf{i} + 1\mathbf{j}$

c) Maximum height = 12.25 m range = 24.5 m

4 a) $t = 0$ $\mathbf{r}_A = 8\mathbf{i} + 4\mathbf{j}$

$\mathbf{r}_B = 0\mathbf{i} + 0\mathbf{j}$

$t = 2$ $\mathbf{r}_A = 12\mathbf{i} - 2\mathbf{j}$

$\mathbf{r}_B = 8\mathbf{i} + 4\mathbf{j}$

$t = 3$ $\mathbf{r}_A = 14\mathbf{i} - 5\mathbf{j}$

$\mathbf{r}_B = 12\mathbf{i} + 6\mathbf{j}$

c) 7.21

5 a) $t = 0$ $\mathbf{r} = 0\mathbf{i} + 2\mathbf{j}$

$t = 0.5$ $\mathbf{r} = 7.5\mathbf{i} + 5.75\mathbf{j}$

$t = 1.0$ $\mathbf{r} = 15\mathbf{i} + 7\mathbf{j}$

$t = 1.5$ $\mathbf{r} = 22.5\mathbf{i} + 5.75\mathbf{j}$

$t = 2.0$ $\mathbf{r} = 30\mathbf{i} + 2\mathbf{j}$

c) $0 \le t \le 2.18$ d) 32.75

e) The spin of the javelin generates lift. Also where do you place a particle on a javelin, at the middle or one end?

6 a) $t = 0$ $\mathbf{r} = 0\mathbf{i} + 0\mathbf{j}$

$t = 1$ $\mathbf{r} = 0.5\mathbf{i} + 0.48\mathbf{j}$

$t = 2$ $\mathbf{r} = 1\mathbf{i} + 1.12\mathbf{j}$

$t = 3$ $\mathbf{r} = 1.5\mathbf{i} + 1.92\mathbf{j}$

c) In the $\mathbf{j}$ direction.

7 a) $t = 0$ $\mathbf{r} = 0\mathbf{i} + 0\mathbf{j}$

$t = 1$ $\mathbf{r} = 20\mathbf{i} + 37\mathbf{j}$

$t = 3$ $\mathbf{r} = 60\mathbf{i} + 63\mathbf{j}$

$t = 3.5$ $\mathbf{r} = 70\mathbf{i} + 63.875\mathbf{j}$

$t = 4$ $\mathbf{r} = 80\mathbf{i} + 64\mathbf{j}$

b) Horizontally

8 a) $2\mathbf{i} + 1\mathbf{j}$

b) $a = 0.2$, $c = 0.3$

c) $b = 0.9$, $d = 0.35$

Exercises 5.1B *(p.62)*

1 a) $t = 0$ $\mathbf{r} = 0\mathbf{i} + 0\mathbf{j}$

$t = 1$ $\mathbf{r} = 4.5\mathbf{i} + 3\mathbf{j}$

$t = 2$ $\mathbf{r} = 10\mathbf{i} + 6\mathbf{j}$

$t = 3$ $\mathbf{r} = 16.5\mathbf{i} + 9\mathbf{j}$

2 a) $t = 0$ $\mathbf{r} = -3\mathbf{i} + 0\mathbf{j}$

$t = 2$ $\mathbf{r} = 1\mathbf{i} + 6\mathbf{j}$

$t = 4$ $\mathbf{r} = 5\mathbf{i} + 12\mathbf{j}$

$t = 6$ $\mathbf{r} = 9\mathbf{i} + 18\mathbf{j}$

c) $t = 1.5$

3 a) $t = 0$ $\mathbf{r} = 0\mathbf{i} + 2\mathbf{j}$
$t = 1$ $\mathbf{r} = 3\mathbf{i} + 9\mathbf{j}$
$t = 2$ $\mathbf{r} = 6\mathbf{i} + 6\mathbf{j}$
$t = 3$ $\mathbf{r} = 9\mathbf{i}\ 7\mathbf{j}$
c) 9.2 m **d)** 2.56 s **e)** 7.67 m
4 a) $t = 0$ $\mathbf{r} = 0\mathbf{i} + 0\mathbf{j}$
$t = 1$ $\mathbf{r} = 3\mathbf{i} + 0.1\mathbf{j}$
$t = 2$ $\mathbf{r} = 6\mathbf{i} + 0.4\mathbf{j}$
$t = 3$ $\mathbf{r} = 9\mathbf{i} + 0.9\mathbf{j}$
$t = 4$ $\mathbf{r} = 12\mathbf{i} + 1.6\mathbf{j}$
c) Arrow in $\mathbf{j}$ direction.
5 c) $0 \le t \le 12$
6 a) B, with the higher path
b) B, 0.94 s
7 a) $t = 2.5$ s
b) Yes
8 b) 3.91 m

Exercises 5.2A *(p.65)*

1 a) $3.83\mathbf{i} + 3.21\mathbf{j}$
b) $-72.08\mathbf{i} + 45.04\mathbf{j}$
c) $-0.42\mathbf{i} - 2.97\mathbf{j}$
3 a) $v_A > v_B$ Velocities have opposite directions.
c) $v_C > v_B$ Velocities have the same direction.
4 a) Car cruising on a straight road.
b) Rocket as it beings to accelerate from a launch pad.
c) Car going round a roundabout.
5 a) A: $5\mathbf{j}$, 5 **B:** $-1\mathbf{i} - 4\mathbf{j}$, 4.12,
C: $4.2\mathbf{i} - 5\mathbf{j}$, 6.53
c) C is fastest and A is slowest.
6 A: 102 m s^{-1}, 078.7°
B: 161.6 m s^{-1}, 158.2°
C: 164.9 m s^{-1}, 194.0°
7 $18.8\mathbf{i} + 6.8\mathbf{j}$
8 40.05 m s^{-1}, 2.9°
9 a) $-2\mathbf{i}$, $1.4\mathbf{i} + 1.4\mathbf{j}$ **b)** 3.53 s
c) $0.36\mathbf{i} - 0.87\mathbf{j}$ **d)** $0\mathbf{i} + 0\mathbf{j}$
10 c) ii) 0, 0, 13.42 m s^{-1}

Exercises 5.2B *(p.67)*

1 a) $0.28\mathbf{i} + 0.96\mathbf{j}$
b) $-5.64\mathbf{i} - 2.05\mathbf{j}$
c) $24.57\mathbf{i} - 17.21\mathbf{j}$
3 a) $3500\mathbf{i} + 6062\mathbf{j}$
b) $7.29\mathbf{i} + 12.63\mathbf{j}$, 14.58 m s^{-1}
c) Greater than or equal to 14.58 m s^{-1}
4 a) The inner plane has a slower speed.
b) 157 m s^{-1}, 236 m s^{-1}
c) Same direction but inner plane has a smaller magnitude.
d) $-78.5\mathbf{i} + 136\mathbf{j}$, $-118\mathbf{i} + 204\mathbf{j}$
5 a) A: 9.43 m s^{-1}, 058.0°,
B: 5 m s^{-1}, 143.1°,
C: 6.32 m s^{-1}, 198.4°
b) $1080\mathbf{i} - 600\mathbf{j}$
c) The ship would probably follow a curved path.
6 a) $33.0\mathbf{i} - 3.2\mathbf{j}$
b) $-17.6\mathbf{i} - 55.6\mathbf{j}$
c) 58.4 m s^{-1}, 197.6°
7 a) 2.01 m s^{-1} **b)** 5.7°
c) $2.01\mathbf{i} + 0.07\mathbf{j}$
8 a) $45.3\mathbf{i} + 21.1\mathbf{j}$, $\mathbf{i}, \mathbf{j}$ horizontal and vertical
b) $50\mathbf{i}$, $\mathbf{i}$ at 25° to horizontal
c) 4.74 s Constant speed assumed.
9 a) $40\mathbf{i}$ **b)** $35\mathbf{i}$ **c)** $49.5\mathbf{i} + 1.74\mathbf{j}$
10 a) 12.8 m s^{-1} at 38.7° to the horizontal **b)** 1.6 s, 16 m
c) For example
$\mathbf{v} = (10 - 1.25t)\mathbf{i} + (8 - 5t)\mathbf{j}$
d) No account of factors other than gravity might change the vertical motion.

Exercises 5.3A *(p.70)*

1 a) $1.4\mathbf{i} + 1.4\mathbf{j}$ **b)** $2.4\mathbf{i} + 4.4\mathbf{j}$
c) $19.1\mathbf{i} + 29.1\mathbf{j}$
2 a) $\mathbf{v} = 180\mathbf{i} - 10t\mathbf{j}$.
$\mathbf{r} = 180t\mathbf{i} + (1.5 - 5t^2)\mathbf{j}$
b) 0.55 s **c)** 180 m s^{-1}
3 a) $200\mathbf{i}$, $80\mathbf{i}$
b) $2600\mathbf{i} + 1200\mathbf{j}$, $160\mathbf{i} + 120\mathbf{j}$
4 a) $10\mathbf{i} + 17.32\mathbf{j}$
b) $\mathbf{r} = 10t\mathbf{i} + (17.32t - 5t^2)\mathbf{i}$
d) 34.6 m
5 a) $\mathbf{r} = t^2\mathbf{i} - 0.75t^2\mathbf{j}$
b) 11.25 m
c) 7.5 m s^{-1}
6 a) $3.83\mathbf{i} + 3.21\mathbf{j}$, $-3.76\mathbf{i} + 1.37\mathbf{j}$
b) $-75.9\mathbf{i} - 1.84\mathbf{j}$.
7 a) $\mathbf{v} = 2\mathbf{i} + 0.5t\mathbf{j}$
b) $\mathbf{v} = 2t\mathbf{i} + 0.25t^2\mathbf{j}$
c) 10.2 m s^{-1}, 40 m
d) Unlikely to be constant due to buoyancy and resistance.
8 a) $23\mathbf{i} + 19\mathbf{j}$
c) $\mathbf{v} = 23\mathbf{i} + (19 - 10t)\mathbf{j}$.
$\mathbf{r} = 23t\mathbf{i} + (19t - 5t^2)\mathbf{j}$
d) 3.86 s, 88.7 m
e) An over estimate
9 a) $4\mathbf{i} + 2\mathbf{j}$, constant acceleration
b) 156.5 m s^{-1}
10 a) 2 seconds
b) $\mathbf{a} = \mathbf{i} - \mathbf{j}$

Exercises 5.3B *(p.72)*

1 a) $1.2\mathbf{i}$
b) $\mathbf{v} = 1.2\mathbf{i} + 0.1t\mathbf{j}$
c) $\mathbf{v} = 1.2\mathbf{i} + 0.15\mathbf{j}$, $\mathbf{r} = 1.8\mathbf{i} + 0.1125\mathbf{j}$ **d)** No
2 a) $7.67\mathbf{i} + 6.43\mathbf{j}$
b) $\mathbf{r} = 7.67t\mathbf{i} + (6.43t - 5t^2 + 2)\mathbf{j}$
c) $t = 1.54$ s, $0 \le t \le 1.54$
d) 11.85 m
3 a) $5.13\mathbf{i} + 14.09\mathbf{j}$, $11.82\mathbf{i} - 2.08\mathbf{j}$ **b)** $13.4\mathbf{i} - 32.3\mathbf{j}$
c) Constant acceleration. A model where acceleration peaks.
4 b) $50\mathbf{i} + 200\mathbf{j}$ **c)** 206 m s^{-1}
d) $0 \le t\ 6. \le 24$ s
5 a) $\pm 25\mathbf{j}$ **b)** $3.4\mathbf{i} + 15\mathbf{j}$ **c)** 77.2°
6 a) $0.5t^2\mathbf{i} - 0.025t^2\mathbf{j}$
b) $0.050\,06t^2$
c) 6.32 s, $6.32\mathbf{i} - 0.316\mathbf{j}$, 6.33 m s^{-1}
7 a) $35.4\mathbf{i} + 24.33\mathbf{j}$
b) $70.9\mathbf{i} + 48.68\mathbf{j}$
c) $165.8\mathbf{i} + 163.14\mathbf{j}$, 167.42 m, 013.0°
8 a) $\mathbf{a} = 0.1\mathbf{i} + 0.4\mathbf{j}$, $\mathbf{v} = 0.1t\mathbf{i} + 0.4t\mathbf{j}$
b) $15\mathbf{i} + 80\mathbf{j}$
9 37.82 m, 38.4 m
10 13.45 m s^{-1}, 54.83 on 24.2°

Consolidation Exercises for Chapter 5 *(p.74)*

1 a) Variable, since direction changes.
b) Variable. Velocity changes, but acceleration need not be parallel to velocity.
2 b) $\alpha = 0.92°$ **d)** 25.6 m

CHAPTER 6
Newton's first and second laws

Exercises 6.1A *(p.78)*

1 784 N
2 780 N, travelling in a straight line.

3 a) 45 kg s^{-1} **b)** 1350 N
c) $R = 40v + 200$, 1400 N
4 At a tangent to the curve.
5 Moves along a tangent at a constant speed.
6 462 N
7 a) 62 N **b)** 62 N
8 a) 3.4 m s^{-1} **b)** 3.9 m s^{-1}
9 704 N
10 283 N

Exercises 6.1B *(p.79)*

1 676.2 N
2 2080 N
3 a) $R = 10 + 9v$ **b)** 55 N
4 At a tangent to the curve.
5 At a tangent to the curve.
6 19 600 N
7 74°
8 13317 N
9 a) 640 N
b) 703 N
10 a) $R = 10gv = 98v$
b) 8 m s^{-1}

Exercises 6.2A *(p.82)*

1 a) 15 000 N **b)** 420 N
c) 1.96 N
2 a) 6.67 m s^{-2} **b)** 6.43 m s^{-2}
c) 26 667 m s^{-2}
3 a) 5715 N **b)** 5704 N
c) 5616 N
4 a) –7.5 m s^{-2} **b)** 60 m **c)** 4 s
5 a) –7.84 m s^{-2} **b)** 19.4 m s^{-1}
c) Air resistance – will increase the initial speed.
6 a) $0.1\mathbf{i} + 1.5\mathbf{j}$ **b)** 0.601 N
7 a) $3.1\mathbf{i} - 20\mathbf{j}$ **b)** $0.6\mathbf{i} - 4\mathbf{j}$
c) 4.04 m s^{-2} at 81.5° below $\mathbf{i}$.
8 b) 250 m s^{-2} **c)** $4\mathbf{i} + 100\mathbf{j}$
9 a) 400 N **b)** 83.6°, 268 N
c) 2 needed.
10 a) 769 N **b)** 4.25 m s^{-1}
c) 3.4 m s^{-1}

Exercises 6.2B *(p.84)*

1 a) 0.0294 N **b)** 15 000 N
c) 12 000 N
2 a) 3.75 m s^{-2} **b)** 1.14 m s^{-2}
c) 10 000 m s^{-2}
3 a) 6867 N **b)** 6720 N **c)** 6860 N
4 a) –2.7 m s^{-2} **b)** 18.52 m
c) 3.7 s
5 a) $10.7\mathbf{i} + 2.6\mathbf{j}$ **b)** $5.35\mathbf{i} + 1.3\mathbf{j}$
c) 5.5 m s^{-2} at 13.7° above $\mathbf{i}$
6 a) $190\mathbf{i} + (R - 710)\mathbf{j}$
c) $-2.53\mathbf{i}$ **d)** 74.25 N
7 a) 101 N **b)** 1.36 m s^{-2}
c) 11.66 m s^{-1}
8 a) $-41.7\mathbf{i}$ **b)** $6.37\mathbf{i} + 5.14\mathbf{j}$
9 1142 N
10 b) $0.125\mathbf{j}$ **c)** $\mathbf{v} = 0.5\mathbf{i} + 0.125t\mathbf{j}$, $\mathbf{r} = 0.5t\mathbf{i} + 0.625t^2\mathbf{j}$
e) $15\mathbf{i} + 31.25\mathbf{j}$

Exercises 6.3A *(p.87)*

1 a) $1750 - T$, $T - 250$
b) $1750 - T = 1000a$, $T - 250 = 700a$ **c)** 866 N, 0.88 m s^{-2}
d) 11.4 s
2 a) i) $-\frac{1}{4}g$ **ii)** $\frac{1}{9}g$ **iii)** 0
b) $\dfrac{(M-m)g}{M+m}$
3 a) 2.8 m s^{-2} **b)** 0.7 m s^{-2}
4 a) $\frac{2}{9}$ m s^{-2} **b)** 4444.4 N
c) 8888.8 N
5 a) 0.75 m s^{-2}
b) 0.17 m s^{-2}, 27.44 N
6 a) 14 m s^{-1} **b)** –4.72 m s^{-2}
c) 20.76 m
7 a) Truck: $(1500 - T\cos 20°)\mathbf{i} + (R_1 - 1500g - T\cos 70°)\mathbf{j}$
Car: $(T\cos 20° - 200)\mathbf{i} + (R_2 + T\cos 70° - 800g)\mathbf{j}$
b) 0.57 m s^{-2} **c)** 698 N
8 a) 4.29 m s^{-2} **b)** 2.06 m s^{-2}
c) 38.7 N

Exercises 6.3B *(p.89)*

1 a) Block: $T - 10$ Mass: $19.6 - T$
b) $a = \dfrac{T-10}{5}$ and $a = \dfrac{19.6-T}{2}$
c) $T = 16.86$ N, $a = 1.37$ m s^{-2}
2 a) 33.7 N, 1.37 m s^{-2}
b) 3.26 N, 3.26 m s^{-2}
c) 17.2 N, 1.2 m s^{-2}
3 a) $49 - T$, $T - 19.6$
b) $\dfrac{49-T}{5}$, $\dfrac{T-19.6}{2}$
c) 28 N, 4.2 m s^{-2}
d) There is no resistance and tension constant in string.
4 a) 1.09 m s^{-2}
b) $\dfrac{2mMg}{(M+m)}$
5 a) 900 N **b)** 2200 N
6 a) 0.517 m s^{-2} **b)** 33.3 N.
7 a) 3.41 m s^{-2}, 40.6 N
8 b) 1 m s^{-2}, 0.5 m s^{-2}

Consolidation Exercises for Chapter 6 *(p.90)*

1 a) 11801 N, 37.2 N **b)** Yes
2 a) 2.308 m s^{-2} **b)** 42.86 m s^{-1}
c) 52.38 m s^{-1}
3 a) 34.8° **b)** 62881 N **c)** 126.5 s
4 a) 1.93 m s^{-1} **b)** 3.1 m s^{-1}
5 b) 7.06 kg
c) The string does not stretch.
6 c) 855 N **d)** Yes
7 a) 0.05 **b)** 10.95 m s^{-1}, 9.1 s
c) 87.4 N
8 a) $\mathbf{A}$: $4ma = 2mg - T$
$\mathbf{B}$: $ma = T - mg$ **b)** $\dfrac{6mg}{5}$

CHAPTER 7

Motion with variable forces and acceleration

Exercises 7.1A *(p.97)*

1 a) $\mathbf{v} = 5\mathbf{i} + (6 - 9.8t)\mathbf{j}$
b) $\mathbf{a} = -9.8\mathbf{j}$, 9.8m s^{-2}, acts straight down.
2 a) $\mathbf{v} = \dfrac{t^2}{10}\mathbf{i} + \dfrac{t^3}{30}\mathbf{j} + 5\mathbf{k}$
$\mathbf{a} = \dfrac{t}{5}\mathbf{i} + \dfrac{t^2}{10}\mathbf{j}$
b) 5 m s^{-1} **c)** Horizontal
3 a) $\mathbf{v} = 16 - 9.8t$ **b)** 15.06 m
c) –9.8 m s^{-2}
4 a) $t = 20$ s
b) 80 m s^{-1}, 1066.7 m
5 26.94 m, $0.86\mathbf{i} - 14\mathbf{j}$
6 a) 20 m s^{-1} **b)** $a = \dfrac{1}{\sqrt{t}}$
c) $a = 0.1$ m s^{-2}
7 a) $v = 20t - 5t^2$ $a = 20 - 10t$
b) $t = 2$, $v = 20$ m s^{-1}, $h = 26.7$ m
c) 46.7 m.
8 a) $v = 5\mathbf{i} + \frac{1}{3}t\mathbf{j}$ $a = \frac{1}{3}\mathbf{j}$
b) $t = 0$, $v = 5$, $t = 4$, $v = \sqrt{5^2 + \left(\frac{4}{3}\right)^2} = 5.17$
c) It is constant at 90° to original direction of motion.
9 a) $v = \dfrac{t^3}{5} - \dfrac{3t^2}{2} + \dfrac{5t}{2}$

$a = \frac{3t^2}{5} - 3t + \frac{5}{2}$

b) -1.25 m s^{-2} **c)** 0 m s^{-1}

10 a) $\mathbf{v} = 2\cos t\mathbf{i} - 2\sin t\mathbf{j} - 0.5\mathbf{k}$

$\mathbf{a} = -2\sin(t)\mathbf{i} - 2\cos t\mathbf{j}$

b) 2.06 m s^{-2}

c) Toward centre of ride and horizontal.

Exercises 7.1B *(p.99)*

1 a) $\mathbf{v} = 12t\mathbf{i} + (10t - 3)\mathbf{j}$

$\mathbf{a} = 12\mathbf{i} + 10\mathbf{j}$

b) $\mathbf{v} = \frac{3t^2}{100}\mathbf{i} + \frac{3t^2}{50}\mathbf{j}$

$\mathbf{a} = \frac{3t}{50}\mathbf{i} + \frac{3t}{25}\mathbf{j}$

c) $\mathbf{v} = (24t^3 - 6t)\mathbf{i} + 4\mathbf{j}$

$\mathbf{a} = (72t^2 - 6)\mathbf{i}$.

2 a) $v = 3t^2 - 8t + 4$ **b)** 1.185 m

c) 0 m s^{-1} **d)** 2.37 m s^{-1}

e) The original is more realistic. In the second model the diver will probably pop out of the water.

3 a) $v = 1.6t$, $a = 1.6$ m s^{-2}

b) Reaches a steady speed.

4 $k = 0.0001$

5 9.8 m s^{-2}, downwards.

6 a) $\mathbf{v} = 5\mathbf{i} + (12 - 0.3t^2)\mathbf{j}$ $\mathbf{a} = -0.6t\mathbf{j}$ **c)** Slope in $-\mathbf{j}$ direction.

7 a) $0 \le t \le 2$

b) $\mathbf{v} = 6.4\mathbf{i} - 8\mathbf{j}$ $\mathbf{a} = 9.6\mathbf{i} - 4\mathbf{j}$

c) No, they are still accelerating downwards.

8 $\mathbf{v} = 0.25\mathbf{i} - 0.0625\mathbf{j}$

$\mathbf{a} = -0.031\,25\mathbf{i} + 0.023\,44\mathbf{j}$

9 a) $\mathbf{v} = 1.04t + 0.66$

$\mathbf{a} = 1.04$ m s^{-2} **b)** Initially the child moves at 0.66 m s^{-1}, probably given a push.

c) $\mu = 0.85$

10 b) $4\mathbf{i} + 0\mathbf{j}$, $4\mathbf{i} + 15\mathbf{j}$

c) $40\mathbf{i} + 100\mathbf{j}$

Exercises 7.2A *(p.103)*

1 a) $\mathbf{r} = 4t\mathbf{i} + (2 + 2t - 5t^2)\mathbf{j}$

b) $\mathbf{r} = (3t + 2t^2)\mathbf{i} + \left(-2 + \frac{t^2}{6}\right)\mathbf{j}$

c) $\mathbf{r} = \left(10 + \frac{t^4}{12}\right)\mathbf{i} + \left(-20 + \frac{t^5}{20}\right)\mathbf{j}$

2 a) $a = 0.5 - \frac{t}{60}$

b) $v = 0.5t - \frac{t^2}{120}$, 7.5 m s^{-1}

c) $s = \frac{t^2}{4} - \frac{t^3}{360}$, 150 m

3 a) $a = 0.4 - \frac{t}{50}$

b) $v = 0.4t - \frac{t^2}{100}$ $t = 20$, $v = 4$ m s^{-1}

c) $s = \frac{t^2}{5} - \frac{t^3}{300}$, $t = 20$, $s = 53.5$ m

4 a) $\mathbf{v} = 10\mathbf{i} + (6t^2 - 36t + 54)\mathbf{j}$

b) $\mathbf{a} = (12t - 36)\mathbf{j}$

c) $\mathbf{r} = 30\mathbf{i} + 54\mathbf{j}$, $\mathbf{v} = 10\mathbf{i} + 0\mathbf{j}$

d) $\mathbf{a} = -9.8\mathbf{j}$, $\mathbf{v} = 10\mathbf{i} - 9.8t\mathbf{j}$

$\mathbf{r} = (10t + 30)\mathbf{i} + (54 - 4.9t^2)\mathbf{j}$

e) Lands 63.2 m from launch point.

5 a) $\mathbf{a} = 12\mathbf{i} + 10\mathbf{j}$

b) $\mathbf{v} = 36\mathbf{i} + 27\mathbf{j}$, $v = 45$ m s^{-1}

c) $6t^2 - 5 = 145$ and $5t^2 - 3t + 8 = 118$. Gives $t = 5$ in both cases.

6 $a = 3 - \frac{kt}{40}$, $v = 3t - \frac{kt^2}{80}$, $k = \frac{9}{2}$

$v = 3t - \frac{9t^2}{160}$, $s = \frac{3t^2}{2}$, $-\frac{3t^3}{160}$, $s = 1200$ m

7 a) $v = (0.2t + 30)\mathbf{i} + 0.05t^2\mathbf{j}$

$\mathbf{r} = (0.1t^2 + 30t)\mathbf{i} + \frac{0.05t^3}{3}\mathbf{j}$

8 a) 2 s **b)** $0 \le t \le 4$ **c)** Use a trigonometric function.

10 a) Opposite to the motion.

b) 20 m s^{-1} **c)** 20 m s^{-1}

e) 12 m s^{-1} **f)** Braking force increased and then relaxed, driver may have felt safe.

Exercises 7.2B *(p.104)*

1 a) $\mathbf{r} = \left(\frac{t^3}{12} + 4t\right)\mathbf{i} + \left(\frac{t^3}{18}\right)\mathbf{j}$

b) $\mathbf{r} = \left(\frac{t^3}{2} + 20t\right)\mathbf{i} + \left(\frac{-t^3}{3} - 18t\right)\mathbf{j}$

c) $\mathbf{r} = \left(4\frac{t^{\frac{5}{2}}}{15} + 10t\right)\mathbf{i} + \left(8\frac{t^{\frac{5}{2}}}{15} - 2t\right)\mathbf{j}$

2 a) 50 m s^{-1} **b)** $333\frac{1}{3}$ m

3 a) 10 m s^{-1} **b)** $133\frac{1}{3}$ m

4 36 s, 1036.8 m

5 a) $\mathbf{v} = \frac{t(40-t)}{10}\mathbf{i} + \frac{t(40-t)}{20}\mathbf{j}$

c) $t = 40$ s **d)** $1067\mathbf{i} + 533\mathbf{j}$

6 a) $\mathbf{v} = \left(8 - \frac{t^2}{40}\right)\mathbf{i} + (2 - 10t)\mathbf{j}$

b) $\mathbf{r} = \left(8t - \frac{t^3}{120}\right)\mathbf{i} + \left(2t - 5t^3\right)\mathbf{j}$

c) $3.2\mathbf{i} + 0\mathbf{j}$

7 a) $136.5\mathbf{i} - 68.3\mathbf{j}$

8 a) $s = 2t - \frac{1}{12}t^3$ **b)** 2.82 s

c) For approximately $0 \le t \le 5$.

d) An alternative could be based on a trigonometric function.

9 a) $\mathbf{v} = \frac{t}{2}\mathbf{i} - \frac{(t-4)^3}{6}\mathbf{j}$

$\mathbf{r} = \frac{t^2}{4}\mathbf{i} - \frac{(t-5)^4}{24}\mathbf{j}$

b) $t = 4$ **c)** 0.4 m

10 a) -9 **b)** 4 m s^{-1}

Consolidation Exercises for Chapter 7 *(p.106)*

1 a) $k = \frac{1}{2}$ $s = \frac{1}{24}t^2(42 - t^2)$

b) Total distance is 36.75 m.

2 a) 20 m s^{-1} **b)** 2.16 N

3 $t = 2$ and $t = 6$

4 b) $2.47\mathbf{j}$

5 a) $s = \frac{20}{3}t^{\frac{3}{2}}$, 18.9 m

b) 3.52 m s^{-2}

c) Infinite acceleration

6 b) i) $F = 8$ N, $R = 29.4$ N: $\mu = 0.272$

CHAPTER 8
Projectiles

Exercises 8.1A *(p.115)*

1 a) $3.76\mathbf{i} + 1.37\mathbf{j}$ **b)** $-9.8\mathbf{j}$

c) $\mathbf{v} = 3.76\mathbf{i} + (1.37 - 9.8t)\mathbf{j}$

d) $\mathbf{r} = 3.76t\mathbf{i} + (1.37t - 4.9t^2)\mathbf{j}$

2 a) $4.83\mathbf{i} + 1.29\mathbf{j}$

b) $\mathbf{r} = 4.83t\mathbf{i} + (1.29t - 4.9t^2)\mathbf{j}$

c) 0.264 s **d)** 1.28 m **e)** 0.09 m

3 a) 7.8 m **b)** No, this is probably the speed for a top athlete.

4 a) 22.46 m **b)** No

c) The lighter ball.

5 a) 48.85 m **b)** 25 m

c) Lower one less likely to be affected by air resistance due to shorter time of flight.

6 a) 5.48 m s^{-1} **b)** Minimum speed at 45°, 4.2 m s^{-1}

7 a) 24° or 60° **b)** None

c) More difficult to give a heavier one the same initial velocity.

8 **b)** $\mathbf{r} = 75t\mathbf{i} - 4.9t^2\mathbf{j}$
c) $225\mathbf{i} - 44.1\mathbf{j}$ **d)** 80.6 m s^{-1}

9 **a)** $\mathbf{r} = 2.05t\mathbf{i} + (2 + 5.64t - 4.92t)\mathbf{j}$ **b)** 3.62 m
c) 1.44 s, 2.94 m

10 **a)** $\mathbf{r} = 6.30t\mathbf{i} + (1.8 + 4.93t - 4.9t^2)\mathbf{j}$ **b)** 0.63 s **c)** 2.96 m
d) Increase the angle of projection a little.

Exercises 8.1B *(p.117)*

1 **a)** $6.06\mathbf{i} + 3.5\mathbf{j}$ **b)** $-9.8\mathbf{j}$
c) $\mathbf{v} = 6.06\mathbf{i} + (3.5 - 9.8t)\mathbf{j}$
d) $\mathbf{r} = 6.06t\mathbf{i} + (3.5t - 4.9t^2)\mathbf{j}$

2 **a)** $1.88\mathbf{i} + 0.68\mathbf{j}$
b) $\mathbf{r} = 1.88t\mathbf{i} + (0.68t - 4.9t^2)\mathbf{j}$
c) 0.14 s **d)** 0.262 m **e)** 0.024 m

3 **a)** 13.8 m **b)** Yes

4 **a)** 19.89 m **b)** Not unless air resistance is considered.
c) Range reduced as air resistance has a greater effect on a lighter ball.

5 **a)** 2.23 m **c)** 8 m s^{-1}
d) Lower angle, less time of flight for resistance etc. to deflect dart.

6 **a)** 19.13 m s^{-1} **b)** 45°
c) 15.33 m s^{-1}

7 **a)** 0.102 m **b)** 8.5 cm **c)** 6.4 cm

8 **a)** Both have some horizontal component of velocity, if the air resistance is negligible.
b) $-4.9t^2\mathbf{j}$ **c)** 20.85 m s^{-1}

9 **a)** $\mathbf{r} = 4.91t\mathbf{i} + (1.7 + 3.44t - 4.9t^2)\mathbf{j}$ **b)** 1.04 s, 5.09 m
c) A large force would be needed to alter the path of a heavy object significantly.

10 **a)** 32.3 m s^{-1} **b)** 612.5 m
c) Lengths of holes need to be increased by a factor 6.125.

Exercises 8.2A *(p.122)*

1 **a)** $y = x - 0.00306x^2$, 326.5 m
b) $y = 2x - 0.049x^2$, 40.82 m
c) $y = 1.5x - 0.034x^2$, 44.12 m
d) $y = 3 + \frac{2}{3}x - 0.136x^2$, 7.75 m

2 $y = 0.5x - 0.05x^2$

3 **a)** $\mathbf{r} = 39.0t\mathbf{i} + (22.5t - 4.9t^2)\mathbf{j}$
b) $y = 0.577x - 0.00323x^2$
c) 10.26 m

4 **a)** $\mathbf{r} = 15.59t\mathbf{i} + (9t - 4.9t^2)\mathbf{j}$
b) $y = 0.577x - 0.0202x^2$
c) 1.55 m **d)** Very unlikely.

5 **a)** $\mathbf{r} = V\cos70°t\mathbf{i} + (V\sin70°t - 4.9t^2)\mathbf{j}$
b) $y = 2.747x - \dfrac{41.9x^2}{V^2}$
c) 8.48 m s^{-1}

6 **a)** $y = 1.732x - \dfrac{19.6x^2}{v^2}$
b) $v > 7.27$ m s^{-1} **c)** No air resistance very significant. Also size important when near to net.

7 **a)** $y = 2.747x - 0.0465x^2$
b) $y = 0.0875x$ **c)** $x = 57.15$, $y = 5.00$ **d)** 57.37 m

8 **a)** $y = 1 - 0.196x^2$ **b)** $y = -0.577x$ **c)** (4.17, −2.40), when $t = 0.834$ s.

9 **a)** $y = x\tan\alpha - \dfrac{4.9x^2}{30^2\cos^2\alpha}$
b) $\alpha = 20.3°$ or 88.1°
c) 20.3° as he is unlikely to get up the ramp at 88°.

10 **a)** $y = x\tan\alpha - \dfrac{4.9x^2}{20^2\cos^2\alpha}$
b) $\alpha = 3.2°$ or 81.6°
c) $\alpha = 3.2°$ because the time of flight will be less and the ball will land in the required area.

Exercises 8.2B *(p.124)*

1 **a)** $y = x - 0.0109x^2$, 91.7 m
b) $y = 1.3333x - 0.0014x^2$, 979.6 m **c)** $y = 5.66x - 6.47x^2$, 0.875 m **d)** $y = x\tan\beta - \dfrac{4.9x^2}{u^2\cos^2(\beta)}, \dfrac{u^2\sin\beta\cos\beta}{4.9}$

2 $y = \dfrac{x}{\sqrt{3}} - \dfrac{4x^2}{375}$

3 **a)** $\mathbf{r} = 12.02t\mathbf{i} + (12.02t - 4.9t^2)\mathbf{j}$ **b)** $y = x - 0.0339x^2$
c) 3.07 m

4 **a)** $\mathbf{r} = 8.66t\mathbf{i} + (5t - 4.9t^2)\mathbf{j}$
b) $y = 0.577x - 0.065x^2$
c) 25.3 cm **d)** Yes

5 **a)** $\mathbf{r} = V\cos60°t\mathbf{i} + (2 + V\sin60°t - 4.9t^2)\mathbf{j}$

b) $y = 2 + 1.732x - \dfrac{19.6x^2}{v^2}$

c) 6.29 m s^{-1}

6 **a)** $y = 1.197x - \dfrac{11.859x^2}{v^2}$

b) $v \geq 5.86$ m s^{-1} **c)** No, air resistance will be significant.

7 **a)** $y = 1.732x - 0.049x^2$
b) $y = 0.466x$ **c)** $25.84\mathbf{i} + 12.04\mathbf{j}$ **d)** 28.5 m

8 **a)** $y = -0.0054x^2$
b) $y = -0.2679x - 0.536$
c) $x = 51.55$, $y = -14.35$.

9 **a)** $y = x\tan\alpha - \dfrac{0.00162x^2}{\cos^2\alpha}$
b) $40\tan\alpha - 2.59\tan^2\alpha - 2.59 = 0$ **c)** 3.72° or 86.5°
d) 3.72°!

10 **a)** $y = x\tan\alpha - \dfrac{5x^2}{v^2\cos^2\alpha}$
b) $0 = 10\tan\alpha$
$-\dfrac{500\tan^2\alpha}{v^2} - \dfrac{500}{v^2} + 1$

Consolidation Exercises for Chapter 8 *(p.126)*

1 **a)** 2.04 s **b)** 45.7 m s^{-1} at 40.7° to horizontal.

2 **a)** 15.3 m s^{-1} **b)** 15.3 m s^{-1}

3 190 m s^{-1}, greater than 190 m s^{-1}

4 **a)** 32.1 m s^{-1} **b)** 2.66 s
c) 8.7 m
No air resistance, no wind, etc.

5 1.834 m. Air resistance etc. Horizontal distance, greater speed gives greater resistance.

6 **b)** 10° or 78° **c)** 2 s

7 **a)** 11, 19.05 **b)** $0.2 \leq t \leq 2$
c) 19

8 **b)** 7.23 m s^{-1}
c) 78.5°, 35.4°, preferred angle 78.5°

CHAPTER 9

Momentum and collisions

Exercises 9.1A *(p.132)*

1 **a)** 4 kg m s^{-1}
b) 4 000 000 kg m s^{-1}
c) 9.5 kg m s^{-1}

3 Momentum of 16 g greater in magnitude, both have the same direction.

4 Constant magnitude, but variable direction.

5 **a)** $2724\mathbf{i} + 5346\mathbf{j}$
b) $-57851\mathbf{i} - 68944\mathbf{j}$
c) $72.4\mathbf{i} - 26.3\mathbf{j}$.

6 **a)** $1.046\mathbf{i} - 0.092\mathbf{j}$

b) $48.3\mathbf{i} + 43.5\mathbf{j}$
c) $672\mathbf{i} - 106\mathbf{j}$.

Exercises 9.1B *(p.133)*

1 a) 476 kg m s^{-1}
b) 200 000 kg m s^{-1}
c) 0.012 kg m s^{-1}
3 No. It may have constant magnitude, but will vary in direction.
4 Momentum has constant magnitude.
5 a) $592\mathbf{i} + 215\mathbf{j}$
b) $-3\,420\,201\mathbf{i} - 9\,396\,926\mathbf{j}$
c) $-44.8\mathbf{i} - 254.1\mathbf{j}$
6 a) $3.45\mathbf{i} + 2.89\mathbf{j}$ **b)** $0.434\mathbf{i} + 0.517\mathbf{j}$ **c)** $70.15\mathbf{i} + 40.5\mathbf{j}$

Exercises 9.2A *(p.136)*

1 a) 0.4125 Ns **b)** 0.275 Ns
2 a) 750 Ns **b)** 625 N
3 a) 7500 Ns **b)** 3750 N
4 b) 544.5 Ns **c)** 453.75 N
5 a) −180 Ns **b)** −0.4 m s^{-1}
6 a) $\mathbf{u} = 1.379\mathbf{i} - 1.157\mathbf{j}$, $\mathbf{v} = 1.5\mathbf{i} + 0.87\mathbf{j}$ **b)** $\mathbf{I} = 0.024\mathbf{i} + 0.406\mathbf{j}$ **c)** 0.049 s
7 a) $-55.2\mathbf{i} + 89.3\mathbf{j}$
8 b) $459\mathbf{i} - 321\mathbf{j}$ **d)** If change takes 0.5 s, then $T = 1120$ N
9 a) $\mathbf{u} = 3.60\mathbf{i} + 3.02\mathbf{j} - 1.71\mathbf{k}$, $\mathbf{v} = 0\mathbf{i} - 5.20\mathbf{j} + 3\mathbf{k}$
b) $\mathbf{I} = -0.108\mathbf{i} - 0.247\mathbf{j} + 0.1413\mathbf{k}$ **c)** 0.609 N
10 a) $\mathbf{u} = 0.62\mathbf{i} + 1.69\mathbf{j}$, $\mathbf{v} = -0.5\mathbf{i} + 0.866\mathbf{j}$ **b)** $\mathbf{I} = -358.4\mathbf{i} - 263.7\mathbf{j}$
c) $1.24\mathbf{i} + 2.91\mathbf{j}$

Exercises 9.2B *(p.137)*

1 a) 0.24 Ns **b)** 0.48 m s^{-1}
2 a) −0.24, 0.12, defining down as positive **b)** 0.36 Ns **c)** 0.72 N
3 a) 3 Ns **b)** 5 N **c)** Increase to a maximum and decrease.
4 a) 375 Ns **b)** Extend t and reduce F. **c)** 208.3 N **d)** 1250 N
5 6842 N, 10833 N
6 a) $\mathbf{u} = 0.410\mathbf{i} - 1.128\mathbf{j}$, $\mathbf{v} = 0.396\mathbf{i} + 0.801\mathbf{j}$
b) $0.0108\mathbf{i} + 0.2315\mathbf{j}$ **c)** 0.290 N
d) Increase and decrease.
7 a) $4.70\mathbf{i} - 1.71\mathbf{j}$, $4.70\mathbf{i} + 1.71\mathbf{j}$
b) $0.2736\mathbf{j}$, $-0.2736\mathbf{j}$ **c)** $0.4\mathbf{j}$
d) 7.3 m s^{-1}

8 a) 8.85 m s^{-1} **b)** $-8.85\mathbf{j}$, $7.97\mathbf{i}$
c) $398.5\mathbf{i} + 442.5\mathbf{j}$
d) $498\mathbf{i} + 553\mathbf{j}$.
9 a) $-159\mathbf{i} + 903\mathbf{j}$
b) 305.6 N, 4.63 m s^{-2}
10 a) $m(-0.103\mathbf{i} + 1.181\mathbf{j})$
b) $0.103\mathbf{i} - 1.181\mathbf{j}$
c) 1.19 m s^{-1} at 84.7° to $\mathbf{i}$

Exercises 9.3A *(p.143)*

1 a) 0.252 Ns **b)** 2.52 N **c)** 2.52 N
2 a) 0.828 Ns **b)** 8 m s^{-1}
3 a) 5437.5 Ns **b)** 2.03 m s^{-1}
4 a) 0.318 Ns **b)** 7.25 m s^{-1}
c) 7.02 m s^{-1}
5 a) 667 Ns **b)** 13.3 m s^{-1}
c) $w = \frac{\pi}{2}$, $A = \frac{500\pi}{3}$
6 a) $0 \le t \le \frac{1}{15}, \frac{1}{15}s$
b) 1.24 m s^{-1} **c)** $25\sin(15\pi t) - 1\frac{2}{3}$
d) 0.59 m s^{-1}
7 a) 4.8 m s^{-1} **b)** $F(0) = F(0.4) = 0$. No account of force of gravity.
c) $a = 2.5$, 0.8 m s^{-1}
8 a) $c = mg = 2$
b) $F(0.5) = -2$
c) $a = 105$, $b = 210$

Exercises 9.3B *(p.144)*

1 a) 1.2 Ns **b)** 3T **c)** 0.4 s
d) Probably more curved.
2 a) 2.4 Ns **b)** 3 m s^{-1} **c)** None.
3 a) 800 **b)** $3\frac{1}{3}$ms^{-1}
4 a) 20 400 kg m s^{1}
b) $I = 7.5P$ **c)** 2720 N
5 a) 100 000 Ns **b)** 67 m s^{-1}
6 b) $\frac{A}{5\pi}$ **c)** $A = 8.64$ **d)** 4 m s^{-1}
7 a) $t = 0.3$ s **b)** 8 m s^{-1}
c) 5.06 m s^{-1}
8 Using the assumption $F(t) = 0$ gives $\alpha = \frac{\pi}{T}$ and $A = \frac{m\pi(v-u)}{2T}$

Exercises 9.4A *(p.149)*

1 a) 0.7 kg m s^{-1} **b)** 0.7 kg m s^{-1}
c) 3.6 m s^{-1}
2 a) 18 kg m s^{-1} **b)** 5.25 m s^{-1}
3 6.2 m s^{-1}
4 3 kg
5 8.87 m s^{-1}
6 6 m s^{-1}
7 3 m s^{-1}
8 0
9 $1\frac{2}{3}$ m s^{-1}
10 $\frac{2}{7}$ m s^{-1}

Exercises 9.4B *(p.151)*

1 a) 0.03 kg m s^{-1}
b) 0.03 kg m s^{-1} **c)** 0.248 m s^{-1}
2 a) −5 kg m s^{-1} **b)** −5 kg m s^{-1}
c) $v = 1$ m s^{-1}
3 0.9 m s^{-1}
4 0.2 m s^{-1}
5 2 kg
6 18 grams
7 30 cm s^{-1}
8 49 m s^{-1}
9 Direction remains unchanged but speed reduced to 0.9 m s^{-1}.
10 3.8 m s^{-1}

Exercises 9.5A *(p.155)*

1 a) $124\mathbf{i}$, $194\mathbf{i} - 112\mathbf{j}$ **b)** $318\mathbf{i} - 112\mathbf{j}$ **c)** 2.85 m s^{-1}, 19.4° to $\mathbf{i}$
2 a) $4\mathbf{i}$, $20\mathbf{i} - 34.6\mathbf{j}$ **b)** 17.5 m s^{-1}
c) Due to the effects of external forces due to movement of wings.
3 a) $2.5\mathbf{i} + 33.8\mathbf{j}$ **b)** $0.035\mathbf{i} + 0.469\mathbf{j}$ **c)** At an angle of 86° above horizontal.
4 a) Final velocities are easy to express. They are at 90° to each other.
b) 3.06 m s^{-1}, 2.57 m s^{-1}
5 a) $-23\,142\mathbf{i} + 1446\mathbf{j}$
b) 19.3 m s^{-1} **c)** The handbrake is on and it may hit the kerb, so its speed is lost quickly.
6 a) $0.421\mathbf{i} + 0.145\mathbf{j}$
b) $5.26\mathbf{i} + 1.81\mathbf{j}$
7 $v = \frac{\sqrt{10}u}{4}$ car 18.4°, motorbike 71.6°.
8 a) $(1.085 \times 10^{6}m + 7.231 \times 10^{-21})\mathbf{i} + (1.879 \times 10^{6}m + 4.175 \times 10^{-21})\mathbf{j}$
b) $m = 2.22 \times 10^{-27}$ kg
9 a) 88.9 Ns
b) 4.85 m s^{-1}
10 5.74 m s^{-1} is the speed at which the gun moves backwards. There is an upward impulse of 19 284 Ns on the gun.

Exercises 9.5B *(p.157)*

1 a) 15 400**i**, 24 513**i** + 20 569**j**
b) 8.8 m s^{-1}, at 27° to **i**
c) Decrease as they skid.
2 a) 1.99 × 106**i** – 5.91 × 10^4**j**
b) 383 m s^{-1} **c)** 1.7°
3 a) 1.73**i** + 133**j**
b) 2 m s^{-1}, 0.75° to **j** **c)** No
4 a) m(3.46 × 106**i** + 2.00 × 106**j**)
b) 3.46 × 10^6, 1.00 × 10^6
5 a) $2m\mathbf{i}$ **b)** 1.03 m s^{-1}, 46.8° to **i** **c)** Momentum conserved, no effects due to rotation of the balls.
6 a) 3.24**i** + 0.94**j**
b) 3.24**i** + 0.94**j**
7 a) $\sqrt{37}mu$ **b)** $\frac{\sqrt{37}u}{4}$
c) 80.5°, 9.5°
8 a) (0.684m + 1.928M)**i** + (1.879m – 2.298M)**j**
b) M = 0.818 m
c) v = 2.76 m s^{-1}
9 a) 68.4 Ns **b)** 4.18 m s^{-1}
10 Speed reduced to 2.91 m s^{-1}.

Exercises 9.6A *(p.163)*

1 a) 0.5 **b)** 6 m s^{-1} **c)** 0.75
d) $10\frac{2}{3}$ m s^{-1}
2 a) 6.26 m s^{-1} **b)** 4.85 m s^{-1}
c) 0.775
3 a) 0.16 m **b)** 2.78 m **c)** 0.632
d) 0.730
4 a) 9.5 m s^{-1}, – 9.5 m s^{-1}
b) 1 m s^{-1}, –1 m s^{-1}
5 a) 3.625 m s^{-1}, 5.625 m s^{-1}
b) The rotation of the balls.
6 a) Moves at 0.05 m s^{-1}.
b) 0.975
7 a) 0, v **b)** $\frac{v}{4}, \frac{3v}{4}$ **c)** $\frac{v}{2}, \frac{v}{2}$
8 a) $v_A = 2.25$, $v_B = 2.75$
b) $v_B = 1.4375$, $v_C = 2.3125$
c) Between A and B.
9 a) $v_A = \frac{u(1-e)}{2}$, $v_B = \frac{u(1+e)}{2}$
b) $v_B = \frac{u(1-e)(1+e)}{4}$, $v_C = \frac{u(1+e)^2}{4}$
10 $\frac{1(1+e^2)}{1-e^2}$

Exercises 9.6B *(p.164)*

1 a) $\frac{2}{3}$ **b)** 0.95 **c)** 5.6 m s^{-1}
d) 23 m s^{-1}
2 a) 9.90 m s^{-1} **b)** 3.96 m s^{-1}
c) 0.8 m
3 a) 1.96 m **b)** 0.837 **c)** 0.775
d) 0.55 m
4 a) 5.5 m s^{-1}, 9.5 m s^{-1}
b) 6.5 m s^{-1}, 8.5 m s^{-1}
5 1.2 m s^{-1}, 2.6 m s^{-1}
6 a) $-\frac{2}{3}$ m s^{-1}, $1\frac{1}{3}$ m s^{-1}
b) 0.4 m s^{-1}, 0.8 m s^{-1}
7 a) 10 m s^{-1} **b)** $\frac{1}{18}$
8 a) $v_A = 2.6$, $v_B = 3.4$
b) $v_B = 2.42$, $v_C = 2.98$
c) A and B
9 a) $v_A = 1.3u$, $v_B = 1.7u$
b) $v_B = 0.51u$, $v_C = 1.19u$
c) Next A hits, $v_A = 0.747u$, $v_B = 1.063u$, $v_C = 1.19u$
10 a) $\frac{v(e+1)-V(e-1)}{2}$, $\frac{v(e+1)-v(e-1)}{2}$

Consolidation Exercises for Chapter 9 *(p.166)*

1 20 m s^{-1}
2 b) The ball does not bounce back towards him.
3 $\frac{au}{2}$
4 a) $\begin{pmatrix}-2\\-4.7\end{pmatrix}$ kg m s^{-1} **b)** $\begin{pmatrix}2\\4.7\end{pmatrix}$
5 $\frac{4}{3}u$, $4mu$
6 b) 0.036 Ns
7 ii) 11.4 m s^{-1} **iii)** 8 m s^{-1}
iv) 0.75
8 b) 0.8 kg

CHAPTER 10

Energy, work and power

Exercises 10.1A *(p.171)*

1 a) Electric motor in a washing machine. Dynamo on a bicycle.
b) Pinball machine.
2 In a power station. From brakes on a car.
3 Chemical to mechanical during run up. Mechanical to potential going up during jump. Potential to mechanical going down during jump.
4 Potential converted to mechanical.
5 Potential energy is converted to mechanical energy, which then is converted to potential energy stored in the elastic.

Exercises 10.1B *(p.171)*

1 a) When a ball goes up in the air. **b)** When a person walks or runs.
2 At extremities maximum potential and zero kinetic. At lowest point minimum potential and maximum kinetic.
3 Chemical to kinetic during run. Kinetic to potential during vault. Also some to elastic energy in pole as it bends, and then back to kinetic.
4 The kinetic energy of the wind is transferred to the kinetic energy of the sails, which is then transferred to electrical energy.
5 Kinetic energy is converted to elastic potential energy and heat. The elastic potential energy is then returned to kinetic energy.

Exercises 10.2A *(p.172)*

1 a) 405 000 J **b)** 0.375 J
c) 0.3 J **d)** 20.48 J
2 1575 J
3 a) 250 J **b)** 250 J **c)** 0.92 m s^{-1}
4 a) 0.192 J **b)** 3.58 m s^{-1}
c) Heat. Squash balls warm up during games.
5 a) 56.6 m s^{-1} **b)** 36.5 m s^{-1}
c) 2.5 kg
6 12.4 m s^{-1}
7 a) 0.8 J **b)** 0.8 J
8 a) 2.24 m s^{-1}, 3.16 m s^{-1}
b) 145 J, 290 J **c)** 580 J
9 a) 6.26 m s^{-1} **b)** 2.94 J
c) All the potential energy is converted to kinetic energy.
10 a) 0.784 J **b)** 0.627 J **c)** 20%
d) 20% loss of energy results in a 20% change in height.

Exercises 10.2B *(p.174)*

1 a) 562.5 J **b)** 2.5×10^6 J **c)** 0.005 76 J **d)** 3.2×10^5 J
2 1.8×10^6 J
3 a) 0.768 J **b)** Some lost to heat, but the bulk transferred to elastic potential energy and back to kinetic energy. **c)** 7.155 m s^{-1}
4 a) 0.9 J **b)** 0.9 J **c)** 6.708 m s^{-1}
5 a) 126 m s^{-1} **b)** 89 m s^{-1} **c)** 5 grams
6 3.16 m s^{-1}. Some energy lost due to air resistance.
7 a) 12 500 J **b)** 12 500 J **c)** Some energy will be lost due to air resistance.
8 a) 6.93 m s^{-1}, 9.80 m s^{-1} **b)** 28 900 J, 57 600 J **c)** 115 200 J
9 a) 24.25 m s^{-1}, 20 580 J **b)** 20 580 J **c)** 17 080 J
10 a) 0.8 m, 0.45 m **b)** 0.8J, 0.45J

Exercises 10.3A *(p.178)*

1 a) 2400 J **b)** 1.85 m s^{-1}
2 a) 4000 J **b)** 9.88 m s^{-1} **c)** No resistance factors have been considered.
3 a) 0.96 J **b)** 5.24 m s^{-1} **c)** 12 N
4 a) 75 000 J **b)** 5.25 10^7 J **c)** 2.1 J
5 a) 6.32 m s^{-1} **b)** 0.569 m s^{-1} **c)** 5 m s^{-1}
6 a) 12 618 J **b)** 12 348 J **c)** 270 J **d)** 0.88 m s^{-1}
7 a) 25 J **b)** 9.8 J **c)** 15.2 J **d)** 3.9 m s^{-1}
8 a) 4000 J **b)** 3000 J
9 a) 7056 J **b)** 12 701 J
10 137 m s^{-1} The force has been assumed to be constant and both the doorpost and the barrier assumed to have the same constitution.

Exercises 10.3B *(p.179)*

1 a) 150 J **b)** 6.12 m s^{-1} **c)** No, speed reached is excessive.
2 a) 4.8×10^5 J **b)** 15.5 m s^{-1} **c)** 3.52×10^5 J
3 a) 729 J **b)** 4.85 m s^{-1} **c)** 73 520 N
4 a) 1423.5 J **b)** 456 J **c)** 243 J
5 a) 96 000 J, 12.73 m s^{-1} **b)** 1×10^5 J, 20.25 m s^{-1} **c)** 6 J, 15.49 m s^{-1}
6 a) 7.4×10^4 J **b)** 7.35×10^4 J **c)** 500 J **d)** 1.15 m s^{-1}
7 a) 2940 J **b)** 3020 J **c)** 5390 J
8 a) 1×10^5 J **b)** 4.0×10^4 J **c)** 20.86 m s^{-1} **d)** The resistance forces will probably increase as the speed increases, so the solution will be an over-estimate.
9 a) 729 J, 4.85 m s^{-1} **b)** 1033 J
10 a) Constant force of 83 N **b)** 129 m s^{-1} **c)** It is likely that the force exerted by the sand will depend on the speed of the bullet. This may allow a greater speed to be used.

Exercises 10.4A *(p.183)*

1 a) 130 J **b)** 22.6 J **c)** 453 J **d)** 4290 J
2 a) 256 238 J **b)** 2.7×10^6 J **c)** 70 m s^{-1} **d)** 67 m s^{-1}
3 a) 85 J **b)** 22.5 J **c)** 62.6 J
4 a) 6938 J, 2400 J **b)** 11.5 m s^{-1}
5 a) 68 400 J **b)** 97 714 J, 29 314 J **c)** 1300 N **d)** The road is horizontal.
6 a) 3688 J **b)** 10.12 m s^{-1} **c)** 2788 J **d)** 16.1 m
7 a) 1 237 500 J **b)** $433T$ **c)** 2858 N
8 a) 37.5 J **b)** 100 J, 2011 J **c)** 2149 J **d)** 1091 N
9 a) 42 690 J **b)** 13.07 m s^{-1} **c)** 8098 J
10 a) The normal reaction is less. **b)** Slope : $25.2m$ J Level : 0.J $F = \mu R$ Slope : $12m$ J Horizontal : $7.84m$ J **c)** 5.67 m s^{-1} **d)** 8.90 m s^{-1}

Exercises 10.4B *(p.185)*

1 a) 0 J **b)** 61.7 J **c)** 1015 J **d)** 752 J
2 a) 31 652 J **b)** 251 m, 270 m
3 a) 150 J **b)** 1334 J **c)** 1184 J
4 a) 46 620 J **b)** 75 881 J **c)** 292.6 N **d)** 23.8 m s^{-1}, a constant resistance force and no driving force.
5 a) 25.6 J **b)** 360 J **c)** 74.2 N **d)** The rotation. Yes as a rotating object must have kinetic energy.
6 a) 1534 J **b)** 597 J **c)** 39.8 N **d)** 23.6 m
7 a) 28 128 J **b)** 1276 m from first position. **c)** 68 128 J, 1276 m
8 a) 8.32 m s^{-1} **b)** 5.27°
9 a) 278 J **b)** 72.6 N
10 Constant resistance of $2.13m$ N parallel to the slope, 1.78 m.

Exercises 10.5A *(p.189)*

1 a) 1960 J **b)** 0.0196 J **c)** 735 J **d)** 0.49 J
2 a) 14 m s^{-1} **b)** 1.4 m s^{-1} **c)** 17.1 m s^{-1} **d)** 3.1 m s^{-1}
3 a) 2430 J **b)** 7.67 m s^{-1} **c)** 8.85 m s^{-1} **d)** No loss of energy, due to air resistance or the jerk.
4 a) 7.46 m s^{-1} **b)** 13 400 J **c)** 6.84 m
5 a) 9.8 J **b)** 19.8 m s^{-1} **c)** 16.7 m
6 a) 695 800 J **b)** 694 913 J
7 5.58 J, 14.23 m
8 a) 98 J **b)** 48.7 m **c)** 98 J
9 a) $0.2 \times 9.8(8t - 4.9t^2)$ **b)** $8 - 15.68t + 9.604t^2$ **c)** 8 J
10 a) $5Mg$ **b)** $\sqrt{2gx}$ **c)** $x = \dfrac{5M}{m}$

Exercises 10.5B *(p.190)*

1 a) 196 J **b)** 14 m s^{-1}
2 a) 448.8 J **b)** 17.3 m s^{-1}
3 a) 2940 J **b)** 7.67 m s^{-1}
4 a) 24 500 J, 14 m s^{-1} **b)** 9.90 m s^{-1} **c)** The same.
5 a) 78.4 J **b)** 12.52 m s^{-1} **c)** 13.28 m s^{-1}
6 a) 40.8 J **b)** 40.8 J, 10.1 m s^{-1}
7 a) 156 800 J **b)** 250 m **c)** 282 240 J
8 a) 6.56 m s^{-1} **b)** 9.28 m s^{-1} **c)** Identical
9 a) 30.6 m, 20 m s^{-1} **b)** $a = 20$, $b = 24.5$
10 a) $v = \sqrt{2gx}$ **b)** $980m$ J

Exercises 10.6A *(p.194)*

1 a) 51000 W **b)** 250 W **c)** 541 W
2 R = 1250 N
3 $3\frac{1}{3}$ m s^{-1}
4 a) 36 N **b)** 432 W **c)** 12.75 m s^{-1}
5 a) 0.125 m s^{-2} **b)** 187 500 W **c)** Acceleration assumed constant. No account taken of working against resistances.
6 a) 22050 J **b)** 19.6 m s^{-2} **c)** $a = \frac{29.4}{v} - 9.8$
7 a) 25.9 **b)** 32.9 m s^{-1}
8 0.56 m s^{-1}
9 a) 25.65 N **b)** 25.65 N **c)** 76.95 J **d)** 7.695 N **e)** No, this would reduce the formal force by a significant proportion.
10 a) 20 **b)** 854 N **c)** 33 m s^{-1}

Exercises 10.6B *(p.195)*

1 a) 250 000 W **b)** 72 W **c)** 3136 W
2 a) 511.6 N **b)** 40.95 m s^{-1}
3 a) 50 W **b)** Double to 100 W **c)** Person can work at 100 W, and exert force of 40 N.
4 a) 39.0625 kg s^{-1} **b)** 33.56 increase of 4.8%
5 a) 113.4 N **b)** 453.6 W
6 a) 20.5 N **b)** 41 W
7 a) 23520 W **b)** 3.8 m s^{-1}
8 a) 37.5, 1.875 **b)** 40 m s^{-1}, 31.75 m s^{-1} **c)** Proportional to speed squared.
9 a) 1285.7 N **b)** Up, 33.2 m s^{-1}, Down, 57.2 m s^{-1} **c)** No as not travelling close to 42 m s^{-1}.
10 a) 3.7 **b)** Up, 7.2 m s^{-1}

Consolidation Exercises for Chapter 10 *(p.196)*

1 a) 7000 N **b)** 0.7 m s^{-2}
2 16 761 J The forces are friction acting on the driven tyres.
3 450 N, 0.5625 m s^{-2}
4 i) 9800 N, 9746 N **ii)** 3024 N **iii)** 50406 W **iv)** 95 km h^{-1}
5 a) 220 **b)** 17500 J
6 a) i) 60 kW **ii)** 63600 W
b) 62250 W
7 a) 10990 J **b)** 122 N **c)** 32.1 m
8 k = 25, P = 18.75, 27.4 m s^{-1}

CHAPTER 11
Energy and work: Variable forces and scalar products

Exercises 11.1A *(p.203)*

1 a) $163\frac{1}{3}$ Nm^{-1} **b)** $32\frac{2}{3}$ N **c)** 6 cm
2 a) 174.2 Nm^{-1} **b)** 6.75 cm
3 a) 3920 N **b)** 0.9 cm
4 a) 13.1 Nm^{-1} **b)** $0.132 \le T \le 0.652$
5 a) 2.45 N **b)** 3.2 cm **c)** 2.4 cm
6 a) 8 N **b)** 15.145 cm, 4.855 cm
7 a) $0.6 - x$ **b)** 0.51 cm **c)** 112 m s^{-2}
8 a) 21.63 N **b)** 158 Nm^{-1}

Exercises 11.1B *(p.204)*

1 a) 588 Nm^{-1} **b)** 117.6 N
2 a) 327 Nm^{-1} **b)** $\frac{2}{3}$ kg
3 a) 327 Nm^{-1} **b)** 2.4 cm
4 a) 19.6 Nm^{-1} **b)** $0.392 \le T \le 1.568$
5 a) 7.84 N **b)** 2.5 cm **c)** 2 cm
7 $\frac{l}{2}\left(1+\frac{mg}{\lambda}\right)$
8 a) 3 Nm^{-1} **b)** 107 m s^{-2}

Exercises 11.2A *(p.208)*

1 a) 400000 J **b)** 28.3 m s^{-1}
2 a) –2400 J **b)** 14.9 m s^{-1}
3 b) 882 J
4 a) 0.08 J **b)** 2.83 m s^{-1}
5 a) 0.144 J **b)** 18000 Nm^{-1}
6 a) 4500000 J **b)** 17.3 m s^{-1}
7 a) $m = 50 - 4x$ **b)** 1960 J
8 a) 17067 J **b)** 1067 m s^{-1}
9 a) 20 J **b)** 29.1 m s^{-1}
10 a) 4.21×10^9 J **b)** 4100 m s^{-1} **c)** 2.14×10^3 N

Exercises 11.2B *(p.210)*

1 a) 9.875 J **b)** 0.628 m s^{-1}
2 a) –4120 J **b)** 19.7 m s^{-1}
3 16 537 500 J
4 a) $2\sqrt{gl}$ **b)** $2\sqrt{gl}$
5 a) 2.808 J **b)** 224 640
6 a) 350 000 J **b)** 26.5 m s^{-1}
7 b) 1078 J
8 a) 625 000 J **b)** 4.44
9 $\sqrt{\frac{2c\ln(2)}{m} + V_0^2}$
10 a) 3.8×10^{13} J **b)** 6.18×10^5 m s^{-1}

Exercises 11.3A *(p.213)*

1 a) 0.0882 J **b)** 0.0882 J **c)** 1764 Nm^{-1}
2 a) 37.5 J **b)** 37.5 J **c)** 19 cm **d)** Brakes probably applied.
3 a) 0.075 J **b)** 1.73 m s^{-1} **c)** 0.495 m **d)** Some energy retained in the plunger.
4 a) 25 480 J **b)** 25 480 J **c)** 127.4 Nm^{-1} **d)** 24.6 m **e)** 21.02 m
5 a) 0.588 J **b)** 4.85 m s^{-1}, 1.2 m
6 a) 1000 N **b)** 0.1 m
7 a) 3.038 J **b)** 15 190 Nm^{-1}
8 More energy in the parallel springs.
9 a) 1176 J **b)** $1176 - 588x$ **c)** 6.33 m s^{-1}
10 $\frac{61l}{40}$ above the floor.

Exercises 11.3B *(p.215)*

1 50 J
2 10 m s^{-1}
3 a) 0.0625 J **b)** 5 m s^{-1} **c)** 2.55 m
4 2.43 m
5 $x\sqrt{\frac{k}{m}}$ **a)** $x\sqrt{\frac{k}{2m}}$ **b)** $x\sqrt{\frac{k}{2m}}$ **c)** $x\sqrt{\frac{k}{m}}$
6 0.894 m s^{-1}
7 a) 2.12 m s^{-1} **b)** 1.5 m
8 0.25 m
9 a) 2 mgl **b)** 0 **c)** $\sqrt{gl}$
10 0.14 m

Exercises 11.4A *(p.218)*

1 a) –13 J **b)** 488 J **c)** –126 J **d)** 70 J
2 a) 5.19 m s^{-1} **b)** 4.58 m s^{-1} **c)** 2.87 m s^{-1}
3 a) 784 J **b)** 12.87 m s^{-1}
4 a) $8P - 8g$ **b)** g **c)** 10.425 N
5 a) –9.8b **b)** 15.3 **c)** Only height gained important.

6 **a)** 41 N
b) Because perpendicular to the surface. **c)** 226 J
d) 9.51 m s^{-1}
7 $15\mathbf{i} + 3\mathbf{j}$
8 $5\mathbf{i} + 10\mathbf{j}$
9 **a)** 5.25 J **b)** 11.2 N **c)** 76.4°
10 $6\mathbf{i} + 3\mathbf{j}$

Exercises 11.4B *(p.219)*

1 **a)** 5 J **b)** 5 J **c)** 15 J **d)** 0
2 **a)** 4.36 m s^{-1} **b)** 4.69 m s^{-1}
c) 4 m s^{-1}
3 **a)** 98 J **b)** 6.57 m s^{-1}
4 **a)** $\sqrt{2P} - 19.6$ **b)** 22.3 N
5 **a)** $-784b$ **b)** 5.1
6 **a)** 22.4 N **b)** Perpendicular to direction of motion. **c)** 4.08 m s^{-1}
7 $\frac{147}{29}\mathbf{i} + \frac{13}{29}\mathbf{j} + \frac{1}{29}\mathbf{k}$
8 **a)** 30 J **b)** 50 N **c)** 78.5°
9 $9\mathbf{i} + 4\frac{1}{2}\mathbf{k}$
10 53.1°

Consolidation Exercises for Chapter 11 *(p.221)*

1 **a)** 90 J **b)** 270 J
2 **a)** 18 J **b)** 7.81 m s^{-1}
c) Reduced due to friction.
3 No air resistance **i)** 5.53 m s^{-1}
ii) 1.861 m
4 **a)** 11 760 N **b)** 27.4 m
c) 20 m s^{-1}
5 $\lambda, \lambda a, \sqrt{\frac{2\lambda a}{m}}$

CHAPTER 12
Circular motion at constant speed

Exercises 12.1A *(p.232)*

1 **a)** 7.27×10^{-5} rad s^{-1}
b) 465 m s^{-1}
2 **a)** 0.157 s **b)** 40 rad s^{-1}
3

Engine flywheel	52.36	0.12	–
Record	4.71	1.33	–
Fairground ride	–	6	10
Planet	8.73×10^{-5}	–	8.3×10^{-6}

4 50 rad s^{-1}, 40 rad s^{-1}
5 **a)** 12 m s^{-1} **b)** 3 rad s^{-1}
c) 2.09 s **d)** 28.65 revolutions
6 8 rad s^{-1}
7 **a)** $\mathbf{r} = 5\sin\left(\frac{\pi t}{5}\right)\mathbf{i} - 5\cos\left(\frac{\pi t}{5}\right)\mathbf{j}$

$\mathbf{v} = \pi\cos\left(\frac{\pi t}{5}\right)\mathbf{i} + \pi\sin\left(\frac{\pi t}{5}\right)\mathbf{j}$

b) $\mathbf{r} = 5\sin\left(\frac{\pi t}{5}\right)\mathbf{i}$

$+ (5 - 5\cos\left(\frac{\pi t}{5}\right)\mathbf{j}$

$\mathbf{v} = \pi\cos\left(\frac{\pi t}{5}\right)\mathbf{i} + \pi\sin\left(\frac{\pi t}{5}\right)\mathbf{j}$

8 997 m s^{-1}
9 3.083×10^3 m s^{-1}
10 25.3 m s^{-1}

Exercises 12.1B *(p.233)*

1 **a)** 1.26 rad s^{-1} **b)** 6.28 m s^{-1}
2 **a)** 2.51 m **b)** 0.502 s
c) 12.52 rad s^{-1}
3 **a)** 524 rad s^{-1}, 0.012 s
b) 3.46 rad s^{-1}, 1.82 s
c) 20.9 rad s^{-1}, 0.3 s
d) 1.05 rad s^{-1}, 6 s
4 **a)** $\mathbf{v} = -6\sin(2t)\,\mathbf{i} + 6\cos(2t)\mathbf{j}$
b) $\mathbf{v} = 6$ m s^{-1} **c)** π s
5 **a)** 133.3 rad s^{-1} **b)** 1273 rpm
6 **a)** 80 rad s^{-1} **b)** 12 m s^{-1}
7 **a)** $\mathbf{r}_L = 0.14\sin\left(\frac{\pi t}{1800}\right)\mathbf{i}$

$+ 0.14\cos\left(\frac{\pi t}{1800}\right)\mathbf{j}$

$\mathbf{r}_S = 0.08\sin\left(\frac{\pi t}{21600}\right)\mathbf{i}$

$+ 0.08\sin\left(\frac{\pi t}{21600}\right)\mathbf{j}$

b) $\mathbf{v}_L = \frac{0.14\pi}{1800}\cos\left(\frac{\pi t}{1800}\right)\mathbf{i}$

$- \frac{0.14\pi}{1800}\sin\left(\frac{\pi t}{1800}\right)\mathbf{j}$

$\mathbf{v}_S = \frac{0.08\pi}{21600}\cos\left(\frac{\pi t}{21600}\right)\mathbf{i}$

$- \frac{0.08\pi}{21600}\cos\left(\frac{\pi t}{21600}\right)\mathbf{j}$

8 **a)** $\mathbf{v} = -5\pi\sin\left(\frac{\pi t}{10}\right)\mathbf{i} +$

$5\pi\cos\left(\frac{\pi t}{10}\right)\mathbf{j}$ **b)** 117 202 J

9 **a)** $\mathbf{v} = -4\sin(20t)\,\mathbf{i} + 4\cos(20t)\mathbf{j}$
b) $1.22\,\mathbf{i} - 3.81\mathbf{j}$ **c)** $2.25\,\mathbf{i} - 7.68\mathbf{j}$, moves at constant speed when the string breaks.

10 **a)** 7.85 m s^{-1} **b)** 0.55 s
c) 8.4 m s^{-1} at 20.9° to tangent.

Exercises 12.2A *(p.238)*

1 **a)** 1.8 m s^{-2} **b)** 2 m s^{-2}
c) 1.794 m s^{-2}
2 As close to outside as possible.
3 **b)** 10 584 N **c)** 8.82 m s^{-2}
d) 21 m s^{-1} **e)** No
4 **b)** 8.58 m s^{-2} **c)** 0.0011 rad s^{-1}
d) 5603 s **b)** 202 914 N
c) 52 518 N, 2.63 m s^{-2}
d) 36.2 m s^{-1} **e)** The same speed.
6 **a)** 7.27×10^{-5} rad s^{-1}
b) $5.29 \times 10^{-9}mr$
c) $\frac{GM_EM}{r^2}$ **d)** 4.23×10^7 m
7 **a)** Yes **b)** 10.8 m s^{-1}
8 **a)** 18.4 m s^{-1} **b)** 23.7 m s^{-1}
9 **b)** 3.37×10^{-2} m s^{-2}
c) $3.37 \times 10^{-2}m$ N **d)** Yes
10 **c)** 6.26 m s^{-1}

Exercises 12.2B *(p.240)*

1 **a)** 140 m s^{-2} **b)** 80 m s^{-2}
c) 1.25 m s^{-2}
2 **a)** 1667 N **b)** 432 N **c)** 9995 N
3 **a)** 3290 m s^{-2} **b)** 3290 N
4 **a)** $\mathbf{a} = -24\pi^2\cos(2\pi t\,\mathbf{i} - 24\pi^2\sin(2\pi t)\mathbf{j}$ **b)** $24\pi^2$
5 **a)** 411.6 N **b)** 2.05 rad s^{-1}
c) No **d)** Possible if children do not hold on.
6 **a)** 11 805 N **b)** 8.3 m s^{-1}
c) Could be greater as 8.3 m s^{-1} is about 19 mph.
7 38.34 m s^{-1}
8 5.6 s
9 **a)** 0.139 N, 1.57 m s^{-1}
b) Causes the ball to roll.
10 4.6 rad s^{-1}

Exercises 12.3A *(p.245)*

1 **b)** $T\cos\theta = mg$, $T\sin\theta = \frac{mv^2}{r}$
c) 76.2°
2 **a)** 520 N **b)** 1.32 rad s^{-1}
c) 4.8 s **d)** Inextensible and light.
e) Tension increased, time not changed.
3 **a)** 1.67 rad s^{-1} **b)** 18.79°
c) 1.04 N
4 1.68 m s^{-1}
5 **b)** 9.8 N **c)** $r = 0.54$ m, $l = 0.9$ m
6 **a)** 71 m s^{-2} **b)** 2.14 N, 13.34 N

7 **c)** 59.8° **d)** 4.2 rpm
e) No tension in lower rod.

8 **b)** $T = m\sqrt{g^2 + \dfrac{r^4}{r^2}}$

Exercises 12.3B *(p.246)*

1 **a)** 0.41 m **b)** 31.3 N
c) 2.95 rad s^{-1}
2 **a)** 1967 N **b)** 0.67 N
3 3.45 m s^{-1}
4 **a)** 0.18 m s^{-1}, 4.87 N
b) 0.925 m s^{-1}, 10.69 N
5 **b)** 0.566 N **c)** 5.66 m s^{-2}
d) 0.578 **e)** Yes
6 **a)** $T = \dfrac{mv^2}{l\sin^2\alpha}$ **b)** $\tan\alpha = \dfrac{v^2}{r}$
7 **a)** $\dfrac{Mg}{\sqrt{2}}$ **b)** $\omega = \sqrt{\dfrac{Mg}{2\sqrt{2}mr}}$
c) $M = \sqrt{\dfrac{3}{2}}m$ **d)** 4.43 rad s^{-1}
8 12 N, 132 N

Consolidation Exercises for Chapter 12 *(p.248)*

1 **a)** 0.066 N **b)** 1.33 m s^{-1}
2 3.83 m s^{-1}
3 **a)** $\mathbf{v} = -2\sin(2t)\,\mathbf{i} + 2\cos(2t)\,\mathbf{j} - 6\mathbf{k}$
$m\mathbf{v} = -6\sin(2t)\mathbf{i} + 6\cos(2t)\,\mathbf{j}$
$-18\mathbf{k}$ **b)** $\mathbf{F} = -12\cos(2t)\,\mathbf{i}$
$-12\sin(2t)\,\mathbf{j}$
4 **b)** 368 N **d)** It does not stretch and it is light. **e)** 4.03 s
5 $-0.4\pi(\cos(\pi t)\mathbf{i} + \sin(\pi t)\mathbf{j})$
6 **a)** 6.4×10^{12} m^3 s^{-2}
b) 1.46×10^3 m s^{-1}

CHAPTER 13
Circular motion with variable speed

Exercises 13.1A *(p.256)*

3 8.78 m s^{-2}
4 **a)** $1\mathbf{e}_t - 4v^2\mathbf{e}_r$ **b)** 0.5 m s^{-2}
c) 3.54 s, 1.77 m s^{-1}, 6.30 m s^{-2}
d) $4v^2$
5 **b)** Normal reaction becomes zero. **c)** 48.1°
d) Straighten out slide at $\theta = 48°$.
6 7.78 m s^{-2}
7 $\dfrac{(100-3t)^2}{500} - 3$

8 $v = \sqrt{392(1-\cos\theta)}$
$a = \sqrt{g^2\cos^2\theta + \dfrac{392^2(1\cos\theta)^2}{20^2}}$
9 1960 N
10 **a)** 0.82 m above the horizontal.
b) 441 N **c)** 260 N
d) Conservation of energy, no, due to losses of energy.

Exercises 13.1B *(p.258)*

3 14.94 m s^{-2}
4 **a)** $10\mathbf{e}_t - 6v^2\mathbf{e}_r$ **b)** $\frac{1}{3}$ m s^{-1}
c) 13.73 s, 4.57 m s^{-1}, 4.2 m s^{-2}
d) $6v^2$
5 **a)** 3.96 m s^{-1} **b)** 21°
6 17.15 m s^{-1}, 1.96 m s^{-2}
7 **a)** 6.26 m s^{-1}, 11.71 m s^{-1}
b) $g\cos\theta$ **c)** $g(0.8 + 2\sin\theta)$
8 **b)** Legs probably bend.
10 0.866 m

Consolidation Exercises for Chapter 13 *(p.260)*

1 **b)** $\dfrac{3a}{16}$
2 **b)** $\dfrac{7rg}{12}$ **c)** $\dfrac{7rg}{4}$
3 $T = 0.233 + 0.590\cos\theta$, 113°
Parabolic trajectory while string slack.

CHAPTER 14
Simple harmonic motion

Exercises 14.1A *(p.269)*

1 **a)** $-20x$ **b)** $-x$ **c)** $-375x$
d) $-50x$
2 **a)** $-60x$ N **b)** $-600x$ m s^{-2}
c) $-90x$ N, $-900x$ m s^{-2}
3 **a)** 2 **b)** 10
4 **a)** 0.314 s **b)** 0.993 s
c) 0.444 s **d)** 0.126 s
5 **a)** $-160x$ **b)** 0.497 s
c) $0.03\cos(\sqrt{160}t)$
d) $-0.03\cos(\sqrt{160}t)$
6 Parallel system has half the period.
7 **a)** 0.444 s **b)** $-kx$
c) 0.444 s **d)** None
8 **a)** $\sqrt{200}$ **b)** $\dfrac{\pi}{2}$ **c)** $-\dfrac{2}{\sqrt{200}}$

9 **a)** $\dfrac{\pi}{2}$ s **b)** 0.9925 m
10 **a)** $-\dfrac{\pi x}{ml}$ **b)** $2\pi\sqrt{\dfrac{ml}{\lambda}}$ or $2\pi\sqrt{\dfrac{m}{k}}$

Exercises 14.1B *(p.271)*

1 **a)** $-30x$ **b)** $-2x$ **c)** $-40x$ **d)** $-10x$
2 **a)** $-80x$ **b)** $-400x$
c) $-120x$, $-600x$
3 **a)** 5 **b)** 9
4 **a)** 0.314 s **b)** 0.444 s
c) 0.795 s **d)** 0.889 s
5 **a)** $-300x$ **b)** 0.363 s
c) $x = 0.05\cos(\sqrt{300}t)$
d) $x = -0.05\cos(\sqrt{300}t)$
6 Period of series system 3 × period of parallel system.
7 118 N m^{-1}
8 **a)** Different amplitudes, same period. **b)** 0.314 s
9 **c)** 1.42 s **d)** 1.07 m
10 **a)** $mg + \dfrac{\lambda x}{l}$ **b)** $-\dfrac{\lambda x}{l}$ **c)** $2\pi\sqrt{\dfrac{ml}{\lambda}}$

Exercises 14.2A *(p.275)*

1 **a)** 0.0625 J **b)** 1.12 m s^{-1}
c) 0.25 J, 2.24 m s^{-1}
2 0.157 m s^{-1}
3 0.314 m s^{-1}
4 5.97 cm
5 **a)** 0.314 s **b)** 0.8 m s^{-1}
6 **a)** $\dfrac{500\pi}{3}$ **b)** 31.4 m s^{-1}
c) 17.4 m s^{-1}
7 **a)** 0.9 cm **b)** 0.047 m s^{-1}
c) 0.052 m s^{-1}
8 **a)** $a\omega\sin\omega t$
9 2.05 cm, 111.8
10 **a)** 0.35 cm **b)** 0.11 s

Exercises 14.2B *(p.276)*

1 **a)** 2.32 m s^{-1} **b)** 3.10 m s^{-1}
2 1.57 m s^{-1}
3 0.314 m s^{-1}
4 1.51 m s^{-1}
5 **a)** 1.73 cm, 0.36 s **b)** 0.3 m s^{-1}
6 **a)** 15π **b)** 3.53 m s^{-1}
c) 2.63 m s^{-1}
7 **a)** 3.2 m **b)** 3.87 m s^{-1}
c) 3.23 m s^{-1}
8 $a\omega\cos\omega t$
9 6.15 cm, 0.084 s
10 **a)** 8.26 cm **b)** 0.259 s

Exercises 14.3A *(p.280)*

1 0.568 s
2 0.062 m
3 **a)** 0.635 s **b)** 1.57 s
4 **a)** 0.993 s **b)** 0.245 m
c) $g > 9.8$ m s^{-2}
5 **a)** 1.42 s **b)** 0.644
c) 0.644, 2.21
6 **a)** 0.201, 3.13 **b)** 0.524, 2.21
c) 0.175, 3.5 **d)** 0.541, 3.74
Most appropriate (c)
Least appropriate (d)
7 **a)** No **b)** 106 cm
c) Shorten by 7 cm
d) Rod used rather than a string.
8 **a)** 1.81 **b)** 0.025 m **c)** 0.129
d) $\frac{7}{30}, -\frac{\pi}{2}$
d) i) Faster **ii)** Greater
iii) The same
9 **a) i)** 0.013 m
ii) 3.13, 0.161, $-\frac{\pi}{2}$
b) i) 0.0046 m
ii) 2.21, 0.0678, $\frac{\pi}{2}$
c) i) 0.0005 m
ii) 3.74, 0.038, $-\frac{\pi}{2}$
d) i) 0.002 m
ii) 1.4, 0.0283, $-\frac{\pi}{2}$
10 **a)** 875 J **b)** 1.33 s

Exercises 14.3B *(p.282)*

1 0.695 s
2 0.140 m
3 **a)** 0.852 s **b)** 2.11 s
4 **a)** 0.99 m **b) i)** 0.602 s
ii) 10.92 m
5 **a)** 1.61 s **b)** 0.505
c) $A = 0.505$, $\omega = 3.91$
6 **a)** 0.384 **b)** 4.01 s **c)** 1.57
d) 2.005 s **e)** θ Larger
7 **a)** 0.834 m **b)** Increase by 16 cm.
8 **a)** 1.57 **b)** 0.885 m s^{-1}
c) 0.142 **d)** $-\frac{\pi}{2}$
9 3.17 s
10 **b)** 1.4 m s^{-1} **c)** No resistance.

Consolidation Exercises for Chapter 14 *(p.284)*

1 $2\pi\sqrt{5}$
2 $g_A = 1.001g_B$, difference in altitude.
3 **a)** $\frac{L}{2}$ **c)** $h \le \frac{L}{2}$ **d)** $\frac{\pi L}{g\sqrt{2}}$
e) $\frac{1}{3}\sqrt{2gL}$
4 **a)** 0.5 m **b)** $\frac{\pi}{15}$ s **c)** 15 m s^{-1}
d) 0.31 s 360 N, 14.4 J
5 $\frac{d^2x}{dt^2} = \frac{-gx}{a}$, $x = \frac{a}{2}\cos\left(\sqrt{\frac{g}{a}}t\right)$
friction between rod and fluid.

CHAPTER 15
Dimensions and dimensional analysis

Exercises 15.1A *(p.292)*

2 MLT^{-2}
3 M^{-1}L^{3}T^{-2}
4 T^{-1}
5 **a)** Yes **b)** Yes **c)** No
6 M L^{-1}T^{-2}
7 $\frac{1}{2}$, $a + 2c = \frac{1}{2}$
8 **a)** ML^{2}T^{-3}
b) $a = 1$, $b = 2$, $c = -3$
9 Yes
10 **a)** T^{-1}
b) Factor that includes mass.
c) $f = \sqrt{\frac{T}{mL}}$ where m is the mass of the string.

Exercises 15.1B *(p.293)*

2 **a)** ML^{2}T^{-2} **b)** MLT^{-2}
3 **a)** MT^{-1} **b)** ML$^{\frac{1}{2}}$T$^{-\frac{3}{2}}$
4 T^{-1}
5 **a)** Dimensionless
b) Not dependent on length and hence area in any way.
6 **a)** Yes **b)** No **c)** Yes
7 ML^{2}T^{-3}
8 **a)** T^{-1}
b) $X = \frac{1}{2}ma^2\frac{d\omega}{dt}$, $X = \frac{ma^2\omega}{t}$
9 **b)** $a = 1$, $b = -1$
10 $L = \frac{k}{f}\sqrt{\frac{p}{d}}$

Consolidation Exercises for Chapter 15 *(p.294)*

1 $c = 2$, $\alpha = 0.0125$
2 $a = \frac{3}{2}$, $b = \frac{1}{2}$, $c = -\frac{1}{2}$
3 $a = -\frac{1}{2}$, $b = \frac{1}{2}$, $c = \frac{1}{2}$
4 **a)** M^{-1}L^{3}T^{-2}
b) $a = \frac{3}{2}$, $b = \frac{1}{2}$, $c = \frac{1}{2}$
5 **ii)** $x + y = 3$ **iii)** $R = \frac{Cr^4}{a}$
iv) 0.2126 m^3 s^{-1}

CHAPTER 16
Monments, couples and equilibrium

Exercises 16.1A *(p.301)*

1 **a)** 2,59 Nm **b)** –300 Nm
c) –64 Nm **d)** –150 Nm
2 **a)** –35 Nm, 40 Nm
b) 5 Nm, rotates anticlockwise
3 **a)** –60 Nm **b)** –74 Nm
c) 20 Nm **d)** 380 Nm
4 **a)** 38.3 N **b)** 46 Nm **c)** No
5 **a)** –96.4 Nm **b)** 90.0 Nm
c) 13.0 Nm **d)** 16.0 Nm
6 –233 N
7 **a)** 3 m **b)** 1.75 m
c) 0.58 m **d)** 2 m
8 **a)** 60° **b)** 70.5°

Exercises 16.1B *(p.302)*

1 **a)** 90 Nm **b)** 5.76 Nm
c) –63 Nm **d)** 0
2 **a)** 1.2 m
b) –2 Nm, Bob goes down.
3 **a)** 5 Nm **b)** 4 Nm
c) – 0.12 Nm **d)** 49 Nm
4 **a)** 1.2 Nm
b) Only if the force continues to act along a tangent.
5 0.8 m
6 **a)** 36.7 Nm **b)** 2.21 Nm
c) 41.1 Nm **d)** –120 Nm
7 **a)** 2.83 m **b)** 1.79 m
8 **a)** 33.3° **b)** 152.5°

Exercises 16.2A *(p.306)*

1 **c)** $130\frac{2}{3}$ N
2 **b)** is true.
3 **a)** A, 700 N; B, 1000 N
b) A, 1190; B, 1490 N
4 A, 68.6 N; B, 78.4 N
5 **a)** Nothing
b) 23 Nm, –23 Nm, 0 Nm
6 **a)** 245 N **b)** Same as above.
7 **a)** 539 N, 762 N **b)** 539**i** + 0**j**
8 **b)** 318.5 N **c)** 1103 N

d) 551.7 N
e) 0.577, Yes reasonable.
9 a) 90.9 N
b) R = 112 N, F = 34.1 N
c) 0.31
10 a) R = 686 N, $F = \dfrac{5g(6x+1)}{\tan 50°}$
b) 1.08 m

Exercises 16.2B *(p.308)*

1 M = 882 N, N = 98 N
2 a) 3 **b)** Greater
3 a) 180 N and 320 N
b) 425 N and 565 N
4 52.5 kg
5 b) $8\frac{1}{3}$ kg
6 a) 588 N **b)** 1568 N
c) 509 N **d)** 1444 N at 79.8° to the horizontal.
7 b) 349 N **c)** 349 N, 376 N
8 a) 48 N **b)** 31 N, 110 N
c) 114 N, 74.3° above horizontal.
9 a) 196 N **b)** 339.5 N **c)** 0.58
10 0.5

Exercises 16.3A *(p.311)*

1 a) No **b)** Yes, 40 Nm **c)** No
d) Yes, 90 Nm **e)** Yes, 80 Nm
f) No
2 a) 40 Nm
b) 40 Nm about any point.
3 a) On AD and BC
b) On AB and CD **c)** In opposite directions along any line passing through the square.
4 a) 10 N downwards **b)** Same place as other 10 N force.
5 a) F = 6, G = 1 N **b)** 45 Nm
6 a) 125 N **b)** 125 N, 11.25 Nm
7 3350 N, 855 Nm
8 a) 600 N, 200 Nm
b) 1.35 m from A

Exercises 16.3B *(p.312)*

1 a) No **b)** Yes, 10 Nm
c) No **d)** No
2 a) 20 Nm **b)** Always 20 Nm.
3 a) On AB and DC.
b) On AD and BC.
4 a) 4.24 N, 45° to DC, 2.75 Nm
5 a) 1000 N, 0.2 m from A.
b) 1000 N, 50 Nm
6 9.8×10^{-4} Nm
7 a) A: Reaction 0, Moment Fd
B: Reaction $2F$, Moment $2Fl$
b) A: Because the reaction force is less.
8 b) 148 N

Consolidation Exercises for Chapter 16 *(p.313)*

1 c) $(R - T\cos 60°)\mathbf{i} + (F + T\sin 60°\ mg)\mathbf{j}$ **e)** Decrease
2 a) 94 N **b)** 0.47 **c)** 134 N
3 a) ii) No horizontal force to balance reaction from wall.
b) ii) $32g$ N
4 a) The ring will not slide as no component of tension parallel to rod.
c) $\dfrac{9W}{5}$
5 a) 3500 N **b)** A: 3500 N, B: 3087 N; T = 1746 N, B: 1293 N

CHAPTER 17
Centre of mass

Exercises 17.1A *(p.321)*

1 a) $\dfrac{28}{17}\mathbf{i} + \dfrac{11}{17}\mathbf{j}$ **b)** $\dfrac{1}{47}\mathbf{i} + \dfrac{7}{30}\mathbf{j}$
c) $-0.055\mathbf{i} + 0.049\mathbf{j}$
d) $-0.762\mathbf{i} - 1.015\mathbf{j}$
2 a) $0.0643\mathbf{i} - 0.0214\mathbf{j}$
b) Circle of radius 6.78 cm.
3 a) 4.39 cm
b) 3.06 cm from the centre.
4 a) 0.4 m from 500 g mass.
b) 120 grams
5 a) $\dfrac{(0.4m+1.6)\mathbf{i} + (0.4m+1.6)\mathbf{j}}{13+m}$
b) 900 grams
6 a) 0.312 m
b) Move closer to the centre.
c) Place point mass at centre.
d) 0.118 m from the centre.
7 a) 0.91 m and 7.09 m
b) 1.41 m from bottom
8 a) $m = \frac{2}{3}$kg, $M = \frac{5}{3}$kg
b) No, as this would require a zero mass at O.

Exercises 17.1B *(p.323)*

1 a) $2.5\mathbf{i} + \mathbf{j}$ **b)** $-0.357\mathbf{i} - \mathbf{j}$
c) $0.25\mathbf{i} + 0.286\mathbf{j}$
d) $0.3125\mathbf{i} + 0.3125\mathbf{j}$
2 a) $0.2\mathbf{i} + 0.267\mathbf{j}$ **b)** 53.2°
3 a) 1.66 cm **b)** 2.08 cm
4 a) $0.12\mathbf{i} + 0.08\mathbf{j}$ **b)** $1\frac{1}{3}$ kg
5 a) $\dfrac{(3.5m)\mathbf{i} + (16.48 + 1.73m)\mathbf{j}}{17+m}$
b) 5.5 kg **c)** 1.16 m
6 a) 3.2 cm
b) Moves towards centre.
c) Yes as the centre of mass is so close to the centre already.
7 10 kg
8 $0.0267\mathbf{i} + 0.02\mathbf{j}$, with $\mathbf{i}$ along AC and $\mathbf{j}$ along AB.

Exercises 17.2A *(p.326)*

1 $1\frac{2}{3}$ cm
2 37.7 cm
3 a) $0\mathbf{i} + 0\mathbf{j}$, $0.0146\mathbf{i} + 0.0354\mathbf{j}$, $0.05\mathbf{i} + 0.05\mathbf{j}$, $0.0854\mathbf{i} + 0.0354\mathbf{j}$, $0.1\mathbf{i}$
4 15.4 cm from front.
5 a) $0.796\mathbf{i} + 0.499\mathbf{j}$
b) $0.719\mathbf{i} + 0.793\mathbf{j}$
6 a) 21.4 cm **b)** 26.7 cm, 6.4 cm
7 a) $0.844\mathbf{i} + 0.247\mathbf{j}$, $0.877\mathbf{i} + 0.110\mathbf{j}$ **b)** 0.147 m **c)** 104 J
d) Probably not perpendicular in both positions.
8 a) 0.193 m **b)** $0.193\mathbf{i} + 2.895\mathbf{j}$
9 a) 3.42 kg **b)** 8.74 cm
c) 4.89 cm **d)** 30.3°
10 $0.29H$

Exercises 17.2B *(p.327)*

1 5.2 cm
2 9.3 cm
3 a) i) $0.225\mathbf{i} + 1.065\mathbf{j}$
ii) $0.915\mathbf{i} + 0.706\mathbf{j}$
iii) $1.25\mathbf{i} + 0.126\mathbf{j}$
4 a) $0.1\mathbf{i} + 0.075\mathbf{j}$ **b)** 36.9°
5 a) 0.25 m **b)** 275 N
6 a) 60.8 cm
b) i) 65.2 cm **ii)** 67.2 cm
7 a) 1.12 cm, 1.62 m
b) 0.5 m **c)** 980 J
8 9.45°
9 $0.158\mathbf{i} + 0.07\mathbf{j}$ with $\mathbf{i}$ along DC and $\mathbf{j}$ along DA.
10 $\dfrac{1}{(t+16)}\left(\dfrac{5t^2}{64\pi} + 9600\right)$

Exercises 17.3A *(p.330)*

1 $\left(\frac{14}{15}, 0\right)$
2 (0.75, 0)
3 8.89 cm from longer side.

4 (2.4, 0)
5 b) 12.7 cm from base.
6 a) $\frac{3h}{8}$
7 a) $\left(\frac{5}{3}, 0\right)$ **b)** $\left(\frac{105}{62}, 0\right)$ **c)** $\left(\frac{5a}{6}, 0\right)$
8 (1, 0)
9 $\frac{7a}{40}$ from base
10 7.31

Exercises 17.3B *(p.331)*

1 $\left(\frac{7}{9}, 0\right)$
2 $\left(\frac{4}{5}, 0\right)$
3 3.55 from base
4 a) $\frac{10}{7}$ **b)** $\left(\frac{10}{7}, 0\right)$
5 4.89
6 $\frac{3h}{13}$
7 b) $\frac{a\left(b-a-b\ln\left(\frac{a}{b}\right)\right)}{b-a}$
8 0.83
9 b) 4.95
10 3.17

Consolidation Exercises for Chapter 17 *(p.332)*

1 $0.1\mathbf{i} + 0.1\mathbf{j}$
2 a) $1.625\mathbf{i} + 1.125\mathbf{j}$ with $\mathbf{i}$ along PQ, $\mathbf{j}$ along PS. **b)** 40.9°
3 a) $\frac{9}{4}\mathbf{i} + 1\mathbf{j}$ **b)** $r = 0.56$ therefore impossible **c)** 0.399 m, $\frac{3}{2}$
5 b) 8 cm **c)** 4.7 cm

CHAPTER 18
Conditions for sliding and toppling

Exercises 18.1A *(p.340)*

1 a) 294 N **b)** 285.8 N **c)** Topples
2 b) 490 N **c)** Topples first because 490 < 693. **d)** $\mu < 0.5$
3 a) $\mu < 1$ **b)** None **c)** None
4 a) Slides $\alpha = 21.8°$
Topples $\alpha = 45°$ Slides
b) Slides $\alpha = 11.3°$
Topples $\alpha = 78.7°$ Slides
c) Slides $\alpha = 45°$
Topples $\alpha = 43.4°$ Slides
d) Slides $\alpha = 42°$
Topples $\alpha = 14°$ Topples

5 b) $\frac{mgd}{6h}$ **d)** $m < \frac{d}{6h}$
e) Results suggest full can should be used. However, ability of cans to recover once hit could be important.
6 a) $15\mathbf{i} + 10\mathbf{j}$
b) Slides when $P = 7.84$ N.
c) Slides when $P = 7.84$ N.
7 Slides when $\alpha = 38.6°$.
8 a) $\alpha = 30°$ **b)** $\mu > \tan 30°$
c) $\alpha = 22.5°$, $\mu > \tan 22.5°$
9 $F = \frac{mg}{\sqrt{2}}$, $\mu > 1$
10 a) $\frac{\mu mg}{\cos\alpha + \mu\sin\alpha}$
b) $\frac{mg}{2\sin(\alpha+60°)}$ **c)** $\frac{1}{\sqrt{3}}$

Exercises 18.1B *(p.341)*

1 a) 274.4 N **b)** 130.7 N
c) Topples
2 b) 1176 N **c)** Topple
3 a) 122.5 N **b)** 196μ N
c) $\mu < 0.625$
4 a) Slides $\alpha = 21.8°$
Topples $\alpha = 26.6°$ Slides
b) Slides $\alpha = 16.7°$
Topples $\alpha = 56.3°$ Slides
c) Slides $\alpha = 50.2°$
Topples $\alpha = 45°$ Topples
d) Slides $\alpha = 38.7°$
Topples $\alpha = 69.4°$ Slides
5 a) 25.68 N **b)** 13.07 N
c) Topples **d)** No
6 a) 470 N **b)** 426μ **c)** $\mu < 1.1$
7 a) tan 40° = 0.84 > 0.2 so boxes can slide. **b)** $b = a \tan 50°$
8 a) F **b)** $B\ F$
9 $P = \frac{2mg}{5}$, $\mu \geq \frac{8}{31}$
10 $\mu < \frac{b}{2a - b\tan\alpha}$

Consolidation Exercises for Chapter 18 *(p.343)*

1 7.3 cm, $\mu \geq 0.36$
2 a) 0.5 m from base **b)** 31°
c) 19.2 kg
4 0.37

CHAPTER 19
Frameworks

Exercises 19.1A *(p.350)*

1 b) 2.5 m **c)** 3750 N
d) 1250 N **e)** $T_{AC} = -2500$ N, $T_{AB} = 2165$ N **f)** $T_{BC} = -4330$ N
2 a) 2347 N **b)** 347 N
c) $T_{AC} = 540$ N, $T_{AB} = -414$ N
d) $T_{BC} = -2383$ N
e) AB: compression
BC: compression
AC: tension
3 a) To give a zero vertical resultant
b) $Q = 400$ N
c) $T_{AB} = 231$ N, compression,
$T_{AC} = 305$ N, compression,
$T_{BC} = 231$ N, tension
4 a) Horizontal bar in tension, others exert thrusts.
b) 1800 N in angled bars, 3118 N in horizontal bar.
6 a) A 1825 N downwards.
B 6825 N upwards
b) $T_{AB} = 1800$ N, thrust
$T_{BC} = 7100$ N, thrust
$T_{BD} = 0$ N
$T_{AD} = 2600$ N, tension
$T_{CD} = 2600$ N, tension
7 a) $R = S = 100$ N
b) $T_{AB} = T_{CD}$ etc.
c) $T_{AB} = T_{CD} = -115$ N
$T_{AE} = T_{DE} = 58$ N
$T_{BE} = T_{CE} = 0$
$T_{BC} = -58$ N
8 a) 1000 N
b) $Q = 500$ N $R = 1000$ N
c) $T_{AD} = -1000$ N, thrust
$T_{AE} = 1414$ N, tension
$T_{BC} = -707$ N, thrust
$T_{BE} = -500$ N, thrust
$T_{CD} = -500$ N, thrust
$T_{CE} = 500$ N, tension
$T_{DE} = 707$ N, tension
d) AE
9 AB 462 N, tension
AF 462 N, tension
BF 462 N, thrust
BC 0 N
BD 462 N, tension
DF 462 N, tension
DC 50 N, tension
DE 50 N, tension
EF 0 N

10 $Q =$ 1500 N, vertical force at E of 2500 N
AB 5196 N, tension
AE 3000 N, thrust
BC 1732 N, tension
BD 2000 N, thrust
BE 2000 N, thrust
CD 2000 N, thrust
DE 3464 N, thrust
AF 3000 N, tension
EF 2598 N, thrust

Exercises 19.2B *(p.352)*

1 a) 800 N **b)** 400 N
c) T_{CA} = 1170 N, tension
T_{BC} = 1099 N, compression
d) T_{AD} = 1170 N, tension
2 a) Light rods and pin-joints
b) R = 283 N P = 217 N
c) AB, 596 N, tension AC, 634 N, thrust BC, 778 N, tension
3 a) 2500 N
b) AD, 1444 N, compression
AC, 2887 N, tension
BC, 2887 N, tension
BD, 1444 N, compression
CD, 0 N
c) The member CD is redundant.
4 b) 8052 N upward at B, 2052 N down at A.
c) AB, 2931 N, compression
BC, 8569 N, compression
AC, 3578 N, tension
5 a) R = 41.4 N **b)** P = –241 N
Q = –100 N
c) AB, 100 N, tension
AD, 0 N
BC, 141 N, tension
BD, 141 N, compression
CD, 141 N, compression
6 a) P = 4000 N, Q = 2000 N, R = 2000 N
b) AB, 2828 N, compression
BC, 2828 N, compression
BD, 4000 N, tension
CD, 2828 N, compression
AD, 2000 N, compression
7 $R = S$ = 25 N
AB, 25.9 N, compression
AE, 6.7 N, tension
BC, 13.4 N, compression
BE, 25.9 N, tension
CD, 25.9 N, compression
CE, 25.9 N, tension
DE, 6.7 N, tension
8 AB, 146 N, tension
AF, 426 N, compression
BC, 146 N, tension
BD, 426 N, compression
BE, 0 N
BF, 426 N, compression
CD, 426 N, compression
DE, 426 N, compression
EF, 426 N, compression
9 245 N
AB 0 N
AC 245 N
AD 0 N
BC 0 N
CD –160 N
10 a) $R = \frac{4P}{3}$
b) AB, 127 N, tension
AF, 90 N, compression
BC, 42.4 N, tension
BF, 84.9 N, tension
CD, 42.4 N, tension
CF, 60 N, compression
DE, 42.4 N, tension
DF, 0 N
EF, 30 N, compression

Consolidation Exercises for Chapter 19 *(p.353)*

1 a) Light, smooth pin-jointed rods **b)** 1132 N
d) AB, 566 N, compression
AC, 1132 N, compression
BC, 1132 N,tension
e) 1497 N, 40.9° above horizontal
2 AB Tension, 100 N
BC Tension
BD Tension
BE Compression
CD Compression
DE Compression
3 a) A, 60 N; B, 80 N
b) AB, 69.3 N
AE, 34.64 N
BC, 40.4 N
4 a) 500 N
b) 610 N at 35° to horizontal
c) AB, 150 N, compression
AD, 495 N, compression
BC, 212 N, compression
BD, 250 N, tension
CD, 150 N, tension

CHAPTER 20
Differential equations in mechanics

Exercises 20.1A *(p.361)*

1 a) $\frac{-v}{2}$ **b)** $10e^{-0.5t}$
c) $20(1 - e^{-0.5t})$.
2 b) $60(1 - e^{-\frac{1}{60}gt})$ **c)** 4.24 s
3 b) $-2.5v^2$ **c)** $7.4e^{-2.5x}$ **d)** 1.72 m
4 a) 4 m s^{-1} **b)** $4 + 6e^{-\frac{1}{4}gt}$
c) $4t + \frac{24}{g}(1 - e^{-\frac{1}{4}gt})$
d) 4 m s^{-1}, 42.45 m
5 a) A **b)** B **c)** D **d)** C
6 9.94 m
7 a) 19.8 m s^{-1}
b) $\sqrt{(392 - 100x^2 + 19.6x)}$
c) 19.82 m s^{-1}, 2.08 m
8 a) $33\frac{1}{3}$ m s^{-1}, free-fall
b) 31.7 m s^{-1}
9 a) $\frac{1}{v} = kt + \frac{1}{20}$, $v = 20e^{-kx}$
b) $\frac{3}{200}$, 92.4 m
10 Distance tends to gT.

Exercises 20.1B *(p.363)*

1 a) $-500v$ **b)** $80e^{-500t}$
c) $\frac{4}{25}(1 - e^{-0.5t})$ **d)** $\frac{4}{25}$ m
2 a) $-16v$ **b)** 1000 m
3 a) Zero **b)** $-300v$
d) $10(1 - e^{-\frac{t}{4}})$ **e)** 2.8 s
4 a) Surface area of hull, depth of hull etc.
5 a) B **b)** A **c)** C **d)** D
6 5.65 m s^{-1}
7 a) 1.71 m
b) Air resistance, friction, $v\frac{dv}{dx} = -200x - 10kv - \mu g$
8 a) $\frac{2 - kv^2}{v}$, $\frac{2 - Ae^{-2kt}}{k}$ **b)** $\frac{1}{72}$
9 5.7 m
10 86 m

Consolidation Exercises for Chapter 20 *(p.364)*

1 b) $\frac{9-8(1-e^{-40t})}{40}$

2 $\frac{3}{400}$ s

3 b) 11.76 **c)** $50(1 - e^{-0.196t})$
d) 15.3 s

5 147.8 m

CHAPTER 21
Relative motion

Exercises 21.1A *(p.371)*

1 a) $0.3\mathbf{i} + 1.2\mathbf{j}$ **b)** 5 s **c)** 1.5 m

2 a) $2\mathbf{i} + 1.2\mathbf{j}$, 6 m
b) 53.1° to bank, 6.25 s

3 b) The air speed may vary with height.

4 a) $87\mathbf{i} + 70\mathbf{j}$
b) 72 km

5 $-20\mathbf{i} - 10\mathbf{j}$

6 a) $-5\mathbf{i} + 6\mathbf{j}$ **b)** $-5t\mathbf{i} + 6t\mathbf{j}$
c) $t\sqrt{61}$

7 a) $4\mathbf{i} + 3\mathbf{j}$ **c)** 71 m

8 a) $64.6\mathbf{i} - 35.4\mathbf{j}$
b) $64.6t\mathbf{i} + (5000 - 35.4t)\mathbf{j}$
c) 4385 m

9 343.6°, 322 s

10 304 m, yes if good visibility.

Exercises 21.1B *(p.373)*

1 a) $5\mathbf{i} + 12\mathbf{j}$ **b)** 260 m

2 a) 2.6 m **b)** 19.5° to bank
c) At right angles to bank.

3 a) 0.79 m s^{-1} **b)** 1.29 m s^{-1}

4 a) 5.29 m s^{-1}, 139° **b)** 2400 m

5 a) $-8\mathbf{i} - 4\mathbf{j}$ **b)** 8.94 m s^{-1}

6 a) $1.23\mathbf{i} - 4.71\mathbf{j}$
b) $1.23t\mathbf{i} - 4.71t\mathbf{j}$
c) $4.87t$

7 a) $2\mathbf{i} - 8\mathbf{j}$, $(2t - 50)\mathbf{i} + (87 - 8t)\mathbf{j}$
b) $d = \sqrt{(2t-50)^2 + (87-8t)^2}$
c) 56 m

8 a) $-6.6\mathbf{i} - 143.2\mathbf{j}$, $-6.6t\mathbf{i} + (5000 - 143.2t)\mathbf{j}$ **b)** 230 m

9 14 m s^{-1}

10 275.4°

Consolidation Exercises for Chapter 21 *(p.374)*

2 a) $5\mathbf{i} - 10\mathbf{j}$ **c)** 75 s

3 125 s, 100 s, 300 m

4 a) 53.1° to the bank **b)** 5 s
c) Not possible to match speed of current.

5 131° to bank, 33 s

6 After $\frac{2}{25}$ hour at 12.05 p.m.

Acknowledgements

We are grateful to the following Examination Groups for permission to reproduce questions from their past examination papers and from specimen papers. Full details are given with each question.
The Examination Groups accept no responsibility whatsoever for the accuracy or method of working in the answers given, which are solely the responsibility of the author and publishers.

Associated Examining Board (*AEB*)
Northern Examinations and Assessment Board (*NEAB*);
also School Mathematics Project, 16–19 (*SMP 16–19*)
Oxford and Cambridge Schools Examination Board (*OCSEB*);
also Mathematics in Education and Industry (*MEI*)
Scottish Examination Board (*SEB*)
University of Cambridge Local Examinations Syndicate (*UCLES*)
University of London Examinations and Assessment Council (*ULEAC*)
University of Oxford Delegacy of Local Examinations (*Oxford*);
also Nuffield Advanced Mathematics (*Nuffield*)
Welsh Joint Examinations Council (*WJEC*)

We are grateful to the following for their permission to reproduce copyright photographs: p.2 John Birdsall; p.45 Andrew Lambert; p.237 Allsport/Pascal Rondeau; p. 347 Arcaid/Martine Hamilton-Knight.